CONVERSION FACTORS

Length

$$1 \text{ m} = 3.281 \text{ ft} = 39.37 \text{ in} = 6.214 \times 10^{-4} \text{ mi}$$
$$0.304 \text{ m} = 1 \text{ ft} = 12 \text{ in} = 1.894 \times 10^{-4} \text{ mi}$$
$$0.0254 \text{ m} = \tfrac{1}{12} \text{ ft} = 1 \text{ in} = 1.578 \times 10^{-5} \text{ mi}$$
$$1609 \text{ m} = 5280 \text{ ft} = 63{,}360 \text{ in} = 1 \text{ mi}$$

Mass

Mass of 1 lb = 2.2046 kg

$$1 \text{ u} = 1.66056 \times 10^{-27} \text{ kg}$$

Force

$$1 \text{ N} = 0.2248 \text{ lb}$$
$$1 \text{ lb} = 4.448 \text{ N}$$

Energy, work, and heat

$$1 \text{ J} = 0.2389 \text{ cal} = 9.481 \times 10^{-4} \text{ Btu} = 2.778 \times 10^{-7} \text{ kWh}$$
$$4.186 \text{ J} = 1 \text{ cal} = 3.968 \times 10^{-3} \text{ Btu} = 1.163 \times 10^{-6} \text{ kWh}$$
$$1055 \text{ J} = 252 \text{ cal} = 1 \text{ Btu} = 2.930 \times 10^{-4} \text{ kWh}$$
$$3.6 \times 10^{3} \text{ J} = 8.601 \times 10^{5} \text{ cal} = 341.3 \text{ Btu} = 1 \text{ kWh}$$

$$1 \text{ J} = 0.7376 \text{ ft} \cdot \text{lb}$$
$$1.356 \text{ J} = 1 \text{ ft} \cdot \text{lb}$$

$$1 \text{ eV} = 1.602 \times 10^{-19} \text{ J}$$
$$(1 \text{ u})c^2 = 1.49244 \times 10^{-10} \text{ J}$$

Power

$$1 \text{ hp} = 745.7 \text{ W} = 550 \text{ ft} \cdot \text{lb/s}$$

Pressure

$$1 \text{ N/m}^2 = 9.87 \times 10^{-6} \text{ atm} = 7.5 \times 10^{-4} \text{ cm Hg} = 1.45 \times 10^{-4} \text{ lb/in}^2$$
$$1.01 \times 10^5 \text{ N/m}^2 = 1 \text{ atm} = 76 \text{ cm Hg} = 14.7 \text{ lb/in}^2$$
$$1333 \text{ N/m}^2 = 0.0132 \text{ atm} = 1 \text{ cm Hg} = 0.1943 \text{ lb/in}^2$$
$$6895 \text{ N/m}^2 = 0.0681 \text{ atm} = 5.171 \text{ cm Hg} = 1 \text{ lb/in}^2$$

Time

$$1 \text{ day} = 8.64 \times 10^4 \text{ s}$$
$$1 \text{ year} = 3.15 \times 10^7 \text{ s}$$

Volume

$$1 \text{ m}^3 = 10^3 \text{ liters} = 35.3 \text{ ft}^3 = 6.1 \times 10^4 \text{ in}^3$$

Velocity

$$1 \text{ m/s} = 3.28 \text{ ft/s} = 2.24 \text{ mi/h}$$

PRINCIPLES OF PHYSICS

PRINCIPLES OF PHYSICS

JAMES P. HURLEY   **CLAUDE GARROD**   University of California, Davis

HOUGHTON MIFFLIN COMPANY   BOSTON

Dallas   Geneva, Illinois   Hopewell, New Jersey   Palo Alto   London

This book is dedicated to our children:

Dave and Gladys

Emily, David, and Susanna

Printed in the U.S.A.

Library of Congress Catalog Card Number: 77-75475

ISBN: 0-395-25036-6

CONTENTS

PREFACE xv

NOTE TO THE INSTRUCTOR xvii

CHAPTER 1 THE TOOLS OF PHYSICS 1
1.1 Mathematical Description of Physical Observables: Coordinate Systems 2
1.2 Mathematical Description of Physical Observables: Vectors 5

CHAPTER 2 KINEMATICS 15
2.1 Velocity 15
2.2 Acceleration 17
*2.3 Kinematics of Uniform Acceleration Employing Calculus 21
2.4 Motion in Two Dimensions 22
2.5 Two-Dimensional Motion in Rectangular Coordinates and Projectile Motion 27
2.6 Kinematics of Rotation in a Plane 30
2.7 Relative Motion 32

CHAPTER 3 DYNAMICS 39
3.1 Mass and Momentum 39
3.2 Conservation of Momentum 42
3.3 Physical Significance of Momentum 47
3.4 Inertial Frames 49
3.5 Force and the Second Law of Dynamics 49
3.6 Pseudoforces 55
3.7 The Second Law of Dynamics and Determinism 57
3.8 Physical Significance of the Second Law of Dynamics 59
3.9 The First Law of Dynamics and the Law of Action and Reaction 59
3.10 Law of Addition of Forces 60
3.11 Newton's Laws of Motion 61

CHAPTER 4 APPLICATIONS OF THE LAWS OF MECHANICS 65
4.1 Problems in Dynamics 65
4.2 Problems in Statics 69
4.3 Frictional Force 70

CHAPTER 5 CONSERVATION OF ENERGY 77
5.1 Principle of Work and Kinetic Energy 79
5.2 Calculating Work 80
5.3 Applications of the Principle of Work and Kinetic Energy 82
5.4 Dependence of Work on the Path 84
5.5 Conservative and Nonconservative Forces 85

5.6   Potential Energy 86
5.7   Principle of Work and Energy 91
5.8   Other Conservative Forces and Their Associated Potential Energy 93
*5.9   Deriving the Force from the Potential Energy 98
5.10 Physical Significance of Kinetic and Potential Energy 99
5.11 Principle of Minimum Potential Energy 102
5.12 Power 103

*CHAPTER 6 ENERGY CONSERVATION IN A SYSTEM OF MANY PARTICLES 109
6.1 Energy Conservation and Metabolism 110
6.2 Oxygen Consumption and Metabolism 112

*CHAPTER 7 DYNAMICS OF SYSTEMS OF MANY PARTICLES 117
7.1   Center of Mass 118
7.2   Equation of Motion for the Center of Mass 120
7.3   Torque 121
7.4   Direction of the Torque 123
7.5   Angular Momentum 124
7.6   Equation for Rotational Motion of a System of Particles 124
7.7   Conditions for Applicability of the Torque—Angular Momentum Equation 127
7.8   Rigid Bodies 130
7.9   Moment of Inertia 134
7.10 Parallel Axis Theorem 135
7.11 Dynamics of Rigid-Body Rotation 139
7.12 Center of Gravity 143
7.13 Work and Energy Principle for Rotating Rigid Bodies 148
7.14 Kinetic Energy for General Motion of a Rigid Body 151

*CHAPTER 8 EQUILIBRIUM OF A RIGID BODY 157
8.1 Equilibrium of Rigid Bodies 157
*8.2 Statics and Anatomy 160

CHAPTER 9 HARMONIC MOTION 167
9.1 Motivation 168
9.2 Kinematics of Harmonic Motion 170
9.3 Amplitude, Period, and Frequency 171

*CHAPTER 10 SIZE AND FUNCTION 179
10.1 Surface-to-Volume Effects 179
10.2 Structural Strength 183
10.3 Running 184

10.4 Jumping 186
10.5 Flying 186
10.6 Conclusion 187

CHAPTER 11 FLUIDS 191
11.1 Scalar and Vector Fields 193
11.2 Pascal's Law 195
11.3 The Hydrostatic Paradox 198
11.4 Archimedes' Principle 200
11.5 Fluids in Motion 201

*CHAPTER 12 SURFACE TENSION 211
12.1 Surface Energy and Surface Tension 212
12.2 Laplace's Law 215
12.3 Capillarity 220

CHAPTER 13 KINETIC THEORY 229
13.1 Ideal-Gas Law 230
13.2 Temperature 232
13.3 Avogadro's Number 235

*CHAPTER 14 THERMODYNAMICS 241
14.1 Law of Large Numbers and Equilibrium States 241
14.2 The First Law of Thermodynamics 245
14.3 Physical Basis for the Second Law of Thermodynamics 246
14.4 The Second Law of Thermodynamics 247
14.5 Mathematical Statement of the Second Law of Dynamics 248
14.6 Applications 248

*CHAPTER 15 MEASUREMENT OF ENTROPY 255
15.1 Changes in Entropy 255
15.2 Some Properties of P and T 257
15.3 The Pressure 259
15.4 Heat 260
15.5 Isothermal and Adiabatic Processes 261
15.6 Measurement of Entropy 262

*CHAPTER 16 GIBBS AND HELMHOLTZ FREE ENERGIES 269
16.1 Gibbs Free Energy 269
16.2 Helmholtz Free Energy 271

*CHAPTER 17 VARIATION OF MOLAR NUMBERS 273

17.1 Chemical Potential 274

17.2 Particle Exchange Across a Surface 276

17.3 Chemical Potential of Ideal Gases and Dilute Solutes 277

17.4 Applications 278

*CHAPTER 18 LAW OF INCREASING ENTROPY 289

18.1 Applications 291

18.2 Entropy and Disorder 295

18.3 The Second Law of Thermodynamics and the Law of Evolution 296

18.4 Disorder and Velocity Distribution 299

18.5 The Second Law of Thermodynamics and the Fate of the Universe 301

18.6 The Evolution of the Species, the Death of the Individual, and the
     Second Law of Thermodynamics 302

18.7 The Second Law of Thermodynamics and Upper Limits on Engine and
     Refrigerator Efficiencies 303

CHAPTER 19 THERMODYNAMICS AND PROPERTIES OF MATTER 313

19.1 Equation of State 313

19.2 Phase Transitions 316

19.3 Specific Heat 318

19.4 Heats of Fusion and Vaporization 321

19.5 Evaporation and Boiling 322

CHAPTER 20 HEAT TRANSFER 329

20.1 Conduction 330

20.2 Radiation 333

*20.3 Heat Transfer and Physiology 336

*20.4 Countercurrent Heat Exchanger 338

CHAPTER 21 ELECTROSTATICS 343

21.1 Coulomb's Law 343

21.2 Electric Fields 347

21.3 Calculating Electric Fields 349

21.4 The Electric Field on the Axis of a Charged Ring 350

21.5 The Electric Field of an Infinite Plane of Charge 351

21.6 The Electric Field of Two Large Parallel Planes with Equal and Opposite
     Charges 353

CHAPTER 22 ELECTRIC FIELD LINES AND ELECTRIC POTENTIAL 357

22.1 Electric Field Lines 357

22.2 Electric Potential 358
22.3 Deriving Electric Fields from Potential Fields 361
22.4 Motion of Charges in Electric Fields 363
22.5 Equipotential Surfaces 365
22.6 Conductors 366

CHAPTER 23 GAUSS'S LAW 373
23.1 Gauss's Law in Fluid Dynamics 374
23.2 Gauss's Law for Electric Fields 377
23.3 Applications of Gauss's Law 378
23.4 A Second Integral Equation for the Electric Field 382
23.5 Solving Electrostatics Problems 383

CHAPTER 24 CAPACITANCE 387
24.1 Capacitance: An Intrinsic Property 388
24.2 Example of the Intrinsic Nature of Capacitance 389
24.3 Capacitors Connected in Series and in Parallel 391
24.4 Dielectrics 397

CHAPTER 25 ELECTRIC CURRENT AND RESISTANCE 405
25.1 Electric Current 405
25.2 Electric Resistance 406
25.3 Resistance and Resistivity 407
25.4 Energy Dissipated in a Resistor 408

CHAPTER 26 DC CIRCUITS 411
26.1 Batteries: emf and Internal Resistance 411
26.2 Resistors Connected in Series 413
26.3 Resistors Connected in Parallel 413
26.4 Kirchhoff's Rules 415

CHAPTER 27 THE MAGNETIC FORCE 421
27.1 The Magnetic Field 421
27.2 The Trajectory of a Charged Particle in a Uniform Magnetic Field 423
27.3 Motion of an Electric Charge in Nonuniform Magnetic Fields 426
27.4 Magnetic Force on an Electric Current 428

CHAPTER 28 THE MAGNETIC FIELD 435
28.1 Magnetic Field of a Point Charge 435
28.2 Qualitative Picture of the Magnetic Field 436
28.3 Calculating the Magnetic Field 439

28.4 Two Parallel Currents 440
28.5 The Integral Form of the Magnetic Field Laws 441
28.6 Calculating Magnetic Fields Using Ampere's Law 444
28.7 Magnetic Materials 446

CHAPTER 29 ELECTRODYNAMICS: DISPLACEMENT CURRENT AND ELECTROMAGNETIC INDUCTION 451
29.1 The Displacement Current 452
29.2 Electromagnetic Induction 457
29.3 Lenz's Law 461
29.4 Inductance 463

CHAPTER 30 ENERGY AND ELECTROMAGNETIC WAVES 473
30.1 Energy in an Electromagnetic Field 473
30.2 Electric Field Energy 474
30.3 Magnetic Field Energy 477
30.4 Electromagnetic Waves 478
30.5 The Velocity of Light 488

CHAPTER 31 WAVE MOTION 495
31.1 String Waves 495
31.2 Wave Functions 497
31.3 Reflection of Waves 503
31.4 Periodic Waves 506
31.5 Harmonic Waves 506
31.6 Standing Waves 508
31.7 Wave Speed of String Waves 511
31.8 Wave Energetics 512
31.9 Partial Reflection of Waves 514
31.10 The Wave Equation 517

CHAPTER 32 SOUND WAVES 527
32.1 Wave Fronts 527
32.2 Plane Waves 528
32.3 Sound Wave Velocity 528
32.4 Acoustic Energy Flux 529
32.5 Impedance-Matching Devices 530
32.6 Waves on Membranes 532
32.7 Normal Mode Vibrations of a Membrane 533
32.8 Forced Oscillations of a System 534

*32.9   The Physics of Hearing 534
*32.10 Frequency Determination by the Ear 536
*32.11 Analysis of Complex Sounds 539

CHAPTER 33 WAVES IN TWO AND THREE DIMENSIONS 547
33.1 Spherical Waves 547
33.2 Interference 549
33.3 The Electromagnetic Spectrum 553
33.4 Diffraction 555
33.5 Single-Slit Diffraction 557
33.6 Refraction 561
33.7 The Law of Reflection 564
33.8 Total Internal Reflection 565

CHAPTER 34 OPTICS 571
34.1 Single-Surface Lenses 571
34.2 The Double-Surface Lens 576
*34.3 The Human Eye 580
*34.4 Magnifying Instruments 583
*34.5 Ultimate Resolution of an Optical System 587
*34.6 Resolution of a Microscope 591

CHAPTER 35 THE THEORY OF RELATIVITY 603
31.1   The Newtonian Principle of Relativity 604
35.2   The Effect of Electromagnetic Theory on the Newtonian Relativity
       Principle 605
35.3   Attempts to Measure the Velocity of the Ether 607
35.4   Einstein's Theory of Relativity 611
35.5   The Lorentz Transformation 614
35.6   Time Dilation 618
35.7   Length Contraction 619
35.8   The Geometry of Spacetime 620
35.9   The Relativistic Addition of Velocity Law 622
35.10 Three-Dimensional Velocity Addition 624
35.11 Momentum Conservation 626
35.12 Relativistic Velocity and Momentum 628
35.13 Transformation Laws of Relativistic Velocity and Momentum 629
35.14 Conservation of Four-Momentum 630
35.15 "Does the Inertia of a Body Depend upon its Energy Content?" 633
35.16 Particles of Zero Mass 636

CHAPTER 36 THE ORIGINS OF QUANTUM PHYSICS 643

36.1   Classical Physics 643
36.2   Cavity Radiation 644
36.3   The Photoelectric Effect 647
36.4   The Compton Effect 648
36.5   Alpha-Particle Scattering 651
36.6   Energy Quantization in Atoms 653
36.7   The Franck-Hertz Experiment 655
36.8   The Hydrogen Atom 656
36.9   De Broglie Waves 658
36.10   The Davisson-Germer Experiment 661

CHAPTER 37 THE SCHRÖDINGER WAVE EQUATION 667

37.1   Diffraction of Light and Electrons 667
37.2   Quantum Theory 668
37.3   The Relationship between Quantum and Classical Physics 671
37.4   The Schrödinger Equation 673
37.5   The Physical Meaning of Wave Function 675
37.6   Standing-Wave States and Discrete Energy Values 677
37.7   The Emission and Absorption Spectrum 679
37.8   Normalization of the Wave Function 682
37.9   The Heisenberg Uncertainty Relation 683
37.10   The Classical Limit of Quantum Mechanics 685
37.11   The Schrödinger Equation with a Potential 686
37.12   Quantum States of a Particle in a Potential 687
37.13   The Harmonic Oscillator 688

*CHAPTER 38 QUANTUM THEORY OF ATOMIC STRUCTURE 697

38.1   The Schrödinger Equation in Two and Three Dimensions 697
38.2   Translational and Rotational Invariance in Classical and
        Quantum Theory 698
38.3   Quantization of Angular Momentum 700
38.4   Three-Dimensional Quantum States in a Symmetrical Potential 701
38.5   The Hydrogen Atom 704
38.6   The Zeemann Effect 706
38.7   Electron Spin 708
38.8   The Exclusion Principle 710
38.9   The Periodic Table of the Elements 710
38.10   The Chemical Bond 712

*CHAPTER 39 NUCLEAR PHYSICS 721

39.1 The Nuclear Force 721
39.2 Nuclear Fluid 723
39.3 Binding Energies of Nuclei 724
39.4 Nuclear-Decay Modes 728
39.5 The Law of Radioactive Decay 731
39.6 Nuclear Stability 732
39.7 Nuclear Shells 736
39.8 Nuclear Fission and Fusion 737

APPENDIX A 749

A.1 Mathematical Formulas 749
A.2 Properties of Logarithms and Exponentials 750
A.3 Partial Derivatives 751
A.4 Complex Numbers 751

APPENDIX B 753

Trigonometric Functions 753
Common Logarithms 754

APPENDIX C 755

Astronomical Data 755

ANSWERS TO ODD-NUMBERED PROBLEMS 757

INDEX 765

PREFACE

Our goals in writing this text are twofold. We have sought first to state the fundamental laws of physics with clarity and precision and second to illustrate their application with interesting examples.

In stating the fundamental laws, we have stressed the requirement that all quantities must have independent operational definitions. If students are presented with the equation $A = BC$ but are only given instructions for assigning numbers to B and C, then that equation is only a definition of the quantity A and is not a physical law. By ignoring this simple, logical principle the study of physics can easily deteriorate from a study of the physical universe to mathematical games played on physics exams. Thus in this book mass is not merely *the amount of matter,* force *the strength of a push or pull,* and temperature *the reading on a thermometer.* We hope to point out clearly the basic physical idea that is enunciated in each law and to illustrate how the law makes quantitative predictions of real physical observables.

The second goal is perhaps the more challenging. How can we communicate to students the excitement that comes in discovering the logical chain that connects observed phenomena with fundamental principles? How can we take advantage of students' innate desire to see into the nature of things and to become aware of relationships among the processes at work around them? We have tried as often as possible to illustrate the laws of physics with real problems involving real objects. The traditional topics such as inclined planes, calorimeters, and vibrating strings are discussed because they allow the novice to gain experience in applying theory to simple systems. But we have then used the same physical principles to answer such intriguing questions as: Why can an ant carry many times its body weight with relative ease? Why does a brick chimney buckle as it falls? Why does water spill from an inverted coffee cup but not from an inverted straw sealed at the top? Is evolution a violation of the second law of thermodynamics? Is the inevitability of death a manifestation of the second law? How does the ear distinguish sounds of different frequencies? If the tides are caused by the gravitational pull of the moon, why are there two high tides daily? Since nature's laws conserve energy, are federal energy conservation laws redundant?

We have extended the usual range of applications to include a number of problems of interest to the biologist. We have done so for three reasons. First, the biological sciences have reached a level of fundamental understanding where interaction between physics and biology is becoming increasingly fruitful. Second, the reader is thereby led to appreciate the pervasive character of the laws of physics, which apply to animate as well as inanimate systems. Finally, it is our experience that students find the application of the laws of physics to living systems fascinating.

We are grateful to the many reviewers who read the original manuscript.

J. P. H.
C. G.

NOTE TO THE INSTRUCTOR

In our opinion there is more material in this text than can be covered in a single year. We have isolated into separate chapters and sections material that may be omitted without loss in continuity, and there are only minor references to this material in later chapters. The topics we suggest as possible condidates for omission are indicated with an asterisk. For example, Chapter 7, "Dynamics of Systems of Many Particles," might be omitted in courses intended primarily for life science students. Chapter 6, which deals mainly with the relationship between energy conservation and metabolism, might be omitted in a course directed to engineers.

Although Chapters 14 through 18 are marked with an asterisk, these chapters must be taken in sequence. With the exception of Chapter 16, "Gibbs and Helmholtz Free Energies," each is a prerequisite for those that follow. It is possible, however, to stop the sequence at any point since only occasional references are made to this material in subsequent chapters.

The thermodynamics presented in Chapters 14 through 18 is basically that of J. Willard Gibbs. This development has been most clearly elucidated in Callen's *Thermodynamics*. We owe a considerable debt to this exposition. It permits a very intuitive development of the subject in marked contrast with the more conventional, but more abstract, Carathéodory approach. It makes it possible to develop thermodynamics to the same degree that mechanics, and electromagnetic theory are traditionally developed. At a time when the energy crisis is of such concern, it is important for students to understand that it is in fact an entropy crisis and not an energy crisis.

Although Chapter 12, "Surface Tension," is marked with an asterisk, Section 12.1 will be required background for Chapter 39, "Nuclear Physics."

PRINCIPLES OF PHYSICS

1
THE
TOOLS
OF
PHYSICS

BEFORE WE CAN get into the basic laws of physics we must consider the language and tools of physics. In physics we attempt to make quantitative predictions on the outcome of well-defined experiments. Sometimes the process is inverted and we attempt to understand the nature of the elements involved in some process given the *results* of the experiment. In either case a quantitative description of the system is required. We must be able to assign numbers to physical observables. This is *the measurement process*. There is basically only one measuring process: counting. We measure the distance between two objects by counting the number of times a standard of length fits into this distance. We measure the time interval between two events by counting, for example, the number of times a pendulum oscillates between the events. We measure the weight of an object by counting the standard units of weight necessary to balance the object. We shall see that we can measure the thermodynamic entropy of a state by observing how much ice melts in an ice-water reservoir in a certain process. We find out how much ice melts by weighing the ice, and we have just seen that weight is determined by counting.

Since the measurement process is one of counting multiples of some chosen standard, it is reasonable to ask how many standards we need. If we need a standard for each observable, we will need a large Bureau of Standards indeed. As a matter of fact, we need only four standards: a standard of length, a standard of mass, a standard of time, and a standard of electric charge. This is an extraordinary fact. It means that if one is equipped with a set of these four standards and the ability to count, one can (in principle) assign a numerical value to any observable, be it distance, velocity, viscosity, temperature, pressure, etc. We will discover why this is so as we proceed through this course.

Unfortunately different sets of standards have come into use. If we leave aside for the moment the unit of electric charge, the three major systems are the meter-kilogram-second (mks) system, the centimeter-gram-second (cgs) system, and the English system. These systems are summarized in Table 1.1.

TABLE 1.1 Standards of length, mass, and time.

	mks	*cgs*	*English*
Length	meter	centimeter	foot
Mass	kilogram	gram	slug[a]
Time	second	second	second

[a]We shall see later how this is related to the pound.

We have not yet defined the concept of mass; this will have to wait until we take up the study of dynamics. Because of their nearly universal use in scientific work, we shall confine ourselves almost exclusively to the mks and cgs units.

1.1 MATHEMATICAL DESCRIPTION OF PHYSICAL OBSERVABLES: COORDINATE SYSTEMS

Suppose we wanted to give a quantitative description of the motion of a projectile in the earth's gravitational field. We would need some convenient way of describing the location of the projectile. In general, to specify the location of some object, we give its location relative to some other object or objects whose positions are known (e.g., two miles due east of Petaluma).

Often we will locate objects by relating them to a coordinate system, and we assume that we know where the coordinate system is (in relation to something else) or else we do not care where it is. (Most people could not care less where our solar system is with respect to the rest of the galaxy, and they still manage to find their socks in the morning.)

A two-dimensional rectangular coordinate system is two perpendicular lines (called axes), each with a positive sense (Fig. 1.1). (We restrict ourselves at present to two dimensions.) To distinguish between the axes, we call the horizontal axis the x axis, or *abscissa,* and the vertical axis the y axis or *ordinate.* We locate a point P by drawing lines from the point parallel to the axes and measuring the distances from the points of interception (called the x and y *intercepts*) to the origin O. These two distances are called the x and y *coordinates* of P. They are positive if the direction *from* the origin *to* the respective intercept is in the chosen positive sense (indicated by the arrows in Fig. 1.1).

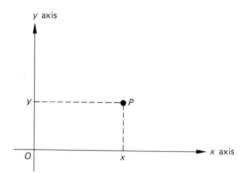

FIGURE 1.1 Rectangular coordinates.

A second way of determining the position of P with respect to the coordinate axes is to find the distance r from P to the origin O and the angle θ between the line joining P to the origin and the x axis (Fig. 1.2). We shall call r and θ the *polar coordinates* and x and y the *rectangular coordinates.*

We measure the angle θ between the two lines in much the same way that we measure length. We choose some standard and define it to be a unit angle (Fig. 1.3). We then determine the magnitude of any specific angle by counting the number of times the unit angle can be fitted into it. For example, the angle between the lines a and b in Fig. 1.4 is 5 units. If the unit angle does not fit in precisely, we may have to divide the unit angle into smaller subdivisions.

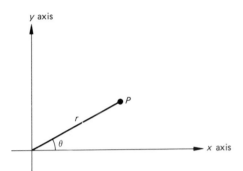

FIGURE 1.2 Polar coordinates.

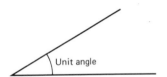

FIGURE 1.3 A unit angle.

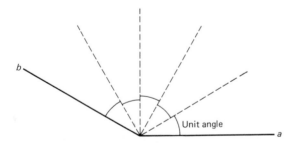

FIGURE 1.4 An angle of five units.

Just as we may construct a ruler by repeatedly laying the unit length on a straightedge and marking each interval, so may we construct a protractor by laying the unit angle about a circle and marking each unit on the perimeter (Fig. 1.5).

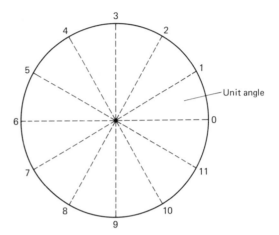

FIGURE 1.5 A protractor.

We are free to choose any unit standard to divide the circumference of the protractor. The three most common units are those that divide a circle into 360 divisions (*degrees*), 2π divisions (*radians*), and 1 division (*revolutions*). If we use the symbol θ_{degrees} to denote the numerical values of the angle between two lines when the reference circle or protractor is divided into 360 equal divisions (Fig. 1.6), then the fraction of the circumference intercepted by two lines on any circle is given by

$$\text{Fraction of circumference intercepted} = \frac{\theta_{\text{degrees}}}{360}$$

Similarly, if we measure the angle in radians, we have

$$\text{Fraction of circumference intercepted} = \frac{\theta_{\text{radians}}}{2\pi}$$

and in revolutions,

$$\text{Fraction of circumference intercepted} = \frac{\theta_{\text{revolutions}}}{1}$$

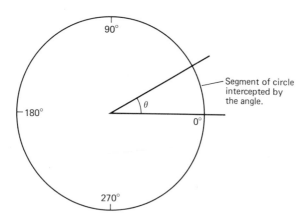

FIGURE 1.6 The angle θ intercepts a fraction of the circumference of the circle.

Since the fraction of circumference intercepted is independent of the unit employed,

$$\text{Fraction of circumference intercepted} = \frac{\theta_{degrees}}{360} = \frac{\theta_{radians}}{2\pi} = \frac{\theta_{revolutions}}{1}$$

This equation allows us to convert from one system to another. For example, if the angle measured in radians is 1 radian,

$$\theta_{radians} = 1$$

then this angle in degrees is given by

$$\frac{\theta_{degrees}}{360} = \frac{\theta_{radians}}{2\pi} = \frac{1}{2\pi}$$

or

$$\theta_{degrees} = \frac{360}{2\pi} = 57.3°$$

Or if $\theta_{degrees} = 90°$, then

$$\theta_{revolutions} = \frac{\theta_{degrees}}{360} = \frac{90}{360} = \frac{1}{4} \text{ revolution}$$

The reason for dividing the circle into 2π divisions can be seen from the following consideration. No matter how the angle is measured (see Fig. 1.7),

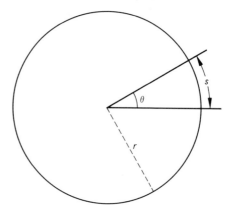

FIGURE 1.7 The fraction of the circumference intercepted is s divided by the circumference.

$$\text{Fraction of circumference intercepted} = \frac{s}{2\pi r} = \frac{\theta}{N}$$

where N is the number of divisions in the protractor and s is the arc length. The value $2\pi r$ is the total circumference. Solving for θ, we have

$$\theta = \frac{N}{2\pi}\left(\frac{s}{r}\right)$$

It is to simplify this relation between the angular separation θ between two lines and the linear distance s along the arc of the circle that N is chosen to be 2π. If we take $N = 2\pi$, then

$$\theta_{\text{radians}} = \frac{s}{r} \qquad (1.1)$$

Unless specified otherwise, we shall use radian measure in the future.

1.2 MATHEMATICAL DESCRIPTION OF PHYSICAL OBSERVABLES: VECTORS

DEFINITION AND NOTATION

A number of physical quantities are best described in terms of a mathematical framework called *vector analysis*. A *vector* is a directed line segment (specified by magnitude and direction) and obeys a law of addition described in the following section. Examples of vectors that we shall encounter later are displacement, velocity, acceleration, and force. Vectors are to be contrasted with *scalars,* which have only magnitude. Examples of scalars are mass, temperature, volume, and pressure.

We shall use boldface type to indicate a vector (**D**) and lightface type (D) to indicate the magnitude of the vector. To denote the direction of a vector, we shall use a vector of unit magnitude, designed by boldface type with a caret above the letter ($\hat{\mathbf{D}}$). If **D** represents a displacement, then $\hat{\mathbf{D}} = \mathbf{D}/D$ represents a unit vector in the direction of the displacement. We may therefore represent a vector **D** by the equation

$$\mathbf{D} = D\hat{\mathbf{D}}$$

which says that the vector **D** has a magnitude of D and a direction specified by $\hat{\mathbf{D}}$.

ADDITION OF VECTORS

We add vectors as we add displacements. If an object is displaced from P to Q and then from Q to R, the net effect of these two displacements is a displacement from P to R. If **D**$_1$ is a vector from P to Q, **D**$_2$ a vector from Q to R, and **D** a vector from P to R, then

$$\mathbf{D} = \mathbf{D}_1 + \mathbf{D}_2$$

This equation is illustrated in Fig. 1.8.

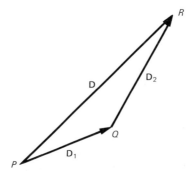

FIGURE 1.8 The displacement **D**$_1$ plus the displacement **D**$_2$ is equivalent to the displacement **D**.

QUESTION

What is the sum of the two perpendicular vectors $\mathbf{D}_1$ and $\mathbf{D}_2$ where $\mathbf{D}_1$ is 1 unit in length and $\mathbf{D}_2$ is 2 units?

ANSWER

Placing the vectors tail to tip

we see that the magnitude of **D** is given by

$$D = \sqrt{D_1{}^2 + D_2{}^2} = \sqrt{1^2 + 2^2} = 2.2$$

The angle θ is given by

$$\tan \theta = 2/1$$

or

$$\theta = 63°$$

Therefore the sum of the vectors $\mathbf{D}_1$ and $\mathbf{D}_2$ is a vector **D** that has a magnitude of 2.2 and makes an angle of 63° with respect to $\mathbf{D}_1$.

QUESTION

Does it matter in what order we add two vectors?

ANSWER

No. We can easily see this by constructing a parallelogram from the two vectors. We see from Fig. 1.9 that we can obtain the sum **D** by placing the tail of $\mathbf{D}_2$ at the tip of $\mathbf{D}_1$ or the tail of $\mathbf{D}_1$ at the tip of $\mathbf{D}_2$.

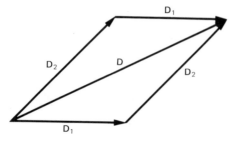

FIGURE 1.9   Parallelogram formed by the vectors $\mathbf{D}_1$ and $\mathbf{D}_2$.

Thus we add two vectors by placing them tail to tip, in any order, and the sum is a vector from the tail of the first to the tip of the second. If we have four vectors (Fig. 1.10), then we place the four vectors tail to tip, and the sum $\mathbf{D} = \mathbf{D}_1 + \mathbf{D}_2 + \mathbf{D}_3 + \mathbf{D}_4$ is a vector from the tail of the first to the tip of the last (Fig. 1.11). We shall call this "tail-to-tip" rule the *polygon method* of vector addition.

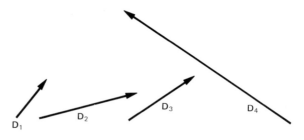

FIGURE 1.10   Four vectors.

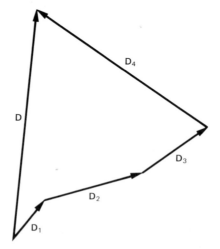

FIGURE 1.11   The sum or resultant **D** is a vector from the tail of the first vector to the tip of the last vector.

QUESTION

We have defined the sum of vectors by the polygon method of placing the vectors tail to tip (in any order), with the sum being a vector from the tail of the first to the tip of the last. How might we define the difference between two vectors? What is **A** − **B**?

ANSWER

While we cannot derive definitions, we can ask what the definition must be if we require that

$$\mathbf{A} - \mathbf{B} = \mathbf{A} + (-\mathbf{B}) \tag{1.2}$$

where −**B** is defined to be a vector equal in magnitude to **B** but opposite in direction. Since we know how to add vectors, we know what **A** + (−**B**) must be. Therefore Eq. 1.2 is in fact a definition of vector subtraction. Graphically we may accomplish subtraction of vectors in the following way. Given two vectors **A** and **B**, we obtain the difference by placing the vectors *tip to tip*, and the difference is a vector from the tail of **A** to the tail of **B**. It is clear from Fig. 1.12 that Eq. 1.2 is satisfied by this definition.

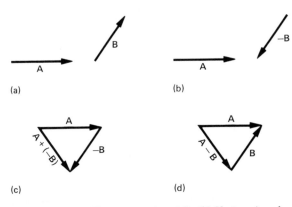

(a)

(b)

(c)

(d)

FIGURE 1.12 (a) Two vectors **A** and **B**. (b) Vectors **A** and −**B**. (c) The sum **A** + (−**B**). (d) The difference **A** − **B**.

This polygon method of vector addition defines the sum uniquely, but it is difficult to use in practice. You can see, for example, that the geometry of Fig. 1.11 is rather complicated, and it would be difficult to determine the sum *quantitatively*. To answer the need for a method of vector addition that is easier to use, we introduce the concept of *resolution into components*. If **D** is a vector and we have some *xy* reference coordinate system, we define the *x* and *y* components of **D** by projecting **D** onto the *x* and *y* axes, as shown in Fig. 1.13 or somewhat more simply in Fig. 1.14.

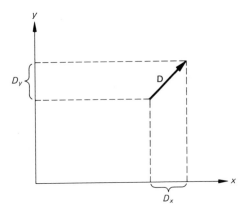

FIGURE 1.13 Resolution of **D** into components D_x and D_y.

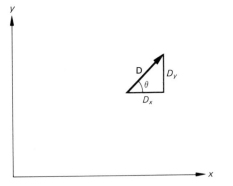

FIGURE 1.14 Resolution of **D** into components D_x and D_y.

We see from Fig. 1.14 that if D is the magnitude of $\mathbf{D}$, then

$$D_x = D \cos \theta$$

$$D_y = D \sin \theta$$

We may also reverse the process: Given D_x and D_y, we can determine D and θ from the equations

$$D^2 = D_x^2 + D_y^2$$

and

$$\tan \theta = \frac{D_y}{D_x}$$

QUESTION

What are the x and y components of the vector $\mathbf{D}$ shown in Fig. 1.15? The magnitude of $\mathbf{D}$ is 10 units.

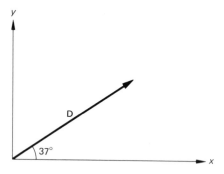

FIGURE 1.15 A vector 10 units in length making an angle of 37° with the x axis.

ANSWER

$$D_x = D \cos 37° = 10(\tfrac{4}{5}) = 8$$

$$D_y = D \sin 37° = 10(\tfrac{3}{5}) = 6$$

Vector addition is now easy. If we have three vectors to add, we place them tail to tip as before, but now we introduce a coordinate system and determine the x and y components of each vector. From Fig. 1.16 we see that if

$$\mathbf{D} = \mathbf{D}_1 + \mathbf{D}_2 + \mathbf{D}_3$$

then

$$D_x = D_{1x} + D_{2x} + D_{3x}$$

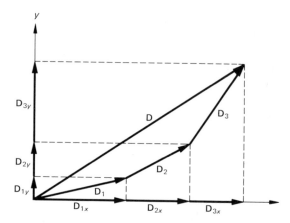

FIGURE 1.16 Sum of three vectors by the method of resolutions into components.

and

$$D_y + D_{1y} + D_{2y} + D_{3y}$$

Knowing the x and y components of $\mathbf{D}$ is equivalent to knowing $\mathbf{D}$ itself.

QUESTION

A man and a woman begin a journey from the same point. The man walks 1 mile due east, 3 miles 30° north of east, and 7 miles 45° south of east. How should the woman walk to make the shortest trip to meet the man?

ANSWER

Let us represent the man's journey as a sum of the three vector displacements $\mathbf{D}_1$, $\mathbf{D}_2$, and $\mathbf{D}_3$ (see Fig. 1.17). The resultant displacement $\mathbf{D}$ is the sum $\mathbf{D}_1 + \mathbf{D}_2 + \mathbf{D}_3$. The x component of the displacement is the sum of the x components of $\mathbf{D}_1 + \mathbf{D}_2 + \mathbf{D}_3$. Therefore

$$D_x = D_{1x} + D_{2x} + D_{3x}$$
$$= 1 + 3 \cos 30° + 7 \cos 45° = 8.5 \text{ mi}$$

Similarly

$$D_y = D_{1y} + D_{2y} + D_{3y}$$
$$= 0 + 3 \sin 30° - 7 \sin 45° = -3.4 \text{ mi}$$

The magnitude of the resultant is

$$D = \sqrt{D_x{}^2 + D_y{}^2} = 9.2 \text{ mi}$$

and the direction is determined from

$$\tan \theta = \frac{D_y}{D_x} = \frac{-3.4}{8.5} \quad \text{or} \quad \theta = -24.2°$$

Thus the equivalent displacement that would take the woman to the same point is a 9.1-mi journey at an angle of 24.2° south of east.

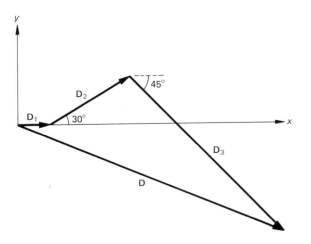

FIGURE 1.17 Journeys of the man and the woman.

We shall find it convenient to represent a vector in terms of its components as a vector relationship. If we let the vector $\mathbf{D}_x$ be a vector in the x direction whose magnitude is D_x and $\mathbf{D}_y$ a vector in the y direction with magnitude D_y, then it is clear from Fig. 1.18 that

$$\mathbf{D} = \mathbf{D}_x + \mathbf{D}_y$$

The sum of the two component vectors placed tail to tip is a vector from the tail of the first to the tip of the second, which is the vector $\mathbf{D}$.

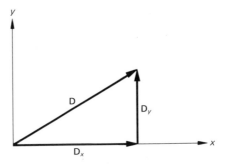

FIGURE 1.18 The vector $\mathbf{D}$ is the sum of $\mathbf{D}_x$ and $\mathbf{D}_y$.

MULTIPLICATION OF A VECTOR BY A SCALAR

We define multiplication of a vector by a scalar as follows: For a positive scalar c and a vector $\mathbf{D}$, the product $c\mathbf{D}$ is a vector with the same direction as $\mathbf{D}$ and a magnitude of c times the magnitude of $\mathbf{D}$. If c is negative, then the direction of $c\mathbf{D}$ is opposite to $\mathbf{D}$. This rule for multiplication is determined by the condition that $2\mathbf{D} = \mathbf{D} + \mathbf{D}$, $3\mathbf{D} = \mathbf{D} + \mathbf{D} + \mathbf{D}$, etc. (see Fig. 1.19).

FIGURE 1.19   Multiplication of a vector by a scalar.

THE SCALAR PRODUCT

Having defined addition of vectors and multiplication of a vector by a scalar, we now wish to define two types of multiplication between vectors. The first is called *scalar multiplication,* or the *dot product.* The scalar or dot product of two vectors **A** and **B** is defined as the product of the magnitudes of **A** and **B** multiplied by the cosine of the angle between the two vectors; that is,

$$\mathbf{A} \cdot \mathbf{B} = AB \cos \theta$$

This is not something to be understood; it is a definition and its justification is its usefulness.

QUESTION

What is the scalar product of the vectors **A** and **B** illustrated in Fig. 1.20? The magnitude of **A** is 10, and the magnitude of **B** is 5.

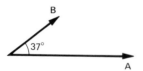

FIGURE 1.20   Two vectors with an angle of 37° between them.

ANSWER

From the definition, the scalar product is

$AB \cos \theta = (10)(5)\cos 37° = 40$

Some of the properties of the scalar product that we will use are:

1. $\mathbf{A} \cdot \mathbf{B} = \mathbf{B} \cdot \mathbf{A}$

2. $\mathbf{A} \cdot \mathbf{A} = A^2$

3. $\mathbf{A} \cdot \mathbf{B} = 0$ if **A** is perpendicular to **B**

4. $\mathbf{A} \cdot \mathbf{B} = $ (component of **A** in the direction of **B**)B
   $= A$(component of **B** in the direction of **A**)

5. $\mathbf{A} \cdot (\mathbf{B} + \mathbf{C}) = \mathbf{A} \cdot \mathbf{B} + \mathbf{A} \cdot \mathbf{C}$

The fourth property above follows very simply from the observation that

$$\mathbf{A} \cdot \mathbf{B} = AB \cos \theta = (A \cos \theta)B$$
$$= A(B \cos \theta)$$

and $A \cos \theta$ is the component of **A** in the direction of **B** and $B \cos \theta$ is the component of **B** in the direction of **A**. We shall find this interpretation of the scalar product very useful in the following discussion. It is also helpful in verifying property 5 above; this is left as a problem (see problem 1.B.8).

Just as we found that sometimes we can sum vectors more easily by resolving them into components, so can we sometimes obtain the scalar product more easily by using components. For example, if we write

$$\mathbf{A} = \mathbf{A}_x + \mathbf{A}_y$$
$$\mathbf{B} = \mathbf{B}_x + \mathbf{B}_y$$

then

$$\mathbf{A} \cdot \mathbf{B} = (\mathbf{A}_x + \mathbf{A}_y) \cdot (\mathbf{B}_x + \mathbf{B}_y)$$
$$= A_x B_x + A_y B_y$$

where we have employed properties 3 and 5 above. Once again a vector operation has been reduced to a simple algebraic operation by the method of resolution into components.

THE VECTOR PRODUCT

Another useful definition is the *vector* or *cross product* between vectors. The vector product between two vectors **A** and **B** is defined as a vector whose magnitude is the product of the magnitudes of **A** and **B** and the sine of the angle between them. Thus

Magnitude of **A** × **B** = $AB \sin \theta$ (1.3)

The direction of the vector product is perpendicular to both **A** and **B**, and the sense is given by the following *right-hand rule:* If the fingers of the right hand are aligned with **A** and rotated into **B** (through the smaller angle), the thumb then points in the direction of the cross product (Fig. 1.21).

We can see from the definition of the vector product that

1. **A** × **B** = −**B** × **A**

2. **A** × **A** = 0

3. **A** × **B** = 0 if **A** is parallel to **B**

4. **A** × (**B** + **C**) = **A** × **B** + **A** × **C**

QUESTION

Two vectors **A** and **B**, defined in Fig. 1.22, lie in the plane of the page. What are the magnitude and direction of the vectors **C** = **A** × **B** and **D** = **B** × **A**?

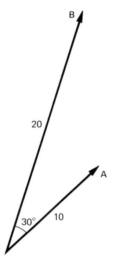

FIGURE 1.22   The vector **A** is 10 units in length. The vector **B** is 20 units in length and makes an angle of 30° with respect to **A**.

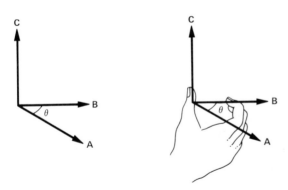

FIGURE 1.21   If **C** = **A** × **B**, then $C = AB \sin \theta$ and the direction of **C** is perpendicular to both **A** and **B**. To decide between the two possible choices (up or down) for this perpendicular, one uses the right-hand rule. The fingers of the right hand are laid along *A* and then curled into *B*. The thumb points in the direction of **C**.

ANSWER

From the definition (Eq. 1.3),

Magnitude of $\mathbf{C} = AB \sin \theta$

$$= (10)(20)\sin 30° = 100$$

and the direction, obtained from the right-hand rule, is out of the page.
   Also

Magnitude of $\mathbf{D} = BA \sin \theta = 100$

and the direction of $\mathbf{D}$ is into the page.

SUMMARY

All physical measurements are fundamentally counting processes: the counting of multiples of some chosen standard. The number of *fundamental* standards is four; these are length, mass, time, and electric charge. One can assign any physical observable a number using these standards and the ability to count. In the mks system of units, the fundamental standards for length, mass, and time are the meter, the kilogram, and the second, respectively.

   We may give a quantitative description of the position of an object by specifying its coordinates—either its rectangular coordinates (x, y) or its polar coordinates (r, θ).

   Angles measure the fraction of the circumference of a circle intercepted by two straight lines. The numerical value of the angle depends on the number of divisions of the circle chosen as the unit. The fraction of circumference intercepted, however, is always the same, so

$$\text{Fraction of circumference intercepted} = \frac{s}{2\pi r} = \frac{\theta_{\text{degrees}}}{360}$$

$$= \frac{\theta_{\text{radians}}}{2\pi} = \frac{\theta_{\text{revolutions}}}{1}$$

In radian measure we have the simple relation

$$\theta_{\text{radians}} = \frac{s}{r}$$

   A vector is a directed line segment that has both magnitude and direction. We add vectors in the same way that we add displacements. We can obtain a vector sum by placing vectors tail to tip, and the sum is then a vector from the tail of the first vector to the tip of the last. We may also add vectors by resolving them into components. Here the components of the sum are the sum of the components. The two ways to multiply two vectors are the scalar or dot product and the vector or cross product. The scalar product of two vectors $\mathbf{A}$ and $\mathbf{B}$ is given by

$$\mathbf{A} \cdot \mathbf{B} = AB \cos \theta$$

For the cross product, if

$$\mathbf{C} = \mathbf{A} \times \mathbf{B}$$

then

$$C = AB \sin \theta$$

and the direction of $\mathbf{C}$ is determined by the right-hand rule.

PROBLEMS

1.A.1 A man walks along a circular path of radius 2 mi for a distance of 3 mi. If he had taken a straight line path between the same two points, how far would he have walked?

1.A.2 Find an equation for the area of the segment of the circle in the figure.

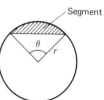

1.A.3 Surveyors use another measure for angles, called the *grad*. One revolution is 400 grads. If $\theta_{grad} = 40$, what are θ_{deg}, θ_{rad}, and θ_{rev}?

1.A.4 Suppose you are at the corner of 1st Street and 1st Avenue and want to go to 4th Street and 5th Avenue. What is the shortest distance you must walk? The blocks are square and 500 ft on a side.

1.A.5 Find the x and y components of a vector 10 units in length which makes an angle of (a) 30°, (b) 120°, (c) 300° with respect to the x axis.

1.A.6 A vector 3 units in length has an x component of 1 unit. (a) What angle does the vector make with the x axis? (b) What is the y component of the vector?

1.A.7
   (a) Find the x and y components of the sum of the three vectors in the figure.
   (b) Find the magnitude and direction of the sum.

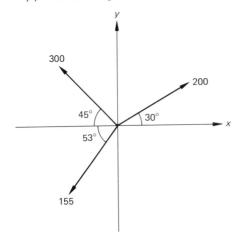

1.A.8 A man walks 3 mi due east, 4 mi 30° north of east, and 5 mi due north. How far and in what direction must he walk to return to his starting point by the most direct route?

1.B.1 Using the polygon method of vector addition and the laws of geometry, show that

$$a(\mathbf{A} + \mathbf{B}) = a\mathbf{A} + a\mathbf{B}$$

where **A** and **B** are arbitrary vectors and a is a scalar.

1.B.2 From the polygon rule for vector addition, prove that

$$(\mathbf{A} + \mathbf{B}) + \mathbf{C} = \mathbf{A} + (\mathbf{B} + \mathbf{C})$$

On the left-hand side of this equation, **A** and **B** are summed first and then **C** is added. On the right-hand side **B** and **C** are summed and then **A** is added.

1.B.3 Let A and B be the endpoints of a straight line. Let P be the midpoint of the line AB. Let O be an arbitrary point. Show that if **a** is a vector from O to A, **b** a vector from O to B, and **p** a vector from O to P, then

$$\mathbf{p} = \tfrac{1}{2}(\mathbf{a} + \mathbf{b})$$

1.B.4 The magnitude of **A** is 2 m and that of **B** 4 m. If the sum is a vector 5 m in length, what is the angle between **A** and **B**?

1.B.5 A man walks 5 m directly up a 37° incline, then angles off at 30° with respect to this path, and walks an additional distance of 8 m to reach the top of the incline. What is the height of the incline?

1.B.6 Show that the area of a triangle is $\tfrac{1}{2}$ mag (**A** × **B**), where **A** and **B** are any two sides of the triangle.

1.B.7 Using the results of Problem 1.B.6, derive the law of sines for a triangle; i.e., show that

$$\frac{A}{\sin a} = \frac{B}{\sin b} = \frac{C}{\sin c}$$

where a, b, and c are the angles opposite sides A, B, and C, respectively.

1.B.8 Show that $\mathbf{A} \cdot (\mathbf{B} + \mathbf{C}) = \mathbf{A} \cdot \mathbf{B} + \mathbf{A} \cdot \mathbf{C}$.

1.B.9 Show that $\mathbf{A} \times (\mathbf{B} + \mathbf{C}) = \mathbf{A} \times \mathbf{B} + \mathbf{A} \times \mathbf{C}$.

1.B.10 Given the sines and cosines of 30° and 45°, find the sine and cosine of 15°.

1.B.11 Using the two unit vectors $\hat{\mathbf{A}}$ and $\hat{\mathbf{B}}$ in the figure show that
(a) $\cos 2\theta = \cos^2 \theta - \sin^2 \theta$
(b) $\sin 2\theta = 2 \sin \theta \cos \theta$

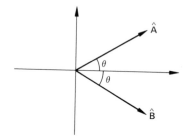

1.C.1 Show that the diagonals of a parallelogram bisect each other. The results of problem 1.B.3 and the following figure may be helpful.

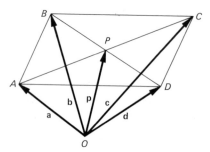

1.C.2 Show that the medians of a triangle intersect at a point that divides the medians in the ratio 1:2. (*Hint:* See problem 1.C.1.)

2
KINEMATICS

KINEMATICS IS THE study of motion without regard to the forces that cause the motion. Suppose we are told that a body slides down an inclined plane with a given acceleration and we want to know how long it takes to reach the bottom. This is a problem in kinematics. On the other hand, we might be given the angle of the incline, the weight of the sliding body, and the coefficient of friction, and be asked what happens. This would be a problem in *dynamics*. We would have to examine the forces acting on the body and determine their effect on the motion. Dynamics is in general the study of forces and their effects.

2.1 VELOCITY

Let us begin with the simplest problem in kinematics — the motion of an object along a straight line. Let x denote the distance between the object and some fixed point on the line (which we shall refer to as the origin). Let t represent the time.

We can conveniently represent the motion of the object by plotting x as a function of t on a rectangular coordinate system, as in Fig. 2.1. This has the virtue of giving us a visual representation of the entire history of the motion. Such visual or geometrical representations have many advantages.

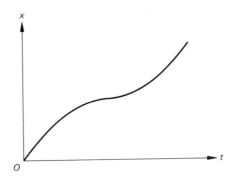

FIGURE 2.1 Space-time plot of particle motion.

QUESTION

A monk climbs to the top of a mountain along a path. He begins at the bottom at 6 A.M. and arrives at the top at 6 P.M. He does not climb at a constant rate. He spends the night at the peak in meditation and begins his descent the next morning at 6 A.M., arriving at the bottom at 6 P.M. He descends on the same path he used for the ascent. Prove that there exists at least one point on the path that the monk occupies at the same time of day on the ascent and descent.

ANSWER

We may solve this problem very simply using the geometrical properties of the lines representing the ascent and descent in a space-time diagram (Fig. 2.2). Although the monk does not move on a straight line, we may let x represent the distance along the path. It is clear from Fig. 2.2 that the two lines representing the ascent and descent must cross at least once (or any odd number of times). Remember: No history is possible that represents motion backward in time.

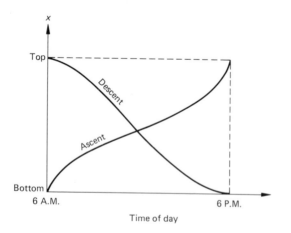

FIGURE 2.2 Space-time histories of the ascent and descent of the monk on the mountain.

In the above example we have illustrated how we may employ the geometrical properties of the space-time plot. The most useful geometrical property of the xt plot is that the slope at any point represent the *instantaneous velocity*. The term "instantaneous velocity" might at first seem self-contradictory. We know, for instance, that we cannot take a snapshot of an object and determine its velocity. The picture can tell us only where the object was at the time the picture was taken. If we take a second picture at a later time, we will have a measure of the rate at which the position of the object is changing in time, and this is what velocity is. However, we cannot associate this velocity with either the first picture or the second one, but with some kind of average between the two. This procedure does not therefore define an instantaneous velocity.

To see how we may define instantaneous velocity, let us consider a similar problem. Suppose we want to define the *slopes* of the two hills represented in Fig. 2.3. The slope is a measure of steepness. For Fig. 2.3(a) we may take as a quantitative measure of steepness the ratio $y/x = \tan \theta$. For Fig. 2.3(b) it is a little more difficult because the steepness varies from point to point. We may define the average steepness between any two points 1 and 2 as the ratio $\Delta y/\Delta x$. The closer point 2 is to point 1, i.e., the smaller Δx and Δy are, the closer we get to a measure of the steepness at point 1. The limit of $\Delta y/\Delta x$ as Δx (and therefore Δy) becomes smaller and smaller is defined as the *slope* of the curve at point 1. We might write

$$\text{steepness} = \underset{\Delta x \to 0}{\text{limit}} \frac{\Delta y}{\Delta x} = \text{slope}$$

From the definition of a derivative,

$$\text{steepness} = \underset{\Delta x \to 0}{\text{limit}} \frac{\Delta y}{\Delta x} = \frac{dy}{dx} \tag{2.1}$$

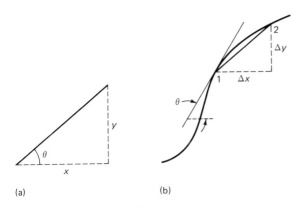

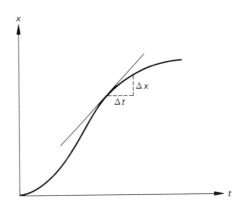

FIGURE 2.3 The slope is defined as the tangent.

FIGURE 2.4 The velocity is the slope of the tangent to the xt curve.

We may also see from the figure that

$$\text{slope} = \tan \theta \tag{2.2}$$

where θ is the angle between the line tangent to the curve and the horizontal.

We may define the instantaneous velocity at a point in a similar way. The average velocity is given by

$$v_{av} = \frac{\Delta x}{\Delta t}$$

where Δx is the displacement during the time Δt. The instantaneous velocity is

$$v = \lim_{\Delta t \to 0} \frac{\Delta x}{\Delta t} = \frac{dx}{dt} \tag{2.3}$$

This limit is clearly the slope of the tangent to the xt curve (Fig. 2.4). The slope of the tangent represents the steepness of the xt curve and hence the rate at which the ordinate (x) changes as the abscissa (t) changes. In geometrical terms,

$$v = \text{slope of } xt \text{ curve} \tag{2.4}$$

2.2 ACCELERATION

We shall also find it useful to consider some geometrical features of a plot of velocity as a function of time. In Fig. 2.5 we have plotted v vs t for the history represented in Fig. 2.4. We would like to explore two geometrical features of this figure.

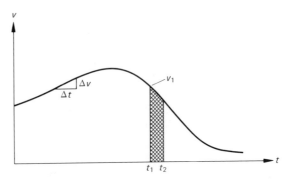

FIGURE 2.5 Plot of v vs t for the particle whose history is represented in Fig. 2.4.

Just as the slope of the x vs t curve gives the rate at which x increases with t, so the slope of the v vs t curve gives the rate at which the velocity v increases with t. The time rate of change of the velocity is defined as the acceleration a,

$$a = \text{slope of } v \text{ vs } t \text{ curve} \qquad (2.5)$$

or, as before, we also have an analytic definition of the acceleration,

$$a = \lim_{\Delta t \to 0} \frac{\Delta v}{\Delta t} = \frac{dv}{dt} \qquad (2.6)$$

Next let us consider the area under the v vs t curve between two neighboring times t_1 and t_2. Let $\Delta t = t_2 - t_1$. The area of the crosshatched region in Fig. 2.5 is approximately that of a rectangle of height v_1 and width Δt. Thus

$$\text{area} = v_1 \, \Delta t$$

But the velocity is given approximately by

$$v_1 = \frac{\Delta x}{\Delta t}$$

so that

$$\text{area} = \Delta x \qquad (2.7)$$

Thus the area under the curve between times t_1 and t_2 is the displacement Δx of the object during this time. If we consider the area under the v vs t curve between *any* two times (not necessarily close together), we can break this area down into a number of smaller areas such as the crosshatched region in Fig. 2.5. The total area is the sum of the smaller areas. In this way we see that the total area is the total displacement.

The area under the v vs t curve between any two points in time is the displacement during this time interval.

QUESTION

What are the position and velocity of an object at $t = 8$ s and $t = 16$ s if the v vs t curve is that of Fig. 2.6 and the object starts at $x = 0$?

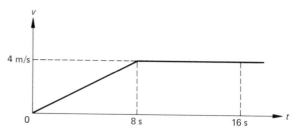

FIGURE 2.6   The v vs t curve for an object that accelerates uniformly to a velocity of 4 m/s and then continues with a constant velocity.

ANSWER

Since the displacement is the area under the curve, after 8 s the displacement is the area of the triangle. This is $\frac{1}{2}(8 \text{ m})(4 \text{ m/s}) = 16$ m. After 16 s the area is the area of the triangle (16 m) plus the area of the rectangle, which is (8 s)(4 m/s) = 32 m. The total displacement after 16 s is therefore 48 m.

QUESTION

What is the displacement after 8 s for a body whose v vs t curve is that of Fig. 2.7?

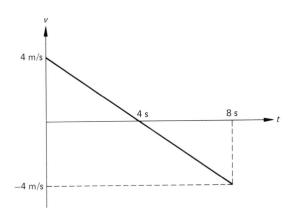

FIGURE 2.7   The v vs t curve for a body that has a constant negative acceleration.

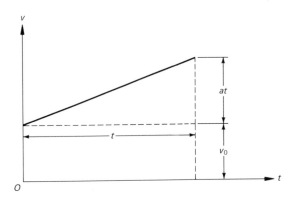

FIGURE 2.8   The v vs t curve for an object moving with constant acceleration.

ANSWER

The displacement is the sum of the areas of the two triangles, and these two areas are equal in magnitude but of opposite sign. (The height of the second triangle is -4 m/s.) The total area is therefore zero, so the net displacement is zero.

Let us apply our result to an important special case. An object moves along a straight line with constant acceleration. At $t = 0$ the object is at $x = 0$ with a velocity v_0. We would like to obtain an expression for the coordinate of the object x as a function of time. In Fig. 2.8 we illustrate the v vs t curve for an object whose acceleration is constant. If the acceleration is constant, then the slope of the v vs t curve must be constant.

The area under this straight line is the displacement x, and we may divide this area into two parts, the triangle and the rectangle. The height of the triangle is at, since the slope of the v vs t curve must be a and the base of the triangle is t. Clearly $at/t = a$. We then have

area of triangle $= \frac{1}{2}(t)(at) = \frac{1}{2}at^2$

area of rectangle $= v_0 t$

so that

$$x = \text{total area} = \tfrac{1}{2}at^2 + v_0 t \qquad (2.8)$$

This is a very useful result, but we must remember that it applies *only when the acceleration is constant.*

We also see from Fig. 2.8 that v is the sum of v_0 and at. Therefore we may write the velocity as a function of time as follows:

$$v = at + v_0 \qquad (2.9)$$

It is often useful to have an expression for v as a function of x. Eliminating t between Eqs. 2.8 and 2.9, we have

$$v^2 = 2ax + v_0{}^2 \qquad (2.10)$$

These last three equations will be used repeatedly in the following discussion. They answer any question one might pose concerning the motion of an object along a straight line with constant acceleration. *We must remember that these results apply only when the acceleration is constant.* They constitute a complete solution to this problem in kinematics.

QUESTION

The superintendent of an apartment building is informed that someone in the building is dropping water bags on passing pedestrians. She discovers that the bags pass the length of her window on the first floor in 0.11 s. Her window is 6 ft long, and the distance between floors is 10 ft. Where should she look for the culprit? (Assume that the bags fall with a constant acceleration of 32 ft/s².)

ANSWER

Let us take as the origin of the x axis the point from which the bags are dropped and choose the downward direction as positive (Fig. 2.9). Let x_1 denote the coordinate of the top of the first floor window and $x_1 + \Delta x$ the coordinate of the bottom of the first floor window. Clearly Δx is 6 ft. Let t_1 and $t_1 + \Delta t$ be the times at which a water bag occupies the positions x_1 and $x_1 + \Delta x$, respectively. Therefore $\Delta t = 0.11$ s. Since the bags are dropped and not thrown, we have in general

$$x = \tfrac{1}{2}at^2$$

where $a = 32$ ft/s². At the top of the first floor window we have

$$x_1 = \tfrac{1}{2}at_1^2$$

and at the bottom of the window,

$$x_1 + \Delta x = \tfrac{1}{2}a(t_1 + \Delta t)^2$$

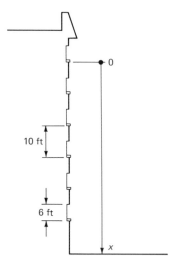

FIGURE 2.9   A bag is dropped from 0 and takes 0.11 s to pass by the first floor window.

Subtracting, we obtain

$$\Delta x = \tfrac{1}{2}a(t_1 + \Delta t)^2 - \tfrac{1}{2}at_1^2$$

$$= at_1\Delta t + \tfrac{1}{2}a\,\Delta t^2$$

Solving for t_1 gives

$$t_1 = \frac{\Delta x - \tfrac{1}{2}a\,\Delta t^2}{a\,\Delta t}$$

$$= \frac{6\text{ ft} - \tfrac{1}{2}(32\text{ ft/s}^2)(0.11\text{ s})^2}{(32\text{ ft/s})(0.11\text{ s})}$$

$$= 1.65\text{ s}$$

so that

$$x_1 = \tfrac{1}{2}at_1^2$$

$$= \tfrac{1}{2}(32\text{ ft/s}^2)(1.65\text{ s})^2$$

$$= 44\text{ ft}$$

The water bags are therefore dropped from a point 44 ft above the top of the first floor window, or 44 ft + 6 ft = 50 ft above the bottom of the first floor window. Since there are 10 ft between floors, the culprit is to be found on the sixth floor.

*2.3 KINEMATICS OF UNIFORM ACCELERATION EMPLOYING CALCULUS[1]

We have seen in Sec. 2.2 that when the acceleration is constant,

$$x = \tfrac{1}{2}at^2 + v_0 t$$

We obtained this result by calculating the area under the v vs t curve, and this was particularly easy in this case because the curve was a straight line. If the curve were not a straight line, we would be hard pressed to find the area. Worse yet, the acceleration might be given by the dynamics as a function of x and not as a function of t. How then can we determine the area under the v vs t curve?

For these more difficult cases we must employ the methods of calculus. We will not take up the general problem here, but will illustrate how we would approach the simple problem of showing that $x = \tfrac{1}{2}at^2 + v_0 t$ when a is a constant using calculus as a tool.

First let us consider the simple task of verifying that the relation

$$x = \tfrac{1}{2}at^2 + v_0 t \qquad \text{(2.11)}$$

is compatible with

$$a = \frac{dv}{dt} = \frac{d}{dt}\frac{dx}{dt} = \frac{d^2 x}{dt^2}$$

_____

[1] Starred sections are optional and may be omitted if calculus is being taken concurrently.

If a is a constant, we may differentiate Eq. 2.11 with respect to t, and we have

$$\frac{dx}{dt} = at + v_0$$

A second differentiation gives

$$\frac{d^2 x}{dt^2} = a$$

as it must.

Next let us perform the calculation in reverse. Now

$$\frac{d^2 x}{dt^2} = \frac{dv}{dt} = a$$

where a is a constant. Multiplying both sides by dt and integrating, we have

$$\int dv = \int a\, dt$$

or

$$v = at + c$$

where c is a constant of integration. To determine c, we set $t = 0$. Now dx/dt at $t = 0$ is just v_0, so

$$c = v_0$$

We have then

$$v = \frac{dx}{dt} = at + v_0$$

Multiplying both sides again by dt and integrating gives

$$\int dx = \int at\, dt + \int v_0\, dt$$

or

$$x = \tfrac{1}{2}at^2 + v_0 t + c'$$

where c' is a new constant of integration. To determine c', we again let $t = 0$. If x_0 is the value of x at $t = 0$, we have

$$x_0 = c'$$

so that

$$x = \tfrac{1}{2}at^2 + v_0 t + x_0$$

which is Eq. 2.8 except that we have generalized the result somewhat by allowing x_0 to be different from zero. Earlier we required that the clock read zero when the object is at the origin.

What we have accomplished is the integration of a differential equation in the special case in which the acceleration is a constant. The general task in kinematics is similar, that is, to integrate the equation

$$\frac{d^2 x}{dt^2} = a$$

where the acceleration a has been determined by the dynamics of the problem to be a certain function of time or position or velocity or some combination of the three.

2.4 MOTION IN TWO DIMENSIONS

In Sec. 2.3 we restricted the motion to a straight line. Let us now consider motion in two dimensions. Here we will not find it convenient to describe the motion by plotting position and velocity as functions of time. Instead we shall draw a path and in some way specify the time at which the particle occupies each point on the path. We shall define several kinematic quantities such as speed, velocity, and acceleration, as in one-dimensional motion along a straight line, and then discuss relations among these quantities.

Let us first define the speed of a particle along a curve. We shall let Δs be the distance the particle travels along the path in the time Δt. The instantaneous speed at point P (Fig. 2.10) is defined as the ratio $\Delta s / \Delta t$ (the rate at which distance is being traversed on the curve) in the limit as Δt becomes very small; hence

$$\text{speed} = \lim_{\Delta t \to 0} \frac{\Delta s}{\Delta t} = \frac{ds}{dt} \qquad (2.12)$$

The speed is a scalar quantity since we have defined only a magnitude and not a direction.

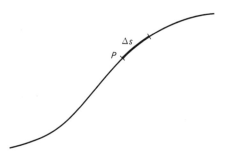

FIGURE 2.10 Distance traveled along a path.

Next we wish to define a velocity. We shall let $\mathbf{r}_1$ be a vector from some origin O to the position of the particle at time t_1 (Fig. 2.11), and shall call this the position vector at time t_1. Similarly, we shall let $\mathbf{r}_2$ be the position vector at time t_2. We then let $\Delta \mathbf{r}$ denote the vector $\mathbf{r}_2 - \mathbf{r}_1$ and let $\Delta t = t_2 - t_1$. We define the velocity as the time rate of change of the position vector; that is,

$$\mathbf{v} = \lim_{\Delta t \to 0} \frac{\Delta \mathbf{r}}{\Delta t} = \frac{d\mathbf{r}}{dt} \qquad (2.13)$$

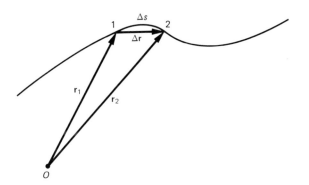

FIGURE 2.11 Position vectors $\mathbf{r}_1$ and $\mathbf{r}_2$ at times t_1 and t_2.

Forget for the moment any preconceived ideas you may have about velocity and just regard Eq. 2.13 as a formal definition. Let us investigate the nature of the thing we have defined. First we notice that Eq. 2.13 is a vector equation, and as such it tells us that the magnitude of the thing on the left must equal the magnitude of the thing on the right and the direction of the thing on the left must be the same as the direction of the thing on the right. Let us consider the magnitude of the velocity. We will use the symbol v to indicate the magnitude of the vector $\mathbf{v}$. From Eq. 2.13,

$$\text{magnitude of } \mathbf{v} = v = \lim_{\Delta t \to 0} \frac{\text{magnitude of } \Delta \mathbf{r}}{\Delta t}$$

From Fig. 2.11 it is clear that as Δt gets smaller and smaller (i.e., as point 2 approaches point 1), the magnitude of $\Delta \mathbf{r}$ approaches Δs. We have then

$$v = \lim_{\Delta t \to 0} \frac{\Delta s}{\Delta t} = \frac{ds}{dt}$$

But we see from Eq. 2.12 that this is just the speed. We have shown then that the magnitude of this newly defined quantity (**v**) is the speed, and we may write

$$\text{magnitude of } \mathbf{v} = v = \text{speed}$$

Let us now look at the other property of **v**, namely its direction. From the definition, Eq. 2.13, we have

$$\text{direction of } \mathbf{v} = \text{direction of } \Delta \mathbf{r}$$

We see from Fig. 2.11 that, as point 2 approaches point 1, $\Delta \mathbf{r}$ points more and more in the direction of the tangent line to the curve at point 1. Thus the vector **v** is parallel to the tangent to the path.

From these considerations we see that the definition of **v** is quite useful, because it gives us two important pieces of information. The magnitude of **v** gives the speed of the object, and the direction of **v** gives the direction of motion.

The nomenclature we have just employed is regretfully self-contradictory. The velocity as we have defined it is a vector quantity. The speed on the other hand is a scalar quantity; it has magnitude but not direction. Furthermore, it is necessarily positive. What we have called the velocity in Sec. 2.1 cannot be a velocity because it is not a vector, and it is not a speed because it can be positive or negative depending on whether the x coordinate is increasing or decreasing. This difficulty will be resolved later when we show that the velocity of Sec. 2.1 is the x component of the velocity vector.

We define the acceleration in a manner analogous to the way we defined the velocity. We write

$$\mathbf{a} = \lim_{\Delta t \to 0} \frac{\Delta \mathbf{v}}{\Delta t} = \frac{d\mathbf{v}}{dt} \tag{2.14}$$

where $\Delta \mathbf{v}$ is the change in the velocity *vector* during the time Δt.

Once again Eq. 2.14 is a vector equation that gives us two pieces of information, but in this case it is not as easy to see what the information is. First it is important to observe that the velocity can change in two ways: Its magnitude can change and its direction can change. Both these factors contribute to the acceleration. A particle moving along a straight line with continuously increasing speed is accelerating because the magnitude of the velocity is changing, and a particle moving along a curve with constant speed is accelerating because the direction of the velocity is changing.

Let us consider separately these two special cases: (1) motion along a straight line with changing speed, and (2) motion in a circular path with constant speed. We shall consider last (3) the general motion of a body in a curved path.

1. *Motion along a straight line with changing speed*   By definition,

$$\mathbf{a} = \lim_{\Delta t \to 0} \frac{\Delta \mathbf{v}}{\Delta t} = \frac{d\mathbf{v}}{dt}$$

Let $\Delta \mathbf{v} = \mathbf{v}_2 - \mathbf{v}_1$, where $\mathbf{v}_1$ is the velocity at time t_1 and $\mathbf{v}_2$ is the velocity at time t_2 (Fig. 2.12). Let $\Delta t = t_2 - t_1$.

FIGURE 2.12   A body moves along a straight line. Its velocity is $\mathbf{v}_1$ at point 1 and $\mathbf{v}_2$ at point 2.

The change in velocity is given in Fig. 2.13 as the difference between $\mathbf{v}_2$ and $\mathbf{v}_1$. We see that this is directed along the straight line, either in the same direction as the velocity or in the opposite direction, depending on whether v_2 is greater or less than v_1, that is, whether the body is speeding up or slowing down. The direction of $\Delta \mathbf{v}$ and hence the acceleration is always tangent to the path.

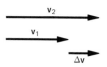

FIGURE 2.13   The change in velocity $\Delta \mathbf{v}$ is the difference $\mathbf{v}_2 - \mathbf{v}_1$.

The magnitude of the acceleration is the magnitude of $\Delta \mathbf{v}$ divided by Δt in the limit as Δt approaches zero. But

$$\text{magnitude of } \Delta \mathbf{v} = v_2 - v_1$$

that is, the difference in the speeds. Let us write for the difference in the speeds Δv, so that

$$\Delta v = v_2 - v_1$$

We then have

$$\text{magnitude of } \mathbf{a} = \lim_{\Delta t \to 0} \frac{\text{magnitude of } \Delta \mathbf{v}}{\Delta t}$$

$$= \lim_{\Delta t \to 0} \frac{\Delta v}{\Delta t} = \frac{dv}{dt}$$

In summary: In the case of a body moving along a straight line with changing speed the acceleration is tangent to the straight line and the magnitude of the acceleration is the time rate of change of the speed. For reasons that will be clear shortly, let us write

$$a_T = \frac{dv}{dt} \tag{2.15}$$

and

$$\text{component of } \mathbf{a} \text{ perpendicular to the path} = 0 \tag{2.16}$$

The symbol a_T refers to the component of the acceleration that is tangent to the path. In this special case the component that is tangent to the path is the only component.

2. *Motion in a circular path with constant speed*   This case represents the other extreme: The speed of the body is constant but the direction of the velocity is continually changing. Once again we have from the definition

$$\mathbf{a} = \lim_{\Delta t \to 0} \frac{\Delta \mathbf{v}}{\Delta t} \tag{2.17}$$

The change in the velocity vector $\Delta \mathbf{v}$ is indicated in Fig. 2.14, and we shall consider first its magnitude. Since the triangles in Fig. 2.14(b) and 2.14(c) are similar,

$$\frac{\text{magnitude of } \Delta \mathbf{v}}{v} = \frac{\Delta s}{R} \qquad (2.18)$$

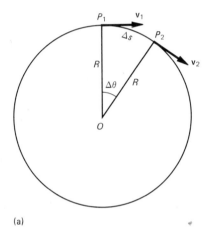

(a)

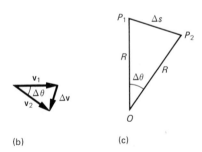

(b)   (c)

FIGURE 2.14 (a) As the body moves from P_1 to P_2 the velocity changes from $\mathbf{v}_1$ to $\mathbf{v}_2$. The length Δs is the distance traveled along the arc of the circle of radius R. (b) Both $\mathbf{v}_1$ and $\mathbf{v}_2$ have the same magnitude, and the angle between them is the same as the angle between the radii to points P_1 and P_2, namely $\Delta \theta$. (c) The triangle formed by the three points O, P_1, and P_2 is an isosceles triangle and is similar to the triangle in part (b). The arc length Δs is approximately equal to the length of the cord if $\Delta \theta$ is small.

where v is the magnitude of $\mathbf{v}_1$ or $\mathbf{v}_2$. (Remember that the magnitudes of $\mathbf{v}_1$ and $\mathbf{v}_2$ are equal.) Therefore, from Eqs. 2.17 and 2.18,

$$\text{magnitude of } \mathbf{a} = \lim_{\Delta t \to 0} \frac{\text{magnitude of } \Delta \mathbf{v}}{\Delta t}$$

$$= \lim_{\Delta t \to 0} \frac{v}{R} \frac{\Delta s}{\Delta t}$$

But

$$\lim_{\Delta t \to 0} \frac{\Delta s}{\Delta t} = v$$

so that

$$\text{magnitude of } \mathbf{a} = \frac{v^2}{R}$$

Let us next consider the direction of $\mathbf{a}$. From the definition we see that

direction of $\mathbf{a}$ = direction of $\Delta \mathbf{v}$

in the limit as Δt approaches zero. But from Fig. 2.14(b) we see that in the limit as Δt approaches zero, $\Delta \mathbf{v}$ becomes perpendicular to both $\mathbf{v}_1$ and $\mathbf{v}_2$, so that $\Delta \mathbf{v}$ is directed toward the center of the circle. We shall call this the radial or centripetal direction. We may write then that the radial (or centripetal) component of the acceleration is given by

$$a_R = \frac{v^2}{R} \qquad (2.19)$$

Since the component of the acceleration tangent to the path is zero,

$$a_T = 0 \qquad (2.20)$$

These results are to be contrasted with the acceleration of a body moving along a straight line, where we found that the component of the acceleration perpendicular to the path was zero ($a_R = 0$) and that $a_T = dv/dt$ (see Eqs. 2.15 and 2.16).

3. *General motion of a body in a curved path in two dimensions* Let us now generalize the situation in two ways. We shall consider motion along any plane curve (i.e., any curve that lies in a plane) and allow the speed of the particle to vary along the curve. We shall not derive the results, although this is not difficult, but we shall be able to verify that the expressions obtained reduce to those of cases 1 and 2 above in these special cases.

First we must define the local radius of curvature of a plane curve; to do this we consider two neighboring points P_1 and P_2 on a plane curve (see Fig. 2.15). At each point we draw a perpendicular to the curve (or better, a perpendicular to the tangent to the curve at that point), and we let C be the point of inersection of these two lines. We define

$$R = P_1 C$$

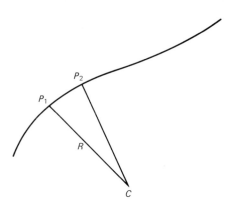

FIGURE 2.15 The lines P_1C and P_2C are perpendicular to the tangent to the curve at P_1 and P_2, respectively. The point C is the point of intersection of these two lines. The radius of curvature R is equal to the distance between P_1 and C in the limit as P_2 approaches P_1.

in the limit as P_2 approaches P_1 as the radius of curvature at P_1. $P_1 C$ is the distance between the two points P_1 and C. From this definition we see that the radius of curvature of a circle is simply the radius of the circle and that the radius of curvature of a straight line is infinite (see Fig. 2.16).

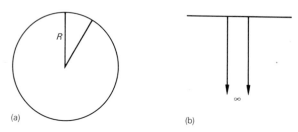

FIGURE 2.16 (a) The radius of curvature of a circle is equal to the radius of the circle. (b) The radius of curvature of a straight line is infinite.

We may now state the general result for the acceleration of a body on a plane curve. The acceleration in general is given by

$$\mathbf{a} = \frac{d\mathbf{v}}{dt}$$

The tangential and radial components of the acceleration at a point are given by

$$a_T = \frac{dv}{dt} \tag{2.21}$$

$$a_R = \frac{v^2}{R} \tag{2.22}$$

where v is the local speed and R is the local radius of curvature.

These results confirm our earlier observation that two factors contribute to the acceleration: (1) the change in the magnitude of the velocity (the speed) and (2) the change in the direction of the velocity. The change in speed is responsible for the tangential component of the acceleration, and the change in direction is responsible for the radial component of the acceleration. The net acceleration of a body, then, is the vector sum of the tangential and radial components (see Fig. 2.17).

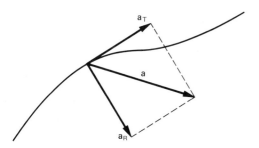

FIGURE 2.17   The net acceleration **a** is the vector sum of $\mathbf{a}_T$ and $\mathbf{a}_R$. The vector $\mathbf{a}_R$ is directed toward the center of curvature of the path; $\mathbf{a}_T$ is parallel to the tangent to the path and points in the same direction as or the opposite direction of the velocity vector, depending on whether the body is speeding up or slowing down.

We may easily verify the results for cases 1 and 2 above as special cases of the general results of Eqs. 2.21 and 2.22. (1) If the curve is a straight line, $R \rightarrow \infty$, so

$$a_T = \frac{dv}{dt}$$

$$a_R = 0$$

These are Eqs. 2.15 and 2.16. (2) If the speed is constant and the body is moving in a circular path of radius R,

$$a_T = 0$$

$$a_R = \frac{v^2}{R}$$

These are Eqs. 2.20 and 2.19.

QUESTION

If the maximum acceleration that a pilot can sustain in pulling out of a vertical dive is $10g$ (10 times the acceleration of gravity), at what height must the pilot begin to pull out if the speed of the aircraft is 400 mi/h (294 ft/s)? Assume that the path of the airplane is a semicircle during pullout.

ANSWER

For radius R and acceleration $10g$,

$$10g = \frac{v^2}{R}$$

or

$$R = \frac{v^2}{10g} = \frac{(294 \text{ ft/s})^2}{10(32 \text{ ft/s}^2)} = 270 \text{ ft}$$

The pilot must begin to pull out before the airplane drops to an altitude of 270 ft.

2.5 TWO-DIMENSIONAL MOTION IN RECTANGULAR COORDINATES AND PROJECTILE MOTION

We observe experimentally that all bodies in free fall near the earth's surface have the same acceleration (if we neglect air friction), and that the acceleration is directed toward the earth. Let us consider the following problem in kinematics: Given the fact that a projectile moves with a constant downward acceleration and given its initial state (its initial position and velocity), we wish to find its path and the velocity as a function of time along the path.

The discussion of Sec. 2.4 is not very helpful to us in solving this problem. In the first place we do not know the path, so we do not know which is the tangential and which is the radial direction; nor do we know the speed or the radius of curvature. Although the resolution into tangential and radial components was useful in helping us understand the significance of the definition of the acceleration, it is generally useful as a calculational tool only when the path is known (e.g., an automobile rounding a curve).

You might ask why we need any new tools to solve this problem. The definitions of velocity and acceleration are much the same as those for rectilinear motion. Why do we not just make rt and vt plots as before, draw tangents, and find the areas under the curves? For example, we were able to show that we can obtain the total displacement during a time interval t_1 to t_2 by finding the total area under the vt curve between t_1 and t_2. The task is not so simple for a particle moving in two dimensions. We cannot add the displacements $\Delta \mathbf{r} \simeq \mathbf{v} \, \Delta t$ as before. Since $\Delta \mathbf{r}$ is a vector, to obtain the total displacement (the sum of all the little $\Delta \mathbf{r}$'s), we must add the $\Delta \mathbf{r}$'s vectorially. This is difficult. We shall now show how this difficult problem in vector addition can be reduced (in two dimensions) to two simpler problems in scalar addition. We do this by resolving the displacement, velocity, and acceleration into rectangular components. That is,

$$\mathbf{v} = \frac{d\mathbf{r}}{dt}$$

can be resolved into two components,

$$v_x = \frac{dx}{dt} \tag{2.23}$$

$$v_y = \frac{dy}{dt} \tag{2.24}$$

Similarly, we have

$$a_x = \frac{dv_x}{dt} \tag{2.25}$$

$$a_y = \frac{dv_y}{dt} \tag{2.26}$$

If a_x and a_y are constants, we have two problems that are completely equivalent to the one we have already solved. If a_x is a constant, then we have already seen that the solution to Eqs. 2.23 and 2.24 (see Eqs. 2.8, 2.9, and 2.10) is given by

$$x = \tfrac{1}{2}a_x t^2 + v_{0x} t \tag{2.27}$$

$$v_x = a_x t + v_{0x} \tag{2.28}$$

$$v_x^2 = 2a_x x + v_{0x}^2 \tag{2.29}$$

Similarly, if a_y is a constant,

$$y = \tfrac{1}{2}a_y t^2 + v_{0y} t \tag{2.30}$$

$$v_y = a_y t + v_{0y} \tag{2.31}$$

$$v_y^2 = 2a_y y + v_{0y}^2 \tag{2.32}$$

QUESTION

What is the trajectory of a projectile in free fall?

ANSWER

The trajectory of a particle is the path the particle takes, and we must determine an equation for this path. For example, we might determine y as a function of x. If we choose a coordinate system in which the y axis is vertical, then $a_x = 0$ and $a_y = -g$. (We shall in the future use the symbol g to indicate the acceleration of a body in free fall. Its value is 32 ft/s^2, 980 cm/s^2, and 9.8 m/s^2 in the English, cgs, and mks systems of units respectively). The x and y coordinates as functions of time are given by Eqs. 2.27 and 2.30:

$$x = v_{0x} t$$

$$y = -\tfrac{1}{2}g t^2 + v_{0y} t$$

Eliminating t, we have

$$y = -\frac{1}{2}\frac{g}{v_{0x}^2}x^2 + \frac{v_{0y}}{v_{0x}}x \qquad (2.33)$$

This is the equation of a parabola in the xy plane (Fig. 2.18).

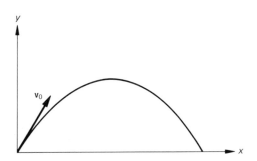

FIGURE 2.18   The path of the projectile is a parabola.

QUESTION

What is the range of a projectile?

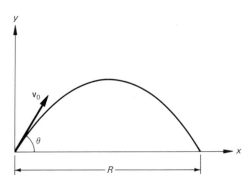

FIGURE 2.19   The range of the projectile is R.

ANSWER

The range is the horizontal displacement when the projectile returns to ground level (Fig. 2.19). Let R be the range. Clearly R is the value of x when $y = 0$. Setting $y = 0$ in Eq. 2.27, we obtain

$$0 = -\frac{1}{2}\frac{g}{v_{0x}^2}x^2 + \frac{v_{0y}}{v_{0x}}x$$

One solution is $x = 0$; this corresponds to the starting position. The other solution is

$$R = \frac{2v_{0x}v_{0y}}{g} \qquad (2.34)$$

where we have set $x = R$.

QUESTION

For a given initial speed v_0, what is the maximum range, and at what angle should the projectile be launched to achieve the maximum range?

ANSWER

Since $v_{0x} = v_0 \cos\theta$ and $v_{0y} = v_0 \sin\theta$, we may write Eq. 2.34 in the form

$$R = \frac{2v_0 \cos\theta\, v_0 \sin\theta}{g}$$

$$= \frac{v_0^2 \sin 2\theta}{g} \qquad (2.35)$$

since $\sin 2\theta = 2\sin\theta\cos\theta$. For a given v_0, R is a maximum when $\sin 2\theta$ is a maximum. The maximum value of the sine function is 1. Thus the maximum range is

$$R_{max} = \frac{v_0^2}{g} \qquad (2.36)$$

Note that the value of θ for which

$$\sin 2\theta = 1$$

is $\theta = 45°$. To achieve the maximum range, therefore, the projectile should be launched at 45°.

2.6 KINEMATICS OF ROTATION IN A PLANE

So far we have considered bodies whose positions could be defined by an x coordinate or by x and y coordinates. Let us now consider a rigid body rotating about a fixed axis, such as the disk in Fig. 2.20 rotating about an axis through O. We should like to find some convenient way of describing the position of this body.

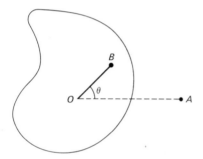

FIGURE 2.20 Cross section of a rigid body rotating about a fixed axis through O.

One method might be to paint a small dot on the body (point B) and to give the coordinates of this dot. A more natural way is to give the angle θ between the line joining the axis of rotation O and the point B and some reference line (OA).

In order to quantitatively describe the rotational motion of the body, we may define an angular velocity ω as the time rate of change of the angle θ:

$$\omega = \frac{d\theta}{dt} \tag{2.37}$$

In a similar way we define an angular acceleration α:

$$\alpha = \frac{d\omega}{dt} \tag{2.38}$$

Comparing Eqs. 2.37 and 2.38 with the equations $v = dx/dt$ and $a = dv/dt$ for motion in a straight line, we see that both sets have the same mathematical form. Only the symbols are different. The results of our previous discussion of motion along a straight line can be transferred by a simple change in notation. The results are tabulated in Table 2.1.

TABLE 2.1 Kinematics of linear and circular motion.

Linear motion	*Circular motion*
x	θ
v	ω
a	α
$v = \dfrac{dx}{dt}$ = slope of xt curve	$\omega = \dfrac{d\theta}{dt}$ = slope of θt curve
$a = \dfrac{dv}{dt}$ = slope of vt curve	$\alpha = \dfrac{d\omega}{dt}$ = slope of ωt curve
$x_2 - x_1$ = area under xt curve between t_1 and t_2	$\theta_2 - \theta_1$ = area under ωt curve between t_1 and t_2
When a = constant: $x = \frac{1}{2}at^2 + v_0 t$ $v = at + v_0$ $v^2 = 2ax + v_0^2$	When α = constant: $\theta = \frac{1}{2}\alpha t^2 + \omega_0 t$ $\omega = \alpha t + \omega_0$ $\omega^2 = 2\alpha\theta + \omega_0^2$

The path of every particle in a rigid body rotating about a fixed axis is a circle (Fig. 2.21). Since θ, ω, and α completely describe the kinematics of rotation, if we know these quantities we should be able to answer any question about the motion of any point in the rotating body. Suppose we

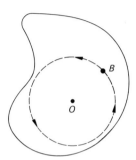

FIGURE 2.21 Each particle in a rigid body moves in a circle.

are given θ, ω, and α, and would like to know the distance traveled (s), the instantaneous speed (v), and the two components of acceleration (a_T and a_R) of some point a distance r from the axis of rotation (Fig. 2.22).

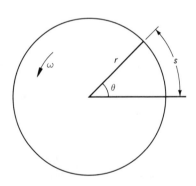

FIGURE 2.22 Motion of a point in a circle.

From the definition of θ in radian measure,

$$\theta = \frac{s}{r} \tag{2.39}$$

or

$$s = r\theta \tag{2.40}$$

Also, by definition the speed is

$$v = \frac{ds}{dt} = r\frac{d\theta}{dt} = r\omega \tag{2.41}$$

since $ds = r\,d\theta$ and $d\theta/dt = \omega$. Finally the tangential and radial accelerations are given by

$$a_T = \frac{dv}{dt} = r\frac{d\omega}{dt} = r\alpha \tag{2.42}$$

and

$$a_R = \frac{v^2}{r} = r\omega^2 \tag{2.43}$$

In sum, given the angular displacement, velocity, and acceleration (θ, ω, and α), the linear displacement, velocity, and acceleration are given by

$$s = r\theta \tag{2.44}$$

$$v = r\omega \tag{2.45}$$

$$a_T = r\alpha \tag{2.46}$$

$$a_R = r\omega^2 \tag{2.47}$$

These last four equations are valid quite generally and do not require that α be constant.

QUESTION

An ultracentrifuge is designed to run at an angular velocity of 10,000 r/min (revolutions per minute). When turned on, it accelerates uniformly for 1 min, at which time it has attained its operating speed. Suppose the distance from the sample to the axis of rotation is 10 cm. Find

(a) the angular acceleration

(b) the number of revolutions before the final angular velocity is reached

(c) the final velocity of the sample

(d) the tangential acceleration of the sample

(e) the final radial acceleration of the sample

ANSWER

(a) Since $\omega = 10,000$ r/min when $t = 1$ min, the angular acceleration is

$$\alpha = \frac{\omega}{t} = \frac{10,000 \text{ r/min}}{1 \text{ min}} = 10,000 \text{ r/min}^2$$

(b) We can obtain the number of revolutions by finding the angle θ when $t = 1$ min:

$$\theta = \tfrac{1}{2}\alpha t^2 = \tfrac{1}{2}(10,000 \text{ r/min}^2)(1 \text{ min})^2 = 5000 \text{ r}$$

(c) The final velocity can be obtained from

$$v = r\omega$$

but when we mix linear and angular quantities it is important that all angular quantities be measured in radian measure. For example,

$$\omega = \left(10,000 \ \frac{\text{r}}{\text{min}}\right)\left(\frac{2\pi \text{ rad}}{\text{r}}\right)\left(\frac{\text{min}}{60 \text{ s}}\right) = 1050 \ \frac{\text{rad}}{\text{s}}$$

and

$$\alpha = \left(10,000 \ \frac{\text{r}}{\text{min}^2}\right)\left(\frac{2\pi \text{ rad}}{\text{r}}\right)\left(\frac{\text{min}}{60 \text{ s}}\right)^2$$

$$= 17 \ \frac{\text{rad}}{\text{s}^2}$$

The velocity is therefore

$$v = r\omega = (10 \text{ cm})(1050 \text{ rad/s}) = 10,500 \text{ cm/s}$$

(d) The tangential acceleration is

$$a_\text{T} = r\alpha = (10 \text{ cm})(17 \text{ rad/s}^2) = 170 \text{ cm/s}^2$$

(e) The final radial acceleration is

$$a_\text{R} = r\omega^2 = (10 \text{ cm})(1050 \text{ rad/s})^2 = 11,000,000 \text{ cm/s}^2$$

or 11,000 times the acceleration of gravity (980 cm/s^2).

2.7 RELATIVE MOTION

When two or more people observe a situation, each person obtains a different view. A bicyclist's view of the rotating wheel of his or her own bicycle is that the axle is fixed and a point on the rim revolves with constant speed about the axle, provided the speed of the bicycle is constant. A pedestrian observing the moving bicycle sees the point of the wheel in contact with the ground at rest and the axle moving with constant speed. As motivation for the discussion that follows, let us pose a somewhat more difficult problem. Suppose the gear ratios, the radius of the pedal arm, and diameter of the rear wheel, and the speed of the bicycle are known. What is the velocity of the pedal at the top of the stroke as observed by the pedestrian? From the observer's point of view the motion of the bicycle parts is rather complicated and this question would be difficult to answer. From the cyclist's point of view the problem is much simpler. Is it possible for the cyclist to observe the motion of the bicycle parts and to convey this information to the pedestrian, and then for the pedestrian to use this information to calculate the motion of the bicycle parts? To put it in another way, if the bicycle were covered with a cloth and the cyclist could talk to the pedestrian, could the pedestrian determine the velocity of the pedal at the top of its stroke relative to the ground?

Problems of this class can be put in a very general framework. Suppose we have an object C and observers A and B, both of whom can view the object but with different points of view, i.e., different frames of reference. Figure 2.23 illustrates the frames of reference and the object. The position of object C relative to observer A is determined by the position vector $\mathbf{r}_{C/A}$. This symbol might be read "position of C relative to A." The positions of C relative to

B and of B relative to A are also defined in the figure. From the law of vector addition we see that

$$\mathbf{r}_{C/A} = \mathbf{r}_{C/B} + \mathbf{r}_{B/A} \tag{2.48}$$

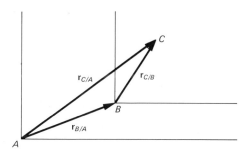

FIGURE 2.23 Two frames of reference with observers at A and B. An object C occupies a position that is determined by position vectors $\mathbf{r}_{C/A}$ and $\mathbf{r}_{C/B}$ relative to observers A and B. The position of observer B relative to observer A is given by $\mathbf{r}_{B/A}$.

(As an aid in writing these equations, note that if the subscripts are written as quotients, they form an equality under multiplication: $C/A = (C/B)(B/A)$. With this crutch we may easily generalize the results to more than two observers. For example, $\mathbf{r}_{D/A} = \mathbf{r}_{D/C} + \mathbf{r}_{C/B} + \mathbf{r}_{B/A}$.) Points A, B, and C may be in relative motion, so the position vectors are changing in time. Since Eq. 2.48 holds for all time, we may differentiate with respect to time, and we obtain

$$\frac{d\mathbf{r}_{C/A}}{dt} = \frac{d\mathbf{r}_{C/B}}{dt} + \frac{d\mathbf{r}_{B/A}}{dt}$$

or

$$\mathbf{v}_{C/A} = \mathbf{v}_{C/B} + \mathbf{v}_{B/A} \tag{2.49}$$

Differentiating again we obtain a relation among the accelerations:

$$\mathbf{a}_{C/A} = \mathbf{a}_{C/B} + \mathbf{a}_{B/A} \tag{2.50}$$

Equations 2.49 and 2.50 give us the desired relations between the motion of the object as viewed by the two observers. For example, if $\mathbf{v}_{p/g}$ is the velocity of the pedal on the bicycle relative to the ground, $\mathbf{v}_{p/c}$ the velocity of the pedal relative to the cyclist, and $\mathbf{v}_{c/g}$ the velocity of the cyclist relative to the ground, then

$$\mathbf{v}_{p/g} = \mathbf{v}_{p/c} + \mathbf{v}_{c/g}$$

(Often the fact that velocities are measured relative to the ground may be understood and the notation abbreviated. For instance, for the above equation we might write

$$\mathbf{v}_{p} = \mathbf{v}_{p/c} + \mathbf{v}_{c}$$

and omit the g subscript.)

QUESTION

It was a dark and stormy night. The rain was falling heavily, but there was no wind. Rather than battle the elements, the highway patrol officer stationed himself in the relative comfort of the turnpike tollbooth. An automobile entered the booth with rain tracts on the side window that made an angle of 76° with the vertical and the driver was given a ticket for speeding. How fast was the driver going? The average terminal velocity of a raindrop is 15 mi/h.

ANSWER

The velocity of the raindrop relative to the ground ($\mathbf{v}_{r/g}$) may be related to the velocity relative to the automobile by

$$\mathbf{v}_{r/g} = \mathbf{v}_{r/a} + \mathbf{v}_{a/g}$$

or more simply

$$\mathbf{v}_{r} = \mathbf{v}_{r/a} + \mathbf{v}_{a}$$

where it is understood that the velocity of the rain and that of the automobile are measured relative to the ground. From the law of vector addition, we may represent the three velocities graphically as shown in Fig. 2.24. From the figure we see that

$$\tan 76° = \frac{v_a}{v_r}$$

so that

$$v_a = v_r \tan 76° = (15 \text{ mi/h})(4.0) = 60 \text{ mi/h}$$

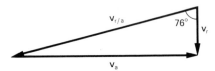

FIGURE 2.24 The rain is falling vertically with a speed of 15 mi/h and leaves a track on the window of the automobile at an angle of 76° with respect to the vertical.

QUESTION

It is raining and windless. A cyclist who has to make a one-mile journey has a rain hat but not a raincoat. The faster he pedals, the faster he gets wet. But also, the faster he pedals, the faster he reaches his destination. If he wishes to minimize the total amount of rain that strikes him broadside, should he ride quickly or slowly?

ANSWER

The cyclist cannot win. In Fig. 2.25 the line AB represents the cyclist. From his point of view the rain is coming at him at an angle. From the figure we see that every drop of rain within the parallelogram will strike the cyclist and every drop outside it will miss him. (Only the two-dimensional projection is illustrated.) Now the area of the parallelogram is the side AB multiplied by the separation between AB and CD measured perpendicularly. But this separation is just the one-mile distance to the cyclist's destination. If the cyclist rides slowly, the parallelogram is steep; if he rides quickly, the parallelogram is less steep. But whatever the pitch, the area is the same: the length AB times one mile.

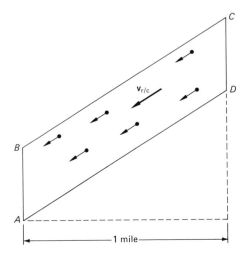

FIGURE 2.25 The cyclist is represented by the line AB. The velocity of the rain relative to the cyclist is parallel to the sides BC and AD of the parallelogram.

SUMMARY

Kinematics is the study of motion without regard to the forces that cause the motion. *Dynamics,* on the other hand, is the study of forces and their effects.

A space-time diagram gives us a graphic representation of the history of the object under study. The slope of the space-time curve at a point represents the velocity at that point.

A velocity-time diagram has two useful geometrical properties: The slope is the acceleration, and the area under the curve between two points in time is the distance traveled during that time interval.

The velocity and acceleration of a body moving along a straight line may also be defined analytically by the equations

$$v = \frac{dx}{dt}, \qquad a = \frac{dv}{dt}$$

If the object is moving in a straight line with *constant acceleration*, then

$$x = \tfrac{1}{2}at^2 + v_0 t$$

$$v = at + v_0$$

$$2ax = v^2 - v_0^2$$

In general, the velocity and acceleration are defined by the equations

$$\mathbf{v} = \frac{d\mathbf{r}}{dt}$$

$$\mathbf{a} = \frac{d\mathbf{v}}{dt}$$

If the object is moving along a curve in two dimensions, the tangential and radial components of the acceleration are related to the speed v and the radius of curvature R by the equations

$$a_T = \frac{dv}{dt}$$

$$a_R = \frac{v^2}{R}$$

It is helpful at times to analyze motion in two dimensions by resolving the position, velocity, and acceleration vectors into their rectangular components. *If a_x = constant and a_y = constant,* then

$$x = \tfrac{1}{2}a_x t^2 + v_{0x} t$$

$$v_x = a_x t + v_{0x}$$

$$2a_x x = v_x^2 - v_{0x}^2$$

with similar equations for the y components.

Angular velocity and acceleration are defined in a manner similar to their linear counterparts:

$$\omega = \frac{d\theta}{dt}$$

$$\alpha = \frac{d\omega}{dt}$$

If α = constant, then

$$\theta = \tfrac{1}{2}\alpha t^2 + \omega_0$$

$$\omega = \alpha t + \omega_0$$

$$2\alpha\theta = \omega^2 - \omega_0^2$$

For motion in a circle of radius r, the angular and linear variables are related by the equations

$$s = r\theta$$

$$v = r\omega$$

$$a_T = r\alpha$$

$$a_R = r\omega^2$$

The above equations apply whether or not the acceleration is constant.

Relative velocities and accelerations are added in the same way that relative position vectors are added. Since

$$\mathbf{r}_A = \mathbf{r}_{A/B} + \mathbf{r}_{B/C} + \mathbf{r}_C$$

we have

$$\mathbf{v}_A = \mathbf{v}_{A/B} + \mathbf{v}_{B/C} + \mathbf{v}_C$$

$$\mathbf{a}_A = \mathbf{a}_{A/B} + \mathbf{a}_{B/C} + \mathbf{a}_C$$

The velocity (acceleration) of an object A is the velocity (acceleration) of A relative to B plus the velocity (acceleration) of B relative to C plus the velocity (acceleration) of C.

PROBLEMS

2.A.1 You can measure your reaction time by attempting to catch a ruler as quickly as possible after it is released. (It must be released by another person.) If your fingers are at the bottom of the ruler at the moment of release and you catch it at the 19-cm mark, what is your reaction time?

2.A.2 Find the following for a body falling from rest:
(a) Its acceleration
(b) The distance it falls in 3 s
(c) Its speed after it has fallen 256 ft
(d) The time required for it to attain a speed of 80 ft/s
(e) The time in which it falls 900 ft

2.A.3 An elevator is ascending at a constant velocity of 10 ft/s. The cable breaks and the elevator hits the bottom of the shaft in 2 s. How high was the elevator when the cable broke?

2.A.4 The crankshaft of an automobile is rotating at an angular velocity of 2000 r/min. What gear ratio must be used to reduce the angular velocity in the drive shaft to 1200 r/min, i.e., what must the ratio of the radii of the gears be?

2.A.5 For the purpose of increasing sedimentation rate, a test tube is placed in an ultracentrifuge whose angular velocity is 30,000 r/min. If the sample is 15 cm from the axis of rotation, what is the acceleration? What multiple of g is this?

2.A.6 A wheel whose radius is 5 cm starts from rest and accelerates uniformly to an angular velocity of 30 r/min after $\frac{3}{4}$ revolutions.
(a) What is the angular acceleration?
(b) What is the angular displacement after 1 s?

(c) Find the magnitude and direction of the acceleration of a point at the bottom of the wheel after 1 s.

2.A.7 A wheel of radius R is rolling on the ground without slipping with a constant speed v.
(a) What is the angular velocity of the wheel?
(b) What is the velocity of the top of the wheel relative to an observer traveling with the wheel at the center of the wheel?
(c) What is the velocity of a point at the top of the wheel relative to the ground?
(d) What is the velocity of a point on the edge of the wheel at the same height as the center?
(e) What is the acceleration of a point at the top of the wheel?
(f) What is the acceleration of a point at the bottom of the wheel?

2.A.8 An automobile enters a 90° circular curve at a speed of 40 mi/h. Its speed increases uniformly to 60 mi/h when the automobile completes the turn. The radius of the curve is 0.25 mi. Compare the radial and tangential acceleration at the end of the turn.

2.B.1 A ball that is thrown vertically downward from the top of a building takes 5 s to reach the ground, where its velocity is 200 ft/s. What is the height of the building?

2.B.2 A juggler performs in a room with a ceiling 10 ft above his hands. If he uses only one hand and it takes him 0.5 s to catch a ball and throw it, how many balls can he keep in the air at one time?

2.B.3 A ball is thrown from a roof top as shown in the figure with an initial velocity v at an angle of 37° above the horizontal. The ball just misses the edge of the building

and lands on the ground a distance s from the base of the building. What are
  (a) the initial velocity of the ball, and
  (b) the distance s?

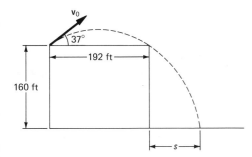

2.B.4 A baseball is hit into center field with an initial velocity of 90 mi/h (132 ft/s) at an angle of 53° with the horizontal. The center fielder, 400 ft from the ball at the moment the ball is struck, begins to run backward with constant speed. What is the minimum speed that will allow him to reach the ball before it strikes the ground?

2.B.5 Two balls are thrown into the air 2 s apart; each reaching a maximum height of 20 ft. How far above the ground are they when they pass?

2.B.6 Three gears that are connected in line have radii 4 cm, 2 cm, and 3 cm (see the figure). If the angular velocity of the first gear is 1000 r/min, what is the angular velocity of the third gear? What are the velocities of the points on the rim where the gears make contact?

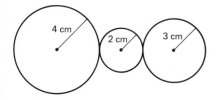

2.B.7 An airplane is flying due east at a speed of 200 mi/h, and the wind is blowing due north at 100 mi/h. In what direction is the airplane pointed?

2.B.8 A truck is traveling due east at a speed of 60 mi/h, and the wind is blowing due north at 20 mi/h. What is the direction of the trail of exhaust smoke?

2.B.9 An airplane's maximum speed in still air is 250 mi/h. If the airplane is flying due east and the wind speed is 150 mi/h due north, what is the maximum ground speed of the plane?

2.B.10 A pilot is flying due east following a highway. By observing the mile markers on the highway, she determines her ground speed to be 200 mi/h. Her air speed indicator reads 250 mi/h, and a pair of socks hung out on the wing to dry make a 45° angle south of west. What is the wind velocity?

2.C.1 There is a room in an Edgar Allan Poe story that has a ceiling that can be slowly lowered to the floor, crushing the victim inside. Suppose the initial height of the ceiling is 10 ft and the ceiling starts down with a velocity of 2 ft/h. A neurotic fly who is pacing up and down one wall begins at the bottom as the ceiling starts down. When the fly reaches the ceiling he turns around and walks back to the floor. He repeats this process until his neurosis is cured. How long does it take and how far does he walk if he maintains a constant speed of 6 ft/min?

2.C.2 A hunter aims a gun directly at a monkey sitting on the limb of a tree. The minute the monkey sees the flash of the gun, it drops from the limb. Will the bullet hit the monkey, pass above it, or pass below it? (*Hint*: Solve the problem relative to the monkey.)

2.C.3 A bullet is fired at a target 500 ft away. The muzzle velocity is 1000 ft/s. At what point above the target should the gun be aimed? (Use small-angle approximations.)

2.C.4 An automobile starts from rest and accelerates with a constant acceleration of $\frac{1}{2}g$. How long could you wait to fire a bullet whose velocity is 600 ft/s and be sure that the bullet hits the vehicle?

2.C.5 Roughly speaking, the moon is in a circular orbit about the earth. The radius of the orbit r is 3.8×10^{15} km, and the period $T_M = 27$ days. The earth is revolving about the sun in a roughly circular orbit. The radius R is 1.5×10^8 km, and the period $T_E = 365$ days.

(a) Show that the acceleration of the moon relative to the sun is given by

$$\mathbf{a}_{M/S} = -4\pi^2 \left(\frac{\mathbf{R}}{T_E^2} + \frac{\mathbf{r}}{T_M^2} \right)$$

where $\mathbf{R}$ is a radius vector from the sun to the earth and $\mathbf{r}$ is a radius vector from the earth to the moon.

(b) Show that

$$\mathbf{a}_{M/S} \cdot \hat{\mathbf{R}} < 0$$

so that the orbit of the moon is always concave toward the sun.

3
DYNAMICS

IN CHAP. 2 we discussed certain relations among kinematic quantities. We showed, for example, that if we are told that an object moves with a given constant acceleration and that its initial velocity is zero, we can find its position and velocity at any future time. But how do we know in general whether a body will move with constant acceleration, and how do we determine what the acceleration will be? In physics we are not presented with mathematical problems; we are presented with situations: A steel block weighing 2 lb rests on a 30% incline. Where will it be in 30 s? We are not given an acceleration; we are given an arrangement of objects. Somehow we must determine how these objects are going to interact and what the outcome of this interaction will be.

In this chapter we shall study the general problem of the interaction between bodies, and we shall propose three laws of motion. These laws are quite general and will not depend on what the interacting bodies are or on the nature of the interaction between them. They have, however, a limited range of application. They will not apply to bodies moving at velocities close to the velocity of light or to individual atomic particles under most circumstances. Such problems must be treated by relativistic or quantum mechanical theories.

3.1 MASS AND MOMENTUM

Often the development of the basic laws of physics has followed a rather tortured path, for nature does not give up her secrets without a struggle. Although this devious process of seeking out essential physical principles from a mass of experimental data is a fascinating study in human fortitude and ingenuity, it is not a path that we would like to follow. On the other hand, the laws of physics should never be divorced from their foundation in experiment. To simply postulate the laws and then deduce their consequences is to put the cart before the horse. We shall take

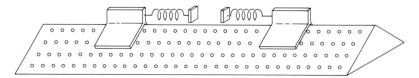

FIGURE 3.1   The cars slide virtually without friction on a layer of air forced through small holes in the track.

a middle ground by postulating a little history; that is, we shall show how the basic laws of mechanics might have been arrived at in an economical fashion through a few simple experiments.

The simplest experiment in dynamics is the two-body collision in a straight line. This is most easily realized on the "air track" illustrated in Fig. 3.1. Two cars slide with very little friction on a cushion of air. Weights may be added to or removed from the cars, and the cars can "interact" by means of bumpers of one sort or another placed on the opposing faces of the cars.

The first experiment we shall perform is to use springs for bumpers and to tie the two cars together tightly with a string, thereby compressing the springs. We now cut the string, and the two cars fly apart with velocities v_1 and v_2. We then repeat the experiment, but this time we tie the two cars together more tightly, so that the springs are compressed even more. Again we sever the string, and the cars fly apart with new velocities v_1' and v_2'. These new velocities are greater than the first ones, but we find that their ratio is the same; that is

$$\frac{v_1}{v_2} = \frac{v_1'}{v_2'}.$$

After several more trials we find that *the ratio of the velocities is always the same,* so that

$$\frac{v_1}{v_2} = \text{constant}$$

for all spring tensions. Let us call the constant k_{12}. We may then write

$$\frac{v_1}{v_2} = k_{12} \tag{3.1}$$

Not only is the ratio of the velocities a constant for a given bumper, but the ratio is a constant (the same constant) for any bumper and any string tension. The ratio of the velocities is a constant for these two cars no matter what device is used between the cars to force them apart.

The results are still rather specialized since we have considered always the same two cars. Let us remove the second car and replace it by a third car. Once again we find that for all springs, regardless of the tension applied to the binding string, the ratio

$$\frac{v_1'}{v_3'} = \text{constant} = k_{13} \tag{3.2}$$

but the constant k_{13} is not the same as k_{12}. It would seem that we must first experiment with each pair of cars to determine the ratio of velocities; then we may predict the ratio for these same two cars, regardless of the thrust that pushes them apart. However, this is not altogether true. If we use cars 2 and 3, we find, as we expect, that

$$\frac{v_2''}{v_3''} = \text{constant} = k_{23} \tag{3.3}$$

but the value of k_{23} can be predicted. It is related to the previously measured constants k_{13} and k_{12} by the equation

$$k_{23} = \frac{k_{13}}{k_{12}} \tag{3.4}$$

It is important to recognize that Eq. 3.4 does not follow from Eqs. 3.1, 3.2, and 3.3; in fact, the velocities in these three experiments are unrelated. The velocity v_1 will not be the same as v'_1, etc.

We can tie all these results together in a neat package in the following way. The constants k_{12}, k_{13}, etc. are *related to the interacting bodies but not to the nature of the inter-action.* Let us say therefore that every body has an intrinsic property called its mass m, such that the ratio of the masses of two interacting bodies, m_2/m_1, is by definition k_{12}, that is,

$$k_{12} = \frac{m_2}{m_1} \tag{3.5}$$

If we perform an experiment to measure k_{12} and find it to be 2, then

$$\frac{m_2}{m_1} = 2$$

But what is m_1 or m_2? All we know is the ratio. Here it is necessary to choose a standard mass, so we select a body of convenient size and call it one unit of mass, say 1 kg (kilogram). If one of the interacting masses is the standard ($m_1 = 1$), then $m_2 = v_1/v_2$. In this way we can determine the mass of any body.

The advantage of setting $k_{12} = m_2/m_1$ is that we may incorporate into a single law both our findings, namely, the result that the ratio of the velocities is a constant and the result that there is a relationship between these constants. If we write

$$k_{12} = \frac{m_2}{m_1}$$

$$k_{13} = \frac{m_3}{m_1}$$

$$k_{23} = \frac{m_3}{m_2}$$

then it follows that

$$k_{23} = \frac{k_{13}}{k_{12}}$$

and this is Eq. 3.4.

We may state the results of our experiments to this point as follows: When two cars at rest on an air track are forced apart by an internal device such as a spring, the ratio of the velocities of the two cars after they have separated, v_1/v_2, is a constant, and this constant is equal to the ratio of the masses of the two bodies, m_2/m_1. Equation 3.1 now becomes

$$\frac{v_1}{v_2} = \frac{m_2}{m_1}$$

or

$$m_1 v_1 = m_2 v_2 \tag{3.6}$$

The mass of a body, defined by Eq. 3.5, is an intrinsic or invariant property of the body, and has all the properties we commonly associate with mass. A massive body has a great deal of inertia, i.e., resistance to change in velocity. If it is at rest, it tends to stay at rest. If it is moving, it tends to stay in motion. To see this, suppose m_1 is the unit standard and m_2 is much greater than 1. Then

$$\frac{v_1}{v_2} = \frac{m_2}{m_1} \gg 1$$

or

$$v_2 \ll v_1$$

so that the massive body has a much smaller velocity than the less massive body. The massive body was at rest before the string was cut and wants to stay at rest, since a body with a large mass has a large inertia.

Often the mass defined by Eq. 3.5 is called the *inertial mass*, for it is a measure of the body's resistance to acceleration. We shall see later that every body also has a *gravitational mass*, which is a measure of the gravitational properties of the body. These two properties, resistance to acceleration and tendency to exert gravitational force, are proportional. If we double *m*, we double the inertia and we double the gravitational force. We shall discuss this property further later in the chapter. For the moment it is important for us to recognize the distinction between inertial mass and gravitational mass and that the mass we have been speaking of is the inertial mass, not the gravitational mass. It is also not to be identified with the weight.

QUESTION

Two cars on an air track are tied together so that a spring between them is compressed. One of the cars has a mass of 1 kg. When the string is cut, the 1-kg car moves away with a speed of 12 m/s, and the other car acquires a speed of 6 m/s. What is the mass of the second car?

ANSWER

From Eq. 3.6,

$$m_1 v_1 = m_2 v_2$$

or

$$m_2 = \frac{m_1 v_1}{v_2} = \frac{(1 \text{ kg})(12 \text{ m/s})}{(6 \text{ m/s})} = 2 \text{ kg}$$

Recognizing that the velocities of the two cars on the air track are in opposite directions, we may write Eq. 3.6 in vector form:

$$m_1 \mathbf{v}_1 = -m_2 \mathbf{v}_2$$

or

$$m_1 \mathbf{v}_1 + m_2 \mathbf{v}_2 = 0 \tag{3.7}$$

We may give Eq. 3.7 a somewhat different interpretation by introducing a new concept: *momentum*. The momentum of a body **p** is defined as the product of its mass and its velocity:

$$\mathbf{p} = m\mathbf{v}$$

so that Eq. 3.7 may be written

$$\mathbf{p}_1 + \mathbf{p}_2 = 0 \tag{3.8}$$

3.2 CONSERVATION OF MOMENTUM

Our results so far have a rather limited application — two cars tied together on an air track. This hardly seems worth bothering about. We would like to consider more general types of interactions between bodies.

Instead of tying the two cars together, suppose we consider a collision between the two cars on the air track. Again the two cars have bumpers of some sort. We would like to know whether there is anything we can say about the velocities (or momenta) of the two cars after impact when we know the velocities (or momenta) before impact.

Let us see what Eq. 3.8 tells us about the momenta after the string is cut, compared to the momenta before the string is cut. Before the string is cut, the momenta are zero. After the cut, the total momentum is $\mathbf{p}_1 + \mathbf{p}_2$. But from Eq. 3.8, $\mathbf{p}_1 + \mathbf{p}_2$ is zero. Therefore the total momentum is the same before and after the string is cut. To be more precise, we should say that the total momentum before the string is cut is the same as the total momentum after the cars have separated. It takes time for the bumpers to expand and the two cars to separate. But what about this separation period? We certainly can observe the momentum of each car during this period. What about the total momentum of the two cars during the separation period? If we perform the experiment and measure the momenta, we find that the total momentum is zero *throughout the separation process*. We find then in this experiment that

total momentum = 0

throughout the experiment.

This suggests how we might generalize these results to include the case in which the two cars collide on the air track. If $\mathbf{p}_1$ is the momentum of the first car at some moment and $\mathbf{p}_2$ is the momentum of the second car at the same time, then

$$\mathbf{p}_1 + \mathbf{p}_2 = \text{constant in time}$$

that is, the total momentum has the same value before, during, and after the collision. The example of the two cars tied together is a special case of the above. There the constant was zero.

We may imagine other kinds of "collisions": two billiard balls on a table, two electrically charged particles interacting in space, two planetary bodies interacting in space, etc. In all such cases of two bodies essentially isolated from their surroundings, we find that the total momentum remains constant in time. (Of course the two cars on the air track were not isolated, but the air track offers so little resistance to movement along its length that we may regard motion parallel to the track as that of two isolated bodies.)

We need not restrict ourselves to two bodies. Indeed we find that in any isolated collection of objects, the total momentum remains constant in time.

Let us summarize our results and dignify them as the first law of dynamics, called the *law of the conservation of momentum*.

The Law of the Conservation of Momentum   *In any isolated system the total momentum remains constant in time. The momentum of any body is proportional to its velocity, the proportionality factor being the mass. The mass is an intrinsic property of the body.*

We may state this result as an equation:

$$\mathbf{p}_1 + \mathbf{p}_2 + \cdots = \text{constant in time}$$

where $\mathbf{p}_1$, $\mathbf{p}_2$, . . . are the momenta of the particles in the isolated system. From the definition of the momentum we may write this equation in another form:

$$m_1\mathbf{v}_1 + m_2\mathbf{v}_2 + \cdots = \text{constant in time}$$

QUESTION

A truck traveling at a speed of 60 mi/h strikes a parked Volkswagen, and the two vehicles stick together. What is their common velocity after impact? The truck has a mass of 3000 kg and the Volkswagen has a mass of 900 kg.

ANSWER

Let m and $\mathbf{v}$ be the mass and velocity of the Volkswagen before the collision and M and $\mathbf{V}$ the mass and velocity of the truck before the collision. The velocities of the Volkswagen and truck after the collision will be denoted by $\mathbf{v}'$ and $\mathbf{V}'$, respectively. Since momentum must be conserved,

$$m\mathbf{v} + M\mathbf{V} = m\mathbf{v}' + M\mathbf{V}'$$

But the two vehicles stick together after the collision, so

$$\mathbf{v}' = \mathbf{V}'$$

And since the Volkswagen is at rest before the collision,

$$\mathbf{v} = 0$$

Therefore

$$M\mathbf{V} = (m + M)\mathbf{v}'$$

or

$$v' = \frac{MV}{m + M} = \frac{(3000 \text{ kg})(60 \text{ mi/h})}{(2000 \text{ kg} + 900 \text{ kg})} = 46 \text{ mi/h}$$

(We should have converted the mass to the English system of units, but the conversion factor would have canceled since the masses appear in both numerator and denominator.)

QUESTION

For the question above, compare the change in momentum and the change in velocity of each vehicle.

ANSWER

Since the total momentum is conserved, the increase in momentum of the Volkswagen must be the same as the decrease in momentum of the truck. The change in speed of the Volkswagen, however, is 46 mi/h and that of the truck is $60 - 46 = 14$ mi/h. Therefore a passenger in the Volkswagen would experience a much greater change in velocity (and momentum) than would a passenger in the truck.

QUESTION

A fullback charges into the line at a speed of 30 ft/s to be met by a 240-lb (109-kg) linebacker traveling at a speed of 25 ft/s in the opposite direction. Both players are brought to rest on impact. What is the mass of the fullback?

ANSWER

This is the reverse of the experiment of the two cars tied together on the air track. Instead of being at rest initially and flying apart, the two bodies run together and come to rest.

Since momentum is conserved, the initial momentum must equal the final momentum:

$$m_1v_1 + m_2v_2 = 0$$

where the subscript 1 refers to the fullback and the subscript 2 refers to the linebacker. Since the velocities are in opposite directions, we have

$$m_1 = \frac{m_2v_2}{v_1} = 109 \text{ kg} \left(\frac{25 \text{ ft/s}}{30 \text{ ft/s}}\right) = 91 \text{ kg}$$

Note that, as promised in Chap. 1, all measurements can be made with a ruler and a clock and standards of mass, length, and time. (Velocity is distance traveled divided by time.) The standard of mass here might be taken as the linebacker.

QUESTION

What is the cause of air resistance or "drag"?

ANSWER

We find that the motion of any object moving through the air is impeded by its interaction with the air. We can understand this effect by applying our momentum principle, but first we must observe that this principle applies only to an isolated system, or a system that behaves like an isolated system at least with respect to motion in one direction, such as the cars on the air track. Let us imagine that we have an object moving through the air on an air track, or perhaps an asteroid moving in space and interacting with the very low-density gas in outer space. Consider as a system the moving object *and* the gas. The combined momentum of the object and the gas must remain constant. But when the object collides with gas molecules, they acquire a component of momentum in the direction of motion of the object. For momentum to be conserved, the object must lose momentum in this direction; that is, the object must slow down.

QUESTION

Why is a canoe propelled forward with each stroke of the paddle?

ANSWER

Take as a system the canoe, the person in it, the paddle, and the water. If the system is at rest, it will remain at rest. If the water is given a momentum in one direction, the canoe must acquire an equal momentum in the opposite direction.

QUESTION

How do rockets and jet planes work?

ANSWER

A rocket works basically by propelling matter out the back end. Consider a rocket of mass M traveling in outer space with a velocity V. If some small mass m is thrown out the back with velocity v_r *relative to the rocket*, the rocket has a net velocity of $V' - v_r$, where V' is the new velocity of the rocket and the mass of the rocket is now $M - m$ (Fig. 3.2).

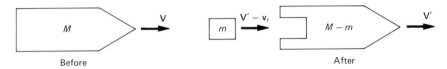

Before   After

FIGURE 3.2 A rocket ejects a mass m out the back with a relative velocity $\mathbf{v}_r$.

Before   After

FIGURE 3.3 Jet plane before intake and after exhaust.

Since the rocket is isolated, the total momentum before the mass is ejected must equal the momentum after the mass is ejected. Therefore

$$MV = (M - m)V' + m(V' - v_r)$$

or

$$M(V' - V) = mv_r$$

so that the increase in velocity of the rocket is

$$V' - V = \frac{mv_r}{M}$$

We see that the more matter that is ejected m, the greater its velocity v_r, and the lighter the rocket M, the greater will be the increase in the velocity of the rocket.

In contrast to the rocket, the jet plane takes in air, heats it, and throws it out the back (Fig. 3.3). The mass of fuel ejected is small in comparison with the mass of air ejected. Let m be the mass of air taken in during some short time. The air is at rest (more or less) before it is drawn into the jet engine. Equating the initial and final momenta gives

$$MV = MV' + m(V' - v_r)$$

The increase in velocity of the jet is then

$$V' - V = \frac{m(v_r - V')}{M}$$

This is not the same as the expression we obtained for the rocket. We see in fact that $V' - V$ may be negative (that is, the jet is slowing down) if v_r is not greater than V'. Therefore the jet must be throwing air out the back at a speed greater than the speed of the jet itself; otherwise the jet is scooping up air in front and ejecting it so that it is traveling in the same direction as the plane. This is not unlike the drag that any body experiences in moving through air: The object causes air around it to acquire a momentum at the expense of the momentum of the body. The trick is to get the air to travel in a direction *opposite* to the direction of jet. Perhaps we can understand this best by considering the limiting case in which $v_r = V'$. Here the jet picks up the air at rest and leaves it at rest ($v_r - V' = 0$), and of course the plane neither speeds up nor slows down ($V' = V$).

QUESTION

What is an upper limit on the velocity of a jet plane?

ANSWER

We have seen that the velocity of the jet plane increases only when the velocity v_r of the air ejected is greater than the velocity of the jet. When the relative velocity of the air ejected is equal to the velocity of the plane, there is no further thrust. Thus the upper limit is v_r.

QUESTION

Two persons are traveling in a small spaceship far from any astronomical body. During the "night" one person jumps ship. How could the deserter be found?

ANSWER

If the initial momentum of the spaceship were known, then the change in momentum of the ship caused by the desertion of one passenger would determine this person's momentum. From Fig. 3.4 it is clear that if we knew the new velocity of the spaceship and the new velocity of the deserter, we could calculate the deserter's position in space as a function of time.

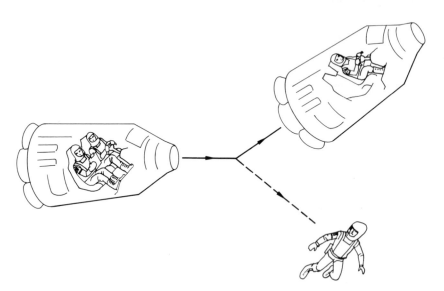

FIGURE 3.4   An astronaut abandons ship in flight.

The procedure outlined above is very useful in analyzing the properties of elementary particles (electrons, protons, neutrons, mesons, etc.). By following the tracks of charged particles in a bubble chamber (Fig. 3.5), we can infer, using our principle of the conservation of momentum, where the uncharged particles that do not leave tracks must have gone.

FIGURE 3.5 A bubble chamber photograph of the decay of a Ξ^- particle. (University of California, Lawrence Berkeley Laboratory)

QUESTION

Suppose, for the purpose of discussion, that we could ignore the other planets in our solar system and consider just the earth revolving about the sun. This is an isolated system, and momentum should be conserved. But this certainly does not appear to be the case. The earth's momentum is changing constantly; it is in fact reversing itself in one-half year. For that matter, what about a falling stone? For all practical purposes we can consider the stone and the earth as an isolated system, yet the stone's momentum increases continuously, and so again momentum does not appear to be conserved. How do we resolve these paradoxes?

ANSWER

This may be difficult to accept, but when a stone is released and moves toward the earth, the earth also moves toward the stone. The earth pulls on the stone, but the stone pulls back. The earth would not have to move very fast to balance the momentum of the stone. If the mass of the stone were 1 kg and its velocity 1 m/s, then the velocity of the earth, which must make the total momentum zero, would be given by

$$M_{earth}V_{earth} = M_{stone}V_{stone}$$

or

$$V_{earth} = \frac{(1 \text{ kg})(1 \text{ m/s})}{6 \times 10^{24} \text{ kg}} = 1.7 \times 10^{-25} \text{ m/s}$$

At this speed, it would take the earth 2×10^{17} yr to move 1 m.

In a similar way, we must include the motion of the sun when considering the total momentum of the earth-sun system. If the earth and sun were the only two bodies in the solar system, they would both revolve about a point between the sun and earth (a point very close to the center of the sun). The total momentum relative to this fixed point would always be zero.

3.3 PHYSICAL SIGNIFICANCE OF MOMENTUM

Momentum is no doubt something that everyone has heard of before. "The momentum has turned in favor of the Petaluma Pets. They have marched from their own twenty-yard line down to the opponents five." The announcer, of course, is not trying to associate momentum with mass times velocity, but is saying that the Pets are hard to stop. Momentum is a measure of how difficult it is to bring something to rest.

FIGURE 3.6   A pellet gun used to measure momentum.

To illustrate this, let us review the process by which we measure something. We measure length by counting how many times a standard of length fits into the distance to be measured. We might measure the momentum of a body in a similar way. We could construct a device that would fire 1-g pellets with a velocity of 1 cm/s and count how many pellets a body must absorb before it is brought to rest (Fig. 3.6). Since momentum is conserved and the final momentum is zero when the body (and the pellets that it has absorbed) has come to rest, the total initial momentum must have been zero; that is, the momentum of the body is equal to and in the opposite direction of the sum of the momenta of the pellets. For example, if it takes 10 pellets to bring a body to rest, then the momentum of the body must have been 10 g·cm/s.

It seems reasonable that the product of mass and velocity is a good measure of a body's momentum. The larger the mass, the larger the momentum. The larger the velocity, the larger the momentum, that is, the harder it is to stop the body.

We do not want to press our physical intuition too far, however. It is *reasonable* that the product of mass and velocity is a measure of the resistance of a body to being stopped, but why isn't mv^2 or $mv + 3m^2v^2$, etc., a measure of momentum? In fact, for a body whose velocity is not negligible compared with the velocity of light, the momentum is given by

$$\mathbf{p} = \frac{m\mathbf{v}}{\sqrt{1 - v^2/c^2}} = m\mathbf{v}(1 + \tfrac{1}{2}v^2/c^2 + \cdots) \qquad (3.9)$$

where c is the velocity of light. If $v \ll c$, then $\mathbf{p} \simeq m\mathbf{v}$.

Of course, there is no physical law in a definition, and we could define momentum as we please. The physical law is that if we define momentum to be $m\mathbf{v}$, then momentum is conserved for an isolated system no matter how the particles interact. Why should this be? Why should anything be conserved? The answer is not clear; no one knows.

The foundations of some conservation principles are clear-cut. For example, the number of clubs in an honest card game is conserved; clubs are neither created nor destroyed. There are units of clubs (ace, two, three, etc.), and these units are indivisible. Mass is conserved in a closed system, for matter is made up of atomic particles that generally cannot be created or destroyed. (We shall find exceptions to this rule when we consider relativistic mechanics.) In both these cases there are discrete elements in the system that can be counted, and the count cannot change. There are, however, no discrete elements of momentum to be counted, and we do not really understand why momentum should be conserved. We must simply accept the fact that it is as empirical. We hope that in time we will gain a more fundamental understanding of this principle and will be able to *explain* or *derive* it. (In fact, it can be shown that the conservation of momentum is related to the uniformity of space, and this is a step toward a more fundamental understanding of this principle.) This will then become the new physical law, and the conservation of momentum will be a derived consequence. We will then no doubt embark on the task of trying to understand or explain the origin of this new law. It is clear that there is no end to this search for the "fundamental" laws of nature.

3.4 INERTIAL FRAMES

We have observed that the momentum of any collection of isolated bodies is constant in time. Let us imagine two observers, a man and a woman, out in space far from any astronomical body. The man holds a billiard ball at arm's length, releases it, and observes that it remains at rest. If he gives it a shove, it moves with constant speed in a straight line. This is a very simple example of the law of the conservation of momentum.

Let us imagine the woman in a state of constant acceleration relative to the man. Her view of the motion of the billiard ball is quite different, and for her the momentum of the billiard ball is not constant in time. Furthermore, she finds that if she releases a baseball, it begins to accelerate. If the baseball is at rest, it does not remain at rest, and if it is moving, it does not move with constant velocity. She does not observe that momentum is conserved either for the man's billiard ball or her baseball.

It appears that the conservation of momentum is not valid for all observers. It is tempting to say that the woman was accelerating and therefore that momentum was not conserved. But how do we decide who was accelerating? All we said was that the woman was accelerating *relative* to the man. If we consider only the geometrical properties of space, it is impossible to say who is accelerating and who is not. We can only make a comparative statement about the woman and the man. If, for example, the man and the woman maintained a fixed relationship and the man and the woman were the *only* existing objects in the universe, it would be meaningless to speak of the motion of either the man or the woman. There would be no translation and no rotation.

This seems to cast a shadow on our first principle of dynamics. Momentum is conserved for some observers but not for others.[1] Moreover, there is no way to tell in

advance for which observers momentum is conserved. However, it is possible to live with these facts! We may still formulate a first law of dynamics:

For any isolated system there exists *a frame of reference in which the total momentum remains constant in time. Such a frame is called an* inertial frame.

Does the mere existence of an inertial frame constitute a physical law, or is this just a philosophical curiosity? Let us put this law to the test: Can it be used to make predictions about untried experiments? Clearly it can. To use the law, we could in principle perform some simple collision experiment of an isolated system in various frames of reference. When we find the frame in which momentum is conserved (we know that one exists from the statement of the first law), we may then assume that momentum will be conserved in any future experiment, and so we can make predictions in this frame of reference. (We shall see later how to make predictions in other noninertial frames.)

Not only does there exist *a* frame of reference in which momentum is conserved, there exist many such frames. It is a simple matter to prove that if momentum is conserved in one frame, then it is conserved in any frame traveling with a constant velocity relative to this frame. This will be left as an exercise.

It is generally assumed that a frame of reference fixed or traveling with a constant velocity with respect to the fixed stars is a frame in which the first law is satisfied.

3.5 FORCE AND THE SECOND LAW OF DYNAMICS

QUESTION

A 1-kg car is moving with a velocity of 1 m/s toward a 2-kg car at rest on an air track. What are the velocities of the two cars after the collision?

---

[1] In his quest for a more fundamental understanding of the concept of inertia, Sir Arthur Eddington has facetiously observed that "every body continues in its state of rest or uniform motion in a straight line, except insofar as it doesn't" [*The Nature of the Physical World* (University of Michigan Press, Ann Arbor, 1958)].

ANSWER

Since momentum is conserved,

$$1 \text{ kg} \cdot \text{m/s} = (1 \text{ kg})v_1' + (2 \text{ kg})v_2'$$

where the initial momentum is $1 \text{ kg} \cdot \text{m/s}$. The velocities v_1' and v_2' are the velocities of the 1- and 2-kg masses after the collision. We have just one equation and two unknowns, v_1' and v_2', so we cannot solve for either one. Momentum conservation is clearly not enough information to enable us to solve this problem. We need to know more, and we need a new dynamical law. We need to know something about the nature of the bumpers between the cars, and we need to know how these compressed bumpers affect the motion of the cars.

From the preceding question we see that there must be more to dynamics than the conservation of momentum if we are to be able to deal with such problems. As a second example, consider two bodies of equal mass moving past each other in opposite directions with equal velocities but not interacting in any way (Fig. 3.7(a)). The two bodies will pass by each other with constant velocity in a straight line (if we neglect gravity). Now consider these same two bodies with an electric charge of the same sign on each. The two bodies repel each other, and they will be deflected by this repulsion (Fig. 3.7(b)). The path is no longer a straight line. However, in both cases momentum is conserved. The initial momentum is zero and remains zero at all times. There are therefore many different paths that the particles can take and still conserve momentum. How do we determine what paths the particles will take?

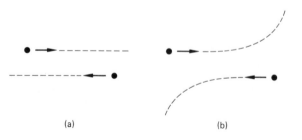

(a)   (b)

FIGURE 3.7 (a) Trajectories of two uncharged bodies, and (b) trajectories of two bodies carrying electric charges of the same sign.

When two particles are interacting, each particle is causing the momentum of the other to change. In the case of the two electric charges in Fig. 3.7(b), the momentum of each charge is continuously changing. *It is reasonable to expect that the change in momentum experienced by either particle depends on the nature of the system at that time.* This seemingly innocuous statement is in fact the second law of dynamics:

The Second Law of Dynamics   *The rate at which the momentum of each particle changes in time depends on the state of the system at that time.*

We might state this in mathematical form:

$$\frac{d\mathbf{p}}{dt} = \frac{\text{some function of the}}{\text{state of the system}} \qquad (3.10)$$

By the "state of the system" we mean those properties necessary to characterize the system, such as the positions of the particles, the velocities of the particles, the masses and electric charges, etc. If we know where all the particles are, what their velocities are, and the physical nature of the particles, in principle we can predict the time rate of change of momentum for each particle. (Of course, it is not necessary to specify the time rate of change of the momentum, or equivalently the product of the mass and acceleration of the particles, for then the law would be empty.)

Before we illustrate this law, let us express it in a somewhat different way. It is customary to speak of the thing that one body does to another body to change its momentum as the *force* it exerts. We say that electric charges exert forces on other electric charges. The earth is exerting a gravitational force on this book. The sun is exerting a gravitational force on the earth. If the force on a body is the thing that is being done to it to make it change its momentum, then certainly a very natural definition of force **F** is that it is equal to the time rate of change of the momentum:

$$\mathbf{F} = \frac{d\mathbf{p}}{dt} \qquad (3.11)$$

or since $\mathbf{p} = m\mathbf{v}$ and $\mathbf{a} = d\mathbf{v}/dt$,

$$\mathbf{F} = m\mathbf{a}$$

It is important to note that *this is a definition of force* and not a physical law. We determine the force acting on a body by observing the time rate of change of its momentum.

We can now state the second law of dynamics in terms of force. From Eqs. 3.10 and 3.11 it follows that

$$\mathbf{F} = \text{some function of the state of the system}$$

This must seem like a strange physical law. The difficulty with the second law is that it is so simple. We can illustrate the nature of this law by considering some specific examples.

Example 1

Suppose we have two interacting electrically charged bodies. For convenience, we hold one fixed and observe the behavior of the second as it moves by. The essence of the second law is that *if we observe the time rate of change of the momentum, we shall find that it is in some way related to the instantaneous state of the system*. It does not depend on the past history. In this example of the two electric charges, we expect to find that the change in momentum of the free charge depends on the quantity of charge on each body, on the distance between the two bodies, perhaps on the masses, perhaps on the velocity — on anything that is associated with the instantaneous state of the system. By starting the free charge at various positions with various velocities and applying various charges to the two bodies, we find that the time rate of change of the momentum at any instant of time is related to the state in the following way:

$$\frac{d\mathbf{p}}{dt} = k\frac{q_1 q_2}{r^2}\hat{\mathbf{r}}$$

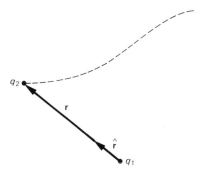

FIGURE 3.8 Two interacting electric charges.

where k is a constant that depends on the medium between the two charges, q_1 and q_2 are the magnitudes of the electric charges, r is the distance between the charges, and $\hat{\mathbf{r}}$ is a vector of unit magnitude directed from the fixed charge to the moving charge. We also find that the change in momentum is along the line of the two charges. From the definition of force it follows that

$$\mathbf{F} = k\frac{q_1 q_2}{r^2}\hat{\mathbf{r}} \qquad (3.12)$$

By freeing the fixed charge and observing the changes in momentum of the two bodies, we find that the force is still given by Eq. 3.12, which is called *Coulomb's law*. (It might have depended on the motion of the other charge. In fact, if the velocities are large, it does.) Coulomb's law is just *one example of the second law of dynamics*. We see in this case how the force depends on the state, the state being determined by q_1, q_2, and r.

Example 2

We find a similar expression for the gravitational force between two bodies of mass m_1 and m_2 at a distance r. The force on either body is attractive and is given by

$$\mathbf{F} = -G\frac{m_1 m_2}{r^2}\hat{\mathbf{r}} \qquad (3.13)$$

where G is the universal gravitational constant and is equal to 6.67×10^{-11} N $\cdot$ m²/kg². The minus sign indicates that the force is attractive. Once again we obtain this expression for the force by observing the time rate of change of the momentum and determining how it depends on the state.

We should notice in Eq. 3.13 that the gravitational force between two bodies depends on the product of their inertial masses. Why should the gravitational force be proportional to the inertial mass? What does gravitational force have to do with inertia? This enigma has not been resolved and is the object of current research.

Often the mass in Eq. 3.13 is referred to as the gravitational mass. As far as can be determined from very careful experiments, these two masses, the inertial mass and the gravitational mass, are equal. (It would be more proper to speak of these masses as proportional, the proportionality constant being absorbed in the universal gravitational constant G.)

A special case of Eq. 3.13 is a mass m_1 near the surface of the earth. The magnitude of the force is

$$F = m_1 \frac{Gm_2}{r^2}$$

where m_2 is the mass of the earth and r is approximately the radius of the earth. Unless the body gets very far from the surface, we do not have to worry about changes in r. Inserting the appropriate values, we find that

$$\frac{Gm_2}{r^2} = 9.8 \text{ m/s}^2 = 980 \text{ cm/s}^2 = 32 \text{ ft/s}^2$$

We will call this quantity g. We have then

$$F = mg$$

Example 3

An object tied to the end of a spring (Fig. 3.9) lies on a smooth table and slides back and forth along a straight line (the axis of the spring). By observing the time rate of change of the momentum, we find that at any instant it is directly proportional to the negative of the displacement x of the spring from its equilibrium position at that time, i.e.,

$$\frac{dp}{dt} \propto -x$$

or

$$\frac{dp}{dt} = -kx$$

where k is the constant of proportionality and is called the spring constant. From the definition of force as the time rate of change of the momentum, we see that

$$F = -kx \tag{3.14}$$

This is called *Hooke*'s *law*, and it is a good approximation for most springs.

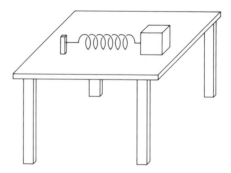

FIGURE 3.9 A mass attached to a spring slides on a smooth table.

In all these cases [$F = k(q_1 q_2 /r^2)$, $F = G(m_1 m_2 /r^2)$, $F = mg$, $F = -kx$] we see that the force depends only on the instantaneous state of the system. All these situations are *examples* of the second law of dynamics, which is the *generalization* that all forces depend only on the instantaneous state of the system.

In each of these examples the force depends only on the positions of the bodies and not on their velocities. (We shall see later that the magnetic force depends on the velocity.) As an example of something that does not depend only on the positions of the bodies, consider the velocity of a body close to the surface of the earth. The velocity does not depend only on the position of the body; it depends on the past history. In Fig. 3.10 we have drawn two trajectories that pass through the same point. The velocities **v** and **v**' are not the same at the point of intersection of the trajectories, but the accelerations and the changes in momenta and the forces are the same. (In this special case the acceleration, time rate of change of momentum, and the force are constant everywhere. In the case of two charges, or the nonuniform gravitational field or the spring, these quantities are not the same everywhere.) No matter how the projectile gets to some point, the force at that point is the same.

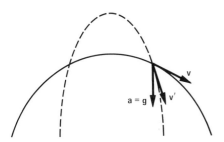

FIGURE 3.10 Two different trajectories in a uniform gravitational field.

The units of force are determined by the definition **F** = m**a**. Force, therefore, is measured in units of a mass times a length divided by a time squared. This combination of units is abbreviated and called respectively a *newton*, dyne, or *pound* in the mks, cgs, and English systems of units. The units are listed in Table 3.1.

TABLE 3.1

System	Fundamental unit	Abbreviation
mks	kg · m/s²	newton
cgs	g · cm/s²	dyne
English	slug · ft/s²	pound

QUESTION

What is the weight of a 2-kg body?

ANSWER

The *weight* is by definition *the force that the earth exerts on the body*. To find this force, let us simply drop the body (Fig. 3.11). The only force acting on the body (if we neglect air resistance) is the weight w. From the definition of force,

$$\mathbf{F} = m\mathbf{a}$$

or since $F = w$ and the observed acceleration is $g(=9.8 \text{ m/s}^2)$,

$$w = mg = (2 \text{ kg})(9.8 \text{ m/s}^2)$$

$$= 19.6 \text{ N}$$

The relation

$$w = mg \tag{3.15}$$

is quite general and will be used repeatedly.

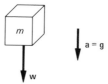

FIGURE 3.11 A freely falling body.

QUESTION

What is the mass of a person whose weight is 200 lb?

ANSWER

The unit of force in the English system of units is the pound. If the person's weight is 200 lb, then the mass (see Eq. 3.15) is given by

$$m = \frac{w}{g} = \frac{200 \text{ lb}}{32 \text{ ft/s}^2} = 6.25 \text{ lb} \cdot \text{s}^2/\text{ft}$$

This unit (lb · s²/ft) is called a *slug*. The mass of the person then is 6.25 slugs.

QUESTION

What is the maximum speed with which an automobile can round a curve whose radius of curvature is 30 m? The maximum frictional force that the road can exert on the automobile is seven-tenths the weight of the automobile (Fig. 3.12).

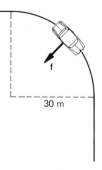

FIGURE 3.12   An automobile rounds a curve whose radius of curvature is 30 m. The maximum force of friction is 0.7 w.

ANSWER

From the definition of force,

$$\mathbf{F} = m\mathbf{a}$$

Consider the component of the equation in the direction perpendicular to the curve, i.e., the radial direction. The force acting in this direction is the force of friction f. The acceleration is a_R. Therefore

$$f = ma_R = \frac{mv^2}{r}$$

from Eq. 2.22. Now

$$f = 0.7w = 0.7mg$$

so that

$$0.7mg = \frac{mv^2}{r}$$

or

$$v = \sqrt{0.7gr}$$
$$= \sqrt{0.7(9.8 \text{ m/s}^2)(30 \text{ m})}$$
$$= 14 \text{ m/s}$$
$$= 31 \text{ mi/h}$$

QUESTION

If you were a race car driver and wanted to round the curve illustrated in Fig. 3.13, what path would you take?

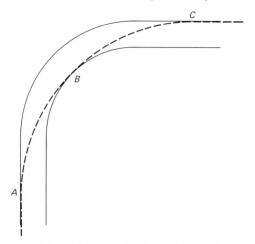

FIGURE 3.13   A 90° bend in the road. The trajectory ABC has the least curvature.

ANSWER

The path that would permit you to maintain the greatest speed is the path with the least curvature (or equivalently the path with the greatest radius of curvature). The circular path $\overline{ABC}$ from outside to inside to outside is the path with least curvature. There are other paths that have less curvature at some points, but they must have greater curvature at other points. On the circular arc $\overline{ABC}$ the speed could be set so that the car is *at all times* on the verge of slipping. At this same speed on another path with greater curvature at *any* point, the car would slip at that point.

Thus you can maximize your speed around the curve by choosing a path of minimum *constant* curvature.

FIGURE 3.14 An unusual poker room.

3.6 PSEUDOFORCES

In the preceding two questions about automobiles rounding curves, it is customary to introduce the notion of *centrifugal force*. This topic is a source of much confusion. There are times, however, when it is useful, and it will be worthwhile for us to discuss the subject, if only to dispel the confusion.

Suppose you are playing poker in a large room and observe the following rather unusual circumstances: The cord holding the overhead light does not hang vertically; the level of liquid in the drinks is not parallel to the surface of the table; a card dropped to the table falls along the line of the lamp cord rather than along a vertical line; and a poker chip placed on edge accelerates as it rolls along the table (Fig. 3.14).

You perform some quantitative measurements and find that the acceleration of the poker chip is a'. If a body is constrained to fall vertically (for example, a bead sliding on a vertical wire), its acceleration is 32 ft/s² (g). The

angle θ that the lamp makes with vertical satisfies the relation

$$\tan \theta = \frac{a'}{g}$$

In short, all the phenomena you observe can be explained if you assume that, in addition to the customary forces, there is force ma' acting on every element of mass in the room ($\mathbf{a}'$ is directed horizontally in the direction in which the poker chip accelerates). You may then apply the standard laws of dynamics, provided you include the ma' force. For example, you would write

$$\mathbf{F} + m\mathbf{a}' = m\mathbf{a}_{m/p} \tag{3.16}$$

where $\mathbf{F}$ is the net force acting on the body, $m\mathbf{a}'$ is this strange new force, and $\mathbf{a}_{m/p}$ is the acceleration of the mass m as observed by the poker player.

Now let us observe this poker game from a different point of view. Actually the room is the back of a large truck (Fig. 3.15), and the truck is accelerating at a constant rate $\mathbf{a}_T$. From the point of view of the ground, the law of motion is

$$\mathbf{F} = m\mathbf{a}_m \tag{3.17}$$

FIGURE 3.15 The poker game is being played in a truck that has acceleration $\mathbf{a}_T$.

where $\mathbf{a}_m$ is the acceleration of the mass m relative to the ground. Now

$$\mathbf{a}_m = \mathbf{a}_{m/p} + \mathbf{a}_p$$

but the poker player is moving with the truck, so

$$\mathbf{a}_p = \mathbf{a}_T$$

Therefore Eq. 3.17 may be written

$$\mathbf{F} = m\mathbf{a}_{m/p} + m\mathbf{a}_T$$

or

$$\mathbf{F} - m\mathbf{a}_T = m\mathbf{a}_{m/p} \qquad\qquad \textbf{(3.18)}$$

Comparing Eqs. 3.16 and 3.18, we see that the results will be compatible if we set

$$\mathbf{a}' = -\mathbf{a}_T$$

Thus the presence of the extra force $m\mathbf{a}'$ in Eq. 3.16 results from an attempt to apply the law of motion in an accelerated frame of reference. If we wish to apply the laws of dynamics in an accelerated frame, we must add a "pseudoforce" (or a "reversed effective force") to the usual forces, and we may measure acceleration and velocity in this accelerated frame.

The most familiar example of such a pseudoforce is the so-called centrifugal force. When we round a curve in an automobile, we "feel" a force directed away from the center of curvature. We also feel a force directed toward the center of curvature. The seat of the automobile exerts this force on us to keep us in place. We say that these two forces balance each other, and this is why we remain in equilibrium *with respect to the automobile.* We are subconsciously applying the laws of motion in the accelerating frame of the automobile. Since we are at rest, we require that the net force be zero, and therefore invent the pseudoforce to satisfy our intuition. There is nothing wrong with this, but we must recognize this force for what it is. It is an inertial term brought from the right-hand side of $\mathbf{F} = m\mathbf{a}$ to the left-hand side and treated as a force. In the inertial term remaining on the right-hand side, the acceleration is measured with respect to the accelerated frame.

To carry this procedure to the extreme, we might observe every particle in a frame in which it is at rest (or better in a frame in which the acceleration is zero). In this frame the real forces $\mathbf{F}$ plus the pseudoforce $(-m\mathbf{a})$ balance to produce equilibrium; that is,

$$\mathbf{F} - m\mathbf{a} = 0$$

Of course this is just another way of looking at the system, and clearly this equation is equivalent to

$$\mathbf{F} = m\mathbf{a}$$

QUESTION

Why do astronauts feel weightless when they are orbiting the earth?

ANSWER

The spaceship is accelerating toward the earth. (Remember that the velocity is not necessarily in the same direction as the acceleration.) If we choose to perform our dynamics relative to the spaceship, we must include in the accelerating frame the pseudoforce $-m a_s$ on any body of mass m, where $\mathbf{a}_s$ is the acceleration of the spaceship. Let us determine this acceleration. The force on the spaceship is

$$F = G\frac{m_s m_e}{r^2}$$

where m_s is the mass of the spaceship and m_e is the mass of the earth. From the law of motion,

$$F = m_s a_s$$

or

$$\frac{Gm_s m_e}{r^2} = m_s a_s$$

so that

$$a_s = \frac{Gm_e}{r^2}$$

If m_a is the mass of the astronaut and $a_{a/s}$ the acceleration of the astronaut relative to the spaceship, then the equation of motion for the astronaut is

$$\mathbf{F}_{\text{other}} + \mathbf{F}_{\text{grav}} - m_a \mathbf{a}_s = m_a \mathbf{a}_{a/s}$$

where $\mathbf{F}_{\text{other}}$ refers to the forces on the astronaut other than the earth's gravitational field. But the gravitational force on the astronaut is

$$\mathbf{F}_{\text{grav}} = G\frac{m_a m_e}{r^2}$$

Clearly $\mathbf{F}_{\text{grav}} - m_a \mathbf{a}_s$ is zero, and the equation of motion becomes

$$\mathbf{F}_{\text{other}} = m_a \mathbf{a}_{a/s}$$

But this is just the equation of motion we would obtain if there were no gravitational field at all. In other words, astronauts feel weightless.

It should be pointed out that this result depends on the fact that gravitational force is directly proportional to inertial mass m.

3.7 THE SECOND LAW OF DYNAMICS AND DETERMINISM

The second law of dynamics tells us that the acceleration depends only on the present state of the system. We would like to show that, as a consequence, any mechanical system is deterministic; that is, given the initial state of the system, the future state is determined. To see this let us consider a particular situation: a planetary body moving about the sun. Suppose we know the position and velocity of the planet at some time, say $t = 0$ (Fig. 3.16). We wish

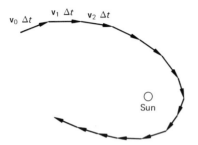

$v_0\,\Delta t$ $v_1\,\Delta t$ $v_2\,\Delta t$

Sun

FIGURE 3.16 Motion of a planetary body about the sun.

to show that we can then determine the entire future of the body.

Let v_0 denote the initial velocity of the body. From the definition of velocity,

$$v = \frac{\Delta r}{\Delta t}$$

in the limit as Δt approaches zero. Therefore, for small Δt we have for the displacement Δr,

$$\Delta r = v\,\Delta t$$

During the short time interval Δt after $t = 0$, the body will undergo a displacement

$$\Delta r_o = v_o\,\Delta t$$

We therefore know where the body is after the short time Δt. But now what? To repeat this process we must know the new velocity v_1 at this new position. This is where our knowledge of the acceleration for *any state* comes in. The acceleration is defined as the time rate of change of the velocity,

$$a = \frac{\Delta v}{\Delta t}$$

so for a short time Δt we may write

$$\Delta v = a_0\,\Delta t$$

where a_0 is the acceleration at $t = 0$. Now the change in velocity is the new velocity minus the old velocity,

$$\Delta v = v_1 - v_0$$

or

$$v_1 = a_0\,\Delta t + v_0 \qquad (3.19)$$

Knowing the acceleration, we can therefore determine the change in velocity. Knowing the change in velocity, we can determine the new velocity. With the new velocity we can repeat the whole process. More briefly, now

$$\Delta r_1 = v_1\,\Delta t$$

where v_1 is given by Eq. 3.19. The change in velocity is

$$\Delta v = v_2 - v_1 = a_1\,\Delta t$$

Remember that we know a_1 since we know the acceleration for all states no matter how the body gets there; hence

$$v_2 = a_1\,\Delta t + v_1$$

Again

$$\Delta r_2 = v_2\,\Delta t$$

and so on.

We see that the key to the success of this method is the fact that we know the acceleration and hence the change in velocity wherever the body goes. Thus the second law of dynamics tells us how to predict the future from the present.

The accuracy of the calculation improves as Δt is chosen smaller and smaller. Of course, the smaller we choose Δt, the more steps we need to project the motion to any given time in the future. The greatest precision is achieved in the limit where Δt approaches zero and the number of steps approaches infinity. It is one of our great intellectual achievements that this process can be carried out. The mathematical discipline that deals with such problems is the calculus.

3.8 PHYSICAL SIGNIFICANCE OF THE SECOND LAW OF DYNAMICS

We have said that the momentum of an isolated particle remains constant in time,

$$\frac{d\mathbf{p}}{dt} = 0$$

If the particle interacts with other particles, then $d\mathbf{p}/dt$ depends on what and where these particles are; that is,

$$\frac{d\mathbf{p}}{dt} = \text{some function of the instantaneous state of the system.}$$

This function is called the force, and this is the essence of the second law of dynamics. This seems like a straightforward statement, perhaps even an obvious one. Not only is it not obvious, it is not entirely correct. The basic significance of the second law is that of *instantaneous action at a distance*. The law states that what a particle feels at some point in time depends on the nature of its environment *at that moment.* On the other hand, if it takes time for signals to propagate, then the time rate of change of the momentum of a particle would depend not on the nature of the environment at that moment, but on where the particles were when they sent the signals that the given particle is now feeling. We shall see later that it does take time for signals to propagate, and we shall find it convenient to describe the interaction between particles in terms of fields generated by the particles.

Now the velocity of propagation of the fundamental forces of nature is very large: 3×10^8 m/s. For many problems this is so large that it might as well be infinite. For the time being we shall assume that particles interact instantaneously, and we shall discuss the finite velocity of signal propagation at great length later.

3.9 THE FIRST LAW OF DYNAMICS AND THE LAW OF ACTION AND REACTION

We would like to give an alternative interpretation to the first law of dynamics in terms of forces. For an isolated system consisting of two interacting particles, the total momentum is a constant of the motion:

$$\mathbf{p}_1 + \mathbf{p}_2 = \text{constant}$$

Consider the time rate of change of both sides of this equation,

$$\frac{d\mathbf{p}_1}{dt} + \frac{d\mathbf{p}_2}{dt} = 0$$

Or, from the definition of force,

$$\mathbf{F}_{2 \text{ on } 1} = \frac{d\mathbf{p}_1}{dt}$$

$$\mathbf{F}_{1 \text{ on } 2} = \frac{d\mathbf{p}_2}{dt}$$

so that

$$\mathbf{F}_{2 \text{ on } 1} + \mathbf{F}_{1 \text{ on } 2} = 0$$

or

$$\mathbf{F}_{2 \text{ on } 1} = -\mathbf{F}_{1 \text{ on } 2} \tag{3.20}$$

Therefore, in order for momentum to be conserved, the force that body 2 exerts on body 1 must be equal in magnitude and opposite in direction to the force that body 1 exerts on body 2. This is often called the *law of action and reaction.*

As an example of this law of action and reaction, consider two charges q_1 and q_2 interacting by means of electrostatic forces. The force of charge 1 on charge 2 is given by Coulomb's law:

$$\mathbf{F}_{1 \text{ on } 2} = k\frac{q_1 q_2}{r_{12}^{2}}\hat{\mathbf{r}}_{12}$$

where $\hat{\mathbf{r}}_{12}$ is a unit vector directed from charge 1 to charge 2 and is attractive if q_1 and q_2 are of opposite sign and repulsive if they are of like sign. The force of charge 2 on charge 1 is given by

$$\mathbf{F}_{2 \text{ on } 1} = k\frac{q_2 q_1}{r_{21}^{2}}\hat{\mathbf{r}}_{21}$$

(Note that we have interchanged the indices.) But r_{12} is equal to r_{21}, since the distance between 1 and 2 is the same as the distance between 2 and 1. The forces are clearly equal in magnitude. We also have $\hat{\mathbf{r}}_{12} = -\hat{\mathbf{r}}_{21}$, so Eq. 3.20 is satisfied by Coulomb's law.

If the force law were

$$\mathbf{F}_{1 \text{ on } 2} = k\frac{q_1 q_2^{2}}{r_{12}^{2}}$$

then the law of action and reaction would not be satisfied and momentum would not be conserved.

It is a shame that we cannot find two bodies that violate the law of action and reaction. If body *A* exerted a larger force on body *B* than body *B* exerted on body *A*, then one might tie these bodies to the front and back bumpers of an automobile and use them to propel the vehicle (Fig. 3.17).

FIGURE 3.17   If $F_{A \text{ on } B}$ is greater than $F_{B \text{ on } A}$, there will be a net force in the forward direction.

3.10 LAW OF ADDITION OF FORCES

If we define force to be $d\mathbf{p}/dt$, or equivalently $m\mathbf{a}$, then what can we say about a book resting on a table top? The momentum is zero (as is the acceleration), and yet a force is acting on the book. The earth is exerting a force on the book, and so is the table top. These two forces must somehow cancel each other. But how do we determine the net effect of two or more forces? We must, of course, return to experiment, for fundamental physical laws can be deduced only from experiment.

Let us consider a series of three experiments. First we allow two particles (labeled 1 and 2) to interact. We find in a certain state that the change in momentum of particle 2 due to particle 1 is

$$\left(\frac{d\mathbf{p}}{dt}\right)_{2 \text{ due to } 1}$$

Next we remove particle 1 and insert another particle, labeled particle 3 (see Fig. 3.18). With particle 2 in the same position (same state) as before, we observe that the rate of change of momentum of particle 2 is now

$$\left(\frac{d\mathbf{p}}{dt}\right)_{2 \text{ due to } 3}$$

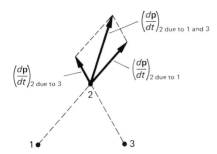

FIGURE 3.18 Two particles interacting with a third particle.

Finally we allow all three particles to interact. Particles 1 and 2 and particles 2 and 3 are in the same state as before. We find that the rate of change of momentum of particle 2 is equal to the vector sum

$$\left(\frac{d\mathbf{p}}{dt}\right)_{2\text{ due to 1 and 3}} = \left(\frac{d\mathbf{p}}{dt}\right)_{2\text{ due to 1}} + \left(\frac{d\mathbf{p}}{dt}\right)_{2\text{ due to 3}}$$

or, from the definition of force,

$$\mathbf{F} \text{ on 2 due to 1 and 3} = \mathbf{F} \text{ on 2 due to 1} + \mathbf{F} \text{ on 2 due to 3}$$

$$(3.21)$$

Thus we can add forces or rates of change of momentum using our rule of vector addition discussed in Chap. 1.

This *law of vector addition of forces*, together with the *law of the conservation of momentum* (or equivalently the *law of action and reaction*) and the *law of motion*, completes the fundamental laws of mechanics.

QUESTION

Let us return to the question posed at the beginning of this section concerning the book at rest on the table top. What is the force of the table on the book?

ANSWER

Two forces are acting on the book: the force of gravity $\mathbf{w}$ and the force of the table $\mathbf{N}$ (Fig. 3.19). Since the book is at rest, the total force must be zero. By the law of vector addition of forces, we have

$$\mathbf{N} + \mathbf{w} = 0$$

These forces are not equal and opposite because of the law of action and reaction, but because the book is at rest. The law of action and reaction states that the force that the table exerts on the book is equal and opposite to the force that the book exerts on the table.

FIGURE 3.19 Forces acting on a book at rest.

3.11 NEWTON'S LAWS OF MOTION

For those who are familiar with Newton's three laws of motion, we should point out that they differ in their order of presentation from the laws as we have stated them. Newton's laws of motion may be stated as follows:

1. Every object continues in its state of rest or uniform motion in a straight line unless acted on by an unbalanced force.

2. The acceleration of a body is directly proportional to the unbalanced force and inversely proportional to its mass.

3. To every action there is an equal and opposite reaction.

Our first law is equivalent to Newton's third law, and our second law is equivalent to Newton's second law. Today Newton's first law is taken as a definition of an inertial frame, but in Newton's day its primary purpose was to dispel the notion that the velocity rather than the acceleration will be zero unless some force is applied. Our third law was an implicit assumption in Newton's second law. The force that Newton speaks of is the unbalanced force, and the unbalanced force is the vector sum of the individual forces.

SUMMARY

The three laws of dynamics are:

1. The law of the conservation of momentum, which states that:

In any isolated system the total momentum remains constant in time. The momentum of any body is proportional to its velocity, the proportionality factor being the mass. The mass is an intrinsic property of the body.

2. The second law of dynamics (or the *law of motion*), may be stated in two ways:
   (a) The time rate of change of momentum of a body depends on the instantaneous state of the system at that moment.
   (b) If we define the force acting on a body to be equal to the time rate of change in momentum that it produces, or equivalently to $m\mathbf{a}$, then the force on a body depends on the instantaneous state of the system at that moment.

3. If more than one force acts on a body, then the resultant force is the vector sum of the individual forces.

The second law of dynamics implies that mechanical systems are deterministic; i.e., knowledge of the present state of a system and the nature of the forces between all elements of the system allows us to calculate the future behavior for all time.

Some examples of the second law of dynamics, i.e., of how forces depend on the state, are as follows.

1. The law of universal gravitation:

$$F = G\frac{m_1 m_2}{r^2}$$

where

$$G = 6.67 \times 10^{-11} \text{ N} \cdot \text{m}^2/\text{kg}^2$$

2. Coulomb's law for electric charges:

$$F = k\frac{q_1 q_2}{r^2}$$

3. A special case of (1) near the surface of the earth:

$$F = mg = w$$

where w is the weight.

4. Hooke's law:

$$F = -kx$$

If the net force on a mass m is $\mathbf{F}_m$, then

$$\mathbf{F}_m = m\mathbf{a}_m$$

where $\mathbf{a}_m$ is the acceleration of the mass. If this body is observed by an observer O who is accelerating with an acceleration $\mathbf{a}_0$, then

$$\mathbf{a}_m = \mathbf{a}_{m/0} + \mathbf{a}_0$$

and we have

$$\mathbf{F}_m + (-m\mathbf{a}_0) = m\mathbf{a}_{m/0}$$

where $\mathbf{a}_{m/0}$ is the acceleration of the mass relative to the accelerating observer. For this observer the equation of motion states that the true force $\mathbf{F}_m$ plus the pseudoforce (or reversed effective force or inertial force) $-m\mathbf{a}_0$ is equal to the mass times the acceleration of the mass relative to this observer.

PROBLEMS

3.A.1 A hunter who is out in the jungle with a pellet gun sees a 90-kg tiger leap out from the brush with a velocity of 20 m/s. This seems like a good opportunity to use the pellet gun to measure the tiger's momentum. If the pellets are 0.1 kg in mass and have a muzzle velocity of 300 m/s, how many must be fired at the tiger to bring it to rest?

3.A.2 Fire hoses are often used for crowd control. Such a hose delivers water at the rate of 30 kg/s with a velocity 20 m/s. If this stream struck you full in the chest (and then ran down to the ground), what force would it exert? Express your answer in pounds.

3.A.3
(a) What is the thrust of a rocket that ejects fuel at a rate of 30 kg/s at a relative velocity of 200 m/s?
(b) If this were a jet engine traveling at a speed of 135 m/s, what would the thrust be?

3.A.4 A 150-lb boy who is standing on an ice-covered handball court throws a 10-lb ball horizontally with a speed of 50 mi/h.
(a) What is the recoil speed of the boy?
(b) The ball strikes the wall and rebounds with the same speed of 50 mi/h. What is the new speed of the boy after he catches the ball?

3.B.1 An astronaut is connected to his spaceship by a tether of length l. Initially he is at rest with respect to the spaceship. He pulls on the tether to return to the ship. How far does each move (along the line of the tether) before they meet? The force is not constant. The mass of the space ship is M and the mass of the astronaut is m.

3.B.2 A proton and an electron are initially at rest 1 m apart. If they are released, where is the proton when it collides with the electron?

3.B.3 Hansel and Gretel are stranded in the middle of a frictionless ice-covered lake 1 mi in diameter (another ruse of their wicked stepmother). They are just out of reach of each other. Hansel still has his trusty bag of bread crumbs (which weighs 5 lb). In his frustration over his predicament, Hansel throws the bag with a speed of 20 mi/h at Gretel, who catches it. Much to their surprise, they each begin to move toward the shore with constant velocity. Each weighs 80 lb.
(a) How long does it take Hansel to reach the shore?
(b) How long does it take Gretel to reach the shore?

3.B.4 A 1-ton open-top railroad car is traveling along a level track with negligible friction at a speed of 40 mi/h. It begins to rain, and the car accumulates rain at the rate of 1 ton/h. If the rain is falling vertically, how long does it take for the speed of the railroad car to be reduced to 20 mi/h?

3.B.5 A plane of mass M in level flight drops a bomb of mass m. What is the vertical acceleration of the plane?

3.B.6 A 10-g bullet with a velocity of 500 ft/s becomes embedded in a stationary block whose mass is 200 g.
(a) Determine the velocity of the bullet and the block after the bullet has come to rest relative to the block.
(b) If the bullet whose initial velocity was 500 ft/s penetrated the block to a depth of 3 cm, to what

depth would a similar bullet traveling at 1000 ft/s become embedded? (Assume that deceleration is constant.)

3.B.7 Show that if momentum is conserved in a certain frame of reference, then in any other frame traveling with a constant velocity relative to this frame, momentum is also conserved.

3.B.8 In a collision between a Volkswagen and a Mack truck, momentum is conserved. The change in momentum of the Volkswagen is equal in magnitude to that of the truck. Since the force is proportional to the change in momentum, the magnitude of the force must be the same for each vehicle. If the force is the same, why is it better to be the driver of the truck?

4
APPLICATIONS OF THE LAWS OF MECHANICS

A PERSON CAN survive a fall of 1000 ft into a snow bank but be killed falling off a roof. The moon creates two high tides daily, not just one. The tides follow a lunar cycle and not a solar cycle even though the gravitational force of the sun on the earth is much larger than that of the moon. A Volkswagen and a Mack truck have the same tendency to skid off the road when rounding a curve. These and many other phenomena can be easily understood by a straightforward application of the three laws of dynamics discussed in the Chap. 3. We shall now discuss some of the techniques used in applying these laws, first to problems in dynamics and then to statics problems.

4.1 PROBLEMS IN DYNAMICS

Although it is difficult (and in some respects undesirable) to devise rules for the solution to all problems in dynamics, we will nevertheless give three steps that may be helpful.

1. Define carefully the *system* to which the law of motion is to be applied.

2. Represent by vectors in a diagram all the external forces acting *on* the system. With the exception of gravity, only those macroscopic objects in *contact* with the system can exert forces on it. (We shall not be concerned with electromagnetic forces in this chapter.)

3. Apply the law of motion $\mathbf{F} = m\mathbf{a}$, where $\mathbf{F}$ is the vector sum of all the external forces acting on the system.

QUESTION

A 200-lb man is standing on a scale in an elevator, and the elevator is accelerating upward with an acceleration of $\frac{1}{2}g$ (Fig. 4.1). What does the scale read?

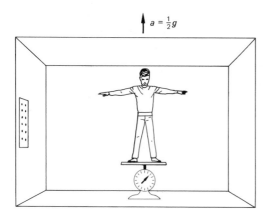

FIGURE 4.1 The effective weight of a man standing in an elevator that has a vertical acceleration of $\frac{1}{2}g$ is one and one-half times his normal weight.

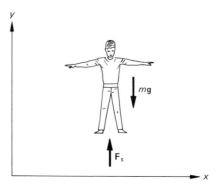

FIGURE 4.2 The man in the elevator and the forces acting on him.

ANSWER

Let us apply our three steps.

1. We may pick any system we please. The one which most readily answers the question asked is the man (Fig. 4.2). The only forces acting *on* the man are the force ($\mathbf{F}_s$) of the scale platform on which he is standing and the force of gravity $m\mathbf{g}$. The floor of the elevator does not exert a force on the man since it is not touching him. The cable exerts no force on the man. Do not include the force that the man exerts on the scale ($-\mathbf{F}_s$ by action and reaction). This is not a force acting on the man but rather a force acting on the scale.

2. $\mathbf{F} = m\mathbf{a}$ is a vector equation. We may write

$$\mathbf{F}_s + m\mathbf{g} = m\mathbf{a}$$

If we choose the y axis vertically upward, the y component of the equation reads

$$F_s - mg = ma$$

Normally when all the vectors point in the same direction, we will immediately write down the scalar equation. *Much confusion will be avoided if we choose the positive sense to be in the same direction as the acceleration.* Since the acceleration of the elevator is up, we measure all vectors up as positive and all vectors down as negative. Solving for F_s we have

$$F_s = mg + ma$$

But $a = \frac{1}{2}g$ (given), so

$$F_s = \tfrac{3}{2}mg$$

Since the man weighs 200 lb ($mg = 200$ lb), the scale will read

$$F_s = \tfrac{3}{2}(200 \text{ lb}) = 300 \text{ lb}$$

QUESTION

There are several accounts of people falling from airplanes into snow banks and surviving. In one well-documented case, the depth of the hole left by a man was measured to be 3.5 ft. If the man weighed 150 lb and his terminal velocity in free fall was 120 mi/h (175 ft/s), what was the average force felt by the man on impact?

ANSWER

Let us assume that the man's acceleration in the snow was constant. From Eq. 2.10,

$$v^2 - v_0^2 = 2ax$$

where v is the velocity after the man has moved a distance x in the snow. If we set $x = 3.5$ ft, then $v = 0$, since the man came to rest at this depth. The acceleration therefore is given by

$$a = \frac{-v_0^2}{2x} = -\frac{(176 \text{ ft/s})^2}{2(3.5 \text{ ft})} = -4425 \text{ ft/s}^2$$

The force necessary to give this acceleration is

$$F = ma = \frac{mg}{g}a = \left(\frac{150 \text{ lb}}{32 \text{ ft/s}^2}\right)(4425 \text{ ft/s}^2)$$

$$= 21{,}000 \text{ lb}$$

Believe it or not!

QUESTION

A 5-lb block slides down the smooth (no friction) incline of Fig. 4.3. If the block starts from rest at the top, what will its velocity be at the bottom?

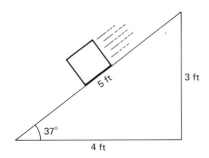

FIGURE 4.3 A block sliding down a frictionless incline.

ANSWER

Take as a system the block at some arbitrary point on the incline. The forces on the block are illustrated in Fig. 4.4. The force $\mathbf{N}$ exerted on the block by the incline is perpendicular to the block because the incline is smooth and there is no friction, that is, no component of force tangent to the surface impeding the motion down the incline. The equation of motion is

$$\mathbf{N} + m\mathbf{g} = m\mathbf{a}$$

Resolving this equation into its components, we have

$$mg \sin 37° = ma_x$$

$$N - mg \cos 37° = ma_y$$

But $a_y = 0$, so

$$N = mg \cos 37° = (5 \text{ lb})(\tfrac{4}{5}) = 4 \text{ lb}$$

We have for the x component

$$a_x = g \sin 37° = (32 \text{ ft/s}^2)(\tfrac{3}{5}) = 19.2 \text{ ft/s}^2$$

This completes the dynamics. The rest is kinematics. Since a_x is constant throughout the motion, we may apply Eq. 2.29 and obtain

$$v_x^2 = 2a_x x + v_{0x}^2$$

But $v_{0x} = 0$, so

$$v_x = \sqrt{2(19.2 \text{ ft/s}^2)(5 \text{ ft})} = 13.8 \text{ ft/s}$$

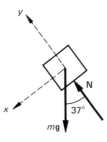

FIGURE 4.4 The forces on the block are the weight $m\mathbf{g}$ and the force of the incline $\mathbf{N}$.

QUESTION

Why are there two high tides per day?

ANSWER

The tides are caused by the interaction between the moon and the earth. Let us first imagine the earth to be fixed in space and the moon rotating about the earth. In this case the moon would attract the water of the earth and there would be a bulge on one side only. We would observe only one high tide.

Certainly the earth is not fixed in space. It is in fact accelerating toward the moon just as the moon is accelerating toward the earth. For simplicity let us imagine the earth falling in a uniform gravitational field (Fig. 4.5). In the accelerating frame of reference of the earth, the pseudoforce of the accelerating frame and the gravitational force of the moon would exactly cancel, just as they did when we considered the astronaut in the spaceship. (If you were in an elevator falling with an acceleration g, then in this frame of reference you would feel no gravitational force.) There would therefore be no tide at all We have gone from bad to worse. At least with the earth fixed we found *one* high tide. Now we have none.

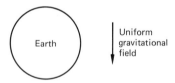

FIGURE 4.5   The earth falling in a uniform gravitational field.

Let us add one more refinement. The gravitational field of the moon is not uniform. It decreases with distance like $1/r^2$.

Let us therefore imagine the earth falling in a *nonuniform* gravitational field. Let the force increase from top to bottom as illustrated in Fig. 4.6. The forces indicated in the figure are not the forces on the earth but the forces that would act on the same object (or different objects of the same mass) at the indicated positions. Thus if we had three independent elements of mass at three points, the lower element (the element closest to the moon) would have the largest acceleration and the upper element the smallest acceleration. If we viewed the motion from the middle particle, we would see the upper and lower mass elements recede from us; that is, in the accelerated frame of the middle particle we would imagine a pseudoforce that acts to draw elements away from us at the top and bottom. From the point of view of the earth's frame of reference, the double tide is caused by this pseudoforce. From the point of view of some external inertial frame, the acceleration of the center of the earth (caused by the moon alone) is intermediate between the acceleration of elements of mass on the near and far side of the earth.

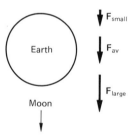

FIGURE 4.6   The earth falling in a nonuniform gravitational field.

ANSWER

We have seen that there would be no tides in a uniform gravitational field. It is the gradient or nonuniformity of the field that is responsible for the tides. Although the sun contributes a larger force, the moon, being closer, produces a larger gradient (Fig. 4.7). The quantitative details are left as an exercise.

We should also observe that the earth's gravitational force has a large gradient at the moon, and that there will therefore be a large tidal force acting on the moon. There is of course no water on the moon, but these forces will have some effect on the internal forces within the moon. The closer the moon is to the earth, the greater these tidal forces become. It is speculated that one of Saturn's moons came too close to Saturn and was broken up by tidal forces.

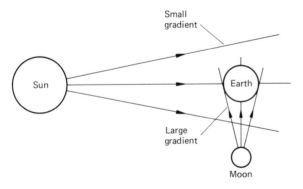

FIGURE 4.7 At the earth the gradient of the sun's gravitational force is much smaller than the gradient of the moon's gravitational force.

4.2 PROBLEMS IN STATICS

A system at rest, or moving with constant velocity, is said to be in static equilibrium. Clearly a necessary condition for static equilibrium is that the net force acting on the system be zero.

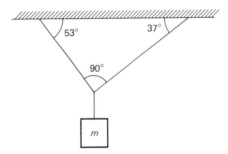

FIGURE 4.8 A block supported by ropes.

ANSWER

We solve problems in statics in much the same way that we solve problems in dynamics. We choose a system, represent the forces acting on the system in a diagram, and finally equate the vector sum of the forces to zero. Let us choose as a system that small particle of rope where all three ropes meet. The forces acting on this point are represented in Fig. 4.9. T_1 and T_2 are the tension in the upper ropes, and $m\mathbf{g}$ is the weight of the hanging block. Since the vector sum of the forces must be zero,

$$\mathbf{T}_1 + \mathbf{T}_2 + m\mathbf{g} = 0$$

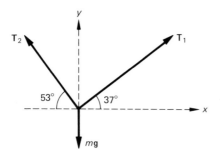

FIGURE 4.9   Three forces are acting on the system.

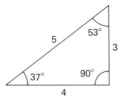

FIGURE 4.10   A 3-4-5 triangle.

Resolving this vector equation into its x and y components, we have

$$T_{1x} + T_{2x} = 0$$

and

$$T_{1y} + T_{2y} - mg = 0$$

or

$$T_1 \cos 37° - T_2 \cos 53° = 0 \tag{4.1}$$

and

$$T_1 \sin 37° + T_2 \sin 53° - mg = 0 \tag{4.2}$$

We can obtain the trigonometric functions from Fig. 4.10. Equations 4.1 and 4.2 become

$$\tfrac{4}{5}T_1 - \tfrac{3}{5}T_2 = 0$$

and

$$\tfrac{3}{5}T_1 + \tfrac{4}{5}T_2 - mg = 0$$

Solving for T_1 and T_2 gives

$$T_1 = \tfrac{3}{5}mg$$

$$T_2 = \tfrac{4}{5}mg$$

4.3 FRICTIONAL FORCE

In the study of dynamic systems we encounter a force that is very difficult to deal with in a realistic way, and this is the force of friction. Friction has its origin in the inter-molecular forces between two objects. Consider, for example, a metal block sliding along a horizontal metal surface. If we magnified the area of contact it might look something like the system illustrated in Fig. 4.11. Determining the nature of the force between these two bodies as a function of the state of the systems is impossible. We do not know the details of either surface: both the shapes of the surfaces and their molecular constituents (e.g., impurities) are unknown. We would nevertheless like to predict how long it will take a block to slide down an inclined plane. To do this we need to know the forces that act on the system as a function of the state of the system. What do we do?

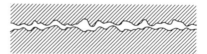

FIGURE 4.11   Magnified view of the region of interaction between two bodies.

We have already encountered a similar situation, namely the force exerted by a stretched spring, and have resolved the problem without discussion. We did not examine the molecular constituents of the spring or the way in which the crystal lattice of the metal changed as the spring was deformed. Yet we wrote an expression for the force:

$$F = -kx$$

We might have arrived at this result in the following way: The restoring force depends on the molecular make-up of the wire and how far the spring is displaced from its equilibrium position. Suppose we performed an experiment with the spring and found an Fx curve something like the one in Fig. 4.12. We know that F is zero at $x = 0$ by definition of $x = 0$. Also, F is negative for positive x and positive for negative x. In the neighborhood of $x = 0$ the straight line tangent to the curve (dashed line) approximates the curve reasonable well, so if x is not too large, we have approximately

$$F = -kx$$

where k is the negative of the slope of the experimental curve at $x = 0$. This gives us a simple relationship between the force and the displacement, and the complexity of the molecular structure is hidden in the spring constant k. That constant must be determined from experiment.

Although the circumstances are a little different, we might approach the problem of the frictional force between two sliding surfaces in the same way. The frictional force between two surfaces depends on the details of the surfaces and the force pressing the surfaces together, which we call the *normal force N*. If we perform an experiment and plot f_k vs N, we might find a curve something like that in Fig. 4.13. Here we use kinetic frictional force f_k which is distinguished from static frictional force f_s, discussed below. Again if N is not too large we have approximately

$$f_k = \mu_k N$$

where μ_k is the slope of the straight line at $N = 0$ and is to be determined by experiment. The complexity of the surface structure is buried in μ_k, the *coefficient of kinetic friction*.

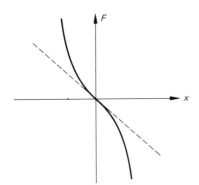

FIGURE 4.12 Restoring force of a spring as a function of the displacement from the equilibrium position.

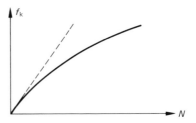

FIGURE 4.13 Frictional force f_k as a function of the normal force N.

We must consider this an empirical relation of question-able accuracy. We may use it to give some qualitative answers to problems in which friction is important.

We have just considered the frictional force between two sliding surfaces. Now we must distinguish this case from the frictional force between two bodies for which the relative motion is zero; this force is the so-called *static friction*. The frictional force acting on a block at rest on a horizontal surface is zero. If an external force is applied to the block and the block does not move, then the force of static friction must exactly balance the applied force (Fig. 4.14):

$$\mathbf{F} + \mathbf{f}_s = 0$$

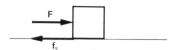

FIGURE 4.14   The force of static friction f_s balances the applied force F and the block does not move.

As the applied force increases, a point is reached at which the static friction is no longer sufficient to counteract the applied force and the block moves. This maximum value of the static frictional force is related to the nature of the surface and the normal force. As before, we assume a relation of the form

$$f_s(\text{max}) = \mu_s N$$

The subscript s denotes static friction. Another way of expressing these results is to say that

$$f_s \leq \mu_s N$$

A 2-kg block rests on a horizontal board. The coefficient of static friction is $\frac{1}{4}$. To what angle may the board be tilted before the block begins to slide?

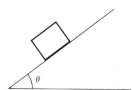

FIGURE 4.15   The block is on the verge of slipping when the board makes an angle θ with the horizontal.

ANSWER

Let us tilt the board until the block slips. We let the angle at which it slips be θ. At an angle infinitesimally smaller than θ the block is on the verge of slipping. Since it is in equilibrium, the sum of the forces must be zero, that is,

$$\mathbf{f}_s + \mathbf{N} + m\mathbf{g} = 0$$

We might take components of the forces to solve the problem, but it is easier to use the polygon rule of vector addition. We note from Fig. 4.16 that

$$\tan \theta = \frac{f_s}{N}$$

FIGURE 4.16   The vector sum $\mathbf{f}_s + \mathbf{N} + m\mathbf{g}$ must be zero. Therefore the three vectors must form a closed triangle.

But the block is on the verge of slipping, so

$f_s = f_s(\text{max}) = \mu_s N$

Therefore

$\tan \theta = \mu_s = \frac{1}{4}$

so that

$\theta = 14°$

We could use this simple experiment to determine the coefficient of static friction.

QUESTION

When you are rounding a curve in an automobile and feel you are in danger of skidding, why should you not apply the brakes?

ANSWER

The maximum static frictional force is $\mu_s N$, where N is equal to the weight of the automobile. If you apply your brakes, then part of this frictional force will be used to slow the car and part of it will be used to keep the car from skidding off the road. The first part is tangent to the path and the second part is centripetal. If you do not apply the brakes, then *all* the frictional force can be used to keep the car from skidding.

SUMMARY

Three steps that are often helpful in solving problems in dynamics and statics are

1. Define carefully the system to which the law of motion is to applied.

2. Represent by vectors in a diagram all the external forces acting *on* the system.

3. Apply the law of motion $\mathbf{F} = m\mathbf{a}$, where $\mathbf{F}$ is the vector sum of all the external forces acting on the system.

The frictional force on a body in motion (kinetic frictional force) is given by

$f_k = \mu_k N$

where N is the normal force between the two surfaces.

The frictional force on a body at rest (static frictional force) cannot exceed $\mu_s N$, that is,

$f_s \leq \mu_s N$

PROBLEMS

4.A.1 What is the minimum acceleration with which a 200 lb person may slide down a rope with a tensile strength of 100 lb without breaking the rope?

4.A.2 What is the acceleration of the 2- and 3-kg masses in the figure? Neglect friction.

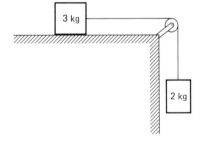

4.A.3 The three blocks shown in the following figure are being pulled along a smooth horizontal surface with an acceleration of 3 m/s². Each block has a mass of 2 kg. Find the tension of the rope at points A, B, and C.

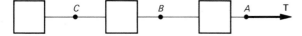

4.A.4 What is the orbital radius of a satellite that circles the earth once every 24 h?

4.A.5 A 100-lb weight is suspended from three cords as shown below. Find the tension in each cord.

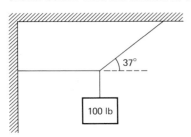

37°

100 lb

4.A.6 A driver is using a rope to pull her automobile out of a rut. Rather than pulling directly on the rope, the driver ties the rope to a tree and makes the rope as taut as possible. She applies a force of 100 lb at the midpoint in a direction perpendicular to the straight line between the automobile and the tree. The rope bends, making an angle of 5° with the straight line. What is the force on the automobile?

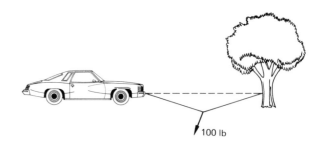

100 lb

4.A.7 If an automobile traveling at a speed v can come to a stop in a distance x, then with what speed v' may the automobile round a curve of radius r?

4.A.8 The dome of the capitol is being resurfaced. If the coefficient of friction between the workers' boots and the surface is $\mu_s = \frac{1}{2}$, to what latitude may they work without a tether? Assume that the dome is a hemisphere.

4.A.9 The lamp cord in Fig. 3.14 hangs at an angle of 10°. What is the acceleration of the truck?

4.B.1 If the 100-kg person in the figure below wants to remain at rest by pulling on the rope, what must the acceleration of the rope be? Neglect the mass of the rope and pulley.

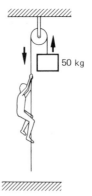

50 kg

4.B.2 A satellite is circling the earth at a distance of 100,000 km from the center of the earth.
  (a) What is the angular velocity of the satellite?
  (b) What is the period of rotation?
  (c) What is the velocity?
  (d) Does v increase or decrease with orbital radius?

4.B.3 A ball hanging from a 5-m-long cord rotates in a horizontal circle, and the cord makes an angle of $37°$ with the vertical. What is the period of rotation?

4.B.4 If the coefficient of static friction between the two blocks shown below is $\mu_s = 0.5$, what is the maximum acceleration that may be given to the system without the upper block slipping?

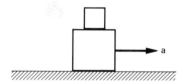

4.B.5 Two bodies of identical material but different masses are sliding along a rough horizontal surface. They are connected by a light rigid rod. What is the tension in the rod? What does this tell you about internal stresses that may be generated when a body slides along a rough surface?

4.B.6 A block starting from rest takes 2 s to slide 5 ft down a $37°$ incline. What is the coefficient of kinetic friction?

4.B.7 The gradient of a function $F(x)$ at a point x may be defined as dF/dx, that is, the rate at which that function changes in the x direction.

  (a) Compare the ratio of the gravitational force exerted on the earth by the moon and the sun; i.e., calculate the ratio $F_{m\,on\,e}/F_{s\,on\,e}$, where $F_{m\,on\,e}$ is the force of the moon on the earth and $F_{s\,on\,e}$ is the force of the sun on the earth.
  (b) Do the same for the ratio of the gradients,

$$\frac{dF_{m\,on\,e}/dr}{dF_{s\,on\,e}/dr}$$

4.B.8
  (a) An automobile is headed directly at a long stone wall with a velocity v. If the coefficient of friction is μ_s, how close can the driver get before he must apply the brakes?
  (b) Instead of applying the brakes, the driver turns in a circular arc to avoid the wall. How close may he get before he must begin his turn?

4.B.9 What force **F** is required to support the 100-lb weight in the block and tackle shown in the following figure? (The tension in the rope is uniform throughout.)

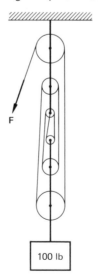

4.C.1 We have observed that a satellite will break up under tidal forces if it gets too close to its parent planet. Estimate how close this would be in the case of the moon. Consider a rock on the far side of the moon (see the figure), and show that it would be on the verge of flying off the moon when

$$r^3 = \frac{3M_e}{M_m}R^3$$

where r is the distance between the center of the moon and the center of the earth, R is the radius of the moon, and M_e and M_m are the mass of the earth and moon, respectively. (Use the approximation $1/(1 + R/r)^2 \simeq 1 - 2R/r$.) Show also that a rock on the near side of the moon would be on the verge of flying off under the same conditions.

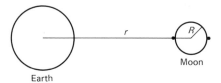

Earth   Moon

4.C.2 Show that μ_s must be greater than or equal to μ_k.

5
CONSERVATION OF ENERGY

We have seen that momentum is conserved for any isolated collection of interacting bodies. If we observe the total momentum of a system at some time, the system must have the same total momentum at any future time. This does not tell us everything we might want to know about the system, but from it we do get some information about a future state without having to solve the dynamic problem for every body in the system. Moreover, we might not be able to do the calculations either because we do not know the nature of the forces or because the system is so complicated that the task is impossible (e.g., the systems of a gas of 10^{23} molecules or a human being). Clearly such conservation theorems are very useful. We shall now discuss another conservation theorem: the conservation of energy theorem. Like the conservation of momentum theorem, the conservation of energy theorem tells us that some function of the state of the system is constant throughout the motion.

Let us consider as an example a projectile. From Eqs. 2.29 and 2.32,

$$v_x^2 = v_{0x}^2 \tag{5.1}$$

$$v_y^2 = -2gy + v_{0y}^2 \tag{5.2}$$

Adding Eqs. 5.1 and 5.2 gives

$$v_x^2 + v_y^2 = -2gy + v_{0x}^2 + v_{0y}^2$$

or, since

$$v^2 = v_x^2 + v_y^2$$

and

$$v_0^2 = v_{0x}^2 + v_{0y}^2$$

we have

$$v^2 + 2gy = v_0^2 \tag{5.3}$$

Thus if we consider the function $v^2 + 2gy$, we see that it has a constant value, in particular the value it had at $y = 0$, which is v_0^2. We may also express Eq. 5.3 in the form

$$\tfrac{1}{2}mv^2 + mgy = \text{constant} \tag{5.4}$$

Later we shall interpret this equation to say that the kinetic energy ($\frac{1}{2}mv^2$) plus the potential energy (*mgy*) is a constant.

Here we have an example of a conservation principle. If this were the way we derived all conservation principles, they would be of little value. We used the known solution of the problem (Eqs. 5.1 and 5.2) to obtain the result. Clearly if we know the solution, we do not need the energy principle. However, we find that there are other situations in which

$$\frac{1}{2}mv^2 + mgy = \text{constant}$$

Examples of these are a bead sliding without friction down a smooth wire (Fig. 5.1), and a ball bouncing elastically down any smooth surface (Fig. 5.2). In these examples the motion of the body is complicated, perhaps too complicated to determine. Yet we are able to say that Eq. 5.4 is satisfied, and we shall see later why this is so.

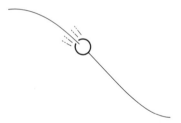

FIGURE 5.1 A bead sliding on a smooth wire.

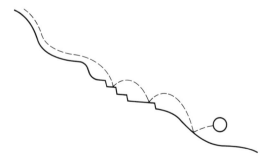

FIGURE 5.2 A ball bouncing down a smooth surface.

We do not want to give the impression that the energy principle always takes this form. For a block sliding back and forth on a smooth table (Fig. 5.3) and attached to a spring, we shall find that

$$\frac{1}{2}mv^2 + \frac{1}{2}kx^2 = \text{constant}$$

where *x* is the displacement of the spring from its equilibrium position and *k* is the spring constant.

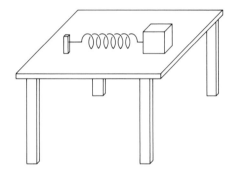

FIGURE 5.3 A block slides along a smooth table and stretches a spring.

Also, in planetary motion we shall find that

$$\frac{1}{2}mv^2 - \frac{GmM}{r} = \text{constant}$$

where *m* is the mass of the planet, *M* the mass of the sun, and *r* the distance between the two.

All these examples share the feature that the sum of a term relating to the motion of the body, called the kinetic energy ($\frac{1}{2}mv^2$), plus a term relating to the position of the body, called the potential energy (*mgy* or $\frac{1}{2}kx^2$ or

$-GmM/r$), is constant throughout the motion. We shall derive the general principle underlying these common features and generalize the results to include a much wider range of phenomena.

5.1 PRINCIPLE OF WORK AND KINETIC ENERGY

Let us focus our attention first on the term we have called the kinetic energy. Let us define the kinetic energy of a body moving with a velocity v to be

$$K = \tfrac{1}{2}mv^2 \tag{5.5}$$

Now let us see whether we can relate the change in the kinetic energy of a body to the forces acting on it. We know from the second law of dynamics that the forces depend in general on the position of the body. We have seen that many forces depend on the configuration of the system. This relation between the forces and the configuration might give us the desired relation between the kinetic energy and the configuration.

First we shall consider the change in kinetic energy between two neighboring points on the trajectory of the body (Fig. 5.4). Let ds be the distance between these points and $\mathbf{F}_T$ and $\mathbf{F}_R$ the tangential and radial components of the net force acting on the body. Now

$$dK = d(\tfrac{1}{2}mv^2) = \tfrac{1}{2}m\,d(v^2) = mv\,dv \tag{5.6}$$

where dv is the change in the speed between the neighboring points. Note that it is not $d\mathbf{v}$, nor is it the magnitude of $d\mathbf{v}$. The change in the magnitude of the velocity is not the same as the magnitude of the change in velocity. For an automobile rounding a curve at 60 mi/h, $dv = 0$, but $d\mathbf{v} \neq 0$.

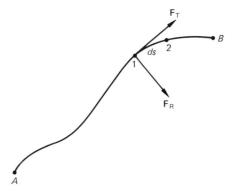

FIGURE 5.4 A body moves along a path between two points A and B subject to a net force **F**. Points 1 and 2 are neighboring points separated by a distance ds.

To find the relationship between dK and the force acting on the body, we must recall that the tangential component of the force is responsible for the change in magnitude of the velocity. (The radial component of the force is responsible for the change in direction.) That is,

$$F_T = m\frac{dv}{dt}$$

or

$$F_T\,dt = m\,dv$$

Substituting this relation into Eq. 5.6, we find

$$dK = F_T v\,dt$$

But

$$v = \frac{ds}{dt}$$

or

$$v\,dt = ds$$

so that

$$dK = F_T\,ds \tag{5.7}$$

or the change in kinetic energy dK between neighboring points is equal to the product of the component of force along the path (F_T) and the distance ds between the neighboring points.

This product $F_T\,ds$ is called *the work* done by the force **F** as the body moves a distance ds. Let us use the symbol dW to denote the work done in this small displacement:

$$dW = F_T\,ds \tag{5.8}$$

From Eqs. 5.7 and 5.8,

$$dK = dW \tag{5.9}$$

which states that the increase in kinetic energy is equal to the work done on the body.

We have derived this result for neighboring points only. Let us consider an object that moves along a finite path from A to B (Fig. 5.5). The path may be broken up into a large number of small segments. If the work done by the force along each segment is equal to the increase in kinetic energy along this segment, it follows that the total work done along the whole path is equal to the net increase in kinetic energy along the path, or

$$W_{A \to B} = K_B - K_A \tag{5.10}$$

where $W_{A \to B}$ is the net work done along the path from A to B, K_B is the kinetic energy at B and K_A is the kinetic energy at A. Often it will be convenient to use an abbreviated notation and to write

$$W = \Delta K \tag{5.11}$$

where W is the work done on the body and ΔK the increase in kinetic energy of the body. We shall call this the *principle of work and kinetic energy*:

The net work done on a body is equal to the increase in kinetic energy of the body.

FIGURE 5.5 The path from A to B may be subdivided into a large number of small segments.

The fundamental units of energy, and therefore of work, can be seen from the definition of kinetic energy. In the mks system they are $kg \cdot m^2/s^2$. This is an awkward combination, so it is abbreviated to the unit called a *joule*. The units of work and energy in the three systems of units commonly used and the abbreviated nomenclature are given in Table 5.1.

TABLE 5.1 Units of energy and work.

System of units	Fundamental units	Abbreviation
mks	$kg \cdot m^2/s^2$	joule
cgs	$g \cdot cm^2/s^2$	erg
English	$slug \cdot ft^2/s^2$	ft · lb

5.2 CALCULATING WORK

If we wish to use Eq. 5.10, we must be able to compute the work done along the path. We must be able to evaluate

$$W = \int F_T\,ds \tag{5.12}$$

First let us observe that we may express the integrand as a vector dot product (see Sec. 1.3). From the definition we have

$$\mathbf{F} \cdot d\mathbf{r} = F \cos\theta\,ds$$

where θ is the angle between **F** and $d\mathbf{r}$ and ds is the magnitude of $d\mathbf{r}$ (Fig. 5.6). But

$$F \cos \theta = F_T$$

so that

$$\mathbf{F} \cdot d\mathbf{r} = F_T \, ds$$

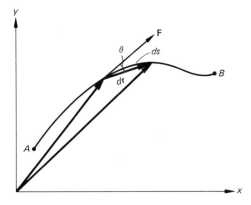

FIGURE 5.6   The angle between the force **F** and the displacement $d\mathbf{r}$ is θ.

We may write then

$$W = \int \mathbf{F} \cdot d\mathbf{r} \qquad\qquad (5.13)$$

No matter how we express the work (Eq. 5.12 or Eq. 5.13), we must perform an integral along a path. Let us consider some simple cases.

Example 1

A body is moved vertically upward a distance of 3 m (Fig. 5.7). What is the work done by a vertical force of 2 N?

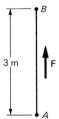

FIGURE 5.7   The force **F** has a magnitude of 2 N. The object moves from A to B, a distance of 3 m.

The work is given by

$$W = \int F_T \, ds$$

Now

$$F_T = F = 2 \text{ N}$$

since the net force is tangent to the path. But F is constant, so we may factor it from the integral

$$W = F_T \int ds = 2 \text{ N} \int ds$$

But

$$\int ds = 3 \text{ m}$$

so

$$W = 6 \text{ N} \cdot \text{m} = 6 \text{ J}$$

Example 2

We may clearly extend the result of Example 1 to any case in which the force is constant. If F_T is constant, then

$$W = \int F_T \, ds = F_T \int ds = F_T s$$

where s is the total distance along the path.

Example 3

What is the work done on a body by a vertical force of 2 N along the diagonal path from A to B in Fig. 5.8?

The work is given by

$$W_{A \to B} = F_T s = F(\cos \theta)s = (2 \text{ N})(\tfrac{3}{5})(5 \text{ m}) = 6 \text{ J}$$

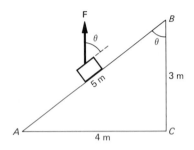

FIGURE 5.8 A vertical force of 2 N is applied to a body as it moves 5 m from A to B.

Example 4

What work is done by the force in Example 3 along the path from A to C to B?

The work from A to C is

$$W_{A \to C} = F_T s = 0$$

since F_T is zero along the line AC. The work from C to B is

$$W_{C \to B} = F_T s = (2 \text{ N})(3 \text{ m}) = 6 \text{ J}$$

We see that the work performed is the same for both paths. Later we shall see that forces for which the work is independent of the path are of very special significance,

and for such forces we shall be able to construct a conservation of energy principle.

Two properties of the work done by a force along a path will be useful to us later. They are:

1. If a path from A to C is divided into two segments AB and BC, where B is some intermediate point on the path (Fig. 5.9), then

$$W_{A \to C} = W_{A \to B} + W_{B \to C} \tag{5.14}$$

FIGURE 5.9 A path from A to C that passes through B.

2. If $W_{A \to B}$ is the work done by a force on an object that is moved along some path from A to B and if $W_{B \to A}$ is the work done by the same force on an object moved along the same path from B to A, then

$$W_{A \to B} = -W_{B \to A} \tag{5.15}$$

since the component of the force tangent to the path from A to B is the negative of the component of the force tangent to the path from B to A.

5.3 APPLICATIONS OF THE PRINCIPLE OF WORK AND KINETIC ENERGY

Before we deal with more complicated forces and paths, let us apply our principle of work and kinetic energy to some simple problems to see how it works.

QUESTION

A stone is released and falls a distance s (Fig. 5.10). What is its velocity?

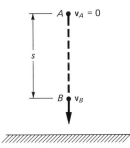

FIGURE 5.10 A stone at rest at A falls a distance s to B.

ANSWER

The work done by the outside forces must equal the increase in kinetic energy. The only outside force is the gravitational force, and the work done by this force is

$$W_{A \to B} = mgs$$

where A is the point of release and B is a point a distance s below A. Now

$$W_{A \to B} = K_B - K_A$$

and

$$K_A = 0$$

Also

$$K_B = \tfrac{1}{2}mv_B^2$$

so that

$$mgs = \tfrac{1}{2}mv_B^2$$

or

$$v_B = \sqrt{2gs}$$

a result obtained earlier in Chap. 2.

QUESTION

A block slides along a frictionless-horizontal surface. What is the relationship between the velocity and the distance traveled?

ANSWER

The forces acting on the block are the normal force $\mathbf{N}$ and the weight $m\mathbf{g}$ (Fig. 5.11). Both these forces are perpendicular to the path and so do no work. Therefore the kinetic energy cannot change and the speed must remain constant.

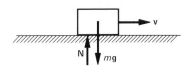

FIGURE 5.11 A block slides without friction on a horizontal surface.

QUESTION

A block slides along a rough table which exerts a constant frictional force f opposing the motion. If the initial velocity of the block is v_A, how far does it travel before it comes to rest (Fig. 5.12).

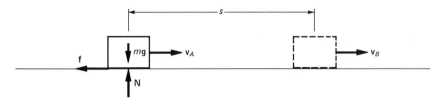

FIGURE 5.12 A block slides along a rough horizontal surface.

ANSWER

Neither the weight nor the normal force does work on the block, but the frictional force does an amount of work given by

$$W_{A \to B} = -fs$$

where s is the distance traveled. Equating the work to the increase in kinetic energy, we have

$$-fs = \tfrac{1}{2}mv_B^2 - \tfrac{1}{2}mv_A^2$$

The block will come to rest when $v_B = 0$. Solving for s we find

$$s = \frac{\tfrac{1}{2}mv_A^2}{f}$$

5.4 DEPENDENCE OF WORK ON THE PATH

The *principle of work and kinetic energy* is not the type of principle we have been seeking. We must compute the work by summing (or integrating) the product of the tangential component of the force and the displacement *along the path*. In general, we do not know the path until we have solved the problem. If we have solved the problem, we do not need an energy principle. (In the examples in Sec. 5.3 the problems were such that we knew the path even before we obtained a solution, e.g., the block was sliding along a table. If on the other hand the problem were to find the velocity of a ball bouncing down a flight of stairs, we would not know the path in advance.)

We would like to show that there are certain types of forces for which we can calculate the work performed as a system evolves from some state A to a new state B *without*

knowing the path! Since the increase in kinetic energy $\Delta K = K_B - K_A$ depends only on the initial and final states and the work performed by such forces is independent of the path, we shall have a principle relating the initial and final states without having to worry about the path taken by the system in going from state A to state B! This can be an enormous simplification. For example, when I burn (oxidize) glucose (state A) in a test tube it turns into CO_2 and H_2O (state B). I can observe the energy released. When I eat glucose (a sugar), it turns into CO_2 and H_2O in the body, and now *I can predict how much energy is released in the body*. The paths of the molecules are completely different in these two situations, but the end states are the same. It is truly amazing that we can say anything at all about a process so extraordinarily complicated.

The energy principle is of profound significance in that it allows us to ignore the details of intermediate states, but we pay a price: it gives us only limited information. If we know the energy of a collection of objects, we only know one number. To completely characterize the collection of objects, we would have to know where each object is and what its velocity is. This requires a great many numbers. Still, one number is better than none.

As we have said, our success in obtaining a path-independent relation between states A and B depends on our ability to show that $W_{A \to B}$ does not depend on the path. This is not always true. It depends on the nature of the forces. Let us consider the simplest force—the uniform gravitational force—and show that the work done by *this* force is independent of the path.

In Fig. 5.13 we show an arbitrary path joining two points A and B. This might be the path of a free projectile, or a bead sliding on a wire, or a ball bouncing down a hill. We wish to compute only the work done by the earth's gravitational force on the body. Never mind any other forces that may be acting. The magnitude of the gravitational force is mg, and the component of this force tangent to the path is

$$F_T = -mg \cos \theta$$

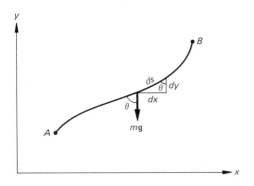

FIGURE 5.13   A mass m moves from A to B along an arbitrary path.

Therefore

$$W_{A \to B} = \int F_T \, ds = -\int mg \cos \theta \, ds$$

But from the figure we see that

$$\cos \theta = \frac{dy}{ds}$$

so that

$$W_{A \to B} = -\int mg \, dy = -mg \int dy$$

But the integral of dy is just the net vertical separation between A and B:

$$\int dy = y_B - y_A$$

Finally

$$W_{A \to B} = -mg(y_B - y_A) \tag{5.16}$$

Thus if we know where A and B are (i.e., we know y_A and y_B), we know the work done by the force of gravity between A and B and we do not have to know the path.

5.5 CONSERVATIVE AND NONCONSERVATIVE FORCES

There must certainly be something special about the force if the work done by the force depends only on the endpoints and not on the path. Not all forces have this property. Consider, for example, the work done by the force of friction on a body sliding on a horizontal surface (Fig. 5.14). Let us assume that the frictional force f is constant and is always directed in the opposite direction of the motion.

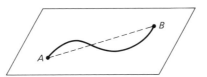

FIGURE 5.14   Two paths joining points A and B on a rough surface.

Then

$$W_{A \to B} = \int F_\tau \, ds = -fs$$

since $F_\tau = -f$ and $\int ds = s$, the total distance traveled along the path. The work is a minimum along a straight-line path and increases with increasing path length. The work obviously depends on the path.

As a second example in which the work depends on the path, suppose we have a force that is directed downward everywhere, such as the gravitational force, but whose strength increases with the distance measured from some vertical axis. In Fig. 5.15 we see that the work done on a body is less along the path AB than along the path $ACDB$ since the force is greater along CD than along AB. The work done along AC and along DB is zero since the component of the force tangent to the path is zero.

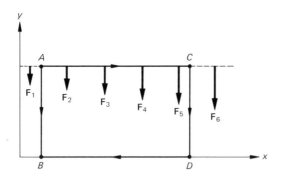

FIGURE 5.15 The force increases with increasing x. Two paths connecting A and B are AB and $ACDB$.

Forces for which the work is independent of the path are called "conservative" forces, and all others are called "nonconservative" forces. The reason we use the terms "conservative" and "nonconservative" is that the total energy (kinetic energy plus potential energy) is conserved for conservative forces and there is no conserved quantity for nonconservative forces. To demonstrate this we must now introduce the notion of potential energy.

5.6 POTENTIAL ENERGY

When we considered a collision between two cars on an air track, we observed that momentum is constant throughout the collision. Suppose we have two cars with the same mass approaching each other at the same speeds. If each is equipped with a good elastic bumper, we find that the speed of each car is the same after the collision as before. Thus in addition to momentum, kinetic energy is also conserved (Fig. 5.16). But what of the time during the collision in which the cars are reversing their velocities? Clearly there comes a time when the kinetic energy of each car is zero, so that kinetic energy is *not conserved* during the collision. Is there some other kind of energy we might add to the kinetic energy so that this total energy might be conserved, i.e., is a constant throughout the motion? Should not there be some sort of energy associated with the compressed bumpers, just as there is energy stored in the spring of a watch? As the kinetic energy decreases during the collision, this spring energy would increase. Perhaps if the forces exerted by the springs are conservative forces, we can find some "energy of compression" which, when added to the kinetic energy, gives us a total energy that is conserved at all times, not just before and after the collision.

To see how this may be done, let us define carefully the *potential energy* of a state.

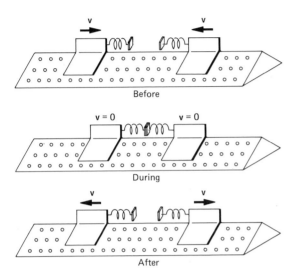

FIGURE 5.16 Collision on an air track between two cars of equal mass.

For every conservative force there exists a potential energy associated with each state which is equal to the work the force does on the system as it goes from this state to the reference state.[1]

This "reference state" serves much the same purpose as the origin of a coordinate axis. We can locate objects in space only by giving them positions relative to other objects. When we say that a body has a coordinate of 3 m, we mean that the object is three meters from the origin. When we speak of the potential energy of a state as being the work done by a force, we must specify where the object came from. Since we will be interested only in *changes* in potential energy, the choice of the reference state is immaterial just as the choice of the origin is immaterial when

we calaculate the distance between two points on a line by finding the *change* in the *x* coordinate.

For example, the gravitational potential energy of a mass at some height above the ground is the work done by the gravitational force on the mass as it moves from this height to the reference state, say some point on the ground. Similarly, the potentital energy of a compressed spring is the work the spring does in expanding to its equilibrium length (the reference state).

For this definition of potential energy to be meaningful, it is important that the force be a conservative force. The definition says nothing about the path by which the system moves to the reference state. If we got different results for different paths, then we would calculate different potentials for the same state. *Potential energy exists only when the force is conservative.*

We can express these ideas very succinctly in mathematical terms. Let $W_{A \to B}$ be the work done by the force as the system goes from state A to state B. States A and B might be two different heights in a uniform gravitational field or two different lengths of a spring. Then

$$W_{A \to B} = \int F_T \, ds$$

where $\int$ denotes a sum calculated along some path (any path) from A to B. If we denote the reference state by O, then the potential energy U_A of a system in state A would be

$$U_A = W_{A \to O} \tag{5.17}$$

The potential energy of state B would be

$$U_B = W_{B \to O} \tag{5.18}$$

---

[1] In some other physics texts the potential energy of a state is defined as the work done *against* the force to take the system *from the reference state to the given state*. Both the force and the direction of the sum along the path are reversed. This doulbe change in sign makes the two definitions equivalent.

We are now in a position to derive a conservation of energy principle. From the principle of work and kinetic energy, we have

$$W_{A \to B} = K_B - K_A \qquad (5.20)$$

If we assume that the forces are conservative, then the path from A to B is immaterial. Let us choose a path that goes through the reference state O (Fig. 5.17). From Eq. 5.14 we may write then

$$W_{A \to B} = W_{A \to 0} + W_{0 \to B}$$

FIGURE 5.17   A path from A to B that passes through the reference state O.

Also, from Eq. 5.15,

$$W_{0 \to B} = -W_{B \to 0}$$

Therefore

$$W_{A \to B} = W_{A \to 0} - W_{B \to 0}$$

or, from the definition of potential energy (Eqs. 5.17 and 5.18),

$$W_{A \to B} = U_A - U_B \qquad (5.21)$$

Combining Eqs. 5.20 and 5.21 gives

$$U_A - U_B = K_B - K_A$$

or

$$K_A + U_A = K_B + U_B \qquad (5.22)$$

for any two states A and B. In other words, the kinetic energy plus the potential energy in any state A is the same as the kinetic energy plus the potential energy in any other state B. The total energy therefore remains constant during the motion. We have proved the principle of *conservation of energy:*

For a system acted on by conservative forces and/or forces that do no work, the total energy remains constant throughout the motion.

It is useful to introduce a symbol to indicate the total energy, kinetic plus potential. If we let E denote the total energy,

$$E = K + U$$

then for Eq. 5.22 we may write

$$E_A = E_B$$

QUESTION

Determine the velocity as a function of position for a simple pendulum (Fig. 5.18).

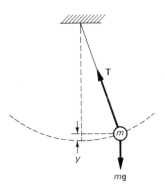

FIGURE 5.18 A pendulum.

ANSWER

Let us assume that the only forces acting on the pendulum bob are gravity and the tension in the string. We shall neglect air resistance. First we note that the tension in the string supporting the bob is always perpendicular to the path and therefore does no work. The gravitational force is a conservative force, and thus we may include its effect in a gravitational potential energy,

$$U = mgy$$

We have chosen the lowest position of the bob as the reference state, and the height y is measured from this level. Since there are no other forces that do work, the total energy is conserved. Therefore

$$\tfrac{1}{2}mv^2 + mgy = \text{constant}$$

We may determine this constant by evaluating the energy at the lowest position ($y = 0$). If we call the velocity there v_0, then

$$\tfrac{1}{2}mv^2 + mgy = \tfrac{1}{2}mv_0^2$$

QUESTION

Determine the velocity as a function of position of a bead sliding without friction on a circular wire (Fig. 5.19).

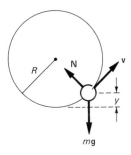

FIGURE 5.19 A bead slides without friction on a circular wire.

ANSWER

This problem is equivalent to that of the pendulum. The only difference is that the tension in the string is replaced by the normal force **N** that the wire exerts on the bead. (If there were a frictional force, there would also be a tangential force that would do work). Once again

$$\tfrac{1}{2}mv^2 + mgy = \tfrac{1}{2}mv_0^2$$

QUESTION

For the above problem of the bead sliding on the smooth wire, what must the minimum velocity of the bead be at the bottom if it is to reach the top of the circle?

ANSWER

If the bead just barely reaches the top, then its velocity there will be zero. Its height above the base will be $2R$, where R is the radius of the circular track. Equating the energy at the bottom to the energy at the top, we have

$$\tfrac{1}{2}mv_0^2 = mg(2R)$$

Therefore the minimum velocity at the bottom must be

$$v_0 = \sqrt{4gR}$$

QUESTION

What is the maximum height that a pole vaulter can clear if he can run at a speed of 30 ft/s carrying the pole? (Note that 30 ft/s is equivalent to 100 yd in 10 s. The record for the 100-yd dash is about 9 s. The record for the pole vault is about 17 ft.)

ANSWER

We shall neglect the energy dissipated as a result of air resistance, and the planting and flexing of the pole. (The fiberglass pole is particularly good at returning stored energy). We shall also neglect the work done by the vaulter's muscles after he leaves the ground. The pole serves as a tool for converting the kinetic energy of the vaulter into gravitational potential energy. We may equate the kinetic energy of the vaulter on the runway to the potential energy at the peak of the vault (Fig. 5.20). (We will neglect the kinetic energy at the peak.) Therefore

$$\tfrac{1}{2}mv^2 = mgy$$

or

$$y = \frac{v^2}{2g} = \frac{(30 \text{ ft/s})^2}{2(32 \text{ ft/s}^2)} = 14 \text{ ft}$$

This is the maximum height that the vaulter could raise his center of gravity. Since his center of gravity is about 3 ft above the ground to begin with, he should be able to clear about 17 ft.

QUESTION

If your best speed on a sled is 30 ft/s on flat ground, to what height could you slide up a hill (Fig. 5.21)?

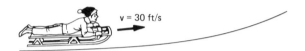

FIGURE 5.21   The speed of the sled at the base of the hill is 30 ft/s.

ANSWER

You could go 14 ft! This problem is equivalent to the problem concerning the pole vaulter. The only difference is the mechanism for converting kinetic energy into potential energy.

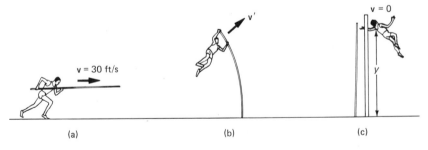

(a)   (b)   (c)

FIGURE 5.20   (a) The pole vaulter approaches the bar at a speed of 30 ft/s, (b) then bends the pole and begins to rise. (c) The pole straightens out again at the top of the vault.

QUESTION

What is the maximum attainable height in the high jump?

ANSWER

Once again 14 ft plus the initial height of the center of gravity! There is a considerable disparity between this figure and the world's record of 7+ ft. The leg is a poor mechanism for converting translational energy into potential energy.

5.7 PRINCIPLE OF WORK AND ENERGY

From the principle of work and kinetic energy we have seen that

$$W_{A \to B} = K_B - K_A$$

That is, the work done by all the forces is equal to the increase in kinetic energy. We have also shown that for conservative forces (Eq. 5.21)

$$W_{A \to B} \text{ (conservative forces)} = U_A - U_B$$

That is, the work done by the conservative forces is equal to the *decrease* in potential energy. For a situation in which both types of forces are present, we may write

$$W_{A \to B} \text{(nonconservative forces)}$$
$$+ W_{A \to B} \text{(conservative forces)} = K_B - K_A$$

or

$$W_{A \to B} \text{(nonconservative forces)}$$
$$= K_B - K_A + U_B - U_A \qquad \textbf{(5.23)}$$

where the work done by the conservative forces has been replaced by the decrease in potential energy. We may call this the principle of *work and energy* and state it as follows:

The work done by the nonconservative forces is equal to the increase in kinetic energy plus the increase in potential energy.

QUESTION

A block slides down a rough inclined plane with a frictional force *f*. What is the velocity of the block at the bottom if it started from rest at the top (Fig. 5.22)?

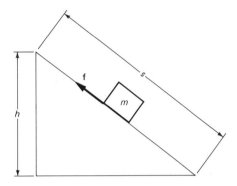

FIGURE 5.22 A block of mass *m* slides from a height *h* to the bottom of an inclined plane. The length of the incline is *s*.

ANSWER

Equating the work done by the frictional force to the increase in kinetic plus potential energy, we have

$$-fs = \tfrac{1}{2}mv^2 - mgh$$

Hence the velocity at the bottom is

$$v = \sqrt{\frac{mgh - fs}{m/2}}$$

QUESTION

How does a swing work?

ANSWER

We may analyze the way in which a child pumps a swing by examining the forces at work, but it is much simpler to apply our principle of work and energy. At the bottom of the arc the child leans back and pulls on the ropes. This bending of the ropes represents work. At the top of the arc the ropes are relaxed. This represents negative work. Since the tension in the ropes is greater at the bottom of the arc than at the top, the positive work done at the bottom is greater than the negative work done at the top. Thus on each cycle there is net work done on the swing and the energy must increase.

As an experiment you might try this with a simple pendulum. Run the string over a fixed support. Now pull on the string when the bob is at the bottom of the arc and allow it to return when the bob is at the top (Fig. 5.23). The amplitude will quickly increase.

FIGURE 5.23   One may increase the amplitude of the pendulum by pulling the string when the tension is greatest and releasing the string when the tension is least.

As we pointed out in the question above, we can analyze the increase in amplitude either by considering the forces at work or by applying the energy principle. In many cases it is very difficult to analyze the forces. Consider for example a bicycle going into a turn. The rider has to lean the bicycle in the direction of the turn. This lowers the center of gravity of the bicycle. Since energy must be conserved, this decrease in potential energy must be compensated for by an increase in kinetic energy. Thus the bicycle must speed up in the turn. It would be difficult to determine what forces are responsible for this acceleration. Something akin to this happens when a coin is spun on a hard, flat surface. As the center of gravity of the coin lowers, the coin revolves faster and faster.

QUESTION

If you want to use a jackscrew to raise the rear end of a 3000-lb automobile, what should be the pitch of the screw if the handle is 1 ft long and you do not want to apply a force in excess of 25 lb?

ANSWER

Let us apply the principle of work and energy to the process in which the handle is given one complete revolution (Fig. 5.24). The work done is the force of 25 lb applied through a distance of $2\pi r$, where r is the length of the handle ($r = 1$ ft). The kinetic energy of the automobile does not change, and therefore the change in energy is the change in potential energy only. The increase in potential energy is mgp, where p is the pitch. If we assume that the weight on the jack is one-half the total weight of the automobile, then mg is 1500 lb. Equating the work to the change in energy yields

$$2\pi r F = mgp$$

or

$$p = \frac{2\pi r F}{mg} = \frac{2\pi(1 \text{ ft})(25 \text{ lb})}{1500 \text{ lb}} = 0.1 \text{ ft}$$

Actually the pitch would have to be less than this because we have neglected friction.

This question could also be approached as a problem in static equilibrium, but this would be more difficult. There are many such statics problems that can be solved more easily with the methods of work and energy.

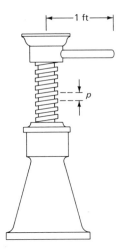

FIGURE 5.24 A jackscrew. The pitch of the screw is the vertical distance between adjacent threads.

5.8 OTHER CONSERVATIVE FORCES AND THEIR ASSOCIATED POTENTIAL ENERGY

We have seen that a uniform gravitational field is conservative; that is, if we calculate the work done by the force of the earth on a body as the body moves along some path between any two points, we obtain the same result regardless of the path. This work is therefore dependent only on the endpoints and, by virtue of our definition of potential energy, is equal to the decrease in potential energy.

Other forces besides the uniform gravitational force are conservative. We shall discuss some of these now.

1. *Nonuniform gravitational force*   The gravitational force of one body on another is

$$\mathbf{F} = -\frac{GmM}{r^2}\hat{\mathbf{r}}$$

(see Eq. 3.13). Let us first consider the work done by this force on a radial path (Fig. 5.25). Let the body of mass m

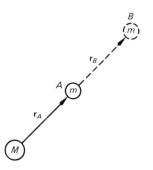

FIGURE 5.25   A body of mass m moves from A to B along the radial lines from the body of mass M.

start from A, a distance r_A from the body of mass M, and move to B at a distance r_B. The work done by the gravitational force is given by

$$W_{A \to B} = \int F_T \, ds = -\int_{r_A}^{r_B} \frac{GmM}{r^2} \, dr$$

$$= GmM\left(\frac{1}{r_B} - \frac{1}{r_A}\right) \tag{5.24}$$

Let us now consider any two points A and B and an arbitrary path between them. As shown in Fig. 5.26, the actual path may be approached with arbitrary precision by the stepped path. Let us compute the work along the stepped path. On those segments for which r is a constant, there is no work done because the force is perpendicular to the path. On the radial segments where θ is constant, the net work done is the sum of the work done on each segment, but this is clearly equal to the work done in

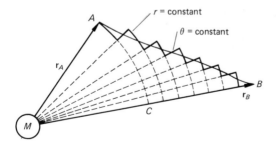

FIGURE 5.26 An arbitrary path joining A and B. The actual path may be approximated by the stepped path, which consists of small segments along which either r or θ is constant, where r and θ are the polar coordinates of the body.

moving the body from C to B, since for every radial segment there corresponds one and only one segment on the line from C to B. The work done along the line CB is

$$W_{C \to B} = GmM\left(\frac{1}{r_B} - \frac{1}{r_C}\right)$$

from Eq. 5.24. But

$$W_{A \to B} = W_{C \to B}$$

and

$$r_C = r_A$$

Therefore

$$W_{A \to B} = GmM\left(\frac{1}{r_B} - \frac{1}{r_A}\right) \tag{5.25}$$

We see that the work is independent of the path and depends only on the position of the endpoints r_A and r_B.

Now our real objective was to calculate the potential energy associated with a state — two masses separated by a distance r. To do this we need only define the reference state and then employ Eq. 5.25 to find the work done in taking the system to the reference state. We shall choose

the reference state to be the state in which the two masses are infinitely far apart. Therefore the potential energy in state A is

$$U_A = W_{A \to \infty}$$

Recall that

$$W_{A \to B} = GmM\left(\frac{1}{r_B} - \frac{1}{r_A}\right)$$

If we let $r_B \to \infty$, we have

$$W_{A \to \infty} = \frac{-GmM}{r_A}$$

Thus

$$U_A = \frac{-GmM}{r_A}$$

or more simply

$$U = \frac{-GmM}{r} \tag{5.26}$$

where U is the potential energy of two masses m and M that are separated by a distance r.

The reason the minus sign occurs in Eq. 5.26 is that if either body is moved to infinity, the gravitational force will be opposite to the direction of motion along the path, so the work done by this force will be negative.

From the manner in which the potential energy was calculated, it would be very natural to assume that U is the potential energy of the body of mass m. After all, we took the integral of $F_T \, ds$, where F_T was the force on m, and it was m, not M, that was moved to infinity. But because of the law of action and reaction, exactly the same amount of work would be required to move M to infinity, if m is kept fixed. What we have calculated is the gravitational potential energy of the *system* of two masses, not the potential energy of either body. A body by itself has no potential energy. Potential energy exists only by virtue of the interrelation between two or more bodies. On the other hand, we may associate a kinetic energy with a single body.

We can perhaps best illustrate this point by considering a collection of three masses m_1, m_2, and m_3, as shown in Fig. 5.27. From the third law of dynamics we know that forces are additive, so the work in moving m_1 to infinity is given by

$$\int \mathbf{F}_{2 \text{ and } 3 \text{ on } 1} \cdot d\mathbf{r} = \int (\mathbf{F}_{2 \text{ on } 1} + \mathbf{F}_{3 \text{ on } 1}) \cdot d\mathbf{r}$$

$$= \int \mathbf{F}_{2 \text{ on } 1} \cdot d\mathbf{r} + \int \mathbf{F}_{3 \text{ on } 1} \cdot d\mathbf{r}$$

$$= \frac{-Gm_1 m_2}{r_{12}} - \frac{Gm_1 m_3}{r_{13}}$$

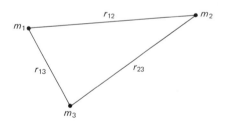

FIGURE 5.27   Three gravitating masses.

Now that m_1 is at infinity, let us move m_2 to infinity. The work done by the force due to m_3 is

$$\int \mathbf{F}_{3 \text{ on } 2} \cdot d\mathbf{r} = \frac{-Gm_2 m_3}{r_{23}}$$

Thus the net work done in putting an infinite distance between the three masses is

$$U = \frac{-Gm_1 m_2}{r_{12}} - \frac{Gm_1 m_3}{r_{13}} - \frac{Gm_2 m_3}{r_{23}} \tag{5.27}$$

and U is the potential energy of the *system* of the three masses.

Not only is U independent of the path, but it is also independent of the process. We leave it as a problem for you to show that we would have obtained the same result (Eq. 5.27) if we have first moved m_3 to infinity and then m_2, or first m_2 and then m_3, or indeed moved all three simultaneously to infinity (i.e., infinitely far from each other). The potentital energy of a system is the work done by the interacting forces as the system is taken to its reference state.

There will be occasions in the material that follows when we may speak of "the potentital energy of a body." We shall do so when only one body moves and all others are fixed. We shall speak of the gravitational potential energy of a body sliding down an inclined plane, or the electrostatic potential energy of a charged particle moving between the charged plates of a capacitor, or the potential energy of a satellite rotating about the earth. In all such cases we are being a bit sloppy and should actually speak of the potential energy of the entire system. Generally speaking, the economy of language makes up for the strain on precision provided, one understands the more fundamental view.

2. *Electrostatic force*   The force between two charges q_1 and q_2 is given by

$$\mathbf{F} = k\frac{q_1 q_2}{r^2}\hat{\mathbf{r}} \tag{5.28}$$

We may derive the potential associated with this force in exactly the same way that we derived the potential for the gravitational force, which has the same form except for a change in notation and a change in sign. (Compare Eqs.

52.8 and 5.23.) The potential energy of the pair of charges is

$$U = k\frac{q_1q_2}{r} \qquad (5.29)$$

3. *The spring* For a spring that obeys Hooke's law,

$$F = -kx \qquad (5.30)$$

If the reference state is taken as the unstretched state, then the work done by this force as the spring moves from x to $x = O$ is

$$U = \int_x^O F\, dx = -\int_x^O kx\, dx = \tfrac{1}{2}kx^2 \qquad (5.31)$$

It may seem that there is only one path and therefore that the question of path independence does not arise. However, there are many ways of taking the spring from one state to another. It might be alternately stretched and compressed several times before being taken to the final state. But no matter how we do it, we always get the same value for the work done.

In Table 5.2 we have listed some forces and their associated potential energies.

TABLE 5.2 Some forces and their associated potential energies.

Force	Potential energy
$\mathbf{F} = \dfrac{-GmM}{r^2}\hat{\mathbf{r}}$	$U = -\dfrac{GmM}{r}$
$\mathbf{F} = k\dfrac{q_1q_2}{r^2}\hat{\mathbf{r}}$	$U = \dfrac{kq_1q_2}{r}$
$\mathbf{F} = -kx$	$U = \tfrac{1}{2}kx^2$
$\mathbf{F} = m\mathbf{g}$	$U = mgy$

QUESTION

A 1-kg mass is dropped from a height of 2 m above a spring with a spring constant of 1000 N/m. What is the maximum displacement of the spring (Fig. 5.28)?

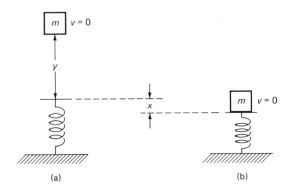

FIGURE 5.28 Initially the block is at rest at a height y above the spring (state A). Later it momentarily comes to rest after compressing the spring a distance x (state B).

ANSWER

If we compute the work done by gravity and the spring by introducing the gravitational and spring potentials, then energy must be conserved because there are no other forces acting on the body. If we choose as reference level for measuring gravitational potential energy the top of the uncompressed spring, then we have

$$mgy = \tfrac{1}{2}kx^2 - mgx$$

where we have equated the energy in state A (see Fig. 5.28) to that in state B. The gravitational potential in state B is negative because the body is at a height $-x$ above the reference level.

Solving for x we have

$$x = \frac{mg}{k} + \frac{\sqrt{m^2g^2 + 2\,mgyk}}{k}$$

$$= \frac{(1 \text{ kg})(9.8 \text{ m/s}^2)}{100 \text{ N/m}}$$

$$+ \frac{\sqrt{(1 \text{ kg})^2(9.8 \text{ m/s}^2)^2 + 2(1 \text{ kg})(9.8 \text{ m/s}^2)(2\text{m})(1000 \text{ N/m})}}{1000 \text{ N/m}}$$

$$= 0.21 \text{ m}$$

QUESTION

What is the escape velocity of a rocket?

ANSWER

The escape velocity is the minimum velocity that a rocket must have at burnout (state A) to be able to reach a point indefinitely far from the earth (state B) (Fig. 5.29). Let us assume that burnout is very close to the surface of the earth and that the rocket moves off to infinity, where its velocity approaches zero. Equating energies in states A and B (see Fig. 5.29) gives

$$\tfrac{1}{2}mv^2 - \frac{GmM}{R} = 0$$

where m is the mass of the rocket, M the mass of the earth, and R the radius of the earth. The kinetic and potential energies are zero at infinity. Solving for v, we have

$$v = \sqrt{\frac{2GM}{R}} = \sqrt{\frac{2(6.67 \times 10^{-11} \text{ N} \cdot \text{m/kg}^2)(6 \times 10^{24} \text{ kg})}{(6.4 \times 10^6 \text{ m})}}$$

$$= 11{,}000 \text{ m/s} = 25{,}000 \text{ mi/h} \qquad (5.32)$$

Note that the direction in which the rocket is fired is irrelevant. (A very small increase in the initial velocity can be achieved by projecting the rocket in the direction of the earth's rotation.) All that matters is v^2. We also see that the mass of the rocket cancels out. Any object traveling in any direction away from the earth at a velocity in excess of 11,000 m/s will escape the earth's gravity.

(We have neglected the effect the sun would have on the rocket. The velocity that a body must have to escape the sun's gravity if it started from a point on the earth's orbit is 42,000 m/s. Therefore a body that escapes from the earth need not escape from the solar system. (See problem 5.C.2.)

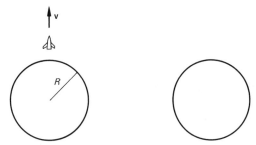

FIGURE 5.29   A rocket leaves the earth with a velocity v (state A) and moves to a great distance, where its velocity is negligible (state B).

QUESTION

The universe is currently in a state of uniform expansion. Will this expansion continue indefinitely?

ANSWER

This question is similar to the preceding question. If the rocket is to continue to move away from the earth indefinitely, it must have a positive total energy. For, if the rocket moved to a very great distance, its potential energy would be nearly zero, but since its kinetic energy is always positive, the total energy would necessarily be positive. If the total energy were negative, it could never recede further than a certain fixed distance. (The planets are examples of this principle; all have a negative total energy and are therefore bound to the sun.)

If the bodies in the universe are to continue to recede from each other, the net energy of the universe must be positive. It is possible to obtain a rough estimate of the energy of the universe, and it appears to be slightly negative. The universe will therefore expand only up to a point and will then begin to contract. It is hypothesized that the universe is in a state of continuous oscillation between two extremes: an incredibly dense state and a very diffuse state.

*5.9 DERIVING THE FORCE FROM THE POTENTIAL ENERGY

In Sec. 5.8 we showed how to obtain the potential energy from the force. We would now like to show how to invert the process, i.e., how to obtain the force from the potential energy.

Suppose we know the potential energy of a body at every point in some region of space:

$$U = U(x, y, z)$$

We know that the work done by the force as the body moves along a path is equal to the negative of the change in potential energy:

$$W = -\Delta U \tag{5.33}$$

Let us consider the work done by the force along three different paths of infinitesimal length. We first consider a path from the point (x, y, x) to the point $(x + dx, y, z)$ (Fig. 5.30). The work done by the force is

$$W = F_x dx$$

where F_x is the component of force tangent to the path and dx is the distance moved.

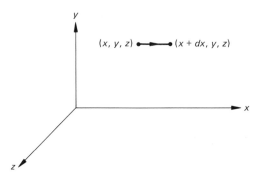

FIGURE 5.30 A body moves from (x, y, z) to $(x + dx, y, z)$.

The change in potential between these two points is

$$dU = U(x + dx, y, z) - U(x, y, z)$$

which, by the definition of the partial derivative, is $(\partial U/\partial x)\, dx$, so that

$$dU = \frac{\partial U}{\partial x} dx$$

Therefore from Eq. 5.33

$$F_x dx = -\frac{\partial U}{\partial x} dx$$

or

$$F_x = -\frac{\partial U}{\partial x} \tag{5.34}$$

Similarly, by taking infinitesimal displacements dy and dz from (x, y, z), we may show that

$$F_y = -\frac{\partial U}{\partial y} \tag{5.35}$$

$$F_z = -\frac{\partial U}{\partial z} \tag{5.36}$$

The quantities $\partial U/\partial x$, $\partial U/\partial y$, and $\partial U/\partial z$ represent the spatial rates of change of the potential energy in three mutually perpendicular directions. This triplet is referred to as the *gradient* of the potential energy. The force therefore is the negative gradient of the potential energy.

We may generalize this result as follows: The component of force in any given direction is the negative of the spatial rate of change of the potential energy in that direction. For example, in a uniform gravitational field

$$F_x(\text{grav}) = -\frac{\partial U(\text{grav})}{\partial x}$$

$$F_y(\text{grav}) = -\frac{\partial U(\text{grav})}{\partial y}$$

$$F_z(\text{grav}) = -\frac{\partial U(\text{grav})}{\partial z}$$

But

$$U(\text{grav}) = mgy$$

so that

$$F_x = 0$$

$$F_y = -mg$$

$$F_z = 0$$

as we expect.

The radial component of the nonuniform gravitational field is given by

$$F_r = -\frac{\partial U}{\partial r} = \frac{\partial}{\partial r}\frac{GmM}{r} = -\frac{GmM}{r^2}$$

The other components of the force are zero since the potential changes only with r.

Lastly, Hooke's law force for a spring is

$$F = -\frac{\partial U}{\partial x} = -\frac{\partial}{\partial x}\left(\frac{1}{2}kx^2\right) = -kx$$

QUESTION

If the potential energy of a body constrained to the x axis is given by

$$U = 2x^3 + 15x^2 + 36x$$

where will it be in equilibrium?

ANSWER

The body will be in equilibrium at points where $F_x = 0$. Now

$$F_x = -\frac{\partial U}{\partial x} = -(6x^2 + 30x + 36)$$

so that

$$6(x^2 + 5x + 6) = 0$$

or

$$6(x + 2)(x + 3) = 0.$$

The body will therefore be in equilibrium at

$$x = -2 \quad \text{and} \quad x = -3$$

5.10 PHYSICAL SIGNIFICANCE OF KINETIC AND POTENTIAL ENERGY

We have pointed out that the kinetic energy is associated with the motion of the system and the potential energy is associated with the configuration of the system. We shall now see exactly what this association is. We shall show that the kinetic and potential energy are the work that a system can do by virtue of its state of motion and its configuration, respectively.

First let us consider a block sliding along a smooth horizontal surface with velocity v. The block strikes a nail and drives it into a wall (Fig. 5.31). We would like to show that the kinetic energy of the block is equal to the work that it does on the nail. From the principle of work and energy,

$$W_{A \to B} = K_B - K_A$$
$$= -\tfrac{1}{2}mv_A{}^2$$

since the velocity in state B is zero. The quantity $W_{A \to B}$ is the work done *on the block* by the nail. This is equal in magnitude but opposite in sign to the work done *on the nail* by the block. Therefore

work done on the nail $= \tfrac{1}{2}mv_A{}^2$

so the kinetic energy of the block is equal to the work that it does on the nail before it comes to rest.

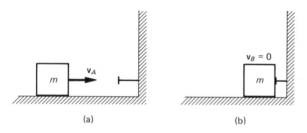

(a)   (b)

FIGURE 5.31 A block is sliding along a frictionless surface with a velocity **v** (state A). The block strikes a nail and drives it into the wall, bringing the block to rest (state B).

Next let us consider a block at rest at a height y_A above the ground. When the block falls it strikes a nail and drives it into the ground (Fig. 5.32). From the principle of work and energy,

$$W_{A \to B} = U_B - U_A = -mgy_A$$

Once again $W_{A \to B}$ is the work done *on the block* by the nail. As before, it follows that the

work done on the nail $= mgy_A$

or the potential energy is equal to the work the block can do in going to its reference state.

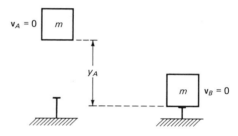

FIGURE 5.32 A block falls from rest (state A) and drives a nail into the ground (state B).

Although we have derived these results for rather special cases, the results are easily generalized. For example, using the above analysis we could show that a spring compressed a distance x is capable of doing work equal to $\tfrac{1}{2}kx^2$.

We should also observe that all our results are predicated on the assumption that the velocity of all objects is much less than the velocity of light. If the velocity of an object is comparable to the velocity of light, then our results must be modified. Furthermore, our interpretation of the energy possessed by a body acquires a new and startling significance.

To see what this modification is, let us begin by summarizing the development of the energy principle for a particle whose velocity is not too great. We start with the law of motion. The time rate of change of the momentum of a particle is equal to a function of the state of the system called the force:

$$\mathbf{F} = \frac{d\mathbf{p}}{dt}$$

The momentum is defined by the relation

$$\mathbf{p} = m\mathbf{v}$$

If the force is conservative, then a potential exists, and for an isolated system

$$K + U = \text{constant}$$

where K is the kinetic energy of the particle, given by $\frac{1}{2}mv^2$, and U is the potential energy.

Now let us consider fast particles, particles whose velocities are not small compared with the velocity of light. We have observed (Eq. 3.9) that the momentum is not $m\mathbf{v}$ but rather

$$\mathbf{p} = \frac{m\mathbf{v}}{\sqrt{1 - v^2/c^2}} \tag{5.37}$$

where c is the velocity of light and has the value 3×10^8 m/s. *If we define the momentum in this way,* we find that for an isolated system the total momentum of all the particles remains constant in time. Note that we may expand the square root in a power series, so that

$$\mathbf{p} = m\mathbf{v}(1 + \tfrac{1}{2}v^2/c^2 + \cdots)$$

If $v \ll c$, then $\mathbf{p} = m\mathbf{v}$ as before.

If we define the momentum by Eq. 5.37, and if the time rate of change of the momentum is equal to a function of the state of the system called the force, and if this force is conservative, then once again

$$K + U = \text{constant}$$

for an isolated system. However, the kinetic energy is not $\frac{1}{2}mv^2$ as before, but becomes instead (see problem 5.C.6)

$$K = \frac{mc^2}{\sqrt{1 - v^2/c^2}} \tag{5.38}$$

If we expand the kinetic energy we find that

$$K = mc^2 + \tfrac{1}{2}mv^2 + \cdots$$

For small velocities (small compared with c),

$$K \simeq mc^2 + \tfrac{1}{2}mv^2 \tag{5.39}$$

We see from Eq. 5.39 that the energy of a single particle is not zero when the velocity is zero. The energy of a body at rest is mc^2. We shall call this the *rest energy*. If mass were always conserved, then this additive constant in the energy would be of no consequence. But the mass is in fact not a conserved quantity; it depends on the energy content. (Equation 5.38 was first proposed by Albert Einstein in a paper titled "Does the Inertia of a Body Depend upon its Energy Content?") To see this, imagine a hydrogen atom in free space at rest with respect to the observer. Then $K = mc^2$, where m is the total mass of the hydrogen atom. Now the atom consists of a proton and an orbiting electron. Suppose we do work on the electron so that it, and therefore the atom as a whole, has a greater energy. The atom is still at rest but it has a greater energy. (The electron is moving but the center of mass of the atom is at rest.) Let the new energy of the atom be $K' = m'c^2$. Since K' is larger than K, it follows that m' is larger than m. The atom increases its mass if its energy content is increased.

This same principle applies to macroscopic bodies as well. When a bar of iron is heated, its energy content is increased and so its mass must increase. When a spring is compressed, its energy content is increased and so its mass is increased. In general, the change in the energy content and the change in the mass are related by the equation

$$\Delta K = \Delta mc^2 \tag{5.40}$$

For macroscopic systems the percentage increase in mass is very, very small. If a spring with a spring constant of 1000 N/m is compressed a distance of 1 m, the increase in mass is $\frac{1}{2}kx^2/c^2 = \frac{1}{2}(1000 \text{ N/m})(1 \text{ m})^2/(3 \times 10^8 \text{ m/s})^2 = 5.6 \times 10^{-15}$ kg. For all practical purposes we may assume that the mass of the spring is constant.

The most familiar example of the relationship between mass and energy content is furnished by radioactive decay of nuclei. If a nucleus at rest breaks up spontaneously into two or more parts, then the mass of the parts must be less than the mass of the original nucleus. To see this we note that before the breakup the energy of the nucleus at rest was mc^2. After the breakup the energy of the fragments is the rest energy $(m_1 c^2 + m_2 c^2 + \cdots)$ of the fragments *plus* the energy associated with their motion. If energy is conserved the rest energy must decrease, that is, $m > m_1 + m_2 + \cdots$.

Another unusual consequence of this new form of the momentum shall be useful to us later. If the velocity is eliminated between Eqs. 5.37 and 5.38, we find

$$K = c\sqrt{m^2 c^2 + p^2} \tag{5.41}$$

5.11 PRINCIPLE OF MINIMUM POTENTIAL ENERGY

We are all familiar with the fact that a ball will eventually settle to the bottom of a bowl. This is also the position at which the gravitational potential energy of the ball is a minimum. Is there a general principle here? Is it true that a minimum potential energy state is a state of static equilibrium? If so, we would have another method of determining the equilibrium state of mechanical systems. Just as we found it easier to analyze the jackscrew by using the energy principle, there may be problems in which it would be easier to look at the potential energy and to search for the state of minimum potential energy than to find that state in which the forces add up to zero.

Suppose we bring a conservative system to rest in some state—for example, a ball held at some point on the side of a bowl. Now we let the system (ball) go and ask whether it will move. Suppose it does move. It will acquire kinetic energy. Since the total energy is conserved and the kinetic energy has increased, the potential energy must have decreased. However, if the state in question is a state of minimum potential energy, the potential energy cannot decrease. Therefore we state that:

The state of minimum potential energy is an equilibrium state for a conservative system.

We shall find an interesting analog of this principle later when we discuss thermodynamics. There we shall see that the state of thermodynamic equilibrium is the state in which a certain function of the thermodynamic state, called the entropy, is a maximum.

QUESTION

What is the equilibrium state of the fluid in the vessel shown in Fig. 5.33?

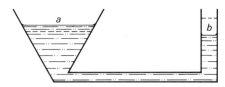

FIGURE 5.33 Two connected vessels containing a fluid.

ANSWER

The potential energy of the fluid has reached a minimum when any further change in the level at a and b produces no increase or decrease in the potential energy. A small decrease in the level at a will produce an increase in the level at b. If a and b are not at the same height, this will mean a change in potential energy. If a and b are at the same height, then a transfer of a small amount of fluid from a to b will not change the potential energy. If a large amount is transferred, then the potential energy will increase whether *a* is lowered and *b* raised or *a* raised and *b* lowered. The equilibrium state therefore is reached when *a* and *b* are at the same levels.

This is exactly what we find when we look for the minimum potential energy of a ball in a trough (see Fig. 5.34). At (1) there is no significant change in potential energy for a small displacement. The potential energy increases away from (1) in either direction for a large displacement. At (2) the potential energy either increases or decreases, depending on the direction of the displacement.

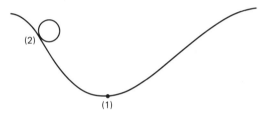

FIGURE 5.34   The potential energy of the ball is a minimum at (1).

5.12 POWER

In many cases it is useful to have some measure of the rate at which work is done. This rate is called the *power P* and is defined by the relation

$$P = \frac{dW}{dt} \tag{5.42}$$

Power has the units of joules per second or watts in the mks system, ergs per second in the cgs system, and foot-pounds per second in the English system. Another common unit of power is horsepower which is equivalent to 746 W or 550 ft·lb/s.

It is possible to express the power in terms of the force and the velocity of the body on which the force acts. Since

$$dW = F_T \, ds$$

then

$$P = F_T \frac{ds}{dt} = F_T v$$

QUESTION

What is the maximum velocity with which a 3000-lb automobile can climb a 45° incline (Fig. 5.35)? (Assume that there is no slipping.) The maximum power output of the engine is 100 hp.

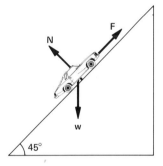

FIGURE 5.35   An automobile climbs a 45° incline with constant velocity.

ANSWER

The power being expended by the automobile is

$$P = Fv$$

where *F* is the driving force. If the automobile is moving up the incline with constant velocity (zero acceleration), the net force tangent to the incline must be zero or

$$F - w \sin 45° = 0$$

so that

$$F = \frac{w}{\sqrt{2}}$$

The power then is

$$P = \frac{w}{\sqrt{2}} v$$

so that the velocity is given by

$$v = \frac{\sqrt{2}P}{w} = \frac{\sqrt{2}(100 \text{ hp})(500 \text{ hp ft} \cdot \text{lb/s})}{3000 \text{ lb}}$$

$$= 26 \text{ ft/s} = 18 \text{ mi/h}$$

SUMMARY

Definitions

Work   The work done on a system by a force as the system moves from state A to state B ($W_{A \to B}$) is the sum along the path of the component of force tangent to the path multiplied by the distance moved:

$$W_{A \to B} = \int F_T \, ds$$

Kinetic energy   The kinetic energy is the energy a body or system possesses by virtue of its state of motion. For a single body of mass m traveling with velocity v, the kinetic energy is given by

$$K = \tfrac{1}{2} mv^2$$

The kinetic energy is also equal to the work that the system can do as a result of its motion in coming to rest.

Potential energy   For every conservative force there exists a potential energy associated with each state which is equal to the work the force does on the system as it goes from this state to the reference state.

Since nothing was said in the definition about the path the system takes to the reference state, it follows that a potential energy exists only when the force is of such a nature that the work done is independent of the path. Such a force is called a conservative force.

Potential energy is the energy a system possesses by virtue of its position or configuration. It is equal to the work the system can do in moving to the reference state by reason of this position or configuration.

Some forms of potential energy are

$$U = mgy \qquad \text{(Uniform gravitational potential)}$$

$$U = -G\frac{mM}{r} \qquad \text{(Universal gravitational potential)}$$

$$U = k\frac{q_1 q_2}{r} \qquad \text{(Electrical potential)}$$

$$U = \tfrac{1}{2} kx^2 \qquad \text{(Spring potential)}$$

Power   Power is defined as the rate at which work is done:

$$P = \frac{dW}{dt}$$

For a single body moving with a velocity v subject to force F, the power expended by this force is

$$P = F_T v$$

where F_T is the component of the force tangent to the path.

There is no physical principle here, merely the definition of a useful concept.

Energy Principles

Principle of work and kinetic energy   The work done on the system by all the outside forces is equal to the increase in kinetic energy of the system:

$$W_{A \to B}(\text{all forces}) = \Delta K = K_B - K_A$$

This principle is quite general and applies to all systems and all forces, conservative and nonconservative.

Principle of work and energy   The work done on the system by the nonconservative forces is equal to the increase in kinetic energy plus the increase in potential energy.

$$W_{A \to B}(\text{nonconservative forces}) = \Delta K + \Delta U$$

$$= K_B - K_A + U_B - U_A$$

$$= \Delta E$$

Principle of minimum potential energy   The state of minimum potential energy is an equilibrium state for a conservative system.

One may determine the force on a body by evaluating the partial derivatives of the potential energy. For example,

$$F_x = -\frac{\partial U}{\partial x}$$

$$F_y = -\frac{\partial U}{\partial y}$$

$$F_z = -\frac{\partial U}{\partial z}$$

PROBLEMS

5.A.1 A ball is thrown vertically into the air from ground level with a speed of 5 m/s.
  (a) With what speed does it strike the ground?
  (b) If the ball were thrown at an angle of 45° above the horizontal with the same speed, with what speed would it strike the ground?

5.A.2 How much work is required to carry a 1-kg package across a room 10 m wide? (The path is horizontal.)

5.A.3 How much work does a 75-kg boy do in one complete chin-up? He pulls himself up 0.25 m and lowers himself 0.25 m.

5.A.4 The 2-kg block in the figure rests on a shelf 1 m above the ground. The block is tied to a spring ($k = 1000$

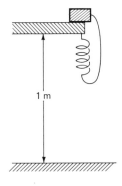

N/m) by a long string. The combined length of string and unstretched spring is 60 cm. If the block is pushed from the shelf, will it strike the ground?

5.A.5 What is the velocity with which the 2-kg block in the following figure will strike the ground if it starts from rest?

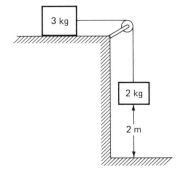

5.A.6 A 2-kg mass is attached to a vertical spring (k = 100 N/m). The spring is unstretched and the mass is at rest. When the mass is released, how far will it fall before it stops.

5.A.7 Show that the escape velocity

$$v_{esc} = \sqrt{\frac{2GM_e}{R_e}}$$

where M_e and R_e are the mass and radius of the earth, respectively, can also be written

$$v_{esc} = \sqrt{2gR_e}$$

5.A.8 A particle moves on a straight line where the potential energy measured in joules is given by $U = 8x^2$. It begins at $x = 1$ m with a kinetic energy of 32 J. Describe the motion. Determine the turning points.

5.A.9 A particle moves on a straight line where the potential energy in joules is given by $U = 8(x^3 + x^2)$.
(a) Where are the equilibrium states?
(b) Are they stable or unstable?
(c) If $U = ax^2y$, what are the x and y components of the force?

5.A.10 The potential energy of a particle is given by

$$U = 3x^2 + 6x + 2y^2 + 4y + 2xy$$

where x and y are measured in meters and U is measured in joules.
(a) Where would the particle be in equilibrium?
(b) How much work must be done on the particle to take it from the origin where it is at rest to the point $x = 1$ m, $y = 1$ m, where it is again brought to rest?

5.B.1 Show that $\mathbf{F} = f(x)\hat{\mathbf{x}}$ is a conservative force for any $f(x)$.

5.B.2 A 5-g bullet enters a 35-g block suspended from a long cord at a speed of 21 m/s and becomes embedded.
(a) Is mechanical energy conserved as the bullet enters the block?
(b) Is momentum conserved as the bullet enters the block?
(c) Is momentum conserved as the bullet and block rise together?
(d) Is energy conserved as the bullet and block rise together?
(e) To what height does the block rise?

5.B.3 In the figure, from what minimum height must the block be released so that it will complete the loop without leaving the track? The loop is a circle of radius 2 ft. (Neglect friction.)

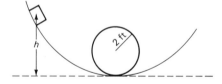

5.B.4 A spring is to be mounted behind the front bumper of a 2000-lb automobile as a cushion. The spring must have the following properties. It must be capable of absorbing the energy of a 30-mi/h head-on collision (a stopping collision), and it must not subject the automobile to more than 30 g's.
(a) What are the spring constant and the compression length?
(b) Explain why a spring would not make a good bumper. Contrast it with a bumper that exerts a constant force over the same distance.

5.B.5 A 170-lb man with an artificial leg seeks to break the world high jump record by raising his center of gravity 5 ft (the equivalent of an 8-ft jump). He inserts a spring

($k = 150$ lb/in) into his artificial leg to store his kinetic energy.

(a) How long must the spring be; that is, how far must the man be able to compress the spring to store enough energy for the feat?

(b) How fast must the man be able to run?

5.B.6 If the velocity of a rocket at burnout is one-half the escape velocity, how far will the rocket rise above the surface of the earth? Assume that the burnout occurs close to the earth's surface.

5.B.7 Determine the force required to support the 100-lb weight in the figure below by the methods of work and energy. (To determine how far the 100 lb is raised when the rope is pulled, imagine that the lower set of three pulleys is raised 1 ft. How much slack does this create in the rope?)

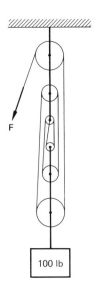

100 lb

5.B.8 We may anticipate some results to be derived later with regard to the equilibrium of rigid bodies. Consider the equilibrium of the system in the following figure. For the system to be in equilibrium, the potential energy must not change when the lever undergoes a small rotation. Show that $wa = Wb$.

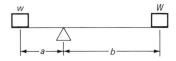

5.B.9 Show that

$$F_x = axy^2$$

$$F_y = 2x^2y$$

is not a conservative force.

5.C.1 If you want to design a golf club to transfer maximum kinetic energy to a golf ball, you must determine how heavy to make the club head. Consider the following idealization. Let the block in the figure represent a club head of mass M. Let m be the mass of the golf ball. Suppose the *kinetic energy* that the golfer can give to the club head is fixed and therefore independent of the choice of M. Show that the velocity of the golf ball will be a maximum if the club head is designed to be of the *same* mass as the golf ball ($M = m$).

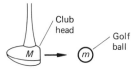

5.C.2 What velocity must a rocket have at burnout to escape both the earth's and the sun's gravitational fields?

5.C.3 Show that in an elastic collision in one dimension the relative velocity with which the particles approach each other before the collision is equal and opposite to the relative velocity with which they recede from each other after the collision.

5.C.4 A block slides without friction on the surface of a sphere. It starts from rest at the top and is displaced slightly. At what latitude does the block leave the sphere?

5.C.5 Show that

$$\frac{\partial F_x}{\partial y} = \frac{\partial F_y}{\partial x}$$

is a necessary condition for a two-dimensional force field to be conservative. (It turns out to be sufficient as well.)

5.C.6 Verify Eq. 5.38 for a particle moving along a straight line (the x axis). You must show that

$$\int \frac{dp}{dt}\, dx = K$$

or equivalently

$$\frac{dK}{dx} = \frac{dp}{dt}$$

where

$$K = mc^2 / \sqrt{1 - v^2/c^2}$$

and

$$p = mv / \sqrt{1 - v^2/c^2}$$

*6
ENERGY CONSERVATION IN A SYSTEM OF MANY PARTICLES

IN CHAP. 5 we discussed principles of work and energy for simple systems, for example, two or three bodies interacting by gravitational forces, a mass on the end of a spring, or a block sliding down an incline. We observed that the work done on the system by outside forces is equal to the change in the total energy of the system. If the system is isolated or the outside forces do no work, then energy is conserved.

In light of these principles, let us examine the motion of a block sliding along a flat horizontal table top. The block will eventually come to rest as a result of friction between the surfaces. When we apply our principle of work and energy to the system consisting of the block and the table, we see an apparent contradiction. The only outside force is that of gravity, and this force does no work on the table since the table does not move, and it does no work on the block since the force is perpendicular to the direction of motion. (The force of friction between the block and the table top is an internal force.) Since there is no work done by outside forces, the total energy ought to be conserved. But the block comes to rest; it looses its kinetic energy. Since the table top is horizontal, the gravitational energy remains constant. We appear to have a system in which no work is done by outside forces and the energy is not conserved.

We might take refuge in the fact that the frictional force between the block and the table is not a conservative force, so the energy principle does not apply. However, when we look at the frictional force on a microscopic level, we find that it arises from the interaction between molecules. This interaction is primarily electrostatic, and we have seen that electrostatic forces are conservative. Therefore, when we look at the system on a microscopic level, we must expect that energy be conserved. This means, however, that we must look at the energy of the system on a microscopic level as well; that is, we must take into account the energy of the molecules that make up the block and the table top. This energy is often called the *internal energy*, and it is to be distinguished from the translational kinetic energy of the block as a whole. If the block is sliding along and eventually comes to rest as a result of frictional force, then the translational kinetic energy of the block is transformed into internal energy in the molecules that make up the block and the table top.

Internal energy may be exhibited in many ways. The molecules may move about more rapidly, thereby increasing their kinetic energy. The molecules may rotate faster or vibrate with a larger amplitude, or some of the electrons may gain energy. Some chemical reactions may also take place. This change in the manner in which the atoms bind together to form molecules represents a change in chemical energy. It is not necessary for us to understand all the details of these various energy forms now. It is sufficient for us to realize that if we look closely enough at any isolated system or any system in which no work is done by outside forces, we shall find that the *total* energy of the system is conserved.

We shall discuss this concept of internal energy at some length in Chap. 14 when we take up thermodynamics, and we shall find that it plays a central role in the development of this subject. However, a number of applications of the energy principle applied to the internal energy states of physical systems follow from the laws of dynamics alone. One in particular that we would like to consider is of considerable importance in all living organisms, and this is the energy requirement for the multitude of metabolic processes that are so vital in sustaining life.

6.1 ENERGY CONSERVATION AND METABOLISM

We have observed that the essential feature of the law of energy conservation for isolated systems is that the total energy has the same value at the final state as it had at the initial state *regardless of the means by which it evolved from one state to the other*. This is very important when we are applying the principle to systems such as plants or animals or even one cubic meter of gas at one atmosphere of pressure. All these systems have so many particles and the overall processes are so complicated that we cannot hope to follow the details.

As an example of the value of not having to know the details, consider the burning of glucose (a sugar) in air. Glucose decomposes into carbon dioxide and water with the release of 2.87×10^6 J of energy per mole.[1]

$$C_6H_{12}O_6 + 6 \ O_2 \rightarrow 6 \ CO_2 + 6 \ H_2O + 2.87 \times 10^6 \ J$$

Another way to break down glucose is to eat and digest it. The process now is not quite so simple. The reaction is facilitated by several enzymes; the energy of two ATP molecules are required to initiate the reaction; and it produces as intermediates 40 ATP molecules from ADP molecules. These ATP molecules will serve as temporary energy storage elements in the body, and later they will be used to drive some other reaction and will release their energy and revert back to ADP. However, when all is said and done, when 6 CO_2 molecules and 6 H_2O molecules have been produced and all the ATP produced has reverted back to ADP, then we know that 2.87×10^6 J of energy must have been released for each mole of glucose. This energy may show up in the form of heat energy, or mechanical work, or osmotic work performed by a nerve cell, or work necessary to make a new DNA molecule, and so on. Since the energy released by the oxidation of glucose is independent of the path, we can perform the oxidation in the test tube and know that the energy released in the body is the same.

This is why a soft drink company can advertise the number of calories in its product. The company does not have to know how the energy is used in the body; it has to know only what happens in the laboratory.

A new unit that is commonly used to denote the energy released in chemical reactions is the calorie (abbreviated cal), which is defined as the amount of energy that must be transferred to 1 g of water to raise its temperature $1 \, °C$.

---

[1] One mole (abbreviated mol) of a substance contains 6.02×10^{23} molecules.

We find that it takes 4.18 J of energy to raise 1 g of water 1°C. Therefore

$$1 \text{ cal} = 4.18 \text{ J} \qquad \text{(6.1)}$$

The quantity that is called a calorie in dietary matters is actually a kilocalorie (abbreviated kcal). A kilocalorie is 1000 calories:

$$1 \text{ kcal} = 1000 \text{ cal} = 1 \text{ dietary calorie}$$

QUESTION

How many calories are released when 1 mol of glucose is oxidized?

ANSWER

The energy released is 2.87×10^6 J or

$$\frac{2.87 \times 10^6 \text{ J}}{4.18 \text{ J/cal}} = 686,000 \text{ cal} = 686 \text{ kcal}$$

The calorie is not just a unit of heat energy. It may be used to denote the magnitude of any form of energy.

QUESTION

How many calories of gravitational potential energy does a 70-kg person possess when standing on a table 1 m high?

ANSWER

The potential energy is

$$E_p = mgh = (70 \text{ kg})(9.8 \text{ m/s}^2)(1 \text{ m}) = 686 \text{ J}$$

$$= \frac{686 \text{ J}}{4.18 \text{ J/cal}} = 164 \text{ cal}$$

QUESTION

In the above example, if the person jumped into a pan containing 100 g of water and all the energy went into heating the water, what would be the increase in temperature of the water?

ANSWER

If 164 cal of energy is added to the water and if 1 cal raises 1 g of water 1°C, the 100 g of water would increase in temperature by 1.64°C.

The fact that glucose is broken down in small discrete stages is of great physiological significance. The source of most of the body's immediate energy needs is a molecule called ATP. This molecule interacts with water to yield a new molecule called ADP plus an inorganic phosphate, and in the process 7.6 kcal of energy are released per mole of reactants. This energy is readily available and is of the right magnitude. It is the primary source of energy for such diverse processes as muscle contraction, transport of ions across cell membranes, and the synthesis of a great many chemical elements needed within the body. The process ATP → ADP + 7.6 kcal can be reversed if 7.6 kcal of energy are supplied. This is where glucose comes in. We have seen that in the breakdown of one mole of glucose 686 kcal of energy are released. It would be very wasteful if 1 mol of glucose were used to supply the 7.6 kcal of energy required to produce 1 mol of ATP from 1 mol of ADP. The degradation of glucose in fact takes place in many small steps, thereby producing many moles of ATP (40 in all). Some of the energy in glucose is obviously given to other energy forms since 686 kcal is greater than 40 (7.6 kcal) = 304 kcal.

By way of analogy, imagine a large ball resting on top of a table. The ball has a great deal of potential energy, and we would like to store this energy in more convenient units. For this purpose we have a number of springs, each equipped with a ratchet so that when a spring is compressed the ratchet will hold it in the compressed state. The energy stored in the compressed spring is much less than the total potential energy in the ball on the table top. If we simply dropped the ball from the table onto one of the springs, thereby compressing it, most of the energy would be lost and only a small fraction would be stored in the single compressed spring.

However, if we bounced the ball down a flight of stairs with a spring on each step, then many springs could be compressed and much more of the energy stored (Fig. 6.1).

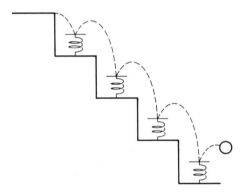

FIGURE 6.1 A ball bounces down a flight of stairs and compresses a spring on each step.

This is analogous to the step-by-step degradation of glucose into carbon dioxide, water, and the 40 ATP molecules. The original energy (686 kcal/mol of glucose) is stored in more usable units of 7.6 kcal/mol of ATP.

QUESTION

We have spoken of the energy stored in glucose and ATP. However, all chemical bonds are attractive or they would not be bonds. Since it *takes* energy to break bonds, how is it possible to *gain* energy from glucose by breaking the bonds?

ANSWER

It surely would take energy to simply break up glucose, but this is not what happens either in the test tube or in the body.

Glucose interacts with other molecules, and in the process some bonds are broken and replaced with other *tighter* bonds. The collection of atoms is proceeding to a lower energy state, and so this decrease in energy is available in some other form: heat, chemical energy, electric energy, etc. So when one speaks of "the energy-rich phosphate bond" in ATP, one must remember that it actually *requires* energy to break this bond, and it is the fact that this bond is replaced by a much stronger bond that makes ATP valuable as an energy source. The same may be said for glucose.

6.2 OXYGEN COMSUMPTION AND METABOLISM

We have seen that 1 mol of glucose reacting with 6 mol of O_2 produces 686 kcal of energy. The body uses other foodstuffs for energy besides glucose; other carbohydrates, proteins, and fats are the principal sources. For each of these it is possible to measure the energy released per mole when the substance combines with oxygen. If we knew the net oxygen consumption of a person or an animal and the relative proportion of foodstuffs being oxidized, we could calculate the total energy being liberated. Measuring the oxygen consumption is a simple matter that is accomplished with an instrument called a respirometer. It is also possible to obtain a good estimate of the kinds of foodstuffs being oxidized by observing the CO_2 exhaled and the amount of nitrogen (a by-product of protein metabolism) in the urine.

The energy released by glucose for each mole of oxygen consumed is $686/6 = 114$ kcal/mol since there are 6 mol of oxygen that combine with every mole of glucose. Table 6.1 gives the average energy released per mole of oxygen for various foods as well as the kilocalories released per gram of foodstuff.

TABLE 6.1 Energy release by some foodstuffs.

Foodstuff	Average kcal/mol of O_2	Average kcal/g
Carbohydrates	110	4.1
Proteins	100	4.2
Fats	105	9.3

QUESTION

How many kilocalories are released per gram of glucose?

ANSWER

There are 686 kcal/mol, and the molecular weight of $C_6H_{12}O_6$ is

$$6(12) + 12(1) + 6(16) = 180 \text{ g/mol}$$

Therefore the number of kilocalories per gram is given by

$$\frac{686 \text{ kcal/mol}}{180 \text{ g/mole}} = 3.8 \text{ kcal/g}$$

which is somewhat below the average of 4.1 kcal/g for the other carbohydrates as a class.

QUESTION

The basal metabolic rate (BMR) is defined as the number of kilocalories required to sustain a person in a resting state under certain standard conditions. If a person inhales and retains 19 mol of oxygen over a 24-h period, what is the individual's BMR? Assume that the person generates 108 kcal/mol of O_2 inhaled.

ANSWER

$$BMR = \left(108 \frac{\text{kcal}}{\text{mol } O_2}\right)\left(19 \frac{\text{mol } O_2}{\text{day}}\right) = 2050 \text{ kcal/day}$$

QUESTION

In the preceding example, if the person fasted for 24 h, how many pounds would be lost? Assume that this individual is feeding on fat reserves and rests for the entire period.

ANSWER

The person's energy needs are 2050 kcal, and fat will generate 9.3 kcal/g. Thus the weight loss is

$$\frac{2050 \text{ kcal}}{9.3 \text{ kcal/g}} = 0.49 \text{ lb}$$

QUESTION

If you wanted to lose 1 lb (450 g) of fat by exercising, how much exercise would be required? Assume that exercise requires 500 kcal/h.

ANSWER

The amount of fat that is burned per hour is given by

$$\frac{500 \text{ kcal/h}}{9.3 \text{ kcal/g}} = 54 \text{ g/h}$$

To lose 450 g you would have to exercise for

$$\frac{450 \text{ g}}{54 \text{ g/h}} = 8.4 \text{ h}$$

This is a difficult way to lose weight. Dieting is much more effective.

SUMMARY

Energy can appear in many forms: kinetic energy, gravitational energy, electrostatic energy (including chemical energy), nuclear energy, and electromagnetic energy. Energy can be transformed from one form to another, but it cannot be gained or lost in an isolated system. The total energy of an isolated system is always conserved.

The fact that energy is conserved and that the energy

of a state depends only on the state and not on the path by which the system got to that state allows us to determine how much energy we obtain from various foodstuffs. If we know the nature of the initial chemical and the nature of the endproducts after metabolism in the body, we may determine how much energy is released in the body by observing the reaction in the test tube. The initial state and the final state are the same in the test tube and in the body. The paths by which the chemicals get to the final state are quite different.

PROBLEMS

6.A.1 Often one sees chemical reactions written in the form

$$4HCl + O_2 \rightleftharpoons 2H_2O + 2Cl_2 + 28.4 \text{ kcal}$$

The equation is somewhat confusing. The double arrow suggests equilibrium, but how can the system be in equilibrium if heat is being generated? The interpretation of this equation is as follows. If we put these compounds into a container and wait a while, the system will reach an equilibrium state in which there are certain proportions of HCl, O_2 H_2O, and Cl_2. The equation gives no information about the relative proportions of each compound. The equilibrium is a dynamic equilibrium. Some HCl and O_2 molecules will be reacting to form H_2O and Cl_2 molecules, but also H_2O and Cl_2 molecules will be reacting to form HCl and O_2 molecules. In the equilibrium state the amount of HCl being formed will be equal to the amount being lost. The same can be said of the O_2, H_2O, and Cl_2 molecules.

The significance of the 28.4-kcal term on the right-hand side of the equation is that if we *add* to the system 4 mol of HCl and 1 mol of O_2 and *remove* 2 mol of H_2O and 2 mol

of Cl_2, then the system will heat up (increase in temperature) unless we remove 28.4 kcal of heat from the system. Roughly speaking, we may say that if the reaction proceeds to the right, heat is generated and the reaction is exothermic. A reaction that proceeds to the left is endothermic.

Consider now the following problem. Suppose we have a mixture of HCl, O_2, H_2O, and Cl_2 in equilibrium, and we add heat to the system. How will the relative abundances of the compounds be affected? Will there be more HCl and O_2 and less H_2O and Cl_2 or vice versa?

6.A.2 Consider the following reactions:
  (a) $H_2 + Cl_2 \rightleftharpoons 2HCl + 44.2 \text{ kcal}$
  (b) $CH_4 + 2O_2 \rightleftharpoons CO_2 + 2H_2O + 212.8 \text{ kcal}$
  (c) $N_2O_4 + 14 \text{ kcal} \rightleftharpoons 2NO_2$
  (d) $2CO_2 + 135.2 \text{ kcal} \rightleftharpoons 2CO + O_2$
  (e) $H_2 + I_2 \rightleftharpoons 2HI + 12.4 \text{ kcal}$
What happens when heat is added to each of these systems?

6.A.3 What is the kinetic energy in calories of a 3-kg block traveling at a speed of 4 m/s?

6.A.4 If your BMR is 2000 kcal/day, what is your power consumption in watts?

6.A.5 To appreciate the significance of the magnitude of the BMR (2000 kcal/day), calculate how high a 100-lb weight could be raised with 2000 kcal of available energy. (There are 4.2 J/cal, 2.2 lb/kg, and 2.54 cm/in)

6.A.6 Does it require more energy to pull apart a hydrogen molecule and a chlorine molecule than is recovered when two hydrogen atoms are joined to two chlorine atoms? ($H_2 + Cl_2 \rightleftharpoons 2HCl + 44.2 \text{ kcal}$.)

6.A.7 If energy is added to a gas of hydrogen, will the percentage of ionized hydrogen atoms increase or decrease?

6.B.1 One mole of gas occupies 22.4 liters (1 ℓ = 1000 cm^3) at 1 atm of pressure of 0°C. One-fifth of these molecules are oxygen. Find an upper limit on how long a person could survive locked in an airtight vault 2 × 2 × 2 m.

6.B.2 The average rate at which solar energy falls on 1 m^2 of the earth's surface perpendicular to the sun's rays is 0.3 kcal/m^2 · s. If all we needed from the sun was energy and we could make 100% utilization of this energy, what is the maximum number of people per square meter that could live at the equator if each person's BMR was 2000 kcal/day?

6.B.3 An air conditioner is rated at 2000 Btu/h, which is equivalent to 504 kcal/h. Roughly speaking, this means that the air conditioner will remove 504 kcal of heat from a room every hour. If the air conditioner runs 50% of the time to maintain a room at 70°F, what fraction of the time would it run if a person whose BMR is 2000 kcal/day entered the room?

6.B.4 Suppose you want to take food to eat after you have climbed a mountain 5000 m high. What minimum number of kilocalories per gram should the foodstuff contain?

*7
DYNAMICS OF SYSTEMS OF MANY PARTICLES

IN CHAP. 3 we formulated the basic laws of particle dynamics. We saw, for example, that if we know the force on a particle, we may determine the acceleration of the particle ($F = ma$). If we know the position and velocity of a particle at some moment and know how the acceleration (or the force) depends on position, we can predict the future motion of the particle. But suppose we are asked to determine how long it would take a wheel to roll down an inclined plane. We might be able to determine the forces acting on the wheel, but what is *the acceleration of the wheel*? There clearly is no single acceleration to be associated with all the particles that make up the wheel. How are we to deal with such problems?

In the past we have dealt with these problems by simply ignoring them. We have treated balls, blocks, people, the earth, and the sun as if they were single particles. They are not, and we shall now attempt to deal with these problems honestly.

Let us look at the wheel as a collection of particles. We know the rules for the dynamics of particles. Is it possible to examine the forces acting on each particle in the wheel, determine the motion of each particle, and thereby determine the motion of the wheel as a whole? This seems like a formidable task. There are many particles in the wheel, and it would take a life time to even count them, much less analyze their motion.

We shall demonstrate in this chapter that it is possible to derive two laws: The first determines the motion of one point in the body (called the *center of mass*), and the second determines the manner in which the body rotates about this point. In the case of a wheel or disk this point is the center of the object (if the mass is distributed symmetrically about the center). If we know how the center of the wheel moves and how the wheel rotates about the center, we have completely determined the motion of the wheel. (Note that if the incline is very slippery, the wheel may be sliding down the incline. Then the rotation of the wheel and the translation of the center of the wheel are independent.)

The remarkable thing about these two laws governing the motion of the center of mass and the rotation about the center of mass is that they *do not involve the complicated internal forces acting among the vast number of individual particles.* They depend on only the *external* forces acting on the body. In the case of the wheel on the incline, we can determine the motion of the center of the wheel and the rotation about the center by considering only the force of gravity on the wheel and the force of the incline on the wheel.

Before we can derive these remarkable theorems from our laws of particle dynamics, we must first define precisely the center of mass of a collection of particles.

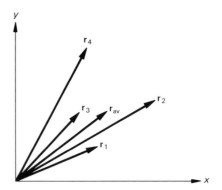

FIGURE 7.1   The average position vector of a collection of particles.

7.1 CENTER OF MASS

As we have mentioned before, we cannot derive definitions. We can, however, make them reasonable or at least intuitively clear. We have said above that the center of mass of a uniform disk is at its center. Whatever "center of mass" means, it would be surprising if it were anywhere else. For a baseball bat, however, the location of the center of mass is not so obvious.

Suppose we want to define an *average position* for a cluster of particles of equal mass. We might define the average position vector in very much the same way that we normally determine averages: We would add the quantities to be averaged and divide by the number of things being averaged (Fig. 7.1). Therefore let us define

$$\mathbf{r}_{av} = \frac{\mathbf{r}_1 + \mathbf{r}_2 + \cdots + \mathbf{r}_n}{n} \tag{7.1}$$

where $\mathbf{r}_1, \mathbf{r}_2, \ldots, \mathbf{r}_n$ are the position vectors of n particles of equal mass.

If two of the particles are coincident, say $\mathbf{r}_1 = \mathbf{r}_2$, then

$$\mathbf{r}_{av} = \frac{2\mathbf{r}_1 + \mathbf{r}_3 + \cdots + \mathbf{r}_n}{n} \tag{7.2}$$

In effect, if we double the mass at some point, we double the contribution from that point.

If we have a collection of unequal masses m_1, m_2, etc., we might imagine them to be a collection of equal mass (multiples of some very small unit of mass) many of which are coincident. The number of multiples of the small unit of mass at any point is proportional to the net mass at that point. We would then obtain for the average position vector for the collection of unequal masses

$$\mathbf{r}_{av} = \frac{m_1\mathbf{r}_1 + m_2\mathbf{r}_2 + \cdots + m_n\mathbf{r}_n}{m_1 + m_2 + \cdots + m_n} \tag{7.3}$$

Note that if $m_1 = m_2 = \cdots = m_n$, then we recover Eq. 7.1.

This average position vector turns out to be a useful quantity. It is normally called the *center of mass* (abbreviated c.m.), so that

$$\mathbf{r}_{c.m.} = \frac{m_1\mathbf{r}_1 + m_2\mathbf{r}_2 + \cdots + m_n\mathbf{r}_n}{m_1 + m_2 + \cdots + m_n} \tag{7.4}$$

It is a simple matter to show that the center of mass of a uniform body that has a center of symmetry is located at the center of symmetry. Suppose the origin of the coordinate system is at the center of symmetry. Then for every mass m_1 at r_1 there is an equal mass at r_2 where $r_2 = -r_1$. Thus the sum $m_1r_1 + m_2r_2 + \cdots$ must be zero, so the center of mass is at the center of symmetry. The center of mass of a homogeneous sphere is at the center of the sphere. The center of mass of a homogeneous parallelepiped is at the center of the parallelepiped.

QUESTION

Where is the center of mass of the two spheres in Fig. 7.2.

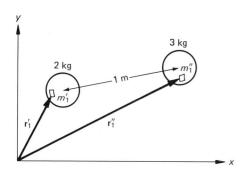

FIGURE 7.2 Two spheres 1 m apart.

ANSWER

The position of the center of mass is given by

$$\mathbf{r}_{\text{c.m.}} = \frac{(m_1'\mathbf{r}_1' + m_2'\mathbf{r}_2' + \cdots) + (m_1''\mathbf{r}_1'' + m_2''\mathbf{r}_2'' + \cdots)}{(m_1' + m_2' + \cdots) + (m_1'' + m_2'' + \cdots)} \qquad (7.5)$$

where we have chosen to group the sum into two parts, the particles associated with the 2-kg sphere (m_1', m_2', etc.) and those associated with the 3-kg sphere (m_1'', m_2'', etc.). Now, from Eq. 7.4

$$m_1'\mathbf{r}_1' + m_2'\mathbf{r}_2' + \cdots = M'\mathbf{r}_{\text{c.m.}}' \qquad (7.6)$$

where M' is the mass of the 2-kg sphere and $\mathbf{r}_{\text{c.m.}}'$ is the position of the center of mass of the 2-kg sphere. Similarly

$$m_1''\mathbf{r}_1'' + m_2''\mathbf{r}_2'' + \cdots = M''\mathbf{r}_{\text{c.m.}}'' \qquad (7.7)$$

where M'' is the mass of the 3-kg sphere and $\mathbf{r}_{\text{c.m.}}''$ is the center of mass of the 3-kg sphere. Therefore

$$\mathbf{r}_{\text{c.m.}} = \frac{M''\mathbf{r}_{\text{c.m.}}' + M''\mathbf{r}_{\text{c.m.}}''}{M' + M''} \qquad (7.8)$$

But this is exactly what we would have obtained if we had treated each sphere as a single mass point located at its center of mass. We leave it as an exercise for you to show that the center of mass is $\frac{3}{5}$m from the 2-kg mass along the line joining the two spheres.

This procedure may be followed quite generally. For example, suppose we want to find the center of mass of the elbow in Fig. 7.3. We may treat the elbow as if it were two adjoining rectangles. The centers of mass of the rectangles are clear. We may then treat the two rectangles as two mass points located at their respective centers of mass. The masses of the rectangles will be proportional to their areas if we assume that they are homogeneous.

FIGURE 7.3 The elbow may be considered two adjoining rectangles.

7.2 EQUATION OF MOTION FOR THE CENTER OF MASS

We promised to obtain both an equation of motion for the center of mass and an equation for the rotation about the center of mass which do not involve the complicated internal structure. Let us look at the equation of motion for each particle and see what we can do about eliminating all the internal forces. These equations are

$$\mathbf{F}_1(\text{ext}) + \mathbf{F}_1(\text{int}) = m_1\mathbf{a}_1$$

$$\mathbf{F}_2(\text{ext}) + \mathbf{F}_2(\text{int}) = m_2\mathbf{a}_2$$

$$\vdots \tag{7.9}$$

$$\mathbf{F}_n(\text{ext}) + \mathbf{F}_n(\text{int}) = m_n\mathbf{a}_n$$

where $\mathbf{F}_1(\text{ext})$ is the sum of all the external forces acting on the particle of mass m_1 and $\mathbf{F}_1(\text{int})$ is the sum of all the internal forces acting on this particle, i.e., all the other particles in the body. Let us add these n equations together. One of the terms will be the sum of all the internal forces on all the particles

$$\mathbf{F}_1(\text{int}) + \mathbf{F}_2(\text{int}) + \cdots + \mathbf{F}_n(\text{int})$$

but by the law of action and reaction this sum must be zero. For every force that one particle exerts on another, there is an equal and opposite force to cancel it. The sum of the n equations then becomes

$$\mathbf{F}_{\text{tot}}(\text{ext}) = m_1\mathbf{a}_1 + m_2\mathbf{a}_2 + \cdots + m_n\mathbf{a}_n \tag{7.10}$$

where $\mathbf{F}_{\text{tot}}(\text{ext})$ is the sum of all the external forces acting on all the particles.

We have accomplished half our objective. We have eliminated the complicated internal forces. But what of the term on the right-hand side of Eq. 7.10? Let us return to Eq. 7.4.

$$M\mathbf{r}_{\text{c.m.}} = m_1\mathbf{r}_1 + m_2\mathbf{r}_2 + \cdots + m_n\mathbf{r}_n$$

This equation holds for all time, so we may differentiate with respect to time to give

$$M\mathbf{v}_{\text{c.m.}} = m_1\mathbf{v}_1 + m_2\mathbf{v}_2 + \cdots + m_n\mathbf{v}_n \tag{7.11}$$

Differentiating again we have

$$M\mathbf{a}_{\text{c.m.}} = m_1\mathbf{a}_1 + m_2\mathbf{a}_2 + \cdots + m_n\mathbf{a}_n \tag{7.12}$$

But the right-hand side of Eq. 7.12 is the same as the right-hand side of Eq. 7.10, so

$$\mathbf{F}_{\text{tot}}(\text{ext}) = M\mathbf{a}_{\text{c.m.}} \tag{7.13}$$

This is the stated objective: a relatively simple equation of motion for the center of mass.

Another equation for a collection of particles that we will have use for later is

$$\mathbf{F}_{\text{tot}}(\text{ext}) = \frac{d\mathbf{p}_{\text{tot}}}{dt} \tag{7.14}$$

which can be obtained immediately from Eq. 7.10 and the fact that the total momentum is given by

$$\mathbf{p}_{\text{tot}} = m_1\mathbf{v}_1 + m_2\mathbf{v}_2 + \cdots + m_n\mathbf{v}_n$$

QUESTION

A bomb that is dropped from an airplane in horizontal flight explodes during its descent. What can you say about the motion of the bomb? (Neglect air resistance.)

ANSWER

From Eq. 7.13,

$$\mathbf{F}(\text{ext}) = M\mathbf{a}_{\text{c.m.}}$$

where $\mathbf{F}(\text{ext}) = M\mathbf{g}$, the weight of the bomb. Therefore

$$\mathbf{a}_{\text{c.m.}} = \mathbf{g}$$

This equation for the acceleration of the center of mass is exactly the same as that for a single mass point, and therefore the trajectory of the center of mass must be a parabola. The motion of the bomb itself will be complicated. It will tumble about the center of mass, and it is impossible to predict the complicated fragmentation when it explodes. However, the center of mass of all the parts will continue in the parabola (Fig. 7.4).

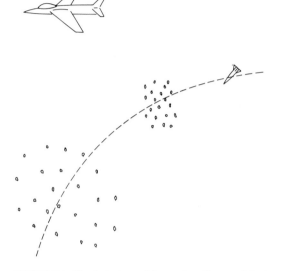

FIGURE 7.4 The trajectory of the center of mass of the bomb is a parabola.

QUESTION

A long string is wrapped around a disk and pulled upward as shown in Fig. 7.5 with a constant force equal to the weight of the disk. The disk is initially at rest. Describe the motion of the center of mass.

FIGURE 7.5 A string is wrapped around a disk, and an upward force equal to the weight of the disk is applied to the end of the string.

ANSWER

Since the net external force acting on the disk is zero, the acceleration of the center of mass must be zero. The center therefore remains fixed, and the disk rotates faster and faster about its center.

QUESTION

A rocket ship is at rest in space in some inertial frame of reference. The rockets are fired and the ship travels in a straight line for a million miles. Find the center of mass of the ship and its exhaust, assuming that it is far from any star.

ANSWER

The center of mass is exactly where it was when the rockets were fired. If there is no external force, $\mathbf{a}_{\text{c.m.}} = 0$.

7.3 TORQUE

Now that we have discovered the law that determines the motion of the center of mass from the external forces alone, we would like to obtain a law that determines the motion of the body about the center of mass. If the body is a rigid body, then the motion of the center of mass and the rotation about the center of mass completely determine the motion of the body. If the body is not a rigid body, then these laws will be of only limited use. If you are struck by shrapnel from a grenade that explodes in flight, it is small consolation to know that you do not lie on the parabolic path followed by the center of mass.

A problem that is very similar to that of the motion of a *free rigid body* rotating about the center of mass is that of a *rigid body rotating about a fixed axis.* This axis may or may not pass through the center of mass. This problem is simpler conceptually and mathematically, and we shall consider it first.

In Fig. 7.6 an external force **F** is applied to a rigid rod constrained to rotate about a fixed axis perpendicular to the page through point O. There are, of course, many internal forces that are not shown in the figure. We will attempt as before to perform the feat of determining the motion of the body (in this case the rotation about the axis rather than the motion of the center of mass) without having to worry about the very complex internal forces.

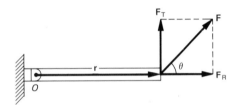

FIGURE 7.6 A rigid rod rotating about a fixed axis perpendicular to the page through point O.

First we would like to define a measure of the ability of the force to rotate the body. Let us resolve the force into three components: one along the radius vector **r**, which we denote by F_R (the radial component); one denoted by F_T (the tangential component) perpendicular to **r**; and one parallel to the axis, denoted by F_P (not shown in Fig. 7.6). The tangential component is not tangent to the body, but tangent to the *circle* in which the point of application of the force will move. Clearly F_R and F_P will have no effect on the rotation of the body. Certainly the rotation will depend on the magnitude of F_T. Also we expect that the further the point of application of the force is from the axis of rotation, the greater its effect in rotating the body. If the force were

applied at the axis, there would be no rotation. Therefore the tendency of the force **F** to produce rotation should depend on F_T and r. Let us define the tendency of the force to cause rotation to be

$$\tau = rF_T \tag{7.15}$$

The quantity τ is called the *moment of force F about the axis O*, which is also referred to as the *torque*.

We may also define torque as the product of the force F and the moment arm l of the force (see Fig. 7.7):

$$\tau = lF \tag{7.16}$$

The moment arm of **F** is the perpendicular distance between the line of action of the force and the axis of rotation. It is clear from Fig. 7.7 that these two definitions are equivalent.

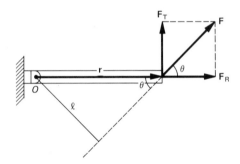

FIGURE 7.7 The moment arm of the force **F** is l.

We see that

$$F_T = F \sin \theta$$

and

$$l = r \sin \theta$$

so that

$$\tau = rF_T = r(F \sin \theta) = (r \sin \theta)F = lF$$

7.4 DIRECTION OF THE TORQUE

As we have defined torque above, it would make no difference if we reversed the direction of the force; $F_T r$ would be the same. It is clear, however, that the direction of the rotation would be changed if **F** were reversed. We would like to incorporate the direction of rotation into the definition of torque. We have seen that the magnitude of the torque is given by

$$\tau = rF \sin \theta$$

But this is just the magnitude of **r** × **F** (see Eq. 1.3). If we define the torque as the vector

$$\tau = \mathbf{r} \times \mathbf{F} \tag{7.17}$$

then the magnitude of τ will give us a measure of the tendency of the force to cause rotation, and the direction of τ will lie along the line of the axis of rotation. Also the sense along this line determines the direction in which the body rotates about the line in the following way. Say **r** × **F** is directed out of the page (as in Fig. 7.8). Then the axis about which the body tends to rotate is perpendicular to the page, and the direction of rotation about this line is determined from a right-hand rule: Place the thumb of the right hand in the direction of τ; then the coiled fingers indicate the direction of rotation.

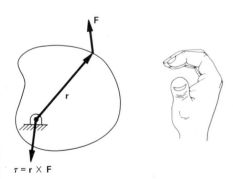

$\tau = \mathbf{r} \times \mathbf{F}$

FIGURE 7.8 The torque is directed out of the page. If the thumb of the right hand is directed out of the page, the fingers coil in the direction of rotation.

This definition of torque may be extended to three dimensions and to bodies that are not constrained to rotate about a fixed axis, i.e., bodies that are tumbling in space. The torque then is defined relative to an origin and not just to a single axis through the origin.

QUESTION

Find the moment of the three forces $\mathbf{F}_1$, $\mathbf{F}_2$, and $\mathbf{F}_3$ in Fig. 7.9 about O.

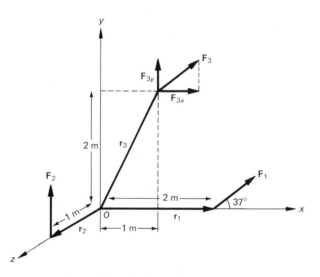

FIGURE 7.9 Three forces $\mathbf{F}_1$, $\mathbf{F}_2$, and $\mathbf{F}_3$.

ANSWER

The moment of $\mathbf{F}_1$ is $\mathbf{r}_1 \times \mathbf{F}_1$, where $\mathbf{r}_1$ is 2 m long and directed along the x axis. The force $\mathbf{F}_1$ makes an angle of 37° with respect to $\mathbf{r}_1$. Therefore

$$\tau_1 = \mathbf{r}_1 \times \mathbf{F}_1 = r_1 F_1 (\sin 37°)\hat{\mathbf{z}} = 1.2 F_1 \hat{\mathbf{z}}$$

The fact that τ_1 is directed along the z axis follows from the fact that the right-hand rule applied to $\mathbf{r}_1 \times \mathbf{F}_1$ produces a vector in the z direction. If you place your right thumb in the z direction, the circulation of your fingers tell you that this torque acting on a rigid body would tend to rotate the body about an axis perpendicular to the page (i.e., about the z axis) in a counterclockwise sense.

The moment of $\mathbf{F}_2$ is given by

$$\tau_2 = \mathbf{r}_2 \times \mathbf{F}_2 = r_2 F_2 \hat{\mathbf{x}} = -F_2 \hat{\mathbf{x}}$$

since $\mathbf{r}_2$ is 1 m long.

The moment of $\mathbf{F}_3$ is $\mathbf{r}_3 \times \mathbf{F}_3$. It is possible to find the angle between $\mathbf{r}_3$ and $\mathbf{F}_3$ (given the components F_{3x} and F_{3y} or the direction of $\mathbf{F}$), but in many cases it is simpler to resolve the vector into its components and determine the moment of the components. The moments of the components are most easily determined by employing the alternative expression for the torque given in Eq. 7.16. The moment arm of F_{3x} is the perpendicular distance between the line of action of the force and the point about which the moment is to be determined. This distance is 2 m. Similarly the moment arm of F_{3y} is 1 m. We must observe, however, that $\mathbf{F}_{3x}$ and $\mathbf{F}_{3y}$ would tend to rotate a rigid body in opposite directions about the z axis. Applying the right-hand rule to obtain the direction of rotation, we see that $\mathbf{F}_{3x}$ would produce a torque along the negative z axis and $\mathbf{F}_{3y}$ a torque along the positive z axis. Therefore

$$\tau_3 = -2F_{3x}\hat{\mathbf{z}} + F_{3y}\hat{\mathbf{z}} = (-2F_{3x} + F_{3y})\hat{\mathbf{z}}.$$

We shall often be dealing with systems in which the forces all lie in a plane. If we take this to be the plane of the page, then the torque will be directed into or out of the page. In such cases we shall find it convenient to choose one direction as positive (either into the page or out of the page—it makes no difference) and immediately write an equation for the *components* of the torque in this direction.

7.5 ANGULAR MOMENTUM

The moment of a force is a torque. We may define the moment of any vector, not just a force. Let us consider the

moment of the momentum vector. If a particle has a momentum $\mathbf{p}$ with a position vector $\mathbf{r}$ relative to some origin, then the moment of momentum is $\mathbf{r} \times \mathbf{p}$. We shall call the moment of the momentum the angular momentum $\mathbf{L}$, so that

$$\mathbf{L} = \mathbf{r} \times \mathbf{p} \tag{7.18}$$

Let us consider the time rate of change of the angular momentum of a particle:

$$\frac{d\mathbf{L}}{dt} = \frac{d}{dt}(\mathbf{r} \times \mathbf{p}) = \frac{d\mathbf{r}}{dt} \times \mathbf{p} + \mathbf{r} \times \frac{d\mathbf{p}}{dt}$$

But the first term on the right-hand side is zero, since

$$\frac{d\mathbf{r}}{dt} \times \mathbf{p} = \frac{d\mathbf{r}}{dt} \times m\mathbf{v} = \mathbf{v} \times m\mathbf{v} = 0$$

This is because $d\mathbf{r}/dt = \mathbf{v}$, and any vector crossed with itself is zero. Hence

$$\frac{d\mathbf{L}}{dt} = \mathbf{r} \times \frac{d\mathbf{p}}{dt} = \mathbf{r} \times \mathbf{F} = \tau \tag{7.19}$$

from the second law of motion, $\mathbf{F} = d\mathbf{p}/dt$, and the definition of torque, $\tau = \mathbf{r} \times \mathbf{F}$. This relationship between the moment of the force and the moment of the momentum

$$\tau = \frac{d\mathbf{L}}{dt} \tag{7.20}$$

is the key to our quest for a relationship that determines the effect of the external force on the rotation of a body.

7.6 EQUATION FOR ROTATIONAL MOTION OF A SYSTEM OF PARTICLES

We must recall that our objective is to find an equation for the rotation of a body that does not involve the internal

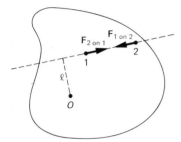

FIGURE 7.10 Two particles in a body and the forces each exerts on the other. The length ℓ is the moment arm of these two forces about O.

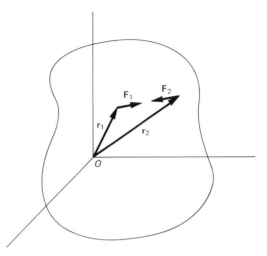

FIGURE 7.11 Two equal and opposite collinear forces acting on two particles at $\mathbf{r}_1$ and $\mathbf{r}_2$.

forces. In Fig. 7.10 we have illustrated two particles in a very flat body and the forces that each exerts on the other. Many other particles are exerting forces on particles 1 and 2, but for the moment we consider only the action and reaction forces between these particles. If we are looking for an equation of motion for the whole body that will not involve these two forces, we must consider only those operations for which they cancel. There are two things we can do. If we simply add the forces, they cancel. We have done this and obtained the equation of motion of the center of mass. Or we can add the moments of these two forces. Clearly the moment arm ℓ is the same for both forces, but one torque is directed out of the page and the other into the page. They cancel. Here then is a way to obtain our second equation.

In the preceding discussion we considered only a very flat body. It is a simple matter to show that the torque resulting from any two equal and opposite forces about *any* point is zero provided the forces are directed along the line joining the two particles. To see this, consider particles 1 and 2 in Fig. 7.11. The net torque about O is

$$\tau = \mathbf{r}_1 \times \mathbf{F}_1 + \mathbf{r}_2 \times \mathbf{F}_2$$

but $\mathbf{F}_2 = -\mathbf{F}_1$, so

$$\tau = \mathbf{r}_1 \times \mathbf{F}_1 - \mathbf{r}_2 \times \mathbf{F}_1 = (\mathbf{r}_1 - \mathbf{r}_2) \times \mathbf{F}_1$$

However $\mathbf{r}_1 - \mathbf{r}_2$ is a vector from particle 2 to particle 1. Since this vector is parallel to $\mathbf{F}_1$, the cross product is zero. We see then that in general the action and reaction forces between particle pairs produce no net torque about any point. Let us examine the equation of motion of all the particles in any system and see whether we can employ this fact to obtain an equation that does not involve the internal forces.

The equations of motion for n particles are

$$\mathbf{F}_1(\text{ext}) + \mathbf{F}_1(\text{int}) = \frac{d\mathbf{p}_1}{dt}$$

$$\mathbf{F}_2(\text{ext}) + \mathbf{F}_2(\text{int}) = \frac{d\mathbf{p}_2}{dt}$$

$$\vdots$$

$$\mathbf{F}_n(\text{ext}) + \mathbf{F}_n(\text{int}) = \frac{d\mathbf{p}_n}{dt}$$

Let us take the moment of each equation and add them up. The moment of all the internal forces will cancel by pairs as we have shown. We are left with the moment of all the external forces or the net external torque $\tau_{\text{tot}}(\text{ext})$. On the right-hand side we obtain

$$\mathbf{r}_1 \times \frac{d\mathbf{p}_1}{dt} + \mathbf{r}_2 \times \frac{d\mathbf{p}_2}{dt} + \cdots + \mathbf{r}_n \times \frac{d\mathbf{p}_n}{dt}$$

But from Eq. 7.19

$$\mathbf{r}_1 \times \frac{d\mathbf{p}_1}{dt} = \frac{d\mathbf{L}_1}{dt}$$

where $\mathbf{L}_1$ is the angular momentum of particle 1. We have then

$$\tau_{\text{tot}}(\text{ext}) = \frac{d\mathbf{L}_1}{dt} + \frac{d\mathbf{L}_2}{dt} + \cdots + \frac{d\mathbf{L}_n}{dt}$$

$$= \frac{d\mathbf{L}_{\text{tot}}}{dt}$$

This is the result we have been searching for. We shall abbreviate the equation and write

$$\tau = \frac{d\mathbf{L}}{dt} \tag{7.21}$$

and it is to be understood that the torque is the net torque of the external forces and that $\mathbf{L}$ is the net angular momentum. This equation is the rotational analog of Eq. 7.14:

$$\mathbf{F} = \frac{d\mathbf{p}}{dt}$$

The net force is equal to the time rate of change of the momentum. From Eq. 7.21 we see that the moment of the forces is equal to the time rate of change of the moment of the momenta, i.e., the time rate of change of the angular momentum.

QUESTION

A particle falls from rest in a uniform gravitational field. What is the acceleration?

ANSWER

Certainly this is not a new problem, and normally we would not apply the torque and angular momentum equation to obtain a solution. It does, however, afford us a very simple problem to practice on.

The particle starts at P. Let O be a point a distance s to the right of P (Fig. 7.12). Now there is a torque on the particle about O given by

$$\tau = mgs$$

the product of the force mg and the moment arm s.

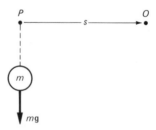

FIGURE 7.12 A particle of mass m falls from point P.

There is also an angular momentum

$$L = mvs$$

the product of the momentum mv and the moment arm of the momentum s. The rate of change of the angular momentum is

$$\frac{dL}{dt} = sm\frac{dv}{dt} = sma$$

Equating τ and dL/dt, we find

$$mgs = sma$$

or

$$a = g$$

as we expect. (Note that the torque remains mgs and the angular momentum mvs at every point on the trajectory. The moment arm s is constant.)

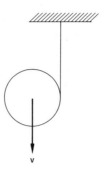

FIGURE 7.13 A cord is wrapped around a disk, then unwinds as the disk falls. The cord hangs vertically.

7.7 CONDITIONS FOR APPLICABILITY OF THE TORQUE–ANGULAR MOMENTUM EQUATION

Let us examine the assumptions made in our derivation of Eq. 7.21 to see the conditions necessary for its application.

The system was any collection of particles interacting through forces that satisfy the law of action and reaction. (We have assumed throughout that the forces between any two particles are not only equal and opposite but are directed along the line joining the two particles. This is the so called "strong form of action and reaction." In the "weak form of action and reaction" forces are equal and opposite but not necessarily directed along the line joining the particles. Our derivation is valid only if we employ the strong form.) The only other assumption we made was that the force on each particle was equal to its mass times its acceleration. Therefore the equation can be applied only in an inertial frame. It turns out that this is a serious handicap. One of our objectives was to find an equation to predict the rotation of a body about its center of mass. Now the center of mass will in general be accelerating. Consider, for example, the problem of a cord wrapped around a disk. The cord is hanging vertically from the ceiling, and the cord unwinds as the disk falls (see Fig. 7.13). We would

like to be able to say something about the torque measured about the center of mass and the time rate of change of the angular momentum about the center of mass. But the center of mass is accelerating, so if we define the moment of a force on the body as $\mathbf{r} \times \mathbf{F}$ and $\mathbf{r}$ is measured from the center of mass, we are performing the calculation in an accelerating frame. It is possible to repeat the derivation, and remarkably we find that it is still true that

$$\tau = \frac{d\mathbf{L}}{dt}$$

even though the torque and angular momentum are measured relative to the accelerating center of mass. We might be more explicit and write

$$\tau_p(\text{ext}) = \frac{d\mathbf{L}_p}{dt} \tag{7.22}$$

where $\tau_p(\text{ext})$ is the sum of the external torques calculated with respect to a point p, and $\mathbf{L}_p$ is the total angular momentum relative to p. The point p is *any point fixed in space or the center of mass.* If p were any other point (e.g., a point fixed to the rim of a rotating wheel), Eq. 7.22 would not be valid.

As a special case of Eq. 7.22, we see that if

$$\tau_{(ext)} = 0$$

then

$$\frac{d\mathbf{L}}{dt} = 0$$

or

$$\mathbf{L} = \text{constant} \tag{7.23}$$

that is, the angular momentum is conserved. If some *component* of τ_{ext} is zero, then the corresponding *component* of the angular momentum is conserved.

This conservation of angular momentum when the torque is zero is the analog of the conservation of linear momentum when the external forces are zero.

QUESTION

Why are Saturn's rings rings (Fig. 7.14)? That is, why has the debris that surrounds the planet taken this shape?

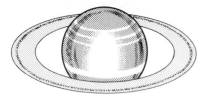

FIGURE 7.14 Saturn's rings.

ANSWER

Let us imagine that initially the debris was more or less uniformly distributed around Saturn, but with some net angular momentum (measured relative to Saturn's center). If we take the debris as the system, then the only significant external force is that of Saturn. This force is directed toward the center of the planet, and hence the external torque on each particle is zero. It follows then that

$$\frac{d\mathbf{L}}{dt} = 0$$

or

$$\mathbf{L} = \text{constant}$$

If the particles are not moving in the same orbit, there will be occasional collisions and mechanical energy will be lost. The particles cannot lose all their energy, for then they would have zero angular momentum and angular momentum would not be conserved. They will therefore lose all the energy they can, consistent with the conservation of angular momentum. It is not difficult to show that the energy reaches a minimum for a given angular momentum when all particles are traveling in the same orbit.

As it turns out, the orbits of all the particles are not the same. Rather they are spread out to form a very flat disk. It is clear that once this state is reached, there would be no further collisions and hence no further energy loss. If there were some mechanism for energy loss, all the particles would come to the same orbit and the disk would shrink to a ring.

QUESTION

Why does a whirlpool develop in a sink as the water drains?

ANSWER

When water runs into a sink it has some angular momentum. Its direction depends on how it enters. If the water is left alone for many hours, this angular momentum will dissipate as a result of friction between the water and the sink. However, if the plug is pulled before the angular momentum has dissipated completely, the angular momentum will remain constant as the water drains if we can neglect the friction between the water and the sink during this short time. Now, the angular momentum of a particle is the moment of its momentum. As the moment arm gets smaller (the water approaches the drain), the momentum must increase if the moment of momentum is to remain constant. Therefore the tangential component of velocity of the water must increase as the water nears the drain.

(If the water is left in the sink until it reaches equilibrium, it will still develop a whirlpool as it runs down the drain. This, however, results from the rotation of the earth, and can be explained by means of a pseudoforce called the *Coriolis force*. The earth is a rotating frame of reference and thus not an inertial frame. We can understand this effect in another way. Suppose we had a circular basin of water with a centrally located drain at the North Pole (Fig. 7.15). Initially the water is at rest with respect to the basin. When the plug is pulled and the water runs down the drain, we find that a whirlpool develops counterclockwise at the North Pole and clockwise at the South Pole. To see why this happens, let us look at the system from an inertial frame, i.e., a frame that is not rotating with the earth but is fixed in space. If we neglect the friction between the basin and the water, angular momentum must be conserved. Initially the water at the edge of the basin is rotating with the earth. It has some angular momentum. After the plug is pulled the water moves toward the drain. If it were to continue to rotate at the same rate as the earth, then its angular momentum would decrease since the moment arm is decreasing. To keep the angular momentum constant, the water must speed up, i.e., it must rotate *faster* than the earth. Since the earth is rotating clockwise—from the point of view of one looking down at the North Pole—the water must be rotating clockwise *relative* to an observer on the earth. The effect is very small and would be difficult to detect in an average sink with an average drain.)

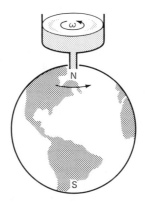

FIGURE 7.15   From the point of view of an inertial frame, the water in a sink at the North Pole rotates with the earth.

QUESTION

Why does a spinning figure skater increase her angular velocity as she draws in her arms (Fig. 7.16)?

Moment arm

FIGURE 7.16   Spinning figure skater.

ANSWER

This question is similar to the one above. We have a spinning mass (hands and arms) that moves toward the axis. The mass must speed up to compensate for the decrease in the moment arm of the momentum.

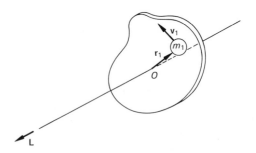

FIGURE 7.17 A planar rigid body is rotating about an axis perpendicular to the plane through O. The mass element m_1 is a distance r_1 from the axis of rotation.

7.8 RIGID BODIES

All the examples of torque and angular momentum discussed above were qualitative in nature. This was because the systems considered were rather complicated, and it would have been difficult to treat them quantitatively. One type of system that is simple enough to treat analytically is the rigid body. We shall develop the results for a rigid body in three stages: (1) a very flat (planar) body rotating about a fixed axis perpendicular to the plane, (2) *any* rigid body rotating about a fixed axis, and (3) a rigid body rotating about an axis through the center of mass which is not fixed in space (e.g. a sphere rolling down an inclined plane).

1. Let us consider first a rigid body rotating about a fixed axis. To simplify the problem we shall assume that the body is very flat so that all the mass lies in a plane (*a planar body*). We would like to determine the angular momentum of the body given its angular velocity ω and the distribution of mass throughout the planar body. The angular momentum of a mass element m_1 traveling with a velocity $\mathbf{v}_1$ at a distance r_1 from the axis of rotation (Fig. 7.17) is given by

$$\mathbf{L}_1 = \mathbf{r}_1 \times \mathbf{p}_1 = m_1 \mathbf{r}_1 \times \mathbf{v}_1$$

Now $\mathbf{r}_1$ and $\mathbf{v}_1$ are mutually perpendicular, so the magnitude of the angular momentum is

$$L_1 = m_1 r_1 v_1$$

and the direction is along the axis of rotation (see Fig. 7.17).

Since the mass element m_1 is rotating in a circle of radius r_1 about O with an angular velocity ω, we have

$$v_1 = r_1 \omega$$

so that

$$L_1 = m_1 r_1^2 \omega$$

The total angular momentum of all the mass elements in the planar body is

$$L = L_1 + L_2 + \cdots L_n$$
$$= (m_1 r_1 + m_2 r_2^2 + \cdots + m_n r_n^2)\omega$$

The quantity in parentheses in this equation is an intrinsic property of the body (unlike the angular velocity, which is variable). It depends on only the mass elements and their distribution about the axis of rotation. It is called the *moment of inertia I*:

$$I = m_1 r_1^2 + m_2 r_2^2 + \cdots + m_n r_n^2$$

We may then write for the angular momentum parallel to the axis of rotation

$$L = I\omega$$

Now the torque and angular momentum are related by the equation

$$\tau = \frac{d\mathbf{L}}{dt}$$

Taking components along the axis of rotation gives

$$\tau = \frac{dL}{dt} = I\frac{d\omega}{dt} = I\alpha$$

where τ is the component of the torque parallel to the axis of rotation.

QUESTION

What is the physical significance of the moment of inertia?

ANSWER

We have discussed moments of forces and moments of momentum. The moment of any vector $\mathbf{B}$ is defined as $\mathbf{r} \times \mathbf{B}$. The moment of inertia is not a cross product of $\mathbf{r}$ with something; it is not even a vector. The name is suggestive but it is not to be confused with moments of vectors.

We can best understand the significance of the moment of inertia by examining the equation that determines rotational motion, namely

$$\tau = I\alpha$$

Solving for α, we find

$$\alpha = \tau/I$$

From this relation we see that I is a measure of the resistance of a body to angular acceleration for a given torque. The larger I is, the smaller is α. In much the same way we have seen that

$$a = \frac{F}{m}$$

and m is a measure of the resistance to linear acceleration for a given force.

We see from the equation

$$I = m_1r_1{}^2 + m_2r_2{}^2 + \cdots + m_nr_n{}^2$$

how the moment of inertia is related to the mass of the elements of the body and their positions relative to the axis of rotation. The greater the inertia of the mass elements and the further they are from the axis, the larger the moment of inertia. It is easy to close the cover of this book but difficult to open the gates of St. Peter's Basilica.

QUESTION

The angular momentum and the torque were calculated about a point in the plane of the body lying on the axis of rotation. Suppose we had chosen a point O' (Fig. 7.18) on the axis of rotation but not in the plane of the body. We know that the torque and angular momentum depend on the location of the origin. How would this new choice of origin affect the equation of motion ($\tau = I\alpha$)?

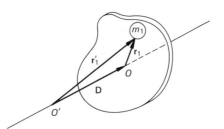

FIGURE 7.18   A planar rigid body is rotating about an axis perpendicular to the plane. The point O lies in the body and on the axis of rotation. The point O' is any other point on the axis. The vector $\mathbf{D}$ represents the position of O relative to O'.

ANSWER

Let **D** represent a vector from O' to O in Fig. 7.18. The angular momentum of a mass element m_1 about the point O' is given by

$$\mathbf{L}_1 = \mathbf{r}_1' \times m_1\mathbf{v}_1$$

Now

$$\mathbf{r}_1' = \mathbf{r}_1 + \mathbf{D}$$

so that

$$\mathbf{L}_1' = \mathbf{r}_1 \times m_1\mathbf{v}_1 + \mathbf{D} \times m_1\mathbf{v}_1$$

The first term is just $\mathbf{L}_1$, the angular momentum calculated with respect to point O and, as we have observed before, $\mathbf{L}_1$ is parallel to the axis of rotation. The second term is perpendicular to the axis of rotation. It follows then that the component of the angular momentum parallel to the axis of rotation is independent of the choice of the origin on the axis of rotation. The same may be said for other mass elements m_2, m_3, . . . in the planar body, and thus the component of the total angular momentum $L_\parallel$ parallel to the axis of rotation is independent of the choice of origin. Therefore

$$L_\parallel' = L_\parallel = I\omega$$

We shall leave as an exercise for you to show that the component of the torque parallel to the axis of rotation is also independent of the choice of origin. It follows that

$$\tau_\parallel = \frac{dL_\parallel}{dt}$$

no matter where the origin is located on the axis of the planar body. We shall find these results useful in determining the rotation of nonplanar bodies.

2. Next let us consider the rotation of an arbitrary rigid body about a fixed axis. We shall divide the body into a large

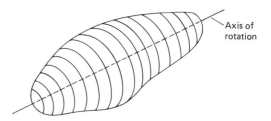

FIGURE 7.19 An arbitrary rigid body divided into a number of slabs.

number of thin slabs (see Fig. 7.19) in the hope of using the results of the preceding discussion dealing with planar bodies. Now we know in general that

$$\tau = \frac{d\mathbf{L}}{dt}$$

where the torque and angular momentum are calculated with respect to a point fixed in space or the center of mass. Let us choose some point O on the fixed axis of rotation. We have seen in the question above that the component of the angular momentum parallel to the axis of rotation is independent of the position of the origin on the axis of rotation. Thus the component of angular momentum parallel to the axis of rotation of each slab is just the product of the moment of inertia of the slab multiplied by ω, that is,

$$L_\parallel(\text{slab}) = I(\text{slab})\omega$$

The total angular momentum component parallel to the axis of rotation is the sum of the angular momentum components of each slab or

$$L_\parallel = I\omega$$

where $L_\parallel$ is the total angular momentum component and I is the total moment of inertia and is given by

$$I = m_1 r_1{}^2 + m_2 r_2{}^2 + \cdots + m_n r_n{}^2$$

Here the sum is over all mass elements in the body and $r_1, r_2, \ldots,$ are the distances of the mass elements *from the axis of rotation*.

Since

$$\tau_{||} = \frac{dL_{||}}{dt}$$

we have

$$\tau_{||} = \frac{d(I\omega)}{dt} = I\alpha$$

so the equation of motion is similar in form to that of a planar body.

3. Finally let us consider a body that has no fixed axis of rotation, e.g., a sphere rolling down a hill, a football thrown into the air, a satellite orbiting the earth.

In general

$$\tau_p = \frac{d\mathbf{L}_p}{dt} \tag{7.24}$$

where p is a point fixed in space or the center of mass. The motion of the body can be described as a translation of the center of mass through space plus a rotation about an axis through the center of mass. Let us assume that the *orientation* of the axis of rotation is constant. Examples of such motion are a sphere rolling straight down an incline, a football thrown in the air spinning about its long axis, a football spinning end over end as in a place kick, a yo-yo, and, to a good approximation, the earth moving about the sun. In each case the axis of rotation moves with the body but it does not change its orientation. Examples of rigid-body motion in which the orientation of the axis of rotation

is not fixed are a sphere rolling down an incline on a curved path, a football thrown in such a way that it is not revolving about an axis of symmetry (a "wobbly" pass), and a precessing gyroscope.

If we restrict ourselves to rigid bodies rotating about an axis of fixed orientation, then we may take the component of Eq. 7.24 parallel to the axis of rotation, and obtain

$$\tau_{||} = \frac{dL_{||}}{dt}$$

but

$$L_{||} = I\omega$$

so that

$$\tau_{||} = I\alpha \tag{7.25}$$

The moment of inertia must be evaluated about the axis of rotation through the center of mass.

Equation 7.13 together with Eq. 7.25 completely describe the motion of a rigid body rotating about an axis of fixed orientation. The first equation describes the motion of the center of mass, and the second describes the rotation about the center of mass.

If the axis of rotation does not have a fixed orientation, then the problem becomes more difficult. It is no longer true that

$$\tau_{||} = \frac{dL_{||}}{dt}$$

even though it is still true that

$$\tau = \frac{d\mathbf{L}}{dt}$$

The reason for this is that the two operations "parallel component of" and d/dt do not commute, that is,

$$\left(\frac{d\mathbf{L}}{dt}\right)_{||} = \text{(parallel component of)} \frac{d\mathbf{L}}{dt}$$

$$\neq \frac{d}{dt} \text{(parallel component of } \mathbf{L}) = \frac{dL_{||}}{dt}$$

To see this let us write

$$L_{\parallel} = \mathbf{L} \cdot \hat{\imath}$$

where $\hat{\imath}$ is a unit vector directed along the axis of rotation (see Sec. 1.2). Also by definition we may write

$$\left(\frac{d\mathbf{L}}{dt}\right)_{\parallel} = \frac{d\mathbf{L}}{dt} \cdot \hat{\imath}$$

Now,

$$\frac{dL_{\parallel}}{dt} = \frac{d}{dt}(\mathbf{L} \cdot \hat{\imath}) = \frac{d\mathbf{L}}{dt} \cdot \hat{\imath} + \mathbf{L} \cdot \frac{d\hat{\imath}}{dt} = \left(\frac{d\mathbf{L}}{dt}\right)_{\parallel} + \mathbf{L} \cdot \frac{d\hat{\imath}}{dt}$$

Unless $d\hat{\imath}/dt = 0$,

$$\frac{dL_{\parallel}}{dt} \neq \left(\frac{d\mathbf{L}}{dt}\right)_{\parallel}$$

Now $d\hat{\imath}/dt$ will be zero only when the direction of the axis of rotation does not change in time.

This is not to say that such problems cannot be solved. It is just that we cannot employ our simple analysis. Unfortunately we must leave such problems to more advanced courses.

Let us summarize our results to this point. For any collection of particles (not necessarily a rigid body)

$$\tau_p = \frac{d\mathbf{L}_p}{dt}$$

where p is any point fixed in space or the center of mass. For a rigid body we have

$$\tau_{\parallel} = I\alpha \tag{7.26}$$

where $\tau_{\parallel}$ is the component of the torque along an axis fixed in space or an axis of rotation through the center of mass

which does not change direction. The moment of inertia about an axis is defined by the equation

$$I = m_1 r_1{}^2 + m_2 r_2{}^2 + \cdots \tag{7.27}$$

where r_1 is the perpendicular distance from the mass element m_1 to the axis. The sum is performed over the entire body.

7.9 MOMENT OF INERTIA

The moment of inertia of a rigid body about an axis is defined by Eq. 7.27:

$$I = m_1 r_1{}^2 + m_2 r_2{}^2 + \cdots + m_n r_n{}^2 \tag{7.28}$$

where r_1 is the perpendicular distance between the mass m_1 and the axis (Fig. 7.20). If the mass distribution is continuous, we may replace the sum by an integral,

$$I = \int r^2 \, dm \tag{7.29}$$

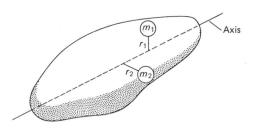

FIGURE 7.20 The moment of inertia about an axis is obtained by adding mr^2 over the entire body, where r is the perpendicular distance between the mass and the axis.

QUESTION

What is the moment of inertia of a hoop of radius R and mass M about an axis through its center and perpendicular to the plane of the hoop (Fig. 7.21)?

FIGURE 7.21 A hoop of mass M and radius R.

ANSWER

The moment of inertia of a hoop or cylinder is particularly simple since all the mass elements are at the same distance from the axis. Now

$$I = m_1 r_1{}^2 + m_2 r_2{}^2 + \cdots + m_n r_n{}^2$$

$$= (m_1 + m_2 + \cdots + m_n)R^2$$

$$= MR^2$$

since $r_1 = r_2 = \cdots = r_n = R$ and $m_1 + m_2 + \cdots + m_n = M$.

QUESTION

What is the moment of inertia of a uniform rod about an axis through one end and perpendicular to the rod (Fig. 7.22)?

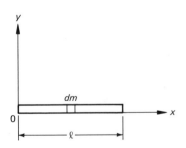

FIGURE 7.22 A rod of length l.

ANSWER

The moment of inertia about the y axis is

$$I = \int x^2 \, dm$$

Let ρ be the mass per unit length of rod. Then

$$dm = \rho \, dx$$

so that

$$I = \int_0^l x^2 \rho \, dx = \tfrac{1}{3}\rho l^3$$

But the rod is uniform, so that

$$M = \rho l$$

where M is the total mass of the rod. Therefore

$$I = \tfrac{1}{3}M l^2$$

The moments of inertia of several simple geometrical forms are given in Table 7.1.

7.10 PARALLEL AXIS THEOREM

There is a useful theorem that allows us to determine the moment of inertia about any axis from a knowledge of the moment of inertia about a parallel axis through the center of mass of the body. For example, we can determine the moment of inertia of a door about an axis through its hinges if we know the moment of inertia of the door about a parallel axis through the center.

We shall prove the theorem for a planar body. One can extend the theorem to nonplanar bodies in a trivial way, by observing that any object can be divided into a large number of planar segments and the moment of inertia of the whole body is just the sum of the moments of inertia of the planar cross sections (see Fig. 7.23).

TABLE 7.1   Some moments of inertia.

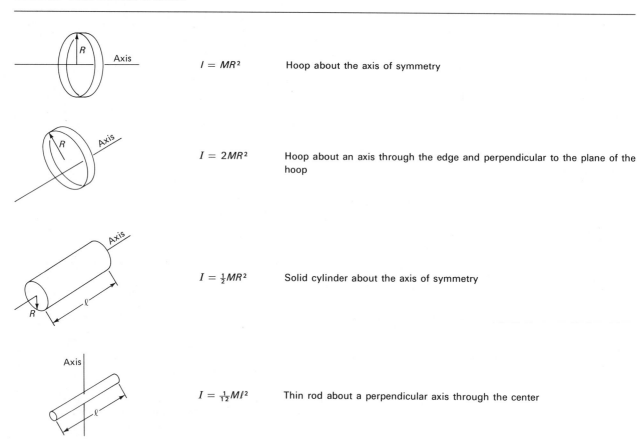

$I = MR^2$ Hoop about the axis of symmetry

$I = 2MR^2$ Hoop about an axis through the edge and perpendicular to the plane of the hoop

$I = \frac{1}{2}MR^2$ Solid cylinder about the axis of symmetry

$I = \frac{1}{12}M\ell^2$ Thin rod about a perpendicular axis through the center

TABLE 7.1 (*continued*)

$I = \frac{1}{3}Ml^2$ Thin rod about a perpendicular axis through one end

$I = \frac{2}{5}MR^2$ Solid sphere about a diameter

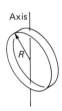

$I = \frac{1}{2}MR^2$ Hoop about a diameter

$I = \frac{3}{2}MR^2$ Hoop about a line tangent to the edge

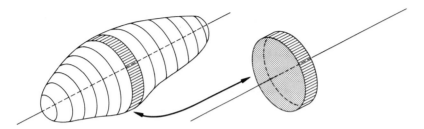

FIGURE 7.23 Any body can be divided into a number of planar sections.

We wish to calculate the moment of inertia of a planar body about an axis through O (Fig. 7.24). From Eq. 7.29,

$$I_O = \int r^2 \, dm$$

But from Fig. 7.24,

$$\mathbf{r} = \mathbf{r}' + \mathbf{r}_{\text{c.m.}}$$

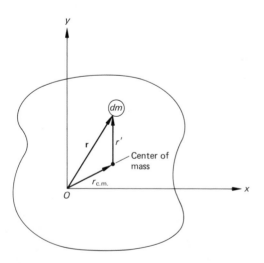

FIGURE 7.24 The element of mass dm has a position vector $\mathbf{r}$ relative to O and $\mathbf{r}'$ relative to the center of mass.

so that

$$r^2 = (r')^2 + 2\mathbf{r}_{\text{c.m.}} \cdot \mathbf{r}' + r_{\text{c.m.}}{}^2$$

We have then

$$I_O = \int (r')^2 \, dm + 2\mathbf{r}_{\text{c.m.}} \cdot \int \mathbf{r}' \, dm + r_{\text{c.m.}}{}^2 \int dm \qquad (7.30)$$

since the vector to the center of mass $\mathbf{r}_{\text{c.m.}}$ is a constant. The first term in Eq. 7.30 is the moment of inertia with respect to an axis through the center of mass,

$$I_{\text{c.m.}} = \int (r')^2 \, dm$$

The second term in Eq. 7.30 is zero since the integral just represents the position of the center of mass *relative to the center of mass*. It is as if we chose a coordinate system with the origin at the center of mass and calculated the center of mass. We would of course find that the coordinates are zero. Therefore

$$\int \mathbf{r}' \, dm = 0$$

In the third term in Eq. 7.30

$$\int dm = M$$

where M is the total mass.

We may therefore write for Eq. 7.30

$$I_O = I_{c.m.} + Mr_{c.m.}^2 \qquad (7.31)$$

The moment of inertia about an axis through O is therefore equal to the moment of inertia about a parallel axis through the center of mass plus the moment of inertia about O of a mass M located at the center of mass. In Eq. 7.31, $r_{c.m.}$ is the distance from the center of mass to point O.

In an arbitrary rigid body the moment of inertia about an axis through a point O is equal to the moment of inertia about a parallel axis through the center of mass plus MR^2, where R is the distance between the two axes,

$$I_O = I_{c.m.} + MR^2$$

QUESTION

What is the moment of inertia of a sphere about an axis tangent to the surface?

ANSWER

The moment of inertia about an axis through O (see Fig. 7.25) is given by

$$I_O = I_{c.m.} + MR^2$$

$$= \tfrac{2}{5}MR^2 + MR^2 = \tfrac{7}{5}MR^2$$

where we have used the result from Table 7.1 that $I_{c.m.} = \tfrac{2}{5}MR^2$ for a sphere.

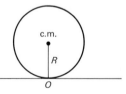

FIGURE 7.25 A sphere of radius R. The point O is on the surface of the sphere.

QUESTION

Two points O and O' are equidistant from the center of mass (Fig. 7.26). What is the relationship between the moments of inertia about two parallel axes through these points?

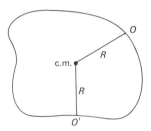

FIGURE 7.26 Points O and O' are each a distance R from the center of mass.

ANSWER

The moments of inertia must be the same. We know from the parallel axis theorem that

$$I_O = I_{c.m.} + MR^2$$

and

$$I_{O'} = I_{c.m.} + MR^2$$

Therefore

$$I_{O'} = I_O$$

7.11 DYNAMICS OF RIGID-BODY ROTATION

We now have all the equipment we need to deal with rigid-body dynamics.

QUESTION

A light cord is wrapped around a homogeneous disk of radius $R = 0.2$ m, and the disk is free to rotate about an axis through its center. The mass of the disk is $M = 3$ kg. A block of mass $m = 2$ kg is tied to the other end of the cord. Initially the block is at rest and the cord is vertical (Fig. 7.27). After the block is released, what is the acceleration of the block, the tension in the cord, and the force that the axle exerts on the disk?

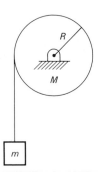

FIGURE 7.27   A block of mass m hangs from a light cord wrapped around a disk of mass M.

ANSWER

First let us consider the block as a system. The forces acting on the block are $m\mathbf{g}$ and the tension in the cord $\mathbf{T}$ (see Fig. 7.28). The equation of motion is

$$m\mathbf{g} + \mathbf{T} = m\mathbf{a}$$

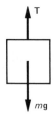

FIGURE 7.28   The forces acting on the block are $\mathbf{T}$ and $m\mathbf{g}$.

Taking components in the downward direction, we have

$$mg - T = ma \tag{7.32}$$

This is one equation with two unknowns.

Take as a second system the disk and a portion of the hanging cord (Fig. 7.29). The equation of motion for the center of mass of the disk is

$$M\mathbf{g} + \mathbf{F} + \mathbf{T} = M\mathbf{a}_{\text{c.m.}} \tag{7.33}$$

where $\mathbf{F}$ is the force that the axle exerts on the disk. Since the center of the disk is fixed, $\mathbf{a}_{\text{c.m.}} = 0$, so that

$$M\mathbf{g} + \mathbf{F} + \mathbf{T} = 0$$

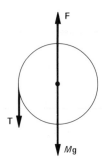

FIGURE 7.29   The forces acting on the disk are its weight $m\mathbf{g}$, the tension in the cord $\mathbf{T}$, and the force of the axle $\mathbf{F}$.

(Note that $M\mathbf{g}$ and $\mathbf{T}$ have no horizontal components, so $\mathbf{F}$ must have no horizontal component.) Taking components in the downward direction, we have

$$Mg - F + T = 0 \tag{7.34}$$

Now we have two equations with three unknowns.

We may obtain a third equation by considering the moments of the external torques on the disk. The moments of **F** and **M**g are zero. The moment of **T** is RT, so that

$$\tau = RT \tag{7.35}$$

Now

$$\tau = I\alpha \tag{7.36}$$

where $I = \frac{1}{2}MR^2$ for a disk. Eliminating τ between Eqs. 7.35 and 7.36, we have

$$RT = I\alpha \tag{7.37}$$

This gives us another equation but also another unknown, α. However, there is a kinematic relationship between a and α. The tangential acceleration of a point on the rim of the wheel must be the same as the tangential acceleration of the cord. Now

$$a_T = R\alpha \tag{7.38}$$

But the tangential acceleration of the cord is the same as the acceleration of the block; that is,

$$a_T = a \tag{7.39}$$

so that

$$a = R\alpha \tag{7.40}$$

In Eqs. 7.32, 7.34, 7.37, and 7.40 we have four equations for four unknowns: a, α, T, and F. Substituting the numerical values we find

$$a = 5.6 \text{ m/s}^2$$

$$T = 8.4 \text{ N}$$

and

$$F = 37.8 \text{ N}$$

There are always a variety of ways of solving any problem in dynamics. The solution above is perhaps the most straightforward for this particular problem. However, it may be worthwhile to approach the problem another way to get a different perspective of the fundamental laws of dynamics applied to a collection of bodies.

Suppose we were asked to determine only the motion of the block and disk and not to find the tension in the cord or the force at the supporting axle. Let us apply our principle of torque and angular momentum to both the disk and the block as a combined system (see Fig. 7.30). Consider the net torque on the system about the center of the disk at O. The only outside force that has a torque about O is mg. The torque is mgR:

$$\tau = mgR$$

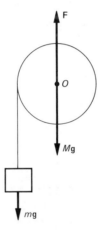

FIGURE 7.30 The system is the disk and the block.

The torque must equal the time rate of change of the angular momentum or equivalently the moment of momentum of the system. The moment of momentum L is composed of two parts, the moment of momentum of the disk, which is $I\omega$, and the moment of momentum of the block about O, which is the momentum of the block mv multiplied by its moment arm R. Therefore

$$L = I\omega + mvR$$

Now

$$\tau = \frac{dL}{dt} = I\alpha + maR \qquad (7.41)$$

since $d\omega/dt = \alpha$ and $dv/dt = a$. But $\tau = mgR$, so that

$$mgR = I\alpha + maR \qquad (7.42)$$

(We could have obtained this equation by eliminating T between Eqs. 7.32 and 7.37.) Noting that

$$a = R\alpha$$

we may solve these last two equations for either a or α. For example,

$$a = \frac{m}{m + I/R^2}g$$

Since $I = \frac{1}{2}MR^2$, we have

$$a = \frac{m}{m + M/2}g$$

As a check, suppose $M = 0$. Then $a = g$ as it should, for then the disk has no inertia. Also if $m = 0$, then $a = 0$ as expected.

QUESTION

A bowling ball is thrown with zero initial angular velocity. The coefficient of friction between the ball and the alley is μ. How long is it before the ball begins to roll without slipping on the alley?

ANSWER

The external forces on the ball are illustrated in Fig. 7.31. Let us apply our two dynamical principles to this rigid body:

$$\mathbf{F}_{\text{tot}}(\text{ext}) = m\mathbf{a}_{\text{c.m.}} \qquad (7.43)$$

and

$$\tau_{\text{tot}}(\text{ext}) = I\alpha \qquad (7.44)$$

where the torques are computed relative to the center of mass.

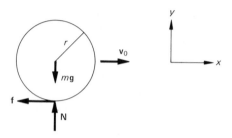

FIGURE 7.31   A ball of radius r is traveling with an initial velocity v_0 and $\omega_0 = 0$. Because of frictional torque, the ball begins to rotate and eventually it rolls without slipping.

From the x component of Eq. 7.43 we have

$$-f = ma_{\text{c.m.}} \qquad (7.45)$$

and from Eq. 7.44 we have

$$fr = I\alpha \qquad (7.46)$$

When the velocity of the center of mass is equal to $r\omega$, the ball is rolling without slipping. The velocity of the center of mass is given by

$$v_{\text{c.m.}} = a_{\text{c.m.}}t + v_0$$

and

$$\omega = \alpha t$$

Letting $v_{\text{c.m.}} = r\omega$, we have

$$a_{\text{c.m.}}t + v_0 = r\alpha t$$

or

$$t = \frac{v_0}{r\alpha - a_{\text{c.m.}}}$$

Substituting $a_{\text{c.m.}}$ and α from Eqs. 7.45 and 7.46, we have

$$t = \frac{v_0}{fr^2/I + f/m}$$

But

$$f = \mu N = \mu mg$$

so that

$$t = \frac{v_0}{\mu mgr^2/I + \mu mg/m} = \frac{2}{7}\frac{v_0}{\mu g}$$

where we have set

$$I = \tfrac{2}{5}mr^2$$

For comparison, a *sliding* block would come to *rest* in a time of $v_0/\mu g$.

7.12 CENTER OF GRAVITY

We have seen earlier in this chapter that we can cope with the problem of a very large number of internal forces by adding all the forces or adding all the torques. Suppose, however, that there are a very large number of *external* forces. Such a situation arises for any body in a gravitational field. Each particle of mass m experiences a force mg. How do we deal with the problem of adding so many forces and the torques associated with these forces? We have in fact been dealing with this problem right along by placing the gravitational force at the center of the body. We must now justify this procedure.

We would like to show that the multitude of gravitational forces can be replaced by a single force acting at a point called the *center of gravity* (abbreviated c.g.). For a rigid body the center of gravity is an invariant point, that is, the location of this point does not depend on the orientation of the body or on the strength of the local gravitational force (provided it can be considered locally uniform). We shall see that the center of gravity is the same as the center of mass, even though these concepts are defined quite differently.

What do we mean when we say that we may replace the gravitational forces on all the particles by a single force at the center of gravity? We do two things with forces: we

add them and we add their moments. If that is all we do, then we might say that the real forces on all the particles and the single force applied at the center of gravity are *equivalent* if (1) the sum of all the individual forces is equal to the single force, and (2) the moment of all the individual forces about any point is the same as the moment of the single force about that point. We are using the word "equivalent" in a technical sense. The single force is equivalent to the real forces if it satisfies conditions 1 and 2. An example of equivalent systems of forces is shown in Fig. 7.32. It is obvious that the net force is the same in both systems (namely 4 lb) and that the net torque is the same about the center of the rod (namely zero). We leave it as an exercise for you to show that the torque about *any* point is the same in both cases. The fact that the net force and the torque are the same for both systems means that both will have the same acceleration of the center of mass and rate of change of the total angular momentum about the center of mass. This does not mean that the systems will behave the same. If the object is a rigid rod, the motion is the same. If the object is a rope, the motion is not. In the case of a rigid body the motion of the center of mass and the rotation about it uniquely determine the motion, so that if two systems of forces applied to a rigid body are equivalent, they will behave the same. Even for rigid bodies, however, the *internal forces* will be different for different but "equivalent" systems of forces, as the two people in Fig. 7.33 will testify.

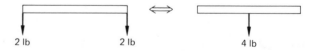

FIGURE 7.32 These two systems are subject to "equivalent" forces.

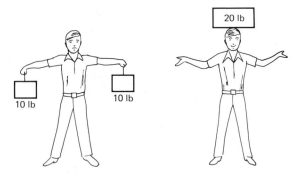

FIGURE 7.33 In these two systems, although the forces are equivalent, the internal forces are different. In one case the strain is in the shoulders, and in the other it is in the neck.

Let us determine the position and magnitude of the equivalent force for a uniform gravitational field. In Fig. 7.34 we have illustrated two identical bodies. One is subject to the actual gravitational forces and the other to a

single force at the center of gravity. First, the net force must be the same. The net force in Fig. 7.34(a) is

$$m_1\mathbf{g} + m_2\mathbf{g} + \cdots + m_n\mathbf{g} = M\mathbf{g}$$

where

$$M = m_1 + m_2 + \cdots + m_n$$

is the total mass. Therefore the single force at the center of gravity must be $M\mathbf{g}$. Second, the moment about O must be the same in Fig. 7.34(a) and (b). The net moment in Fig. 7.34(a) calculated about the origin O is

$$\tau_a = \mathbf{r}_1 \times m_1\mathbf{g} + \mathbf{r}_2 \times m_2\mathbf{g} + \cdots + \mathbf{r}_n \times m_n\mathbf{g}$$

$$= (m_1\mathbf{r}_1 + m_2\mathbf{r}_2 + \cdots + m_n\mathbf{r}_n) \times \mathbf{g}$$

The net moment in Fig. 7.34(b) is

$$\tau_b = \mathbf{r}_{c.g.} \times M\mathbf{g} = M\mathbf{r}_{c.g.} \times \mathbf{g}$$

These will be equal (for all orientations of the body or equivalently for all orientations of $\mathbf{g}$) if and only if

$$M\mathbf{r}_{c.g.} = m_1\mathbf{r}_1 + m_2\mathbf{r}_2 + \cdots + m_n\mathbf{r}_n$$

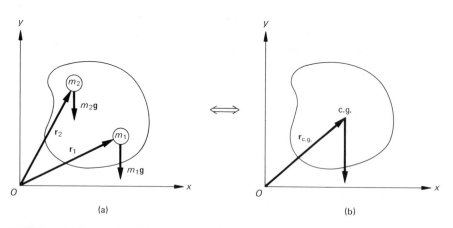

(a)　　　　　　　　　　　　　　(b)

FIGURE 7.34 The two sets of forces are equivalent if the net force is the same and the net moment about O is the same.

or

$$r_{c.g.} = \frac{m_1 r_1 + m_2 r_2 + \cdots + m_n r_n}{M} \qquad (7.47)$$

But this is just the definition of the position of the center of mass (see Eq. 7.4). Therefore

$$r_{c.m.} = r_{c.g.} \qquad (7.48)$$

Note that the value of g does not enter into Eq. 7.47, so that the center of gravity would be the same at Chicago or Pike's Peak or the moon.

The very fact that a center of gravity exists has some interesting consequences. Let us suspend a body by a string from some point A on its surface (see Fig. 7.35). We project the line of the string as shown. Now let us suspend the body from some other point B on its surface and project this line as shown. If the body is flat, these two lines will intersect at a point. It is by no means obvious that if we suspend the body from a third point C, the extension of the string will pass through the same point. However,

if the body is in equilibrium in each of the above configurations, then the extension of the string must pass through the center of gravity for it is only in this way that the net torque on the system can be zero, and the torque must be zero or else the body would rotate.

As a second physical consequence of the existence of the center of gravity, we observe that there is one point (the center of gravity) from which the body could be supported and be in equilibrium in *any* orientation. This point may or may not be accessible.

To show that the existence of the center of gravity is not trivial, it suffices to show that a unique center of gravity does not exist in a nonuniform gravitational field. The earth's gravitational field decreases with increasing distance like $1/r^2$. Suppose we hang a uniform rod from its center in such a field (Fig. 7.36). The rod will be in equilibrium in a horizontal or vertical position. However, at any other angle the lower elements of mass will be heavier than the higher elements, and the rod will not be in equilibrium.

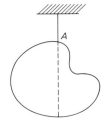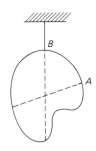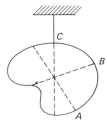

FIGURE 7.35   A flat body is suspended successively at three different points on its edge.

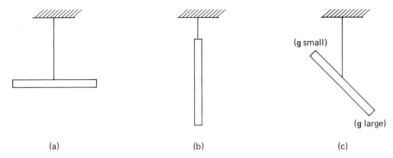

(a)   (b)   (c)

FIGURE 7.36 A thin rod suspended at its center will be in equilibrium in (a) and (b) but not in (c) if the gravitational field decreases with increasing height.

QUESTION

A rod pinned at one end is initially at rest in a vertical position (Fig. 7.37). It begins to tip. What is the equation of motion?

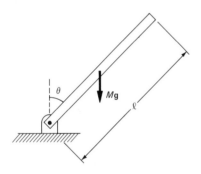

FIGURE 7.37 A rod of length l pinned at one end.

ANSWER

The external forces acting on the rod are the forces at the pin and the gravitational forces on all the mass elements. We have just shown that the latter forces may be replaced by a single force at the center of gravity. The net torque of the external forces about the axis of rotation is therefore given by

$$\tau = Mg\frac{l}{2}\sin\theta$$

But

$$\tau = I\alpha$$

where

$$I = \tfrac{1}{3}Ml^2$$

so that

$$\alpha = \frac{Mg(l/2)\sin\theta}{\tfrac{1}{3}Ml^2} = \frac{3}{2}\frac{g\sin\theta}{l}$$

We see that α increases uniformly as θ increases from 0° to 90°.

QUESTION

A falling brick chimney is illustrated in Fig. 7.38. Why does the chimney buckle as it falls?

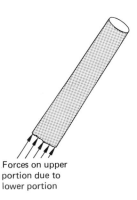

Forces on upper
portion due to
lower portion

FIGURE 7.39 A segment of the upper portion of the chimney of Fig. 7.38.

FIGURE 7.38 A chimney under demolition buckles as it falls. (Permission of Courier-Journal and Louisville Times)

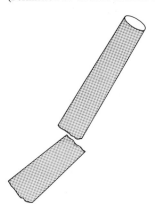

the angular acceleration of *any portion of the rod*, increases with θ. Now we may apply our equation for the general motion of a rigid body to the upper portion of the rod or the chimney (Fig. 7.39). In particular,

$$\tau_{c.m.} = I_{c.m.}\alpha$$

where the torque and moment of inertia are calculated with respect to the center of mass of the upper portion of the rod. Since α increases with θ, the torque of all the forces about the center of mass of the upper portion must increase as θ increases. Now the gravitational force on this segment acts effectively at the center of mass and hence causes no torque. The increasing torque necessary to produce the increasing angular acceleration can only be caused by forces acting on the upper portion of the rod as a result of the lower portion.

The radial components of these forces are shown in Fig. 7.39. The forces will be larger on the upper side and smaller on the lower side. This will create a clockwise torque, which is responsible for the observed clockwise angular acceleration. As θ increases and hence α increases, the forces on the lower edge will eventually be reversed and a *tension* will be created in the rod at this point. Now for a rod this creates no problem, but for a brick chimney this tension will undoubtedly cause a

ANSWER

Let us return for a moment to the falling rod in the preceding question and consider some portion of the upper end. We know that the angular acceleration of the rod, and therefore fracture. Brick structures are capable of sustaining great compressive forces, but only very small tensile forces. We would expect the chimney to fracture very shortly after this tensile force is created in the lower edge of the chimney.

(We have ignored the tangential force that acts between the upper and lower segments. This force does contribute a torque about the center of mass of the upper segment, but for a cross section whose distance from the base is less than one-third the length of the chimney, this tangential force *opposes* the rotational acceleration. This means that the torque resulting from the radial force shown in Fig. 7.39 must be even larger below this point. It can be shown that the point at which the fracture is most likely to occur in a uniform chimney is in fact one-third of the length of the chimney measured from the base.)

7.13 WORK AND ENERGY PRINCIPLE FOR ROTATING RIGID BODIES

We have seen that the work done on a system is equal to the change in the total energy of the system. If a telephone pole breaks at the base and falls over, no work is done by outside forces as it falls. (The system is the pole and the earth, so the weight of the pole is not an outside force.) Therefore the total energy should be conserved. We should be able to determine the velocity of the pole just before it hits the ground by equating the energy of the pole just before it hits the ground to the energy of the pole just after it breaks. It is a simple matter to determine the change in potential energy, but how do we determine the change in kinetic energy? Since the pole is rotating about its base, different points on the pole are moving with different velocities.

Let us consider the general problem of calculating the kinetic energy of a rigid body rotating about a fixed axis. As before, we divide the body into slabs by planes that are perpendicular to the axis of rotation. The total kinetic energy will be the sum of the kinetic energies of all the

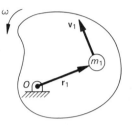

FIGURE 7.40 A planar rigid body is rotating about O with an angular velocity ω.

slabs. One such slab is shown in Fig. 7.40. The axis of rotation of this planar body is perpendicular to the page and intersects the body at O. The instantaneous angular velocity of the body is ω. The kinetic energy is given by

$$K = \tfrac{1}{2}m_1 v_1^2 + \tfrac{1}{2}m_2 v_2^2 + \cdots$$

From Fig. 7.40 and Eq. 2.41 we have

$$v_1 = r_1 \omega$$

where v_1 is the speed of a particle at a distance r_1 from the axis of rotation. Therefore

$$K = \tfrac{1}{2}m_1 r_1^2 \omega^2 + \tfrac{1}{2}m_2 r_2^2 \omega^2 + \cdots$$

Note that ω is the same for all mass elements in a rigid body. This is an essential simplifying feature. For a nonrigid body ω is not the same for all mass elements nor is r fixed for any given mass element.

Factoring we obtain

$$K = \tfrac{1}{2}(m_1 r_1^2 + m_2 r_2^2 + \cdots)\omega^2$$

But the moment of inertia about the axis of rotation is

$$I_0 = m_1 r_1^2 + m_2 r_2^2 + \cdots$$

so that

$$K = \tfrac{1}{2}I_0\omega^2 \qquad (7.49)$$

This is the rotational analog of

$$K = \tfrac{1}{2}mv^2$$

for a mass element in translation.

If we add the kinetic energy of the other slabs in the rigid body to this kinetic energy, we obtain a similar expression except that I_0 is now the moment of inertia of the entire body about the axis of rotation.

QUESTION

A telephone lineman is working on top of a telephone pole when it breaks at the base (Fig. 7.41). Should the lineman jump off or ride the pole to the ground? (Neglect the weight of the person in comparison to the pole and assume that the pole is uniform.)

FIGURE 7.41 A lineman atop a pole.

ANSWER

Let us calculate the velocity of the top of the pole just before it strikes the ground. We do this by equating the total energy of the pole just before it begins to fall (state A) to the total energy just before it strikes the ground (state B). Now

$$E_A = K_A + U_A$$

The kinetic energy is zero and the potential energy is

$$U_A = mg(l/2)$$

where m is the mass and l the length of the pole. Also

$$E_B = K_B + U_B$$

The potential energy just before the pole strikes the ground is zero and the kinetic energy is

$$K_B = \tfrac{1}{2}I_0\omega^2$$

where I_0 is the moment of inertia of the pole calculated about the fixed point of rotation at the base. Equating E_A and E_B, we have

$$mg(l/2) = \tfrac{1}{2}I_0\omega^2$$

Now from Table 7.1

$$I_0 = \tfrac{1}{3}ml^2$$

so that

$$\omega = \sqrt{3g/l}$$

and the velocity of the top of the pole (v_{top}) is given by

$$v_{\text{top}} = l\omega = \sqrt{3gl}$$

We have seen that the velocity in free fall of a body that falls from rest through a distance l is given by (see Eq. 2.10)

$$v = \sqrt{2gl}$$

Since this is less than the velocity of the top of the pole just before the pole strikes the ground, the lineman is better off jumping from the pole before it begins to fall (Fig. 7.42).

FIGURE 7.42 A telephone pole of length l breaks at the base.

where we have employed the fact that

$$v_{c.m.} = R\omega$$

The kinetic energy of a body rotating about a fixed axis can therefore be determined as either a kinetic energy of pure rotation about the fixed axis ($\frac{1}{2}I_0\omega^2$) or as the kinetic energy of rotation about the center of mass ($\frac{1}{2}I_{c.m.}\omega^2$) plus the kinetic energy of translation associated with the center of mass ($\frac{1}{2}Mv_{c.m.}^2$).

QUESTION

If the angular velocity of a rigid body about a fixed axis is ω, what is the angular velocity of the body about the center of mass?

ANSWER

Actually a rigid body has only one angular velocity, and this is the rate at which the angular displacement of any line fixed in the body changes with time. Every line fixed in the body rotates in exactly the same way (Fig. 7.43). If the angular velocity about a point were defined as the rate of change of angular displacement of a line through that point, it would be a simple task to prove that the angular velocity about all points is the same. (This is left as an exercise.)

There is an alternate form of Eq. 7.49 for which we shall find a more general application. If the axis of rotation O does not pass through the center of mass, then from the parallel axis theorem

$$I_0 = I_{c.m.} + MR^2 \tag{7.50}$$

where R is the perpendicular distance between the axis through O and the parallel axis through the center of mass. We may then write Eq. 7.49 in the form

$$K = \frac{1}{2}(I_{c.m.} + MR^2)\omega^2$$
$$= \frac{1}{2}I_{c.m.}\omega^2 + \frac{1}{2}MR^2\omega^2$$
$$= \frac{1}{2}I_{c.m.}\omega^2 + \frac{1}{2}Mv_{c.m.}^2 \tag{7.51}$$

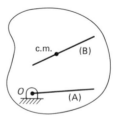

FIGURE 7.43 A rigid body rotates about a fixed axis through O. The lines A and B are fixed in the body and pass through O and the center of mass, respectively. If line A rotates through some angle, then line B must rotate through the same angle.

7.14 KINETIC ENERGY FOR GENERAL MOTION OF A RIGID BODY

In Sec. 7.13 we considered the special case of a rigid body rotating about a fixed axis. Now suppose there is no fixed axis. What, for example, is the kinetic energy of the disk in Fig. 7.44? There is no fixed axis of rotation, but there is an *instantaneous* axis of rotation through O. This point on the disk is clearly at rest since there is no relative motion between the cord and the disk at this point, and the cord is at rest. This instantaneous axis of rotation is continuously changing, but that is no problem. We may still calculate the kinetic energy by using either Eq. 7.49 or Eq. 7.51.

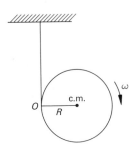

FIGURE 7.44 A cord that is wrapped around a disk is tied to the ceiling and unwinds as the disk falls.

Is there always an instantaneous axis of rotation? Is there an axis about which any rigid body may be considered to be rotating instantaneously? The answer is yes. To find the instantaneous center we need to know only the direction of the velocity of two points in the body (provided the velocities are not parallel). In Fig. 7.45, A and B are two points in a planar body. (The argument can be extended to an arbitrary rigid body.) The dashed lines perpendicular to $\mathbf{v}_A$ and $\mathbf{v}_B$ intersect at C. Now the velocity of C must be perpendicular to the line AC, for if it were not, then the velocity of C relative to A would have a component along the line AC. This would mean that C is moving toward (or

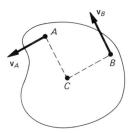

FIGURE 7.45 Two arbitrary points A and B have instantaneous velocities $\mathbf{v}_A$ and $\mathbf{v}_B$. The dashed lines are perpendicular to $\mathbf{v}_A$ and $\mathbf{v}_B$ and intersect at C. The point C is the instantaneous center of rotation.

away from) A. But the body is rigid, so the relative distance between any two points cannot change in time; that is, all points must maintain a fixed separation. Therefore the velocity of C must be perpendicular to the line AC. For the same reason the velocity of C must be perpendicular to the line BC. The point C cannot have a component of velocity that would cause the distance between B and C to change. But the velocity C cannot be perpendicular to AC and BC at the same time. The only conclusion is that C is at rest. The point C is the instantaneous center of zero velocity.

QUESTION

A ladder is standing against a vertical wall. It begins to slip. Where is the instantaneous center of rotation?

ANSWER

We have seen that the instantaneous center will lie on a line perpendicular to the velocity vector of any point. Now we know the direction of the velocity at the points at which the ladder meets the wall (B) and the floor (A). Lines perpendicular to

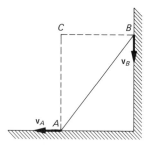

FIGURE 7.46 A ladder sliding down a vertical wall.

$\mathbf{v}_A$ and $\mathbf{v}_B$ intersect at C (Fig. 7.46). The ladder is instantaneously rotating about C.

We have shown that a planar body has an instantaneous center of zero velocity, i.e., an instantaneous axis of rotation. One may also show that any rigid body (not necessarily planar) has an instantaneous axis of rotation. Generally we do not know where the instantaneous axis is, but the very fact that it exists allows us to write for the kinetic energy (see Eq. 7.49)

$$K = \tfrac{1}{2}I_0\omega^2 \tag{7.49}$$

where I_0 is the moment of inertia about the instantaneous axis of rotation. Employing the parallel axis theorem, we may also write Eq. 7.49 in the form (see Eq. 7.51)

$$K = \tfrac{1}{2}I_{c.m.}\omega^2 + \tfrac{1}{2}Mv_{c.m.}{}^2 \tag{7.52}$$

where $I_{c.m.}$ is the moment of inertia about an axis through the center of mass. The kinetic energy of any rigid body that has an instantaneous axis of rotation is expressed by Eq. 7.49. But every rigid body has an instantaneous axis of rotation. *Therefore Eq. 7.52 is a general expression for the kinetic energy of any rigid body.*

QUESTION

A light cord is wrapped several times around a uniform disk of mass M and radius R. The cord is tied to the ceiling, and initially the disk is at rest and the cord is vertical. What is the velocity of the center of the disk after it has fallen a distance h?

ANSWER

Let us equate the energy in the initial state A to the energy in the state B in which the disk has fallen a distance h. Now

$$E_A = K_A + U_A$$

but $K_A = 0$ and

$$U_A = Mgh$$

if we choose the reference level to be the position of the center of the disk when it has fallen a distance h. Also

$$E_B = K_B + U_B$$

but $U_B = 0$ and

$$K_B = \tfrac{1}{2}I_{c.m.}\omega^2 + \tfrac{1}{2}Mv_{c.m.}{}^2$$

where

$$I_{c.m.} = \tfrac{1}{2}MR^2$$

and

$$v_{c.m.} = R\omega$$

Equating the energies we have

$$Mgh = \tfrac{1}{2}I_{c.m.}\omega^2 + \tfrac{1}{2}Mv_{c.m.}{}^2$$

Solving for $v_{c.m.}$, we find

$$v_{c.m.} = \sqrt{\tfrac{4}{3}gh}$$

This is less than the velocity in free fall ($\sqrt{2gh}$) because the cord is exerting an upward force on the disk.

SUMMARY

In this chapter we have defined several new quantities.

Center of mass The center of mass of a collection of particles is a weighted average of the position vectors of the particles:

$$\mathbf{r}_{c.m.} = \frac{m_1\mathbf{r}_1 + m_2\mathbf{r}_2 + \cdots + m_n\mathbf{r}_n}{M}$$

Angular momentum The angular momentum of a particle with respect to a point p is the moment of the momentum of the particle relative to this point:

$$\mathbf{L}_p = \mathbf{r} \times m\mathbf{v}$$

where $\mathbf{r}$ is a vector from p to the particle.

Torque The torque of a force $\mathbf{F}$ about a point p is the moment of this force relative to this point.

$$\tau_p = \mathbf{r} \times \mathbf{F}$$

where $\mathbf{r}$ is a vector from p to the point of application of the force.

Moment of inertia The moment of inertia of a rigid body about an axis through O is given by

$$I_0 = m_1 r_1^2 + m_2 r_2^2 + \cdots + m_m r_n^2$$

where r_1 is the perpendicular distance from the particle of mass m_1 to the axis.

Center of gravity If we replace the gravitational forces on all the particles in a body by a single force equal to $M\mathbf{g}$ located at the center of gravity, then the magnitude of this single force will equal the net gravitational force on all the particles, and the moment of this single force about any point will equal the sum of the moments of the gravitational forces on all the particles of the body about the same point. The center of gravity is located at

$$\mathbf{r}_{c.g.} = \frac{m_1\mathbf{r}_1 + m_2\mathbf{r}_2 + \cdots + m_n\mathbf{r}_n}{M}$$

It is coincident with the center of mass.

We have also derived the following laws for systems of particles:

1. The net external force acting on a system of particles is equal to the total mass multiplied by the acceleration of the center of mass:

$$\mathbf{F} = M\mathbf{a}_{c.m.}$$

2. The net external torque calculated about a point p is equal to the time rate of change in the angular momentum about the same point provided p is a point fixed in space or the center of mass:

$$\tau_p = \frac{d\mathbf{L}_p}{dt}$$

3. If p is a point on a fixed axis of rotation or a point on an axis through the center of mass that does not change its orientation, then for a *rigid body*

$$\tau_{\parallel} = I\alpha$$

where $\tau_{\parallel}$ is the component of the external torque parallel to this axis and I is the moment of inertia about this axis.

4. The parallel axis theorem states that the moment of inertia about any axis is equal to the moment of inertia about a parallel axis through the center of mass plus MR^2, where R is the distance between the two axes:

$$I_0 = I_{c.m.} + MR^2$$

5. The kinetic energy of a rigid body rotating about a fixed axis through O is given by

$$K = \tfrac{1}{2}I_0\omega^2$$

6. The kinetic energy of a rigid body with no fixed axis of rotation is given by

$$K = \tfrac{1}{2}I_{c.m.}\omega^2 + \tfrac{1}{2}Mv_{c.m.}^2$$

PROBLEMS

7.A.1 A force of 10 N making an angle of 37° with the x axis is applied at the point $x = 3$ m, $y = 8$ m.
   (a) What is the torque about the origin?
   (b) What is the torque about the point $x = 8$ m and $y = 3$ m?

7.A.2 Let L be an arbitrary line on a flat disk. Suppose the disk is rotated through an angle θ about some point (not necessarily a point on L). Show that L is rotated through the same angle.

7.A.3 A 5-g bullet is fired into a 10-g block and becomes embedded. The block is constrained to move in a smooth horizontal circular track 4 ft in diameter. The block and bullet rotate in the tract with an angular velocity of 1600 r/min. What is the minimum velocity the bullet had before impact?

7.A.4 Two disks are free to rotate on parallel, concentric axes, as shown in the figure. The first disk has a radius of 10 cm and a mass of 30 g, and rotates with an angular velocity of 100 r/min. The second disk has a radius of 5 cm and a mass of 12 g, and is intially at rest. The two disks are brought together and eventually rotate with a common angular velocity.
   (a) What is this angular velocity?
   (b) How much energy was lost in the "collision"?

7.A.5 A 3-g mass is located at $x = 2$ cm and $y = 4$ cm. A 12-g mass is located at $x = 1$ cm and $y = 8$ cm. A 9-g mass is located at $x = -4$ cm and $y = -10$ cm. Where is the center of mass?

7.A.6 A hoop, a disk, and a sphere are at rest at the top of an inclined plane. All are of the same radius and mass. All three begin at the same time to roll down the plane. In what order do they reach the bottom?

7.A.7 We have shown that when a uniform vertical rod of length l is released, the upper end will strike the ground with a velocity $v = \sqrt{3gl}$ if it is free to rotate about the lower end. Compare this with the velocity with which the top of a uniform door of height l and width s will strike the ground when released. The door rotates about the base.

7.B.1 A 10-g block hangs from a 30-g pulley by means of a light cord wrapped around the pulley. The radius of the pulley is 8 cm. Assume that the block starts from rest and that the pulley is uniform.
   (a) What is the acceleration of the block?
   (b) What are the tension in the cord and the velocity of the block after the block has fallen 10 cm?

7.B.2 A 10-g mass is tied to one end of a string that passes through a hole in a smooth horizontal table and the other end is held fixed below the table (see the following figure).

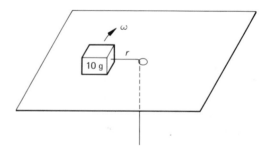

Initially the distance between the hole and the mass is 20 cm, and the mass is rotating about the hole with an angular velocity of 400 r/min. The string is now pulled down, so the distance r between the mass and the hole is shortened.

(a) Determine an expression for the force as a function of r.

(b) How much work is done in pulling the string a distance of 5 cm?

(c) What is the new angular momentum?

(d) What is the new angular velocity?

(e) What is the change in kinetic energy?

7.B.3 Derive the expression for the moment of inertia of a uniform disk of radius R and mass M about its center from the definition of the moment of inertia, $I = \int r^2 \, dm$.

7.B.4 A planar rigid body is pinned at an axis through O. A force $\mathbf{F}$ is applied at some point whose displacement vector relative to O is $\mathbf{r}$.

(a) Determine the work dW done by $\mathbf{F}$ as the body undergoes a rotation through a small angle $d\theta$ about O.

(b) Determine the change in kinetic energy associated with the work calculated in part (a).

(c) From the principle of work and kinetic energy, show that $\tau = I\alpha$.

7.B.5 A hoop rolls down a $37°$ incline without slipping. What is the acceleration of the center of the hoop?

7.B.6 A uniform rod is broken into three segments whose lengths are 3 ft, 4 ft, and 5 ft. When the segments are joined together to form a triangle, where is the center of mass of the triangle?

7.B.7 We have shown that a block placed on an incline will not slip if $\mu > \tan \theta$. Show that if a hoop is placed on an incline and released, it will roll without slipping provided $\mu > \frac{1}{2} \tan \theta$.

7.B.8 A bowling ball of radius 12 cm is thrown without rotation onto a horizontal surface. The initial velocity of the ball is 10 m/s. How many revolutions does the ball make before it begins to roll without slipping if the coefficient of sliding friction is 0.5?

7.B.9 What is the moment of inertia of three identical rods forming an equilateral triangle about an axis through one vertex and perpendicular to the triangle? The length of each rod is l and the mass m.

7.C.1

(a) Given the fact that the medians of a triangle meet at a point that divides each median in the ratio $1:2$, show that the center of mass of the *isosceles* traingle in the figure below is $\frac{1}{3}h$ from the base, where h is the height of the triangle.

(b) From part (a), show that the center of mass of any *right* triangle is $\frac{1}{3}h$ from the base, where h is the height of the triangle (see the figure below).

(c) From part (b), show that the center of mass of *any* triangle is $\frac{1}{3}h$ from the base, where h is the height of the triangle measured from that base, as indicated below.

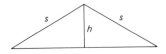

7.C.2 A uniform rod of length l and mass m is standing in a vertical position on a frictionless horizontal surface. It is given a small displacement, and it begins to tip and slip.

(a) Using energy principles, determine the velocity of the top of the rod just before it strikes the surface.

(b) Note that this is the same value we obtained for the rod that is pinned at the bottom, even though the motion is quite different. How might you have deduced this similarity from energy considerations without actually solving the problem?

7.C.3 Two uniform disks A and B are free to rotate on parallel axes, as shown in the following figure. Initially disk A is at rest and disk B is rotating with an angular velocity of 3000 r/min. The rim of A is now allowed to touch the rim of B, and the frictional force between them accelerates A and decelerates B.

(a) If the mass of A is 20 g and the mass of B is 40 g, what is the final angular velocity of B when there is no further slipping? (It is not necessary to assume a constant force of friction or to know the radius of either disk.)

(b) Is angular momentum conserved?

7.C.4 A 20-g stick 20 cm in length is free to slide without friction inside a sphere of radius 30 cm (see the figure). Initially the stick is at rest in a vertical position. What is the velocity of the center of the stick when it reaches its lowest point? Where is the instantaneous axis of rotation throughout the motion?

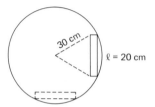

7.C.5 A 5-g cord 200 cm long is wrapped several times around a uniform cylinder whose radius is 10 cm and whose mass is 100 g. Initially the system is at rest, but a small piece of cord is hanging down from the rim and the cylinder begins to rotate and the cord unwinds. What is the angular velocity of the cylinder when the cord is fully unwound?

7.C.6 A man is sitting atop a long pole that is resting on the ground, and the bottom of the pole cannot be moved. He has a large watch on the end of a chain. Can he balance himself on the pole by rotating the watch on the end of the chain?

7.C.7 A drag racer is trying to place her engine in such a way that her acceleration is a maximum. The further the engine is toward the rear of the car, the greater the frictional force at the rear wheels. However, if the engine is too far back, the front end will lift off the ground. Suppose the coefficient of friction between the rear wheels and the ground is $\frac{3}{4}$ and the height of the center of gravity above the ground is 2 ft. How far in front of the rear wheels should the center of gravity be located?

*8
EQUILIBRIUM
OF
A
RIGID
BODY

In chap. 4 we discussed systems in static equilibrium. A system is in static equilibrium if the acceleration of each of its parts is zero; hence any system at rest or moving with a constant velocity is in static equilibrium. We restricted ourselves in Chap. 4 to those problems that we could solve by summing the forces acting on a single body and requiring that this sum be zero. Consider, however, the rigid rod in Fig. 8.1. The net force acting on this rod is zero, but clearly the rod is not in equilibrium: it will rotate. Only one point in this rod does not accelerate, and that is the center of mass.

FIGURE 8.1 A rigid rod subject to equal and opposite forces on its ends.

For a system to be in equilibrium it is not enough that the net external force on the system be zero. The net force on every particle in the system must be zero! In general, it is a difficult task to solve statics problems. There is a special case, however, which is comparatively easy and that is a rigid body.

8.1 EQUILIBRIUM OF RIGID BODIES

We know in general that

$$\mathbf{F}_{tot} = m\mathbf{a}_{c.m.}$$

and

$$\tau_{tot} = \frac{d\mathbf{L}}{dt}$$

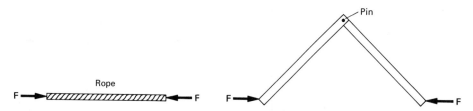

FIGURE 8.2 Two systems for which $a_{c.m.} = 0$ and $L = 0$ but which are not in static equilibrium.

These equations apply to any collection of particles, not just to a rigid body. Suppose, however, that the system is a rigid body and suppose that $F_{tot} = 0$ and $\tau_{tot} = 0$. The acceleration of the center of mass must then be zero, and the time rate of change of the angular momentum must also be zero. If the system is initially at rest and is a rigid body, it must remain at rest; i.e., it must be in static equilibrium. This is not true of a nonrigid body. See, for example, the two systems in Fig. 8.2.

We have then two conditions for static equilibrium of a rigid body. These are

$$F_{tot} = 0$$

$$\tau_{tot} = 0$$

where F_{tot} and τ_{tot} are, respectively, the net force and net torque resulting from the external forces. The torque may be computed about *any* point. These conditions are necessary and sufficient for static equilibrium.

QUESTION

A uniform beam 10 ft long and weighing 10 lb is hinged at one end to a vertical wall. The beam is supported in a horizontal position by a rope tied to the free end. The rope is attached to the wall and makes an angle of 45° with the vertical (Fig. 8.3). What is the tension in the rope and the force of the hinge on the beam?

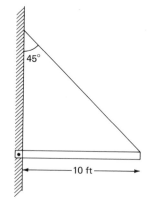

FIGURE 8.3 The beam is supported by a hinge at one end and a rope at the other.

ANSWER

Let the beam be the system. The forces on the beam are illustrated in Fig. 8.4. The sum of the forces in the horizontal direction must be zero, so that

$$F_x - T \cos 45° = 0$$

and in the vertical direction

$$F_y + T \sin 45° - w = 0$$

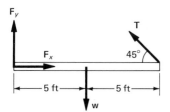

FIGURE 8.4 The forces on the beam are the force of the hinge **F**, the tension in the rope **T**, and the weight of the beam **w**.

The moment of the forces about any point must be zero. We choose to take moments about the right end of the beam, and we find that

$$5w - 10F_y = 0$$

Since w is known to be 10 lb, we have three equations and three unknowns. Solving for T, F_x, and F_y, we have

$$T = 5\sqrt{2} \text{ lb}$$

$$F_x = 5 \text{ lb}$$

$$F_y = 5 \text{ lb}$$

Notice that if we had taken moments about the left end or about the middle of the beam the results would have been the same. We leave it as an exercise for you to verify that the moments are zero about these points, given the solution obtained here.

We see that the x and y components of **F** are equal, so **F** makes a 45° angle with respect to the horizontal. We might have seen this without obtaining a complete solution in the following way. Consider the moment of the forces **T** and **w** about the point of intersection of the lines of action of these two forces (see Fig. 8.5). Since the net torque about this point must be zero, it follows that the line of action of **F** must also pass through this point. In fact, we may say that whenever three forces act on a body in equilibrium, they must intersect at a common point.

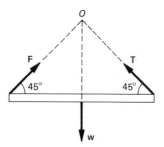

FIGURE 8.5 The lines of action of **w** and **T** intersect at O. The line of action of **F** must also pass through O.

QUESTION

A horizontal force of 2000 lb is applied at A to the steel frame in Fig. 8.6. All the beams are pinned at the joints. Point E is fixed

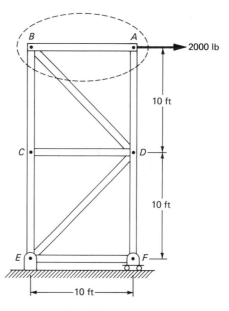

FIGURE 8.6 A steel frame.

and point F is free to slide horizontally. If we neglect the weight of the beams, what is the tension in the beam BC?

ANSWER

We choose as a system everything inside the dashed line in Fig. 8.6. We must replace all the things removed by the forces they exert. These forces are shown in Fig. 8.7. Note that all forces within the beams act along the length of the beams. This is true only so long as we can neglect the weight of the beams. (The student should consider an isolated beam and verify this statement.)

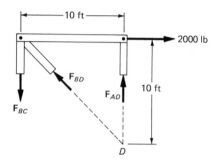

FIGURE 8.7   Forces acting on a portion of the steel frame.

We were asked to determine the force $\mathbf{F}_{BC}$ and not $\mathbf{F}_{BD}$ or $\mathbf{F}_{AD}$. Let us therefore take moments about point D, the intersection of the line of action of $\mathbf{F}_{BD}$ and $\mathbf{F}_{AD}$. We have

$$(10 \text{ ft})(F_{BC}) - (10 \text{ ft})(2000 \text{ lb}) = 0$$

so that

$$F_{BC} = 2000 \text{ lb}$$

(The reason for the nature of the constraints at E and F is that if both E and F were pinned to a fixed position, the system would be overdetermined. When the system is overdetermined, it is not possible to determine all the forces uniquely. A simple example is a single rod fixed at both ends. There is no way to determine the stress in the rod.)

*8.2   STATICS AND ANATOMY

A number of interesting examples of the application of the laws of statics to anatomy exist. In many of these it is fascinating to see how evolutionary pressures have developed some wonderful machines.

QUESTION

Figure 8.8 is a schematic diagram illustrating the function of the flexor and extensor muscles in the human arm. What tension must be developed in the flexor (extensor) muscle to exert a force of 60 lb at the hand? (Assume that the tendons are roughly vertical.)

ANSWER

If $F(E)$ is the tension in the flexor (extensor), we find, taking moments about the joint,

$$(2 \text{ in})(F) - (10 \text{ in})(60 \text{ lb}) = 0$$

and

$$-(2 \text{ in})(E) + (10 \text{ in})(60 \text{ lb}) = 0$$

so that

$$F = E = 300 \text{ lb}$$

The required tension is the same for either muscle.

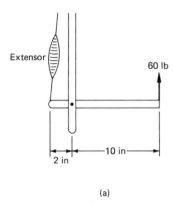

(a)

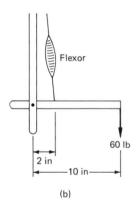

(b)

FIGURE 8.8 A simplified model of the human arm. The extensor (a) opens the arm, and the flexor (b) closes the arm.

QUESTION

What tension is required in the flexor muscle illustrated in Fig. 8.9 to raise a 60-lb weight?

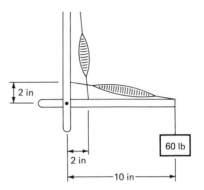

FIGURE 8.9 The 60-lb weight might be supported by either of two flexor muscles.

ANSWER

We have already seen what tension is required by the muscle attached to the upper portion of the upper arm (the biceps). To see what tension is required in the other portion of the arm, let us choose as a system the forearm and see what forces are required for equilibrium (Fig. 8.10). We let **F** be the force at the joint and **T** the tension in the muscle or tendon. Taking moments about the joint and approximating the moment arm of **T** about the joint to be 2 in, we have

$$-(60 \text{ lb})(10 \text{ in}) + T(2 \text{ in}) = 0$$

or

$$T = 300 \text{ lb}$$

so that both muscles are equally effective.

If both are equally effective, then why is the flexor on the upper arm (the biceps) rather than on the lower arm? There is a clear advantage in having the flexor close to the shoulder. It reduces the moment of inertia of the forearm about the elbow and the moment of inertia of the entire arm about the shoulder. Since $\alpha = \tau/I$, this reduction of the moment of inertia will increase the angular acceleration for a given torque.

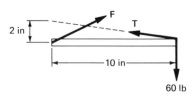

FIGURE 8.10   Forces acting on the forearm.

QUESTION

In the examples considered above, the forearm was horizontal and the upper arm vertical. Suppose the forearm made an angle θ with the horizontal (Fig. 8.11). How would the force required in the flexor change with θ for a given load of 60 lb?

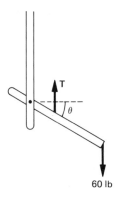

FIGURE 8.11   The forearm makes an angle θ with the horizontal.

ANSWER

The moment arm of both the load and the tension in the tendon are changed by the same factor ($\cos \theta$), so the tension will be the same as before. Thus the effort to hold a weight will be the same for all angles. Note that this is not the case when the upper arm is not vertical.

QUESTION

How does the moment arm of the extensor vary with the angle between the forearm and the upper arm in the geometry shown in Fig. 8.12?

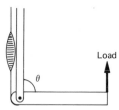

FIGURE 8.12   The tendon connecting the extensor to the forearm passes around a nearly circular arc. The point of pivot is approximately the center of curvature of this arc.

ANSWER

The moment arm of the extensor is equal to the radius of the circular arc and is independent of angle. This geometry is much more effective than that of Fig. 8.8 and in fact more closely represents the configuration in the human arm. This is one of nature's clever solutions to a physics problem.

QUESTION

What is the function of the patella (kneecap) in the knee?

ANSWER

The patella provides protection, but it also serves a mechanical function. The quadraceps tendon engulfs the patella and attaches to the tibia. Increasing the distance between this tendon and the axis of rotation increases the moment arm, and thus increases the torque (Fig. 8.13). If the patella becomes arthritic it may be removed, but the leg is then not as strong.

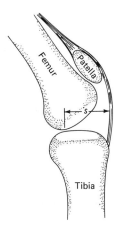

FIGURE 8.13 The patella increases the moment arm *s* of the tendon from the quadraceps to the tibia.

SUMMARY

Necessary but not sufficient conditions for equilibrium of an arbitrary body are

$$F_{tot} = 0$$

and

$$\tau_{tot} = 0$$

If the body is a rigid body, these conditions are necessary and sufficient.

PROBLEMS

8.A.1 The mobile in the following figure is in equilibrium. Neglecting the weights of the rods, determine the weights of balls *A*, *B*, and *C*.

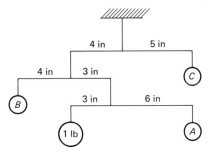

8.A.2 In the following figure, how strong must the cord be if it is to support the 10-lb shelf in a horizontal position?

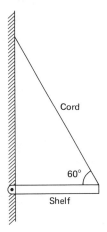

8.A.3 The Swiss Family Robinson has decided to adopt their meter stick as a standard of both length and weight. An object that is equal in weight to the meter stick weighs *one stick*. The Swiss Family Robinson purchases a sack of potatoes from the friendly natives at a price of 10¢ a stick. The meter stick balances on a knife edge at the 2-cm mark when the sack is hung from the 0-cm mark. What did they pay for the potatoes?

8.A.4 What weight W will balance the 2-lb weight in the figure below? Neglect the weight of the rods.

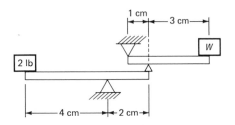

8.A.5 A 40-ft board weighing 30 lb extends 4 ft beyond the edge of a horizontal floor. How close to the overhanging end may a 150-lb person walk?

8.A.6 A 20-ft scaffold board is supported by two horizontal pipes 5 ft from either end. If the board weighs 20 lb, how far beyond the supports may a 200-lb person walk before the board begins to tip?

8.A.7 The human arm pictured below is supporting a 50-lb weight. What is the tension in the deltoid muscle? What force does the scapula exert on the humerus?

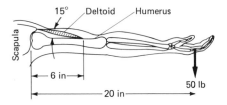

8.B.1 A 200-lb individual climbs a 10-ft stepladder (see the figure). The crossbar is located halfway up the ladder and will break if subjected to a tension greater than 50 lb. How far up the ladder may the person climb? (Neglect friction.)

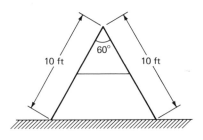

8.B.2 A 10,000-lb truck crosses a bridge as shown in the following figure. What is the stress in the beam at A when the truck reaches the midpoint? (Neglect the weight of the beams.)

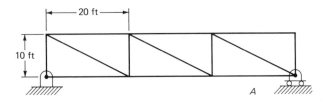

8.B.3 Neglecting the weight of the struts, determine the compressive force in the strut GH in the bridge shown below. A load of 2000 lb is applied at B, and each strut is 20 ft long. (Consider the system within the dashed line and select the point about which moments are taken with some care.)

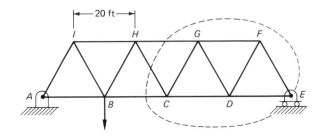

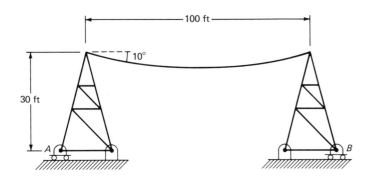

8.B.4 The two towers in the above figure support a 200-lb cable. What is the tension in the cable at the midpoint? How heavy must the towers be if they are not tied down at *A* and *B*?

8.B.5 A large uniform gate weighing 20 lb is attached to a gatepost, as shown in the figure. Determine the horizontal and vertical components of the force at the hinge and the tension in the cord.

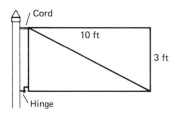

8.B.6 The garden roller shown below weighs 50 lb and is to be pulled over a 3-in curb by a horizontal force. If the radius of the roller is 20 in, what force is required?

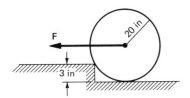

8.B.7 A 150-lb person is standing on the toes of *one* foot (see the figure). What is the tension in the Achilles' tendon? What is the stress in the tibia?

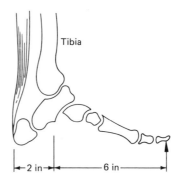

8.C.1

(a) A plank of length *l* is supported from below by an identical plank. The center of the upper plant is displaced a distance $l/2$ to the right of the center of the lower plank. Since the center of gravity of the upper plank lies directly above the edge of the plank below, the upper will be on the verge of tipping. Where should a third plank be placed below the two above so that the upper two planks will be on the verge of tipping on the third plank?

(b) Continue this process and show that a stack of arbitrarily horizontal displacement is possible (see the figure).

(c) Suppose you wanted to cross a 10-ft-wide gorge by dropping down from the end of the top board. How many boards would you need if the boards were 10-ft long and equal to your weight?

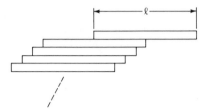

8.C.2 A uniform block $10 \times 10 \times 30$ cm high rests on a horizontal surface. The coefficient of static friction is $\frac{1}{4}$. A horizontal force perpendicular to one face of the block is applied 15 cm from the base and gradually increased until the block moves. Will the block tip or slip?

9
HARMONIC
MOTION

IN THIS CHAPTER we shall discuss a very special type of motion called *simple harmonic motion* (or *harmonic motion*). We may define harmonic motion as motion for which the acceleration is proportional to the displacement from the equilibrium position but opposite in direction, so that if the displacement is positive the acceleration is negative. We may write

$$a \propto -x$$

or

$$a = -(\text{constant})x$$

where the constant is positive. We shall find that a convenient notation for this constant is ω^2. Use of this notation will eliminate a lot of square roots later on. We may write, then,

$$a = -\omega^2 x \tag{9.1}$$

or equivalently

$$\frac{dv}{dt} = -\omega^2 x \tag{9.2}$$

As an example of a system in which the motion is harmonic, consider a mass on the end of a spring which obeys Hooke's law,

$$F = -kx \tag{9.3}$$

Since

$$F = ma$$

it follows that

$$a = -\frac{k}{m}x$$

so that a is equal to some negative number times x. In this case $\omega^2 = k/m$.

We can see roughly what the motion will be without actually solving Eq. 9.2. This equation tells us that the time rate of change of the velocity is negative for positive x and positive for negative x. Since the velocity is the slope of the space-time curve, the slope must be continuously decreasing for positive x and continuously increasing for negative x. A curve that has this property is illustrated in Fig. 9.1. From A to B the slope continuously decreases, and from B to C the slope continuously increases. The curve is everywhere concave toward the time axis. A curve that does not have this property is illustrated in Fig. 9.2.

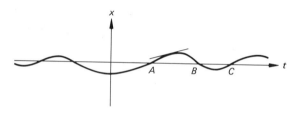

FIGURE 9.1 A space-time curve that is everywhere concave toward the time axis.

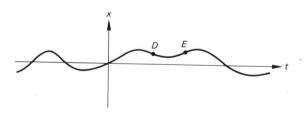

FIGURE 9.2 A space-time curve that is not everywhere concave toward the time axis. Between D and E the curve is convex toward the time axis.

The motion described in Fig. 9.1 is more or less what we would expect physically from a restoring force $F = -kx$. The force is always directed toward the origin, and the further the body gets from the origin, the greater the restoring force. The body should therefore oscillate about the origin. We shall see that *harmonic motion* is a very *special type* of oscillatory motion; not all oscillatory motion is harmonic motion.

9.1 MOTIVATION

One might ask why we give special attention to simple harmonic motion, since it is just a very special form of oscillatory motion. The answer to this question lies in the fact that for oscillations of small amplitude (the maximum displacement from the equilibrium position), we may, with a high degree of accuracy, approximate any oscillatory motion by harmonic motion.

To demonstrate this fact, let us consider the motion of a body subject to any restoring force, not just a restoring force proportional to the displacement. Such a force must be negative for positive x and positive for negative x (Fig. 9.3). We have seen in Chap. 4 that for small values of x, the tangent to the Fx curve gives a good approximation to the force. The equation of the tangent is

$$F = -c'x$$

where c' is a constant (the slope at $x = 0$). But this is equivalent to the condition for harmonic motion. Therefore, for small displacements any oscillatory motion is very nearly harmonic motion.

There are many examples of systems, mechanical and nonmechanical, which exhibit harmonic motion. Let us consider two examples.

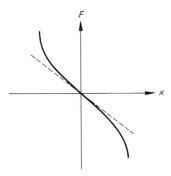

FIGURE 9.3 A nonlinear restoring force. The dashed line has the same slope as the solid line at the origin.

1. A mass m is suspended from a spring whose spring constant is k (Fig. 9.4). If the spring is extended a distance x beyond its equilibrium length, the spring will exert a force $-kx$. There will also be a gravitational force equal to the weight mg. The total force is given by

$$F = -kx + mg$$

and since this force is equal to ma,

$$ma = -kx + mg$$

or

$$a = \frac{-k}{m}x + g$$

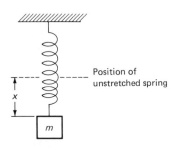

Position of
unstretched spring

FIGURE 9.4 A mass suspended from a spring.

Now this does not appear to be of the form required for harmonic motion. The acceleration is equal to a negative constant times the displacement plus an additional constant. However, if we make a simple change of variable, in effect change the origin of the x axis, we obtain the standard form for harmonic motion. We let

$$X = x - \frac{mg}{k}$$

Then

$$a = -\frac{k}{m}X \qquad\qquad (9.4)$$

and the motion will be harmonic. The mass will oscillate about the point $X = 0$ or $x = mg/k$. This is the point at which the net force on the mass is zero, so this is the equilibrium point of the suspended mass.

As we shall see, the motion as predicted by Eq. 9.4 is one in which the system would oscillate indefinitely. Of course, we would expect such a system eventually to come to rest at $x = mg/k$. We have neglected all damping effects such as air resistance and energy dissipated in stretching and compressing the spring. Such effects can be included and would predict a gradual decrease in the amplitude of oscillation. We shall assume that the damping effects are small.

2. Many systems are controlled by servomechanisms that exhibit harmonic motion. A servomechanism is a device, such as a room thermostat, that reacts with the system to control the behavior. Also a number of processes in living systems are self-regulating, and in general such self-regulating systems are called *homeostatic systems*. For example, if the body produces some chemical element x

and wants to maintain the element at some definite concentration, then a regulating mechanism is required. If there is too much x, then the body may generate a neutralizing agent y that will inhibit the production of x. This interaction of x and y under certain conditions will exhibit harmonic behavior; that is, x and y will oscillate about mean values. Normally some damping effect is also present, and this allows both x and y to settle down to steady-state values.

Some of these mechanisms can be quite complicated, involving two, three, or more interacting elements and often delayed reaction effects. We will not attempt to discuss these very complicated control problems. To understand them, however, the student must have a firm grasp of the simplest type of oscillatory behavior; harmonic motion.

9.2 KINEMATICS OF HARMONIC MOTION

We have said that the motion of a system is simple harmonic motion if

$$a = -\omega^2 x \tag{9.5}$$

If we examine the total force acting on the system and find that it is proportional to the displacement from some equilibrium position, then we know that Eq. 9.5 is satisfied and the motion is harmonic. This is where dynamics ends and kinematics begins. Now that we have this relation between acceleration and displacement, we would like to know such things as:

1. What is x as a function of time?

2. What are the velocity and acceleration as functions of time?

3. What are the period, frequency, and amplitude of the motion?

These are all questions involving the motion of the system, given the law of dynamics that governs the motion. The task of answering these questions for harmonic motion is very similar to the problem we faced for a body whose acceleration is governed by the equation

$$a = \text{constant}$$

We were able to show that

$$x = \tfrac{1}{2}at^2 + v_0 t$$

and that

$$v = at + v_0$$

by integrating the differential equation $d^2x/dt^2 = a$, where a is a constant.

Now we would like to integrate the differential equation

$$\frac{d^2x}{dt^2} = -\omega^2 x \tag{9.6}$$

We shall do this by guessing the solution (the correct solution) and verifying that it satisfies the differential equation. Needless to say, this is not very elegant, but without some exposure to differential equations it will have to do. Let us show that

$$x = A \cos(\omega t + \phi) \tag{9.7}$$

is a solution of Eq. 9.6, where A and ϕ are two arbitrary constants of integration. (Equation 9.6 is a second-order differential equation, and there should be two constants of integration.) Differentiating Eq. 9.7 once we find

$$v = \frac{dx}{dt} = -A\omega \sin(\omega t + \phi) \tag{9.8}$$

Differentiating a second time yields

$$a = \frac{d^2 x}{dt^2} = -A\omega^2 \cos(\omega t + \phi) \qquad (9.9)$$

But

$$x = A \cos(\omega t + \phi)$$

so that

$$a = -\omega^2 x$$

as it should. Therefore the solution to our problem in kinematics, $a = -\omega^2 x$, is

$$x = A \cos(\omega t + \phi) \qquad (9.10)$$

$$v = -A\omega \sin(\omega t + \phi) \qquad (9.11)$$

$$a = -A\omega^2 \cos(\omega t + \phi) \qquad (9.12)$$

and

$$x^2 + \frac{v^2}{\omega^2} = A^2 \qquad (9.13)$$

where we obtained Eq. 9.13 by eliminating time between Eqs. 9.10 and 9.11.

The constants A and ϕ depend on how the oscillation was started. We shall see presently that A is the amplitude of the oscillation, that is, the maximum displacement from the equilibrium position, and that ϕ depends on the choice for the zero of time.

When we plot the space-time trajectory (Eq. 9.10), we see that it satisfies the general condition required of it, namely, it is always concave toward the time axis and oscillates about $x = 0$ (Fig. 9.5).

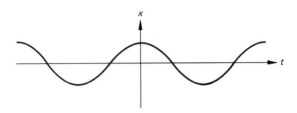

FIGURE 9.5 Space-time trajectory for harmonic motion with $\phi = 0$.

9.3 AMPLITUDE, PERIOD, AND FREQUENCY

We have seen that the solution of the equation of motion

$$a = -\omega^2 x$$

is

$$x = A \cos(\omega t + \phi) \qquad (9.14)$$

This equation is characterized by three quantities, A, ω, and ϕ, and we would now like to relate these constants to parameters of physical significance. First let us consider ϕ. The quantity ϕ is called the *phase,* and its value depends on when we start our clock. If we write Eq. 9.14 in the form

$$x = A \cos \omega\left(t + \frac{\phi}{\omega}\right)$$

we see that ϕ/ω is a constant added to t. With a suitable choice of the zero of time, we can see that $\phi/\omega = 0$. Wait until $x = +A$ and start the clock. The clock reads $t = 0$, and so $x = A \cos \phi$. If $x = A$ at $t = 0$, then $\cos \phi = 1$ or $\phi = 0$. If we choose to start the clock when $x = 0$ and the velocity is positive, then ϕ is determined by the equations

$$0 = A \cos \phi$$

and

$$v = -A\omega \sin \phi > 0$$

One choice of ϕ that satisfies these equations is

$$\phi = -\pi/2$$

Other solutions are $-\pi/2 + 2n\pi$, where n is any integer. If $\phi = -\pi/2$, then

$$x = A \sin \omega t$$

$$v = A\omega \cos \omega t$$

$$a = -A\omega^2 \sin \omega t$$

(Note that a is still equal to $-\omega^2 x$.)

Next let us look at the parameter A. Since the cosine function ranges between $+1$ and -1, we see from Eq. 9.14 that the value of x ranges between $+A$ and $-A$. Thus A represents the maximum displacement from the equilibrium position. The quantity A is called the *amplitude*.

Finally let us consider the significance of ω. We observe that if the coordinate of the body (x_0) at time t_0 is given by

$$x_0 = A \cos(\omega t_0 + \phi)$$

then x at time $t_1 = t_0 + 2\pi/\omega$ is given by

$$x_1 = A \cos(\omega t_1 + \phi) = A \cos(\omega t_0 + 2\pi + \phi)$$

$$= A \cos(\omega t_0 + \phi) = x_0$$

We see then that if we wait a time $2\pi/\omega$, the body returns to its original state. (Note that v and a also return to their original values.) Thus the motion repeats itself after a time $2\pi/\omega$, that is, the motion is periodic. If we define the *period* T to be the time required for the system to repeat itself, then

$$T = \frac{2\pi}{\omega} \tag{9.15}$$

It is important to note that the period depends on ω only and not on the amplitude A or the phase ϕ. The amplitude and phase depend on the initial conditions. The value of ω depends on the dynamic properties of the system, e.g., mass, spring constant, etc. This is the reason a spring-driven watch works as well as it does. As the driving spring runs down, the amplitude of oscillation of the control spring decreases. If the period depended on the amplitude, the watch would keep very poor time. It is only to the extent that the spring is not strictly a harmonic oscillator that it slows down as it runs down.

Another parameter of some importance in periodic motion is the *frequency,* the number of oscillations per second. But the number of oscillations per second is just the reciprocal of the number of seconds per oscillation. If we denote the frequency by f, then

$$f = \frac{1}{T} \tag{9.16}$$

The unit of frequency is s^{-1}, which is also called a hertz, abbreviated Hz.

We may now summarize the solution of problems in harmonic motion: We determine from the law of motion the relation between the acceleration and the displacement. If it takes the form

$$a = -\omega^2 x \tag{9.17}$$

where ω is some constant depending on the parameters of the problem, then

$$T = \frac{2\pi}{\omega} \tag{9.18}$$

$$x = A \cos(\omega t + \phi) \tag{9.19}$$

$$v = -A\omega \sin(\omega t + \phi) \tag{9.20}$$

$$a = -A\omega^2 \cos(\omega t + \phi) \tag{9.21}$$

$$x^2 + \frac{x^2}{\omega^2} = A^2 \tag{9.22}$$

QUESTION

A 50-g mass suspended from a spring causes a deflection of 1 cm. What mass should be attached to the spring to give it a natural period of oscillation of 1 s?

ANSWER

If a 50-g mass causes a deflection of 1 cm, then the spring constant must satisfy the relation

$$F = -kx$$

where

$$F = -mg = (0.05 \text{ kg})(9.8 \text{ m/s}^2)$$

$$= -0.49 \text{ N}$$

and

$$x = 0.01 \text{ m}$$

Therefore

$$k = \frac{0.49 \text{ N}}{0.01 \text{ m}} = 49 \text{ N/m}$$

If in a mass m is now suspended from the spring and allowed to oscillate, we have (see Eq. 9.4)

$$ma = -kX$$

or

$$a = -\frac{k}{m}X$$

where X is the displacement from the equilibrium position. Now from Eq. 9.15

$$T = \frac{2\pi}{\omega}$$

where ω^2 is the constant of proportionality between a and x (that is, $a = -\omega^2 x$), and we see that

$$\omega = \sqrt{k/m}$$

so that

$$T = 2\pi\sqrt{\frac{m}{k}}$$

Solving for m and setting $T = 1$ s, we obtain the mass that gives a 1-s period,

$$m = \frac{T^2}{4\pi^2}k = \frac{(1 \text{ s})^2(49 \text{ N/m})}{4\pi^2} = 1.2 \text{ kg}$$

QUESTION

Why should the oscillatory motion of a mass on a straight line be described in terms of trigonometric functions? And where does the π come from in $T = 2\pi/\omega$? The quantity π is the ratio of the circumference of a circle to its diameter. What does a circle have to do with motion on a straight line?

ANSWER

We can see the origin of these trigonometric functions and of the π's from the following consideration. Let P be a point that moves with constant angular velocity ω on a circle of radius A. Let θ be the angular displacement of P (Fig. 9.6). Now from Table 2.1

$$\theta = \omega t$$

and the projection of the point P on the x axis is

$$x = A \cos \theta = A \cos \omega t$$

where A is the radius of the circle. Since this is Eq. 9.19 with ϕ equal to zero, this is the equation of a point moving in simple harmonic motion. Furthermore,

$$v_P = A\omega$$

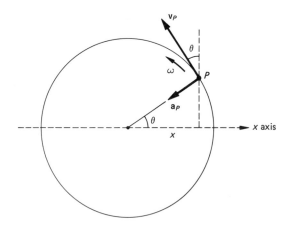

FIGURE 9.6 The point P moves with constant angular velocity ω on a circle of radius A.

and its projection on the x axis is

$$\frac{dx}{dt} = -v_P \sin \theta = -v_P \sin \omega t = -A\omega \sin \omega t$$

which is just Eq. 9.20. Finally the x component of a_P is

$$\frac{d^2x}{dt^2} = (a_P)_x = -a_P \cos \theta$$

But from Eq. 2.47

$$a_P = A\omega^2$$

so that

$$\frac{d^2x}{dt^2} = -A\omega^2 \cos \omega t$$

which is Eq. 9.21.

The period of the projection point on the x axis is simply the time it takes P to make one complete revolution. Now

$$\theta = \omega t$$

and setting $\theta = 2\pi$ and $t = T$, we have

$$2\pi = \omega T$$

or

$$T = \frac{2\pi}{\omega}$$

which is Eq. 9.18.

So we see that we can deduce all our results for simple harmonic motion by following the projection on a diameter of a point moving with constant angular velocity ω on a circle of radius A.

We have been considering the harmonic motion of a system moving along a straight line. We would also like to consider such problems as a mass supported by a string (a simple pendulum), where the mass moves along an arc of a circle rather than a straight line. Before we contrast the two problems, let us determine the equation of motion for the simple pendulum.

Let θ be the angle the string makes with the vertical (Fig. 9.7). The component of the forces perpendicular to the string (in the direction of increasing θ) is $-mg \sin \theta$. (The tension T is perpendicular to this direction and has no component perpendicular to the string.) This tangential force must be equal to the mass times the component of the acceleration in the tangential direction $a_T = r\alpha$. Therefore

$$mr\alpha = -mg \sin \theta$$

Now

$$\sin \theta = \frac{x}{r}$$

and

$$\theta = \frac{s}{r}$$

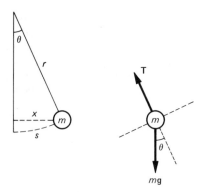

FIGURE 9.7 A simple pendulum.

But for small displacements from equilibrium

$$x \simeq s$$

so that

$$\sin \theta \simeq \theta$$

We have then approximately

$$\alpha = -\frac{g}{r}\theta$$

This equation is of exactly the same form as Eq. 9.17 except that instead of a linear displacement we have an angular displacement and instead of a linear acceleration we have an angular acceleration. All our previous equations apply equally well to this problem. In particular, since

$$\alpha = -\frac{g}{r}\theta$$

it follows that

$$T = 2\pi\sqrt{r/g} \qquad\qquad (9.23)$$

Thus the longer the pendulum string r, the longer the period; and the greater g (the gravity pull per unit mass), the shorter the period. Note that the period is independent of the mass. This might seem strange at first. One would

expect that the greater the inertia, the greater the period (as in the case of the mass on the spring). However, in this case, as the inertia increases, the restoring force (proportional to mg) also increases.

QUESTION

Suppose you wanted to design a grandfather's clock so that each swing of the pendulum would take exactly 1 s. How long should you make the pendulum? (Assume that it is a simple pendulum.)

ANSWER

For each swing to take 1 s, the period (the time required for the pendulum to swing back and forth) must be 2 s. Solving Eq. 9.23 for r, we find

$$r = \frac{T^2}{4\pi^2}g = \frac{(2\text{ s})^2(9.8\text{ m/s}^2)}{4\pi^2} = 0.99\text{ m}$$

SUMMARY

The motion of a system is *simple harmonic* motion if

$$a = -\omega^2 x$$

where ω is a constant that depends on the physical situation. The motion of any system oscillating with small amplitude about a stable equilibrium state may be approximated as simple harmonic motion.

The kinematics of a system executing simple harmonic motion is given by the relations

$$x = A\cos(\omega t + \phi)$$

$$v = -A\omega\sin(\omega t + \phi)$$

$$a = -A\omega^2\cos(\omega t + \phi)$$

where A is the amplitude of the motion and represents the maximum displacement from the equilibrium position.

The period T is related to ω by the equation

$$T = \frac{2\pi}{\omega}$$

For a mass m suspended from a spring with spring constant k,

$$\omega = \sqrt{k/m}$$

For a simple pendulum,

$$\omega = \sqrt{g/r}$$

PROBLEMS

9.A.1 If the amplitude of a simple harmonic oscillator is 10 cm and the maximum velocity is 20 cm/s, what is the frequency?

9.A.2 A body undergoing simple harmonic motion has an amplitude of 10 cm and a period of 2π s. What is the displacement when the velocity is 4 cm/s?

9.A.3 A body oscillates according to the equation

$$x = 8 \cos(2t + 5)$$

where x is measured in centimeters and t in seconds. Determine (a) the period, (b) the frequency, (c) the amplitude, (d) the phase, (e) the maximum velocity, and (f) the maximum acceleration.

9.A.4 When a 10-kg mass is hung on a spring and slowly lowered, it stretches the spring 10 cm. The mass is then pulled down an additional 5 cm and released. Determine (a) the period, (b) the amplitude, (c) the maximum velocity,

(d) the position at which the velocity is one-half the maximum, and (e) the time it takes the mass to move from $x = 0$ to $x = A/2$.

9.B.1 A body is hung on an unstretched spring and released. The body falls 10 cm before it stops and starts back up. What is the period?

9.B.2 A horse is trotting with a time between gaits of 0.75 s. Assuming that the vertical motion of the saddle can be approximated as simple harmonic motion, determine the maximum vertical amplitude of the saddle for which it is unnecessary for the rider to hold on to avoid leaving the saddle.

9.B.3 A uniform rigid rod of length l and mass m is suspended from its upper end. Determine the period for small amplitude oscillations.

9.B.4 A disk of radius r is cut in half along a diameter and pinned at the center of curvature (see the figure). What is the period for small amplitude oscillations?

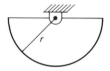

9.B.5 We have shown that the motion of a simple pendulum is simple harmonic motion if the amplitude is small. To get an idea of the error involved for large-amplitude oscillation, consider a simple pendulum whose angular amplitude is $\pi/2$; that is, it oscillates in a semicircle.

(a) Determine the velocity at the bottom of the arc exactly in terms of the mass and the length of the pendulum.

(b) Assuming that the motion is simple harmonic motion with amplitude $\pi/2$, determine the maximum velocity of the pendulum.

(c) Determine the ratio of the velocities calculated in parts (a) and (b).

9.B.6 Two unstretched springs (k_1 = 1 N/m, k_2 = 3 N/m) are attached to a 16-kg mass as shown in the following figure. Determine the period of oscillation.

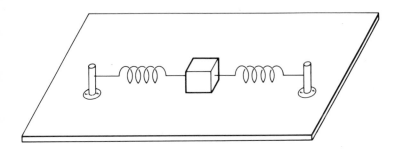

9.C.1 A 2-g body moves along a straight line with a potential energy $U(x) = 5x^3 + 4x^2$ where x is measured in centimeters. What is the period for small-amplitude oscillations about $x = 0$?

9.C.2 A "seconds pendulum" keeps accurate time at a location where g = 980 cm/s². The pendulum is moved to a new location and loses 3 min/day. Determine g at the new location. (A seconds pendulum has a period of 2 s.)

9.C.3 A board of negligible mass is pinned at a point three-fourths of the distance from the left end. A spring (k = 3 N/m) is attached to the left end and a 3-kg mass rests on the right end.

(a) Determine the period.
(b) What is the maximum amplitude of the right end if the 3-kg mass does not leave the board?

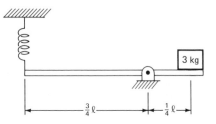

9.C.4 A rod of length l oscillates without friction in the bottom of a bowl of radius r. The mass of the rod is m. Show that for small-amplitude oscillations the motion is simple harmonic motion and determine the period.

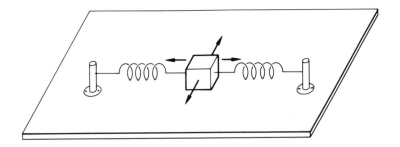

9.C.5 Two identical springs whose equilibrium lengths are 10 cm are stretched an additional 5 cm and joined. A mass is attached at the union (see the figure). The mass rests on a horizontal surface and can oscillate along the line of the springs with a period $T_\parallel$ or perpendicular to the line of the springs with a period $T_\perp$. Determine the ratio $T_\parallel / T_\perp$ of the periods.

*10
SIZE
AND
FUNCTION

We hear a great deal about the physical feats of small animals. Ants, for example, can lift three to four times their body weight. If a 200-lb man could lift three times his body weight, he could raise 600 lb. A flea can jump 200 times its own height. If a 6-ft man were capable of this feat, he could jump 1200 ft; he could leap over buildings in a single bound. (Superman is a flea with a cape and a thyroid problem.)

A great many questions may be asked concerning the relationship between size and function. How does size affect speed, strength, endurance, resistance to water and heat loss, terminal speed in free fall, flight, etc.? To answer these questions, we must examine how the particular function depends on the size of the animal or object being studied. For example, to see how speed depends on size, we must study all the factors that determine speed, such as the inertia of the moving parts, the strength of the muscles, and the distance through which the muscles contract. It is not obvious how all these factors will relate. A larger animal has stronger muscles that would tend to make it faster, but its legs have more inertia, which slows it down. We shall see later when we examine this problem quantitatively that these two effects exactly compensate each other and that similarly constructed animals of all sizes should have roughly the same maximum speed.

10.1 SURFACE-TO-VOLUME EFFECTS

The simplest class of examples that demonstrate the relationship between size and function are surface-to-volume effects, i.e., effects that depend on the ratio of the surface area to the volume. As we shall see, a large number of phenomena depend on this ratio. First let us examine how this ratio depends on size.

The solid with the simplest geometry is the cube. The area of a cube of side L is given by

$$A = 6L^2$$

since the cube has six faces of area L^2. The volume of a cube is

$$V = L^3$$

The ratio of area to volume is therefore

$$\frac{A}{V} = \frac{6L^2}{L^3} = \frac{6}{L}$$

For a cube, then, the surface-to-volume ratio decreases with increasing size. The volume increases faster than the surface area.

For a sphere,

$$\frac{A}{V} = \frac{4\pi r^2}{\frac{4}{3}\pi r^3} = \frac{3}{r}$$

and once again the surface-to-volume ratio is inversely proportional to the size of the object.

As a third example consider a cylinder, for which

$$\frac{A}{V} = \frac{2\pi rL + 2\pi r^2}{\pi r^2 L} = \frac{2(1 + L/r)}{L}$$

where L is the length of the cylinder and r is its radius. Again the ratio is inversely proportional to a characteristic length. (The factor L/r is a constant. If all the dimensions are increased proportionately, then the ratio L/r does not change. If L is doubled and r is doubled, then L/r is unchanged.)

One can show that in *all* cases

$$\frac{A}{V} = \frac{k}{L}$$

where k is a constant that depends only on the shape and not on the size and L is some characteristic length. For a cube $k = 6$, for a sphere $k = 3$, and for a cylinder $k = 2(1 + L/r)$. (The constant k will have different values for different choices of the characteristic length.)

QUESTION

Both objects in Fig. 10.1 have the same shape, and $L' = 2L$. If A/V is 10 for the object on the left, what is A'/V' for the object on the right?

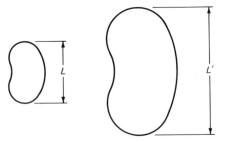

FIGURE 10.1   Two bodies of similar shape but different size.

ANSWER

Since

$$\frac{A}{V} = \frac{k}{L} = 10$$

for the object on the left, then

$$k = 10L$$

Now

$$\frac{A'}{V'} = \frac{k}{L'}$$

where k is the same, since both objects are the same shape. Therefore

$$\frac{A'}{V'} = \frac{10L}{L'} = \frac{10L}{2L} = 5$$

so the surface-to-volume ratio is halved if the size is doubled.

We shall now employ this principle,

$$\frac{\text{surface area}}{\text{volume}} \propto \frac{1}{\text{characteristic length}}$$

to explain some characteristics of living systems.

1. Why are insects cold-blooded? Larger animals have found that a well-regulated warm-blooded circulatory system is much more efficient. Since all chemical reactions take place more rapidly at higher temperatures, a warm-blooded animal can react more quickly. Why are all living creatures not warm-blooded? A clue to the answer to this question can be found in the eating habits of large and small warm-blooded animals. Large animals eat only a small fraction of their body weight in one day, while small animals may consume as much as ten times their body weight in the same period. Most of this food energy is used to maintain a constant body temperature, so it seems that small animals have a difficult time maintaining a high body temperature. The reason for this is clear: A small animal has a very large surface-to-volume ratio. Since heat is lost only across the skin, the heat lost will be proportional to the surface area, while the heat generated by an animal through chemical reactions in the body is proportional to the volume. That is,

$$\frac{\text{heat lost}}{\text{heat generated}} \propto \frac{\text{surface}}{\text{volume}} \propto \frac{1}{\text{size}}$$

Therefore very large animals will have difficulty dissipating heat, and very small animals will have difficulty retaining heat. Elephants spend a lot of time at the water hole, and they also cover themselves with mud and dust to insulate themselves from the sun. At the other extreme, the shrew consumes ten times its body weight in one day. (A human will consume one-fiftieth of his or her body weight in a day.) Insects have given up the task of maintaining a constant body temperature.

To gain some appreciation of the problem faced by the elephant, imagine yourself with a group of people on a hot day in Central Africa. Suppose you all pack yourselves together, and the total volume of the pack is the same as that of the elephant. The surface-to-volume ratio of the group has been considerably reduced, since those on the interior of the pack have no exposed surface and those on the perimeter have only a fraction of their body surfaces exposed (Fig. 10.2). Misery!

2. We find a remarkable uniformity in the size of biological cells. The cells of small animals and plants are pretty much the same as those of larger animals and plants. To understand this we must examine the energy requirements of a cell. Every living cell requires nourishment. This nourishment must pass across the cell walls, and it is then consumed within the volume of the cell. If the cell were to increase in size, the amount of nourishment that could be taken in would increase in proportion to the area, but the amount of nourishment required would increase in proportion to the volume:

nutrients absorbed $\propto$ surface area

nutrients required $\propto$ volume

We have seen that with increasing size the volume increases faster than the surface area, so there would in-

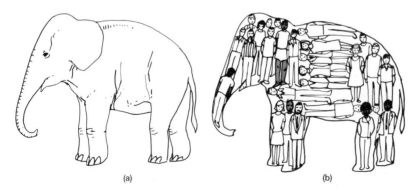

FIGURE 10.2 (a) An elephant and (b) a number of people packed into a space with the same volume as the elephant.

evitably come a point at which the supply could not equal the demand.

Certain cells (nerve cells) are very long — 1 m or more — and seem to violate our rule that all cells are roughly of the same size. We have shown that for a cylinder the ratio of surface area to volume is $2(1/L + 1/r)$. As L becomes much larger than r, the surface-to-volume ratio becomes more or less independent of L and approaches a constant value equal to $2/r$, so that in these cells the surface-to-volume ratio is not much different than that of a sphere, that is, $3/r$. Although these cells are very long, they nevertheless have the same surface-to-volume ratio as most other cells.

If a living organism is to increase much beyond the size of a single cell, it must develop some means of feeding itself other than diffusion across its surface. In animals this was accomplished by the development of a circulatory system that transports nutrients throughout the body. The problem then becomes one of getting this nourishment into the blood, by some means other than diffusion through the skin. This is accomplished through eating and the use of the lungs. These are not diffusion processes but convection processes. Of course, after the nourishment has been taken in, the nourishment is extracted by diffusion processes. In the lungs, digestive tract, and kidneys, the effective surface area is greatly expanded by folding and filigreeing.

This same sort of convection process is achieved in plants by means of a number of small tubes called xylem, which carry sap. The pumping of the sap is accomplished by means of osmotic and capillary forces.

3. A bug does not get hurt when it falls from a bush. This is because of its large surface-to-volume ratio and, as we shall see in the next section, because self-destructive forces are less important in small objects.

The frictional drag f resulting from air resistance is proportional to the area A and the square of the velocity v.

Hence

$$f = k_1 A v^2$$

The weight of the bug is proportional to its volume, so

$$w = k_2 V$$

As the bug falls, its velocity increases until the drag just balances the weight,

$$f = w$$

or

$$k_1 A v^2 = k_2 V$$

Therefore the terminal velocity is

$$v = \frac{\sqrt{k_2/k_1}}{\sqrt{A/V}}$$

The terminal velocity is inversely proportional to the square root of the surface-to-volume ratio. A flea therefore has a low terminal velocity and a human has a high terminal velocity.

   This also explains why on steep grades you see signs requiring trucks to use low gear. A truck has a low surface-to-volume ratio. Its terminal velocity may be terminal.

10.2 STRUCTURAL STRENGTH

We generally tend to associate strength with size. An elephant is stronger than a mouse, an adult is stronger than a child, and a 2 × 4 is stronger than a match. Yet some paradoxes do exist. A toy automobile falling from a height of 10 ft suffers no damage, but a real automobile falling the same distance would more than likely suffer some damage. Large buildings have collapsed under their own weight, while a small shed may last until it rots. That is not to say that a toy automobile is stronger than a real one or that a shed is stronger than a massive building. When an automobile falls over a cliff and strikes the ground, certain *self-destructive* forces are at work. If the vehicle lands on its front bumper, the entire weight of the automobile is supported there and the bumper will collapse. The strength of the bumper is proportional to the cross-sectional area of its supports. The weight of the automobile is proportional to its volume.

$$\frac{\text{strength}}{\text{weight}} \propto \frac{\text{area}}{\text{volume}} \propto \frac{1}{\text{size}}$$

The smaller the automobile, the greater the strength-to-weight ratio.

QUESTION

A Cadillac is twice as long as a Toyota; i.e. if L_{Cad} is the overall length of the Cadillac and L_{Toy} is the length of the Toyota, then

$$L_{\text{Cad}} = 2L_{\text{Toy}}$$

If the diameter of the front axle of the Toyota is 2 in, what should the diameter of the front axle on the Cadillac be?

ANSWER

If the strength-to-weight ratio is to be the same, then

$$\frac{A_{\text{Toy}}}{V_{\text{Toy}}} = \frac{A_{\text{Cad}}}{V_{\text{Cad}}}$$

where A is the cross-sectional area of the axle and $V \propto L^3$ is the volume. Solving for A_{Cad}, we have

$$A_{\text{Cad}} = \frac{V_{\text{Cad}}}{V_{\text{Toy}}} A_{\text{Toy}} = \frac{L_{\text{Cad}}^3}{L_{\text{Toy}}^3} A_{\text{Toy}} = 8 A_{\text{Toy}}$$

Since the cross-sectional area of the axle is proportional to the square of the diameter,

$$D^2_{\text{Cad}} = 8D^2_{\text{Toy}}$$

or

$$D_{\text{Cad}} = \sqrt{8}D_{\text{Toy}} = \sqrt{8}(2 \text{ in}) = 5.7 \text{ in}$$

So the diameter of the axle (and other components of the suspension) must be increased almost three times ($\sqrt{8} = 2.8$) if the length is doubled. The suspension must be increased *disproportionately.*

These same considerations apply to the suspension system in animals as well. The legs of an elephant and those of a hippopotamus are disproportionately thick, but an ant can get along quite well with very thin legs (very thin in proportion to its size).

Not only must the legs of larger animals be disproportionately large, but the entire skeletal system must also. The fractional weight of the skeleton of a mouse is 8%, a dog 13%, and a human 17%. On the other hand, the fractional weights of the skeletons of the porpoise and whale are similar because weight it not a burden in the sea.

The structure of plants and trees is also governed by the rule that strength is proportional to cross-sectional area and weight burden is proportional to volume. If an oak tree grew as large as a redwood, all its limbs would break off. As the limbs became larger and larger, the weight would increase as L^3 and the strength as L^2. We observe in fact that redwoods have branches that are proportionately much smaller than those of oak trees.

Fruit trees cannot bear fruit that is very large unless the branches are very thick. The branch strength is propor- tional to L^2, and the weight of the fruit to L^3. This is why melons and pumpkins grow on the ground.

10.3 RUNNING

A remarkable similarity exists in the maximum speeds of animals over a wide range of sizes (see Table 10.1). The maximum speeds of the rabbit, dog, and horse all fall within the range of 35–42 mi/h. (There are of course a number of exceptions, e.g., the turtle and the cheetah.) Is there some way we might understand this similarity?

TABLE 10.1   Maximum speeds of various animals.

Animal	Speed (mi/h)	Animal	Speed (mi/h)
Human	25	Wild donkey	41
Whippet	34	Giraffe	30
Greyhound	37	African elephant	35
Horse	42	Wolf	41
Ostrich	51	Fox	45
Rabbit	41	Cheetah	65
Gazelle	62		

The principal limiting factor in running is the work that must be done in accelerating the limbs. We shall analyze a simple model that illustrates the important features. A more realistic study is rather complicated, but the conclu- sion is the same.

Let us represent the motion of a single leg by a mass that is accelerated from rest to some velocity v_{max} and then brought to rest again when the mass is returned to the ground (Fig. 10.3). This maximum velocity will be proportional to the mean velocity of the animal. Work is done by the muscles in accelerating the mass to its maximum velocity, and from our principle of work and energy,

$$W_{A \to B} = \tfrac{1}{2} m v_{max}^2 \qquad (10.1)$$

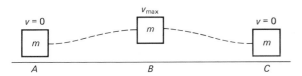

FIGURE 10.3 The block of mass m represents the leg. Initially it is at rest on the ground (A). It is accelerated to a maximum velocity (B), and then brought to rest again as it is returned to the ground (C).

This kinetic energy is not returned to the muscles when the mass is decelerated and brought to rest again at C (see Fig. 10.3); instead the energy is dissipated in the form of heat.

Now the work done by a muscle is the product of the force the muscle exerts and the distance through which it acts:

$$W_{A \to B} = (\text{force})(\text{distance})$$

This distance is not the distance the mass moves but the distance through which the muscle contracts. Equating the work to the increase in kinetic energy, we have

$$W_{A \to B} = \tfrac{1}{2} m v_{max}^2$$

so that

$$\tfrac{1}{2} m v_{max}^2 = (\text{force})(\text{distance})$$

or

$$v_{max}^2 = \frac{2(\text{force})(\text{distance})}{m}$$

The force exerted by the muscle is proportional to the cross-sectional area of the muscle, i.e., the number of fibers in the muscle (not the length of the fibers in the muscle but the number of pulling fibers). If we let L be some characteristic length in the animal, then

$$\text{force} \propto L^2$$

The distance through which the force acts is proportional to L:

$$\text{distance} \propto L$$

and the mass m being moved is proportional to L^3. Therefore

$$v_{max}^2 = \frac{2(\text{force})(\text{distance})}{m} \propto \frac{(L^2)(L)}{L^3}$$

so that v_{max} is *independent of the size.*

Since the maximum velocity of the mass is proportional to the velocity of the animal, we conclude that the speed of the animal is independent of its size and depends only on its construction. Two animals of similar construction but different proportions should have the same maximum speed. This is not to say that a turtle and a cheetah or a human being and a horse will have the same maximum speed because they are not of proportional construction. However, a fox and a greyhound should have the same maximum speed even though the greyhound is two or three times larger than the fox.

One might ask why animals have not developed with *disproportionately* large muscles. They could increase their velocity by just increasing the muscle size without increasing in the same proportion the total body weight. Unfortunately, the body is a finely tuned, well-integrated system. If the muscles were larger, the tendons would all have to be larger or they would break under the additional strain. As it is, we find in professional athletes constant tearing of cartilage, tendons, and ligaments. These people are using their muscles to their fullest potential. Also, if the muscles were larger, they would require a greater blood supply to bring in nutrients and carry away waste. Hence the lungs would have to be larger to supply oxygen to the blood, the kidneys larger to purify the blood, the bones larger to support the additional weight, and so on. Before long a great deal of additional weight would have been added to support the additional muscle.

10.4 JUMPING

We can examine in the same manner the ability to jump. In the standing high jump the muscles in the legs contract, giving the body an upward velocity, and this kinetic energy is converted into potential energy (mgh) at the maximum height of the jump. The work done by the muscles is given by

$$\text{work} = (\text{force})(\text{distance})$$

where the force is the force exerted by the muscles and the distance is the distance through which the muscles contract. If L is some characteristic length of the body and the force of the muscle is proportional to its cross-sectional area, we have

$$\text{force} \propto L^2$$

and

$$\text{distance} \propto L$$

so that

$$\text{work} \propto L^3$$

Equating the work to the change in energy yields

$$\text{work} = mgh$$

Solving for h, we have

$$h = \frac{\text{work}}{mg} \propto \frac{L^3}{L^3}$$

since the mass is proportional to L^3. Therefore the height of the jump is independent of L, so two animals of similar construction but different proportion should be able to jump the same height. It is not surprising then that a grasshopper is able to jump many many times its height. Rabbits and deer and human beings can all raise their centers of gravity about 4 ft. Even a flea can jump several feet; it could probably jump much higher, but air resistance is more important for so small an insect.

10.5 FLYING

We observe that birds do not grow as large as most land animals, and therefore expect that the power required for flight increases faster with size than the power available. We will not go into the theory here, but it can be shown that

$$\text{power required for flight} \propto L^{7/2}$$

or

$$\text{power required for flight} = k_1 L^{7/2}$$

On the other hand,

$$\text{power available} \propto L^2$$

or

power available $= k_2 L^2$

We have no way of determining what the proportionality factors are, but for any k_1 and k_2 there will exist an L such that

$k_1 L^{7/2} > k_2 L^2$

In particular, if

$$L > \left(\frac{k_2}{k_1}\right)^{2/3}$$

then the power available will be less than the power required.

10.6 CONCLUSION

We can deal with many other problems by applying these scaling techniques. In fact, many evolutionary forces are nature's way of solving problems associated with size.

QUESTION

We have spoken of the seemingly remarkable ability of ants to lift several times their own weight. What is really surprising is that they cannot lift even greater weights.

Suppose there were an animal the size of an ant ($\frac{1}{4}$ in in length) but with the structure of a man. What weight could the animal lift? Assume that a 6-ft man can left a weight equal to his body weight, say 200 lb.

ANSWER

Let W_L be the maximum weight a man can lift and W_B be the body weight of the man. Now

$W_L \propto$ strength $\propto L^2$

where L is some characteristic length (say the man's height). Therefore

$W_L = k_1 L^2$

Also

$W_B \propto L^3$

or

$W_B = k_2 L^3$

Therefore

$$\frac{W_L}{W_B} = \frac{k_1 L^2}{k_2 L^3} = \frac{k_1/k_2}{L}$$

When $L = 6$ ft, $W_L/W_B = 1$; that is, a 6-ft man can lift a weight equal to his body weight. Therefore

$$1 = \frac{k_1/k_2}{6}$$

or

$k_1/k_2 = 6$

When $L = \frac{1}{4}$ in $= \frac{1}{48}$ ft,

$$\frac{W_L}{W_B} = \frac{k_1/k_2}{L} = \frac{6}{\frac{1}{48}} = 288$$

or

$W_L = 288 W_B$

In other words, the weight that may be lifted is 288 times the body weight! Thus the strength of the ant pales by comparison. Of course, if a man were $\frac{1}{4}$-in tall, he would freeze to death almost immediately because of his large surface-to-volume ratio. If he became cold-blooded like the ant, he would lose much of his muscular efficiency and probably would be no stronger than the ant.

QUESTION

Assuming that the weight per unit cross section of bone is the same for a horse and a dog, show that a horse four times as large as a dog would have legs eight times as thick.

ANSWER

Given that the weight supported per unit cross section of bone is the same for both the dog and the horse, we have

$$\frac{W}{A} = \text{constant}$$

where W is the weight of either animal and A is the cross-sectional area of bone. If D is the bone diameter, then $A = \pi D^2/4$, so that

$$\frac{W}{D^2} = \text{constant},$$

or the ratio W/D^2 is the same for the dog and the horse. That is,

$$\frac{W_H}{D_H{}^2} = \frac{W_D}{D_D{}^2}$$

or

$$\frac{D_H{}^2}{D_D{}^2} = \frac{W_H}{W_D} = \frac{L_H{}^3}{L_D{}^3}$$

since the weight is proportional to L^3, where L is the size. Now $L_H = 4L_D$, so that

$$\frac{D_H{}^2}{D_D{}^2} = 4^3 = 64$$

or

$$D_H = 8D_D$$

QUESTION

If a 6-ft person dies of thirst in the desert in three days, how long would a 3-ft person last?

ANSWER

The rate of water loss is proportional to the body surface area, so

$$\frac{\text{water loss}}{\text{time}} = k_1 L^2$$

or

$$\text{time} = \frac{\text{water loss}}{k_1 L^2}$$

Now the total water that is available is proportional to the total volume, or

$$\text{Water loss} = k_2 L^3$$

Hence

$$\text{time} = \frac{k_2 L^3}{k_1 L^2} = \frac{k_2}{k_1} L$$

Thus the time required to lose the water available is directly proportional to the size: Double the size and the time is doubled; halve the size and the time is halved. Therefore, the 3-ft person would last half as long as the 6-ft person, or one and one-half days.

SUMMARY

The ratio of the surface area A to the volume V of a body satisfies the relation

$$\frac{A}{V} = \frac{k}{L}$$

where k depends on the shape but not the size of the body and L is a characteristic length. Therefore the surface-to-volume ratio varies with $1/L$ for a fixed shape.

This inverse proportionality explains why insects are cold-blooded, why most living cells are roughly the same size, why small objects have small terminal velocities while large objects have large terminal velocities, why large animals have less difficulty with water loss than small animals, and so on.

The strength-to-weight ratio of a structure is proportional to the ratio of the cross-sectional area (not the surface area) to the volume. This ratio in turn in inversely proportional to the size, i.e.,

$$\frac{strength}{weight} \propto \frac{1}{size}$$

This explains why the supports for large buildings, cars, and animals must be disproportionately large.

The maximum speed of an animal depends on the ratio of the mass to be accelerated (the legs and feet) and the work done by the muscles in the legs. The work of the muscles in turn is proportional to the product of the cross-sectional area of the muscle (a measure of the strength of the muscle) and the distance through which it contracts. The mass is proportional to the cube of the size, and the work of the muscles is also proportional to the cube of the size. Thus the maximum speed is independent of size.

The jumping ability of an animal is proportional to the ratio of the work done by the muscles to the weight to be lifted. Again this ratio is independent of size.

Flying ability is proportional to the ratio of the power available ($\propto L^2$) to the power required ($\propto L^{7/2}$) or to $L^2/L^{7/2} = 1/L^{3/2}$. Thus flying ability decreases with increasing size.

PROBLEMS

10.A.1 Show that the surface-to-volume ratio for a parallelepiped of sides a, b, and c can be expressed as the ratio

of a factor that depends on the shape to a factor that depends on the size.

10.A.2 It is advantageous for an animal foraging in deep snow to have large feet. If the foot of a 4-ft-long animal is 3 in long, how long should the foot of a 2-ft-long animal be if it is to have the same facility for walking in deep snow?

10.A.3 If the terminal velocity of a 6-ft person in free fall is 100 mi/h, what is the terminal velocity of (a) a 3-ft child in free fall, and (b) a 4-in bird?

10.A.4 If all the dimensions of a muscle were to increase by a factor of 2, by what factor should the diameter of the tendons associated with this muscle increase?

10.A.5 If a human 6 ft tall can lift 200 lb, what weight could a similarly proportioned person 3 ft tall lift?

10.B.1 If a 6-ft person has a tibia that is 1.5 in in diameter, what would be the diameter of the tibia of a 2-ft person if the strength-to-weight ratio is to be the same?

10.B.2 When a tree bends in the wind, there is a danger of the trunk's breaking at the base.
   (a) Letting the height of the tree be L and the diameter of the trunk D, determine the torque exerted at the base of the tree when the tree tilts through an angle θ. (This is of course an oversimplification. The diameter of a tree is not uniform, and the bending angle is not constant.)
   (b) How would the resistance of the trunk to breaking depend on the diameter D?
   (c) If a tree were to double its length, by what factor should the diameter increase for the tree to have the same resistance to breaking?

10.B.3 The lungs must supply the oxygen needs of the body by diffusion across a surface.

(a) Ignoring the differences in heat radiation, determine the relative surface area of the lungs of two animals one twice the size of the other.

(b) Taking heat radiation into account, determine the relative surface area of the lungs of two warm-blooded animals, one twice the size of the other.

11
FLUIDS

THE STATES OF matter are generally divided into three classes or phases: solid, liquid, and gaseous. Although the distinctions among the states are not always clear-cut, there are some things we can say that are of general validity.

The molecules of a solid are closely packed into a crystalline structure. Because of thermal energy (random motion of molecules in a material not at absolute zero temperature), the molecules will vibrate about their lattice sites in the crystal but will not diffuse through the solid.

The basic reason that molecules (or atoms) form crystalline structures is that the intermolecular (interatomic) forces are attractive at long range and repulsive at short range. If the temperature is not too high (so that the thermal kinetic energy is not too great), then the mutual attraction of the molecules (atoms) predominates and the molecules (atoms) are pulled together. When they get very close, the repulsive force takes over to keep the particles from coalescing. When a number of particles come together, the mutual attractive forces bind the system into a regular structure.

If the solid is heated, a point is reached at which the particles in the crystal have sufficient thermal kinetic energy to break away from their lattice sites but still not enough energy to entirely free themselves from their bonds with neighboring molecules. When this happens the substance loses its rigid crystal structure and enters the liquid state, where the free flow of particles becomes possible. The particles are still closely packed, so the compressibility is still low and the density is still high and both are comparable to those of the solid state. If the liquid is heated further, the particles eventually have sufficient thermal energy to break the bonds with their neighbors and become almost free. In this state the molecules are far apart, and the density is low and the compressibility high. This is the gaseous state.

In summary, then, the state of the substance and its physical properties depend on the relative importance of the thermal kinetic energy and the interaction potential energy. When the interaction energy dominates, the substance is in the solid state. When the thermal energy dominates, the substance is in the gaseous state. When the interaction and thermal energies are comparable, the substance is in the liquid state.

We have observed that it is only in the liquid and gaseous states that the particles are not bound to fixed sites; that is, liquids and gases can flow. *We shall call any substance that can flow a fluid.* A fluid can therefore be a liquid or a gas.

Another way to define a fluid is to say that it is *a substance that cannot sustain a shear when it is in equilibrium.* By a shear we mean a tangential force acting across any surface in the substance. In Fig. 11.1 we have illustrated a substance that is divided into two parts by an imaginary plane. The molecules on one side of the plane exert forces on the molecules on the other side over the entire face of the plane. We may resolve these forces into two components, one normal to the surface and the other tangential to the surface. The tangential component is referred to as a shearing force, and such a force is possible in a solid in equilibrium. Consider a bar resting on a table top, with a portion of the bar extending beyond the edge of the table (Fig. 11.2). If the overhanging portion of the bar is to be in equilibrium, then the sum of the forces acting on this portion must be zero. The only way for the weight of the bar to be balanced is for there to be a shearing force acting across the interface.

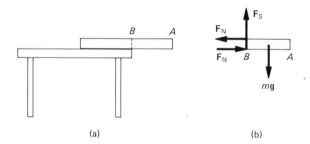

FIGURE 11.2   (a) A bar resting on a table top. (b) The forces that act on the overhanging segment AB of the bar are its weight $m\mathbf{g}$, a shearing force $\mathbf{F}_S$, and normal forces $\mathbf{F}_N$.

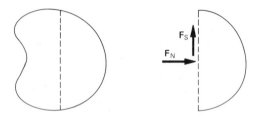

FIGURE 11.1   A substance divided by an imaginary plane. The molecules on one side of the plane exert forces on the molecules on the other side. These forces may be divided into two classes, normal forces $\mathbf{F}_N$ and shearing forces $\mathbf{F}_S$.

Since shearing forces are impossible in a fluid in equilibrium, it is impossible to hang a fluid over the edge of a table. This is not exactly a startling revelation, but we shall see presently how one may apply this same sort of argument to resolve the hydrostatic paradox and derive Pascal's law and Archimedes' principle. To clearly understand the behavior of fluids, you must recognize this fundamental property of a fluid at rest: *All fluid forces across a surface must be normal to that surface.* This is not true, however, of fluids in motion. Shearing forces in a moving fluid are responsible for viscous effects.

11.1 SCALAR AND VECTOR FIELDS

A description of the state of a fluid is much more difficult than a description of the state of a single particle or a solid. We cannot simply give the mass of the fluid and let it go at that. We must give the density over the entire fluid, and to describe the flow, we must give the velocity everywhere throughout the fluid. The motion of the fluid is determined by the pressure variation within the fluid, so the pressure must be given throughout the fluid. When we specify a quantity, e.g., the density or pressure or velocity, throughout a domain, we are defining a *field*.

A field is the unique assignment of a number to every point in a domain.

Generally this number will vary continuously from point to point.

THE DENSITY FIELD

The *density* at a point in a fluid is defined as the ratio of the mass to the volume within some arbitrarily small surface surrounding the point (Fig. 11.3):

$$\rho = \frac{m}{v} \tag{11.1}$$

In a liquid the density is fairly uniform over the entire volume; in a gas the density can vary widely from place to place. From our earlier definition we see that the density of a fluid is a field: it has a numerical value at every point in the domain occupied by the fluid (Fig. 11.4).

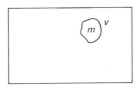

FIGURE 11.3 The density at a point in a fluid is the mass per unit volume in the neighborhood of that point.

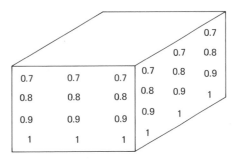

FIGURE 11.4 The density field in a fluid is an example of a field. The density at a point in the fluid is a measure of the mass per unit volume in the immediate neighborhood of the point. The density will in general vary continuously throughout the fluid.

Of course we may define densities for solids as well as for fluids. The densities of some common substances are given in Table 11.1.

THE PRESSURE FIELD

Another field needed to describe a fluid is the pressure field, and the pressure exists at all points throughout the fluid, not just on the walls containing the fluid. *Pressure* is

TABLE 11.1   Densities of some common substances.

Solids	Density (g/cm³)	Liquids	Density (g/cm³)	Gases	Density at 0°C, 760 mm Hg (g/cm³)
Aluminum	2.70	Alcohol	0.79	Air	0.001293
Brass	8.44–8.70	Ether	0.74	Ammonia	0.000771
Copper	8.93	Glycerol	1.26	Carbon dioxide	0.001977
Cork	0.22–0.26	Mercury	13.596	Helium	0.000178
Glass, common	2.4–2.8	Oil, olive	0.92	Hydrogen	0.000090
Glass, flint	2.9–5.9	Oil, paraffin	0.8	Oxygen	0.001429
Gold	19.3	Water	1.00		
Ice	0.917				
Iron	7.9				
Lead	11.34				
Nickel	8.8				
Osmium	22.5				
Platinum	21.37				
Silver	10.49				
Tungsten	19.3				
Uranium	18.7				
Wood, cedar	0.31–0.49				
Wood, ebony	0.98				
Wood, elm	0.54–0.60				
Wood, white pine	0.35–0.50				
Zinc	6.9				

defined as the force per unit area that the fluid molecules on one side of a small surface exert on the fluid molecules on the other side of the surface (Fig. 11.5):

$$P = \frac{\Delta F}{\Delta A} \qquad (11.2)$$

The orientation of this surface is of no consequence, since the magnitude of the force will be the same for all orientations of the surface.

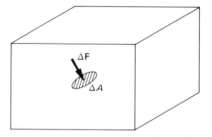

FIGURE 11.5   The area ΔA is some small area in the neighborhood of a point at which the pressure is to be determined. The force ΔF is the normal force that molecules on one side of the surface exert on molecules on the other side.

THE VELOCITY FIELD

The motion of a fluid is determined by a velocity field, which is simply the velocity of the elements of the fluid (Fig. 11.6). This field, however, is more than just an assignment of a number to every point in some domain. It is the assignment of a *vector* to every point, so this field is a *vector field*. The density and pressure fields are *scalar fields*.

FIGURE 11.7 A liquid in a vertical circular cylinder.

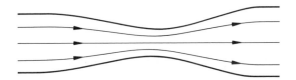

FIGURE 11.6 Fluid flow in a pipe. The velocity of the fluid changes from point to point, so the fluid motion is described by a velocity field.

We shall have occasion to introduce a number of different fields later when we take up electromagnetic theory, thermodynamics, and quantum mechanics.

11.2 PASCAL'S LAW

We shall now use our definition of a fluid as a substance that cannot sustain a shear when in equilibrium to determine how the pressure varies in an incompressible liquid of constant density. First let us consider the simple case of a liquid in a container with vertical walls (Fig. 11.7). We let h be the depth of the fluid. Since the fluid is in equilibrium, the sum of all the external forces must be zero. Let us isolate the fluid and replace the things removed by the

forces that they exert (Fig. 11.8). The force acting on the top is $P_0 A$, the atmospheric force per unit area multiplied by the area A. Similarly the force on the bottom is PA. As shown in the figure, the walls exert horizontal forces, and, as always, the weight w of the fluid exerts a downward force. Since the total vertical force must be zero,

$$PA = P_0 A + w$$

where

$$w = mg$$

and if the fluid is incompressible so that the density is uniform,

$$m = \rho V = \rho Ah$$

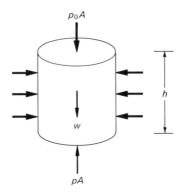

FIGURE 11.8 Forces acting on the fluid.

Therefore

$$PA = P_0A + \rho Ahg$$

or

$$P = P_0 + \rho gh \qquad\qquad (11.3)$$

We used implicitly the fact that a fluid cannot sustain a shear when we assumed that all the forces were perpendicular to the surfaces of the fluid.

We may calculate the pressure at any depth in a fluid in a vessel of any shape in a similar way (Fig. 11.9).

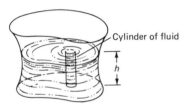

FIGURE 11.9 A fluid is contained in a vessel of arbitrary shape.

Consider the forces acting on a cylinder of fluid of height h and area A. As before, we find that the vertical forces balance if

$$P = P_0 + \rho gh$$

Suppose the point in the fluid does not lie directly below the surface. Let us find the pressure at a point C in the liquid in the vessel illustrated in Fig. 11.10. The pressure at point B is given by Eq. 11.3:

$$P_B = P_0 + \rho gh$$

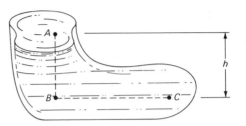

FIGURE 11.10 A vessel in which some of the liquid does not lie directly below the surface.

We can see that the pressure at C is the same as the pressure at B by considering the horizontal forces acting on the cylinder in Fig. 11.11. Equating the horizontal forces to zero, we have

$$P_BA - P_CA = 0$$

or

$$P_B = P_C$$

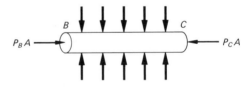

FIGURE 11.11 A horizontal cylinder of fluid.

Therefore

$$P_C = P_0 + \rho gh$$

We may now state Pascal's law:

The pressure at a depth h in a constant-density fluid in equilibrium is equal to the pressure at the surface plus ρhg

We may generalize this law in the following way:

The difference in pressure between any two points in a constant-density fluid in equilibrium is ρgh, where h is the vertical separation between the points.

The derivation of this generalization follows in a straightforward way from Pascal's law.

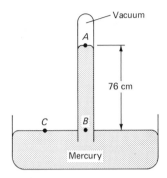

FIGURE 11.12   In an evacuated tube, mercury will rise to a height of 76 cm.

QUESTION

What is the minimum pressure that must be created by a giraffe's heart to raise blood to its head? The distance between the giraffe's heart and head is 200 cm.

ANSWER

The pressure difference is

$$\rho gh = \left(1\,\frac{g}{cm^3}\right)\left(980\,\frac{cm}{s^2}\right)(200\ cm) = 196{,}000\ cm^2/dyn$$

When the giraffe lowers its head, there will be a substantial increase in pressure, and special anatomical precautions have been taken to prevent damage.

QUESTION

Compare the pressure difference of 196,000 dyn/cm² for the giraffe with standard atmospheric pressure. Standard atmospheric pressure will push mercury 76 cm up an evacuated tube. The density of mercury is 13.6 g/cm³.

ANSWER

The pressure at C in Fig. 11.12 is standard atmospheric pressure (we call this *one atmosphere*). Since B and C are at the same height, the pressure at B is the same as that at C. The pressure at A is zero since the pressure in a vacuum is zero. The pressure difference between A and B is given by our theorem:

$$P_B - P_A = \rho gh = \left(13.6\,\frac{g}{cm^3}\right)\left(980\,\frac{cm}{s^2}\right)(76\ cm)$$

$$= 1.01 \times 10^6\ dyn/cm^2$$

Now

$$P_B = P_C = 1\ atm$$

and

$$P_A = 0$$

Therefore

$$1\ atm = P_C = P_B = 1.01 \times 10^6\ dyn/cm^2$$

Therefore a pressure of 196,000 dyn/cm² is equivalent to

$$196{,}000\ dyn/cm^2\,\frac{1\ atm}{1.01 \times 10^6\ dyn/cm^2} = 0.194\ atm$$

Hence the pressure in the heart of a giraffe must be at least 0.194 atm above atmospheric pressure. This would create a pressure in the blood vessels in the head of at least 1 atm if there were no loss in pressure as a result of viscosity. If the pressure in the head were to fall below atmospheric pressure, the blood vessels would collapse.

QUESTION

Determine the readings on pressure gauges A and B in Fig. 11.13. Atmospheric pressure is 10^6 dyn/cm².

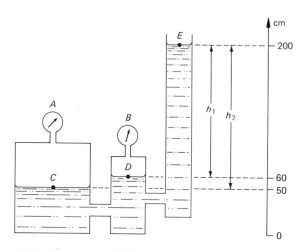

FIGURE 11.13   A vessel filled with a mixture of water and air.

ANSWER

The pressure at C is the same as the pressure at A, since the density of air is small and $\rho_{air}gh$ is negligible. The pressure at D is given by

$$P_D = P_E + \rho g h_1 = 10^6 \frac{\text{dyn}}{\text{cm}^2} + \left(1 \frac{\text{g}}{\text{cm}^3}\right)\left(980 \frac{\text{cm}}{\text{s}^2}\right)(140 \text{ cm})$$

$$= 10^6 \frac{\text{dyn}}{\text{cm}^2} + 0.137 \times 10^6 \frac{\text{dyn}}{\text{cm}^2} = 1.137 \times 10^6 \frac{\text{dyn}}{\text{cm}^2}$$

Similarly

$$P_C = P_E + \rho g h_2 = 10^6 \frac{\text{dyn}}{\text{cm}^2} + \left(1 \frac{\text{g}}{\text{cm}^3}\right)\left(980 \frac{\text{cm}}{\text{s}^2}\right)(150 \text{ cm})$$

$$= 1.147 \times 10^6 \text{ dyn/cm}^2$$

11.3 THE HYDROSTATIC PARADOX

In Fig. 11.14 the height of the liquid is the same in each vessel. In which of these is the pressure greatest at the bottom? We might expect the pressure in the bottom of vessel (a) to be greater than that at the bottom of (b). Vessel (a) might be a huge storage 40-ft-high tank containing thousands of gallons of water, and vessel (b) might be a 40-ft-high straw containing only a cupful of water. Or again it might seem obvious that the pressure at the bottom of vessel (c) would be greater than the pressure at the bottom of vessel (d), since they both contain the same quantity of water, but in (d) the weight of the water is supported by a much larger base than in (c).

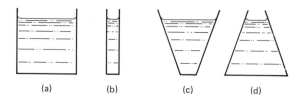

FIGURE 11.14   Four vessels containing fluid at the same height.

If it is true that the pressure at a given depth in a container depends on the size or shape of the container, one might think twice about diving into the ocean. One might be crushed under a few feet of water.

One way to solve this problem is to take an experimental approach. Suppose the vessels are joined at their bases by a tube (Fig. 11.15). If the pressure were greater at the bottom of any one of these vessels, then water would flow into the connecting tube from the higher-pressure points to the lower-pressure points. As a result the level of

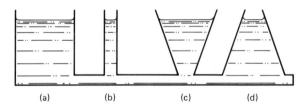

FIGURE 11.15   The four vessels are joined at their bases by a single tube.

the fluid would change in the vessels. But we have seen (Sec. 5.11) that the level will be the same in connected vessels. *Therefore the pressure must be the same at the bottom of all four vessels.*

We may also solve the problem by applying Pascal's law. The pressure at a given depth depends *only* on the depth and not on the shape of the vessel.

The reason the pressure at the base of vessel (c) is not greater than that at the base of vessel (d) is that the slanting walls in (c) help to support the liquid, but in (d) they have the opposite effect.

QUESTION

The base of a piston is attached by a string to one end of a beam balance. The cylinder enclosing the piston is attached to some rigid support that is independent of the balance, as shown in Fig. 11.16. When the cylinder is filled with water to a height h, a weight W_1 must be placed on the other end of the beam to achieve balance. However, if the liquid is frozen and freed from the sides of the cylinder, a weight W_2 much less than W_1 will keep the beam in balance. Why?

ANSWER

Suppose the shape of the cylinder in Fig. 11.16 is changed to one of uniform cross section, as illustrated in Fig. 11.17. The force acting on the piston will be unchanged since the pressure of the water on the piston is the same as in Fig. 11.16. The

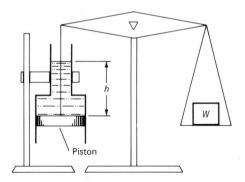

FIGURE 11.16   The weight W is balanced by the water pressure on the piston.

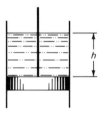

FIGURE 11.17   A cylinder of uniform cross-sectional area.

pressure depends only on the depth of the fluid. Since the walls exert horizontal forces only, it is clear that the net force on the piston in Fig. 11.17 must be equal to the weight of the water. Now the weight of the water in the cylinder of nonuniform cross section in Fig. 11.16 is less than that of the cylinder of uniform cross section in Fig. 11.17. Therefore the force acting on the piston in Fig. 11.16 must be *greater* than the weight of the water above it. However, if the water is frozen and becomes detached from the walls, the piston will support only the weight of the ice. The force acting on the piston will therefore decrease if the water is frozen.

Another way to understand this result is to observe that the horizontal portion of surface of the cylinder in Fig. 11.16 exerts a downward force on the fluid which is transmitted to the fluid. If the water is frozen and detached from the walls, this force is absent.

QUESTION

The beam in Fig. 11.18 is balanced by the weight W. Is the equilibrium stable or unstable?

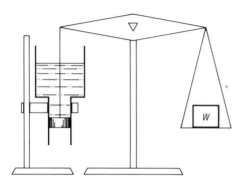

FIGURE 11.18 The weight W is balanced by the pressure of the water on the piston.

ANSWER

If the piston moves down a small distance, the height of the water above the piston increases, and hence the pressure on the piston increases. As a result the force acting on the piston increases, and the piston will be forced down further. The equilibrium is therefore unstable.

11.4 ARCHIMEDES' PRINCIPLE

It is a familiar fact that any body immersed in a fluid is buoyed up by the fluid. Since pressure increases with increasing depth of fluid, the force exerted on the bottom of an object immersed in a fluid will be greater than the force exerted on the top, and this difference in force is responsible for buoyant force. Let us see whether we can determine

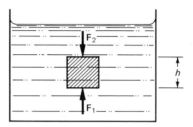

FIGURE 11.19 A rectangular block immersed in a fluid. The forces F_1 and F_2 are exerted on the bottom and top of the block by the fluid.

this buoyant force quantitatively. Consider first a parallel-epiped entirely immersed in a fluid. Let the area of the top and bottom surfaces be A, and let h be the height of the block (Fig. 11.19). The buoyant force will be given by

$$\text{buoyant force} = F_1 - F_2 = P_1 A - P_2 A = (P_1 - P_2)A$$
$$= \rho g h A$$

since the difference in pressure is $\rho g h$, where ρ is the *density of the fluid*. Now Ah is the volume V of the block,

$$Ah = V$$

and

$$\rho V = m$$

where m is the mass of the fluid displaced by the block. Therefore

$$\text{buoyant force} = mg$$

or

buoyant force = weight of fluid displaced

This is Archimedes' principle.

Our derivation applies only to the rectangular block, but the result is of general validity. To see this, consider a solid body of arbitrary shape immersed in a fluid. Consider also a similar vessel without the immersed body. We may imagine that a similar *body* has been immersed in the fluid, but that the *body* is composed of fluid (see Fig. 11.20). Now the buoyant force on this *fluid* body must be equal to the weight of the fluid body since the whole system is in equilibrium. But the buoyant force on the *fluid body* in Fig. 11.20(b) must be the same as that on the solid body in Fig. 11.20(a), since the force of the outside fluid does not depend on the composition of the immersed body.

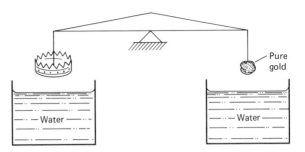

FIGURE 11.21 If the crown and the gold balance out of water, they must have the same mass. If they still balance underwater, they must have the same volume.

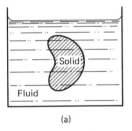 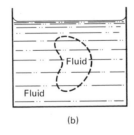

(a)   (b)

FIGURE 11.20 A body of arbitrary shape immersed in a fluid.

We see therefore that

buoyant force = weight of fluid displaced

for a body of arbitrary shape.

Archimedes supposedly discovered this principle when confronted with the problem of determining whether the king's crown was pure gold. If the crown and an equal weight of pure gold are balanced on a scale and both immersed in water, they will still balance the scale (Fig. 11.21). If the scale does not balance, we know that the volume of displaced fluid is not the same for the crown and the same mass of pure gold. If they both have the same mass but displace different volumes of water, then the materials have different densities and cannot be the same substance.

Although we have considered only the case of a totally submerged body, Archimedes' principle is valid for partially submerged bodies as well.

QUESTION

How might we determine what load we could safely put in a rowboat?

ANSWER

We could fill the boat with water on land. Then, if we neglect the thickness of the hull, the weight of the water (less the weight of the boat) is the weight of water that would be displaced if the boat were on the verge of sinking, and must therefore equal the maximum load the boat can support.

11.5 FLUIDS IN MOTION

The general subject of fluids in motion is very complicated, and we shall consider only a special case: streamline flow of an incompressible, inviscid fluid in a steady state. By *streamline flow* we mean that it is possible to describe the

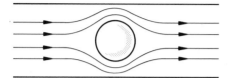

FIGURE 11.22   Streamline flow.

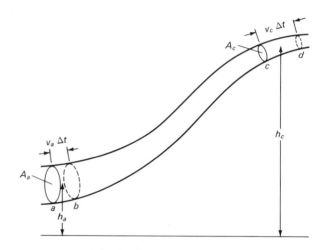

FIGURE 11.24   A bundle of streamlines.

flow pattern by a number of lines along which the fluid particles move. The tangent to a streamline at any point gives the direction of the fluid velocity at this point (Fig. 11.22). In turbulent flow it is not possible to construct such a set of stream lines (Fig. 11.23). By *incompressible* we mean that the density is constant both in time and throughout the fluid. By *inviscid* we mean that the effects of viscosity are negligible. And by *steady state* we mean that every portion of the fluid passing a given point will have the same velocity (magnitude and direction) as all the fluid elements that preceded it and all that will follow. Although the fluid is moving, the nature of the motion does not change with time.

FIGURE 11.23   Turbulent flow.

A very powerful theorem, called *Bernoulli's theorem,* applies to streamline flow. It may be stated in the form of a conservation principle; that is, some function is constant along a streamline. We shall prove that the quantity

$$P + \tfrac{1}{2}\rho v^2 + \rho g h = \text{constant} \tag{11.4}$$

along a streamline.

To derive this theorem, we consider a narrow bundle of streamlines (Fig. 11.24). Let us focus our attention on the fluid enclosed in this bundle between the two cross sections at a and c in Fig. 11.24 and follow its motion. During a short time dt the fluid at a will move a distance $v_a\,dt$ to b, and the fluid at c will move a distance $v_c\,dt$ to d. We would like to say two things about the system consisting of the fluid bounded by the bundle of streamlines and lying between the two cross sections initially at a and c. First, the mass of fluid remains unchanged and, second, the work done on the fluid during the time interval dt is equal to the total change in energy of the fluid.

Let us consider the mathematical consequences of these two statements. If the mass between a and c is equal to the mass between b and d, then the mass between a and b must equal the mass between c and d. Let us call the mass dm. Now the mass between a and b is the density times the volume between a and b, so

$$dm = \rho A_a v_a\,dt \tag{11.5}$$

Similarly

$$dm = \rho A_c v_c\,dt \tag{11.6}$$

so that

$$A_a v_a = A_c v_c \qquad (11.7)$$

But a and c are two arbitrary points on the streamline. Therefore

$$Av = \text{constant along a streamline} \qquad (11.8)$$

This is the mathematical expression of the conservation of mass in a moving fluid.

A familiar example of Eq. 11.8 is the fact that the column of water falling from a faucet becomes narrower and narrower until it finally breaks up into drops (Fig. 11.25). The fact that it breaks up into drops is a consequence of surface tension effects that we shall discuss later, but the narrowing of the water column follows from Eq. 11.8 and the fact that the falling water accelerates. As the velocity increases, the cross-sectional area A must decrease in order to keep the product Av constant.

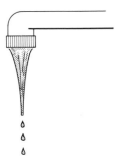

FIGURE 11.25   A narrowing column of water flowing from a faucet.

QUESTION

Suppose the water coming from a faucet whose diameter is 2 cm has a velocity of 5 cm/s. What is the velocity at a point in the water column at which the diameter is 1 cm?

ANSWER

Since

$$Av = \text{constant}$$

we have

$$\frac{\pi D^2}{4} v = \text{constant}$$

or

$$D_1{}^2 v_1 = D_2{}^2 v_2$$

where the subscript 1 refers to the point at which the diameter is 2 cm and the subscript 2 refers to the point at which the diameter is 1 cm. Inserting the given information, we obtain

$$(2 \text{ cm})^2 (5 \text{ cm/s}) = (1 \text{ cm})^2 v_2$$

so that

$$v_2 = 20 \text{ cm/s}$$

Let us now consider the application of the energy principle. The system is the fluid between a and c, and during a time dt this sytem moves to the region between b and d. The work done by the outside forces (the rest of the fluid in contact with our system) must equal the change in kinetic plus potential energies. No work is done on the sides made up of the bundle of streamlines, since the fluid forces must be perpendicular to this surface, not because the fluid can sustain no shearing force (this only applies to fluids at rest), but because we have assured that the fluid is inviscid and there are no shearing forces in an inviscid fluid, whether it is in motion or not. The only work is that done by the pressure of the outside fluid on the two cross sections at a and c. The work is the forces $p_a A_a$ and $-P_c A_c$ times the distances $v_a \, dt$ and $v_c \, dt$, respectively (Fig. 11.26). Therefore

$$\text{work} = P_a A_a v_a \Delta t - P_c A_c v_c \Delta t \qquad (11.9)$$

The change in total energy during dt is the energy in the fluid between b and d minus the energy in the fluid between

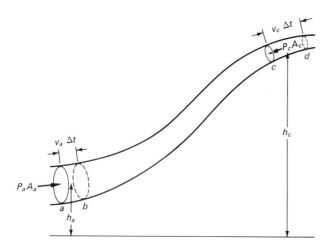

FIGURE 11.26 Work done on a fluid is the product of force times distance.

a and c. Since the energy of the fluid between b and c is common to each term, the net change in energy is given by

$$\begin{array}{c} \text{change} \\ \text{in energy} \end{array} = \text{(energy of the fluid between } c \text{ and } d\text{)}$$
$$- \text{(energy of the fluid between } a \text{ and } b\text{)}$$

Now

$$\begin{array}{c} \text{energy of the fluid} \\ \text{between } a \text{ and } b \end{array} = \tfrac{1}{2} dm \; v_a{}^2 + dm \; gh_a$$

Similarly

$$\begin{array}{c} \text{energy of the fluid} \\ \text{between } c \text{ and } d \end{array} = \tfrac{1}{2} dm \; v_c{}^2 + dm \; gh_c$$

Equating the work performed to the total change in energy, we have

$$P_a A_a v_a \, dt - P_c A_c v_c \, dt$$
$$= \tfrac{1}{2} \, dm \; v_c{}^2 + dm \; gh_c - \tfrac{1}{2} \, dm \; v_a{}^2 - dm \; gh_a \quad \text{(11.10)}$$

Multiplying both sides of Eq. 11.10 by ρ yields

$$P_a (\rho A_a v_a \, dt) - P_c (\rho A_c v_c \, dt)$$
$$= \tfrac{1}{2} \rho v_c{}^2 \, (dm) - \rho g h_c \, (dm) - \tfrac{1}{2} \rho v_a{}^2 \, (dm) - \rho g h_a \, (dm)$$

where the terms in parentheses are all equal (see Eqs. 11.5 and 11.6). Canceling these common terms and transposing the a terms to the left and the c terms to the right, we have

$$P_a + \tfrac{1}{2} \rho v_a{}^2 + \rho g h_a = P_c + \tfrac{1}{2} \rho v_c{}^2 + \rho g h_c \quad \text{(11.11)}$$

Now a and c are any two points on the narrow bundle of streamlines, so we may say that

$$P + \tfrac{1}{2} \rho v^2 + \rho g h = \text{constant along a streamline} \quad \text{(11.12)}$$

This is Bernoulli's equation.

Let us consider two special cases of Bernoulli's equation. If $v = 0$, then

$$P + \rho g h = \text{constant}$$

so that if we consider any two points in a fluid at rest,

$$P_1 + \rho g h_1 = P_2 + \rho g h_2$$

or

$$P_1 - P_2 = \rho g (h_2 - h_1)$$

Thus the pressure difference between two points is equal to ρg times the difference in height between these two points. This is just Pascal's law.

As a second special case of Bernoulli's equation, let $h_1 = h_2$. We find then that

$$P + \tfrac{1}{2} \rho v^2 = \text{constant}$$

along a streamline. *As the velocity increases, the pressure must therefore decrease in order for the sum to remain constant.*

As a simple application of this principle, consider the flow of a liquid into a constriction in a long cylindrical tube (Fig. 11.27). The velocity of the liquid must increase ($Av = \text{constant}$) as the fluid enters the constriction. Where the velocity is high, the pressure is low, so the pressure will be lower in the constriction than it is elsewhere. We

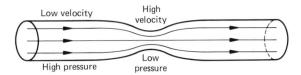

FIGURE 11.27 A liquid passing through a constriction in a tube.

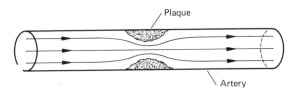

FIGURE 11.28 Blood flow constricted by the accumulation of plaque in an artery.

might have expected just the opposite. It would seem that the pressure should be greater in the constriction, but if the velocity of the fluid is greater in the constriction, the fluid must have accelerated. To accelerate fluid in a given direction it is necessary to have a net force in that direction, i.e., a high pressure behind the fluid and a low pressure in front. Therefore the fluid pressure must decrease in the direction of the acceleration.

An interesting example of Bernoulli's principle can be found in the flutter that sometimes occurs in arteriosclerosis. When a constriction is formed in an artery because of the growth of plaque, the blood pressure is lower in the constriction (Fig. 11.28) than elsewhere. When the constriction is sufficiently narrow, the pressure may fall below the pressure in the surrounding tissue (roughly atmospheric pressure), and as a result the artery may close at the constriction. But as soon as the artery closes, the flow stops and the pressure reduction caused by the Bernoulli effect is no longer present. The pressure then builds up very quickly and the artery opens again. As the blood begins to flow the pressure drops in the constriction, and again the artery closes. The process is repeated again and again very rapidly, and the flow in the artery pulsates. These pulsations are easily detected. This phenomenon is easily demonstrated with a piece of flexible rubber tubing. When water is forced through the tubing and the tubing is gradually pinched more and more tightly at a certain point, eventually the tubing begins to flutter violently (Fig. 11.29).

Reed instruments, such as the clarinet, oboe, and bassoon, operate on exactly the same principle.

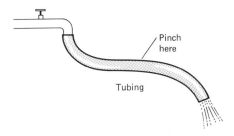

FIGURE 11.29 The flow of water through rubber tubing pulsates when the tubing is constricted.

QUESTION
How does an aspirator work?

ANSWER
When the bulb is squeezed (see Fig. 11.30), air is forced through the constriction at c. The pressure is lowered below atmospheric pressure, and this causes the liquid to rise in the tube and mix with the air in the constriction.

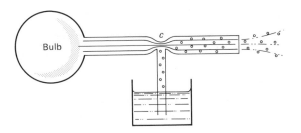

FIGURE 11.30 An aspirator.

Which way will the baseball shown in Fig. 11.31 curve?

FIGURE 11.31 A spinning baseball.

ANSWER

In the frame of reference in which the air is at rest and the ball moving, the air flow is not in a steady state. However, if we observe the ball from a frame of reference moving with the ball, then the ball is spinning about a fixed position and the air is streaming past the ball in a steady state (Fig. 11.32). Because of the direction of rotation of the ball, it will support the flow of air over the top and impede the flow past the bottom. The air velocity will therefore be greater at top than at bottom, and thus the pressure will be lower at the top. The ball will rise (if we neglect gravity).

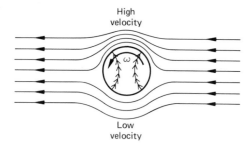

FIGURE 11.32 Air flowing past a spinning ball.

QUESTION

If air is blown into the center hole of a spool of thread and a card is laid flat against the other end, will the card be blown away (Fig. 11.33)?

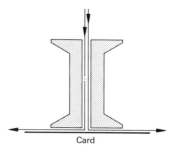

FIGURE 11.33 Air passes through the center of the spool against the card.

ANSWER

No. The air flow between the spool and the card will cause a drop in pressure, and this will force the card against the spool.

QUESTION

What is the velocity with which water will flow from a hole in the bottom of a tank filled with water to a height h (Fig. 11.34)?

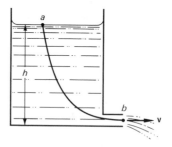

FIGURE 11.34 Water flows from a hole in the bottom of a tank.

ANSWER

Let us apply Bernoulli's principle to a point on the surface (a) and a point on the same streamline at the point (b) where the water leaves the hole. We have

$$P_a + \rho g h = P_b + \tfrac{1}{2}\rho v_b^2$$

since the velocity of the water at a is practically zero. Now P_a and P_b are both atmospheric pressure, since the water is exposed to the atmosphere at both points. Therefore

$$\rho gh = \tfrac{1}{2}\rho v_b{}^2$$

or

$$v_b = \sqrt{2gh}$$

This is just what the velocity would be if the water were to fall freely through a distance h.

QUESTION

How is *lift* achieved on an airplane wing?

ANSWER

Consider the flow of air past the wing (Fig. 11.35). We see that the air flowing over the top of the wing will be moving faster at point a than at point b. The air in region A in effect flows through a constriction and speeds up as it does so. This is another example of Eq. 11.8. The fluid velocity in region B does not increase. Since the velocity at a is greater than the velocity at b, the pressure at a must be less than the pressure at b. (We should not be applying Bernoulli's equation to two points that are not on the same streamline, but if we connect a and b to points on their respective streamlines far upstream where the pressure is the same on all streamlines, we can compare the pressure difference between a and b.) Since the pressure on the top of the wing is less than the pressure on the bottom, there will be a net lift. *Flight* is achieved with a Frisbee in the same way.

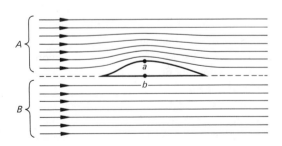

FIGURE 11.35 Air flow past an airplane wing.

SUMMARY

A liquid is a substance that cannot sustain shearing forces when in equilibrium. From this it follows that fluid forces across a surface must be normal to that surface when the fluid is in equilibrium.

Three fields are useful in describing a fluid: the density field ρ, the pressure field P, and the velocity field $\mathbf{v}$, where

$$\rho = \frac{\Delta m}{\Delta v}$$

and

$$P = \frac{\Delta F}{\Delta A}$$

In general a field is an assignment of a scalar or a vector to every point in a domain.

Pascal's law   The pressure difference between any two points in a fluid at rest with constant density is ρgh, where h is the vertical separation between the points.

Archimedes' principle   The buoyant force on a body wholly or partially immersed in a fluid is equal to the weight of the fluid displaced.

The streamlines of a fluid in motion allow a graphic representation of the velocity field. The velocity of the fluid at any point is parallel to the tangent to the streamline through that point.

Bernoulli's theorem   In streamline flow of an inviscid, incompressible fluid in a steady state, the quantity

$$P + \tfrac{1}{2}\rho v^2 + \rho gh$$

is a constant along any streamline.

Mass conservation   If we follow a bundle of streamlines, the product

$$vA = constant$$

along the bundle, where A is the cross-sectional area of the bundle of streamlines.

PROBLEMS

11.A.1 If the gauge pressure in each of the four tires of a 2000-lb automobile is 22 lb/in², what is the minimum area of rubber on the road? (The gauge pressure is the pressure above atmospheric pressure.)

11.A.2 A certain wood floats three-quarters submerged. What volume of this wood is required to build a raft to support a 150-lb child?

11.A.3 A uniform cylinder has a density of 0.75 g/cm³. When the cylinder is placed in water, what fraction of the volume is submerged?

11.A.4 A large glass bulb is balanced on a scale by a small lead weight (see the figure below). After a time the glass bulb begins to rise. Is the barometric pressure rising or falling?

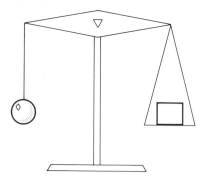

11.A.5 A tube contains mercury to a height of 10 cm and above the mercury a layer of water 15 cm thick. What is the gauge pressure at the bottom of the tube?

11.A.6 A submarine is disabled in 100 ft of water. The area of the escape hatch is 1 ft². If the pressure in the submarine is 1 atm, what force is necessary to open the hatch? (The weight density of water is 62.4 lb/ft³.)

11.A.7 Some fish balance themselves at a given depth in the water by inflating or deflating a float bladder. Is such a mechanism stable or unstable with respect to vertical displacements?

11.A.8 The basic principle of a hydraulic press is illustrated in the following figure. What force F is required to lift the weight W?

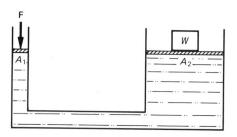

11.A.9 A garden hose projects a vertical stream of water 20 ft into the air. If the cross-sectional area of the water jet just after it emerges from the nozzle is 0.2 cm², what is the velocity of the water within the hose where the cross-sectional area is 2 cm²?

11.B.1 Assume that the left ventricle of the heart is analogous to a simple piston and chamber, and that 70 cm³ of blood are ejected during each stroke against an average pressure of 105 mm Hg.
  (a) Calculate the work done in a single contraction.
  (b) Assuming a pulse of 60 beats per minute, calculate the power generated.
  (c) Calculate the total work performed during a 70-yr life span.
  (d) If this work in part (c) were applied to lifting a one-million-pound building, how high would it be raised?

11.B.2 Your host at a party presents you with a cocktail. The level of the liquid is exactly level with the edge of the glass, and there are ice cubes floating in the drink. Do you have to worry about the drink overflowing as the ice melts? (It's a weak drink, and you may assume that the density of the liquid is equal to the density of water.)

11.B.3 To determine the density of an ore sample, a geologist finds that 80 g will balance the sample when weighed in air and 75 g will balance it when the ore sample is submerged in water. Determine the density of the ore sample.

11.B.4 To determine the density of a liquid, an investigator weighs a rock in air, in water, and in the liquid. The results are 10 lb, 9 lb, and 8 lb, respectively. What is the density of the liquid?

11.B.5 A tank filled with water rests on a scale that reads 10 lb (see the figure). A stone is suspended from a scale that reads 5 lb. When the stone is lowered into the water tank, the scale supporting the stone reads 4 lb. What does the scale supporting the water tank read?

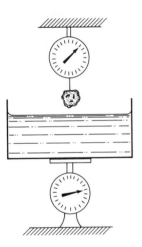

11.B.6 A large coffee pot has a vertical glass tube attached to the outlet pipe to show the level of the coffee in the pot, as shown in the following figure. When the spigot is opened and the coffee begins to flow, the level in the glass tube falls.

(a) If the initial height of the coffee in the tube is 20 cm and it falls to 10 cm, at what rate is coffee flowing from the pot?

(b) The cross-sectional area of the outlet pipe is 3 cm². What is the cross-sectional area of the stream of coffee flowing from the spigot?

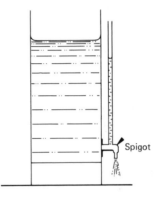

Spigot

11.B.7 To determine the flow rate in a pipe, an investigator makes a small constriction and then places a transparent vertical tube at a point where the tube diameter is 2 in and another where the diameter is 1.9 in (see the figure). The heights of fluid in the tubes are 6 ft and 4 ft. What is the velocity of the fluid in the section of pipe where the diameter is 2 in?

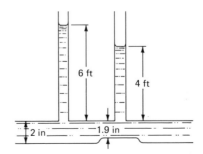

6 ft

4 ft

2 in   1.9 in

11.B.8 Water falls from a faucet with a cross-sectional area of 0.5 cm² with a velocity of 3 cm/s. As the water falls from the faucet the stream becomes increasingly narrow until it breaks up into drops. If the radius of the drops is 0.2 cm, what is the time interval between their formation?

11.B.9 Water is supplied to a city from a large tower 100 ft high. What is the city water pressure? Neglecting friction, determine how high a hose could project a stream of water into the air.

11.B.10 The velocity of ocean waves depends only on the wavelength, the density of water, and g. (The amplitude of the waves is too small in comparison with the ocean depth for the depth to be important.) From dimensional considerations, determine the dependence of the wave velocity on the wavelength. Explain why it is that if the surf is good, it will probably get worse. (Long waves are better than short waves for surfing and both are generated by a storm far at sea.)

11.C.1 A cylinder weighted at one end floats in an upright position in water. The total mass of the cylinder is 20 g and its radius is 5 cm. If the cylinder is displaced vertically from its equilibrium position, it will oscillate.

(a) Neglecting the viscosity of the water and assuming that the force of the water on the cylinder is equal to the weight of the water displaced, show that the motion is simple harmonic motion and determine the period.

(b) Why is the answer given for part (a) incorrect? As a clue, consider a person jumping from a bridge and landing on her stomach in the water. At the moment that half the body is submerged, is it true that the net force on the body is equal to the weight of water displaced?

11.C.2 Show that the hydrostatic force on the face of a vertical dam may be replaced by a single *equivalent* force whose magnitude is equal to the pressure at $h/2$ multiplied by the wetted surface area, where h is the water depth. The point of application of the *equivalent* force is $h/3$ measured from the bottom. The solution to this problem can be greatly simplified by contrasting it with the problem of determining the center of gravity of a right triangle.

11.C.3 Water in a large chamber is maintained at a constant pressure of 10 atm and ejected from a nozzle whose cross-sectional area is 4 cm². What is the thrust of this "engine"?

11.C.4 Water streams from a faucet 5 cm in diameter with a downward velocity of 10 cm/s. Determine the shape of the stream of water.

*12
SURFACE
TENSION

SURFACE TENSION, AS the name implies, has to do with forces associated with a surface interface between two substances. This force is relatively small and becomes important only when the surface-to-volume ratio is large, as for example in the lungs. The primary function of the lungs is to bring oxygen to the blood and to carry away carbon dioxide. This exchange of gases takes place by passive diffusion, which is a very slow process. Therefore, obtaining a significant gas exchange between the lungs and the blood requires a very large contact surface, about 70 m². (The body surface area is of the order of 2 m².) The structure of the lungs is illustrated in Fig. 12.1. The gas exchange

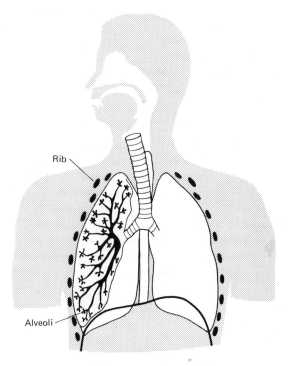

FIGURE 12.1(a)   The human respiratory system.

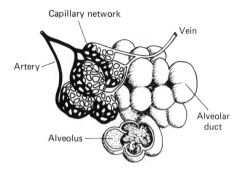

FIGURE 12.1(b)   Detail of part (a).

takes place across the alveolar membrane into the capillary network carrying blood from the heart. These alveoli expand and contract as we breathe. A very thin film covers the surface of each alveolus, and the function of this film can only be understood through a very careful study of the nature of surface tension.

12.1 SURFACE ENERGY AND SURFACE TENSION

To understand the physical basis of surface tension, let us consider a liquid drop (Fig. 12.2). The drop is held together by cohesive forces between the molecules. If the drop is uniform throughout, then a molecule in the interior experiences no net force since it is pulled as much in one direction as in another. However, a molecule on the surface experiences a net force toward the interior of the drop. Certainly the drop can not exist in equilibrium in this state. If there is a net force on a particle, the particle will accelerate in the direction of the force. All the molecules near the surface will move inward forming a more compact

FIGURE 12.2   A liquid drop.

surface distribution. Molecular forces are attractive only up to a point, but there is a repulsive central core, so the molecules can be pushed only so close.

The molecules near the surface therefore exist in a somewhat different state than those in the interior. Because of the very short-range character of the attractive forces ($\sim 10^{-6}$ cm), this distinctive surface is very thin. (The reason such a phenomenon as surface tension does not exist for a planetary body—a very large drop also held together by cohesive forces—is that gravitational force has a *very long-range* character.)

Hence we may divide the energy of the drop into two parts: the volume energy E_V of the interior molecules and the surface energy E_A of the thin layer of more compact molecules near the surface. If E_{tot} is the total energy,

$$E_{tot} = E_V + E_A$$

The shape of the drop will not affect E_V, since the energy of the interior molecule depends only on their immediate surroundings and not on what the rest of the drop looks like (Fig. 12.3). This again is because of the very short-range character of the forces.

Similarly the surface energy E_A depends only on the extent of the surface and not on its shape. For this reason we can write

$$E_A = \sigma A \tag{12.1}$$

where A is the total surface area and σ is the energy per unit surface area. We can now understand why a liquid

FIGURE 12.3   The energy of the molecules in the volume ΔV does not depend on the surroundings.

drop assumes a spherical shape (if we neglect gravity). Whenever some mechanism exists by which a system can lose energy, it will do so. A ball placed in a bowl will roll about in the bowl and, as a result of friction, will eventually come to rest at the bottom of the bowl. The ball assumes a state of minimum energy. (See also Sec. 5.11.) Similarly, if a liquid drop is free to change its shape, it will do so if it can find a shape with a lower energy. The total energy of a drop is

$$E_{\text{tot}} = E_V + \sigma A \tag{12.2}$$

Since the volume energy and the energy per unit area σ are fixed (if we assume that the temperature is constant), the only thing that can change is the area A. Thus the drop will assume a shape with a minimum area, and the shape that has minimum area for a given volume is a sphere (Fig. 12.4).

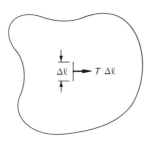

FIGURE 12.5 The force that the fluid exerts across a line segment of length Δl on the surface is $T\,\Delta l$.

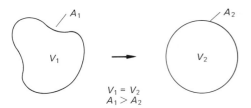

FIGURE 12.4 A sphere has the smallest surface area for a given volume.

This chapter is titled *surface tension,* and so far we have spoken only of surface energy. We would like to show that these two concepts are in fact the same. Surface tension is defined in a very pragmatic way: If a line is drawn on the surface of some interface, then one can determine the equilibrium state by assuming that the molecules on one side of the line exert a force T per unit length of line on the molecules on the other side (Fig. 12.5). The force is perpendicular to the line and tangent to the surface. In any problem in which we have to take into account the compacting of the molecules near the surface, we may take either the surface-energy or the surface-tension point of

view. For example, we could have shown that a liquid drop in equilibrium will assume a spherical shape (if we neglect gravity) by examining the surface-tension forces rather than by minimizing the energy. Suppose the shape of the drop were something like that shown in Fig. 12.6. Let us imagine the drop cut by a horizontal plane, and let us consider as a system in equilibrium the liquid above the plane. For the system to be in equilibrium, the sum of the forces acting on this system must be zero. The forces acting on the system are the surface-tension forces around the perimeter and the forces acting across the plane. If the pressure in the drop is P and the area of the plane A, this latter force is PA and is directed vertically. Because of the asymmetry of the perimeter, it is clear that there will be a net horizontal

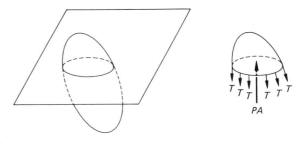

FIGURE 12.6 A liquid drop of irregular shape.

force to the right as a result of the surface-tension forces. There is no other force to balance this, and therefore with this shape the drop could not be in equilibrium. We have not proved that the shape must be a sphere, but this can be done.

QUESTION

As a means of measuring the surface tension of a liquid, a wire frame is immersed in the liquid and then slowly drawn upward to a height h (Fig. 12.7). A two-sided thin film of liquid is spread out over the frame. If the force necessary to hold the frame at rest is F, what is the surface tension of the liquid?

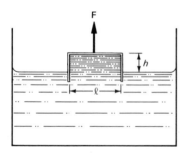

FIGURE 12.7 A wire frame is drawn out of the liquid to a height h above the level of the liquid.

ANSWER

Let us approach this question from the point of view of (a) surface energy and (2) surface tension.

1. Consider the work necessary to raise the frame upward a distance h. This is given by

$$\text{work} = Fh$$

(We have assumed F to be constant, and we will see why this is legitimate presently. We could have avoided the assumption by increasing h by an amount dh and examining only the work done in this small displacement.) We may equate this work to the increase in energy of the system, which is the increase in surface energy and the increase in gravitational energy of the elevated film of water and wire. This latter energy can be made

negligibly small by use of a very thin wire. The increase in surface energy is

$$\text{increase in surface energy} = \sigma 2\ell h$$

since σ is the energy per unit area and the total area of the two-sided film is $2\ell h$. From the principle of work and energy,

$$Fh = \sigma 2\ell h$$

or

$$\sigma = \frac{F}{2\ell}$$

By measuring F and ℓ we may determine σ.

2. Consider the balance of forces on the horizontal segment of wire and a portion of the elevated fluid. The forces acting are F and the surface-tension forces of ℓT on either side (Fig. 12.8). In equilibrium these forces must balance, so that

$$F = 2\ell T$$

or

$$T = \frac{F}{2\ell}$$

We see that T is independent of h. The tension in the surface film does not increase when the surface is stretched, as is the case of a spring or an elastic surface such as rubber.

We see in this example that

$$T = \sigma$$

This, as we have said, is a general relation, which we shall now prove.

FIGURE 12.8 Cross-sectional view of the wire and a small portion of the liquid film.

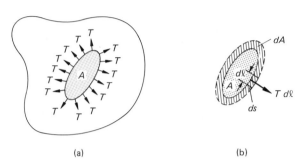

FIGURE 12.9 The surface A is some portion of a closed surface. We imagine this surface increased in area by an amount dA.

We would now like to derive this general equivalence of surface energy σ and the surface tension T. Imagine some arbitrary portion of a surface of area A (Fig. 12.9). Let us consider the work necessary to increase the area by an amount dA, as shown in Fig. 12.9(b). The force on a small line segment on the perimeter of length dl is $T\,dl$, and the distance this force moves is ds. Therefore the work done on dl is given by

work on $dl = T\,dl\,ds$

and the work done around the entire perimeter is given by

total work = sum of $(T\,dl\,ds) = T$ sum of $(dl\,ds)$

where the sum is performed around the entire perimeter. Since T is constant, it may be factored outside the sum. Now

sum of $(dl\,ds) = dA$

so that

total work = $T\,dA$

From the principle of work and energy, this work must equal the increase in energy (surface energy), so that

$T\,dA = \sigma\,dA$

since σ is the energy per unit area and dA is the increase in area. Canceling the common factor dA, we have

$$T = \sigma \tag{12.3}$$

so that surface tension and surface energy are equivalent.

12.2 LAPLACE'S LAW

As we have pointed out, surface tension plays an important role in the functioning of the lungs. The exchange of gases between the lungs and the blood takes place across the surface of small alveolar sacs surrounded by a capillary network. These alveoli expand and contract as we breathe. The alveolar sacs are quite small, so they have a large surface-to-volume ratio. We expect therefore that surface-tension effects should be important, and can gain much information about these effects in the alveoli by studying surface-tension effects on a spherical drop.

Let us consider first a uniform spherical drop, e.g., a drop of water. We would like to show that surface tension causes the pressure inside the drop to be larger than the pressure outside, and to derive a quantitative relation (Laplace's law) between this pressure difference, the surface tension, and the radius of the drop. To determine this relation, let us consider as a system just the upper half of the drop and study the forces acting there (Fig. 12.10).

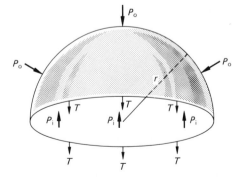

FIGURE 12.10 Forces acting on the upper hemisphere of the liquid drop. P_0 and P_i are the outside and inside pressures and T is the surface tension.

These are the force caused by the atmosphere outside the drop where the pressure is P_o, the force resulting from the pressure P_i in the lower hemisphere of the drop, and the surface tension force T around the circular perimeter of the hemisphere. We consider first the force resulting from the inner pressure P_i. If r is the radius of the drop, then the force is just the pressure times the area of the circle:

Force resulting from $P_i = \pi r^2 P_i$

We consider next the force caused by the outside pressure P_o. This force is in different directions on different portions of the hemispherical surface, so we must add these forces together vectorially. This is difficult to do; in fact, we must perform a vector integration over the spherical surface. We may circumvent the calculation in the following way. Imagine a hemisphere subject only to a *uniform* pressure P_o (Fig. 12.11). Clearly the net external force must be zero. The force on the plane circular surface is directed vertically upward and is given by

force on plane circular surface $= P_o \pi r^2$

Since the hemisphere must be in equilibrium, the remaining force on the hemispherical portion of the surface must be equal in magnitude and opposite in direction. Therefore

force on hemispherical surface $= P_o \pi r^2$

Returning to the hemispherical drop in Fig. 12.10, we find that the surface-tension force around the perimeter is given by

surface-tension force $= 2 \pi r T$

Equating the upward forces to the net downward force yields

$$\pi r^2 P_i = \pi r^2 P_o + 2 \pi r T$$

or

$$P_i - P_o = \frac{2T}{r} \qquad \text{(Spherical drop)} \qquad (12.4)$$

so that the pressure difference between the inside and the outside is directly proportional to the surface tension and inversely proportional to the radius. This is *Laplace's law.*

If the system is a bubble rather than a liquid drop, there is an additional contribution to the surface tension because there are two surfaces in a bubble, an inner surface and an outer one. So in a bubble

$$P_i - P_o = \frac{4T}{r} \qquad \text{(Bubble)} \qquad (12.5)$$

One further example of some interest is that of a cylinder. By methods entirely similar to those dicussed above, we find

$$P_i - P_o = \frac{T}{r} \qquad \text{(Cylinder)} \qquad (12.6)$$

QUESTION

Let us imagine two bubbles of equal size on either end of a rigid cylinder (Fig. 12.12). Is this a stable or an unstable system?

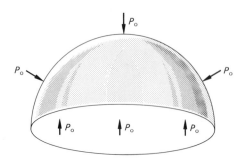

FIGURE 12.11 A hemisphere subject to a uniform outside pressure of P_0.

FIGURE 12.12 Two bubbles of equal size on either end of a solid cylinder.

ANSWER

Suppose the bubble on the left were to become slightly smaller and the bubble on the right slightly larger. From Laplace's law the pressure inside will increase if the radius decreases. (The outside pressure and the surface tension are constants.) Thus the pressure in the left bubble would increase, and conversely, the pressure in the right bubble would decrease. This would create a pressure gradient within the connecting cylinder that would cause air to flow from left to right. This would further decrease the size of the bubble on the left and increase the size of the bubble on the right. Thus the system is unstable.

This presents a problem in understanding the functioning of the alveoli in the lungs. Since all the alveoli are interconnected, why are they not unstable? Equally serious is the fact that the pressure difference necessary to inflate the alveoli should be very large when r (the radius of the alveolar sac) is very small. The resolution of these difficulties lies in the facts that: (1) the alveoli are not simple fluid surfaces like a bubble, but have an underlying elastic membrane, and (2) there is a very thin coating of fluid called a *surfactant* which reduces the surface tension enormously. It is very difficult to inflate a collapsed alveolus when this film is missing. Infants are occasionally born without this surfactant, and this defect is called *hyaline membrane disease*. The surface tension in the lungs of such infants is high, and many of the alveoli are in a collapsed state and cannot be inflated.

Another interesting application of Laplace's law is the formation of bubbles in boiling water. We have seen that the discontinuity in pressure across a surface is inversely proportional to the radius of the drop. The same equation

$$\Delta P = \frac{2\sigma}{r}$$

applies to bubbles in a liquid. A pressure difference of $2\sigma/r$ is required to sustain a bubble of radius r in equilibrium. This being the case, how do we account for the generation of large numbers of bubbles in boiling water? If these bubbles began with radius zero, the pressure difference would have to be infinite. However, these bubbles are in fact not new bubbles at all, but pinched-off portions of gas bubbles already existing in small cracks in the vessel. It is almost impossible to prepare a surface so that it will be free of such air pockets, and an almost infinite number of bubbles can be generated out of such an air pocket. Certainly there is not an infinite amount of air, but even the smallest quantity can serve as a nucleus. *Water vapor* fills the small nucleus of air and causes the bubble to grow. The bubble is filled almost entirely with this vapor and only slightly with air.

This inverse dependence of the pressure difference across a surface on the radius of curvature also explains the phenomenon of a stream of water falling from a faucet breaking up into small drops (Fig. 12.13). We have already seen that a falling column of water must become thinner and thinner as it continues to accelerate. (See Eq. 11.8.) Now let us imagine what would happen if there were a small fluctuation in the diameter of the column (see Fig.

FIGURE 12.13 A column of falling water eventually breaks up into small drops because of an instability created by surface tension.

FIGURE 12.14 A cylindrical column of water with a small fluctuation in the radius.

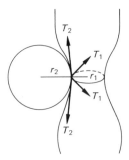

FIGURE 12.15 The two principal radii are r_1 and r_2.

12.14). Let us examine the column in a frame of reference falling with the water. In this frame the water is at rest and we may ignore the force of gravity. Where the column is constricted, the radius of curvature is less, so the liquid has a higher pressure in these regions. This higher pressure forces the liquid from the constricted areas into the larger protions of the column. Where the column is thin, it becomes thinner, and where it is wide, it becomes wider. Eventually the column will break up into a series of drops.

One might ask why this occurs only after the column has become very thin and not in a fat column. Our earlier formulas for the pressure difference across a drop ($\Delta P = 2\sigma/r$) or across a cylinder of fluid ($\Delta P = \sigma/r$) are not really applicable to the deformed cylinder in Fig. 12.14. At the point of construction there are *two* radii of curvature (see Fig. 12.15), one associated with the circle in the plane of the page (r_2) and the other associated with the cross section of the cylinder (r_1). It can be shown that the difference in pressure is given by

$$P_i - P_o = \sigma\left(\frac{1}{r_1} + \frac{1}{r_2}\right)$$

where P_i is the pressure inside the column of water and P_o the pressure outside. This is the appropriate modification of Laplace's law. The two radii are positive or negative depending on whether the center of curvature is inside the fluid or outside: Inside is positive, outside is negative. For a sphere both radii are equal and positive, and we have

$$P_i - P_o = \frac{2\sigma}{r} \qquad \text{(Sphere)}$$

For a cylinder one of the radii is infinite, so if the radius of the cylinder is r, then

$$P_i - P_o = \sigma\left(\frac{1}{r} + \frac{1}{\infty}\right) = \frac{\sigma}{r} \qquad \text{(Cylinder)}$$

as before. In Fig. 12.15, r_2 is negative and r_1 is positive. So long as $|r_2|$ is less than $|r_1|$, the inside pressure is lowered, and thus when a constriction forms, fluid is pushed *into* the lower-pressure constricted region and it is filled out again. Such a system is stable. But when $|r_2|$ becomes greater than $|r_1|$, then the inside pressure is raised and fluid is forced out of the constricted area, and this causes a further pinching of this area. Such a system is unstable and breakup occurs. This explains why we see breakup only after the column has become rather narrow.

If the column is fat and a small constriction occurs, $|r_2| < |r_1|$.

QUESTION

If the surface tension of water at 20°C is 72.8 dyn/cm, what is the pressure discontinuity across a water drop the size of a typical biological cell? ($r = 10^{-4}$ cm.)

ANSWER

From Laplace's law,

$$P_i - P_o = \frac{\sigma}{r} = \frac{72.8 \text{ dyn/cm}}{1 \times 10^{-4}} = 7.28 \times 10^5 \frac{\text{dyn}}{\text{cm}^2}$$

$$= 7.28 \times 10^5 \frac{\text{dyn}}{\text{cm}^2} \frac{1 \text{ atm}}{1.01 \times 10^6 \text{ dyn/cm}^2}$$

$$= 0.72 \text{ atm}$$

This is not the pressure discontinuity of a cell within the body. Within the body a cell is surrounded by another fluid, and the presence of this fluid greatly affects the surface tension. In fact, if we allow the extreme example of a water drop in water, the surface tension is zero. Furthermore, the cell is not a uniform fluid held together solely by cohesive forces; it has internal structure as well as an outer membrane surface.

QUESTION

When a cup filled with water is turned upside down, the water runs out. When a narrow tube partially filled with water is inverted, the water does not spill out (Fig. 12.16). Why?

FIGURE 12.16 Water will run out of the cup but not out of the narrow tube.

ANSWER

Certainly the water could be sustained in equilibrium in the cup by the pressure of the air, since 1 atm of pressure will support a column of water 32 ft in height irrespective of its diameter. Therefore both the cup of water and the tube of water could be in equilibrium. Let us consider the stability of the equilibrium.

It would seem that both systems are unstable. There exists a lower state of potential energy: both liquids on the floor. However, to get out of either vessel, the liquid has to increase its surface area. Perhaps the increase in energy associated with the increase in surface area is greater than the decrease in gravitational potential energy. To investigate this question quantitatively, let us consider a square tube of side a partially filled with water. Imagine a perturbation of the surface as shown in Fig. 12.17. The water drops down in a small triangular wedge in one-half of the tube and, to conserve mass, a triangular cavity forms on the other half. Let us calculate the change in

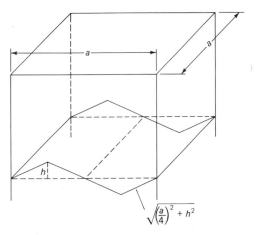

FIGURE 12.17 A small triangular perturbation of the surface. Each triangular wedge is of height h and length a.

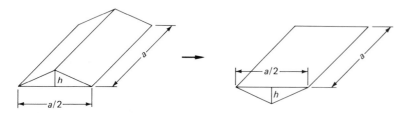

FIGURE 12.18 A triangular wedge of fluid is inverted.

surface energy and the change in gravitational potential energy ΔE_s is given by

$$\Delta E_s = \sigma \Delta A$$

where

$$\Delta A = 4a\sqrt{\left(\frac{a}{4}\right)^2 + h^2} - a^2$$

$$= a^2\sqrt{1 + \left(\frac{4h}{a}\right)^2} - a^2$$

$$\simeq a^2\left[1 + \frac{1}{2}\left(\frac{4h}{a}\right)^2\right] - a^2 = 8h^2$$

and where we have assumed that $h \ll a$ and have used the approximation $\sqrt{1 + \epsilon} \simeq 1 + \frac{1}{2}\epsilon$ if ϵ is small.

Next consider the change in gravitational potential. The effect of the perturbation is to take a triangular wedge of fluid and invert it. The center of mass of a triangle is at a point one-third the distance from the base. Thus a mass given by

$$m = \frac{1}{4}\rho a^2 h$$

has been effectively dropped a distance of $\frac{2}{3}h$. Therefore the change in gravitational potential energy ΔE_g is

$$\Delta E_g = -\left(\frac{1}{4}\rho a^2 h\right)(g)\left(\frac{2}{3}h\right)$$

The net change in energy is

$$\Delta E_s + \Delta E_g = 8h^2\sigma - \frac{1}{6}g\rho a^2 h^2$$

$$= \left(8\sigma - \frac{1}{6}g\rho a^2\right)h^2$$

We see that if a is small the change in energy is positive, and therefore the system is stable; if a is large the change in energy is negative, and the system is unstable. The borderline between stability and instability occurs when

$$8\sigma - \frac{1}{6}g\rho a^2 = 0$$

or when[1]

$$a = \sqrt{\frac{48\sigma}{g\rho}} = 6.93\sqrt{\frac{\sigma}{g\rho}}$$

If we let $\sigma = 72$ dyn/cm and $\rho = 1$ g/cm^3, we find that

$$a = 1.88 \text{ cm}$$

Hence we would expect that water would spill out of a tube whose edge is greater than 1.88 cm but would not spill out of a tube less than 1.88 cm on a side.

12.3 CAPILLARITY

So far we have been speaking of the surface tension of fluids, yet any surface is actually an interface between *two* substances. We cannot just speak of the surface tension of water or mercury but rather must discuss the

---

[1] If we had considered a tube of circular cross section and looked at the most general possible perturbation, we would have found the critical diameter to be $a = 6.28\sqrt{\sigma/\rho g}$.

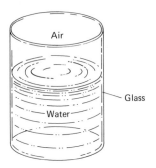

FIGURE 12.19 A glass of water.

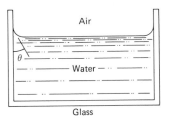

FIGURE 12.21 The angle θ is the angle of contact.

surface tension of a water-air interface or a mercury-glass interface. In many problems of interest we deal with a junction of three interfaces on a line. For example, in a glass of water in air the three interfaces are glass-water, glass-air, and water-air (Fig. 12.19). Let us suppose that these three surfaces intersect perpendicularly, i.e., the surface of the water is flat. The forces acting on a unit length of the line where the three interfaces meet (the water line in the glass) are shown in Fig. 12.20, where σ_{ga} is the surface tension of the glass-air interface, σ_{wa} the surface tension of the water-air interface, and σ_{gw} the surface tension of the glass-water interface. Now there cannot

be any net component of force along the surface of the solid since the liquid is free to move in this direction, but there may be a net component of force perpendicular to the surface of the solid. If we measure σ_{ga} and σ_{gw}, we find that $\sigma_{ga} > \sigma_{gw}$. Therefore the surface of the water must rise and make an angle θ with the glass (called the angle of contact) (Fig. 12.21). This angle is determined by the balance of forces parallel to the surface of the solid. From Fig. 12.22 we see that the net force in the vertical direction will be zero if

$$\sigma_{ga} = \sigma_{gw} + \sigma_{wa}\cos\theta$$

or

$$\cos\theta = \frac{\sigma_{ga} - \sigma_{gw}}{\sigma_{wa}} \tag{12.7}$$

The molecular basis for the fact that $\sigma_{ga} > \sigma_{gw}$ is that the water molecules are attracted to molecules in the glass. We saw that surface tension (or surface energy) was

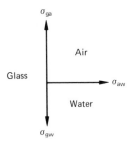

FIGURE 12.20 Forces acting at the intersection of the air, glass, and water.

FIGURE 12.22 Surface tension forces.

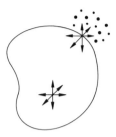

FIGURE 12.23 When molecules across the surface attract molecules within the surface, the compacting of the surface molecules is reduced.

caused by the fact that molecules near the surface are drawn down to the interior by attractive forces and that this causes a constriction near the surface. If other molecules across the surface are *pushing* on the surface molecules, the compacting of the surface is *enhanced* and the surface energy is larger. On the other hand, if the molecules across the surface are *pulling* the surface molecules, then the surface tension is *diminished* (Fig. 12.23). Mercury molecules are repelled by glass, so the surface tension of the glass-mercury interface is very large and the angle of contact is obtuse (Fig. 12.24). Water molecules are attracted to the glass surface, so the surface tension is lower than the surface tension of the glass-air interface.

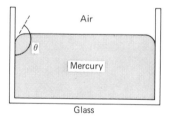

FIGURE 12.24 The angle of contact θ between mercury and glass is obtuse.

Often the surface tensions of the various interfaces are not known, but the angle of contact can be measured. Table 12.1 lists some common angles of contact. These values can change appreciably if the surface is not clean. They also vary with temperature.

TABLE 12.1 Angles of contact of some interfaces.

Interface	θ
Water-glass	~0°
Water-paraffin	110°
Mercury-glass	148°
Turpentine-glass	17°
Kerosene-glass	26°

From the angles of contact and the surface tension of the liquid-air interface, it is possible to determine the distance of the rise or fall of a liquid in a small tube of radius R when the tube is placed in a larger container of liquid. Assuming for the moment that the angle of contact is acute, let us consider the forces acting on the fluid in the tube which lies above the fluid level outside the tube (see Fig. 12.25). The forces acting are those caused by the surface tension, the pressure of the atmosphere on top, the pressure of the water at the base (which is also atmospheric pressure), the pressure of the glass walls, and the weight of the fluid. The pressure forces on the top and bottom cancel. (For an exercise prove this.) The vertical force caused by surface tension is $\sigma 2\pi R \cos \theta$, and the weight of the liquid is $\rho \pi R^2 h g$. If the liquid is in equilibrium, these forces must balance, so that

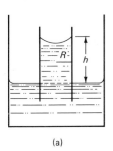 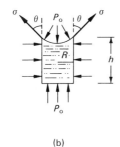

(a)   (b)

FIGURE 12.25 The rise or fall of a fluid in a narrow tube is due to capillary action. The forces acting on the elevated fluid in the tube are shown in part (b).

$$\sigma 2\pi R \cos\theta = \rho\pi R^2 hg$$

or

$$h = \frac{2\sigma \cos\theta}{\rho Rg} \qquad (12.8)$$

For water in a glass tube, $\sigma = 72$ dyn/cm. If the radius of the tube is $R = 0.01$ cm, then

$$h = \frac{2(72 \text{ dyn/cm})(1)}{(1 \text{ g/cm}^3)(0.01 \text{ cm})(980 \text{ cm/s}^2)} = 14.7 \text{ cm}$$

We may approach this problem from a different point of view. We have seen that there is a discontinuity in pressure across a curved surface (Eq. 12.4). In Fig. 12.26 the pressure drop between points 1 and 2 is $2\sigma/r$, where r is the radius of curvature of the surface. The pressure rise between points 2 and 3 is ρgh. The pressures at points 1, 3, and 4 are all atmospheric pressure. Therefore the pressure drop between 1 and 2 must equal the pressure rise between 2 and 3, since the pressure difference between 1 and 3 is zero. Therefore

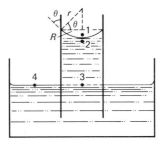

FIGURE 12.26 Liquid in a capillary tube.

$$\frac{2\sigma}{r} = \rho gh$$

or

$$h = \frac{2\sigma}{\rho gr}$$

From Fig. 12.26,

$$\cos\theta = \frac{R}{r}$$

so that

$$h = \frac{2\sigma \cos\theta}{\rho gR}$$

as before.

This formula applies equally well to the case in which θ is obtuse. The value of $\cos\theta$ will be negative and the height h of the fluid will be negative; i.e., the fluid will be depressed.

QUESTION

Water rises in plants and trees in small tubes called xylem. What is the maximum distance water can rise if the radius of the xylem is 10^{-4} cm? Assume that the contact angle is zero and that the surface tension is 70 dyn/cm.

ANSWER

From Eq. 12.8,

$$h = \frac{2\sigma \cos \theta}{\rho g R} = \frac{2(70 \text{ cm/dyn})}{\left(1 \frac{g}{cm^3}\right)\left(980 \frac{cm}{s^2}\right)(10^{-4} \text{ cm})}$$

$$= 1429 \text{ cm} = 47 \text{ ft}$$

Thus capillary action is not sufficient to drive water to the tops of giant redwood trees. It is still not fully understood how this is accomplished.

It is also instructive to obtain this equation for the rise of fluid in a capillary tube by using the energy principle. We know that a system is in equilibrium when it is in a minimum energy state. Consider the surface energy E_s of a fluid in a capillary tube (Fig. 12.27). This energy is given by

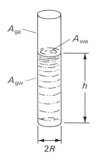

FIGURE 12.27 The areas of the three interfaces are A_{ga} (glass-air interface), A_{gw} (glass-water interface), and A_{wa} (water-air interface).

$$E_s = \sigma_{ga}A_{ga} + \sigma_{gw}A_{gw} + \sigma_{wa}A_{wa}$$

The three interfacial areas can be expressed as

$$A_{gw} = 2\pi Rh$$

$$A_{ga} = A_{cap} - 2\pi Rh$$

and

$$A_{wa} = \text{constant}$$

where A_{cap} is the total inside area of the capillary tube. Although A_{gw} and A_{ga} depend on the height h of the fluid rise in the capillary, A_{wa} is independent of h.

The other energy associated with the fluid in the capillary is the gravitational potential energy:

$$E_g = mg\frac{h}{2}$$

where $h/2$ is the height of the center of gravity of the fluid in the capillary. The mass m of this fluid is given by

$$m = \rho\pi R^2 h$$

so that

$$E_g = \tfrac{1}{2}\rho g\pi R^2 h^2$$

The total energy, surface plus gravitational potential, is

$$E_{tot} = \sigma_{ga}(A_{cap} - 2\pi Rh) + \sigma_{gw}2\pi Rh + \sigma_{wa}A_{wa}$$
$$+ \tfrac{1}{2}\rho g\pi R^2 h^2$$
$$= \tfrac{1}{2}\rho g\pi R^2 h^2 - 2\pi Rh(\sigma_{ga} - \sigma_{gw}) + \text{constants}$$

The column of fluid will seek that level that makes this energy a minimum. To find this minimum, let us differentiate E_{tot} with respect to h and set the derivative equal to zero. Thus

$$\frac{dE_{tot}}{dh} = \rho g \pi R^2 h - 2\pi R(\sigma_{ga} - \sigma_{gw}) = 0$$

Solving for h, we find

$$h = \frac{2(\sigma_{ga} - \sigma_{gw})}{\rho g R}$$

But from Eq. 12.7,

$$\sigma_{ga} - \sigma_{gw} = \sigma_{wa}\cos\theta$$

so that

$$h = \frac{2\sigma_{wa}\cos\theta}{\rho g R}$$

which is just Eq. 12.8.

SUMMARY

Because of the short-range character of the intermolecular forces in a liquid, the molecules in a thin layer on the surface exist in a more compacted state than those in the interior. The *surface energy* σ is the energy per unit area of these compacted molecules on the surface.

If a line is drawn on the surface of a fluid, it is possible to determine the equilibrium state by assuming that the surface on one side of the line exerts a force T (called the *surface tension*) per unit length of line on the surface on the other side. The surface tension is equal to the surface energy:

$$T = \sigma$$

Laplace's law states that

$$P_i - P_o = \frac{2T}{r} \qquad \text{(Spherical drop)}$$

$$P_i - P_o = \frac{4T}{r} \qquad \text{(Bubble)}$$

$$P_i - P_o = \frac{T}{r} \qquad \text{(Cylinder)}$$

The angle of contact between water and glass is given by

$$\cos\theta = \frac{\sigma_{ga} - \sigma_{gw}}{\sigma_{wa}}$$

where the subscripts g, a, and w refer to glass, air, and water, respectively.

The capillary rise of water in a glass tube is determined by the relation

$$h = \frac{2\sigma_{wa}\cos\theta}{\rho R g}$$

PROBLEMS

12.A.1 How high will water rise in a glass tube whose diameter is 0.01 cm?

12.A.2 What is the heaviest metal disk 2 cm in radius that can float on water? The disk is coated with paraffin and the contact angle is 110°.

12.A.3 The diameter of the veins varies significantly throughout the body. Compare the stress in the walls of two blood vessels whose diameters differ by a factor of 2. Assume that the blood pressure is the same in each.

12.A.4 A 1-in-diameter water pipe from the city water supply is reduced to $\frac{3}{4}$ in as it enters the house. If the thickness of the house pipe is $\frac{1}{16}$ in, what should be the thickness of the city water pipe?

12.A.5 A small air bubble trapped in a vessel of water has a radius of 10^{-2} cm. What is the difference in pressure between the air in the bubble and the water outside?

12.B.1 Given that the critical diameter for an inverted tube of water can depend only on ρ, g, and σ, determine the functional relationship from dimensional considerations alone.

12.B.2 A glass rod 1 mm in radius is dipped into water end first. A scale supporting the rod registers an initial increase in tension. As the rod is lowered further into the water, the scale reading begins to fall. At what depth will the scale reading be the same as it was when the rod was supported in air?

12.B.3 If a water bug were twice as large as it is, how much longer would its legs have to be?

12.B.4 Two identical capillary tubes are dipped in water. One is vertical and the other makes an angle θ with a vertical. Compare the height to which water will rise in each tube.

12.B.5 A light rod floats on the surface of a rectangular vessel 5 cm on a side and filled with water. The rod is 5 cm long and divides the surface in half. Benzene is poured into the vessel on one side of the rod. If the surface tension of the benzene-air interface is 30 dyn/cm and the surface tension of the benzene-water interface is 35 dyn/cm, what is the force on the rod?

12.B.6 Two parallel glass plates separated by a distance s are dipped into water. If the plates are vertical, to what height will water rise between the plates?

12.B.7 Two barometric tubes are used to measure atmospheric pressure. One is of radius $r = 0.1$ cm and the other is large enough so that surface-tension effects may be neglected. What is the difference in height in the two tubes? The density of mercury is 13.6 g/cm³, the surface tension 465 dyn/cm, and the contact angle 148°.

12.B.8 In the figure the water rises to a height h in the capillary on the left. Can water be held at height h in the second capillary if the radii of the two capillaries are the same at this height?

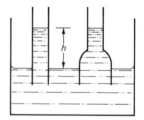

12.B.9 An inventor comes to you with the following invention: When a capillary tube is inserted in a bucket of water, the water rises in the tube to a height of 10 cm. Suppose the capillary is only 5 cm long. Then the water will spill over the top, and this falling stream of water may be used to turn a paddle wheel which in turn can drive a small generator producing electricity. This is a perpetual-motion machine that can be used to generate power with no expenditure of fuel.

Should you invest in this scheme?

12.C.1 If a bent capillary of radius r is dipped into water (see the figure), to what height will water rise in the capillary? (This is a good example of how an energy principle can greatly simplify a statics problem.)

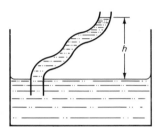

12.C.2 Two large glass plates are touching at one edge and are separated by a wedge at the opposite edge. The plates are placed vertically in water. Show that the air-water interface between the plates is an equilateral hyperbola.

13
KINETIC
THEORY

WE CAN UNDERSTAND a number of properties of macroscopic systems by recognizing that matter is made up of a large number of atoms and applying the laws of mechanics to their motion.

QUESTION

Why does a liquid cool when it evaporates?

ANSWER

We have seen that the molecules of a liquid are bound together by cohesive forces. Therefore energy is required for a molecule in the liquid to break these bonds and escape. The faster, more energetic molecules have a better chance of escaping than the slower molecules. The faster molecules escape and the slower ones are left behind. As we shall see, the temperature of a body is proportional to the mean kinetic energy per molecule; as this energy falls, the temperature falls. This is the principle means by which we cool ourselves in hot weather.

QUESTION

Why does a gas cool on expansion and heat on compression?

ANSWER

Once again we may understand this phenomenon from a microscopic point of view. Suppose the gas is being compressed (Fig. 13.1). We know that a particle rebounding elastically from a stationary wall does not change its energy. However, if the wall is moving toward the particle, the particle will have a greater energy after it is struck by the moving wall. (Conversely, if the wall is moving away from the particle, the particle will rebound with a decrease in energy.) Thus each particle striking the advancing wall will pick up energy, and, as before, an increase in the mean kinetic energy means an increase in temperature.

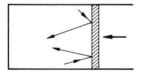

FIGURE 13.1   A gas being compressed.

QUESTION

Why is it that summer winds blowing off a plateau down into a valley heat the valley?

ANSWER

The potential energy stored in the air on the plateau is converted into kinetic energy when it falls into the valley. This increase in the mean kinetic energy per molecule means an increase in temperature.

In all the above examples, we were able to study what happens to individual molecules and, by applying the basic laws of mechanics, to explain the observed result. This is not always possible. If the interaction between the molecules is strong or the thing being measured depends on the detailed interrelation of the molecules, then it would clearly be impossible to determine the motion of so large a number of interacting molecules and to attempt a molecular interpretation.

13.1 IDEAL-GAS LAW

One system that lends itself particularly well to a detailed mechanistic approach is a gas of noninteracting molecules, an *ideal gas.* No real gas has this property that the molecules do not interact, but most gases at low density may be approximated as ideal gases. If the density is low, the molecules are far enough apart that they seldom feel their neighbors. We shall try to relate the pressure that such a gas exerts on the walls of its container to the translational kinetic energy of the molecules within the container.

For simplicity we shall choose a container with a simple geometry — a cube of side L (Fig. 13.2). Let us look at each particle individually. If a particle strikes a wall elastically (i.e., it loses no energy in colliding with the wall), its normal component of velocity will be reversed and its tangential component of velocity will be unchanged. Suppose the wall is perpendicular to the x axis. If the components of velocity before the particle strikes the wall are v_x, v_y, and v_z, then the components after the particle strikes the wall will be $-v_x$, v_y, and v_z. The change in the components of momentum will be

$$\Delta p_x = m(-v_x) - mv_x = -2mx_x$$

$$\Delta p_y = mv_y - mv_y = 0$$

$$\Delta p_z = mv_z - mv_z = 0$$

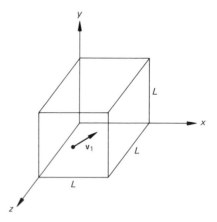

FIGURE 13.2   A single particle with velocity $\mathbf{v}_1$ in a cube of side L.

If there is a change in the x component of momentum of this particle, then there must be a force on the particle in the x direction, determined by the relation

$$\mathbf{F} = \frac{d\mathbf{p}}{dt}$$

The force on the wall is, by the law of action and reaction, equal and opposite to the force on the particle. Let F_{av} be the average force on the wall. The contribution to this force by each particle that strikes the wall is

$$F_{av} = \frac{\text{change in momentum per collision}}{\text{time between collisions with the wall}}$$

Now

change in momentum per collision $= 2mv_x$

and

time between collisions with the wall $= \dfrac{2L}{v_x}$

This is the time it takes a particle to move across the cube and back (a distance of $2L$) if its x component of velocity is v_x. Therefore

$$F_{av} = \frac{2mv_x}{2L/v_x} = \frac{mv_x^2}{L}$$

This is the average force on the wall caused by just one particle, so the total force is the sum of the forces of all the molecules. Let us label the molecules with masses m_1, m_2, etc., and velocities $\mathbf{v}_1$, $\mathbf{v}_2$, etc. The total force is then

$$F_{tot} = \frac{m_1 v_{1x}^2}{L} + \frac{m_2 v_{2x}^2}{L} + \cdots$$

The pressure on this side of the box is the force divided by the area (L^2):

$$P = \frac{F_{tot}}{L^2} = \frac{m_1 v_{1x}^2 + m_2 v_{2x}^2 + \cdots}{L^3}$$

This is the pressure on the wall whose normal is in the x direction. The pressure on a wall whose normal is in the y direction would be

$$P = \frac{m_1 v_{1y}^2 + m_2 v_{2y}^2 + \cdots}{L^3}$$

and the pressure on a wall whose normal is in the z direction,

$$P = \frac{m_1 v_{1z}^2 + m_2 v_{2z}^2 + \cdots}{L^3}$$

If the particles are moving about in the box in such a way that there is no preferential direction, then the force on all the walls in the cube must be the same. We shall asume that the pressure P in the three equations above is the same. If we add the three equations together, we have

$$3P = \frac{m_1(v_{1x}^2 + v_{1y}^2 + v_{1z}^2)}{L^3}$$

$$+ \frac{m_2(v_{2x}^2 + v_{2y}^2 + v_{2z}^2)}{L^3} + \cdots$$

$$= \frac{m_1 v_1^2 + m_2 v_2^2 + \cdots}{L^3}$$

since

$$v_1^2 = v_{1x}^2 + v_{1y}^2 + v_{1z}^2$$

But the numerator is just twice the total kinetic energy of the molecules in the box,

$$K = \tfrac{1}{2}m_1 v_1^2 + \tfrac{1}{2}m_2 v_2^2 + \cdots$$

Thus

$$3P = \frac{2K}{L^3}$$

Noting that L^3 is the volume V, we may write

$$PV = \tfrac{2}{3}K \tag{13.1}$$

This result is quite general and applies to all ideal gases. It makes no special assumption about the distribution of velocities except that there is no preferred direction. (If there is a preferred direction, then the pressure will not be the same on all walls of the cube.) It must be observed, however, that K is the *translational kinetic energy only*. It does not include whatever energy we might associate with the internal structure of the molecules, such as rotational energy, vibrational energy, energy in the chemical bonds, etc. There are many different kinds of ideal gases, i.e., gases for which we can neglect the interaction between molecules. Many of their properties will differ, but in all ideal gases

$$PV = \tfrac{2}{3}K$$

Suppose an ideal gas is composed of two distinct types of molecules, A and B. Equation 13.1 is still valid, and we can write

$$P_{A+B}V = \tfrac{2}{3}K_{A+B}$$

But we know that the kinetic energy of a collection of particles is just the sum of the kinetic energies of the individuals so

$$K_{A+B} = K_A + K_B$$

We have then

$$P_{A+B} = \frac{\tfrac{2}{3}K_A}{V} + \frac{\tfrac{2}{3}K_B}{V}$$

But $\tfrac{2}{3}K_A/V$ is the pressure if only molecules of type A are present, and the same applies to $\tfrac{2}{3}K_B/V$. It follows then that

$$P_{A+B} = P_A + P_B$$

or the pressure of a collection of two groups of molecules is the sum of the partial pressures of each alone. This is called *Dalton's law of partial pressures*.

13.2 TEMPERATURE

We may write Eq. 13.1 in a somewhat different form by defining the average kinetic energy per molecule (ϵ). This is the total translational kinetic energy divided by the total number of molecules N:

$$\text{average energy per molecule} = \epsilon = \frac{K}{N}$$

Equation 13.1 may then be written

$$PV = \tfrac{2}{3}N\epsilon \tag{13.2}$$

This average kinetic energy per molecule has an interesting property. If we have a mixture of several different types of molecules, then we find the following:

The average translational kinetic energy is the same for each type of molecule for any system in thermodynamic equilibrium.

This is not to say that the kinetic energy of each molecule will be the same, only that the average energy of one type will be the same as the average energy of any other type in the same system.

We cannot prove this remarkable fact here. It has its origins in the laws of probability and statistics and cannot be derived from the laws of mechanics. We shall discuss how systems with large numbers of molecules are governed by probabilistic laws in Chap. 14.

QUESTION

If a box containing equal numbers of H_2 and O_2 molecules is in thermal equilibrium and a very small hole is made in the surface, which will leak out faster, the oxygen or the hydrogen?

ANSWER

The hydrogen. Since all molecules have the same average kinetic energy,

$$\tfrac{1}{2}m_{H_2}v_{H_2}{}^2 = \tfrac{1}{2}m_{O_2}v_{O_2}{}^2$$

or

$$\frac{v_{H_2}}{v_{O_2}} = \sqrt{\frac{m_{O_2}}{m_{H_2}}} = \sqrt{\frac{32}{2}} = 4$$

So the velocity of the hydrogen molecule is on the average four times that of the oxygen molecule. Since the hydrogen is moving about in the box more rapidly than the oxygen, it is more likely to strike the hole and escape.

QUESTION

Why is there so little hydrogen in the atmosphere?

ANSWER

As we saw in the preceding question, the hydrogen molecules are traveling faster than the heavier oxygen molecules. Now the velocity required for a molecule to escape the pull of the earth's gravity is independent of its mass. Thus any molecule with a velocity greater than the escape velocity (that does not collide with other molecules) will be lost. A hydrogen molecule is more likely to exceed this escape velocity than heavier molecules.

This property that the average kinetic energy is the same for all types of molecules for a system in thermal equilibrium applies not only to ideal gases but to any gas, any liquid, any solid, or any mixture of solid, liquid, and/or gas.

As an example, consider a closed container that is partially filled with water. Above the surface of the water is a mixture of air and water vapor. The average kinetic energy of the water molecules in the vapor is the same as that of the oxygen or nitrogen molecules in the air. Since the whole system is in thermal equilibrium, the water molecules in the liquid must also have the same average kinetic energy as those in the gas above. At first this appears very strange; it does not seem that the molecules in the liquid could have the same freedom to move about as the molecules in the gas. In fact, they do not have the same freedom of movement, but this simply means that the molecules in the liquid do not go very far before they collide with other water molecules. Even more startling is the fact that the molecules in the solid container holding the liquid and gas also have the same average kinetic energy as do the molecules in the liquid and the gas. In a solid the molecules are bound in place by strong cohesive forces. How can they have the same average kinetic energy? The answer is that they simply jiggle about in a very localized region, i.e., they oscillate rapidly about their equilibrium position in the solid lattice structure.

Consider next what happens when we bring two bodies together and allow them to reach thermal equilibrium. If the mean kinetic energy per molecule is the same before they are brought together, it will be the same after. If there is an energy transfer from one body to the other, then the mean energies per particle cannot be the same for all molecules in both bodies. If one has a higher mean energy per particle, it will give up some of its energy to the other until the mean energy per particle is the same for both.

We see that the mean kinetic energy per molecule has all the properties we normally identify with temperature.

1. It has a unique value for every body in thermal equilibrium.

2. Two bodies with the same mean kinetic energy per molecule before contact will have the same mean kinetic energy after contact.

3. When one body has a higher mean kinetic energy per molecule than another, there will be an energy flow from the body with the higher mean kinetic energy to the body with the lower mean kinetic energy if the bodies are brought together and allowed to exchange energy.

Since we have no prior definition of temperature, we are free to choose a definition now.

The temperature of any body is directly proportional to the mean kinetic energy per molecule:

$$T = (constant)\epsilon$$

The constant defines the temperature scale and may be chosen at will. The easiest choice would be to set it equal to one so that the temperature would be the mean kinetic energy itself. Unfortunately, physics, like all sciences, is a victim of its historical development. Before the true nature of temperature was understood, the scale was chosen so that the difference in temperature between an ice-water bath and a steam-water bath at 1 atm of pressure is 100 units. This is the *Kelvin* scale. To make matters worse, the symbol for the constant of proportionality is chosen to be $(\frac{3}{2}k)^{-1}$, where k is called the *Boltzmann constant*. The reason for the factor of $\frac{3}{2}$ here is to eliminate a factor of $\frac{3}{2}$ elsewhere. Thus

$$\epsilon = \tfrac{3}{2}kT \qquad\qquad (13.3)$$

and the numerical value of k is 1.38×10^{-23} J/K, where K represents Kelvin degrees. With this choice of k the difference in temperature between ice-water and steam-water is 100 Kelvin degrees (100 K).

(We have a certain freedom in assigning dimensions to k and T. If we choose to assign the dimensions of energy to k, then we see from Eq. 13.3 that T is dimensionless. When we say that T is, say, 200 Kelvin degrees, the term "Kelvin degrees" is not a dimension but a notation to indicate how the temperature is defined. This is analogous to the units for angular measure. All angles are dimensionless, but we write $\pi/2$ radians or $90°$ or $\frac{1}{4}$ revolution. The term "radian" or "degree" or "revolution" indicates how the unit angle is defined.)

To use this definition of temperature in any quantitative theory, we must be able to devise some experimental procedure by which we can assign numerical values to T. How, for example, do we measure the temperature of a bar of steel? We cannot very well look at the molecules in the steel and measure the average kinetic energy. For an ideal gas we know that (Eq. 13.1)

$$PV = \tfrac{2}{3}K = \tfrac{2}{3}N\epsilon$$

and from Eq. 13.3 we have

$$PV = NkT \qquad \text{(Ideal-gas law)}$$

so that

$$T = \frac{PV}{Nk}$$

All the quantities on the right-hand side are measurable, so we can calculate T. (We may measure N by weighing the gas and dividing by the weight per molecule.)

It would seem then that we can only determine the temperatures of ideal gases. However, we have observed that any composite system in thermal equilibrium has a unique value for the mean kinetic energy per molecule. If we place a small vessel containing an ideal gas in thermal contact with the sample whose temperature is to be determined (say the bar of steel), and the ideal gas is small compared with the sample, the gas will not significantly affect the sample. The composite system will come to thermal equilibrium, and the mean kinetic energy per molecule will be the same for molecules in the sample and molecules in the ideal gas. To put it another way, the temperatures will be equal. Since we can measure the temperature of the ideal gas, we can determine the temperature of the sample, say the steel bar. In effect we have used the ideal gas as a thermometer (Fig. 13.3). In principle we can determine the temperature of any body by using an ideal-gas thermometer.

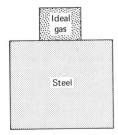

FIGURE 13.3 An ideal gas used as a thermometer.

We must point out that there is really no such thing as a true ideal gas, but most gases behave like ideal gases at very low densities. In Chap. 14 we shall redefine temperature in such a way that no recourse is made to special substances such as ideal gases. More important, we shall not have to use unproved results such as the equality of mean kinetic energies per molecule.

13.3 AVOGADRO'S NUMBER

An awkward problem arises in the practical application of the ideal-gas law,

$$PV = NkT$$

where

$$k = 1.38 \times 10^{-23} \text{ J/K}$$

The number k is very small, and it is multiplied by N, the number of molecules, which is generally very large. The product of these two numbers generally falls within reasonable limits. It would be convenient to choose a new unit for measuring N so that it would not be so large. This may

seem strange at first. How can we define a new way of counting? Actually we do it all the time: A dozen is 12 units; a gross is 12 dozen or 144 units. The new unit of measure we choose for N is 6.02×10^{23}, which is the number of molecules in one mole of a substance. This basic unit is called *Avogadro's number* (N_o):

$$N_o = \text{number of molecules per mole} = 6.02 \times 10^{23}$$

This number was not chosen arbitrarily; it is defined more or less[1] as the number of hydrogen atoms necessary to make up one gram. The gram-molecular weight of a pure substance is defined as the mass in grams of one mole of the substance.

gram-molecular weight = mass in grams of one mole

QUESTION

What are the gram-molecular weights of H, H_2, O_2, and C?

ANSWER

From the definition, the gram-molecular weight of H is 1. Since one mole of H_2 has N_o molecules and thus $2N_o$ atoms, the gram-molecular weight is 2. Oxygen has an atomic number of 16 (that is, one oxygen atom is roughly 16 times as heavy as one hydrogen atom), so the gram-molecular weight of O_2 is 32. The gram-molecular weight of C is 12.

Now if we had N eggs, the number of dozens would be just the total number divided by the number per dozen:

$$\text{number of dozens} = \frac{N}{12}$$

In exactly the same way, the number of moles of molecules n is just the number of molecules N divided by the number of molecules per mole N_o. Thus

$$n = \frac{N}{N_o} \tag{13.4}$$

[1] For technical reasons the standard is actually carbon.

The quantity n is called the *number of moles* or sometimes the *molar number.*

Another way to determine n for a sample is to weigh the sample to find the mass and then to divide the total mass by the mass per mole:

$$n = \frac{\text{mass}}{\text{mass per mole}}$$

But the mass per mole is the molecular weight, so

$$n = \frac{\text{mass}}{\text{molecular weight}} \qquad (13.5)$$

QUESTION

How many molecules are there in an O_2 gas whose mass is 4 g?

ANSWER

From Eq. 13.5,

$$n = \frac{\text{mass}}{\text{molecular weight}} = \frac{4\ \text{g}}{32\ \text{g/mol}} = \frac{1}{8}\ \text{mol}$$

and from Eq. 13.4,

$$N = nN_o = (\tfrac{1}{8}\ \text{mol})(6.02 \times 10^{23}\ \text{molecules/mol})$$

$$= 7.5 \times 10^{22}\ \text{molecules}$$

We may now write the ideal-gas law in the form

$$PV = NkT = nN_o kT$$

Since N_o and k are universal constants, the product is a constant. This product is called *universal gas constant R*:

$$R = N_o k = (6.02 \times 10^{23}\ \text{mol}^{-1})(1.38 \times 10^{-23}\ \text{J/K}) \qquad (13.6)$$

$$= 8.3\ \text{J/K} \cdot \text{mol}$$

The ideal-gas law becomes

$$PV = nRT \qquad (13.7)$$

where n is the number of moles.

QUESTION

Compare the volume occupied by 1 mol of H_2 gas with that occupied by 1 mol of O_2 gas at the same pressure and temperature.

ANSWER

$$V = \frac{nRT}{P}$$

Since n, T, and P are the same for both gases, the volume will be the same for both gases. Thus all ideal gases occupy the same volume if n, T, and P are the same.

QUESTION

In the preceding question we saw that all ideal gases with the same P, T and n occupy the same volume. It is found that an ideal-gas thermometer containing 1 mol of gas occupies a volume of 22.4 liter (1 liter $= 1000\ \text{cm}^3 = 10^{-3}\ \text{m}^3$) when placed in thermal contact with an ice-water bath (Fig. 13.4). What is the temperature of an ice-water bath in equilibrium at atmospheric pressure?

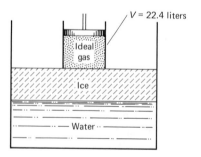

FIGURE 13.4 One mole of any ideal gas occupies 22.4 liters at atmospheric pressure when in thermal equilibrium with an ice-water bath.

ANSWER

Since 1 atm $= 1.01 \times 10^5$ N/m², then

$$T = \frac{PV}{nR} = \frac{(1.01 \times 10^5 \text{ N/m}^2)(22.4 \times 10^{-3} \text{ m}^3)}{(1 \text{ mole})(8.3 \text{ J/K})}$$

$$= 273 \text{ K}$$

QUESTION

The Celsius scale is defined so that ice-water is 0°C and steam-water 100°C at 1 atm and a difference of 1°C is equal to a difference of 1 K. What is body temperature (37°C) in Kelvin degrees?

ANSWER

From the definition of the Celsius scale we see that the two temperatures must differ only by a constant, that is, $T_K = T_C +$ constant (Fig. 13.5). But when $T_C = 0$ we know that $T_K = 273$ K, so the constant must be 273. Therefore the Kelvin equivalent of 37°C is

$$T_K = 37°C + 273 = 310 \text{ K}$$

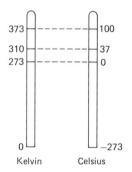

FIGURE 13.5   The Kelvin and Celsius temperature scales.

QUESTION

A scuba diver rises to the surface from a depth of 34 ft. How much air must be expelled from the diver's lungs? (Assume that the air remains at a constant temperature.)

ANSWER

A diver must expel air from the lungs as he or she rises or the lungs will burst. The diver wants to achieve a balanced pressure:

air pressure pushing outward on the lungs

$\qquad$ = water pressure pushing in on the chest

Imagine a cylinder with a piston at one end. Let there be a volume V of gas within, and let the water pressure be P. If T is a constant, then as the cylinder rises PV must remain constant:

$$PV = P_0(V + \Delta V)$$

where P_0 is the surface pressure (1 atm) and ΔV is the increase in volume (Fig. 13.6). Now from hydrostatics we recall that

$$P = P_0 + \rho g h$$

Therefore

$$(P_0 + \rho g h)V = P_0(V + \Delta V)$$

or

$$\frac{\Delta V}{V} = \frac{\rho g h}{P_0}$$

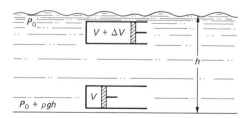

FIGURE 13.6   A cylinder of gas rises from a depth h to the surface, where the volume has increased by ΔV.

Since 1 atm will support 34 ft of water, it follows that $P_o = \rho g h_o (h_o = 34 \text{ ft})$. Therefore

$$\frac{\Delta V}{V} = 1$$

or

$$\Delta V = V$$

so that the volume of air expelled would equal the volume of the lungs itself. This is of course the volume that the air occupies at 1 atm of pressure. The fraction of the air expelled is one-half. From a depth of 68 ft the lungs would have to expel a volume of air twice the volume of the lungs, or two-thirds of the initial air must be expelled.

QUESTION

A jar is partially filled with liquid at 100°C. Assuming that the air above the liquid is 100°C when the lid is put on, determine the force holding the lid on after the jar has cooled to room temperature (20°C). The area of the lid is 7 in².

ANSWER

Since the volume of the air is more or less constant,

$$\frac{P_i}{T_i} = \frac{P_f}{T_f}$$

where i and f refer to "initial" and "final." The lid was put on when the air was at atmospheric pressure and the temperature was 373 K. Therefore

$$P_f = \frac{T_f}{T_i} P_i = \frac{293}{373} \text{ atm}$$

The difference in pressure across the lid after it has cooled is the outside pressure of 1 atm minus the inside pressure of 293/373 atm or 0.21 atm. Since 1 atm is 14.7 lb/in², the net force is

$$\text{Force} = (0.21 \text{ atm})\left(14.7 \frac{\text{lb/in}^2}{\text{atm}}\right)(7 \text{ in}^2) = 22 \text{ lb}$$

SUMMARY

An ideal gas is a collection of molecules in which the interaction between molecules may be neglected. Many different kinds of ideal gases exist, but for all

$$PV = \tfrac{2}{3}K$$

where K is the total kinetic energy of translation of the molecules.

In an ideal gas consisting of two or more types of molecules, the total pressure is the sum of the partial pressures:

$$P = P_A + P_B + \cdots$$

The temperature of any body is proportional to the mean translational kinetic energy per particle:

$$\epsilon = \tfrac{3}{2}kT$$

Avogadro's number ($N_o = 6.02 \times 10^{23}$) is a convenient device for measuring the number of molecules in a macroscopic system. The number of molecules in one mole of a substance is equal to Avogadro's number. A system containing N molecules will have n moles, where

$$n = \frac{N}{N_o}$$

The equation of state for an ideal gas is

$$PV = nRT = NkT$$

PROBLEMS

13.A.1 A vessel contains 1 mol of molecules at a temperature T. What effect, if any, does the composition of the gas (say H_2 or O_2 or a mixture) have on the pressure?

13.A.2 The mass of a hydrogen gas is 10 g. How many molecules of H_2 are there in the gas?

13.A.3 What is the mean velocity of O_2 molecules in the air at a temperature of 300 K? How long would it take a molecule traveling at this speed to go from New York City to Davis, California—a distance of 5000 km?

13.A.4 If the mean velocity of O_2 molecules in the air is 480 m/s, what is the mean velocity of N_2 molecules?

13.A.5 A bubble of air rises from the bottom of a lake 100 ft deep. The temperature at the bottom is 10°C and the temperature at the top is 25°C. Determine the ratio of the bubble volume at the surface to the bubble volume at the bottom.

13.A.6 A sample of O_2 occupies a volume of 0.02 m³. The temperature is 300 K and the pressure is 10^5 N/m².
(a) How many moles of O_2 are there?
(b) How many molecules of O_2 are there?
(c) What is the mass of the gas?

13.A.7 What is the average velocity of an atom in a lead block at 27°C? The atomic weight of lead is 207.

13.A.8 A certain gas is contained in a pressurized cylinder at a pressure of 5 atm. What fraction of the gas is usable? (When the pressure has dropped to 1 atm the gas will stop coming out.)

13.B.1 In principle all physical measurements can be made with a ruler, a watch, and standards of length, time, mass, and electric charge. To measure the temperature of a lake, an investigator performs the following experiment. A gas that obeys the ideal-gas law is placed in a box with a piston for a lid. The box is dipped in an ice-water bath and the volume is measured (with the ruler) to be 1 m³. The lid is free to move, so the pressure in the gas is atmospheric pressure. The box is next dipped in boiling water, and the new volume (again at atmospheric pressure) is observed to be 1.37 m³. The box is finally dipped in the lake, and when thermal equilibrium is reached the volume is 1.1 m³. What is the temperature of the lake? (Assume only that the temperature difference between ice-water and boiling water is 100°.)

13.B.2 At a temperature of 310 K the partial pressure of O_2 in a vessel is $\frac{1}{2}$ atm and the partial pressure of N_2 is also $\frac{1}{2}$ atm. What are the mass densities of O_2 and N_2?

13.B.3 An ordinary tire pressure gauge gives the true pressure in the tire minus 1 atm. (Thus if the pressure in the tire is atmospheric pressure, as it would be in a flat tire, you get a reading of zero). If the tire is inflated to a gauge pressure of 30 lb/in² (1 lb/in² = 6.8×10^3 N/m²) above atmospheric pressure at a temperature of 20°C and then driven on a highway until the temperature of the tire is 30°C, what will be the pressure in the tire as measured by a pressure gauge? Assume that the tire cannot expand.

13.B.4 A large balloon with a volume of 1000 ft³ is to be filled with hydrogen at a final pressure of 1 atm. The hydrogen is supplied in cylinders, each of which has a volume of 3 ft³ at a gauge pressure of 100 lb/in². How many cylinders are needed?

13.B.5 The Ajax Gas Company sells a certain gas for $8 a bottle in 4 m³ bottles at an absolute pressure of 4 atm at 300 K. The Acme Gas Company sells the same gas at $6 a bottle in 5 m³ bottles at an absolute pressure of 3 atm at 300 K. Bearing in mind that when the pressure in a bottle reaches 1 atm no more gas can be extracted, decide which company you would buy from. How many usable moles per dollar do you get from each company?

13.B.6 If 2 g of O_2, 3 g of N_2, and 1.5 g of CO_2 were mixed in a 4-liter bottle at $37°C$, what would their partial pressures be?

13.B.7 How many joules of kinetic energy are contained in the air in a typical room? (Let $P = 1$ atm and the dimensions of the room be $3 \times 3 \times 3$ m.) If all this energy could be converted into electricity and used to light a 100 W bulb, how long would the bulb burn?

13.C.1 Estimate the time it would take a gas to leak from a hole in a box into a vacuum. Assume that all particles travel with the mean velocity and that at any moment one-sixth of the particles are traveling in the six mutually perpendicular directions. Let the area of the hole be A, the volume of the box V, and the average velocity v. (There is an important effect that we have ignored. The faster particles are first out the hole, leaving behind the slower particles. The temperature will therefore continually drop, so the fraction of particles leaving per unit time will continually decrease.)

*14
THERMODYNAMICS

IN CHAP. 13 we were able to deduce certain properties of simple systems by applying the laws of dynamics, for example, $PV = \frac{2}{3}K$. To do this we had to make a number of assumptions regarding the interaction between particles and the random distribution of velocities. What do we do when these assumptions are invalid? What if the system is a liquid or a solid, in which case the interaction between molecules is strong? Is there anything at all that we can say about such systems? We cannot very well apply the laws of dynamics to such complicated systems of interacting molecules. Even if we could, there would be no reasonable way of either recording or utilizing such an overwhelming amount of data. In a typical macroscopic system there would be of the order of 10^{23} molecules. That is an awful lot of molecules to keep track of.

14.1 LAW OF LARGE NUMBERS AND EQUILIBRIUM STATES

Remarkable as it may seem, a great deal can be said about the equilibrium states of such systems—systems with large numbers of particles—no matter what the particle interactions are. The physical ideas involved are simply that *energy is conserved in any closed system* and what we shall call, for lack of a better name, the *law of large numbers*.

Before we discuss these ideas, we must first point out the most striking feature of thermodynamics, which is how inappropriate the name is. There would be no subject of thermodynamics at all were it not for the extraordinary phenomenon of the existence of equilibrium states. Thermodynamics deals only with systems in an equilibrium state. A better name for the subject would be *thermostatics*.

Fundamental to the subject is a clear understanding of the meaning and physical basis for the phenomenon of equilibrium states. By an equilibrium state we mean that the macroscopic properties of the system (e.g., pressure, temperature, density, etc.) do not change in time. However, if we look at the system on a small scale, we find molecules moving about very rapidly, colliding with one another, and in general behaving in a very complicated way. The microscopic features are not at all constant in time.

This phenomenon of systems with large numbers of particles exhibiting this steady-state behavior for macroscopic observables is too easily taken for granted. Familiarity breeds oversights. We are accustomed to systems *running down* and reaching an equilibrium state. For instance, if several balls are put in a bowl, they will bounce around for a while and eventually settle to the bottom of the bowl in equilibrium. This is not the phenomenon we are speaking of. The balls are losing energy (mechanical energy) through friction, and the collection is approaching a state of minimum mechanical energy in which all the balls are at rest in the bottom of the bowl.

Contrast this with the following example. All the molecules of air in a room are confined to a corner and then released. For a short time there are violent air currents, but eventually they die out and the molecules distribute themselves uniformly about the room and an equilibrium state is reached. Why? The system does not lose energy; it does not *run down* the way the balls in the bowl do. All the gas molecules do not just settle to the bottom of the room; they continue to move about vigorously. Why does the system come to a steady state? Why does the gas not just blow back and forth about the room instead of diffusing uniformly over the room, with uniform pressure and uniform temperature? It is important to realize that something peculiar is happening here and that this needs to be explained.

To understand the basis of the phenomenon, let us consider a box containing a large number of coins. Some of the coins are heads and some are tails. We would like to define the *microstate* and the *macrostate* of the system. We specify the microstate when we identify each coin and state whether it is heads or tails. We say, for example, that coin 1 is heads, coin 2 tails, coin 3 tails, etc. We then know the state of each coin. On the other hand, we specify the macrostate by stating the number of heads and tails in the box. We do not say *which* coins are heads and *which* are tails, only that so many are heads and so many are tails. In this example the microstate is characterized by N quantities (if there are N coins), but the macrostate is characterized by just one: the number of heads (the rest of course are tails).

We shake the box periodically and observe the fraction of heads and tails, i.e., the macrostate. If there are just two coins in the box, we do not observe a steady macrostate. Sometimes we see two heads, sometimes two tails, and sometimes one head and one tail. The macrostate is not constant in time. However, if $N = 50$, we find that most of the time we observe roughly half heads and half tails, and the macrostate is fairly steady. The microstate, on the other hand, continuously changes and never approaches a steady state.

In this simple example it is easy to understand the behavior of the system, the key to which lies in the *laws of probability*, not in the *laws of dynamics*. The coins reach the same steady state no matter how the box is shaken, and the result has nothing to do with the dynamics of a tumbling coin. It can be shown that the number of ways of obtaining n heads and $N - n$ tails if there are N coins is

$$\text{number of microstates} = \frac{N!}{n! \, (N - n)!} \tag{14.1}$$

For example, if we have three coins, the list of all micro-states for which there is one head and two tails is given in Table 14.1. There are three microstates for this particular macrostate. Note that

$$\frac{3!}{1!(3-1)!} = 3$$

TABLE 14.1 Microstates associated with the macrostate in which one coin is heads and the other two are tails.

Coin	Microstate		
1	H	T	T
2	T	H	T
3	T	T	H

TABLE 14.2 Number of microstates associated with the indicated macrostate, for $N = 50$.

Macrostate, number of heads, n	Number of microstates
25	1.2×10^{14}
5 (or 45)	2×10^6
1 (or 49)	50
0 (or 50)	1

In Table 14.2 we list the number of microstates corre-sponding to the indicated macrostates for $N = 50$. We see that there are $1.2 \times 10^{14}/2 \times 10^6 = 60$ million times as many microstates corresponding to 25 heads and 25 tails as there are microstates corresponding to 5 heads and 45 tails (or 45 heads and 5 tails). If one micro-states is as likely as another, we would expect to observe half heads and half tails 60 million times as often as 5 heads and 45 tails. Every time the box is shaken, the coins assume a new microstate (i.e., which coins are heads and which are tails), and associated with each microstate is a macrostate (i.e., how many are heads and how many are tails). The macrostate that is observed most often is that

with the most microstates. Since the macrostate in which half the coins are heads and half are tails has the most microstates, it will be observed most often. Those macro-states in which *nearly* half the coins are heads and *nearly* half are tails will occur often, and those macrostates in which the number of heads differs significantly from the number of tails will occur very seldom. The more coins there are, the less often one observes a macrostate that differs significantly from the macrostate that has the most microstates associated with it. This macrostate with the most microstates will then be the steady state, i.e., the state that is observed almost all the time.

The steady state is that macrostate that has the most microstates associated with it.

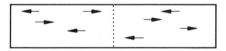

FIGURE 14.1 Particles randomly distributed in a long narrow box.

As a second example, and one more directly related to thermodynamic systems, consider a number N of particles moving back and forth at different rates inside a long narrow box (Fig. 14.1). Let all the particles begin in the left half of the box. We shall assume that after a sufficiently long time any given particle is as likely to be found in the left half of the box as in the right half. The microstate of this system is characterized by specifying which half of the box each particle is in. (We could ask for more detailed information on the microstate, but we shall not.) The macrostate is characterized by specifying how many particles are in the left half and how many are in the right half. The number of microstates associated with the macrostate having n particles on the left and the rest, $N - n$, on the right, is given by

$$\text{number of microstates} = \frac{N!}{n!(N - n)!} \qquad \textbf{(14.2)}$$

(The reason this expression is the same as that for the coins is that if we exchange the words "heads" and "tails" for "left half" and "right half" in the above discussion, we are clearly asking the same question that we asked before.) From our previous discussion we see that the macrostate with the most microstates associated with it is that in which half the particles are on the left and half are on the right.

If we divided a box into a large number of equal compartments, we would find that the macrostate with the most associated microstates is that for which there are roughly equal numbers of particles in each compartment (Fig. 14.2). This is why we find the air molecules uniformly distributed over a room in thermal equilibrium. It is not because of any law of dynamics that drives the system toward uniform density; it is purely a question of probabilities. The uniform state is overwhelmingly the most probable state.

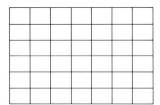

FIGURE 14.2   A box partitioned into equal compartments.

The examples discussed above suggest an explanation for the phenomenon that a system approaches a steady state.

For most systems there is one macrostate *that is associated with a great many microstates, while other macrostates are associated with comparatively few microstates. As the system evolves from one microstate to another, it will more often than not be found in the macrostate with the most microstates. Since the system is most often found in the* same *macrostate, it will be in a* steady state *(or equilibrium state). The equilibrium state is the macrostate with the most microstates associated with it.*

As the number of particles in the system becomes larger and larger, the fraction of those microstates whose associated macrostate differs significantly from the equilibrium state becomes smaller and smaller. For most systems of thermodynamic interest, so many particles are involved (10^{24}) that there is almost no chance of finding the system in one of those unusual macrostates. This is the extraordinary *law of large numbers* that we spoke of earlier: Most states available to a system with many particles look alike when only the gross macroscopic features are observed.

This observation lies at the very foundation of the subject of thermodynamics. In fact it is almost all that need be said. As we shall see, the entire subject of thermodynamics grows out of the mathematical formulation of the fact that most microstates available to a system are associated with the same macrostate.

Before we formalize the basic laws, a word must be said about what is meant by *states available to a system.* If the system is a gas and is contained in a box, then states in which particles are outside the box are not available to the system. If the gas is isolated and has an energy E, then states of different energy are not available. If no particles are created or destroyed, then the number of elementary particles must remain constant in a closed system. Aside from these three constraints, any state is possible for such a system. Any position and velocity of any particle is possible so long as the particles are all in the box and the total energy is E.

14.2 THE FIRST LAW OF THERMODYNAMICS

We may now state the basic laws of thermodynamics. Just two laws are of interest to us. (There is a third law, but it has limited application.)

First law of thermodynamics   There exist equilibrium states of macroscopic systems that are completely characterized macroscopically by the internal energy E, the volume V, and the number of moles n_1, n_2, . . . of the various constituents.

Let us consider a simple system with only one type of particle, and let us try to see the consequences of the first law of thermodynamics. If it is to be a physical law, then we must be able to make predictions from it; otherwise it is a definition, or even less, a non sequitur. Let us imagine a system characterized macroscopically by a certain energy E and volume V (we shall hold the n's fixed). We now change the energy and volume very slowly (quasistatically), but eventually return the system to the original energy and volume (Fig. 14.3). The first law tells us that the system must return to the same macrostate; i.e., *all* the macroscopic variables must return to their original values. If the energy and volume were the *only* macroscopic observables, this would be an empty statement. But there are many other observables: pressure, temperature, specific heat, color, smell, etc. *The first law of thermodynamics tells us that all macroscopic observables are unique functions of E, V, and the n's.* It does not tell us how they depend on E, V, and the n's, only that they do.

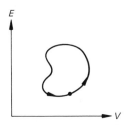

FIGURE 14.3   A process in which the system returns to its initial state.

This law is an expression of our earlier observation that an isolated system will in time reach an equilibrium state. This equilibrium state is unique. If the system is isolated, its energy is fixed. If the system is confined, its volume is fixed. For a given composition the state is then uniquely defined by the energy, volume, and number of moles in the system.

For this law to be of use it is important that we be able to measure the energy. Since all thermodynamic experiments involve change from one state to another, we only need to be able to measure changes in energy between states. We shall assume that it is always possible to do this by taking the system from one state to the next by doing only mechanical work, e.g., compressing a piston or stirring the system. We do not have to be able to take the system from state A to state B by doing mechanical work. It would be sufficient to be able to take the system from B to A or from B to C and from A to C by doing only mechanical work. In any case we would know the energy difference between states A and B.

The first law of thermodynamics is sometimes interpreted as a reformulation of the conservation of energy principle to include internal energy and/or heat energy. As we said earlier, all thermodynamic laws arise out of the laws of probability and not out of the laws of mechanics. Surely we may use the principle of energy conservation to determine the internal energy, but the law itself is not an energy-conservation principle. The law is a consequence of the fact that an isolated system (its energy is fixed) in a box (its volume is fixed) composed of certain numbers of particles (its molar numbers are given) will come to a unique equilibrium state. Therefore the energy, volume, and molar numbers uniquely characterize the thermodynamic equilibrium state.

14.3 PHYSICAL BASIS FOR THE SECOND LAW OF THERMODYNAMICS

The first law of thermodynamics is an expression of the fact that isolated, confined systems come to a *unique* equilibrium state. This law does not seem to incorporate explicitly the probabilistic basis for this behavior, namely: The steady state is that macrostate with the *most microstates associated with it.* Suppose we had a choice between two possible steady states, e.g., returning to the coins in the box, one-quarter heads and three-quarters tails or half heads and half tails. The state with half heads and half tails has more microstates. In fact, it has more microstates than any other macrostate and is therefore the equilibrium state.

Is there some way of formulating this idea as a general physical law? Can we postulate a second law of thermodynamics that will express the fact that all physical systems, if left alone, will approach an equilibrium state *that is the most probable state,* i.e., the macrostate with the most microstates associated with it?

Let us define a thermodynamic quantity S (called the *entropy)* which is a measure of the number of microstates (and thus the probability) associated with any possible macrostate. This entropy will be a function of the nature of the macrostate, so different macrostates will have different entropies. If the system can exist in two or more macrostates (e.g., half the air molecules in the room in the left half and half in the right, or one-quarter in the left half and three-quarters in the right half, etc.), that macrostate with the largest entropy will have the largest number of microstates associated with it and thus will be the most probable macrostate. Therefore the observed macrostate or the state of thermodynamic equilibrium will be that state with the greatest entropy.

All we have done is to substitute the word "entropy" for "number of microstates" and observe that the entropy (or number of microstates) depends on the macrostate (i.e., is some function of the macrostate). To say that S is a function of the macrostate is, by the first law of thermodynamics, to say that S is a function of E, V, and n, for these are the quantities that completely characterize the macrostate.

After one last observation we will be ready to state the second law of thermodynamics. We would like to choose the association between the entropy and the number of microstates in such a way that the entropy is additive. By this we mean that if the system is composed of two or more parts, then the entropy of the whole is the sum of the entropies of the parts. One can show that if we define the entropy as the logarithm of the number of microstates, the entropy will then be additive. This seems unnatural. Why not define the entropy to be just the number of microstates? The answer lies in the fact that if N_1 is the number of microstates for one part of the system and N_2 the number of microstates for the second part, then the total number of microstates for the whole system is not $N_1 + N_2$ but $N_1 N_2$. This is because of the way in which we count states. Suppose, for example, that system 1 can be in two states, A and B, and system 2 in three states, C, D, and E. Then the possible states for the combined system are (A, C), (A, D), (A, E), (B, C), (B, D), and (B, E). There are six possible states in all, that is, 2×3, not $2 + 3$.

If the total number of states is $N_1 N_2$, then $\log N_1 N_2 = \log N_1 + \log N_2$, so that if

$$S_{12} = \log N_1 N_2$$

is the entropy of the whole system,

$$S_1 = \log N_1$$

is the entropy of the first part, and

$$S_2 = \log N_2$$

is the entropy of the second part, then

$$S_{12} = S_1 + S_2$$

The actual definition of entropy in terms of the logarithm of the number of microstates is not necessary for our purposes. All we need to know is that it is possible to define entropy in such a way that it is additive.

FIGURE 14.4   Two bodies with initial energies E_1 and E_2 brought into thermal contact. The final steady-state energies are E_1' and E_2'.

14.4 THE SECOND LAW OF THERMODYNAMICS

We may now put these general ideas together into a statement of the second law of thermodynamics.

For all macroscopic systems there exists a function (called the entropy S) that is defined for all equilibrium states i.e., all possible values of the energy, volume, and molar numbers. The entropy of a composite system is the sum of the entropies of the components. In the absence of an internal constraint the values assumed by the energies, volumes, and molar numbers of the components in a closed system are those that maximize the total entropy.

This is one of the most far-reaching of the fundamental laws of physics, and we shall spend some time in developing its meaning and applications.

First we should put the second law of thermodynamics to the test: Can it make predictions? Let us consider one of the most elementary thermodynamic problems—two bodies with energies E_1 and E_2. Their volumes and molar numbers are V_1, V_2, and n_1, n_2, respectively. The two bodies are brought together and allowed to exchange energy (Fig. 14.4). What is the final steady state? How will the energy be shared between them? (The volumes and molar numbers do not change.) If we know the entropy of each system as a function of energy, all we have to do is find that distribution of energy between the two bodies such that

$$S_{tot} = S_1 + S_2$$

is a maximum subject to the constraint

$$E_{tot} = E_1 + E_2 = E_1' + E_2'$$

(The quantities E_1' and E_2' are the energies of each body after equilibrium is reached.) Physically we are searching for the most probable state or, equivalently, the state that has the most microstates associated with it, or the state that has the largest entropy. Therefore we can use the second law of thermodynamics to *predict* the outcome.

The application of the second law of thermodynamics to this problem is clear-cut, provided we know the entropy of each body as a function of energy. There's the rub. How do we determine the entropy? For that matter, how is it defined? We cannot say that it is the number of microstates associated with the given macrostate because we do not know how to calculate or measure the number of microstates. This conception of the entropy in terms of the number of microstates is vital for an intuitive understanding of the physical basis of the second law, but it is not properly a part of thermodynamics. (In statistical mechanics it is possible to actually calculate the number of microstates for a given macrostate for simple systems, but thermodynamics should be a self-contained subject.)

The second law of thermodynamics is both a definition and a physical law. The entropy is defined as that function *that is a maximum in the equilibrium state,* and the physical law is that *this function depends only on the thermodynamic state,* i.e., on E, V, and the n's. This may not seem like much of a definition, but we shall see that it is enough. We shall show that, given a meter stick and a clock, we shall be able to devise an experiment that allows us to assign a numerical value to the entropy of every thermodynamic state using only the defining property. This is not an easy task, and before we attempt it let us consider some more examples of how to use the principle. It is easier to see what the entropy is after we have seen how it works.

14.5 MATHEMATICAL STATEMENT OF THE SECOND LAW OF THERMODYNAMICS

Mathematics is the language of physics, so we should be able to formulate our verbal statement of the second law of thermodynamics in mathematical terms. There are three basic ideas.

1. Entropy depends only on the thermodynamics state, i.e., on E, V, and the n's. Therefore

$$S = S(E, V, n_1, n_2, \ldots) \tag{14.3}$$

which can be read: S is a function of E, V, n_1, n_2, etc.

2. Entropy is additive; therefore

$$S_{tot} = S_1 + S_2 \tag{14.4}$$

which can be read: The total entropy is the sum of the entropies of the parts.

3. The entropy is largest for that state that is the equilibrium state. Therefore

$$S_{\text{equilibrium state}} \geq S_{\text{other states}}$$

which can be read: The entropy of the equilbrium state is greater than or equal to the entropy of any other state compatible with the constraints. (We shall clarify what we mean by constraints shortly.) It might happen that this other state is itself the equilbrium state or a state very close to it, so we allow for this possibility by saying "greater than or equal to." We may also write

$$S_{\text{equilibrium state}} - S_{\text{other state}} \geq 0$$

or

$$\Delta S \geq 0 \tag{14.5}$$

where ΔS is the difference in entropy between the equilibrium state and some other state. In the future we shall use the maximization principle in the form of Eq. 14.5.

These three euations, 14.3, 14.4, and 14.5, contain all the information in our verbal statement of the second law of thermodynamics.

14.6 APPLICATIONS

We shall put aside for the moment the question of how entropy is measured and shall apply the results to some simple problems.

ENERGY EXCHANGE

Let us return to the problem, discussed earlier, of two bodies exchanging energy and let us consider first two identical gases in boxes of equal volume. Initially the gases are separated and have energies E_1 and E_2 ($E_1 < E_2$). To determine the state of thermodynamic equilibrium, we must find the state that has the maximum entropy. Let us compare the entropy of the initial state with the state in which the energy of the first gas is $E_1 + dE$ and the energy of the second gas is $E_2 - dE$. Note that the total energy remains $E_1 + E_2$. From Fig. 14.5 we see that the slope of the entropy curve is greater at E_1 than at E_2, so the increase in the entropy of the first gas will be greater than the decrease in the entropy of the second gas. Therefore

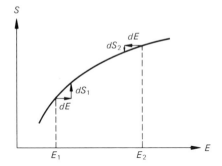

FIGURE 14.5 Entropy as a function of energy.

the total entropy will increase. If we continue in this manner to compare the total entropy as we increase the energy of the first gas and decrease the energy of the second gas by an equal amount, we find a continuous increase in entropy until each system has the same energy. If the energy of the first gas is increased further, the total entropy will decrease. Thus the state with the maximum entropy is that in which *the energy is shared equally between the two gases.*

Note that we have used all the principles in the statement of the second law of thermodynamics or, equivalently, Eq. 14.3, 14.4, and 14.5. We used Eq. 14.5 when we said that the entropy in the equilibrium state is a maximum, Eq. 14.4 when we said that the total entropy was the entropy of the first gas plus the second, and Eq. 14.3 when we determined the entropy of the various possible macrostates given the energy, volume, and molar numbers of the components. (The volume and molar numbers are hidden in the graph. For different volume and molar numbers the graph would be different.)

In this very simple example we have proved from the second law of thermodynamics that two similar systems placed in thermal contact will exchange energy until the

total energy is shared equally. This is just what we might have guessed. This result is analogous to the example we discussed earlier of particles (rather than energy) being shared equally between two equal halves of a box.

This is not to say, however, that whenever any two systems are placed in thermal contact the total energy will be shared equally between the two systems. In the example above, the two systems are identical. If one were contained in a larger box than the other or had more molecules or consisted of a different substance, the systems would not share the energy equally. If the two entropy-vs-energy curves are different (Fig. 14.6), the systems will exchange energy until the *slopes* of the two curves are the same. When the two slopes are the same, dS_1 will be equal and opposite to dS_2, and so dS will be zero, that is, S will be a maximum. But since the two curves are different, the energies at which the slopes are the same are different. When the slopes are equal, any further increase in energy of system 1 above E'_1 and decrease in energy of system 2 below E'_2 causes a decrease in the total entropy.

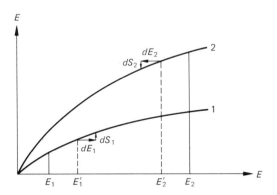

FIGURE 14.6 The slope at E'_1 is the same as that at E'_2.

We shall see later that equality of the two slopes is equivalent to equality of temperature.

QUESTION

Consider two different systems of fixed volume and molar numbers. Suppose the slopes of the entropy-vs-energy curves for fixed V and n are

slope of S_1 vs $E_1 = \dfrac{2}{E_1}$ (V_1 and n_1 fixed)

slope of S_2 vs $E_2 = \dfrac{1}{E_2}$ (V_2 and n_2 fixed)

Initially the two systems are separated, and $E_1 = 10$ J and $E_2 = 40$ J. What is the final energy of each system after the systems are brought in contact and allowed to exchange energy?

ANSWER

At equilibrium the two slopes must be equal, so that

$$\frac{2}{E_1} = \frac{1}{E_2}$$

or

$$E_1 = 2E_2$$

But since the total energy is conserved,

$$E_1 + E_2 = \text{constant} = 10 \text{ J} + 40 \text{ J} = 50 \text{ J}$$

Solving for E_1 and E_2, we find

$$E_1 = 33.3 \text{ J}$$

$$E_2 = 16.7 \text{ J}$$

QUESTION

Two systems with entropies

$$S_1 = \ln(\alpha E_1^2)$$

$$S_2 = \ln(\beta E_2)$$

where α and β may depend on the molar numbers and volumes, are brought into thermal contact and allowed to exchange energy. Before thermal contact the energies are $E_1 = 10$ J and $E_2 = 40$ J. What are the energies of the systems after thermal contact?

ANSWER

We must maximize the total entropy S given by

$$S = S_1 + S_2 = \ln(\alpha E_1^2) + \ln(\beta E_2)$$

considered as a function of E_1 and E_2 but subject to the constraint that $E_1 + E_2 = \text{constant} = 50$ J. Now we determine the maxima of a function by finding those places where the function does not change, in this case where $dS = 0$. Consider a change in entropy associated with an exchange of energy:

$$dS = \frac{\partial S_1}{\partial E_1}dE_1 + \frac{\partial S_2}{\partial E_2}dE_2 = \frac{2}{E_1}dE_1 + \frac{dE_2}{E_2}$$

But if the total energy is to remain constant,

$$d(E_1 + E_2) = 0$$

or

$$dE_2 = -dE_1$$

so that we may write

$$dS = \frac{2}{E_1}dE_1 - \frac{dE_1}{E_2} = \left(\frac{2}{E_1} - \frac{1}{E_2}\right)dE_1$$

Now S is a maximum where $dS = 0$ or where

$$\frac{2}{E_1} - \frac{1}{E_2} = 0$$

or

$$E_1 = 2E_2$$

But

$$E_1 + E_2 = 50 \text{ J}$$

Solving for E_1 and E_2, we find

$$E_1 = 33.3 \text{ J}$$

$$E_2 = 16.7 \text{ J}$$

This question is in fact identical to the preceding question. The only difference is that in one case the rates of change of the entropies with energy are given and in the other the entropies themselves are given and the rates of change determined.

VOLUME EXCHANGE

As a second example of how the second law of thermodynamics works, consider two gases in a large box and separated by a piston (Fig. 14.7). The piston can move and

FIGURE 14.7 Two systems separated by a movable, thermally conducting piston.

is thermally conducting. This is similar to the preceding example except that now the two systems can exchange both energy and volume. If we know the entropy of both gases as functions of volume and energy, we can find the equilibrium state by finding those values of the volume and energy of each compartment that give the largest value for the total entropy.

QUESTION

Two systems at equal temperatures are separated by a fixed piston. The initial volumes are V_1 and V_2 and the SV curves are as shown in Fig. 14.8. Which way will it move when it is released?

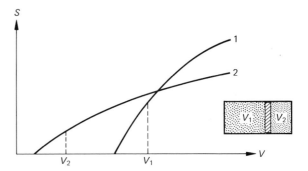

FIGURE 14.8 Entropy-volume curves for fixed energy and molar numbers.

ANSWER

If the piston moves to the right, V_1 will increase and V_2 will decrease. Since the slope of the SV curve is greater at V_1 for system 1 than that of system 2 at V_2, the entropy of system 1 will increase more than the entropy of system 2 will decrease. Thus the total entropy will increase. If the piston were to move to the left, the total entropy would decrease. Since the total entropy must increase (approaching its maximum value), the piston will move to the right. (Caution: It is not possible to determine the final equilibrium position of the piston from these curves. When the piston begins to move, energy will be exchanged between the two systems even if they were initially at the same temperature. The above curves are SV curves for fixed n and E. To determine the final state, we would have to know S as a function of E also.)

PARTICLE EXCHANGE

We might also imagine a membrane between the two compartments that allows particles of one type to pass but not others. Now the two gases can exchange one kind of molecule, and since these molecules carry energy, the gases can exchange energy as well. The steady state will be reached when the entropy reaches a maximum with respect to exchange of energy and exchange of that one type of molecule that can diffuse through the membrane.

In all of these examples the total system is composed of two (or more) parts. These parts can exchange energy, volume, and/or particles, depending on the constraints. The condition for thermodynamic equilibrium is that if some process is possible, that process will continue until the total entropy has reached a maximum value. In each case there are constraints: particles are confined inside boxes, energy was conserved, dividing walls may or may not move, particles may or may not pass through walls, etc. It was within the limits of these constraints that the entropy assumes a maximum value. If one or more of the constraints is relaxed, the entropy can increase further. *Every process* in a closed system will proceed in the direction of increasing entropy. This applies to living as well as nonliving systems.

SUMMARY

Every macroscopic system will eventually come to an equilibrium state, which is that macrostate that has the most microstates associated with it. The existence of an equilibrium state is a consequence of the fact that one macrostate is overwhelmingly more probable than other macrostates. The two fundamental laws of thermodynamics are as follows:

1. There exist steady states of macroscopic systems that are completely characterized macroscopically by the internal energy E, the volume V, and the number of moles $n_1, n_2, \ldots$ of the various constituents.

2. For all macroscopic systems there exists a function (called the entropy S) that is defined for all equilibrium states, i.e., all possible values of the energy, volume, and molar numbers. The entropy of a composite system is the sum of the entropies of the components. In the absence of an internal constraint, the values assumed by the energies, volumes, and molar numbers of the components in a closed system are those that maximize the total entropy.

The physical basis for the first law of thermodynamics is that steady states exist and are determined by the constraints on the states available to the system. In an isolated system these constraints are that energy is conserved, volume is fixed, and the molar numbers are fixed.

The physical basis for the second law of thermodynamics is that the equilibrium state is the most probable state. Entropy is a measure of probability.

PROBLEMS

14.A.1 Two systems A and B are identical except that A has an initial energy of 2 J and B has an initial energy of 7 J. Determine the energy of the steady state if the states are allowed to exchange energy.

14.A.2 The entropy as a function of energy for two systems A and B is given in the figure. Initially $E_A = 6$ and $E_B = 1$. Estimate the energies of the equilibrium state.

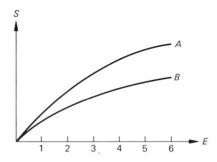

14.B.1 When two dice are thrown, what is the most likely macrostate? The macrostate is the sum of the two numbers showing. The microstate is the number showing on the first die and the number showing on the second die. Thus if the macrostate is 4, there are three associated microstates: die 1 = 1, die 2 = 3; die 1 = 2, die 2 = 2; die 1 = 3, die 2 = 1.

14.B.2 With three dice, how many microstates are there in which the sum is (a) 7? (b) 9?

14.B.3 Two systems at the same temperature are separated by a fixed piston. The entropies as a function of volume are given in the following figure. If the initial volumes are $V_1 = 6$ m³ and $V_2 = 1$ m³ and the piston is released, which way will it move?

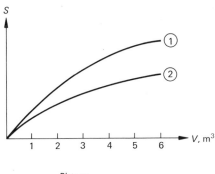

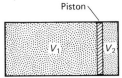

14.B.4 The entropies of two substances A and B are given by

$$S_A = 2 \ln (E_A V_A{}^2)$$

and

$$S_B = 4 \sqrt{E_B V_B}$$

Initially

$$E_A = 4 \text{ J}$$

$$V_A = 1 \text{ m}^3$$

$$E_B = 2 \text{ J}$$

$$V_B = 1 \text{ m}^3$$

If the two systems are brought into thermal contact, what is the final state? (The volumes V_A and V_B are fixed.)

14.B.5 Two systems at the same temperature are separated by a fixed permeable membrane. The entropies as a function of molar number are given in the figure. Initially $n_1 = 1$ mol/m³ and $n_2 = 5$ mol/m³. In which direction is the net diffusion?

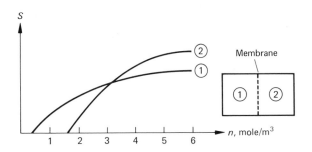

14.C.1 Six numbered balls are distributed among three boxes. How many microstates are there for the macrostate in which

(a) each box has two balls?

(b) the first box has three balls, the second two balls, and the third one ball?

14.C.2 The entropy of two identical systems A and B is given in the figure as a function of energy.

(a) Show that there exists no stable steady state if the systems are allowed to exchange energy.

(b) Show that the slope of the entrope-vs-energy curve must be a uniformly decreasing function of energy.

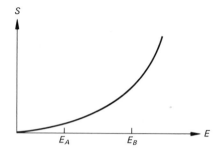

*15
MEASUREMENT
OF
ENTROPY

IT IS CLEAR how useful it would be to have a function such as entropy. To determine the equilibrium state of any system, we only need to find that state that maximizes the entropy. But how do we go about finding the thing that is to be maximized if all we know about it is that it is a maximum (and is additive)? This seems like a vicious circle.

By analogy we might have begun our study of static equilibrium in a gravitational field by postulating the existence of a function U (called the potential energy), defined for all configurations, which is a minimum when the system is in equilibrium or, equivalently, we might have said that the potential energy of the equilibrium state is less than the potential energy of any other state compatible with the constraints. By performing some simple experiments, we may show that the only additive function that satisfies these requirements and the experimental observations in a uniform gravitational field is $U = mgy$ for a single mass or $U = \Sigma\ mgy$ for a system. (We may add any constant to U or multiply by any positive constant without changing the result.) In an entirely similar way, we shall now demonstrate how the entropy may be determined from experiment.

15.1 CHANGES IN ENTROPY

If we are to determine the entropy of a system from the fact that it is a maximum in the equilibrium state, then we shall be interested in *changes* in the entropy when those variables on which it depends—namely the energy, volume, and molar numbers—change. We are interested in the changes in the entropy because it is by observing changes in a function that we determine where the maximum is.

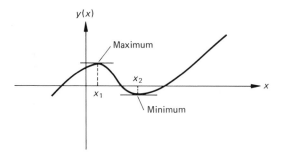

FIGURE 15.1   Maxima and minima of a function of x.

For example, if $y(x)$ is some function of x, then to find the maxima or minima of $y(x)$ we search for those places (those values of x) such that $y(x)$ does not change significantly for small changes in x. In Fig. 15.1, $y(x)$ is a maximum at x_1 and a minimum at x_2. At these points the slope is zero; i.e., any small change in x leaves y practically unchanged.

Consider first the change in entropy associated with a small change in energy. (We are holding fixed the volume and molar numbers.) If the change in entropy is dS, then

$$dS = S(E + dE, V, n) - S(E, V, n)$$

But by definition

$$\lim_{\Delta E \to 0} \frac{S(E + \Delta E, V, n) - S(E, V, n)}{\Delta E} = \frac{\partial S(E, V, n)}{\partial E}$$

so that

$$dS = \frac{\partial S}{\partial E} dE$$

If the volume changes and the energy and molar numbers are fixed, then

$$dS = \frac{\partial S}{\partial V} dV$$

If both the energy and the volume change (n = constant), then

$$dS = \frac{\partial S}{\partial E} dE + \frac{\partial S}{\partial V} dV$$

These partial derivations will obviously play a significant role in our search for the maximum of the entropy function, and we find it convenient to introduce a shorthand notation for these derivatives. Let us define T by the equation

$$\frac{1}{T} = \frac{\partial S}{\partial E} \tag{15.1}$$

and P by the equation

$$\frac{P}{T} = \frac{\partial S}{\partial V} \tag{15.2}$$

Not only is this a convenient shorthand, but as we shall see the functions P and T are indeed what the notation suggests: the pressure and temperature. This, however, remains to be seen, and for the moment we will consider them only as abbreviations for partial derivations.

With these definitions we may write

$$dS = \frac{1}{T} dE + \frac{P}{T} dV = \frac{dE + P\, dV}{T} \tag{15.3}$$

(There will be additional terms later when we allow for a change in molar numbers.)

Suppose we want to determine the change in entropy associated with a change in energy dE and a change in volume dV. From Eq. 15.3 we see that we could calculate dS if we could in some way measure P and T. We can certainly observe the change in volume dV, and we have discussed earlier how we can measure changes in energy. All the quantities on the right-hand side of the equation can be measured, so we can determine dS associated with this change of state.

If we can calculate or measure the change in entropy dS between any two neighboring states, we can then determine the difference in entropy between *any* two states A and B (not necessarily neighboring states) by simply summing the change in entropy between all neighboring states for some process that takes the system from state A to state B. To visualize this, let us represent the state of the system as a point in a coordinate system in which the abscissa is the volume V and the ordinate is the energy E (Fig. 15.2). (We shall assume that the molar numbers are fixed.) From the first law of thermodynamics we know that the macrostate is uniquely characterized by the values of the energy, volume, and molar numbers. Since we are assuming that the molar numbers are specified and fixed, then a point in the EV coordinate system completely specifies the system. This EV coordinate system plays the same role in thermodynamics that the xy coordinate system plays in mechanics (in two dimensions). Just as we com-

puted changes in potential energy ΔU between states in mechanics by summing changes in potential between neighboring states along some path in the xy space, so we may compute ΔS between any two states by summing (integrating) changes in entropy dS between neighboring states along some path in the EV space. If we arbitrarily choose some state as the reference state, then we may determine the entropy everywhere relative to this reference state.

However, all this hinges on our ability to evaluate $dS = (dE + P/dV)/T$. Let us therefore investigate the properties of P and T.

15.2 SOME PROPERTIES OF *P* AND *T*

If P and T are to be the pressure and temperature, they must have certain properties. For example, if two systems can exchange energy, their temperatures must be equal. And if two systems separated by a movable piston can exchange energy and volume, then their pressures must be equal. Let us examine these two situations.

1. Two systems are separated by a heat-conducting wall (Fig. 15.3). Since the total system is in equilibrium, any change in the total entropy must be zero. Therefore

$$dS_{\text{tot}} = 0$$

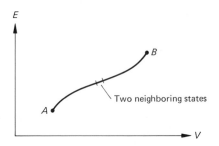

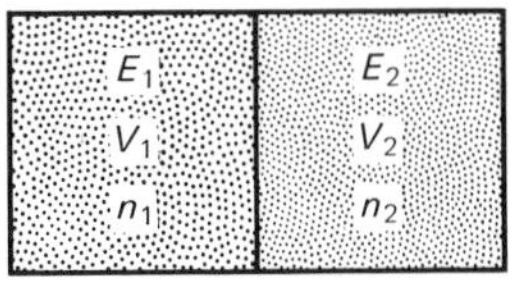

FIGURE 15.2   The state of a system may be represented by a point in an energy-volume coordinate system. A process that takes the system from one state A to another state B is represented by a line in this coordinate system.

FIGURE 15.3   Two systems that can exchange energy.

Now

$$S_{tot} = S_1(E_1, V_1, n_1) + S_2(E_2, V_2, n_2)$$

where S_1 is the entropy of the subsystem on the left and S_2 the entropy of the subsystem on the right (see Fig. 15.3). Let us consider a small transfer of energy from one system to the other. The change in the entropy is given by

$$dS_{tot} = dS_1 + dS_2 = \frac{\partial S_1}{\partial E_1} dE_1 + \frac{\partial S_2}{\partial E_2} dE_2$$

and from the definition of temperature (Eq. 15.1), we may write

$$dS_{tot} = \frac{1}{T_1} dE_1 + \frac{1}{T_2} dE_2$$

Since the total energy is conserved,

$$d(E_1 + E_2) = 0$$

or

$$dE_2 = -dE_1$$

so that

$$dS_{tot} = \left(\frac{1}{T_1} - \frac{1}{T_2} \right) dE_1$$

But $dS_{tot} = 0$, so that

$$\frac{1}{T_1} - \frac{1}{T_2} = 0$$

or

$$T_1 = T_2$$

Of course, we considered this very problem in Chap. 14; the *only* difference here is that we have introduced a new symbol for $\partial S/\partial E$ or the slope of the entropy-vs-energy curve, i.e., we have set $1/T = \partial S/\partial E$.

2. Two subsystems are separated by a movable, heat-conducting wall (Fig. 15.4). Since the total system is in equilibrium, the entropy must be a maximum; i.e., the change in the entropy must be zero for any change of state compatible with the constraints. Now

$$S_{tot} = S_1(E_1, V_1, n_1) + S_2(E_2, V_2, n_2)$$

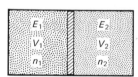

FIGURE 15.4   Two systems that can exchange volume.

Let us consider a small transfer of energy and volume. Since the two systems are separated by a movable, heat-conducting wall, they can exchange both energy and volume. Therefore

$$dS_{tot} = dS_1 + dS_2$$

$$= \frac{\partial S_1}{\partial E_1} dE_1 + \frac{\partial S_1}{\partial V_1} dV_1 + \frac{\partial S_2}{\partial E_2} dE_2 + \frac{\partial S_2}{\partial V_2} dV_2$$

or from Eqs. 15.1 and 15.2,

$$dS_{tot} = \frac{1}{T_1}dE_1 + \frac{P_1}{T_1}dV_1 + \frac{1}{T_2}dE_2 + \frac{P_2}{T_2}dV_2$$

But

$$dE_2 = -dE_1$$

and

$$dV_2 = -dV_1$$

so that

$$dS_{tot} = \left(\frac{1}{T_1} - \frac{1}{T_2}\right)dE_1 + \left(\frac{P_1}{T_1} - \frac{P_2}{T_2}\right)dV_1$$

Since dE_1 and dV_1 are arbitrary, dS_{tot} will be zero for *any* choice of dE_1 and dV_1 if and only if

$$T_1 = T_2$$

and

$$P_1 = P_2$$

The fact that the values of T as we have defined it are equal for two systems that can exchange energy does not necessarily mean that T is what we are accustomed to calling the temperature (for example, the quantity T that appears in the ideal-gas law $PV = nRT$). This T could be the square or the cube of the temperature and still have the property of being equal for two systems that can exchange energy. The same may be said for our quantity P. All that we can say at this point is that P and T are possible candidates for the pressure and temperature.

15.3 THE PRESSURE

To determine the significance of the quantity P, we will consider an experiment in which the pressure is easily identified and will show that P is this pressure. Let us consider a substance contained inside an insulated vessel with a piston of area A on top (Fig. 15.5). A weight W rests on the piston. From our earlier definition of pressure,

$$\text{pressure} = \frac{\text{force}}{\text{area}} = \frac{W}{A}$$

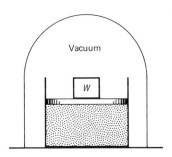

Vacuum

W

FIGURE 15.5 A gas confined to a cylinder.

When the system is in equilibrium any change in the entropy associated with a small displacement in the position of the piston must be zero. Therefore

$$dS = \frac{\partial S}{\partial E}dE + \frac{\partial S}{\partial V}dV = \frac{1}{T}dE + \frac{P}{T}dV = 0$$

or

$$dE + P\,dV = 0$$

where dV is the increase in the volume and dE is the increase in the energy of the system resulting from this increase in volume. Of course, if dV is positive, dE must be negative since the system will lose energy in pushing up the weight. If the weight moves up a distance ds, then the work done by the gas is given by

$$\text{work done by the gas} = W\,ds$$

Now this work done by the expanding gas is at the expense of the internal energy of the gas. There will be a decrease in the energy of the gas equal to $W\,ds$ or, since dE is the increase,

$$dE = -W\,ds$$

But

$$dE + P\,dV = 0$$

so that

$$P = -\frac{dE}{dV} = \frac{W\,ds}{dV} = \frac{W}{A}$$

since

$$dV = A\,ds$$

But

$$\frac{W}{A} = \text{pressure}$$

It follows then that

$$P = \text{pressure}$$

and we have succeeded in establishing the identification. This thought experiment is crucial to our quest for a way to measure the entropy.

15.4 HEAT

Before we proceed further with the problem of measuring the entropy, we will find it convenient to define a new quantity: heat transfer.

There are many ways in which we can increase the internal energy of a system: We can stir the system, rub the system with a rough object, compress the system, or place the system in contact with a hotter body. In this last instance we speak of *heat energy* being transferred from the hot body to the cold one. If we denote this kind of energy by a new symbol dQ and assume that the system

is not being rubbed or stirred, then the increase in the energy in any process is given by

$$dE = dQ - P\,dV \tag{15.4}$$

This equation states that the increase in the internal energy of the system dE is equal to the heat energy added to the system dQ minus the decrease in energy of the system as a result of expansion $P\,dV$. If dV is negative, then the system is contracting and will be gaining energy.

This equation should be regarded as a definition of dQ. We recognize two different mechanisms for increasing the energy of a body: doing mechanical work $(-P\,dV)$ and adding heat (dQ).

We may rewrite the expression for the change in the entropy associated with some process (not involving a change in the molar numbers). From Eq. 15.3 and 15.4,

$$dS = \frac{dE + P\,dV}{T} = \frac{dQ}{T} \tag{15.5}$$

This gives us a new point of view in calculating entropy changes. It may be easier to determine the heat added to the system than to determine the increase in the total energy and $P\,dV$.

A very important distinction must be made between dQ and dE or dV or dS. The quantity E represents the internal energy, V the volume, and S the entropy of the system. Each of these quantities has definite values for a system in any thermodynamic equilibrium state. We cannot, however, associate a quantity Q with a thermodynamic equilibrium state; there is no such thing as the amount of heat in a system. The quantity dQ does not represent the change in some function Q, but rather the amount of heat added to the system in some process. On the other hand, dS represents both the change in the entropy function (and is a true differential of a function in the sense of differential calculus) and at the same time the change in the entropy between two thermodynamic equilibrium states.

Perhaps we can make this distinction clearer by considering a similar situation in mechanics. From the principle of work and energy,

$$dW = dK + dU$$

where dW is the work done on a body, dK the increase in kinetic energy and dU the increase in potential energy of the body. Now K and U are defined for any dynamic state of the body, and we know K and U if we know the position and velocity of the body. But what is W? The quantity W itself has no meaning, but dW is well defined even though it is not the differential of some function.

15.5 ISOTHERMAL AND ADIABATIC PROCESSES

A *thermodynamic process* is a transition of a system with a large number of particles through a series of thermodynamic equilibrium states. One may approach this in a limiting procedure by making only very small changes in the state of the system, e.g., increasing the volume slightly or adding a small amount of heat energy. The system changes from one thermodynamic equilibrium state to a new thermodynamic equilibrium state in a *quasistatic* manner.

A process that is not a thermodynamic process is a transition in which the system does not pass through a sequence of equilibrium states. When a gas is confined to a box and suddenly the box is opened, the particles will rush out into the room. When a hot body is brought in contact with a cold body, the two bodies will eventually reach a common temperature, but the process is not a slow progression through a series of equilibrium states. These processes are not thermodynamic processes.

Two thermodynamic processes of special interest are isothermal processes and adiabatic processes. An *isothermal process* is one in which the temperature of the system remains constant. We may achieve this in practice by placing the system in thermal contact with a very large system — a system so large that it would not be affected by the addition or loss of a small amount of heat. Such a system is called a *thermal reservoir*. When a gas confined to a cylinder is placed in contact with a thermal reservoir and the volume of the gas is allowed to increase slowly, the thermodynamic state of the gas will change, but the temperature of the gas will remain constant.

An *adiabatic process* is one in which no heat is taken from or added to the system. When a gas in a cylinder is isolated from its surroundings and the gas is allowed to expand slowly, the thermodynamic state will change, but no heat is taken from or added to the system.

In summary, if

$$dT = 0$$

in a thermodynamic process, then the process is isothermal. If

$$dQ = 0$$

in a thermodynamic process, then the process is adiabatic. Since $dS = 0$ if $dQ = 0$, it follows that an adiabatic process is also an *isentropic* process.

We may represent these two processes by curves in our *EV* space. In Fig. 15.6 the solid lines represent isothermal processes and the dashed lines adiabatic processes. More precisely, we should say that a process in which a system is changed very slowly so that it moves along one of these lines is an isothermal process or an adiabatic process.

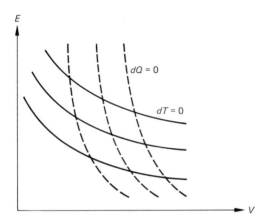

FIGURE 15.6 Each solid line represents a set of thermodynamic equilibrium states that are all at the same temperature. The temperature is a different constant on each line. The dashed lines represent a series of thermodynamic equilibrium states that all have the same entropy. The entropy is a different constant on each line.

QUESTION

Why is it that wherever an adiabat intersects an isotherm, the adiabat is steeper?

ANSWER

Imagine a gas in a cylinder equipped with a piston. The cylinder is isolated from its surroundings, and the piston is pushed in a short distance, thus decreasing the volume. This is an adiabatic process. Now the energy of the system will increase because work has been done on the gas. We also expect the temperature of the system to rise, since when energy is added to the system it will become hotter.

Consider next an isothermal process. Place the cylinder on a thermal reservoir and slowly decrease the volume. The energy of the system will increase as before but not as rapidly. Were it not for the reservoir, the temperature would rise. For the temperature to remain constant, some energy must be transferred from the system to the reservoir. Therefore part of the work done in pushing in the piston will go into the reservoir and part of it will go into the gas. In the adiabatic process *all* the work was transformed into the internal energy of the gas.

Therefore the increase in energy resulting from a decrease in volume will be larger for an adiabatic process than for an isothermal process, and hence the adiabats are steeper than the isotherms.

Of course many other thermodynamic processes besides adiabatic and isothermal processes exist. A process in which the volume is held fixed is an *isovolumic* process. A process in which the pressure is held fixed is an *isobaric* process. And there are processes in which nothing is held fixed.

15.6 MEASUREMENT OF ENTROPY

Let us consider now the fundamental problem of measuring the entropy associated with an arbitrary thermodynamic state. We have mentioned that we need to determine only changes in entropy between states, for it is in this way that we determine the state in which the entropy is a maximum. Therefore we may assign an arbitrary value to the entropy of some reference state and measure the entropy of all other states relative to this state. (This is not unlike the arbitrary manner in which we chose the reference state in defining potential energy. There too it is only changes in potential energy that are important.) The choice of the reference state is of no particular importance. Let us call the temperature of the reference state T_0. Now we can assign any value we please (other than zero) to T_0; this just defines the temperature scale. The reference state is indicated in Fig. 15.7 as the state with energy E_0 and volume V_0. Our task is to determine the entropy of the arbitrary state in Fig. 15.7 relative to the reference state $(S - S_0)$, given that along any path (thermodynamic process)

$$dS = \frac{dE + P\,dV}{T} = \frac{dQ}{T}$$

and that the temperature of the reference state is T_0. We can measure dE, P, dV, and dQ along any path, but we cannot measure T. We would like to find a path joining the arbitrary state to the reference state on which we do not need to measure T.

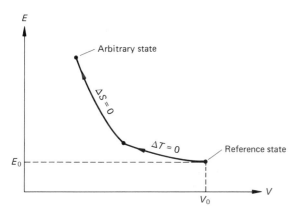

FIGURE 15.7 An arbitrary state linked to the reference state by an adiabat and an isotherm.

Since we cannot yet measure T, the only paths on which we can measure changes in entropy are the isotherm that passes through the reference state and any adiabat. Along the isotherm $T = T_0$ we have

$$\Delta S = \int dS = \int \frac{dQ}{T} = \frac{\Delta Q}{T_0}$$

where ΔQ is the heat added to the system from the reservoir. And along any adiabat

$$\Delta S = 0$$

Now we see in Fig. 15.7 that the isotherm that passes through the reference state intersects the adiabat that passes through the arbitrary state. (Because the adiabats

are steeper than the isotherms, we can be sure that there will always be a point of intersection.) We may measure the entropy difference between the arbitrary state and the reference state by measuring dS along these intersecting segments. In particular,

$$S - S_0 = \frac{\Delta Q}{T_0}$$

where S is the entropy of the arbitrary state, S_0 the entropy of the reference state, T_0 the temperature of the reference state, and ΔQ the heat added to the system from the reservoir during the isothermal stage.

We may simplify the task of measuring the heat taken from the reservoir. If we choose the reference state to be the state in which the temperature is that of an ice-water bath at a pressure of 1 atm, then 1 g of ice will form for every 336 J of energy removed, or 1 g of ice will melt for every 336 J of energy added to the reservoir. To determine ΔQ we simply measure the change in the mass of ice in grams and multiply by 336 J.

QUESTION

A gas is confined to a vessel with a sliding piston. The energy is 0.3×10^5 J above the energy of the reference state, the volume is $\frac{1}{2}$ m³, and the volume of the reference state is 1 m³. The piston is slowly moved out (but no heat is added) until the temperature falls to the temperature of the reference state, i.e., the temperature of an ice-water bath at a pressure of 1 atm. We shall define this reference temperature arbitrarily to be 100°A (A for arbitrary). The gas is now placed on a large ice-water bath and the piston again moved out until the volume equals 1 m³, the volume of the reference state. During this last process it is observed that 70 g of ice melt. What is the entropy of the initial state relative to the reference state?

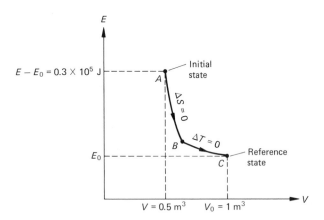

FIGURE 15.8 A two-stage process that takes the system from the initial state A to the reference state C.

ANSWER

The process may be represented on an EV diagram (Fig. 15.8). Initially the system is at A and the volume increased but no heat added until the temperature falls to that of the ice-water bath. This is state B. The entropy change from A to B is zero since $dS = dQ/T$ and $dQ = 0$. Along the isotherm from B to C the entropy change is

$$\Delta S = S_0 - S = \int_S^{S_0} dS = \int \frac{dQ}{T} = \frac{\Delta Q}{T_0}$$

where dQ is the heat added to the gas and T_0 is the temperature of the isotherm. Now T_0 is a constant equal to $100°A$ and $\int dQ = \Delta Q$ is the *net* heat added to the gas, which must be equal to the net heat removed from the ice-water bath. It takes 336 J to freeze 1 g of water or to melt 1 g of ice, and since 70 g of ice melted, the heat added *to the gas* is given by

$$\Delta Q = -(70 \text{ g})(336 \text{ J/g}) = -0.24 \times 10^5 \text{ J}$$

so that

$$S_0 - S = \frac{-0.24 \times 10^5 \text{ J}}{100} = -240 \text{ J}$$

Therefore the entropy of the state in which $E = 0.3 \times 10^5$ J $+ E_0$ and $V = 0.5$ m³ differs from the entropy of the reference state in which $E = E_0$ and $V = 1$ m³ by 240 J. We could in principle find the entropy of any state in this manner.

(Note that the unit of entropy is the joule or erg or foot-pound or calorie. This is unfortunate since entropy is a measure not of energy but of probability. This choice of unit is an accident of the historical development of the subject. If temperature had been defined in such a way that it had the units of energy, then entropy would be dimensionless.)

Now that we can measure the entropy of every state, we can calculate the temperature. All we need to do is calculate $\partial S / \partial E$.

QUESTION

Suppose we find from experiment that the entropy as a function of E and V is given by the relation

$$S = k\sqrt{EV}$$

where $k = 2$ J$^{1/2}$m$^{-3/2}$ and E and V are measured in joules and cubic meters, respectively. What is the temperature at $E = 12$ J and $V = 1$ m³?

ANSWER

The temperature is defined by the relation

$$\frac{1}{T} = \frac{\partial S}{\partial E}$$

Now

$$\frac{\partial S}{\partial E} = \frac{1}{2}k\sqrt{\frac{V}{E}} = \frac{1}{2}(2 \text{ J}^{1/2}\text{m}^{-3/2})\sqrt{\frac{1 \text{ m}^3}{12 \text{ J}}} = 0.29$$

Therefore

$$T = \frac{1}{0.29} = 3.4$$

We still have not shown that this temperature is the same as that used in the ideal-gas law, and this is not easy to do at this level. Suffice it to say that if we choose T_0 for an ice-water bath a 1 atm of pressure to be 273 and measure the entropy of an ideal gas, it can be shown that the two temperatures are in fact identical. From now on we shall fix our temperature scale by choosing $T = 273$ K to be the temperature of an ice-water bath, and this is the Kelvin scale.

by T. We can adjust our temperature scale as we please, and so we may choose R as we please. If R is taken to be the universal gas constant, then T is the conventional absolute temperature measured in Kelvin degrees.

Note that although the relation

$$PV = nRT$$

is the same for all ideal gases, it is not true that other thermodynamic relations will be the same. These other relations depend on the function $f(E, n)$.

QUESTION

The most general form of the entropy for an ideal gas can be shown to be

$$S = nR \ln\{V[f(E, n)]\}$$

where R is a constant and $f(E, n)$ is some function that depends on the nature of the ideal gas. (Not all ideal gases are alike. An ideal gas is a collection of noninteracting molecules, but there are many different kinds of noninteracting molecules.) What is the pressure as a function of temperature, volume, and molar number?

ANSWER

The pressure is given by

$$\frac{P}{T} = \frac{\partial S}{\partial V}$$

But

$$\frac{\partial S}{\partial V} = \frac{nR}{V}$$

so that

$$\frac{P}{T} = \frac{nR}{V}$$

or

$$PV = nRT$$

which is the ideal-gas law provided we choose R to be the universal gas constant. In the above equation R is multiplied

QUESTION

The entropy of a particular ideal gas is given by

$$S = nR \ln(VE^{3/2}) \tag{15.6}$$

(Plus a function of n, which plays no role in this question.) What is the relation between the energy and the temperature?

ANSWER

The temperature is given by

$$\frac{1}{T} = \frac{\partial S}{\partial E} = \frac{3}{2}nRE^{-1}$$

or

$$E = \frac{3}{2}nRT$$

which is the relation we obtained earlier for an ideal-gas composed of point particles, i.e., particles that had no internal energy states such as rotational or vibrational energy.

QUESTION

In the preceding question we assumed an entropy of the form

$$S = nR \ln(VE^{3/2}) \tag{15.7}$$

Suppose we double the size of the system. Clearly n, V, and E will all double, but we see from the above equation that S *will not double*. However, the entropy must be an additive function; i.e., the entropy of two identical systems must be twice the entropy of either system. Thus if we double the size of the system, the entropy should double. How might the entropy function be modified so that it doubles when we double E, V, and n and yet the system has the same properties as the ideal gas, namely

$$PV = nRT \tag{15.8}$$

and

$$E = \tfrac{3}{2}nRT? \tag{15.9}$$

ANSWER

We obtained Eqs. 15.8 and 15.9 by considering $\partial S/\partial V$ and $\partial S/\partial E$. If we add to S a function of n alone, neither of these derivatives will be affected. Let us search for a function of n which, when added to the entropy, makes the entropy additive. Consider

$$S = nR \ln \frac{V}{n}\left(\frac{E}{n}\right)^{3/2}$$

$$= nR \ln VE^{3/2} - nR \ln n^{5/2}$$

It is clear that if E, V, and n are doubled, S will double and the entropy will be the same as in Eq. 15.7, except for an added function of n (namely $-nR \ln n^{5/2}$).

The most general entropy for an ideal gas which satisfies the additivity condition is

$$S = nR \ln \left[\frac{V}{n} f\left(\frac{E}{n}\right)\right]$$

where $f(E/n)$ is an arbitrary function. The nature of the function f depends on the intrinsic properties of the molecules in the ideal gas. For example, $f(E/n)$ might be $(E/n)^{5/2}$ for a gas for which the molecular constituents can take some of the total internal energy of the gas. For such a gas we would find that $E = \tfrac{5}{2}nRT$ rather than $\tfrac{3}{2}nRT$. Since $\tfrac{3}{2}nRT$ is the internal energy associated with the translational motion of the mole-

cules, it follows that the extra nRT must be associated with the intrinsic energy of the molecules, e.g., rotational or vibrational energy.

SUMMARY

We must measure entropy using only its defining property: It is that function that is a maximum in the equilibrium state.

Changes in entropy with respect to energy and volume define the pressure and temperature:

$$\frac{\partial S}{\partial E} = \frac{1}{T}$$

$$\frac{\partial S}{\partial V} = \frac{P}{T}$$

The change in entropy then becomes

$$dS = \frac{dE + P\,dV}{T}$$

and P can be identified with the pressure and measured.

If the energy of a system can increase only when work is done on the system ($-P\,dV$) or heat is added to the system (dQ), then

$$dE = dQ - P\,dV$$

With this definition of heat energy we may write

$$dS = \frac{dQ}{T}$$

Choosing the temperature of the standard state (an ice-water bath), we can measure dS between the standard state and an arbitrary state by following an adiabat and an isotherm from the given state to the standard state. In this process we need to measure only changes in energy, volume, and pressure.

PROBLEMS

15.A.1 A gas is contained in a thermally insulated cylinder at a pressure of 1 atm. The gas is compressed by a small amount so that its change in volume dV is 0.1 m³. Determine the change in the internal energy E of the gas.

15.A.2 If the entropy of one mole of a substance is found to be

$$S = R \ln EV^3,$$

what is the entropy of n moles of the substance?

15.A.3 If $S = 3R \ln(VE^2)$, what are the pressure and temperature when $V = 1$ m³ and $E = 2$ J?

15.A.4 Show that a substance whose entropy is given by $S = 4R \ln V^2 E^5$ is an ideal gas. Determine the number of moles in the gas.

15.B.1 A system with an entropy given by $S = 4R \ln VE^{3/2}$ is in a state in which $E = 8$ J and $V = 1$ m³. A small quantity of heat $dQ = 0.1$ J is added to the system, and the volume increases by 0.2 m³. What is the change in energy? (Assume that all changes are small enough that they may be approximated by differentials.)

15.B.2 A substance with an entropy given by $S = 8\sqrt{EV}$ is at a temperature of 300 K, and its volume is 3 m². Determine the pressure.

15.B.3 Suppose $S = 4\sqrt{EV}$
 (a) What is the pressure when the temperature is 7 K?
 (b) What is the temperature when the pressure is 8 N/m²?

15.B.4 If $P = (n/V)RT$ and $E = \frac{7}{2}nRT$, what is $S(E, V, n)$?

15.B.5 The pressure and energy of a system are observed as functions of temperature and volume and found to have the following dependence: $P = 4T/V^2$ and $E = 3T^2$. Determine the entropy as a function of energy and volume.

15.B.6 Suppose $S = 3 \ln(VE^3)$.
 (a) What is S as a function of T and V?
 (b) What is S as a function of T and P?

15.B.7 If $S = 2\sqrt{EV}$, how much heat is required to raise the temperature 1 K at a constant volume of 1 m³ and an initial temperature of 300 K?

15.B.8 The entropies of two systems A and B are given by $S_A = 4 \ln(V_A + E_A)$ and $S_B = 2 \ln V_B E_B^2$. Initially the systems are in states in which $T_A = 0.5$, $V_A = 0.25$, $T_B = 0.2$, and $V_B = 0.4$. If the systems are brought into thermal contact, what is the final temperature?

15.B.9 If the entropy is given by $S = 2E^{1/2}V^{1/4}$, what is the internal energy when the temperature is 100 K and the pressure 10^5 N/m²?

15.C.1 It was observed in problem 14.C.2 that the slope of the entropy-vs-energy curve must be a uniformly decreasing function of energy. Curves A and B in the figure have this property. However, curve A has a region of negative slope, which implies negative temperatures in this region. Suppose such a system were to come into thermal contact with a system that does not have a region of negative slope (curve B). Show that the final equilibrium temperature of both systems would be positive no matter what the initial states are.

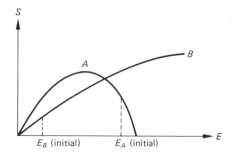

15.C.2 A gas with entropy $S = 2R \ln VE^{3/2}$ is confined to a cylinder that is in a vacuum (see the figure). The temperature of the gas is 300 K. The cross-sectional area of the cylinder is 1 m², and the height is 1 m. Initially the lid of the cylinder is fixed in place by pegs, but then a weight of 10^3 N is placed on the lid and the pegs removed. How far and in what direction will the lid slide?

15.C.3 The entropies of two systems are given by $S_A = 0.4 \ln(E_A + V_A)$ and $S_B = 2 \ln V_B E_B{}^2$. The initial temperatures and pressures are given by $T_A = 0.1$, $P_A = 2$, $T_B = 0.2$, and $P_B = 4$. Determine the final temperatures and pressures if the two systems are allowed to freely exchange energy and volume.

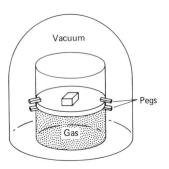

*16

GIBBS
AND
HELMHOLTZ
FREE
ENERGIES

SO FAR WE have dealt only with isolated systems, and we determined the equilibrium state of an isolated system by finding the state with the maximum entropy. There are many examples of thermodynamic systems in which the system being studied is not isolated. For example, reactions taking place in an open test tube are interacting with the atmosphere by exchanging heat and volume. The atmosphere acts as a large reservoir that holds the pressure and temperature fixed. The same may be said of reactions taking place in an animal, and here the reservoir is the rest of the body. We may also consider reactions taking place in a closed vessel whose temperature is held fixed by exchanges of energy with a large external reservoir. Here the volume and temperature, rather than the pressure and temperature, are held fixed.

Is there some way we may determine from the second law of thermodynamics the equilibrium states of such systems even though they are not isolated? Certainly we may consider the composite system of the reservoir and the system of interest and require that the combined entropy be a maximum in the equilbrium state, but this would seem to require a knowledge of the entropy of the reservoir. Generally we do not know this, and it does not seem to make much difference what the reservoir is anyhow. Remarkably, we shall be able to show that it is possible to find the equilibrium states of such systems by determining the minimum of a new thermodynamic function which depends only on the system of interest and on the temperature and pressure of the reservoir. It will make no difference what the reservoir is. The only thermodynamic properties of the reservoir that we shall require are its temperature and pressure (or in the case of reactions taking place at fixed volume, only the temperature).

16.1 GIBBS FREE ENERGY

Let us consider first the problem of determining the equilibrium state of a system whose temperature and pressure are controlled by a reservoir. The system may have several

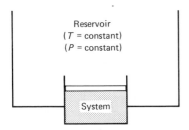

Reservoir
(T = constant)
(P = constant)

System

FIGURE 16.1   A system whose pressure and temperature are held fixed by a large reservoir.

components, but it is represented in Fig. 16.1 by a single component.

The equilibrium state of the total system, reservoir plus interacting system, is determined by the condition that the total entropy be a maximum:

$$dS_{tot} = 0$$

Now the total entropy is the sum of the entropies of the reservoir and the system,

$$S_{tot} = S_{res} + S_{sys}$$

so that

$$dS_{res} + dS_{sys} = 0$$

Remember now that dS represents the difference in entropy between the equilibrium state and any neighboring state that is compatible with the constraints. The change in entropy of any system can be written

$$dS = \frac{dQ}{T}$$

so for the reservoir

$$dS_{res} = \frac{dQ_{res}}{T}$$

where dQ_{res} is the heat that is added to the reservoir to produce the change in state. This heat could only have come from the system, so that

$$dQ_{res} = -dQ_{sys}$$

where $-dQ_{sys}$ is the heat *removed* from the system. From the definition of heat energy,

$$dQ_{sys} = dE_{sys} + P \, dV_{sys}$$

Putting these last five equations together, we see that equilibrium is reached when

$$T \, dS_{sys} - dE_{sys} - P \, dV_{sys} = 0 \qquad (16.1)$$

and we see that each term refers *to the system alone.* We can determine the equilibrium state from a knowledge of the entropy, internal energy, and volume of the system. The nature of the reservoir enters only through its temperature and pressure.

If we define a new function G, called the *Gibbs free energy,* by the relation

$$G = E + PV - TS \qquad (16.2)$$

the change in G when P and T are fixed is

$$dG = dE + P \, dV - T \, dS$$

Comparing dG with Eq. 16.1, we see that if P and T are fixed the condition for equilibrium becomes very simply

$$dG_{sys} = 0 \qquad (16.3)$$

It can easily be shown that an increase in total entropy corresponds to a decrease in the Gibbs free energy. (If we had written $dS_{tot} \geq 0$, we would have obtained $dG_{sys} \leq 0$.) Therefore the condition for thermodynamic equilibrium of an *isolated* system is that the *entropy* be a maximum, and the condition for a *system in contact with a constant-temperature, constant-pressure bath* is that the *Gibbs free energy* be a minimum. Just as no process is possible in an isolated system that decreases the entropy, so no process is possible in a system whose pressure and temperature are fixed that increases the Gibbs free energy. These are equivalent statements.

We shall reserve applications of this principle until we have discussed systems in which the molar numbers are not fixed.

16.2 HELMHOLTZ FREE ENERGY

Consider next the example of a system whose temperature and volume are held fixed (Fig. 16.2). Is there some thermodynamic function of the system alone whose maximum or minimum determines the equilibrium state? Let us proceed as before and apply the second law of thermodynamics to the total system:

$$dS_{tot} = dS_{res} + dS_{sys} = 0$$

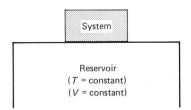

FIGURE 16.2   A system whose volume and temperature are fixed.

Again

$$dS_{res} = \frac{dQ_{res}}{T} = \frac{-dQ_{sys}}{T}$$

Also

$$dQ_{sys} = dE_{sys} + P\, dV_{sys} = dE_{sys}$$

since now the volume of the system is fixed. Therefore

$$\frac{-dE_{sys}}{T} + dS_{sys} = 0$$

or

$$T\, dS_{sys} - dE_{sys} = 0$$

when the system is in equilibrium. If we define a function

$$F = E - TS$$

called the *Helmholtz free energy,* we have, if $dT = 0$,

$$dF = dE - T\, dS$$

so that in equilibrium

$$dF_{sys} = 0$$

Thus the condition that

$$dS_{tot} = 0$$

is completely equivalent to

$$dF_{sys} = 0$$

if $dT = dV = 0$.

To summarize, we have shown that the universal condition that the total entropy be a maximum for equilibrium for an isolated system is completely equivalent to the condition that the Gibbs or Helmholtz free energies be a minimum for the system in contact with a reservoir, depending on whether the temperature and pressure are fixed, or whether the temperature and volume are fixed.

SUMMARY

The Gibbs free energy of a system is defined by the equation

$$G = E + PV - TS$$

The Gibbs free energy is a minimum in the steady state for any system whose temperature and pressure are fixed.

The Helmholtz free energy is defined by the equation

$$F = E - TS$$

The Helmholtz free energy is a minimum in the steady state for any system whose temperature and volume are fixed.

PROBLEMS

16.A.1 If $S = nR \ln VE^{3/2}$, what are F and G as functions of E, V, and n?

16.A.2 A vessel of fixed volume and temperature is divided into two equal compartments by a membrane that is permeable to a certain molecule. The Helmholtz free energy at this temperature and volume is given by $F = -2n_1^{1/2} - 6 \ln n_2$, when n_1 and n_2 are the molar numbers of this molecule on each side of the partition. If $n_1 + n_2 = 8$ mol, what are n_1 and n_2 in thermodynamic equilibrium?

16.B.1 If $S = (nEV)^{1/3}$, what are $F(n, V, T)$ and $G(n, P, T)$?

16.B.2 Show that F and G are minima when S_{tot} is a maximum.

16.B.3 If $S = nR \ln VE^{3/2}$, what are F as a function of n, V and T, and G as a function of n, P, and T?

16.B.4 A vessel is divided into two parts by a movable partition. Each compartment contains an ideal gas with 2 mol on one side and 4 mol on the other. The total Helmholtz free energy is $F = 3RT - 2RT \ln V_1(2T)^{3/2} + 6RT - 4RT \ln V_2(4T)^{3/2}$. The temperature is held fixed, as is the total volume of 6 m³. What is the equilibrium state, i.e., the position of the partition?

*17 VARIATION OF MOLAR NUMBERS

SO FAR WE have restricted ourselves to situations in which the molar concentrations do not change. In a great many problems, however, this is not the case.

1. Whenever chemical reactions are possible, the molar numbers of the chemical elements or compounds can vary.

2. Whenever particles can diffuse across a membrane, the concentration on each side can vary.

3. Whenever the substance can exist in two (or more) phases simultaneously (e.g., ice-liquid or liquid-steam), the number of moles of each phase can vary.

4. Whenever a gas and a liquid exist in thermal equilibrium, the number of moles of gas absorbed into the liquid can vary. This also applies to the adsorption of gas onto the surface of a solid.

5. Whenever we have a gas in a gravitational field, the number of moles at any given height can vary. The same condition applies to solutes dissolved in a liquid, in a gravitational field, or in a centrifuge, or to ions in an electric field.

6. Even for a pure substance, say a hydrogen gas, many different components are possible. There will be H atoms in the lowest energy state, H atoms in various excited energy states, H_2 molecules in the various energy states, H^+, and electrons. The number of moles of each is variable.

In all these cases, and many more, the number of moles of the various constituents will adjust themselves so that the entropy is a maximum. If we can find the entropy as a function of the molar concentrations, we shall be able to predict the distribution of the various constituents. For example, in a hydrogen and oxygen gas we shall be able to predict the relative proportions of H_2, O_2, and H_2O molecules. If we can determine the entropy for all possible distributions of a gas in a gravitational field, we will obtain the equilibrium distribution by searching for that distribution that has the greatest entropy, and thus we will be able to determine the density of the gas as a function of height in a gravitational field.

17.1 CHEMICAL POTENTIAL

Since we shall be looking for the distribution of molar numbers that maximize the entropy and since this maximum is attained when any further change in molar numbers produces no further change in entropy, we must consider how the entropy changes as we change molar numbers. Suppose the number of moles of the first constituent is n_1, the number of the second n_2, etc. (For example, for a gas of hydrogen molecules, oxygen molecules, and water molecules in equilibrium, n_1 might be the number of moles of H_2, n_2 the number of moles of O_2, and n_3 the number of moles of H_2O.) If n_1 is changed by the amount dn_1, then the change in the entropy dS associated with this change in n_1 alone is given by

$$dS = \frac{\partial S}{\partial n_1} dn_1$$

If we permit both n_1 and n_2 to change, then the change in the entropy is given by

$$dS = \frac{\partial S}{\partial n_1} dn_1 + \frac{\partial S}{\partial n_2} dn_2$$

For the general case in which all the variables on which the entropy depends can change, the change in the entropy is given by

$$dS = \frac{\partial S}{\partial E} dE + \frac{\partial S}{\partial V} dV + \frac{\partial S}{\partial n_1} dn_1 + \frac{\partial S}{\partial n_2} dn_2$$

$$+ \frac{\partial S}{\partial n_3} dn_3 + \cdots \qquad (17.1)$$

We have previously defined

$$\frac{\partial S}{\partial E} = \frac{1}{T}$$

$$\frac{\partial S}{\partial V} = \frac{P}{T}$$

We now add to these definitions the *chemical potential*, which is defined by

$$\frac{\partial S}{\partial n_1} = -\frac{\mu_1}{T} \qquad (17.2)$$

This defines the chemical potential of the first species. The chemical potentials of the second species, third species, etc., are defined in a similar way.

With these definitions we may write for Eq. 17.1

$$ds = \frac{dE}{T} + \frac{P}{T} dV - \frac{\mu_1}{T} dn_1 - \frac{\mu_2}{T} dn_2 \cdots$$

or

$$dS = \frac{dE + P\,dV - \mu_1\,dn_1 - \mu_2\,dn_2 \cdots}{T} \qquad (17.3)$$

For the moment we must regard the chemical potential as a definition related to the derivative of the entropy with respect to n. This is exactly how the temperature and pressure were first introduced. It was only after we examined some familiar experiments that we were able to understand the significance of P and T. We will pursue this same course to give some meaning to the concept of chemical potential.

First we would like to show that the chemical potential of a particular species is related to the change in internal energy associated with an increase in the number of moles of that species in the system. We have seen earlier, when we could ignore changes in molar numbers, that heat was defined by the relation

$$dQ = dE + P\,dV$$

or

$$dE = dQ - P\,dV \qquad (17.4)$$

that is, the increase in the total internal energy is equal to the heat energy added minus the energy expended by the system in expansion. This does not include the possibility of energy being released or stored in chemical reactions

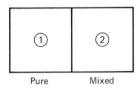

Pure   Mixed

FIGURE 17.1 Mixed and pure systems in thermal equilibrium.

or, more generally, of any change in composition. To see how this equation is modified when the molar numbers can change, consider two systems in thermal contact, that is, two systems that can exchange heat energy (Fig. 17.1). One of the systems contains a pure substance (so the number of moles cannot change), and the other consists of two or more different species whose abundance may or may not change. When the total system is in equilibrium, the total entropy must be a maximum with respect to all fluctuation compatible with the constraints. Therefore

$$dS_{tot} = dS_1 + dS_2 = 0$$

Consider a small change compatible with the constraints— an exchange of heat between the two systems. Now we have seen that for a pure system

$$dS_1 = \frac{dE_1 + P_1 dV_1}{T_1} = \frac{dQ_1}{T_1}$$

Therefore

$$\frac{dQ_1}{T_1} + dS_2 = 0$$

but

$$dQ_1 = -dQ_2$$

that is, the heat added to system 1 is equal to the heat taken from system 2. Also $T_1 = T_2$ since the systems are in thermal contact. We have then

$$dS_2 = \frac{-dQ_1}{T_1} = \frac{dQ_2}{T_1} = \frac{dQ_2}{T_2}$$

or

$$dS_2 = \frac{dQ_2}{T_2}$$

exactly as for the pure system. We see then that the relation

$$dS = \frac{dQ}{T}$$

is valid quite generally; i.e., the change in entropy is always equal to the heat added divided by the temperature, even in cases in which there are variations of molar numbers.

Let us combine the relation

$$dS = \frac{dQ}{T}$$

with our previous expression for the change in entropy for a general system in terms of the change in the energy, volume, and molar numbers,

$$dS = \frac{dE + P\,dV - \mu_1\,dn_1 - \theta_2\,dn_2}{T}$$

Eliminating dS, we have

$$dQ = dE + P\,dV - \mu_1\,dn_1 - \mu_2\,dn_2 - \cdots$$

or

$$dE = dQ - P\,dV + \mu_1\,dn_1 + \mu_2\,dn_2 + \cdots \tag{17.5}$$

This is the generalization of Eq. 17.4 for the case in which the molar numbers might change. It states that the increase in the total internal energy is equal to the heat added minus the work done by the system in expansion plus the additional terms $\mu_1\,dn_1 + \mu_2\,dn_2 + \cdots$. These extra terms must be the contribution to the increase in total energy associated with the change in molar numbers by the amount dn_1, dn_2, etc. The chemical potential μ_1 is therefore the change in energy per mole of the first reactant. If dn_1 moles of the first reactant are produced, then the change in the internal energy will be $\mu_1\,dn_1$.

17.2 PARTICLE EXCHANGE ACROSS A SURFACE

Earlier we consider examples of systems that interact by exchanging energy and volume. We saw that two systems that can exchange energy will ultimately reach the same temperature, and two systems that can exchange volume will reach the same pressure. We shall now consider the case of two systems that can exchange particles (Fig. 17.2). This exchange may take place across a liquid-vapor interface, a geometrical surface separating two domains, a biological membrane such as a cell wall separating the inside of the cell from the outside, blood vessel walls which allow nutrients and waste to pass, etc. This surface or membrane may allow some types of particles to pass but not others. Perhaps small molecules may permeate but not large molecules. The entropy of each part of the system will be a function of the molecular composition of that part, and therefore the entropy will depend on the relative proportions of those molecules that can cross the surface. By maximizing the total entropy, we shall try to determine the condition for thermodynamic equilibrium. This condition should determine the relative abundance of the permeating particle on either side of the surface.

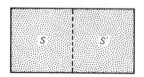

FIGURE 17.2 Particle exchange between two systems.

Let us assume that the entire system is in equilibrium and see what distribution maximizes the entropy. Let the entropy of the left side of the system be S and that of the right side S'. Let n_1 and n_1' be the number of moles on the left and the right of some molecule that is free to cross the surface. The total entropy is

$$S_{tot} = S(E, V, n_1, n_2 \ldots) + S'(E', V', n_1', n_2' \ldots)$$

where n_2, n_3, etc., are the number of moles of other constituents that may or may not be free to cross the surface. Consider a small change in n_1 and n_1'. The associated change in entropy is

$$dS_{tot} = dS + dS' = \frac{\partial S}{\partial n_1}dn_1 + \frac{\partial S'}{\partial n_1'}dn_1'$$

$$= -\frac{\mu_1}{T}dn_1 - \frac{\mu_1'}{T'}dn_1' \qquad (17.6)$$

where dn_1 is the increase in the number of moles on the left and dn_1' is the increase in the number of moles on the right. Clearly

$$dn_1 + dn_1' = 0$$

since no molecules are lost. We also know that the temperatures of both systems must be equal,

$$T = T'$$

and therefore we may write Eq. 17.6 as

$$dS_{tot} = \frac{\mu_1 - \mu_1'}{T}dn_1'$$

Since the total entropy is a maximum, the change caused by this small variation must be zero. Thus

$$S_{tot} = 0$$

so that

$$\mu_1 = \mu_1'$$

or the chemical potential of this molecule is the same on both sides. When we have a membrane (or surface) across which a type of molecule (or a number of types) can pass, the chemical potential of that molecule must be the same on both sides of the membrane.

We have seen that whenever two systems can exchange *energy,* their *temperatures* will be equal; and whenever two systems can exchange *volume* (i.e., there is movable piston between them), their *pressures* must be equal. We have just shown that whenever two systems can exchange *particles,* their *chemical potentials* must be equal.

17.3 CHEMICAL POTENTIAL OF IDEAL GASES AND DILUTE SOLUTES

So far we have been able to deduce our results from first principles because we have dealt with results valid for general systems. We have just shown, for example, that two components of a system that can exchange particles must have the same chemical potential. But this result does not tell us how that chemical potential depends on the thermodynamic state of the system. Different substances will have different chemical potentials, just as different substances will have different entropies or pressures or temperatures or specific heats. All these properties can in principle be deduced from statistical mechanics, where methods are available for counting the number of microstates for a given macrostate in terms of the fundamental properties of the interacting molecules. With these methods it is possible to determine the dependence of the chemical potential on the concentration for an ideal gas and a dilute solute. We will quote some of the results below.

IDEAL GAS

For an ideal gas (i.e., a gas for which we can neglect the interaction between molecules) the chemical potential is of the form

$$\mu = RT \ln c + \mu_0 \qquad (17.7)$$

where $c (= n/V)$ is the concentration in moles per cubic meter and μ_0 is a function of temperature and the nature of the ideal gas. The quantity μ_0 will not be the same function for a H_2 gas and an O_2 gas, but in all cases it will depend only on temperature and not on the pressure or the concentration. Tables for μ_0 at various temperatures for different gases exist. If we know the temperature (e.g., a chemical reaction at room temperature, or body temperature), we can find μ_0.

There is a certain problem with dimensions in Eq. 17.7, and we may avoid this difficulty by writing the equation in the form

$$\mu = RT \ln \left(\frac{c}{1 \ \mathrm{mol/m^3}} \right) + \mu_0$$

Since we may only take logarithms of dimensionless quantities, it follows that in the application of this equation the concentration must be expressed in moles per cubic meter. In the future the $1 \ \mathrm{mol/m^3}$ is to be understood in the logarithm.

QUESTION

What is the chemical potential for an ideal gas whose entropy is given by

$$S = nR \ \ln \left[\frac{V}{n} \left(\frac{E}{n} \right)^{5/2} \right]$$

ANSWER

As we have seen, this is the entropy of some ideal gas. We may obtain the chemical potential from the definition (Eq. 17.2):

$$\frac{\mu}{T} = -\frac{\partial S}{\partial n} = -\frac{\partial}{\partial n} nR \ \ln \left[\frac{V}{n} \left(\frac{E}{n} \right)^{5/2} \right]$$

$$= -R \ \ln \left[\frac{V}{n} \left(\frac{E}{n} \right)^{5/2} \right] + \frac{7}{2} R \qquad (17.8)$$

We also have, from the definition of temperature,

$$\frac{1}{T} = \frac{\partial S}{\partial E} = \frac{\frac{5}{2}nR}{E}$$

or

$$E = \frac{5}{2}nRT$$

Substituting this value of E into Eq. 17.8, we have

$$\frac{\mu}{T} = R \ln\frac{n}{V} - R \ln(\tfrac{5}{2}RT)^{5/2} + \tfrac{7}{2}R$$

which is of the form

$$\mu = RT \ln c + \mu_0(T)$$

where $c = n/V$. In this instance

$$\mu_0(T) = -RT \ln(\tfrac{5}{2}RT)^{5/2} + \tfrac{7}{2}RT$$

DILUTE SOLUTE

The chemical potential of a dilute solute is very similar to that of an ideal gas. It is given by

$$\mu = RT \ln c + \mu_0$$

where μ_0 is now a function of the pressure, the temperature, and the nature of the solution. It must be emphasized that this is the chemical potential of the solute alone. The solvent will have an entirely different chemical potential.

CHEMICAL POTENTIAL IN A GRAVITATIONAL FIELD

In the above expressions for chemical potential we neglected the effect of gravity. Sometimes this is justifiable and sometimes not. We shall see shortly when we can neglect it.

Since the chemical potential is equal to the energy required to add 1 mol (Eq. 17.5), the chemical potential will be greater at greater heights because it takes more work to raise molecules to this greater height. This additional work is just the potential energy of 1 mol at a height z. Thus if m is the mass of 1 mol (in kilograms),

$$\mu = RT \ln c + mgz + \mu_0 \tag{17.9}$$

for an ideal gas and

$$\mu = RT \ln c + m'gz + \mu_0 \tag{17.10}$$

for a dilute solute, where m' is the *effective* mass of 1 mol, i.e., the real mass minus the mass of solvent displaced. The solvent exerts a buoyant effect on the solute. This effective mass may be positive or negative.

17.4 APPLICATIONS

We shall consider just a few applications of the use of chemical potential to determine concentrations in systems in thermodynamic equilibrium.

SEDIMENTATION IN GASES: THE BAROMETRIC EQUATION

The settling of particles in solution to the bottom is called *sedimentation*. If the particles in suspension are very heavy, most will accumulate very close to the bottom. If they are lighter, they will distribute themselves more uniformly, but will still be denser at the bottom than at the top.

Even in gases we find greater densities at lower levels. The earth's atmosphere is a prime example. The density of the atmosphere decreases steadily with increasing height. In fact, if the earth were scaled to the size of a basketball, 99% of the atmosphere would lie in a film 0.02 in thick (Fig. 17.3).

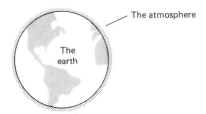

The atmosphere

The earth

FIGURE 17.3 The earth's atmosphere forms a very thin layer around the earth.

Let us see whether we can determine quantitatively how the density of a gas varies with height. Consider a gas in a closed vessel (Fig. 17.4), and imagine the vessel divided into horizontal layers. The gas in each layer can be considered a thermodynamic system. Each layer can exchange both particles and energy with adjacent layers. We have seen that two or more systems that can exchange energy must have the same temperature when equilibrium is reached. We have also seen that two or more systems that can exchange particles must have the same chemical potential. Thus the chemical potential must be the same for all layers in the gas. Therefore from Eq. 17.9

$$\mu = RT \ln c + mgz + \mu_0 = \text{constant}$$

But μ_0 is also a constant, so

$$RT \ln c + mgz = \text{constant}$$

where c is the molar density of the gas in the layer at the height z. This equation tells us that the quantity $RT \ln c + mgz$ has the same value at all points throughout the vessel.

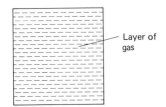

Layer of gas

FIGURE 17.4 A vessel of gas divided into horizontal layers.

If we call c_0 the concentration at $z = 0$ (that is, at the bottom of the vessel), then

$$RT \ln c + mgz = RT \ln c_0 + 0$$

or, solving for c,

$$c = c_0 \exp\left(-\frac{mgz}{RT}\right)$$

This is often called the *barometric equation*. We see that the concentration decreases exponentially with height, and the rate at which it decreases depends on the magnitude of the term multiplying z, that is, on the magnitude of mg/RT. If this quantity is very large, then the exponential will be small and the concentration will be small even for small values of z. In Table 17.1 we have listed values of c/c_0 for different gases at various atmospheric heights and at a temperature of 300 K.

TABLE 17.1 Values of c/c_0 for H_2 and O_2 at $T = 300$ K and various heights.

Gas	Height		
	1 m	*10 km*	*30 km*
H_2	0.999992	0.92	0.79
O_2	0.999871	0.28	0.023

QUESTION

Calculate the fractional decrease in density of O_2 at 300 K at a height of 10 km.

ANSWER

For O_2, $m = 32$ g/mol or 0.032 kg/mol, so that

$$\frac{c}{c_0} = \exp\left(-\frac{mgz}{RT}\right) = \exp\left[-\frac{(0.032 \text{ kg})(9.8 \text{ m/s}^2)(10,000 \text{ m})}{(8.3 \text{ J})(300)}\right]$$

$$= e^{-1.26} = 0.28$$

or the density at 10 km is 28% that at ground level.

Actually the earth's atmosphere is not at uniform temperature, so this calculation is only approximate.

We see from Table 17.1 that H_2 decreases in density much less rapidly than does O_2. We would expect to find a greater percentage of the lighter elements at high altitudes than the heavy elements. This is in fact the case.

We have seen that the larger the quantity mg/RT, the more rapidly the density decreases with height. There are three ways to make this quantity large: (1) by making m large, (2) by making g large, and (3) by making T small. If m is large, we find a rapid decrease in density with height. This is to be expected, since the heavier the particles, the more they tend to accumulate at the bottom.

If T is small, we also find a tendency for particles to accumulate at the bottom. When we cool a gas in a vessel, the particles tend to settle. If T is large, the thermal agitation is large and the particles tend to spread out.

It would seem that we have no control over g, the gravitational constant. We can, however, artificially increase g by placing the vessel in a centrifuge. The centrifugal force per unit mass is the *effective g*. If the settling action of gravity is too small to observe, the *ultracentrifuge* can greatly enhance the effect.

It is also true, of course, that g will be different on different astronomical bodies. A light planet will have difficulty holding an atmosphere.

SEDIMENTATION IN SOLUTIONS

In the preceding section we considered the sedimentation of gases. Of greater importance is the sedimentation of dissolved solutes in solution. Sedimentation techniques are used extensively to determine molecular weights and thus to identify molecules.

The primary difference between sedimentation in solutions and in gases is the buoyant effect of the solvent. If the solute is less dense than the solvent, it will tend to

rise to the top rather than to settle to the bottom. The concentration as a function of height is given by

$$\frac{c}{c_0} = \exp\left(-\frac{m'gz}{RT}\right) \tag{17.11}$$

where m' is the effective mass per mole of the solute, i.e., the true mass per mole minus the mass of solvent displaced. If the solution is centrifuged, then

$$\frac{c}{c_0} = \exp\left(-\frac{m'g'z}{RT}\right) \tag{17.12}$$

where g' is the effective gravitational constant.

REACTION EQUILIBRIUM

As a last example of how the molar numbers of the various constituents of a system adjust themselves to maximize the total entropy, let us consider reaction equilibrium. We shall begin with a very simple problem: the dissociation of molecular hydrogen H_2 into atomic hydrogen H. We shall then generalize the technique so that it may be applied to a wide class of problems in biology, engineering, geology, meteorology, physics, and other sciences.

Let us consider a box of hydrogen atoms. These atoms can interact to form hydrogen molecules, and when thermodynamic equilibrium is attained, the total number of atoms and molecules will be constant. As always, the steady state is dynamic on a microscopic level: there will be atoms combining to form molecules and molecules breaking up to form atoms. This dynamic equilibrium is represented by what is called a *reaction equation:*

$$2H \rightleftharpoons H_2 \tag{17.13}$$

The entropy of the system will depend on the number of atoms (n_H) and the number of molecules (n_{H_2}). (These are the molar numbers.) Let us maximize the entropy with respect to change in n_H and n_{H_2}. Now

$$dS = \frac{\partial S}{\partial n_H}dn_H + \frac{\partial S}{\partial n_{H_2}}dn_{H_2}$$

From the definition of chemical potential,

$$dS = -\frac{\mu_H}{T}\,dn_H - \frac{\mu_{H_2}}{T}\,dn_{H_2}$$

Since the entropy is a maximum in the equilibrium state, the entropy will neither increase nor decrease with this change in n_H and n_{H_2}. Therefore $dS = 0$, so that

$$\mu_H\,dn_H + \mu_{H_2}\,dn_{H_2} = 0 \qquad (17.14)$$

Now dn_H and dn_{H_2} are not independent. We must consider only those variations in the entropy that are compatible with the constraints. Since for every hydrogen molecule lost there is a gain of two hydrogen atoms,

$$dn_H = -2\,dn_{H_2}$$

(For purposes of deriving the more general relation to follow, it is more convenient to express this constraint in the following way: If there are M moles reacting, then

$$dn_H = 2M$$

and

$$dn_{H_2} = -M$$

Note that the multiplier of M is the coefficient in the reaction equation written in the form $2H - H_2 \rightleftharpoons 0$.)
We may now write Eq. 17.14 as

$$-2\mu_H + \mu_{H_2} = 0$$

or

$$2\mu_H = \mu_{H_2}$$

This is the *fundamental equation* for reaction equilibrium between H and H_2. It determines uniquely the relative abundances of hydrogen atoms and molecules in thermodynamic equilibrium.

Before we can apply the fundamental equation for reaction equilibrium, we must know how the chemical potentials depend on the thermodynamic state. In much the

same way, the condition that determines the equilibrium state of a mass suspended from the end of a spring is that the net force on the mass must be zero. To apply this condition we must know how these forces depend on the state.

If the reacting elements are ideal gases, we know (Eq. 17.7) that

$$\mu = RT \ln c + \mu_0$$

where μ_0 is some function of the temperature to be determined from tables.

QUESTION

At a temperature of 3000 K the chemical potential of 1 mol/m³ of H is 11,000 J/mol and the chemical potential of 1 mol/m³ of H_2 is 0 J/mol. (We have a certain amount of freedom in setting numerical values for the chemical potential. We are always adding or subtracting chemical potentials, just as we always add or subtract electric or gravitational potentials. We may always add a constant to the potential without affecting the result. In this instance the chemical potential of 1 mol/m³ of H_2 is chosen to be zero.) A 1 m³ vessel containing hydrogen is raised to a temperature of 3000 K, and it is observed that there are 2 mol of H_2 in the steady state. How many moles of atomic hydrogen are in the vessel in the steady state?

ANSWER

The reaction equation is

$$2H \rightleftharpoons H_2$$

and therefore the fundamental equation for reaction equilibrium is

$$2\mu_H = \mu_{H_2}$$

Assuming that both substances are ideal gases at this temperature, we have

$$\mu_H = RT \ln c_H + 11{,}000 \text{ J/mol}$$

and

$$\mu_{H_2} = RT \ln c_{H_2}$$

(Note that when $c_H = 1$ mol/m³, $\mu_H = 11{,}000$ J, and when $c_{H_2} = 1$ mol/m³, $\mu_{H_2} = 0$.)

We see then that

$$2RT \ln c_H + 22{,}000 \text{ J} = RT \ln c_{H_2}$$

or

$$RT \ln \frac{c_H^2}{c_{H_2}} + 22{,}000 \text{ J} = 0$$

or

$$\frac{c_H^2}{c_{H_2}} = \exp\left(-\frac{22{,}000 \text{ J}}{RT}\right) = \exp\left[-\frac{22{,}000 \text{ J}}{(8.3 \text{ J})(3000)}\right]$$

$$= 0.41 \text{ mol/m}^3$$

Now we were given that the concentration of H_2 is 2 mol/m³, so

$$c_H = \sqrt{(2 \text{ mol/m}^3)(0.41 \text{ mol/m}^3)} = 0.91 \text{ mol/m}^3$$

It is a straightforward matter to extend the preceding analysis to cover more general reactions. For example, if the reaction equation is

$$\nu_A A = \nu_B B$$

where A and B are two elements and ν_A and ν_B are integers, then the fundamental equation for reaction equilibrium is

$$\nu_A \mu_A = \nu_B \mu_B \tag{17.15}$$

QUESTION

We shall see later that atoms and molecules can exist in many different discrete states. The simplest atom, the hydrogen atom, has many states for the electron. There is a lowest energy state, called the ground state, in which the electron is tightly bound to the proton. Hydrogen can also exist in a slightly higher energy state, called the first excited state, in which the electron is less tightly bound. Many other excited states with even higher energy exist. The atom can make transitions from one state to another (giving off electromagnetic radiation), and such a transition can be considered a chemical

reaction. Thus if we denote the hydrogen atom in the ground state by H and the hydrogen atom in some excited state by H*, then we have the reaction

$$H \rightleftharpoons H^*$$

(We should include in this equation the electromagnetic radiation, but it can be shown that the associated chemical potential is zero and therefore can be omitted.) Many molecules of biological interest undergo similar transitions, called conformational deformation. A large molecule A changes its shape to some new state A*. There is a discrete difference in energy between these two shapes. In equilibrium we have

$$A \rightleftharpoons A^*$$

Let the energy difference between H and H* (or A and A*) be ϵ. What is the relative proportion in equilibrium?

ANSWER

We have seen that the chemical potential is equal to the work required to add one mole (see Sec. 17.1). If it takes an energy of μ_H to add one mole of H, then it must take $\mu_H + N_0\epsilon$ to add one mole of H*, since it takes an energy ϵ to raise one atom of H to its excited state H*. Thus if

$$\mu_H = RT \ln c_H + \mu_0$$

then

$$\mu_{H^*} = RT \ln c_{H^*} + N_0\epsilon + \mu_0$$

Now the fundamental condition for thermodynamic equilibrium for the reaction

$$H \rightleftharpoons H^*$$

is (see Eq. 17.15)

$$\mu_H = \mu_{H^*}$$

Therefore

$$RT \ln c_H + \mu_0 = RT \ln c_{H^*} + N_0\epsilon + \mu_0$$

or, solving for the relative concentrations, we have

$$\frac{c_{H^*}}{c_H} = \exp\left(-\frac{N_0\epsilon}{RT}\right) = \exp\left(-\frac{\epsilon}{kT}\right) \tag{17.16}$$

since $R = N_0 k$ (see Eq. 13.6).

QUESTION

The molecule glucose can exist in two different conformations, called the α and β conformations. The energy required to convert one mole of β glucose to α glucose is 1700 J/mol. What are the relative concentrations at 25°C?

ANSWER

From the preceding question we have for the reaction

α glucose $\rightleftharpoons$ β glucose

that the relative concentrations are given by

$$\frac{c_\alpha}{c_\beta} = \exp\left(-\frac{N_0\epsilon}{RT}\right) = \exp\left[-\frac{1700 \text{ J}}{(8.3 \text{ J})(273 + 25)}\right] = 0.5$$

If the reaction equation were a little more complicated, such as

$$\nu_{A_1}A_1 + \nu_{A_2}A_2 \rightleftharpoons \nu_{B_1}B_1 + \nu_{B_2}B_2$$

then the fundamental equation for reaction equilibrium would be

$$\nu_{A_1}\mu_{A_1} + \nu_{A_2}\mu_{A_2} = \nu_{B_1}\mu_{B_1} + \nu_{B_3}\mu_{B_3} \tag{17.17}$$

QUESTION

What is the fundamental reaction equilibrium equation for the reaction

$$2H_2 + O_2 \rightleftharpoons 2H_2O$$

ANSWER

From Eq. 17.17,

$$2\mu_{H_2} + \mu_{O_2} = 2\mu_{H_2O}$$

We may apply these results for reaction equilibrium to a wide variety of problems. We have already considered chemical equilibrium, molecular dissociation, and excita-

tion equilibrium. We may add to this list the following examples.

1. *Ionization* The ionization of hydrogen may be represented by the reaction equation

$$H \rightleftharpoons H^+ + e^-$$

where H^+ represents a proton and e^- an electron. Thermodynamic equilibrium is attained when

$$\mu_H = \mu_{H^+} + \mu_{e^-}$$

2. *Adsorption* Gas molecules in a vessel may attach themselves to the walls of the vessel, and this process is called *adsorption*. When the state of thermodynamic equilibrium is reached, there will be as many molecules attaching themselves to the walls as leaving the walls. If we let M(gas) represent a molecule in the gas and M(adsorbed) represent an adsorbed molecule, then the process may be represented by the reaction equation

$$M(gas) \rightleftharpoons M(adsorbed)$$

The reaction equilibrium equation is

$$\mu(gas) = \mu(adsorbed)$$

3. *Phase equilibrium* At the right temperature and pressure, we would find in a closed box of water molecules that some water molecules will bind together to form a liquid and some will remain separated in the gaseous phase above the liquid. In the steady state there is a dynamic equilibrium. Some water molecules in the gas will be combining with the liquid, and some molecules in the liquid will be evaporating into the gas above. We may represent this process by the reaction equation

$$H_2O(gas) \rightleftharpoons H_2O(liquid)$$

Equilibrium is attained when

$$\mu_{H_2O}(gas) = \mu_{H_2O}(liquid)$$

4. *Sedimentation*   We may also represent the process of sedimentation, studied earlier, by a reaction equation. If we denote the particles in a given layer of gas by P_1 and the particles in an adjacent layer by P_2, then the process of particles moving from one layer to another can be represented by

$$P_1 \rightleftarrows P_2$$

Equilibrium is attained when

$$\mu_1 = \mu_2$$

Since this is true for all adjacent layers, it follows that

$$\mu = \text{constant}$$

This is the equation we used in Sec. 17.4 to derive the barometric equation.

5. *Particle exchange across a surface or membrane*   Let us imagine a surface or membrane that separates a vessel into two parts, a left side and a right side. Let us denote particles on the left side by P_L and particles on the right side by P_R. If the particles can pass from one side to the other, we may represent this process by the reaction equation

$$P_L \rightleftarrows P_R$$

The steady state is attaned when

$$\mu_L = \mu_R$$

This result was derived earlier in Sec. 17.2.

SUMMARY

The chemical potential is defined by the relation

$$\frac{\partial S}{\partial n} = \frac{-\mu}{T}$$

If the energy of a system can increase only when work is done ($-P\,dV$), heat dQ is added and particles are added, then

$$dE = dQ - P\,dV + \mu\,dn$$

so that μ is the increase in energy per mole added to the system.

The chemical potential of any molecule that is free to permeate a membrane or surface must be the same on both sides of the membrane or surface.

The chemical potential of an ideal gas is given by

$$\mu = RT\ln c + \mu_0$$

where μ_0 is a function of the temperature alone.

The chemical potential of a dilute solute is given by

$$\mu = RT\ln c + \mu_0$$

where μ_0 is a function of the temperature and pressure of the solvent.

In a gravitational field we must add mgz to the chemical potential, where m is the effective mass in kilograms of one mole of the gas or solute.

The concentration of gas or dilute solute varies with height in a gravitational field according to the relation

$$c = c_0 \exp\left(-\frac{m'g'z}{RT}\right)$$

If the reaction equation is

$$\nu_{A_1}A_1 + \nu_{A_2}A_2 \rightleftarrows \nu_{B_1}B_1 + \nu_{B_2}B_2$$

then the fundamental equation for reaction equilibrium is

$$\nu_{A_1}\mu_{A_1} + \nu_{A_2}\mu_{A_2} = \nu_{B_1}\mu_{B_1} + \nu_{B_2}\mu_{B_2}$$

PROBLEMS

17.A.1 In the earth's gravitational field a 0.1% variation in the concentration of a sediment is observed over a distance of 10 cm. Suppose you want to put the sample in a centrifuge to obtain a 10% variation in concentration over a distance of 10 cm. What is the g' of the centrifuge?

17.A.2 The concentration of oxygen in the blood is about 1 mol/m³. Oxygen combines in the blood with myoglobin (Mb) to form oxymyoglobin (MbO_2) according to the reaction

$$Mb + O_2 \rightleftharpoons MbO_2$$

The chemical potential of 1 mol/m³ of Mb and O_2 is zero. The chemical potential of 1 mol/m₃ of MbO_2 is 5900 J/mol at body temperature (310 K). What fraction of the myoglobin molecules are oxygenated?

17.A.3 The energy of the first excited state of hydrogen is 1.63×10^{-18} J (10.2 eV) above the energy of the ground state. At a temperature of 6000 K (the temperature of the sun), what is the ratio of hydrogen atoms in this excited state to hydrogen atoms in the ground state?

17.A.4 What is the fundamental equation for reaction equilibrium for the following reactions?
(a) $H_2 + I_2 \rightleftharpoons 2HI$
(b) $2H_2 + O_2 \rightleftharpoons 2H_2O$
(c) $N_2 + 3H_2 \rightleftharpoons 2NH_3$
(d) $2NH_3 \rightleftharpoons N_2 + 3H_2$

17.B.1
(a) Determine the height at which the atmospheric density is one-tenth the density at sea level. (The mean molecular weight of the atmosphere is 29 g/mol.) Let the temperature be 300 K.
(b) What fraction is this of the earth's radius?
(c) If g on the moon is 1.7 m/s² and the moon had an atmosphere similar in composition and mean temperature to the earth's, at what height would the density fall to one-tenth the density at ground level?

(d) In the above calculations g was assumed to be constant throughout the atmosphere. Is this a reasonable approximation?

17.B.2 A dilute solute is centrifuged so that the concentraion at a height of 1 cm is 50% of the concentration at the bottom. The temperature is 300 K. To what temperature should the solution be cooled so that the concentration at 1 cm will decrease to 30% of that at the bottom?

17.B.3 What are the concentrations of N_2 at 1 m, 10 km, and 30 km relative to the sea-level concentrations in an isothermal atmosphere at 300 K?

17.B.4 Suppose you want to centrifuge a sample so that particles in suspension whose effective molecular weight is 64 kg/mol will come out as a sediment. Your objective is to obtain a concentration at 1 mm that is one-tenth the concentration at the bottom. If the temperature of the solution is 330 K, at what effective g' should you run the centrifuge?

17.B.5 A preliminary determination of the effective molecular weight of an unknown solute yields a value of 200 g/mol. The experiment was run at such a high g that the bulk of the sediment was confined to the bottom of the test tube. It would be preferable to spread the sediment out over a larger distance. If the temperature is 300 K, at what effective g should the centrifuge be run to obtain a decrease in density of 10% in 1 cm?

17.B.6 Determine the molecular weight of a substance whose density is 1.5 g/cm³ and whose concentration at a height of 5 cm in a test tube is 20% of the concentration at the bottom. The test tube has been in a centrifuge with an effective g' of 10,000g. The temperature is 300 K.

17.B.7 Three gases are in equilibrium at a temperature of 310 K. They react according to the equation $3A + 2B \rightleftharpoons C$. The chemical potentials are $\mu_A = RT \ln c_A + 1270$ J/mol, $\mu_B = RT \ln c_B + 2700$ J/mol and $\mu_C = RT \ln c_C + 6670$ J/mol. Determine the concentration of C if the concentration of A is 0.1 mol/m³ and the concentration of B is 0.2 mol/m³.

17.B.8 A 1-m³ vessel contains n_0 mol of HI at a low temperature in the crystalline state. The vessel is heated to 750 K, at which temperature the crystal vaporizes into H_2, I_2, and HI molecules. These react according to the reaction equation

$$H_2 + I_2 \rightleftharpoons 2HI$$

At 750 K, $\mu_{0,H_2} = \mu_{0,I_2} = 0$ and $\mu_{0,HI} = -12{,}000$ J/mol. It is observed that n_{HI} in the equilibrium state is 0.5 mol. What is n_0, the original number of moles of HI in the crystal?

17.B.9 The chemical potential of two substances A and B are given by $\mu_A = RT \ln c_A + 3000$ J/mol and $\mu_B = RT \ln c_B + 8000$ J/mol. Given that the substances react according to the equation $A \rightleftharpoons B$, what are the equilibrium concentrations at 350 K if the initial concentrations are $c_A = 1$ mol/m³ and c_B is zero.

17.B.10 At a temperature of 700 K the reaction $H_2 + I_2 \rightleftharpoons 2HI$ is at equilibrium when the densities (in moles per cubic meter) of the three chemical species are

H_2: 3.56×10^{-3}

I_2: 1.25×10^{-3}

HI: 15.6×10^{-3}

If 1-mol of pure HI is placed in an empty closed vessel of volume 1 m³ and heated to 700 K, what are the concentrations of H_2, I_2, and HI when equilibrium is reached?

17.C.1 A vessel whose volume is 2 m³ is divided in two by a semipermeable membrane that will pass H_2 but not I_2 or HI. Suppose there was initially 1 mol of H_2 on the left side of the vessel and 1 mol of I_2 on the right side. If there were no chemical reaction, the equilibrium state would be achieved when there is an equal distribution of H_2 on both sides (that is, $\frac{1}{2}$ mol). But there is a chemical reaction

$$H_2 + I_2 \rightleftharpoons 2HI$$

so that many of the hydrogen atoms on the right will become bound in the form of hydrogen iodide (see the figure). (This is analogous to the manner in which many nutrients are drawn into biological cells. If something enters the cell and then takes part in a reaction, this

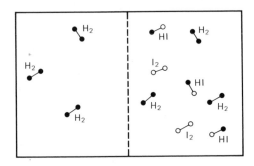

depletes its concentration inside and thus will stimulate further diffusion of the element into the cell.) In the equilibrium state there must be equal concentrations of H_2 on both sides. Determine this concentration. The temperature is 750 K and $\mu_{0,H_2} = \mu_{0,I_2} = 0$ and $\mu_{0,HI} = -12{,}000$ J/mol.

17.C.2 Three gases A, B and C are in equilibrium in a vessel whose volume is 1 m³ at a temperature of 300 K. The gases react according to the equation

$$A + B \rightleftharpoons C$$

and the chemical potentials are of the form

$$\mu_A = RT \ln c_A + \mu_{0,A}$$

$$\mu_B = RT \ln c_B + \mu_{0,B}$$

$$\mu_C = RT \ln c_C + \mu_{0,C}$$

where $\mu_{0,A} = 1270$ J/mol, $\mu_{0,B} = 2700$ J/mol, and $\mu_{0,C} = 6670$ J/mol.

(a) If there was initially 1 mol of C and no A or B, determine the equilibrium concentrations of A, B, and C.

(b) If an additional 0.2 mol of A is added to the vessel (and the temperature kept at 300 K), what are the new concentrations of A, B, and C?

WE HAVE BEEN concerned primarily with the problem of determining the equilibrium state of a thermodynamic system. We have seen that the equilibrium state of an isolated system is determined by the condition that the entropy of the equilbrium state is larger than the entropy of any other state compatible with the constraints, that is,

$$\Delta S = S_{equilibrium} - S_{other} \geq 0$$

The quantity ΔS does not represent a change in entropy associated with some process; it represents a comparison between the entropy of two possible macrostates. (To be sure of this, we must have some way of making the comparison, and we do this by taking the system through a sequence of equilibrium states in a quasistatic manner and evaluating dQ/T or $(dE + P\,dV)/T$ along the path. This path is used as a calculational tool and is not to be confused with the actual time-dependent process by which a system changes from one state to another.)

We would not like to consider the question of the *time evolution* of thermodynamic systems, and would like to show that the second law of thermodynamics can be used to determine the direction of thermodynamic processes in time. What happens, for example, when a hot body and a cold body are brought together? What happens when hydrogen and oxygen are brought together? What happens when a candy bar and a child are brought together (what chemical reactions take place within the candy bar and the human metabolism)?

Consider the following simple example. A vessel is divided into two compartments that are separated by a piston fixed in place by pegs. Both compartments are in equilibrium. The pegs are now pulled out and the piston is free to move (see Fig. 18.1). Which way will it go? Before the pegs are removed the entropy is at a maximum value compatible with the constraints. Once the constraints are removed the entropy is no longer at a maximum, and it will go to a new maximum. What can we say about this new maximum entropy? Will it be larger or smaller than the old maximum?

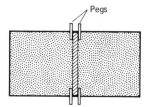

FIGURE 18.1   A vessel divided in two by a constrained piston.

Immediately after the pegs are pulled out the system is in a well-defined macrostate. Let us call the entropy of this macrostate S. (This is the same as the entropy just before the pegs are pulled out, since the macrostate has not changed in this short time.) When the system comes to its final equilibrium state the new entropy is S'. Now from the second law of thermodynamics,

$$\Delta S = S' - S \geq 0$$

since both S and S' refer to the same system with the same constraints (pegs removed). Note that S' is the entropy of the equilibrium state and S is the entropy of another macrostate compatible with the constraints. But S is also the entropy before the pegs are pulled out. Therefore the equilibrium entropy increases (or remains constant) in time when the constraint is removed.

We could have applied this argument to any system and any constraint. We therefore have a new principle, deduced from the second law of thermodynamics, which states that the entropy of any isolated system will increase (or remain constant) in time if a constraint is removed,

$$\Delta S \geq 0$$

where the change ΔS now refers to a change with time. We might say that this is a thermodynamic principle as opposed to our earlier thermostatic principle.

The physical basis behind this principle is the same as that behind the second law itself: Any macroscopic system will seek that macrostate that has the most microstates associated with it. The entropy is a measure of the number of associated microstates. When a constraint is relaxed, *many new macrostates* become available to the system. Some of these will have more associated microstates, and therefore the system will evolve to a state of larger entropy.

As a simple example, consider four particles that initially can be located in only one of two compartments in a box, called compartments A and B. A wall prevents the particles from occupying compartments C and D (see Fig. 18.2). The macrostate with the most associated microstates is that in which there are two particles in compartment A and two in compartment B. Six microstates are associated with this macrostate (count them), and this is the initial equilibrium state.

FIGURE 18.2   When the four particles are constrained to compartments A and B, the steady state is that in which two particles are in A and two particles are in B.

Suppose the dividing wall is removed so that compartments C and D are accessible. Now the macrostate with the most associated microstates is that in which there is one particle in each box (Fig. 18.3). Twenty-four microstates are associated with this macrostate (count them), and this is the new equilibrium state.

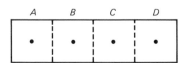

FIGURE 18.3   When the constraint is relaxed, the steady state is that in which one particle is in each compartment.

Since the entropy of a macrostate is a measure of the number of associated microstates and since the number of associated microstates has increased from 6 to 24, it follows that the entropy of the final equilibrium state is greater than the entropy of the initial equilbrium state.

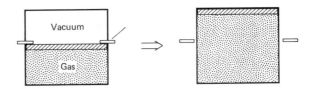

FIGURE 18.4   Free expansion of a gas.

18.1 APPLICATIONS

Let us see what sort of problems we can apply this principle to.

1. As a first example let us consider *free expansion*. A gas is confined in a vessel and restricted to the lower half of the vessel by a piston held in position by pegs. (For simplicity we assume that the piston is of negligible weight.) The pegs are pulled out and the piston is pushed rapidly to the top of the vessel by the gas (see Fig. 18.4). We should be able to demonstrate that the entropy of the final state is greater than the entropy of the initial state, and we must do this using only the second law of thermodynamics. We may not say that there are more microstates for the expanded gas and thus a greater entropy, since we really do not know how to count microstates. (We shall see presently that we have in fact ignored entirely the contribution to the microstates associated with the velocity distribution of the molecules.) This microscopic viewpoint is vital to our understanding of the conceptual basis for thermodynamics; however, if thermodynamics is to be a self-contained discipline, it must stand on the first and second laws alone. We should be able to use these laws to *prove* that the entropy of final state is larger than the entropy of the initial state.

Let us represent the macrostate of the system by a point in an energy-volume coordinate system (as the first law permits). We assume that there are no reactions so the n's are fixed. We cannot represent the actual transition from the initial state to the final state by a line in this energy-volume coordinate system since a line represents a sequence of points and each point represents a thermodynamic state. The system moves to the final state in a rather chaotic manner.

We have two states at the same energy but at different volumes (see Fig. 18.5), and we wish to show that the entropy of the final state is greater than the entropy of the initial state. We know that the entropy is characteristic of the state and independent of how the system got to this state. We also know that the change in entropy dS associated with a small change in energy dE and a small change in volume dV is given by

$$dS = \frac{dE + P\,dV}{T} \tag{18.1}$$

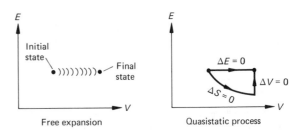

FIGURE 18.5   Representation of the initial and final states in an energy-volume coordinate system. Note that the energy of the initial state is the same as the energy of the final state since no work was done by or on the system.

This gives us the change in entropy between *any* two *neighboring* thermodynamic states. Since the total entropy change $S_f - S_i$ is independent of the process, where the subscripts refer to "final" and "initial," let us invent a process (any process) that takes us quasistatically from one state to the other. If this is done very slowly, we *can* represent the process by a line in the energy-volume coordinate system, and this line represents a sequence of equilibrium states. The change in entropy dS between any two neighboring states is given by Eq. 18.1. The total change in entropy is

$$S_f - S_i = \int dS$$

where $\int dS$ is the sum of the changes in entropy along the path. The simplest path to choose is a straight line connecting the two points. On this line $dE = 0$, so

$$dS = \frac{P \, dV}{T}$$

where P and T will change continuously along the line. Since P, T, and dV are positive, dS is positive. If all the dS's are positive, then $S_f - S_i$ must also be positive, so we have proved that S_f is indeed greater than S_i.

This same transition might be executed in practice by a controlled expansion. Each time the piston is moved up a short distance, heat energy must be added to keep the energy of the gas constant, for the gas is losing energy by pushing on the receding piston. We may compute the entropy change by observing that $dS = dQ/T$ and dQ is positive throughout the process. So once again the entropy increases.

QUESTION

Initially the system was isolated and the pegs were pulled out. No heat was added. If $dS = dQ/T$, how is it that the entropy increased? We proved that the entropy increased when we added heat as the piston moved up slowly. Since we added heat the system cannot be isolated. The calculation for the quasistatic process seems to have nothing to do with the real process.

ANSWER

The formulas

$$dS = \frac{dE + P \, dV}{T}$$

and

$$dS = \frac{dQ}{T}$$

apply to *neighboring states in a quasistatic process*, and not to some chaotic transition. The fact that no heat was added in the sudden free expansion does not imply that $dS = 0$. To calculate entropy changes we note that (1) the entropy depends only on the states, (2) entropy changes can be *calculated* only for *quasistatic processes*, and (3) all quasistatic processes give the same result. Therefore, to determine the entropy change between any two states we may choose any quasistatic process that connects the states. This process does not necessarily resemble the actual process.

---

2. As a second example, and one that will be useful to us later when we discuss the increasing entropy of the universe, let us consider a vessel with a piston for a lid. A weight W is on the piston. Initially the piston is held fixed in position by pegs. When the pegs are removed the piston will move up or down, depending on whether the weight is heavy or light (see Fig. 18.6). We wish to show that in *either* event the entropy increases. Let us devise a quasistatic process that connects the initial and final states. And let us control the movement of the piston so that it moves very slowly to the final state. If we control the movement of the piston, the piston is doing work on us and the system must be losing energy. But in the actual process the total system is isolated, so the net energy must remain constant (internal energy of the gas plus the gravitational energy of the weight). To prevent the system from

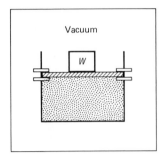

FIGURE 18.6 Expansion or contraction of a confined gas.

losing energy, we must add heat energy to the gas. But if we do this the entropy must increase. In this quasistatic process—controlling the motion of the piston and adding heat energy to make up for the energy lost—the entropy increases continuously.

In this discussion we have been careful not to say whether the piston moves up or down. Whichever way it moves, we must push against it to *control* the movement, and thus work is done on us and we must restore this energy in the form of heat energy. The entropy increases regardless of the direction of motion. If the weight is light, the piston moves up, but if it is heavy, the piston moves down. In either case the entropy increases.

QUESTION

Can an object use its internal energy to spring spontaneously from the floor to a table top?

ANSWER

If we can show that the entropy of the system decreases in the process, then it is impossible for the object to do this. (The system must, of course, be an isolated system. We might take the room and its contents as the system.) To evaluate the entropy change, we must take the system quasistatically to the final state. First we slowly raise the object to the table top.

Since no heat is added, the entropy change in this process is zero. The object, however, does not have the proper internal energy. (When it jumps to the table top spontaneously, it uses internal energy to gain gravitational energy.) To reduce the internal energy to the value for the final state, we must remove heat energy, so the entropy decreases. Thus the entropy of the final state is less than the entropy of the initial state, and the second law is violated. The object cannot spring from the floor to the table top.

QUESTION

It would be very useful to have a filter that would allow a certain type of molecule to pass in one direction but not in the opposite direction. We could use this filter to separate certain molecules from a mixture, or a biological system could use such filters (membranes) to allow nutrients into a cell but not allow them to pass back out. The engineer could use a filter such as this to build an engine that takes its energy from the air or sea water.

To see how this engine might work, imagine a cylinder filled with a gas for which there is a one-way filter. Inserted in the filter is a valve that may be opened or closed (see Fig. 18.7). The cycle would proceed as follows.

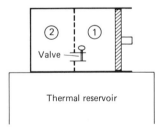

FIGURE 18.7 A four-cycle engine using a one-way filter. The filter will allow particles to pass from left to right but not from right to left.

1. The valve is closed and the molecules crowd into region 1. (The filter allows molecules to pass from left to right, but not from right to left.)

2. The piston is allowed to move out, doing work on some outside body.

3. The valve is opened, and the pressure in region 1 drops.

4. The piston is returned to its original position. (Since the pressure is lower, the work required to return the piston to its position is less than the work done by the piston on expansion.) The valve is closed again and the process repeated. The cylinder rests on a thermal reservoir which keeps the temperature constant. Thus we have created a nonpolluting, four-cycle engine that feeds on the thermal energy of the reservoir. This reservoir might be the ocean.

ANSWER

To see that the existence of such a one-way filter is impossible, imagine a vessel divided into two parts by a filter that allows certain particles to pass in only on direction. Initially the gas fills the vessel uniformly, and in the final state the gas is crowded into one side of the vessel. Is the entropy of the final state greater than that of the initial state? Let us take the system to this same final state by a quasistatic process and measure the entropy change. We may do this by compressing the gas into the restricted region by pushing in on a piston (Fig. 18.8). But in pushing the piston we have done work on the gas, and this increases the internal energy of the gas. Since the gas was

isolated in the original process, its internal energy could not change. Therefore we must extract heat from the gas to restore the original internal energy. In pushing the piston in there is no change in entropy, but the entropy decreases when heat is removed. The entropy of the system decreased in violation of the second law, and therefore it is impossible to make a one-way filter.

QUESTION

In the preceding question, we saw that we could not use a one-way filter to extract heat from a reservoir in a cyclic process and convert this energy into useful work. Is there some other device that would accomplish this objective?

ANSWER

Let us suppose there is such a device. It extracts heat from a reservoir, does work, and returns to its original state ready to repeat the cycle (see Fig. 18.9). Consider the change in entropy of the component parts of this system. First, the work done may be used to raise a weight. The entropy of the weight does not change, since the internal energy, volume, and molar numbers of the weight are not changed. (The internal energy does not include gravitational energy.) The entropy of the device itself does not change, since the device is operating in a cycle, i.e., it returns to its original state after each cycle. The reservoir, however, clearly loses entropy, since heat is removed from the reservoir. The net effect has been a decrease in entropy in violation of the second law of thermodynamics.

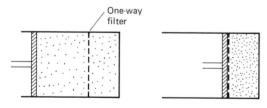

FIGURE 18.8 By pushing in the piston one may compress a gas to a restricted volume.

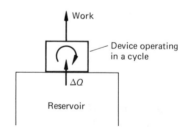

FIGURE 18.9 Heat ΔQ is taken from a reservoir by a device operating in a cycle that does work.

3. As a last example let us consider two bodies, one hot and one cold, brought into thermal contact. Eventually the bodies reach some intermediate temperature. We should be able to show that the entropy increases in this process.

Let us imagine a quasistatic process in which a small amount of heat dQ is taken from the hot body and added to the cold body. The change in entropy of the cold body is

$$\Delta S_{cold} = \int \frac{dQ}{T}$$

where the integral is over those dQ's necessary to raise the temperature of the cold body to the final intermediate temperature. Since the dQ's are positive, this represents an increase in entropy.

The change in entropy of the hot body is

$$\Delta S_{hot} = \int \frac{dQ}{T}$$

where the sum is over those dQ's necessary to lower the temperature of the hot body to the final intermediate temperature. Since heat is removed, the dQ's are negative and the entropy of the hot body decreases. We must show that the decrease in entropy of the hot body is less than the increase in entropy of the cold body.

At any stage of the process the increase in entropy of the cold body is

$$\frac{dQ}{T_{cold}}$$

and the decrease in entropy of the hot body is

$$\frac{dQ}{T_{hot}}$$

Since T_{hot} is at all times greater than T_{cold}, it follows that

$$\frac{dQ}{T_{cold}} > \frac{dQ}{T_{hot}}$$

and thus the entropy increases.

18.2 ENTROPY AND DISORDER

It is often convenient to think of entropy as a measure of disorder. To make this association, we must define order and disorder, not in an artistic or esthetic sense, but in a mathematical way that permits evaluation; i.e., give me a state and I can give the state a grade for neatness.

When we say that a system is orderly, we mean that the elements of the system are arranged in some unusual way. We have seen, however, that one microstate is just as likely as another. For example, one bridge hand is just as likely as another, be it a particular assortment of face cards and number cards of various suits, or a hand of 13 spades from deuce to ace. Even though one hand is as likely as another, we would like to ascribe an orderliness to the 13 spades that we would not give to the motley hand of various suits. If you left the room while the hands were dealt and came back to discover that each player had 13 cards of one suit, you might suspect that the deal was rigged, i.e., that the hands were ordered. If you returned to find that your hand consisted of the 3, 4, 5, 10 of spades, the 2, 9, and jack of hearts, the 7, 9, 10 of diamonds, and the 2, 6, 8, of clubs, you would assume that the cards had been shuffled (and you would pass). If one hand is as likely as the other, why would you suspect that one deal was rigged (ordered) and the other was shuffled (disordered)? Each hand has a probability of occurrence of 1 in 6×10^{11}.

When you put a hand together, you arrange the cards in suits. Thus the hand has an overall structure, namely, how many spades, hearts, diamonds, and clubs. We might call this the macrostate of the hand and the particular distribution of cards the microstate (see Table 18.1). Now there are a great many ways of achieving a 4-3-3-3 distribution (1.7×10^{10} ways) and only one way of achieving 13 spades. So if the hands are dealt at random, we would expect to get a 4-3-3-3 distribution often and a 13-0-0-0 distribution very rarely.

TABLE 18.1   Macrostate for a bridge hand.

	Spades	Hearts	Diamonds	Clubs
Microstate	3, 4, 5, 10	2, 9, jack	7, 9, 10	2, 6, 8
Macrostate	4	3	3	3

You would say that the deck had been well shuffled (disordered) if you received a distribution of high probability, and that the deck had been ordered if you received a distribution of low probability. This concept of disorder is then exactly the same as that associated with entropy. A disordered system is in a macrostate of high probability, that is, a state with a large number of associated microstates. A measure of disorder of a macrostate is the measure of the number of microstates associated with this macrostate. This is precisely the entropy. Therefore entropy and disorder are equivalent, and we now have a quantitative definition of disorder.

The law of increasing entropy may then be paraphrased as follows:

The disorder of an isolated system will either increase or remain constant.

A large number of bricks put together to form a building is a well-ordered system, but a pile of bricks lying in a heap on the ground is a disordered system. There are a great many ways of arranging the bricks so that they form a heap on the ground, but relatively few ways of arranging the bricks so that they form a building.

Two objects at different temperatures are an ordered system, but two objects at the same temperature are a disordered system. A collection of atoms structured into a human being is an ordered system, but the same collection of atoms moving about at random in a box at $98.6\,^{\circ}F$ is a disordered system.

18.3   THE SECOND LAW OF THERMODYNAMICS AND THE LAW OF EVOLUTION

All closed systems proceed from a state of order to a state of disorder. When maximum disorder is reached (compatible with internal constraints), the system is in an equilibrium state and no further change occurs.

But suppose we scatter a deck of cards over the floor of a room, leave, and lock up. On returning the next day we find the cards put together in an elaborate card house. We find this very strange.

Or suppose we are cruising about in a spaceship and observe a certain planet to be in a molten state. On a return visit to this planet many years later we find something entirely different. Now there are cities, and people, bugs and trees. What an extraordinary thing! A disordered system has evolved into a highly ordered system—a clear-cut violation of the second law of thermodynamics. On investigation we find operating on this planet a law called the *law of evolution,* which states that living systems proceed from less-ordered systems to more highly ordered systems better able to function in their environment. Not only has there been *a* violation of the second law of thermodynamics, but there is in *continuous* operation a conflicting law. Trouble. Big trouble!

Are living systems somehow exempt from the second law of thermodynamics? Perhaps, but on our first visit to the planet there were *no* living systems. Living systems evolved from nonliving systems. We might attribute the initial appearance of living matter to extraordinary chance: The planet was dealt a hand of 13 spades. But what about subsequent evolution to systems of greater and greater complexity—the law of evolution?

The answer to these questions lies in the fact that the planet, earth, is not a closed system. We might for all practical purposes consider the sun, the earth, and outer space as a closed system. The second law of thermodynamics requires that the entropy or the disorder of this system increase. In its present state it is far from equilibrium. There is a great deal of order: The sun is hot, the earth cool, and outer space very cold. The system is proceeding toward the disordered state of uniform temperature, i.e., equilibrium. The order we find on the earth is acquired at the expense of a much greater disorder in the whole system. The earth is feeding on the order in the sun—earth—outer space system.

To see the significance of the order in this system, imagine the earth enveloped by a large container whose interior surface is perfectly reflecting so that all radiation is reflected and none escapes. Such a system might remain at a reasonable temperature, say 70°F, with no energy gained or lost. However, all life would disappear. One could not survive in such an isolated environment. The system would continuously lose order until a steady state was reached.

The sun therefore must play an essential role in creating order on earth. However, if this interaction consisted only of the sun radiating energy and the earth absorbing the energy, *the sun would become ordered and the earth disordered.* The sun would be losing heat energy, so its entropy would decrease. The earth would be absorbing heat energy, so its entropy and hence its disorder would increase. If we increased the size of the container mentioned earlier to include the sun *and* the earth, we would become very disordered very quickly. We have seen earlier that whenever two systems have means of exchanging energy, they will reach the same temperature, in this case 11,000°F, the temperature of the sun.

Since the sun-earth system alone is incapable of creating order on earth, the other essential feature must therefore be outer space. The earth absorbs ordered radiation from the sun, extracts some of this order, and reradiates disordered radiation into outer space.

The principal mechanism on earth that is able to use the order in the sun—earth—outer space system is the photosynthesis that takes place when sunlight interacts with the chlorophyll molecule in plants. The principal overall reaction is

$$6CO_2 + 12H_2O + light + chlorophyll \rightarrow$$
$$6O_2 + C_6H_{12}O_6 + 6H_2O + chlorophyll$$

where $C_6H_{12}O_6$ is glucose, a sugar. Glucose is a highly ordered molecule built from a less-ordered collection of carbon dioxide and water. This order is obtained from sunlight. *With some minor exceptions, photosynthesis is the only mechanism on earth that uses the order in the sun—earth—outer space system.* It requires light of rather high energy but certainly not light that is in thermodynamic equilibrium. An example of light in thermodynamic equilibrium is the electromagnetic radiation trapped inside an oven at some temperature above absolute zero. If the light is in thermal equilibrium, it is in a state of maximum disorder and has nothing to offer. On the other hand, light proceeding uniformly in one direction, away from the sun, represents a great deal of order. It does not make any difference which direction as long as the light is not a collection of randomly oriented rays.

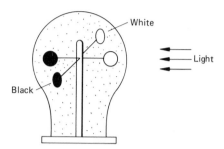

FIGURE 18.10   A radiometer.

An artificial device that is also able to extract order from sunlight is the radiometer. It consists of a set of vanes (usually four) attached to a central shaft that is free to rotate (Fig. 18.10). One surface of each vane is white and the other black. The system is contained inside a partially evacuated glass bulb. The black surfaces are more absorbent than the white and so become hotter. The gas in the neighborhood of the black surface becomes hotter than the gas in the neighborhood of the white surface. Thus the gas molecules striking the black surface will exert a stronger force on that surface than will the gas molecules near the white surface. Therefore the black surface will recede from the light, and the vanes will rotate about the central shaft. As each vane makes one-half revolution, it presents a different shade to the light and so the vanes will continue to rotate.

If the light were randomly directed at the radiometer (e.g., the radiometer were put into an oven), then the entire system would come to a uniform temperature and the vanes would not rotate.

The ability of chlorophyll to extract order from the directed rays of the sun explains how living organisms are able to withstand the continuous degradation of order required by the second law of thermodynamics. They feed on the order generated by the chlorophyll reaction. Not only may living organisms combat the continuous degradation of order, but they may evolve into systems of greater and greater complexity, greater and greater order. However, another fundamental dilemma is associated

with the existence of life on earth: Where did chlorophyll come from? The photosynthetic process is a very complex, well-ordered biological machine for turning out glucose. What is the origin of this machine? How did it evolve out of the early chaos on earth? This is not an easy question to answer, and indeed we do not know the full answer.

There are two paths we might take to explain the origin of chlorophyll. We might imagine some extraordinary event (a violation of the second law) in which the earliest biological molecules were created, or we might search for some other natural process in which order is extracted from directed electromagnetic radiation.

We do not have to search very far to find such mechanisms. They are all about us in the weather. The winds, for example, represent order. If the earth were in thermodynamic equilibrium, there would be no winds. Winds are generated by nonuniform heating of the earth's surface. Even if the earth were not rotating, we would have winds because the earth's surface is not uniformly absorptive. Areas that absorb more sunlight become hotter than areas that absorb less. The air above these hotter areas becomes hotter and hence less dense than the air above the cooler areas. As the hot air rises it is replaced by the cooler surrounding air, and therefore steady air currents are generated. Now air currents generate storms, which in turn generate lightning. A lightning bolt represents a great deal of order. Perhaps the earliest complex biological molecules were generated by the extraction of order from a lightning flash. (Some success along these lines has been achieved in the laboratory, with artificial lightning—electric sparks—used to generate biological molecules.)

Maybe some day when the chain of events that led from the primeval collection of carbon, hydrogen, oxygen, nitrogen, etc., and the occurrence of lightning to the generation of the first photosynthetic process is understood, we will no longer marvel at the existence of life on earth but will view it as an inevitable consequence of the workings of natural laws. Given the right conditions, life is not only possible but the most probable consequence in a system that is continuously gorging itself on the enormous order contained in directed sunlight.

QUESTION

Which came first, plants or animals?

ANSWER

Clearly plants came first, since only plants have a mechanism for feeding on the order in sunlight. A completely carnivorous society is impossible because it has no means of combating its continuous degradation of order. (With our present technology we could in principle use solar batteries, radiometers, or windmills to synthesize the required nutrients, but imagine the enormous scale on which this artificial synthesis would have to be conducted if it were to take the place of photosynthesis!)

18.4 DISORDER AND VELOCITY DISTRIBUTION

Since systems always seek the most disordered state, why do we find so much order in most systems? For example, in hydrogen gas we find most of the atoms paired to form H_2 molecules. In a mixture of hydrogen and oxygen we find two hydrogen atoms coupled with an oxygen atom to form a water molecule. In a gas or dilute solute in a gravitational field we find greater densities at the bottom than at the top. In all these examples we have a great deal of order. Atoms linked in groups have more order than the same atoms scattered about at random. A box with more particles near the bottom than the top has more order than if the same particles were uniformly distributed.

The reason for these apparent violations of the second law of thermodynamics is that we are looking only at the order associated with the relative *positions* of the particles.

To put it another way, in counting microstates associated with a given macrostate, we should not look at only the positions of the particles. In describing the microstate we must also consider the distribution of *velocities* of the particles. The microscopic features of a system are completely described only if we give the positions *and* velocities of all the particles. Many different microstates are associated with different possible velocity distributions. However, the total energy of an isolated system is fixed, and only those velocity distributions for which the total energy has the proper value are possible.

The velocities of the particles are not independent of their positions. If there are attractive forces between particles, then the particles will move faster after they come together. Thus when two hydrogen atoms come together they convert some of their potential energy into kinetic energy. If we compare two possible macrostates of hydrogen gas—one in which the atoms are uniformly distributed and one in which the hydrogen atoms are all paired into hydrogen molecules—with a fixed total energy, we find that the kinetic energy of the H_2 molecules would be greater than the kinetic energy of the H atoms. Therefore more velocity microstates would be available to the H_2 gas, but fewer position microstates would be available to the H_2 gas since all the atoms are ordered by the pairing. The gas will seek some compromise. Some of the atoms will pair up, sacrificing many position microstates but gaining velocity microstates; others will remain separated. The proportion of H atoms and H_2 molecules will depend on the total energy. For large values of the energy, the relative increase in velocity microstates resulting from binding will be relatively small, so most of the atoms will remain separate. For small energies, most of the atoms will pair to form molecules. When we heat hydrogen gas we find fewer molecules and more atoms at the higher temperatures.

We find greater densities at the bottom of a sediment for a reason very similar to why atoms bind together to form molecules. More velocity disorder is gained than positional disorder lost when some of the particles give up gravitational potential energy to gain kinetic energy.

It is tempting to ascribe the breakup of H_2 molecules into H atoms at higher temperature to the fact that higher temperatures mean higher average kinetic energy per molecule, and these high-energy molecules would break up more easily in collisions with other molecules. However this argument is based on dynamic principles rather than on statistical principles, and so has nothing to do with the thermodynamic equilibrium state of a large aggregate of hydrogen molecules and atoms. Equilibrium states of large numbers of particles are determined by statistics. A similar mechanistic argument may be made to explain the increase in sedimentation at lower temperatures. Again this argument is fallacious.

QUESTION

We have observed that the entropy of a system increases when heat is added ($dS = dQ/T$). Suppose we have a box of non-interacting particles—an ideal gas. Neglecting the effect of the gravitational field, we would expect the particles to be uniformly distributed throughout the box when the system has achieved thermodynamic equilibrium. The uniform distribution is the macrostate that has the most microstates associated with it. When the gas is heated the particles will still be uniformly distributed throughout the box in the equilibrium state. Thus it appears that the number of microstates associated with the equilibrium macrostate does not change when heat is added. There is only one macrostate, the uniform distribution, and the number of associated microstates cannot change. But the entropy is a measure of the number of microstates associated with the equilibrium state. How can the entropy increase if the number of microstates cannot change?

ANSWER

The resolution of this paradox lies in the manner in which we count microstates. We must remember that the microstate depends not only on the positions of the particles but also on their velocities. When heat is added to the system the total internal energy of the system increases, and therefore the particle velocities increase. This increase in the particle velocities increases the number of velocity microstates available to the system.

We may demonstrate this result quantitatively with the following crude example. Suppose there are only four particles in the box and the total internal energy of the system is 1 J. To facilitate the problem of counting the number of velocity microstates, let us assume that the individual particles can have only discrete energies of 0 J, 1 J, 2 J, etc. (In classical mechanics this is an artificial constraint. We shall see later, however, when we take up the subject of quantum mechanics, that the energy of a particle in a box is indeed quantized; i.e., the energy can have only certain discrete values.) For the macrostate in which the total energy of the system is 1 J, there is only one macrostate, namely one of the four particles has an energy of 1 J and the other three have zero energy. This macrostate has four associated microstates.

Let us increase the total internal energy of the system by heating it until the total energy is 3 J. Now a number of macrostates are possible, as illustrated in Table 18.2. Since there is more than one possible macrostate, the second law of thermodynamics tells us that the equilibrium state is the macrostate that has the most associated microstates. From the table we see that this is the macrostate in which one particle has an energy of 2 J, one an energy of 1 J, and the remaining two, zero energy.

TABLE 18.2 Possible macrostates for which the total energy is 3 J, and the number of associated microstates.

Macrostate			
3 J	—	—o—	—
2 J	—o—	—	—
1 J	—o—	—	—ooo—
0 J	—oo—	—ooo—	—o—
Number of associated microstates	12	4	4

The principal point of this illustration is to demonstrate that the entropy increases when heat is added. We have seen that when the internal energy of the system is 1 J, there is only one macrostate, and the number of associated microstates is four. When the total energy is increased to 3 J, there are three possible macrostates, and the most probable one is that in which one particle has an energy of 2 J, one particle an energy of 1 J, and the remaining two particles zero energy. The number of associated microstates is 12. Since the entropy of a macrostate is a measure of the number of microstates associated with that macrostate, it follows that the entropy of the equilibrium state (or the most probable state) must increase as the energy increases from 1 J to 3 J. The number of microstates associated with the equilibrium state increases from 4 to 12.

QUESTION

Why is it possible to express the chemical potential of an ideal gas in the form

$$\mu = RT \ln c + \mu_0$$

where the first term, $RT \ln c$, is the same for all ideal gases and the last term, μ_0, is independent of the concentration?

ANSWER

In ideal gases we can neglect entirely the interaction between the elements of the gas. This means that the possible velocity states or states of vibration or rotation of the elements are completely independent of the positions of the molecules. We may therefore count the two types of microstates separately. In counting the microstates associated with the positions of the elements, it makes no difference what kind of molecule it is since we are neglecting its interaction with its neighbors. The first term in the chemical potential is the part that comes from counting the number of different microstates associated with the positions of the molecules that correspond to the given macrostate. The second part depends on the number of velocity states or internal rotations and vibrations of the molecules, and thus depends on what type of molecules are present. Therefore μ_0 will be different for different ideal gases.

18.5 THE SECOND LAW OF THERMODYNAMICS AND THE FATE OF THE UNIVERSE

The so-called *big bang* theory is perhaps the most widely accepted theory of the evolution of the universe. According to this theory all the matter in the universe was initially condensed to an incredibly compact state. This very dense matter exploded and expanded to fill the present bounds of the universe, which is still expanding. Whether it continues to expand indefinitely will depend on the total energy of the universe. If the total energy is positive, then it is possible for the galaxies to recede to indefinitely large distances, where the galaxies would have positive kinetic energy and almost zero potential energy. Thus the total energy would be positive. But if the kinetic energy at the present time of all the galaxies is *less* than the absolute value of the total gravitational potential energy (negative total energy), the universe must at some time in the future stop expanding and begin to contract. Astronomers have estimated that the total energy of the universe is very close to zero. Whether the universe will eventually begin to contract is therefore uncertain. If it does, it will contract back to a very localized region and then expand again, oscillating back and forth. It may in fact have already completed many oscillations.

The universe must obey the second law of thermodynamics, i.e., its entropy must continuously increase. The conclusion seems reasonable that in an expanding universe the system is approaching a more and more disordered state much as a gas does in a free expansion. There are more microstates associated with the expanded system. But what if the universe begins to contract? If it contracts back to a localized region, it would seem to be going to a state of greater order.

Once again we must consider the order in the velocity distribution as well as the order in the configuration. As the universe contracts, it will be gaining order in the positions of the galaxies, but losing more order in the velocity distribution. This is analogous to the example illustrated in Fig. 18.6, where we proved that the gas increases its entropy whether it expands *or* contracts. In that example, if the gas contracts, the weight on the piston gives energy to the gas inside the vessel, and this increases the velocities of the molecules. This increase in velocity allows an increase in microstates associated with the distribution of velocities.

If and when the universe begins to contract, the order gained in the localization of matter will be more than compensated for by the disorder generated in the increasingly random velocity distribution. As the universe contracts, the total disorder will increase.

18.6 THE EVOLUTION OF THE SPECIES, THE DEATH OF THE INDIVIDUAL, AND THE SECOND LAW OF THERMODYNAMICS

Why do we grow old and die? At first the answer seems obvious. Everything about us deteriorates: The car runs down; the house becomes messy; the roads become bumpy. Everything is subject to the ultimate degradation of order demanded by the second law of thermodynamics, and there is no reason to expect the individual to be exempt. And yet the human species is exempt! The species feeds on the order generated by photosynthesis and evolves to states of greater and greater order. But why can't the individual share in this bonanza on which the species thrives? We do very nicely for a time. By eating the order passed down from the plants, we maintain the

order in our bodies. Through photosynthesis the plant forms a glucose molecule, which is eaten by a cow who uses the order in the glucose to create a protein molecule. We then eat the cow and use the order in the protein to form protein molecules of our own. The refuse is passed from our bodies in a disordered state. Why can't we consume as much order as we need to sustain life indefinitely i.e., to maintain our order? *There is nothing in the second law of thermodynamics that requires the death of the individual.*

The death of the individual is, however, vital to the evolution of the species, and each death is in a sense a passing by the species of unneeded waste. Once an individual has reproduced a number of times and produced progeny that may be his or her superior, the species will profit more from the procreation of the superior progeny than from further procreation of the inferior parent, so the parent must die. But observe that it is the laws of evolution at work here, not the second law of thermodynamics. Evolutionary forces have determined the finite life span just as they have developed the eye, the kidney, the reproductive system, and all the other elements vital to the perfection of the species.

The actual mechanism that assures that the life span of an individual is finite is not known. It might be built into the DNA molecule, or it might just be some natural degradative process for which there have been no evolutionary pressures to find a remedy. Indeed a remedy would be harmful to the species.

QUESTION

If energy is conserved, why do we have an energy crisis?

ANSWER

We do not have an energy crisis; we have an entropy crisis. An automobile does not consume energy. After the automobile has burned a gallon of gasoline, the heat energy in the bearings, the heat and chemical energy in the exhaust, etc., are exactly equal to the initial energy of the gallon of gasoline. But the order in the gasoline has been lost. Before the gasoline is burned its chemical constituents are in an ordered state; after burning, however, they are in a disordered state. All systems proceed from a state of order to a state of disorder, and once this order has been lost it cannot be recovered.

The fact that energy is extracted from the gasoline is coincidental. The gasoline burns because of the second law of thermodynamics, a statistical law; that is, because of a tendency to seek a state of lower order not a state of lower chemical energy. The fact that this state of lower order is a state of lower chemical energy is indeed fortunate, but it is an effect not a cause.

Consider a more elementary example. Two large objects are at the same temperature. This is a state of maximum entropy and a state of thermodynamic equilibrium. The system has considerable thermal energy, but there is no way to get it out (see the question on page 294). Suppose we transfer in some way heat energy from one object to the other. Now one object is hot and the other cold, but there is no net change in the energy of the system. However, since the system is now an ordered system, things can happen. The system wants to go to the more disordered state, the state of equal temperature, and in the process perhaps we can extract energy from the system, even though the net energy in the system is precisely what it was before the heat exchange. To see how we might get this system to perform useful work, let us construct the following engine, which consists of a gas in a vessel with a piston for a lid.

1. We place the vessel on the hot body. Now the gas gets hot, expands, and pushes the piston. The piston can do work.

2. We move the vessel over to the cold body. Now the gas cools and we can compress it; but the work necessary to compress the piston is now smaller because the gas is cool, and therefore the pressure in the gas is smaller.

3. Next we put the vessel back on the hot body and repeat the process.

We may repeat these steps until the two large objects come to the same temperature. (There is a continuous transfer of heat from one to the other as the gas is moved back and forth.) When equal temperatures are reached, we can extract no more work from the system, for it has come to a state of maximum entropy, or maximum disorder, at this new energy. (Remember that energy has been removed by the engine.)

Two things are required in most processes; energy and order. To maintain life on earth we need both, and we get them from the sun. For our hearts to beat we must establish in the body a state in which the ordered state is a relaxed heart muscle and the disordered state is a contracted heart muscle. Furthermore, the transition from the ordered state to the disordered state must be capable of supplying the energy required to perform the work of contraction. Once the contraction has been completed, the heart must reach a state that repeats the cycle.

18.7 THE SECOND LAW OF THERMODYNAMICS AND UPPER LIMITS ON ENGINE AND REFRIGERATOR EFFICIENCIES

We have seen in Sec. 18.6 that it is possible to use ordered systems to perform work. All engines operate on this basic principle. In all real engines there is a net loss of order in every cycle of the engine. However, we would like to show that it is possible in principle to construct an engine operating in a cycle in which there is no net loss of order or, equivalently, no net change in entropy either in the engine or in the outside system on which it feeds.

Consider two reservoirs at temperatures T_{hot} and T_{cold}. We would like to operate an engine between these two reservoirs to perform work. Now this is to be a very special engine, *a reversible engine*. Let us define what a reversible engine is.

Every engine runs through a sequence of steps. Imagine that each of these steps is reversed so that the engine is running backward. If the engine behaves in *exactly* the opposite way when it runs backward, then the engine is reversible. If at some state in the forward operation the engine performs work, then in the reverse operation an equal amount of work is done on the engine. If at some stage of the forward operation heat is extracted from a reservoir, then in the reverse operation exactly the same amount of heat must be given to that reservoir. If in the complete cycle of the engine a net amount of work is performed, a net amount of heat energy is given to some reservoirs, and a net amount of heat energy is taken from other reservoirs, and if in the reverse operation the same net amount of work is done on the engine and the reverse cycle completed, then the reservoirs and the engine must return to their original states.

Any engine in which there is friction is not reversible. If there is friction, some of the working parts will heat up. These parts will not cool down when the operation is reversed, and therefore the reverse cycle does not reverse the mode of operation.

If a piston in a vessel containing gas is pulled out with a velocity greater than the velocity of any of the gas molecules confined behind the piston, the gas will not lose any energy. (A particle can lose energy only by bouncing off a receding wall.) However, if the piston is pushed back with the same velocity, it will clearly add energy to the gas. This operation is therefore not reversible.

An example of a process that is reversible is one in which a frictionless piston is pulled out very, very slowly. If the piston is pushed back in the same way, the gas will return to the same state. Even if the vessel is held at constant temperature through heat exchange with a large reservoir, the heat that is removed from the reservoir on the slow expansion will be returned to the reservoir on the slow contraction.

The facts that the reversible process described above must be very slow and that there is no friction are limiting requirements, requirements that may be approached if greater and greater care is taken in the operation of the engine. No real system is reversible, but *it is possible in principle to approach reversibility.*

Given the existence of a reversible engine as a limiting situation, we would now like to demonstrate that no engine is more efficient than a reversible engine, and furthermore that the efficiency e of any reversible engine is given by

$$e_{rev} = 1 - \frac{T_{cold}}{T_{hot}} \tag{18.2}$$

when the engine operates between two reservoirs at temperatures T_{hot} and T_{cold}.

The operation of the engine is illustrated schematically in Fig. 18.11. Here Q_{hot} is the heat energy extracted from the hot reservoir, Q_{cold} the heat energy given to the cold reservoir, and W the work done by the engine in one cycle. First let us observe that energy is conserved, so

$$Q_{hot} = W + Q_{cold} \tag{18.3}$$

The heat energy taken from the hot reservoir goes partly into work and partly into heating the cold reservoir. Observe that Eq. 18.3 would not be true if the engine were not operating in a cycle. Only because the engine returns to its starting point after each cycle do we know that no energy has been stored in the engine. Whatever energy comes in must go out.

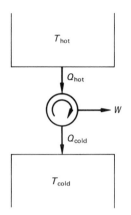

FIGURE 18.11   A reversible engine operating between hot and cold reservoirs.

Next let us consider the entropy change of everything— the reservoirs and the engine—during one cycle. The change in entropy of the hot reservoir is

$$\Delta S_{hot} = -\frac{Q_{hot}}{T_{hot}} \tag{18.4}$$

(We need the minus sign because we have suspended our convention of always denoting the energy added to a system by Q. The alternative would be to have negative heat flowing into the hot reservoir, and this is less desirable than temporarily changing our convention. In Eq. 18.4, $-Q_{hot}$ is the heat added to the hot reservoir.)

The change in entropy of the engine is zero, not because the engine is reversible, but because it is operating in a cycle and so has returned to its original state after completing the cycle. We know from the second law of thermodynamics that the entropy is a unique function of the state, so if the engine returns to the same state it must return to the same entropy.

Finally, the change in entropy of the cold reservoir is

$$\Delta S_{cold} = \frac{Q_{cold}}{T_{cold}} \tag{18.5}$$

If the work done by the engine is used perhaps to raise a weight, then there is no entropy change in the weight. It changes its height only and not its thermodynamic state. The net entropy change is therefore

$$\Delta S_{tot} = \frac{-Q_{hot}}{T_{hot}} + \frac{Q_{cold}}{T_{cold}} \tag{18.6}$$

Since this is an isolated system (if we include the system on which the work is done), the entropy change must be either zero or positive.

Now let us use the fact that the engine is reversible. When the engine is operated in reverse it becomes a refrigerator. Heat energy Q_{cold} is removed from the cold reservoir, heat energy Q_{hot} is given to the hot reservoir, and work W must be performed to accomplish the transfer (Fig. 18.12). Since the engine is reversible, all these quantities

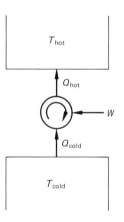

FIGURE 18.12   A refrigerator operating between a hot reservoir and a cold reservoir.

have the same magnitude as before; only the directions have changed. The change in the total entropy is now

$$\Delta S'_{tot} = \frac{Q_{hot}}{T_{hot}} - \frac{Q_{cold}}{T_{cold}} \qquad (18.7)$$

We used S' for the entropy change since we have no reason to expect the same change in entropy as before. In fact we see that

$$\Delta S'_{tot} = -\Delta S_{tot} \qquad (18.8)$$

But $\Delta S'_{tot}$ must be either zero or positive for the same reason that ΔS was either zero or positive: The system is isolated, and in any process the entropy must either increase or remain constant. The only way we can satisfy the requirements

$$\Delta S_{tot} \geq 0$$

$$\Delta S'_{tot} \geq 0$$

and

$$\Delta S'_{tot} = -\Delta S_{tot}$$

is for

$$\Delta S_{tot} = 0$$

and $\qquad\qquad\qquad\qquad\qquad\qquad (18.9)$

$$\Delta S'_{tot} = 0$$

Therefore the entropy does not change. From Eqs. 18.7 and 18.9 we have

$$\Delta S_{tot} = \frac{-Q_{hot}}{T_{hot}} + \frac{Q_{cold}}{T_{cold}} = 0$$

so that

$$\frac{Q_{hot}}{T_{hot}} = \frac{Q_{cold}}{T_{cold}} \qquad (18.10)$$

This is a fundamental result and applies to *any* reversible engine (or refrigerator) operating between two reservoirs.

Let us use this result to obtain an expression for the efficiency. We define the efficiency e of any engine as the ratio of the work output to the heat drawn from the hot reservoir:

$$e = \frac{W}{Q_{hot}} \qquad (18.11)$$

but $W = Q_{hot} - Q_{cold}$, so that

$$e = 1 - \frac{Q_{cold}}{Q_{hot}} \qquad (18.12)$$

This result applies to any engine. For a reversible engine, using Eq. 18.10 we have

$$e_{rev} = 1 - \frac{T_{cold}}{T_{hot}} \qquad (18.13)$$

The efficiency of a reversible engine then depends only on the temperatures of the reservoirs, not on the nature of the reversible cycle or on the working substance used in the engine.

We would also like to show that the efficiency of a reversible engine is greater than or equal to the efficiency of any other engine. Let us suppose that a more efficient engine exists, and let us use this engine to drive a reversible engine so that it is operating as a refrigerator (see Fig. 18.13). If we let e' be the efficiency of the engine and e_{rev} the efficiency of the reversible engine, then

$$e' = \frac{W}{Q'_{hot}}$$

and

$$e_{rev} = \frac{W}{Q_{hot}}$$

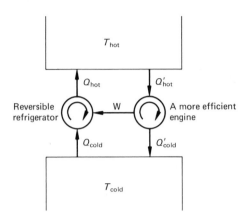

FIGURE 18.13   A reversible refrigerator being driven by a more efficient engine.

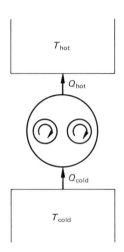

FIGURE 18.14   A device that takes heat from a cold reservoir to a hot reservoir.

But if $e' > e_{rev}$, then

$$Q'_{hot} < Q_{hot}$$

Therefore a net amount of heat energy is being given to the hot reservoir, and this energy can only come from the cold reservoir. We have created a device operating in a cycle whose sole effect is to take heat from a cold reservoir and give it up to a hot reservoir. The net effect is illustrated in Fig. 18.14. We shall leave it as an exercise for you to show that such a device is impossible. (See problem 18.B.4 and also the question on page 294.) (We have used the fact that the refrigerator is reversible when we have assumed that the ratio W/Q_{hot} does not change when the engine is reversed and used as a refrigerator.)

We might observe parenthetically that we have also shown that the ratio W/Q_{hot} for *any* engine is less than or equal to W/Q_{hot} for *any* refrigerator.

QUESTION

We said earlier that we need order to produce work. Now the entropy of a reversible engine and that of the reservoirs do not change during a cycle. But if there is no change in entropy there is no change in order, since entropy is a measure of order (or rather disorder). How is it that we can obtain energy from a system without affecting the order? Furthermore it is clear that after a time the two reservoirs will reach the same temperature since the engine is constantly taking heat energy from the hot reservoir and giving heat energy to the cold reservoir. It would appear that we certainly have lost order. We started with hot and cold reservoirs and we end up with two reservoirs at the same temperature. When this happens it is no longer possible to obtain energy from the system. And yet *the net entropy is unchanged,* so the system is just as ordered as it was initially. But initially the system could do work, and after the reservoirs have reached the same temperature the system cannot do work. What exactly are the requirements for obtaining energy from a system?

ANSWER

Let us first consider the question of how it is that the order in the reservoirs does not change even though initially the two reservoirs were at different temperatures and after the reversible engine has run for some time they will have reached the same temperature. This certainly seems to represent a loss of order.

We must remember that work has been done by the engine, so heat energy has been extracted from the reservoirs and converted into mechanical work. When we extract heat from a body the order in that body increases. The molecules move more slowly, so fewer energy states are available for a given macrostate. Now the combined system of the two reservoirs has lost heat energy, so the order associated with the internal motion of the molecules has increased. This is compensated for exactly by the decrease in order associated with the decrease in the temperature difference, and thus the net order remains unchanged.

It is difficult to see precisely how the order does not change when a reversible engine is run between two reservoirs at different temperatures. As a simpler example, imagine a gas confined inside a vessel with a piston at one end, and let the cylinder be in a vacuum. Now if the piston is fixed in place and the gas in the cylinder is at a uniform temperature, there is no way to extract work from this system. However, if we free the piston (relax a constraint, making available higher entropy states), the gas can expand and in so doing it can do work. If it expands slowly *there is no change in entropy of the system.* The entropy change would be given by $dS = dQ/T = (dE + P\,dV)/T$, and dQ is zero or, equivalently, $dE = -P\,dV$. Now this system is doing work and it appears to be going to a more disordered state, the state of larger volume, which is a less localized state. A localized collection of particles is ordered, but a nonlocalized (or less localized) system is less ordered. However, as we observed, the entropy and therefore the order do not change ($dS = 0$). The order does not change because energy is being taken from the system ($dE < 0$) at the same time that the gas is expanding ($P\,dV > 0$) in such a

way that the disordering associated with the increase in position states is compensated for exactly by the ordering resulting from the decrease in velocity states available. If we let the gas expand rapidly, the energy decrease would be smaller (the slower particles cannot follow the piston and so cannot lose energy by bouncing off the receding piston). Therefore the increase in positional disorder would be larger than the decrease in velocity order, and consequently there would be a net decrease in order in this irreversible process.

The system wants to increase its disorder (to expand), but if we are clever and can use this expansion to perform work at a rate that keeps the entropy constant, the order remains unchanged. The clever device that does this is a reversible engine. (The reversible engine in this case is very simple. It is just the gas-filled cylinder with a frictionless piston for a lid.)

Therefore the requirement for extracting work from a system is that higher entropy states be available to the system *at the given energy* (for example, two reservoirs, one hot and one cold, or a gas in a cylinder surrounded by a vacuum). As we take advantage of the tendency of the system to go to higher entropy states, it is possible for us to devise a process to extract work at a rate that maintains the order. A reversible process is such a process, and it is the most efficient use of energy.

When the system has reached a condition in which higher entropy states are no longer available *at that energy,* we can no longer obtain work from the system. This occurs when the two reservoirs reach the same temperature, or when the gas in the cylinder expands to the point at which the pressure in the gas falls to zero.

QUESTION

Why do all bodies have the same lower limit of temperature, and why is this lower limit zero?

ANSWER

We have seen that the most efficient engine is a reversible engine. In exactly the same way, we may show that the most efficient refrigerator is a reversible refrigerator. If Q_{cold} is the heat energy taken from the body to be cooled at a temperature T_{cold}, Q_{hot} the heat energy given to the hot reservoir at a temperature T_{hot}, and W the work necessary to perform the transfer, then in one cycle of a reversible refrigerator the net entropy change is zero, so that

$$\frac{Q_{hot}}{T_{hot}} - \frac{Q_{cold}}{T_{cold}} = 0 \tag{18.14}$$

And since energy is conserved,

$$Q_{cold} + W = Q_{hot} \tag{18.15}$$

Now the efficiency of a refrigerator, unlike that of an engine, is determined by the amount of heat that is taken from the body to be cooled and the work W needed to complete a cycle. We define the efficiency of a refrigerator by

$$e = \frac{Q_{cold}}{W} \tag{18.16}$$

From Eqs. 18.14 and 18.15 we can determine the efficiency of a reversible refrigerator to be

$$e_{rev} = \frac{T_{cold}}{T_{hot} - T_{cold}} \tag{18.17}$$

This is the upper limit on the efficiency of any refrigerator operating between these temperatures.

We see that the efficiency of a refrigerator approaches zero as the temperature of the cold body approaches zero. It follows then that $T = 0$ is a lower limit of temperature. Whatever device we use to cool a body, the efficiency of that device becomes lower and lower as the temperature of the body it is trying to cool approaches zero. Thus the absolute zero of temperature of any body is not so much a consequence of the properties of that body as a result of the efficiency of the device used to cool the body. In making a cooling device we are limited not by technical factors but by the second law of thermodynamics. If one could construct a device that is more efficient than that predicted by Eq. 18.17, then one could devise a process that would violate the second law of thermodynamics.

SUMMARY

An isolated system with given constraints will come to an equilibrium state, which is the macrostate with the greatest entropy. If one or more constraints are relaxed, many new macrostates become accessible. *One* of the accessible macrostates is the *initial* macrostate. The new equilibrium state will again be that macrostate with the greatest entropy. Unless the macrostate remains unchanged when the constraint is removed, the entropy of the new steady state must be greater than the entropy of the initial state, since both states are accessible macrostates. Therefore, unless there is no change in the macrostate, the entropy of an isolated system will increase whenever a constraint is relaxed.

The disorder associated with a given macrostate is a measure of the number of microstates associated with this macrostate. Disorder and entropy are proportional.

In searching for the macrostate with the most associated microstates, we must remember that the microstate is characterized by the positions *and velocities* of all the particles. It is possible, as in sedimentation, to give up some position microstates and gain more velocity microstates.

No engine is more efficient than a reversible engine. In a reversible engine operating between two reservoirs at temperatures T_{hot} and T_{cold},

$$\frac{Q_{hot}}{T_{hot}} = \frac{Q_{cold}}{T_{cold}}$$

and the efficiency is

$$e_{rev} = 1 - \frac{T_{cold}}{T_{hot}}$$

The efficiency of a reversible refrigerator is

$$e_{rev} = \frac{T_{cold}}{T_{hot} - T_{cold}}$$

To cool a body we must do work to extract heat energy. The efficiency of any device used to extract heat approaches zero as the temperature of the body being cooled approaches zero. Thus $T = 0$ is a limiting temperature that can be approached but not attained.

PROBLEMS

18.A.1 A steam turbine is driven by steam from a boiler at 230°C, and the steam is then exhausted into a condenser at 35°C. What is the maximum efficiency of this engine?

18.A.2 An engine is operating between two reservoirs at 400 K and 300 K. On each cycle 3 J of work are performed.
  (a) What is the minimum amount of heat that must be extracted from the high-temperature reservoir?
  (b) What is the minimum amount of heat that must be given to the low-temperature reservoir?

18.A.3 A reversible engine operates between an ice-water bath and a steam-water bath at atmospheric pressure.
  (a) If 10 J of work are performed on every cycle, how much heat is drawn from the steam-water bath?
  (b) How much heat energy is lost in the ice-water bath?

18.B.1 Let us define the entropy of any macrostate to be the common logarithm of the number of microstates associated with that macrostate [$S = \log$ (number of microstates)]. Initially a set of four particles are confined in a state of thermodynamic equilibrium in the two compartments on the left in the figure. (The macrostate and hence the thermodynamic state are characterized by the number of particles in each compartment.) The plug is pulled and the four particles have access to all four compartments. Eventually the system becomes a new thermodynamic equilibrium state. What is the change in entropy, i.e., the final entropy minus the initial entropy?

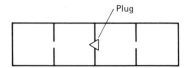

18.B.2 Suppose we have a box containing an ideal gas in a steady state. The molecules in the gas will have some distribution of energies. If we plot the number of molecules as a function of energy, we might get something like the curves in the figure. If both curves represent a system with a given energy, then we must get the same value when we add up the energies of all the molecules for either distribution. Now we may consider the distribution of energies to be a macrostate of the system. We would determine the associated microstates by specifying which molecule has what energy. Clearly many microstates exist for each macrostate, and the most likely macrostate will be that

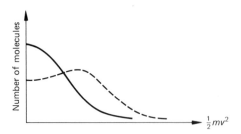

with the most microstates. It would be difficult for us to carry out the determination of the most likely macrostate since the mathematics is a little complicated. Let us examine a simple example. Suppose the molecules can have only certain discrete energies, O, ϵ, 2ϵ, 3ϵ, etc. Let there be just four molecules and let the total energy be 3ϵ. Find the most likely macrostate, i.e., the distribution of particles in the various energy states that has the most microstates associated with it.

18.B.3 Three balls of equal mass m are distributed among four boxes at heights of $0m$, $1m$, $2m$, and $3m$ above the ground (see the figure). If the total potential energy is $3mg$, what is the most likely macrostate?

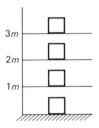

18.B.4 Prove that it is impossible to create a device operating in a cycle whose sole effect is to remove heat energy from a low-temperature body and give heat energy to a high-temperature body.

18.B.5 If you leave your refrigerator door open on a hot summer day, will the room be heated or cooled? Compare the heating or cooling of the room with the refrigerator door open with the heating or cooling of the room with the door closed.

18.B.6 If the maximum efficiency of an engine operating between two reservoirs is $\frac{3}{4}$, what is the maximum efficiency of a refrigerator operating between the same two reservoirs?

18.B.7 Obtain the general relationship between the efficiency of a reversible engine and that of a reversible refrigerator operating between the same two reservoirs.

18.B.8 A reversible engine draws heat from a reservoir at 400 K and gives heat to a second reversible engine at 350 K; this engine in turn gives heat to a reservoir at 300 K. What is the efficiency of this two-stage engine?

18.B.9 The "heat pump" commonly used for home heating and air conditioning is essentially an air conditioner that can be run forward or backward. A heat pump (operating as a heater) and a radiant heater are illustrated in the figure. The radiant heater is 100% efficient; all the electric power is converted into heat. Assuming that the heat pump has the maximum efficiency permitted by the second law of thermodynamics, compare the heat delivered to the room by each device when the outside temperature is 0°C and the inside temperature is 25°C.

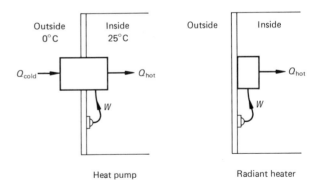

Heat pump   Radiant heater

18.C.1 A vacuum radiometer rotates in the direction opposite to the gas radiometer discussed in the text. Explain.

18.C.2 A system contains three distinguishable particles which can exist only in discrete energy states of 0 J, 1 J, 2 J, 3 J (see the figure below).

(a) If the total energy of the system is 1 J, what is the distribution of particles in the discrete states that has the most associated microstates? (How many?)

(b) If the system is now heated so that the energy is 3 J, what is the macrostate with the most associated microstates? (How many?)

(c) Since the entropy is a measure of the number of microstates associated with a given macrostate, show that the entropy in part (b) is greater than the entropy in part (a).

(This problem illustrates how the entropy or disorder of a system increases as the total energy available to the system increases.)

19
THERMODYNAMICS AND PROPERTIES OF MATTER

19.1 EQUATION OF STATE

FOR EVERY SUBSTANCE we can find a unique relation between the pressure, temperature, volume, and number of moles of the substance. Therefore, given T, V, and n, we should be able to predict the value of P, since it is not an independent variable.

A simple example of this relation is the ideal-gas law:

$$P = \frac{nRT}{V}$$

The relation might be more complicated. For example, we might have something like

$$P = \frac{nRT}{V} + \frac{cnRT^2}{V}$$

where c is some numerical constant.

Such an equation is called an *equation of state*. We may not even be able to write down the equation of state as an equation! Instead the relationship might be expressed only in the form of tables. From these tables we can determine P by interpolation if n, V, and T are given. Of course, this relation is different for each substance.

The existence of such a relationship follows from the second law of thermodynamics and the definitions of T and P. We have defined

$$\frac{1}{T} = \frac{\partial S}{\partial E} \tag{19.1}$$

and

$$\frac{P}{T} = \frac{\partial S}{\partial V} \tag{19.2}$$

From the second law of thermodynamics the entropy S is a unique function of n, V, and E, and thus $\partial S/\partial E$ and $\partial S/\partial V$ are unique functions of n, V, and E. Therefore, from Eq. 19.1 we know that T is a unique function of n, V, and E:

$$T = \text{function of } n, V, \text{ and } E \tag{19.3}$$

Similarly from Eq. 19.2,

$$\frac{P}{T} = \text{function of } n, V, \text{ and } E \qquad \text{(19.4)}$$

The two functions in Eqs. 19.3 and 19.4 will be different.

We can in principle eliminate E between Eqs. 19.3 and 19.4. This will give us some relation between P, n, V, and T, and this of course is just the equation of state. The determination of the equation of state hinges on our knowing $\partial S/\partial E$ and $\partial S/\partial V$, which in turn depends on our knowing S as a function of E, V, and n. We can say in general that we can determine all thermodynamic properties of a system if we know S as a function of E, V, and n. (On the other hand, it is not true that we can determine all thermodynamic properties from the equation of state.)

QUESTION

If S is given by

$$S = 6nR \ln\left(\frac{E + bV}{n}\right)\left(\frac{V}{n}\right)$$

where b is a constant, what is the equation of state?

ANSWER

Let us calculate $\partial S/\partial E$ and $\partial S/\partial V$ and eliminate E. Now

$$\frac{1}{T} = \frac{\partial S}{\partial E} = \frac{6nR}{E + bV}$$

and

$$\frac{P}{T} = \frac{\partial S}{\partial V} = \frac{6nRb}{E + bV} + \frac{6nR}{V}$$

Eliminating E, we have

$$\frac{P}{T} = \frac{b}{T} + \frac{6nR}{V}$$

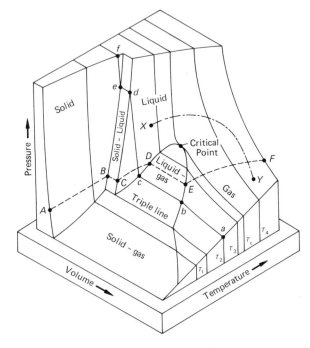

FIGURE 19.1 PVT surface for a substance that expands on melting.

The relation between P, T, n, and V is most easily visualized on what is called a PVT surface. Such a surface is shown in Fig. 19.1 for one mole of a substance that expands on melting (such as CO_2). This diagram contains a vast amount of information. First, it is important to understand that for one mole of this substance the state representing the substance must lie on this surface. It is impossible to have values of P, T, and V that lie above or below the surface. The state point is constrained to move on this surface, just as the state point must satisfy the equation of state. The surface is simply a plot of the equation of state.

We see that the surface is rather complicated. Some portions of the surface are designated *solid,* some *liquid,* and some *gas.* These designations refer to the phase of the substance for those values of *P, T,* and *V.* Some regions are a mixture of two phases in thermal equilibrium. The solid lines represent constant-temperature processes and the dashed lines constant-pressure processes. Let us follow these lines and see what happens to the substance.

CONSTANT-PRESSURE PROCESSES

Consider the line *ABCDEF*. All points on this line are at the same pressure (the line is an isobar). We might imagine that the system is exposed to the atmosphere and that heat is gradually being added. We begin in the solid state at *A,* and as the temperature rises the volume gradually increases. (Most solids expand when heated.) When we reach *B* the solid begins to undergo a phase transition. This is a most unusual process. During this transition we find that as we continue to add heat at constant pressure, more and more of the bonds that hold the molecules in the solid phase are broken, and those molecules freed from the solid enter the liquid phase. During this phase transition from *B* to *C* the temperature remains constant. Not only is *BC* an isobar, it is also an isotherm. For this particular substance the volume increases on melting. We shall discuss the nature of a phase transition in more detail shortly.

At any point between *B* and *C* a certain portion of the substance is in the solid phase and the rest is in the liquid phase. Since this substance expands on melting, the liquid is less dense than the solid so the liquid floats on the solid. At *C* all the solid has melted, and as more heat is added the temperature and volume of the liquid continue to increase until a second phase transition begins to take place at *D*. Once again as heat is added the temperature

does not change. All the heat energy goes into breaking the bonds that hold the molecules in the liquid phase, and none of the heat goes into increasing the temperature. At *E* all of the substance is in the gaseous state, and as more heat is added the gas expands and the temperature increases.

We can see from Fig. 19.1 that if we start at a lower pressure, we can go from the solid state directly into the gaseous state. This process is called *sublimation.* We also see that at a critical pressure and temperature all three phases may exist simultaneously. This occurs on what is called *the triple line.*

Note from the figure that it is possible for a substance to go from the liquid phase to the gaseous phase without undergoing a phase transition. As the substance proceeds from *X* to *Y,* it changes very gradually from a liquid to a gas. In this region of transition it is impossible to characterize the substance as either a liquid or a gas. The *critical point* defines a temperature above which it is impossible for the gas to undergo an abrupt phase transition.

CONSTANT-TEMPERATURE PROCESSES

Let us briefly follow a constant-temperature process (*abcdef*) in Fig. 19.1. We might imagine that the vessel containing the substance is in contact with a large thermal reservoir and that an ever-increasing pressure compresses the substance. We start with a gas at *a,* and as the pressure increases the gas contracts until the molecules are so close together that they begin to bind together in the liquid state. From *b* to *c* the gas undergoes a phase transition; both the temperature and pressure are constant as more and more of the gas condenses into the liquid. From *c* to *d* a very large increase in the pressure produces only a small decrease in the volume. Liquids as a rule are difficult to compress. At *d* the molecules have been forced so close together that they form stronger bonds and undergo a second phase transition into the solid state.

This simple picture of the state surface does not really do justice to the great diversity of forms possible in the solid state. In Fig. 19.3 the solid portion of the *PVT* surface is shown for water. The different phases of ice denote different crystalline forms; as the ice goes from one form to another, it undergoes a phase transition.

19.2 PHASE TRANSITIONS

A phase transition is a very remarkable physical process, and we would have difficulty accepting it if it were not such a familiar phenomenon. The phenomenon of substances existing in various states is quite reasonable. At low temperatures the molecules organize into a regular crystalline structure. As the substance is heated the molecules begin to oscillate more and more vigorously and eventually break their bonds in the lattice. At this point the substance enters the liquid state (unless conditions are right for sublimation, the direct transition from solid to gas). At still higher temperatures the molecules break their bonds in the liquid (vaporize) and the substance enters the gaseous phase.

The precise manner in which the phase changes occur is truly amazing. If we begin to heat ice in a container at a constant pressure of 1 atm, the ice continues to rise in temperature until it reaches 0°C, at which point it begins to melt. The temperature, however, remains constant during this melting process. Why does not some of the heat energy go into melting the ice and some into raising the temperature? The answer must lie in the fact that there are many more microstates associated with the phase change than with the increased temperature states. The second law of thermodynamics tells us that a system will always exhibit those macroscopic features that have the most associated microstates. As we continue to add heat (still holding the pressure at 1 atm), all the ice eventually melts and the temperature again begins to rise until it reaches 100°C, where a second phase change takes place in an entirely similar fashion. The vaporization takes place at constant temperature.

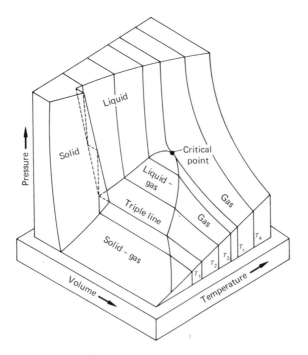

FIGURE 19.2 *PVT* surface for a substance that expands on freezing.

The state diagram for a substance that contracts on melting, such as water, is illustrated in Fig. 19.2. We see that for such a substance the transition region separating the solid from the liquid phase slopes the other way (i.e., at constant volume the temperature decreases with increasing pressure). We also see that for such a substance an increase in pressure can cause the solid to melt. A thin film of water forms under the blade of an ice skate, where the pressure is very great, for this reason. The film lubricates the motion of the skate.

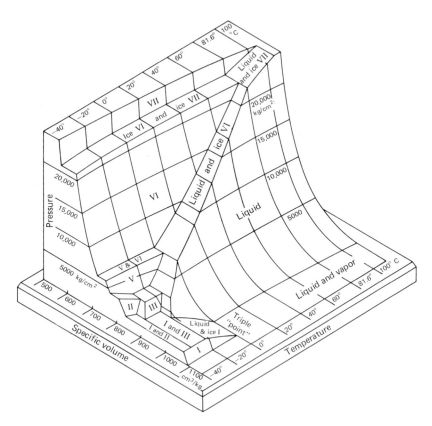

FIGURE 19.3 *PVT* surface showing the various phases of ice.

To appreciate the peculiarity of this phase transition, imagine a box of coins that are free to slide around in the bottom of the box without friction. Initially all the coins are heads. As energy is added the coins slide about more and more vigorously, but all remain heads. Suddenly at a certain point the coins cease to speed up, and as more energy is added they begin to flip over one by one until they are all tails. They have made a transition from the heads phase to the tails phase with *constant translational energy*. If we now add more energy the coins begin to speed up once again. This, of course, is not what happens in a phase transition, but this example does indicate the unusual nature of such a transition.

In the transition of water from the solid to the liquid phase, the ice does not simply heat up until each molecule is on the verge of bursting its bonds in the crystal structure, for then all the ice would melt at once when we added just a small amount of energy. Instead we find that it takes 344 J to melt each gram of ice (at 1 atm pressure).

The transition from the liquid to the vapor phase may take place without the abrupt phase transition. In the process indicated in Fig. 19.1 where the substance goes from *X* to *Y*, the substance gets hotter and hotter and bonds are continuously being broken. There is no point at which the temperature remains constant while the phase change occurs.

19.3 SPECIFIC HEAT

When measuring the transfer of heat between bodies we often find it more convenient to use the unit called a *calorie* than to use the mks unit of energy, the joule. Recall that the *calorie* (abbreviated cal) is defined as that amount of energy that must be added to 1 g of liquid water at 1 atm pressure to raise the temperature 1°C. The kilocalorie (kcal) is 1000 cal.

Experiment shows that the amount of work that must be done to raise 1 g of water 1°C is 4.186 J. Therefore

1 cal = 4.186 J

When the accuracy demands only two significant figures, we shall use the value 4.2 J.

QUESTION

Suppose you want to boil an egg by heating the water through friction. How much work must you do to bring the water to 100°C? The water is initially at 30°C and has a mass of 100 g.

ANSWER

If it takes 1 cal to heat 1 g of water 1°C, then it takes 70 cal to heat 1 g from 30°C to 100°C. The heat required to bring 100 g from 30°C to 100°C would be 7000 cal or

$$(7000 \text{ cal})\left(4.186 \frac{\text{J}}{\text{cal}}\right) = 29{,}300 \text{ J}$$

QUESTION

Suppose that in the preceding example you supply the energy by rubbing the vessel with a rough object. The frictional force is 1 lb = 4.4 N. If the rough object is moved 20 cm in one stroke (back and forth), how many strokes are required? (Neglect the heat energy generated in the rough object.)

ANSWER

Work = 29,300 J

work = (force)(distance)

distance = n(0.20 m)

where n = number of strokes. Since the force is 4.4 N,

(4.4 N)(n)(0.20 m) = 29,300 J

or

$$n = \frac{29{,}300 \text{ J}}{(4.4 \text{ N})(0.20 \text{ m})} = 33{,}300 \text{ strokes}$$

Clearly the use of friction to increase heat is not a practical method for cooking!

Different amounts of heat energy are required to raise the temperatures of different substances a given amount. The *specific heat c* of a substance is defined as the amount of heat required to raise a unit mass 1°C. The numerical value of the specific heat depends on the units used to measure heat and mass. For example, the specific heat of water may be written

$$c = 1 \frac{\text{cal}}{\text{g} \cdot {}^\circ\text{C}} = 4.2 \frac{\text{J}}{\text{g} \cdot {}^\circ\text{C}} = 4.2 \times 10^3 \frac{\text{J}}{\text{kg} \cdot {}^\circ\text{C}}$$

The specific heat depends not only on the substance, but also on its temperature, pressure, and phase. The specific heat of water at 1°C is not quite the same as that of water at 99°C. The specific heat of ice between −10°C and 0°C at atmospheric pressure is about 0.5 cal/g · °C, and the specific heat of steam at the same pressure near 100°C is 0.48 cal/g · °C.

Table 19.1 gives the specific heats of some common substances at 1 atm pressure. Note that water has the highest specific heat of all the substances listed. This makes it an effective coolant in water-cooled engines, since it will absorb more heat per unit increase in temperature. It also makes it a suitable environment for biological reactions in warm-blooded animals. Temperature regulation is simpler when small amounts of heat do not give rise to large changes in temperature.

TABLE 19.1 Specific heats of some common substances.

Substance	Specific heat (cal/g · °C)	Substance	Specific heat (cal/g · °C)
Alcohol	0.60	Iron	0.11
Aluminum	0.21	Lead	0.03
Brass	0.09	Mercury	0.03
Chloroform	0.23	Silver	0.06
Copper	0.09	Tin	0.06
Glass (crown)	0.16	Steam (100°C)	0.48
Ice (0°C)	0.50	Glycerine	0.58
Body tissue	0.83	Water	1.00

The specific heat depends not only on the substance, temperature, pressure, and phase of the substance, but also on the process. For example, consider water at room temperature. If the water were exposed to the atmosphere and heated, we would find that it takes 1 cal to raise the temperature 1°C. This is a constant-pressure process. If, on the other hand, a vessel were filled and a cap put on it, we would find that it takes a different amount of heat to raise the water 1°C. This is a constant-volume process. We could therefore define two different specific heats, say c_P and c_V—the specific heat at constant pressure and the specific heat at constant volume. The specific heats listed in Table 19.1 are all at constant pressure and, unless otherwise specified, this will be the specific heat used in the future.

QUESTION

Why is c_P always greater than c_V?

ANSWER

Suppose we take 1 g of the substance at a pressure P_0 and find that it takes x cal to raise the temperature from 300 K to 301 K with the volume held fixed. It follows then that $c_V = x$. If we now allow the system to expand until the pressure returns once again to P_0, the temperature will drop as a result of the expansion. Suppose we have to add an additional y cal to restore the temperature to 301 K. It follows that $c_P = x + y$, since the net effect has been a 1° rise in temperature with no change in pressure. Since $c_V = x$ and $c_P = x + y$, it follows that $c_P > c_V$.

From the definition of specific heat as the amount of energy necessary to raise a unit mass of the substance 1°C, we see that if the temperature of a substance of mass m is increased by an amount ΔT, then the heat energy required is given by

$$Q = cm \, \Delta T$$

QUESTION

A woman puts on a 500-g silver bracelet at 0°C. How many calories of heat must she expend to raise it to her body temperature (37°C)?

ANSWER

$$Q = cm \, \Delta T = (0.06 \text{ cal/g} \cdot \text{°C})(500 \text{ g})(37\text{°C})$$

$$= 1110 \text{ cal} = 1.11 \text{ kcal}$$

QUESTION

One gram of body fat is capable of releasing approximately 9.3 kcal of energy. How many pounds of fat would the lady in the preceding example burn up in heating the bracelet?

ANSWER

$$\frac{1.11 \text{ kcal}}{9.3 \text{ kcal/g}} = 0.12 \text{ g} = 0.00027 \text{ lb}$$

This is not a very effective way to lose weight!

QUESTION

A 10-kg lead weight is hung from a pulley that turns a paddle wheel immersed in 0.01 kg of water (see Fig. 19.4). The weight falls a distance of 1 m and strikes the floor with a velocity of 2 m/s. Assuming that no heat is transmitted to the floor, determine the increase in temperature of the lead and of the water.

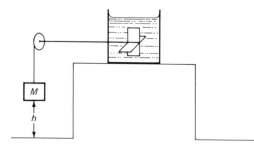

FIGURE 19.4 A lead weight falls to the floor and turns the paddle wheel immersed in water.

ANSWER

The increase in energy of the water must equal the decrease in energy of the lead weight just before it strikes the floor, or

$$\Delta E = Mgh - \tfrac{1}{2}Mv^2$$

The increase in energy of the water gives an increase in temperature given by

$$\Delta E = cm \, \Delta T.$$

Therefore

$$cm \, \Delta T = Mgh - \tfrac{1}{2}Mv^2$$

so that

$$\Delta T = \frac{Mgh - \tfrac{1}{2}Mv^2}{cm} = \frac{(10)(9.8)(1) \text{ J} - \tfrac{1}{2}(10)(2)^2 \text{ J}}{(1 \text{ cal/g} \cdot {}^\circ\text{C})(10 \text{ g})(4.2 \text{ J/cal})}$$

$$= 1.9{}^\circ\text{C}$$

The increase in internal energy of the lead weight after it strikes the floor must equal the decrease in kinetic energy of the weight just before it strikes the floor, or

$$\Delta E_{\text{lead}} = \tfrac{1}{2}Mv^2$$

Therefore

$$\Delta T = \frac{\tfrac{1}{2}Mv^2}{cM} = \frac{\tfrac{1}{2}(10)(2)^2 \text{ J}}{(0.03 \text{ cal/g} \cdot {}^\circ\text{C})(10,000 \text{ g})(4.2 \text{ J/cal})}$$

$$= 0.02{}^\circ\text{C}$$

QUESTION

A 5-g mass of iron at 40°C is placed in a vessel containing 10 g of water at 10°C. What is the final state? What is the net change in entropy? (Neglect the heat taken up by the vessel.)

ANSWER

Since energy is conserved, the heat energy given up by the iron must equal the heat absorbed by the water. The heat gained by the water is

$$\Delta Q_{\text{H}_2\text{O}} = c_{\text{H}_2\text{O}} m_{\text{H}_2\text{O}} (T - T_{0,\text{H}_2\text{O}})$$

$$= (1 \text{ cal/g} \cdot {}^\circ\text{C})(10 \text{ g})(T - 10)$$

$$= 10(T - 10) \text{ cal}$$

where T is the final temperature of the water in Celsius degrees.

The heat given up by the iron is

$$\Delta Q_{iron} = c_{iron} m_{iron}(40 - T)$$

$$= (0.11 \text{ cal/g} \cdot {}^\circ\text{C})(5 \text{ g})(40 - T) = 2.75(40 - T) \text{ cal}$$

(Note that ΔQ_{H_2O} is the heat energy *gained* by the water and ΔQ_{iron} is the heat energy *lost* by the iron.) Equating the energies gained and lost, we have

$$10(T - 10) \text{ cal} = 2.75(40 - T) \text{ cal}$$

and solving for T, we find that

$$T = 16.5°\text{C}$$

The increase in entropy of the water is

$$(\Delta S)_{H_2O} = \int dS = \int \frac{dQ}{T} = \int \frac{c_{H_2O} m_{H_2O} \, dT}{T}$$

$$= \left(1 \frac{\text{cal}}{\text{g} \cdot {}^\circ\text{C}}\right)(10 \text{ g}) \int_{273+10}^{273+16.5} \frac{dT}{T}$$

$$= 10 \ln\left(\frac{289.5}{283}\right) \text{ cal} = 0.23 \text{ cal}$$

The increase in entropy of the iron is

$$(\Delta S)_{iron} = \int dS = \int \frac{dQ}{T} = c_{iron} m_{iron} \int \frac{dT}{T}$$

$$= \left(0.11 \frac{\text{cal}}{\text{g} \cdot {}^\circ\text{C}}\right)(5 \text{ g}) \int_{273+40}^{273+16.5} \frac{dT}{T}$$

$$= 0.55 \ln\left(\frac{289.5}{313}\right) \text{ cal} = -0.04 \text{ cal}$$

The net change in entropy is

$$(\Delta S)_{tot} = (\Delta S)_{H_2O} + (\Delta S)_{iron} = 0.23 \text{ cal} - 0.04 \text{ cal}$$

$$= 0.19 \text{ cal}$$

and is positive, as it must be for any isolated system.

19.4 HEATS OF FUSION AND VAPORIZATION

We have noted that heat is required to bring about a phase transition. The amount of heat depends on the substance, the pressure, and whether the phase transition is fusion (solid to liquid), vaporization (liquid to gas), or sublimation (solid to gas). For water at a pressure of 1 atm the *heat of fusion* is

$$H_f = 80 \text{ cal/g}$$

and the *heat of vaporization* is

$$H_v = 540 \text{ cal/g}$$

The heats of fusion and vaporization give the amount of heat required to melt or vaporize a unit mass, and for mass m this heat would be

$$Q = mH_f \tag{19.5}$$

or

$$Q = mH_v \tag{19.6}$$

QUESTION

A 75-kg person climbs a flight of stairs that has a vertical height of 12 m.

(a) Calculate the work done by the individual's muscles.

(b) Given that the overall efficiency of the muscular system is 10%, calculate the total chemical energy expended.

(c) Calculate the heat energy generated.

(d) If this heat were not lost from the body, but were distributed about the body (whose average specific heat is 0.83 cal/g · °C), what would be the rise in temperature of the body?

(e) Calculate the amount of perspiration that would have to evaporate to keep the body temperature constant if there were no heat loss by other means.

ANSWER

(a) Work performed $= Mgh = $ (75 kg)(9.8 m/s^2)(12 m) $=$ 8820 J

(b) (Chemical energy)(.10) $=$ work performed $=$ 8820 J. Therefore

chemical energy $=$ 88,200 J

(c) Heat energy $Q = $ (0.90)(total chemical energy)

$$= (0.90)(88,200 \text{ J})$$

$$= 79,380 \text{ J}$$

$$= \frac{79,380 \text{ J}}{4.2 \text{ J/cal}} = 18,900 \text{ cal}$$

(d) $Q = cm \, \Delta T$. Therefore

$$\Delta T = \frac{Q}{cm} = \frac{18,900 \text{ cal}}{(0.83 \text{ cal/g} \cdot {}^\circ\text{C})(75,000 \text{ g})} = 0.3{}^\circ\text{C}$$

(e) $Q = mH_v$ where m is the mass evaporated, so

$$m = \frac{Q}{H_v} = \frac{18,900 \text{ cal}}{540 \text{ cal/g}} = 35 \text{ g}$$

19.5 EVAPORATION AND BOILING

Imagine a large closed vessel containing air on the top and a layer of water on the bottom (see Fig. 19.5). Some of the water molecules near the surface of the liquid have sufficient energy to escape from the liquid and enter the air above. When the vessel is large and the layer of water is thin, all the water evaporates, but when the proportion of water is large, only some of the water evaporates. Not only do molecules leave the liquid to enter the gas above,

FIGURE 19.5   The region above the liquid contains a mixture of air and water. Saturation is reached when equal numbers of water molecules are entering and leaving the liquid.

but also water molecules in the gas reenter the liquid. The more water molecules there are in the gas, the faster the condensation. The system eventually reaches a state of equilibrium, in which the evaporation equals the condensation. When this state is reached, we say that the air is *saturated* with vapor.

Let us compare the phenomenon of evaporation with boiling. As we add heat to water, at a certain point bubbles form in the liquid. These bubbles consist *solely* of water vapor. If there were no gravity and no more heat were added to the water, the bubbles would remain in equilibrium in the liquid. On a microscopic level this equilibrium implies that as many water molecules leave the liquid and enter the bubbles as leave the bubbles and enter the liquid; on a macroscopic level this implies that the pressure of the vapor in the bubbles is equal to the pressure in the liquid. If a bubble formed at some lower temperature, it would immediately collapse, since the pressure of the vapor inside the bubble would be less than the pressure in the water.

Let us imagine water in a cylinder whose pressure can be controlled by a piston. With the pressure set at 0.40 atm we find that boiling takes place at 30°C (Fig. 19.6). The boiling is allowed to continue until a given volume of water vapor is formed, say 1 m^3. In a similar vessel we inject air into the vapor above the liquid until at equilibrium the pressure is 1 atm. This pressure is the sum of the partial

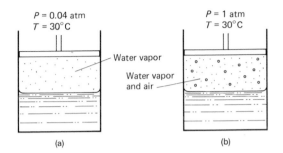

FIGURE 19.6 Saturated water vapor without (a) and with (b) an admixture of air.

pressures of the air and the saturated water vapor. The volume occupied by the mixture is 1 m³. We would like to compare the pressure of the water vapor in the two cases. In both cases the water vapor behaves like an ideal gas, and both vapors are at the same temperature and occupy the same volume. If their densities were the same, then their pressures would be equal. Their densities are determined by the rate at which water molecules leave the surface of the liquid, and this in turn is determined by the state of the liquid, that is, its temperature and pressure. The temperature is the same in both cases, but the pressure is 0.04 atm in the boiling water and 1 atm in the saturated air mixture. Now the properties of water, and of most other liquids for that matter, do not depend strongly on the pressure; certainly the density is not affected significantly by pressure. Water is almost incompressible: For a change in pressure of 100 atm the change in volume of 1 m³ of water is only 0.5 m³. Similarly, the evaporation rate is relatively constant over a wide pressure range. It follows then that the rate at which molecules leave the water surface will be the same at both pressures, but the rate at which water molecules in the air reenter the liquid depends on their density and mean velocity (i.e., the temperature). Since the temperature is 30°C in both vessels, it follows that the density of water molecules above

the liquid must be the same in both vessels. If the density, temperature, and volume of the water molecules above the liquid are the same in both vessels, we see that the vapor pressure must be the same in both. (We assume that the water vapor and the air behave like independent ideal gases.) We may make the following general statement:

The saturated vapor pressure depends only on the temperature and is equal to the pressure at which boiling occurs at that temperature.

In Table 19.2 we give the saturated vapor pressures for water at different temperatures. We see that the saturated vapor pressure of water at 100°C is 1 atm, as we expect.

TABLE 19.2 Saturated vapor pressure of water at different temperatures.

Temperature (°C)	Saturated vapor pressure (atm or N/m² × 10⁵)
0	0.006
5	0.009
10	0.012
15	0.016
20	0.023
30	0.042
40	0.072
50	0.12
60	0.20
70	0.31
80	0.47
90	0.69
100	1.00
120	2.0
140	3.6

QUESTION

The saturated vapor pressure increases with increasing temperature. The density of the saturated vapor is given by

$$\frac{n}{V} = \frac{P}{RT}$$

Although P increases with increasing temperature, so does RT. Does the saturated vapor density increase or decrease with increasing temperature?

ANSWER

The answer to this question is not obvious, and we must look to the data of Table 19.2 to see whether P/RT increases or decreases with temperature. Let us calculate the saturated vapor density at 0°C, 50°C, and 100°C. These are

$$\frac{n}{V} = \frac{0.006 \times 10^5 \text{ N/m}^2}{(8.3 \text{ N} \cdot \text{m})(273 + 0)} = 0.26 \text{ mol/m}^3 \quad \text{at } 0°C$$

$$\frac{n}{V} = \frac{0.12 \times 10^5 \text{ N/m}^2}{(8.3 \text{ N} \cdot \text{m})(273 + 50)} = 4.5 \text{ mol/m}^3 \quad \text{at } 50°C$$

$$\frac{n}{V} = \frac{1.00 \times 10^5 \text{ N/m}^2}{(8.3 \text{ N} \cdot \text{m})(273 + 100)} = 32.3 \text{ mol/m}^3 \quad \text{at } 100°C$$

The saturated vapor density does increase with increasing temperature.

We have assumed that the saturated water vapor behaves like an ideal gas, and this is a good approximation.

QUESTION

Water is allowed to evaporate in a room until saturation is reached. The volume of air and its saturated vapor is 300 m³. If the room temperature is 30°C, what is the total mass of water vapor?

ANSWER

The saturated vapor pressure at this temperature is 4.2×10^3 N/m². Assuming that the vapor satisfies the ideal-gas law, we have

$$P = \frac{n}{V}RT$$

or

$$\frac{n}{V} = \frac{P}{RT} = \frac{4.2 \times 10^3 \text{ N/m}^2}{(8.3 \text{ N} \cdot \text{m})(273 + 30)} = 1.7 \text{ mol/m}^3$$

$$= \left(1.7 \frac{\text{mol}}{\text{m}^3}\right)\left(18 \frac{\text{g}}{\text{mol}}\right)$$

$$= 30 \frac{\text{g}}{\text{m}^3}$$

since the molecular weight of water is 18. Thus the mass in 300 m³ is $(30 \text{ g/m}^3)(300 \text{ m}^3) = 9$ kg. This is a lot of water— 9 liters.

QUESTION

How much water is lost in breathing?

ANSWER

The lungs and air passages are very moist, so air taken in becomes saturated with water vapor. The temperature of the air in the lungs is nearly equal to the body temperature (37°C), and the saturated vapor pressure at 37°C is 0.063×10^5 N/m². The water vapor density is therefore

$$\frac{n}{V} = \frac{P}{RT} = \frac{0.063 \times 10^5 \text{ N/m}^2}{(8.3 \text{ N} \cdot \text{m})(273 + 37)} = 2.4 \text{ mol/m}^3$$

The volume of air inhaled with each breath is on the average 0.5 liters $= 5 \times 10^{-4}$ m^3. Therefore

$$n = (2.4 \text{ mol/m}^3)(5 \times 10^{-4} \text{ m}^3) = 1.2 \times 10^{-3} \text{ mol}$$

Since the molecular weight of water is 18 g/mol, this is equivalent to a water loss with each breath of

$$m = (18 \text{ g/mol})(1.2 \times 10^{-3} \text{ mol}) = 0.02 \text{ g}$$

In this calculation we have made some approximations that make the result larger than it should be. First we have neglected the water vapor in the air inhaled. If the air is saturated (at the room temperature), the amount of additional water that the air can hold at the higher body temperature will be reduced. Second we have assumed that the air is exhaled at 37°C, but actually its temperature is reduced as it passes out. The air passages of the nose and throat are cooled by evaporation, so as the air passes out, it in turn is cooled by contact with the cooler walls. This reduces the saturated vapor pressure and hence the saturated vapor density. Some water will condense as the air is exhaled. This condensation is very important to the desert rat who is trying to conserve water.

On the other hand, a dog, whose principal means of cooling itself is through evaporation from the lungs, wishes to minimize the condensation on the nasal passages and so to increase the total water loss. The dog accomplishes this by breathing in through its nose and exhaling through its mouth, thus minimizing the cooling of the air passages and allowing the exhaled air to carry more water vapor.

QUESTION

If you were immersed in a tub of hot water at 98.6°F (37°C) and your only means of heat dissipation were panting, could you survive? Assume that the body is generating 2000 kcal/day (the basal metabolic rate).

ANSWER

The normal amount of air inhaled is about 6 liters min, and the maximum possible value is about 25 times this amount. Let us assume that the air comes in dry and leaves saturated at 37°C. The water density of the exhaled air is 2.4 mol/m^3 (see the preceding question) or (2.4 mol/m^3)(18 g/mol) = 43 g/m^3. For every gram of water lost, 540 cal of heat are lost. Therefore the heat loss is (43 g/m^3)(540 cal/gm) = 23 kcal/m^3. So for the body to dissipate 2000 kcal/day, the total volume of air intake must be (2000 kcal/day) ÷ 23 kcal/m^3 = 86 m^3/day or 50 liters min. Thus by breathing at 10 times the normal rate you would be able to dissipate the heat generated by the normal body functions (the BMR).

Actually the BMR is measured in a room at 70°F. Since part of this energy expended by the body is used to maintain constant body temperature, one would expect the metabolic rate to decrease in a tub of hot water.

SUMMARY

The equation of state defines a relation between P, V, n, and T for any substance.

During a phase change at constant pressure, the temperature remains constant.

The specific heat of a substance is the energy required to raise a unit mass of the substance 1 °C. It follows that

$$Q = cm \, \Delta T$$

is the heat energy required to raise a mass m through a temperature change ΔT.

The heat of fusion H_f is the heat energy required to melt or solidify a unit mass. The heat of vaporization H_v is the heat energy required to vaporize or liquify a unit mass.

When the rates of evaporation and condensation in the space above a liquid in a closed vessel become equal, the region is said to be saturated. The partial pressure of the vapor is the saturated vapor pressure, which is a function of temperature alone.

PROBLEMS

19.A.1 If $S = \frac{3}{2}\alpha n^{1/3}(EV)^{2/3}$, what is the equation of state?

19.A.2 If $S = 3\alpha(nEV)^{1/3}$, what is the equation of state?

19.A.3 A 10-g mass of lead at $0°C$ is dropped from a height of 20 m into a large ice-water bath.
(a) How much ice melts?
(b) What is the change in entropy?

19.A.4 If a person were to lose all the heat he generated by the body through the evaporation of perspiration, how much water would the person lose per day if the heat generated were 2000 kcal?

19.A.5 What is a BMR of 2000 kcal/day in watts?

19.A.6 In problem 11.B.1 the power output of the heart was estimated to be about 1 W. What fraction is this of the BMR if the BMR is 2000 kcal/day?

19.A.7 At what temperature will water begin to boil on a mountain top where the pressure is 0.69 atm?

19.B.1 Twenty grams of copper at $40°C$ are dropped into a 50-g aluminum vessel containing 100 g of water at $10°C$. What is the equilibrium temperature?

19.B.2 Two beakers containing 10 g of water each are at temperatures of $10°C$ and $50°C$. The beakers are brought into thermal contact and reach a common temperature.
(a) Determine the equilibrium temperature.
(b) Determine the change in entropy of each system.
(c) Is the net change in entropy positive or negative?

19.B.3 Ten grams of water at $10°C$ are mixed with 100 g of water at $40°C$. What is the final temperature?

19.B.4 A 10-g mass of iron at $40°C$ is dropped gently into a large ice-water bath. How much ice melts? What is the net change in entropy?

19.B.5 Suppose you want to design a water heater from which you may draw 1 kg of water per minute (16.7 g/s) continuously. The water comes from the city at $20°C$ and you want hot water at $80°C$. What should be the wattage of the heater?

19.B.6 An automobile weighing 4000 kg and traveling at a speed of 60 mi/h (27 m/s) comes to a stop in 100 ft. Given that the total mass of the brake drums and lining is 2 kg and that the average specific heat is 0.8 cal/g · $°C$, determine the increase in temperature of the brakes, neglecting heat losses.

19.B.7   Water condenses on the outside of a cold glass of water. As the glass warms up the dew on the glass disappears at $10°C$. The glass is in a room at $30°C$. What is the partial pressure of the water vapor in the room?

19.B.8 Water vapor condenses on a cold glass. As the glass warms up the vapor disappears at $20°C$.
(a) What is the density of the water vapor in the room?
(b) If the room temperature is $30°C$, what is the partial pressure of the vapor in the room?

19.C.1 A 50-g copper block is pushed with constant velocity along a flat horizontal surface of aluminum for a distance of 1 m. The force of friction between the two surfaces is 0.25 N. (Assume that both the block and the surface begin at the same temperature.)

  (a) From the information given, could you determine the increase in temperature of the copper block if it were removed from the aluminum surface after traveling a distance of 1 m?

  (b) If the mass of the aluminum were 50 g, could you determine the increase in temperature of the aluminum?

  (c) If both the copper and the aluminum had a mass of 50 g and you waited until they reached a common equilibrium temperature, could you predict the increase in temperature?

19.C.2 A 1-kg block of copper at 500°C is dropped gently into a vessel containing 80 g of ice at 0°C. Determine the final state of the system.

20
HEAT
TRANSFER

WE HAVE SEEN that any system that is not at a uniform temperature will in time approach a uniform temperature if there is some mechanism for the transfer of energy. In the absence of some internal constraint, the equilibrium state is one of uniform temperature.

Energy can be transferred from one part of a system to another by three mechanisms: convection, conduction, and radiation.

CONVECTION

When a hot-water radiator is placed in a room, the air near the radiator is heated, and since this hot air is less dense than the surrounding cool air, it rises. As cool air moves in to take the place of the hot air, it in turn is heated by the radiator. This process is repeated over and over again until the whole room reaches the temperature of the radiator (if we assume no heat loss from the room). The heat energy has been transferred by convection: The hot gas has been physically transported from one point in the room to another. This is also the primary mechanism by which heat is transported in the body. The blood is heated in various organs and then carries this heat to other parts of the body.

CONDUCTION

Imagine a bar of some solid that is thermally insulated from its surroundings. Initially one end of the bar is hotter than the other end, but eventually the bar acquires a uniform temperature. Certainly the hot molecules at one end cannot be transported to the other end by convection. The molecules are bound to their lattice sites in the crystal, i.e., they oscillate about fixed positions. Now the molecules at the hot end are oscillating more vigorously than those at the other end, and it is possible for a molecule that is oscillating vigorously to affect its neighbors. In fact, these intermolecular forces between neighbors bind the molecules into a crystal structure. When a hot molecule is adjacent to a cold molecule, the hot molecule will transfer some of its energy to the cold molecule. This molecule in turn will transfer some of its newly acquired energy to its other neighbors, and so on. Thus heat energy is transferred throughout the solid, although hot molecules are not actually moved to cold regions. In this case the heat is transferred by conduction.

RADIATION

Imagine two bodies, one hot and one cold, placed inside a box whose walls are mirrors. We find that even if the box is evacuated, the two bodies will reach the same temperature. If there is no medium between the two bodies to either convect or conduct the heat, why do they reach the same temperature? Some other mechanism for transmitting energy from one body to the other must exist.

The mechanism is electromagnetic radiation. Matter is composed of electrons, protons, and neutrons. The electrons and protons are electrically charged, and the thermal motion of molecules causes a movement of these charges. We shall find later in our study of electricity that moving charges (actually accelerating charges) create electromagnetic waves. Visible light is an example of electromagnetic waves with a wavelength to which the eye is sensitive. Other examples of electromagnetic waves are x rays, radio waves, infrared rays, and ultraviolet light. These are all electromagnetic waves, but their wavelengths do not stimulate the rods and cones of the eye. These waves carry energy and do not need a medium for their propagation. (In contrast, sound waves do require a medium.) Let us consider, for example, the light radiated by the sun. The space between the earth and the sun is practically a vacuum, and hot summers and cold winters are clear evidence that the sun's rays carry energy.

The hotter a body, the more energy it radiates, so the hot body in the box will radiate more energy than the cold body. Both bodies will absorb radiation, and therefore the radiation serves as a mechanism for the exchange of energy between the two bodies. Eventually thermal equilibrium will be reached and both bodies will come to a common temperature.

(We would like to interrupt our discussion of heat transfer for a moment to remark on a common feature of all these examples. In each case the mechanism is entirely different, but the end result is always the same. The transfer of heat energy always results in a common temperature throughout the system. This is a clear-cut example of the fact that the nature of the equilibrium state is independent of the forces that take the system to this equilibrium state. Any mechanism of interaction will take the system to the same state, and the final state is determined only by the laws of chance. The final state is the macrostate with the most microstates.)

We will now attempt to deal quantitatively with these last two mechanisms of heat transfer. (Convection generally involves the turbulent motion of a fluid and is much too difficult to analyze.)

20.1 CONDUCTION

As an example of thermal conduction, let us consider a rectangular bar of length L and cross-sectional area A (see Fig. 20.1). One end of the bar is held fixed at a temperature T_1 (by immersion in a large constant-temperature reservoir), and the other at a temperature T_2 ($T_2 > T_1$). Let us assume that the temperature at the midpoint of the bar is the mean temperature $(T_1 + T_2)/2$. Thus if $T_1 = 40°C$ and $T_2 = 60°C$, the temperature at the midpoint is $50°C$. If we also assume that the heat Q flowing down

FIGURE 20.1 A rectangular bar with the temperature fixed at both ends.

the bar is proportional to the temperature difference ($\Delta T = T_2 - T_1$), then it follows that the heat flow must in fact be proportional to the temperature gradient $\Delta T/L$ that is, the temperature difference per unit length:

$$Q \propto \frac{\Delta T}{L}$$

To see this we may imagine that the right half of the bar is part of the mechanism that holds the midpoint at a temperature of $(T_2 + T_1)/2$. Now the heat flow down the bar cannot be changed just because we have changed our point of view. Thus the heat flow through a bar of length L with a temperature difference of ΔT must be the same as the heat flow through a rod of length $L/2$ with a temperature difference of $\Delta T/2$ (see Fig. 20.2). The only way Q will remain unchanged if we change L to $L/2$ and ΔT to $\Delta T/2$ is for Q to depend only on the ratio $\Delta T/L$.

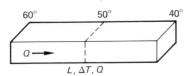

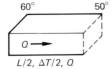

FIGURE 20.2 A bar of half the length and half the temperature difference conducts the same amount of heat.

Since we have assumed that $Q \propto \Delta T$, it follows that

$$Q \propto \frac{\Delta T}{L}$$

We see then how Q depends on ΔT and L, but how does it depend on A, the cross-sectional area? Suppose a second rod, identical in all respects, is placed alongside the first (Fig. 20.3). Clearly this second rod will conduct the same amount of heat. In effect we have just doubled the cross-sectional area of the first rod. Therefore Q must be directly proportional to A, and we have

$$Q \propto A \frac{\Delta T}{L}$$

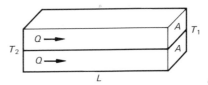

FIGURE 20.3 Two identical rods placed one on top of the other.

The time t is another factor that determines the amount of heat flow. If we double the time, the heat flow will double, and therefore

$$Q \propto At \frac{\Delta T}{L}$$

The only other factor on which the heat conducted depends is the nature of the material. This dependence is complicated, and is generally lumped into a factor of proportionality K, called the *thermal conductivity*. We may then write

$$Q = KAt \frac{\Delta T}{L} \tag{20.1}$$

The factor K will depend on the nature of the substance, somewhat on the temperature, and of course on the units chosen (see Table 20.1).

TABLE 20.1 Thermal conductivities of various substances at room temperature (30°C).

Substance	$K(kcal/m \cdot s \cdot °C)$
Metals	
Silver	10×10^{-2}
Copper	9.2×10^{-2}
Aluminum	5.1×10^{-2}
Brass	0.25×10^{-2}
Iron and steel	0.11×10^{-2}
Other solids	
Ice	5.2×10^{-4}
Concrete	4.1×10^{-4}
Glass	1.9×10^{-4}
Brick	1.7×10^{-4}
Oak	0.38×10^{-4}
Pine	0.28×10^{-4}
Liquids	
Water	1.4×10^{-4}
Insulating materials	
Sawdust	0.14×10^{-4}
Cork	0.1×10^{-4}
Rock and glass wool	0.093×10^{-4}
Kapok	0.083×10^{-4}
Gases	
Hydrogen	0.41×10^{-4}
Air	0.055×10^{-4}

QUESTION

Which would you rather sleep on out-of-doors, an air mattress or a kapok pad of the same thickness?

ANSWER

The kapok would be much superior even though it has a higher thermal conductivity. The thermal conductivity given for air is for still air. In an air mattress there are very large convection currents, and for this reason the air is a very poor insulator. Goose down is good insulator because it fluffs up well, creating a thick layer of air and at the same time preventing the air from circulating.

QUESTION

The heat output of a male husky dog was measured at several outside air temperatures (see Table 20.2). The internal body temperature remained constant at 38°C, and the total surface area was 1.31 m². If the thickness of the skin-fur layer is about 0.05 m and the outside temperature is -32°C, what is the thermal conductivity of the layer? Compare this value with the thermal conductivity of still air.

TABLE 20.2 Heat output by a male husky at various outside air temperatures.

Air temperature (°C)	Heat output (kcal/day)
8	1050
-2	1040
-12	1070
-22	1040
-32	1050
-42	1120
-52	1260

ANSWER

$$K = \frac{QL}{At\Delta T} = \frac{(1050 \text{ kcal/day})(0.05 \text{ m})}{(1.31 \text{ m}^2)(24 \times 60 \times 60 \text{ s/day})(70)}$$

$$= 6.6 \times 10^{-6} \text{ kcal/m} \cdot \text{s}$$

The thermal conductivity of still air is 5.5×10^{-6} kcal/m · s, so the conductivity of fur is only slightly higher. The chief function of the fur is in fact to form a layer of trapped air.

It may be difficult to understand some of the figures in Table 20.2, for in some cases the heat output goes down as the temperature goes down. We shall see later how the effective thickness of the insulating layers of skin can be increased by the reduction of the blood circulation to the skin.

20.2 RADIATION

As we pointed out earlier, in the transfer of energy by radiation, one body emits electromagnetic waves that may be absorbed by another body. These electromagnetic waves carry energy that may be converted into heat.

Radiation emitted by a so-called blackbody has some very interesting characteristics. A *blackbody* will absorb all the electromagnetic waves incident on it and reflect none. (It will, at any temperature above absolute zero, emit electromagnetic waves, but these are to be distinguished from reflected waves.) A body that appears black to the human eye may in fact be quite different from a blackbody because the eye sees only a very limited portion of the electromagnetic spectrum. For example, we do not see infrared rays, ultraviolet rays, x rays, radar waves, or radio waves. These are all electromagnetic waves whose wavelengths are outside the visible range.

We may approach a blackbody in practice by making a small hole in a large box whose inside surface is a good absorber. Any light that impinges on the hole from the outside is reflected by the interior walls many times before it reemerges from the hole. On each reflection much of the radiation is absorbed, so the fraction of light that escapes through the hole is very small and the hole appears black. The laws governing blackbody radiation have been discovered by studying the radiation emitted from a hole in an oven. (The radiation escaping through the hole is not caused by the reflection of incident light, but by the escaping electromagnetic radiation that is in thermodynamic equilibrium within the box. The hole will not appear black if the box is hot. In this case substantial radiation in the visible spectrum will escape through the hole. A blackbody is not something that appears black, but something that is nonreflecting.)

QUESTION

Why is it more difficult to make fine furniture out of oak than out of walnut?

ANSWER

If there is a gap where two pieces of wood are joined, the line of the joint will appear black. (The gap is similar to the hole in a box.) Now oak is a light colored wood and walnut is very dark, so a poor joint will stand out very prominently in oak and will hardly be noticeable in walnut.

It is observed experimentally that when a blackbody is heated to a temperature T, the total energy emitted per unit surface area per unit time (R_B) is proportional to the fourth power of the absolute temperature. Thus

$$R_B = \sigma T^4 \tag{20.2}$$

where σ is a constant (the *Stefan-Boltzmann* constant) and has the value

$$\sigma = 1.36 \times 10^{-11} \text{ kcal/m}^2 \cdot \text{s}$$

$$= 5.67 \times 10^{-8} \text{ W/m}^2$$

This is also the energy emitted per unit time per unit area from a hole in a box at a temperature T.

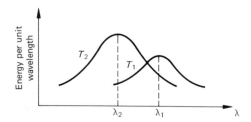

FIGURE 20.4 Radiation energy from a blackbody as a function of wavelength. The temperature T_2 is greater than T_1.

Equation 20.2 represents the energy emitted over the total electromagnetic spectrum. The energy emitted in any given wavelength interval may also be observed through the use of filters. When this energy is plotted for a blackbody at two different temperatures T_1 and T_2 ($T_2 > T_1$) as a function of wavelength λ, curves such as those in Fig. 20.4 result. It is observed experimentally that if λ_1 is the wavelength at which maximum energy is emitted by the blackbody at temperature T_1, and similarly λ_2 is that wavelength for the body at temperature T_2, then

$$\lambda_1 T_1 = \lambda_2 T_2 = \text{constant} = 0.2897 \text{ cm K}$$

Thus we may determine the temperature of the sun, for example, by finding that wavelength at which the maximum energy is emitted (assuming that the sun radiates as a blackbody).

If the radiating body is not a blackbody, we may approximate the radiant energy by introducing an emissivity e and writing

$$R = e\sigma T^4 \tag{20.3}$$

The emissivity lies between 0 and 1, and for a blackbody $e = 1$. The emissivity is not strictly a constant.

We would now like to show that the rate at which a body radiates energy determines the rate at which it absorbs energy. We can see that these two rates are related by observing that any object placed in an oven whose temperature is T will come to thermal equilibrium at this same temperature. This is true regardless of the nature of the object: If we place a blackbody, a white body, a roller skate key, and a Willkie button in an oven, they will all reach the same equilibrium temperature, and this will be the temperature of the oven walls. Now some of these objects are better radiators than others, and they will all have different emissivities. But since all are in equilibrium, they must all be emitting at the same rate at which they are absorbing. Thus the objects that are good absorbers must also be good radiators.

If R_{rad} is the rate at which a body is radiating energy per unit surface area and R_{abs} is the rate at which a body is absorbing energy, then at equilibrium we have

$$R_{rad} = R_{abs}$$

Now $R_{rad} = e\sigma T^4$, so

$$R_{abs} = e\sigma T^4$$

Let us consider a nonequilibrium situation: a body at temperature T_{body} in a room at temperature T_{room} (Fig. 20.5). What is the net rate at which energy is being lost by the body? Now

$$R_{net} = R_{rad} - R_{abs}$$

where R_{rad} is the energy radiated by the body per unit area per unit time and R_{abs} is the energy absorbed per unit area per unit time. The energy absorbed turns out to be practically independent of the temperature of the body. If $T_{body} = T_{room}$, then R_{abs} equals $e\sigma T_{room}^4$. If R_{abs} does not depend on the temperature of the absorber, then

$$R_{abs} = e\sigma T_{room}^4$$

and

$$R_{rad} = e\sigma T_{body}^4$$

Therefore

$$R_{net} = e\sigma(T_{body}^4 - T_{room}^4) \tag{20.4}$$

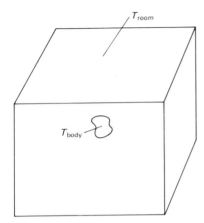

FIGURE 20.5 Energy exchange between a body and a room at different temperatures.

QUESTION

The lizard is a cold-blooded animal that takes its heat from the sun. What is the maximum body temperature attainable by a lizard sitting in the sun? Assume that the lizard loses heat only through radiation.

ANSWER

The average rate at which solar energy falls on 1 m² of surface perpendicular to the sun's rays is 1400 W/m² = 0.3 kcal/m² · s. Suppose the sun is directly overhead. In equilibrium the rate at which energy is absorbed must equal the rate at which energy is radiated. Since the energy is radiated at a rate of

$$R_B = \sigma T^4$$

if we assume that the lizard is a blackbody, then in equilibrium

$$\sigma T^4 = 1400 \text{ W/m}^2$$

or

$$T^4 = \frac{1400 \text{ W/m}^2}{5.67 \times 10^{-8} \text{ W/m}^2} = 2.47 \times 10^{10}$$

or

$$T = 396 \text{ K} = 123°C$$

which of course is much too hot for a lizard. Fortunately the lizard can also lose heat by convection and conduction.

QUESTION

What is the total energy radiated by the sun per second?

ANSWER

Given that the energy flux at the earth is 1400 W/m² and the earth is 1.5×10^{11} m from the sun, the total energy per second passing through a sphere of radius 1.5×10^{11} m is the energy per second per square meter (1400) times the total area of the sphere ($4\pi r^2$), or

$$\text{total energy/sec} = (1400 \text{ W/m}^2)(4\pi)(1.5 \times 10^{11} \text{ m})^2$$

$$= 4.0 \times 10^{26} \text{ W}$$

QUESTION

The radius of the sun is 7×10^8 m. What is its temperature?

ANSWER

The total energy radiated by the sun per second is 4×10^{26} W, which is equal to the energy radiated per second per unit area (σT^4) times the surface area of the sun, that is,

$$4 \times 10^{26} \text{ W} = A\sigma T^4$$

$$= 4\pi(7 \times 10^8 \text{ m})^2(5.67 \times 10^{-8} \text{ W/m}^2)T^4$$

or

$$T^4 = 1.15 \times 10^{15}$$

or

$$T = 5820 \text{ K}$$

QUESTION

Why might a room in which the air temperature is, say, 65°F be quite comfortable in the summer but feel cool in the winter?

ANSWER

We exchange heat not only with the air but also with the walls through radiation. The walls will not necessarily be at the same temperature as the air; they will be cold in the winter and warm in the summer. If the walls are cold, we lose heat to them at a rate of $e\sigma(T_{body}^4 - T_{room}^4)$. If the walls are warm, we gain heat at a rate of $e\sigma(T_{room}^4 - T_{body}^4)$.

QUESTION

Suppose you are stranded in outer space and are radiating into a vacuum. How thick should your sleeping bag be (or how thick should the walls of your capsule be)? The bag is made of down ($K = 0.006$ cal/m · s) and the outer surface area is 2 m². Let your BMR be 2000 kcal/day = 23 cal/s.

ANSWER

When a steady state is reached, the bag must be radiating as much heat as you are generating, that is, 23 cal/s or 97 J/s. Therefore the outer surface temperature of the bag must satisfy the equation (if we assume blackbody radiation)

$A\sigma T_o^4 = 97$ J/s

where T_o is the temperature of the outside surface of the bag. Solving for the outside surface temperature, we find

$T_o = 171$ K

This same amount of heat is being conducted through the walls of the bag. If the temperature of the inside wall is 37°C = 310 K, then the thickness L of the bag must be such that

$KA\dfrac{\Delta T}{L} = 23$ cal/s

or

$$L = \frac{KA\,\Delta T}{23 \text{ cal/s}} = \frac{(0.006 \text{ cal/m}\cdot\text{s})(2 \text{ m}^2)(310 - 171)}{23 \text{ cal/s}}$$

$$= 0.07 \text{ m} = 7 \text{ cm}$$

This is not a particularly thick bag.

*20.3 HEAT TRANSFER AND PHYSIOLOGY

It is interesting to note that horses, dogs, and humans are warm-blooded but that insects are cold-blooded. We have uncovered the physical basis for this difference by the method of scaling discussed in Chap. 10. Now we shall discuss some quantitative features of this phenomenon.

Let us consider the energy required by the body to maintain constant body temperature. When the temperature of the body is greater than that of the surroundings, then energy is lost since heat flows from high temperatures to low temperatures. This heat loss is proportional to the surface area. Let us assume for the moment that heat is generated in the body in proportion to the volume of the body. If we are to maintain an energy balance, then

energy lost in heat

+ energy used in forming body tissue, etc.

= energy intake

Let us consider the ratio of energy intake to body weight:

$$\frac{\text{energy intake}}{\text{body weight}} = \frac{\text{energy lost in heat}}{\text{body weight}}$$

$$+ \frac{\text{energy used in forming body tissue, etc.}}{\text{body weight}} \qquad \text{(20.5)}$$

Now as we consider smaller and smaller animals, the first term on the right-hand side of Eq. 20.5 becomes larger and larger, since the numerator is proportional to

TABLE 20.3 Basal metabolic rates of various animals

Animal	BMR (kcal/day)	Weight (kg)	BMR/body weight (kcal/kg/day)	BMR/surface area (kcal/m²/day)
Mouse	3.82	0.018	212	1185
Guinea pig		0.5		1246
Dog (small)		3.19		1212
Dog (medium)	773	15	51.5	1039
Dog (large)		31.2		1036
Human	2054	64	32.1	1042
Pig	2444	128	19.1	1074
Horse	4833	441	11.3	948

the surface area and the denominator is proportional to the volume. If we let L be a characteristic length of the animal, then the area is proportional to L^2 and the volume is proportional to L^3. The ratio, $L^2/L^3 = 1/L$ approaches infinity as L approaches zero. On the other hand, the energy used in forming body tissue and in other internal functions is proportional to the volume, so the second term on the right-hand side remains constant as L approaches zero. Thus we see that the ratio of energy intake to body weight must become very large in small animals. An insect would have to consume several hundred times its body weight to maintain a high body temperature. Table 20.3 gives the BMRs for various warm-blooded animals. The BMR measures the minimal energy intake necessary to sustain life in a state of complete inactivity. The fact that the BMR per unit surface area per day is very nearly constant over a range of body weights from 0.018 kg to 441 kg suggests that most of the caloric intake under rest conditions is used to maintain constant body temperature.

To support this remark, let us calculate the energy radiated by a person in a room at 20°C. (We assume blackbody radiation.)

Energy radiated/day $= \sigma(T_{body}^4 - T_{room}^4)At$

where A is the surface area of the body (about 1.5 m²) and t is the time (one day). Then

$\sigma(T_{body}^4 - T_{room}^4)At$

$$= 5.67 \times 10^{-8} \frac{J}{s \cdot cm^2 \cdot deg^4}(310^4 - 293^4)$$

$$(1.5 \ m^2)(60)(60)(24)\frac{s}{day}\left(\frac{1}{4.2 \ J/cal}\right)$$

$$= 3.26 \times 10^6 \frac{cal}{day} = 3260 \frac{kcal}{day}$$

Since the BMR is approximately 2000 kcal/day, this is a very large energy loss.

We defend ourselves against such a heat loss in many ways, the most obvious of which is by wearing clothing. The heat lost by radiation is determined by the exposed surface temperature, i.e., the temperature of the clothing which, because of the low conductivity of clothing, is much less than 37°C (98.6°F).

Physiological changes also prevent heat loss. The blood

flow to the skin, i.e., the blood flow in the arterioles, can be greatly reduced (up to 99% in the extremities) by vasoconstriction. Skin temperature is determined primarily by the blood flow since the thermal conductivity of body tissue is low.

Another physiological change, the appearance of goose pimples, is a relic of the time when humans had more body hair. The erect hairs provided better insulation. In addition, since most of the body heat is generated in muscle, these tissues tone up when the body is cold. In extreme cold, shivering increases heat generation. Also, people tend to hunch over when they are cold, thus reducing the radiating surface area.

In cold weather 60—70% of the heat loss is due to radiation; the rest is due to evaporation of moisture from the skin and lungs (only about 2—5% in the case of the lungs) and convection and conduction. In warm weather or during heavy exercise most of the heat loss is due to evaporation, but also the peripheral blood vessels dilate, the skin temperature is raised (the skin appears flushed), and thus radiation is enhanced. In addition, muscle tone relaxes so as to generate less heat.

Variations of radiation from the skin surface are used to diagnose certain tumors through a process called *thermography*. Many tumors cause an anomaly in the circulation and thus cause local temperature differences that can be detected. To estimate the effectiveness of thermography, let us calculate the difference in the radiation from two adjacent portions of skin at temperatures T and $T + \Delta T$.

$$\frac{\Delta R}{R} = \frac{e\sigma(T + \Delta T)^4 - e\sigma T^4}{e\sigma T^4} \simeq \frac{4e\sigma T^3 \, \Delta T}{e\sigma T^4}$$

$$= 4\frac{\Delta T}{T}$$

If we take $\Delta T = 1$ K and $T = 310$ K (body temperature), we find that

$$\frac{\Delta R}{R} = 1.3\%$$

Such variations are easily detected.

20.4 COUNTERCURRENT HEAT EXCHANGE

Countercurrent heat exchange is a device used by many fish, some wading birds, and the male human to minimize heat loss. The whale, for example, is a warm-blooded animal that, because of its small surface-to-volume ratio and thick insulating layer of blubber, is able to live in very cold water. The flippers of the whale, however, are relatively thin and have little insulation. In addition they require large amounts of blood and so are particularly vulnerable to heat loss. To minimize heat loss in the flippers, the artery carrying blood from the heart through the flipper is surrounded by the veins carrying blood back to the heart from the flipper. This proximity of the artery and veins serves a dual purpose: The blood in the veins is cold and that in the artery warm. Heat flowing from the artery to the veins cools the arterial blood and warms the venous blood (Fig. 20.6). The cooled arterial blood loses less heat in the flipper be-

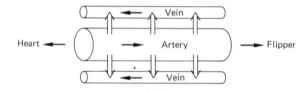

FIGURE 20.6 Countercurrent heat exchange.

cause its temperature is closer to that of the water. Furthermore, the heat that is removed from the artery is not wasted, but is used to reheat the blood returning to the body in the veins. In this manner the heat loss from the blood in the flipper is minimized.

Countercurrent heat exchange is also used in the circulatory system in the male testicles. Sperm cells cannot survive at a temperature as high as normal body temperature. The suspension of the testes in the scrotum outside the body helps reduce the temperature. However, there is a large blood flow to the testes, and the cooling of this blood in the exposed scrotum is not sufficient. To enhance the cooling, the arteries carrying incoming blood are located very close to the veins carrying blood away. The cooler blood in the veins removes heat from the warm blood in the arteries, and thus the arterial blood received by the testes is cooler than it would be without countercurrent heat exchange.

SUMMARY

The three mechanisms of heat transfer are convection, conduction, and radiation. Convection is the transfer of hot material from one place to another. Conduction is the transfer of energy from one particle to another particle by collision. Radiation is the transfer of energy by means of electromagnetic radiation.

The heat energy conducted through a bar of length L and cross-sectional area A during a time t is given by

$$Q = KAt\frac{\Delta T}{L}$$

where ΔT is the temperature difference between the ends. The constant K is called the thermal conductivity.

The energy radiated by a body at a temperature T per unit area is given by

$$R = e\sigma T^4$$

where e is the emissivity and σ the Stefan-Boltzmann constant. For a blackbody $e = 1$. The rate of absorption is practically independent of the body temperature. In thermal equilibrium the rate of absorption is equal to the rate of emission.

PROBLEMS

20.A.1 One wall of a brick house is $3 \times 10 \times 0.2$ m thick. How many calories flow through the wall if the outside temperature is $-20°C$ and the inside temperature is $+20°C$?

20.A.2 A copper ball 6 cm in diameter is coated with lamp black and heated to a temperature of $427°C$.
    (a) How much heat will this body radiate per second?
    (b) If the ball is in a room whose walls are at $27°C$, how much heat will the ball absorb per second?

20.A.3 The rear wall of a fireplace has an effective area of 0.8 m². If the temperature of the wall is $327°C$, how much energy per second is radiated into the room?

20.A.4 A factory roof measures 20×50 m. If the sun's radiation strikes the roof broadside and the energy is converted into mechanical energy with an efficiency of 10%, what power in kilowatts is available?

20.A.5 A blackened solid aluminum sphere of radius 4 cm is placed in an evacuated enclosure whose walls are kept at $70°C$. At what rate must energy be supplied to keep the sphere at a temperature of $100°C$?

20.A.6 Energy from the sun hits an area of black asphalt at a rate of 1400 W/m².

(a) If the incident radiation is normal to the surface, what is the maximum temperature?

(b) If the incident radiation makes an angle of 45° with the surface, what is the maximum temperature?

20.A.7 The temperature of the surface of the sun is 6000 K. At what wavelength is the radiation most intense?

20.B.1 Two bars of the same material are joined together. The length of the first bar is one-half the length of the second bar, and the cross-sectional area of the first is twice that of the second (see the figure below). If the free end of the first bar is immersed in an ice-bath at 0°C and the free end of the second in a steam-bath at 100°C, what is the temperature of the junction?

20.B.2 A long rod that is insulated to prevent heat loss has one end immersed in boiling water and the other end immersed in an ice-water mixture. The rod consists of 100 cm of copper (one end in boiling water), and a length L_2 of aluminum (one end in ice-water). The rod has a cross-sectional area of 5 cm². The temperature of the copper-aluminum junction is 60°C.

(a) How many calories per second of heat energy flow from the boiling water to the ice-water mixture?

(b) What is the length L_2?

20.B.3 A container that has a wall area of 5000 cm² and a wall thickness of 2 cm is filled with water in which there is an electric resistor. The thermal conductivity of the walls is 0.0005 cal/s · cm · °C, and the outer surface of the walls is kept at 0°C. The power delivered to the resistor is 100 W. What is the steady-state temperature of the water in the container? Assume that the temperature of the water is uniform.

20.B.4 If we neglect heat generated internally, the temperature of a planet is determined by the amount of energy it absorbs from the sun and by its emissivity. Assuming that the earth is a blackbody, so that it absorbs 1400 W per square meter of area perpendicular to the sun, determine the mean temperature of the earth.

20.B.5 How does the temperature of a planet depend on its size? (Assume that the planet is a blackbody and that no heat is generated internally.)

20.B.6 How does the temperature of a planet depend on its distance from the sun? (Assume that the planet is a blackbody and that no heat is generated internally.)

20.B.7 Goldilocks, on entering the house of the three bears, found the large bowl of porridge too hot, the medium-size bowl too cold, and the littlest bowl just right. Putting aside the question of breaking and entering, discuss the scientific merits of this tale. How does the rate of cooling vary according to size?

20.B.8

(a) If you were stranded in outer space and your surface temperature were 27°C, at what rate in calories per second would you be radiating energy? Assume that your total mass is 75 kg, your mean specific heat 0.8 cal/g-°C, and your surface area 2 m². Let $e = 1$.

(b) Suppose a drop in temperature of 3°C would cause death and your maximum heat production is 4000 kcal/day $= 46$ cal/s. How long would you last?

20.C.1 The air temperature above a lake is $-20°C$ and the lake is at $0°C$ throughout. How long does it take for a layer of ice 10 cm thick to form? (*Hint:* Suppose that the ice is x cm thick. At what rate is heat conducted from the water to the surface?) The density of ice is 0.917 g/cm^3.

20.C.2 Suppose 700 W of energy are delivered to the filament of an oven. The total surface area of the outside of the oven is 1.5 m^2, the walls are 0.04 m thick, and the thermal conductivity of the walls is 0.4 W/m · °C. Assuming that the outside walls radiate as a blackbody and that radiation is the only means of energy loss, determine the inside and outside wall temperatures when equilibrium is reached.

20.C.3 In the construction of a building it is desirable to minimize the exposed surface area in order to reduce heat exchange.
  (a) Is a rectangular floor plan or a square floor plan of the same area preferable?
  (b) As the area of the floor plan increases, a point is reached at which the exposed surface area can be reduced by use of a two-story structure. Assuming a square floor plan and 8-ft ceilings, determine that area at which it is preferable to use two stories. (Neglect heat exchange with the ground.)

21
ELECTROSTATICS

21.1 COULOMB'S LAW

IN THIS CHAPTER we shall discuss one of the fundamental forces of nature: the electromagnetic force. To appreciate the nature of this force, we shall find it helpful to contrast it with the other fundamental forces. Four fundamental forces exist in nature: gravitational forces, electromagnetic forces, and two types of nuclear forces called *strong* and *weak* interactions. In the interaction between any two bodies the net force is composed of one or more of these four fundamental forces. Because of the differences in the range, strength, and direction of these forces, their relative importance depends on the situation. To make this clear, let us discuss the properties of these fundamental forces.

THE GRAVITATIONAL FORCE

We have already discussed this force in our treatment of dynamics. The force between two bodies of mass m_1 and m_2 separated by a distance r is attractive, and its magnitude is given by

$$F = G \frac{m_1 m_2}{r^2}$$

where $G = 6.67 \times 10^{-11}$ N · m²/kg².

THE ELECTROSTATIC FORCE

We will restrict ourselves to the electrostatic force for the moment, and will consider magnetic forces later. The electrostatic force between two bodies can be either attractive or repulsive. This is an important difference between electrostatic forces and gravitational forces. We find that all electrically charged bodies (i.e., all bodies that exert electrostatic forces on each other) can be divided into two classes. All bodies of the first class repel each other and all bodies of the second class repel each other, but all bodies of the first class are attracted to all bodies of the second class. These two classes are called *positive* and *negative charges.* (The choice of which is called positive and which negative is completely arbitrary.) The dependence of the interaction on the magnitude of the charges (q_1 and q_2) and the distance between the charges is given by

$$F = k \frac{q_1 q_2}{r^2}$$

where k depends on the nature of the medium between the two charges and the units employed to measure the charges. The quantities q_1 and q_2 are taken as positive or negative, depending on whether they belong to the class of positive charges or the class of negative charges. Note that the force is positive (signifying repulsion) when the charges q_1 and q_2 are of like sign and negative (signifying attraction) when they are of unlike sign. This force law is called *Coulomb's law,* after its discoverer Charles A. de Coulomb.

It is by no means obvious that nature should behave in this way. It is conceivable that if body *a* repels body *b* and body *b* repels body *c*, then body *c* could attract body *a*. Fortunately this is not the case, and we are able, through the simple expedient of assigning charged bodies positive and negative numbers, to express the force law in such a convenient manner.

The nature of the interaction between charged bodies has its origins in the nature of the fundamental atomic particles (protons, electrons, neutrons) of which matter is composed. All atomic particles are either electrically neutral or have a positive or negative charge of a given magnitude. By convention, protons are assigned a positive charge and electrons a negative charge; neutrons are neutral.

The value we assign to the quantity of charge depends on the units employed. It is unfortunate that the electron and proton were not discovered before electrical phenomena were observed, for then we might have a more convenient unit of electric charge. The fundamental unit of electric charge in mks units is called the *coulomb* (C). The charge of a proton is 1.6×10^{-19} C. Often the basic element of charge, the charge on the proton or electron, is given the symbol e, so

$$e = 1.6 \times 10^{-19} \text{ C}$$

The coulomb is a new fundamental unit in the mks system, and we should now augment our fundamental standards and call this the mksc system of units, i.e., the meter-kilogram-second-coulomb system of units. This set of four units is now complete; we will not have to add some new unit later. All physical observables can in principle be assigned unique numerical values if we have standards for length, mass, time, and electric charge.

QUESTION

Should there not be a fundamental unit or standard for temperature, which has nothing to do with length, mass, time, or electric charge?

ANSWER

Temperature is in fact dimensionless. It is the derivative of the entropy with respect to energy, but entropy has the same units as energy, and so the derivative is dimensionless. When we write that the temperature is, say 300 K, the K is not a dimension but a notation to indicate how many divisions we choose to make in the temperature interval between an ice-water bath and a steam-water bath at 1 atm pressure. If the number of intervals is 100, then the temperature scale is the Kelvin scale.

With this choice of unit for electric charge, the constant k that appears in Coulomb's law

$$F = k \frac{q_1 q_2}{r^2} \qquad (21.1)$$

has the numerical value

$$k = 9 \times 10^9 \, \text{N} \cdot \text{m}^2/\text{C}^2$$

if the charges are in a vacuum. If the charges are embedded in some medium, k will depend on the nature of the medium.

This force law is often written in another form, in which k is expressed in terms of another constant ϵ_0:

$$k = \frac{1}{4\pi\epsilon_0} \qquad (21.2)$$

so that

$$F = \frac{1}{4\pi\epsilon_0} \frac{q_1 q_2}{r^2}$$

where ϵ_0 is called the *permittivity* of a vacuum and has the numerical value

$$\epsilon_0 = 8.9 \times 10^{-12} \, \text{C}^2/\text{N} \cdot \text{m}^2$$

We shall see later that by choosing this complex form for the proportionality constant we simplify Gauss's law. If we choose k as the proportionality constant, then Gauss's law becomes cumbersome. We shall use either the k or ϵ_0, depending on which is more convenient.

Now that we have a quantitative statement of the force law between charged bodies, it is worthwhile to compare the strength of this electrostatic force with the gravitational force. Consider the case of two electrons. The gravitational force F_g is given by

$$F_g = G \frac{m^2}{r^2}$$

and the electrostatic force F_e by

$$F_e = k \frac{e^2}{r^2}$$

(Remember that $q = e$ if the charge is that of the electron.) Consider the ratio of these two forces:

$$\frac{F_e}{F_g} = \frac{ke^2}{Gm^2}$$

$$= \frac{(9 \times 10^9 \, \text{N} \cdot \text{m}^2/\text{C}^2)(1.6 \times 10^{-19} \, \text{C})^2}{(6.67 \times 10^{-11} \, \text{N} \cdot \text{m}^2/\text{kg}^2)(9.1 \times 10^{-31} \, \text{kg})^2}$$

$$= 4.2 \times 10^{42}$$

so that the electrostatic force is 4.2×10^{42} times stronger than the gravitational force. This is a staggering figure.

As another comparison between the relative strengths of electric and gravitational forces, suppose we removed all the electrons from 1 cm³ of iron and took them to the moon. The electric force between these electrons and the equal number of protons that would be left behind on the earth would be approximately 1700 lb! (See problem 21.A.1.)

QUESTION

Since the electrostatic force is so much stronger than the gravitational force, why are we even aware of gravitational forces?

ANSWER

Two factors are responsible for our awareness of gravitational forces despite the overwhelming strength of electric forces: first, the very fact that the electric force is so large and, second, the fact that electric forces can be either attractive or repulsive, depending on the sign of the charges. We have seen that if we removed the electrons from a piece of iron and took them to the moon, an enormous force of attraction would exist between the electrons and protons even though they were separated by 240,000 mi. In any region, in which there is a significant deviation from charge neutrality, strong forces will attempt to bring about charge neutrality; and when charge neutrality is achieved, the region will no longer be electrically active.

This is not the case with gravitational forces. If two bodies are drawn together because of their gravitational attraction, the composite system is not gravitationally inactive; gravitational forces are always attractive. Gravitational forces act to create a system of *enhanced* gravitational activity. Electric forces act to create a system of *diminished* electrical activity. Like charges repel each other and tend to disperse while unlike charges attract each other and thus create a neutral system. In either case electrical activity is reduced.

Since electric forces are so large, we would expect a collection of particles first to achieve a neutral-charge state, and then perhaps the gravitational forces might be strong enough to pull the particles together into a star or a planetary body. The gravitational force becomes significant when a sufficient number of particles have been drawn together.

QUESTION

Since electric forces are so large that they draw particles together into a neutral-charge state, what keeps us from falling through the floor? The electric forces have been neutralized and that leaves only the gravitational pull of the earth.

ANSWER

Here we run into a problem that we are not ready to deal with, namely the quantum mechanical laws that govern the behavior of matter on an atomic scale. A proton will attract an electron, and we might expect that the electron would be pulled indefinitely close to the proton. As we shall see later, there exist certain discrete states of atoms that cannot be explained by the laws of classical mechanics. In these states there will be a certain distribution of charge. In atomic and molecular systems it is not possible to treat the constituents as particles in the conventional sense, i.e., as objects with size, position, velocity, etc. However, the force responsible for the structure of these atoms and molecules is the electrostatic force. (We shall deal shortly with the forces that hold together the nuclei of these atoms.)

Even though the total electric charge of an atom or molecule may be zero, this does not mean that electric charge is every-

FIGURE 21.1 A positive charge and an equal negative charge separated by a distance *a*. The points *P* and *P'* are in line with the axis of the dipole.

where zero. Consider a pair of equal and opposite charges (see Fig. 21.1). Such a pair is called a *dipole*. The dipole has a *dipole moment p*, which is defined as the product of the charge and the distance separating the charges; that is,

$$p = aq$$

where *a* is the separation between particles and *q* is the magnitude of the charge on either particle. The total charge of a dipole is zero, but we nevertheless expect that the dipole will be electrically active. To see this, suppose that a positive charge is located at *P* (Fig. 21.1). This charge will be repelled by the positive charge $+q$ and attracted by the negative charge $-q$. But the positive charge is closer to *P* than the negative charge, so the force of repulsion will be greater than the force of attraction, and thus there will be a net repulsive force on a charge at *P*.

Now suppose a negative charge is at *P'*. The net force on this charge as a result of the dipole will be attractive, but less attractive than the repulsive force on a positive charge of equal magnitude at *P*. Therefore the dipole on the left of the figure will exert a net repulsive force on a dipole with positive charge at *P* and equal negative charge at *P'*.

In a crude way this answers the question posed. Electrically neutral systems can exert electric forces on other neutral systems as a result of the charge separations. Gravitational pull does not cause us to fall into the earth because electrostatic forces between our feet and the ground cause the ground to repel our feet.

NUCLEAR FORCES

The two types of nuclear forces are called strong and weak interactions. These forces are necessary to hold the nucleus of an atom together. The nucleus is composed of protons and neutrons, and were it not for these nuclear forces, the protons would all fly apart because of their electrostatic repulsion. Since the nucleus does not fly apart, we may conclude that these nuclear forces are stronger than the electrostatic forces.

Fortunately these nuclear forces have a very short range. They do not diminish with distance like $1/r^2$ as do the Coulomb forces and gravitational forces, but they diminish much faster. The nuclear forces are essentially zero for two nuclear particles separated by a distance greater than 3×10^{-15} m. Since the size of the smallest atom is of the order of 10^{-10} m, it follows that we may ignore the nuclear force entirely when dealing with the forces between atoms or molecules. Since the chemical properties of matter are determined by the structure of atoms and molecules, we see that the Coulomb forces together with the quantum mechanical rules for dealing with the dynamics of atomic particles, are entirely responsible for the chemical properties of matter.

21.2 ELECTRIC FIELDS

The force between two charged particles described by Coulomb's law is called an *action-at-a-distance* force, as opposed to a *contact* force, because the two charges exert a force on each other even though they are not touching. In contrast, two billiard balls exert no force on each other until they come in contact. Gravitational force is another example of an action-at-a-distance force.

When we examine more closely the so-called contact forces, we find that they are not contact forces at all. When two billiard balls collide, none of the atomic particles in either ball makes contact with the atomic particles in the other ball. The interaction force is fundamentally the electrostatic force between charged particles and therefore an action-at-a-distance force.

Another way of looking at an action-at-a-distance force is to imagine that one particle creates a field (gravitational or electric) and that the other particle, finding itself in this field, responds by accelerating. One particle creates a field, and the field exerts a force on other particles in this field.

Let us define the *electric field* quantitatively:

The electric field **E** *is a vector field. Its magnitude and direction at a point are given by the force that a positive unit charge would experience at that point.*

Earlier we defined vector fields. We saw, for instance, that the velocity field of a fluid is a vector field; it has magnitude and direction at every point in the fluid. Similarly, the electric field **E** has a magnitude and direction at every point in space. Since this field is the force on a unit charge, we may write

$$\mathbf{E} = \frac{\mathbf{F}}{q'} \tag{21.3}$$

where **F** is the force that a charge q' experiences at a given point in space. Then **E** is the electric field at this point in space. We may think of the charge q' as a test charge used to determine the field. To find the field at a point we place a charge q' at that point, and divide the observed force by the strength of the test charge q'; the result is the field **E**. As an example, let us determine the electric field of a single point charge q at a distance r. We place a test charge q' at r. The force on q' is given by Coulomb's law,

$$F = k\frac{qq'}{r^2}$$

The field then is

$$E = \frac{F}{q'} = k\frac{q}{r^2} \tag{21.4}$$

FIGURE 21.2   The electric field **E** of a single point charge q at a distance r from q.

The field is directed toward or away from q, depending on whether q is positive or negative respectively (Fig. 21.2). Equation 21.4 gives only the magnitude of **E**. We might express the vector **E** as follows:

$$\mathbf{E} = k \frac{q}{r^2} \hat{\mathbf{r}}$$

where $\hat{\mathbf{r}}$ is a unit vector pointing away from the charge q. This vector field **E** exists at every point in space whether or not there is a charge q' at the point to "feel" or "test" the field.

Suppose we have two charges q and q' and we want to know the force that q exerts on q' (see Fig. 21.3). First we must find the field at q' generated by the charge q. This field is given by Eq. 21.4:

$$E = k \frac{q}{r^2}$$

FIGURE 21.3   The force **F** on a charge q' in the field **E** generated by the charge q is q'**E**.

The field E at a point is the force that would act on a *unit* charge at that point, and the force that would act on a charge of strength q' is

$$F = q'E$$

Therefore the force on q' is

$$F = k \frac{qq'}{r^2}$$

This is of course Coulomb's law.

We have presented a field theory of the interaction between charges, not an action-at-a-distance theory. The charge q does not exert a force on the charge q' through a distance, but rather the charge q creates a field **E**. This field exists everywhere in space, in particular at the point occupied by q'. The charge q' feels the field **E** *at the point where it is located.* This is a *local* theory.

You might ask why we bother with a field theory. It gives exactly the same results as the action-at-a-distance theory. For charges at *rest* there is no difference whatever, and there is no real need for a field theory. It is just another way of looking at the same thing. But for *moving* charges there is a very real difference between the field theory and the action-at-a-distance theory. In fact in this case the action-at-a-distance approach to the interaction between charges is no longer feasible.

When you look at a star, electrons in your retina respond to the motion of electrons on that star—not to their motion at the moment of observation, but their motion many years earlier. In the interaction between two or more charged particles (one on the star and one in the retina), if one is moved, it takes time for the other particle or particles to become aware of this motion. The signal takes time to propagate. The velocity of propagation is large, 3×10^8 m/s. All electromagnetic fields propagate with this velocity. Light, of course, is an electromagnetic field.

If we have a system of many charged particles in a state of motion, motion rapid enough so that the finite velocity of electromagnetic signals must be taken into account, it would be difficult to construct a systematic treatment of the interaction between particles from an action-at-a-distance point of view. It is much simpler to construct a physical model in which each charged particle responds not to the state of the system at some earlier time, but to the instantaneous state of an electromagnetic field at the point in space that the particle occupies. To complete this theory we must develop some means of calculating this field from the positions and velocities of the charges that create it. These equations are called *Maxwell's equations,* and we will develop them in the course of our study of electromagnetic theory.

This field theory of the interaction between charged particles represents something of a departure from our second law of dynamics. The second law states that the force on a particle is some function of the instantaneous state of the system, but this is true only if signals propagate with an infinite velocity. The necessary modification of the second law is that the force acting on a charged body depends on the instantaneous state of the field at the point where the charge is located.

This new second law of dynamics, together with Maxwells's equations which predict the nature of the field, complete the dynamics for interacting charged particles. (Coulomb's law is a special case of one of Maxwell's equations for charges at rest. It is a good approximation for charges that are moving slowly.)

Our picture then of the interaction between charged particles on a star and charged particles in the retina of the eye is that the charged particles on the star generate electromagnetic fields (we shall see later that magnetic fields always accompany propagating electric fields) and that these fields propagate toward the earth with a velocity of 3×10^8 m/s, called the velocity of light and denoted by the symbol c. These fields then interact with charged particles in the retina and we *see* the star.

Gravitational signals also propagate with a finite velocity, in fact with the same velocity as electromagnetic signals. To treat gravitational interactions we must therefore use a *gravitational field theory.* For most purposes this gravitational field theory reduces to Newton's law of gravitation, which states that the force is proportional to the product of the masses and inversely proportional to the square of the distance, and this is the analog of Coulomb's law.

21.3 CALCULATING ELECTRIC FIELDS

We have calculated the electric field of a single point charge (Eq. 21.4), and would like to be able to calculate the electric field at any point in space for any distribution of electric charges. Let us begin with a simple case. Two positive charges of 1 C each are separated by a distance of 0.6 m. What is the electric field at a point equidistant from the charges and 0.4 m from the intersection of the lines through the charges (see Fig. 21.4)? Since the electric field is the force on a positive unit test charge and we know from the third law of dynamics that forces add like vectors, the total field must be the vector sum of the fields

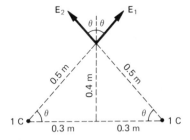

FIGURE 21.4 Two 1-C charges separated by a distance of 0.6 m.

of the two individual charges acting alone. The fields of the individual charges are given by Eq. 21.4:

$$E_1 = k\frac{q}{r^2} = \frac{(9 \times 10^9 \text{ N} \cdot \text{m}^2/\text{C}^2)(1 \text{ C})}{(0.5 \text{ m})^2}$$

and

$$E_2 = k\frac{q}{r^2} = \frac{(9 \times 10^9 \text{ N} \cdot \text{m}^2/\text{C}^2)(1 \text{ C})}{(0.5 \text{ m})^2}$$

where E_1 and E_2 are the *magnitudes* of the electric fields created by the left and right charges respectively. The magnitudes are of course equal. The directions of these fields are indicated in Fig. 21.4. The vector sum of $\mathbf{E}_1$ and $\mathbf{E}_2$ is easily obtained in this case. The horizontal components of $\mathbf{E}_1$ and $\mathbf{E}_2$ are equal and opposite and therefore cancel. The vertical components are the same, so the net electric field E is given by

$$E = E_1 \cos\theta + E_2 \cos\theta = 2E_1 \cos\theta$$

or

$$E = 2\left(9 \times 10^9 \, \frac{\text{N} \cdot \text{m}^2}{\text{C}^2}\right)\frac{1 \text{ C}}{(0.5 \text{ m})^2}\left(\frac{0.4 \text{ m}}{0.5 \text{ m}}\right)$$

$$= 5.8 \times 10^{10} \, \frac{\text{N}}{\text{C}}$$

since $\cos\theta = 0.4$ m/0.5 m. The direction of the net field is vertical.

Note that this electric field is very large, 5.8×10^{10} N/C. This field acting on a charge of 1 C would create a force of 5.8×10^{10} N. One never encounters forces of this magnitude. Macroscopic charges are generally of the order of 10^{-6} C. This is called a microcoulomb and is abbreviated 1 μC.

We can determine the net electric field resulting from any configuration of charge in a similar way. We simply add the Coulomb fields of the individual charges vectorially. If the charges are distributed continuously throughout a volume or over a surface, the vector summation will require the methods of calculus.

21.4 THE ELECTRIC FIELD ON THE AXIS OF A CHARGED RING

To see how we calculate the electric fields of a continuous charge distribution, let us consider the field on the axis of a ring of charge q and radius a (see Fig. 21.5). Let us calculate the electric field at a point P on the axis of the ring and a distance x from the center of the ring created by a small segment of the ring of length ds. Let us call this field dE. It will lie along the line joining ds and P, and if dq is positive, it will point away from ds. The field dE has a component $dE \cos\theta$ along the axis of the ring and a component perpendicular to the axis. If we consider the electric field at P created by an element of charge in the segment ds' on the opposite side of the ring from ds, we see that this field will have the same component along the axis but an equal and opposite component perpendicular to the axis. Therefore the perpendicular components will cancel, and we have to consider only the components along the axis.

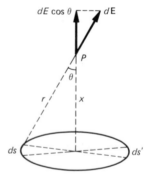

FIGURE 21.5 A ring of radius a carries a total charge q. The point P is on the axis of the ring a distance x from the center. The field $d\mathbf{E}$ is created by the charge on a small segment of the ring of length ds.

The magnitude of $d\mathbf{E}$ is given by

$$dE = k\frac{dq}{r^2}$$

where dq is the charge on the segment ds.

The component of this field along the axis is

$$dE\cos\theta = \frac{k\cos\theta\, dq}{r^2}$$

The net field along the axis is

$$E = \int dE\cos\theta = \int \frac{k\cos\theta\, dq}{r^2}$$

where the integral is performed over the entire ring. Now all the factors in the integrand are constant over the ring and so may be factored from the integral to give

$$E = \frac{k\cos\theta}{r^2}\int dq$$

But $\int dq$ is just the total charge q on the ring, so

$$E = \frac{k\cos\theta\, q}{r^2}$$

Since $r^2 = a^2 + x^2$ and $\cos\theta = x/r$, we may write

$$E = kq\frac{x}{(a^2 + x^2)^{3/2}} \qquad (21.5)$$

We will leave it as an exercise for you to show that this field approaches that of a point charge as x approaches infinity or as a approaches zero.

21.5 THE ELECTRIC FIELD OF AN INFINITE PLANE OF CHARGE

Let us consider a slightly more difficult problem: the electric field at a distance x from an infinite plane of charge carrying a charge σ per unit area. The charge σ is called the *surface charge density*. First let us observe that we may divide the plane into a large number of concentric

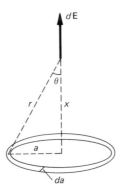

FIGURE 21.6 An infinite plane of charge can be subdivided into an infinite sequence of concentric rings.

rings (see Fig. 21.6), and we have just calculated the electric field on the axis of a ring. If we let $d\mathbf{E}$ represent the electric field created by a ring of charge dq of radius a, of width da, at a distance x from the center, then from Eq. 21.5

$$dE = k\frac{x\, dq}{(a^2 + x^2)^{3/2}}$$

Now since σ is the charge per unit area and $2\pi a\, da$ is the area of the ring,

$$dq = 2\pi a\sigma\, da$$

so that

$$E = \int dE = 2\pi k\sigma x \int_0^\infty \frac{a\, da}{(a^2 + x^2)^{3/2}}$$

$$= \pi k\sigma x \int_0^\infty \frac{da^2}{(a^2 + x^2)^{3/2}}$$

$$= -\pi k\sigma x \left[\frac{2}{\sqrt{a^2 + x^2}}\right]_0^\infty$$

$$= 2\pi k\sigma$$

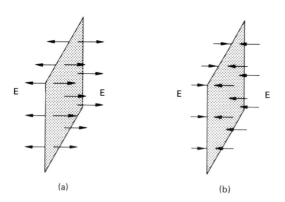

FIGURE 21.7 The electric field of (a) a positive infinite plane and (b) a negative infinite plane.

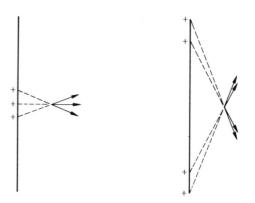

FIGURE 21.8 The electric fields of charges close to the field point are aligned and tend to support each other, while the electric fields of charges far from the field point tend to cancel.

or from Eq. 21.2

$$E = \frac{\sigma}{2\epsilon_0} \tag{21.6}$$

The remarkable thing about this result is that the electric field is independent of the distance x from the plane.

If σ is negative, the field is directed toward the plane rather than away from the plane (see Fig. 21.7).

QUESTION

How is it possible that the electric field strength does not decrease with distance from the plane? Certainly the field strength created by each charge on the plane decreases in inverse proportion to the square of the distance from the plane. Since the field strength of each charge diminishes, why does not the field strength created by the entire body of charges diminish?

ANSWER

We must remember that electric fields add like vectors and not like scalars. The electric field at a point close to the plane is primarily due to the charges close to the point, for these charges create fields that are all closely aligned. The charges

further from the point create fields that are nearly parallel to the plane, and because of the symmetry they nearly cancel (see Fig. 21.8).

As the distance between the field point and the plane increases, more and more of the charges give rise to fields that are aligned, so that even though the magnitude of the field created by each charge decreases with increasing distance from the plane, the alignment of the fields becomes more favorable for increasing the field strength.

QUESTION

What are the force and torque on a dipole that makes an angle θ with respect to a uniform electric field?

ANSWER

The net force on the dipole in Fig. 21.9 is

$$\mathbf{F}_{tot} = q\mathbf{E} - q\mathbf{E} = 0$$

since the electric field is the same at both charges.

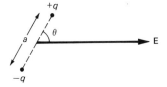

FIGURE 21.9 A dipole makes an angle θ with respect to a uniform electric field.

The net torque, however, is not zero. The torque on the positive charge about the center of the dipole is the product of the force qE and the moment arm $a \sin \theta/2$. Therefore

$$\tau_+ = qE \sin \theta/2$$

Similarly

$$\tau_- = qEa \sin \theta/2$$

Both torques tend to rotate the dipole in the same direction, so the net torque is given by

$$\tau = qEa \sin \theta$$

or, since qa is the dipole moment p,

$$\tau = pE \sin \theta$$

Using the rule for cross products, we may express this result in vector form:

$$\tau = \mathbf{p} \times \mathbf{E} \tag{21.7}$$

21.6 THE ELECTRIC FIELD OF TWO LARGE PARALLEL PLANES WITH EQUAL AND OPPOSITE CHARGES

Next let us consider the electric field created by two large planes carrying equal and opposite charges (see Fig. 12.10). By ''large planes'' we mean planes that are large in comparison to the distance between them. We may determine the electric field for this charge configuration by using the results of Sec. 21.5.

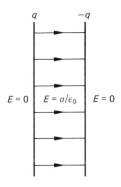

FIGURE 21.10 The electric field created by two large planes carrying equal and opposite charges.

In the region far from the ends of the planes the field of each plane is given by Eq. 21.6. Between the two planes the contributions of the planes add, while outside the planes the fields cancel. Therefore the field between the planes is

$$E = \frac{\sigma}{\epsilon_0} \tag{21.8}$$

and the field outside is

$$E = 0$$

QUESTION

What is the electric field created by the charge distribution illustrated in Fig. 21.11?

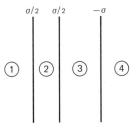

FIGURE 21.11 Three large planes carrying charge densities $\sigma/2$, $\sigma/2$, and $-\sigma$.

ANSWER

The field at any point is just the sum of the fields created by the three planes individually. From Eq. 21.6 the fields in regions 1, 2, 3, and 4 are given by

$$E_1 = 0$$

$$E_2 = \frac{\sigma}{2\epsilon_0}$$

$$E_3 = \frac{\sigma}{\epsilon_0}$$

$$E_4 = 0$$

SUMMARY

Coulomb's law gives the force between two charge particles, and states that

$$\mathbf{F} = k\frac{q_1 q_2}{r^2}\hat{\mathbf{r}}$$

where

$$k = 9 \times 10^9 \text{ N} \cdot \text{m}^2/\text{C}^2$$

if the charges are in a vacuum. Alternatively, we may write

$$\mathbf{F} = \frac{1}{4\pi\epsilon_0}\frac{q_1 q_2}{r^2}\hat{\mathbf{r}}$$

The interaction between charged particles must in general be described by a field theory. The electric field $\mathbf{E}$ is a vector field. Its magnitude and direction at a point are given by the force that a positive unit charge would experience at that point. Thus

$$\mathbf{E} = \mathbf{F}/q$$

The electric field of a point charge q is

$$\mathbf{E} = k\frac{q}{r^2}\hat{\mathbf{r}}$$

The electric field of a collection of charges is the vector sum of the electric fields of the individual charges.

The electric field on either side of an infinite plane of charge is given by

$$E = \sigma/2\epsilon_0$$

where σ is the surface charge density.

PROBLEMS

21.A.1 The density of iron is 7.9 g/cm^3, and the mass of one iron atom is 9.2×10^{-23} g. If all the electrons in 1 cm^3 of iron were taken to the moon (a distance of 3.8×10^8 m), what would be the electric force between the moon and the earth? (There are 26 electrons per iron atom.)

21.A.2 How many electrons must be transferred from one neutral body to another 1 m away to create a force of 1 N?

21.A.3 A 0.5-g particle is supported by a uniform vertical electric field of 100 N/C. What is the charge on the particle?

21.A.4 A 100-g ball hangs from a string, and the other end of the string is attached to a vertical infinite plane. The surface charge density on the plane is $1.5 \ \mu\text{C/m}^2$, and the charge on the ball is $5 \ \mu\text{C}$. What angle does the string make with the plane?

21.A.5 Assuming that the electron in a hydrogen atom obeys the laws of Newtonian dynamics, determine the angular velocity if the electron orbit were a circle of radius 5×10^{-10} m.

21.A.6 A 3-μC charge and a -1-μC charge are separated by a distance of 1 m. Where is the electric field zero?

21.A.7 A 3-μC charge and a 5-μC charge are separated by a distance of 1 m. Where is the electric field zero?

21.A.8 What is the electric field in regions 1, 2, 3, and 4 in the figure? The surface charge densities on the plates are σ, $-\sigma/2$, and σ from left to right.

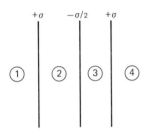

21.A.9 What are the electric fields in regions 1, 2, 3, and 4 in the following figure? The surface charge densities on the plates are $\sigma/2$, σ, and $\sigma/2$ from left to right.

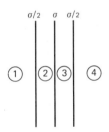

21.B.1 Two 100-g balls carry charges of 0.3 μC and 0.5 μC. The balls are tied to strings 100 cm long, and the strings are attached to a common point. What angle does each string make with the vertical? Use small-angle approximations.

21.B.2 An electron is moving along the axis of a positive ring charge of radius a. Where is the acceleration a maximum?

21.B.3 A proton is projected into a uniform electric field of 1 N/C between two plates with a velocity of 10^4 m/s, as shown in the figure. What is the vertical displacement Δy as the proton emerges from the plates?

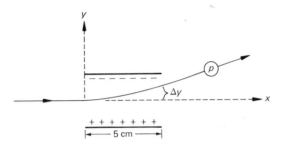

21.B.4 A negative charge moves in a circular orbit about an infinite positive line charge. Show that the speed is the same for all orbits. Determine the speed as a function of the charge per unit length of the line charge λ and the charge and mass of the rotating particle.

21.B.5 Electrons are projected through a small hole into a uniform electric field between two oppositely charged plates, as shown in the figure. It is found that for all velocities below 10^6 m/s the electrons turn in the electric field and reemerge from the hole. For all greater velocities they strike the negatively charged plate. If the distance between the plates is 1 cm, what is the surface charge density on the plates?

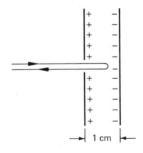

21.B.6 An electron with an initial velocity of 10^6 m/s passes through small holes in two large parallel plates, as shown in the following figure. The separation between the plates is 1 cm. What is the increase in velocity Δv of the emerging electron if the electric field is 1 N/C? Assume that the change in velocity is small compared with the initial velocity ($\Delta v \ll v$).

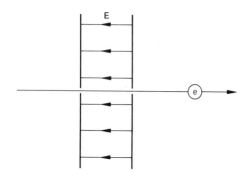

21.B.7 Four charges q, q, $-q$, and $-q$ are placed in sequence on the corners of a square of side a. What is the electric field at the center of the square?

21.B.8 A charge q is uniformly distributed over a semicircle of radius R. What is the electric field at the center?

21.B.9 Determine the electric field of the dipole in the figure on the x and y axes. Assume that a is much less than the distance to the field point.

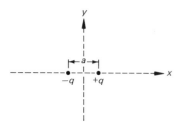

21.B.10 Determine the force between the two dipoles shown in the figure. The dipole moments are p_1 and p_2 and all four charges lie on a line. Assume that r is much greater than the charge separation in each dipole.

21.C.1 What is the force of attraction per unit area between two infinite parallel plates with surface charge densities of $+\sigma$ and $-\sigma$? (*Hint:* The answer is not $\sigma E = \sigma^2/\epsilon_0$.)

21.C.2 Two semi-infinite parallel plates are uniformly charged with equal but opposite surface charge densities σ.

   (a) Compare the electric field at A (see the following figure) with the electric field at B well within the plates.

   (b) What is the direction of the electric field at any point C directly above the edges?

(*Hint:* The field of two infinite parallel plates is the superposition of the electric fields of two semi-infinite parallel plates.)

21.C.3 What force would a 5-μC charge exert on an infinite sheet of charge if the surface charge density σ were 9 μC/m²? (*Hint:* There is an easy way to find the answer.)

22
ELECTRIC
FIELD
LINES
AND
ELECTRIC
POTENTIAL

IN THIS CHAPTER we shall introduce electric field lines and equipotential surfaces. We shall find these helpful in visualizing the electrostatic field. We shall also introduce electric potential, which is related to potential energy in much the same way that the electric field is related to the force. The electric potential is the potential energy that a unit charge would have, just as the electric field is the force that would act on a unit charge.

22.1 ELECTRIC FIELD LINES

As we have seen, the motion of a fluid is described by a velocity field. The velocity field lines or streamlines give a graphic representation of the flow, which is helpful in visualizing the fluid motion. Not only do these lines give a visual picture of the motion of the fluid, but their density tells us something about the magnitude of the velocity. Where the streamlines are close together the velocity is large, but where they are far apart, the velocity is small.

We may use this same visual display to picture the electric field. An *electric field line* is a directed line whose tangent at a point gives the direction of the electric field at that point. Furthermore the closeness of the field lines determines the relative strength of the electric field. (We shall prove this result in Chap. 23.)

The electric field lines associated with a single point charge provide the simplest example of these lines (Fig. 22.1). In this case the electric field lines are radial lines emanating from the charge. Note that the lines are more closely packed near the charge than they are further from the charge. In fact, in this instance it is a simple matter to show that the density of the lines varies in inverse proportion to the square of the distance from the charge. This

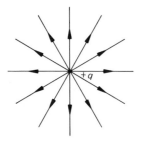

FIGURE 22.1   Electric field lines of a single point charge. (Only the field lines in the plane of the page are shown.)

follows from the fact that the number of lines through every sphere centered at the charge must be the same. The number of lines through the sphere is the product of the density of field lines and the area of the sphere, and this product must be a constant. Since the area of a sphere is proportional to r^2, it follows that the density must be proportional to $1/r^2$ and hence the product is independent of r.

The field lines of other configurations of point charges are illustrated in Fig. 22.2.

Some common features of these electric field lines may be stated as rules:

1. Electric field lines begin only on positive charges and end only on negative charges; electric field lines cannot begin or end in a vacuum. (This applies to the electrostatic field only. In the case of time-varying fields, we shall see that the electric field lines may close on themselves forming a loop.)

2. Electric field lines cannot intersect each other. If they could, there would be a point at which the electric field had two directions simultaneously.

3. Electric field lines always intersect the surface of a conductor perpendicularly. This requirement will be explained later.

22.2 ELECTRIC POTENTIAL

In our study of the gravitational interaction between bodies, we found that the force was directly proportional to the product of the masses and inversely proportional to

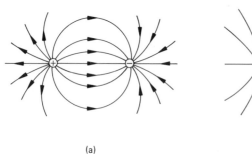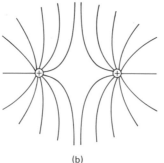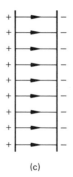

| (a) | (b) | (c) |

FIGURE 22.2   Field lines of (a) a positive charge and an equal negative charge, (b) two equal positive charges, and (c) two oppositely charged planes.

the square of the separation distance. From this force law we could predict the motion, given the initial conditions. We found, however, that certain problems could be solved more easily with the energy principle. For example, we were able to determine the escape velocity of a rocket from the surface of the earth by equating the total energy (kinetic plus potential) at the surface of the earth to the energy at infinity. We did not need to determine the trajectory. (If the rocket is not projected vertically from the surface of the earth, the trajectory is not a straight line.)

We found that the gravitational potential energy of two bodies of mass m and M separated by a distance r is given by $-GmM/r$. We also found that the electrostatic potential energy of two charges q_1 and q_2 is given by (see Eq. 5.29)

$$U = k\frac{q_1 q_2}{r}$$

In describing the interaction potential energy we took the *action-at-a-distance* point of view. We shall now introduce a field-theory picture of potential energy. We would like to associate an *electric potential* with every point in space, and this electric potential is associated with the point, not with a charge at the point. We shall define the electric potential at a point as *the potential energy that a positive unit charge would have at that point* (just as the electric field at a point is the *force* that would act on a positive unit charge at the point). Note that we should not speak of the potential energy of a unit charge, since it is really the system that possesses potential energy. However, often one particle is moving in the field created by a collection of fixed charges, and in such cases it is *convenient* to speak of the potential energy of the moving charge. We should consider this a shorthand for the more accurate concept, which is the potential energy of the system.

Let us put our definition in mathematical language. If U is the potential energy that a charge q' would have at a point in space, then the electric potential V at that point in space is given by

$$V = \frac{U}{q'} \tag{22.1}$$

The potential is the potential energy per unit charge just as the electric field

$$\mathbf{E} = \frac{\mathbf{F}}{q'} \tag{22.2}$$

is the force per unit charge.

The electric potential V is a scalar field that has a numerical value at every point. The units for the electric potential are joules per coulomb, ergs per coulomb, or foot-pounds per coulomb. In the mksc system the joule per coulomb is abbreviated and called a *volt* (abbreviated V).

Recalling our definition of potential energy, we may give V a more physical interpretation by saying that the electric potential created by some configuration of charges (or just a single charge for that matter) at a point is the work that the field created by these charges would do on a positive unit charge as it moves along any path from the point in question to the reference point. In most electrostatic problems the most convenient reference point is a point at infinity.

Let us formulate this statement in mathematical language. If the potential at a point P is the work done by the field on a positive unit charge as it moves from P to infinity (Fig. 22.3), then

$$V_P = \int_P^\infty \mathbf{E} \cdot d\mathbf{s} \tag{22.3}$$

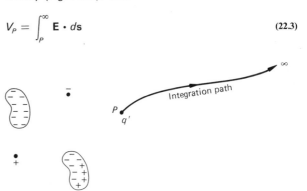

FIGURE 22.3 The potential at P is the work that is done by the electric field on a unit test charge as it moves from P to the reference state at infinity.

where $d\mathbf{s}$ is an element of length on the path from P to infinity.

We have seen in Chap. 5 that the work is independent of the path if the force is of the Coulomb type.

The potential difference between two points P and P' is given by

$$V_P - V_{P'} = \int_P^\infty \mathbf{E} \cdot d\mathbf{s} - \int_{P'}^\infty \mathbf{E} \cdot d\mathbf{s} = \int_P^{P'} \mathbf{E} \cdot d\mathbf{s} \qquad (22.4)$$

and is the work done by the field on a positive unit charge as it moves from P to P' along any path. If P and P' are coincident, then

$$\oint \mathbf{E} \cdot d\mathbf{s} = 0 \qquad (22.5)$$

where the symbol $\oint$ denotes an integral around any closed path that contains the two coincident points. We shall discuss Eq. 22.5 in some detail in Chap. 23.

QUESTION

What is the electric potential at a distance r from a single point charge q?

ANSWER

To determine the electric potential at a point, we place a test charge q' at that point and measure the potential energy U of the test charge. Now the potential energy of two charges q and q' is given by

$$U = k\frac{qq'}{r}$$

(see Eq. 5.29). Therefore

$$V = \frac{U}{q'} = \frac{kq}{r} \qquad (22.6)$$

QUESTION

What is the electric potential at a point 5 m from two charges of 3 μC each? The distance between the two charges is 8 m (see Fig. 22.4).

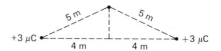

FIGURE 22.4   Two positive charges of 3 μC separated by 8 m.

ANSWER

Since the potential field is a scalar field we may add potentials as we add scalars. Therefore the potential field created by a collection of charges is just the scalar sum of the potentials of the individual charges. The potential at a point equidistant from two equal charges is then twice the potential of either charge at that point, so in this case

$$V = 2k\frac{q}{r} = 2\left(9 \times 10^9 \,\frac{\text{N} \cdot \text{m}^2}{\text{C}^2}\right)\left(\frac{3 \times 10^{-6} \text{ C}}{5 \text{ m}}\right)$$

$$= 1.08 \times 10^4 \text{ V}$$

QUESTION

Suppose that in the preceding question a charge of 2 μC with a mass of 2 kg is located at a point 5 m from the two charges. Initially the charge is at rest. When the charge is released, it will be repelled by the two 3-μC charges and will move away with every-increasing speed. What ultimate speed will it approach?

ANSWER

Since energy is conserved we know that the initial energy just after the charge is released is equal to the final energy when the charge has moved a great distance. Calling these two states A and B, respectively, we have

$$K(A) + U(A) = K(B) + U(B).$$

Now

$$K(A) = 0$$

$$U(A) = qV(A) = (2 \times 10^{-6} \text{ C})(1.08 \times 10^4 \text{ V})$$

$$= 2.2 \times 10^{-2} \text{ J}$$

$$K(B) = \tfrac{1}{2}mv^2 = \tfrac{1}{2}(2 \text{ kg})v^2$$

and

$$U(B) = 0$$

so that

$$0 + 2.2 \times 10^{-2} \text{ J} = \tfrac{1}{2}(2 \text{ kg})v^2 + 0$$

or

$$v = 0.15 \text{ m/s}$$

QUESTION

What is the potential on the axis of a uniformly charged ring at a distance x from the center of the ring?

ANSWER

This calculation is similar to that for the electric field, only easier. Since the potential field is a scalar field, we do not have to add vectors. The potential at a point on the axis is just the scalar sum of the potentials of each element on the ring. The potential of a single element is

$$dV = k\frac{dq}{r}$$

so the net potential is

$$V = \int dV = \frac{k}{r}\int dq = \frac{kq}{r}$$

$$= \frac{kq}{\sqrt{x^2 + a^2}} \tag{22.7}$$

where a is the radius of the ring (see Fig. 21.5).

22.3 DERIVING ELECTRIC FIELDS FROM POTENTIAL FIELDS

We have already seen in our study of potential energy that we may derive the force acting on some element in a system by determining the change in potential energy associated with a displacement. The component of the force in the direction of the displacement is just the negative gradient of the potential energy in that direction. If this is true of an arbitrary object in a general system, it must be true of a unit test charge among a group of electric charges.

To see how we may derive the electric field from the potential field, let us first recall the definitions of **E** and V. The electric field **E** is the force that would act on a unit test charge, and V at a point is the work that would be done on a unit test charge by the electric field **E** if the charge were moved from the given point to infinity.

Suppose a unit test charge is moved from the point x, y, z to the point $x + dx$, y, z. The work done by the electric field is given by

$$W_{x \to x+dx} = E_x \, dx \tag{22.8}$$

that is, the component of the force is the direction of motion E_x times the distance moved dx. Now if $V(x, y, z)$ is the work done by the field on the unit test charge as it moves from x, y, z to infinity and $V(x + dx, y, z)$ is the work done by the field as it moves from $x + dx$, y, z to infinity, then

$$V(x, y, z) - V(x + dx, y, z)$$

is the work done by the field as the unit charge moves from x, y, z to $x + dx$, y, z. Hence

$$W_{x \to x+dx} = V(x, y, z) - V(x + dx, y, z)$$

But from the definition of the partial derivative,

$$W_{x \to x+dx} = -\frac{\partial V}{\partial x} \, dx \tag{22.9}$$

Comparing Eqs. 22.8 and 22.9 we see that

$$E_x \, dx = -\frac{\partial V}{\partial x} \, dx$$

or

$$E_x = -\frac{\partial V}{\partial x} \tag{22.10}$$

This is a special case of Eq. 5.34,

$$F_x = -\frac{\partial U}{\partial x}$$

Dividing by q' we have

$$\frac{F_x}{q'} = -\frac{\partial}{\partial x}\left(\frac{U}{q'}\right)$$

But

$$E_x = \frac{F_x}{q'}$$

and

$$V = \frac{U}{q'}$$

and so we have Eq. 22.10.

We may show in a similar way that

$$E_y = -\frac{\partial V}{\partial y}$$

and

$$E_z = -\frac{\partial V}{\partial z}$$

QUESTION

Given the potential on the axis of a ring charge, what is the electric field on the axis?

ANSWER

We have shown (see Eq. 22.7) that the potential on the axis of a ring charge is given by

$$V = \frac{kq}{\sqrt{x^2 + a^2}}$$

Therefore the field along the axis is given by

$$E_x = -\frac{\partial V}{\partial x} = \frac{kqx}{(x^2 + a^2)^{3/2}}$$

which is the result we obtained in Sec. 21.4.

We can see from the above question that often it may be easier to obtain the electric field by first calculating the electric potential and then differentiating the potential to determine the electric field. To obtain an electric potential created by a system of electric charges we simply add individual potentials by scalar addition, but to obtain an electric field we must perform a vector summation, and this is generally more difficult.

QUESTION

What is the electric potential field associated with a uniform electric field $\mathbf{E}$ in the x direction?

ANSWER

If

$$E = -\frac{\partial V}{\partial x}$$

and E is a constant, then V must be given by

$$V = -Ex$$

plus an arbitrary constant. (The reference point is $x = 0$ if the constant is chosen to be zero.)

QUESTION

What is the potential energy of a dipole $p(p = qa)$ that makes an angle θ with respect to a uniform field $\mathbf{E}$?

ANSWER

Let us choose the coordinate system as shown in Fig. 22.5. We have seen in the preceding question that the potential field of a uniform electric field in the x direction is

$$V = -Ex$$

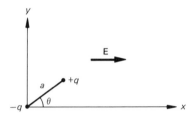

FIGURE 22.5   A dipole makes an angle θ with respect to a uniform electric field.

The potential energy of a charge q in this field is (Eq. 22.1)

$$U = qV$$

The potential energy of the positive charge is

$$U_+ = qV_+$$

where V_+ is the electric potential at the position occupied by the positive charge, and similarly

$$U_- = -qV_-$$

Now V_- is zero since the negative charge is at the origin $(x = 0)$. And

$$V_+ = -Ea \cos \theta$$

since the x coordinate of the positive charge is $a \cos \theta$. Thus the total potential energy is

$$U = -qEa \cos \theta = -pE \cos \theta$$

or in vector form

$$U = -\mathbf{p} \cdot \mathbf{E}$$

24.4 MOTION OF CHARGES IN ELECTRIC FIELDS

Our purpose in constructing the theory of electric fields was to provide a framework for studying the interaction between electric charges. Up to this point we have emphasized the nature of the field generated by various configurations of charges, but we must not lose sight of the effect of these fields on the motion of electric charges. We have already discussed the basic law: The electric field at a point is the force that would act on a positive unit charge at that point. The force then on a charge q will be

$$\mathbf{F} = q\mathbf{E}$$

QUESTION

What is the path of a charge q in the presence of an infinite plane carrying a uniform negative charge σ per unit area?

ANSWER

We must determine the interaction between the charge q and the infinite plane of negative charges. The charges on the plane create a field $\mathbf{E}$ given by $E = \sigma/2\epsilon_0$ and directed toward the plane (see Fig. 22.6). The charge q moves in the field, which exerts a force given by

$$F = qE = \frac{q\sigma}{2\epsilon_0}$$

From the second law of dynamics,

$$F = ma = \frac{q\sigma}{2\epsilon_0}$$

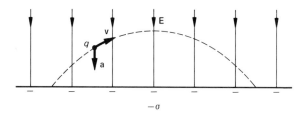

FIGURE 22.6 A charge q will move in a parabolic path in a uniform electric field.

so the acceleration is

$$a = \frac{q\sigma}{2\epsilon_0 m}$$

and is directed toward the plane. We see that the acceleration is constant in both magnitude and direction, so the path must be the same as that of a mass in a uniform gravitational field, namely a parabola.

QUESTION

Two parallel plates carrying charges per unit area of $+\sigma$ and $-\sigma$ are separated by a distance of 1 cm, and the voltage difference is 1 V. What is the surface charge density? If an electron left the negative plate with negligible velocity, with what velocity would it strike the positive plate (see Fig. 22.7)?

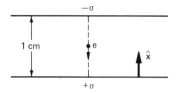

FIGURE 22.7 An electron leaves the negative plate and moves toward the positive plate 1 cm away.

ANSWER

First let us observe that it is not necessary to find σ or the electric field to determine the velocity of the electron just before it strikes the positive plate. All we need to do is use the conservation of energy theorem: The change in total energy of the electron must be zero as it moves from one plate to the other. The change in kinetic energy is

$$\Delta K = \tfrac{1}{2}m_e v^2 - 0 = \tfrac{1}{2}(9.1 \times 10^{-31} \text{ kg})v^2$$

and the change in potential energy is the change in potential energy per unit charge, which is the change in electric potential $\Delta V = 1$ V multiplied by the strength of the charge $-e$. Thus

$$\Delta U = -e\,\Delta V = -1.6 \times 10^{-19} \text{ J}$$

Equating the change in total energy to zero, we have

$$\Delta K + \Delta U = \tfrac{1}{2}(9.1 \times 10^{-31} \text{ kg})v^2 - 1.6 \times 10^{-19} \text{ J} = 0$$

and thus

$$v = 5.9 \times 10^5 \text{ m/s}$$

(A note about the change in electric potential ΔV is required. The change in electric potential is the electric potential at the lower plate minus the electric potential at the upper plate, i.e., the electric potential in the final state minus the electric potential in the initial state. When we said that the potential difference was 1 V we did not say whether this was the potential at the lower plate minus the potential at the upper plate or vice versa. To determine which it is, let us observe that an object at rest in a force field will tend to move from a high potential energy state to a low potential energy state. For example in falling to the ground a mass moves from a high gravitational potential energy state to a low potential energy state. If it begins to move its kinetic energy must be increasing, so for the total energy to remain constant, the potential energy must be decreasing. Thus a *positive* charge q whose potential energy is qV will move from high V to low V, thus decreasing qV. But a negative charge $-q$ whose potential energy is $-qV$ will move from low V to high V, again decreasing its potential energy. Since an electron is repelled by the negative charges on the upper plate and attracted by the positive charges on the lower plate, it will move from the upper plate to the lower plate. In other words, the negative charge will accelerate as it moves from the upper plate to the lower plate. Thus the positive lower plate is at the higher potential and the upper plate at the lower potential. The potential at the lower plate minus the potential at the upper plate is $+1$ V. We may also obtain this same result formally by noting that $E_x = -\partial V/\partial x$, so the electric potential will decrease in the direction of the electric field, i.e., in the direction of the force on a unit positive charge.)

Now let us turn to the first part of the question. What is the surface charge density σ? We have seen (Eq. 21.8) that the electric field between two such plates is given by

$$\mathbf{E} = \frac{\sigma}{\epsilon_0}\hat{\mathbf{x}}$$

This is a constant field. The electric potential associated with a constant electric field is

$$V = -Ex$$

The potential difference between two points a distance Δx apart is

$$\Delta V = -E\,\Delta x$$

Since Δx is the separation between the plates, $\Delta x = 0.01$ m and ΔV is then -1 V. Therefore

$$E = -\frac{\Delta V}{\Delta x} = +\frac{1\text{ V}}{0.01\text{ m}} = +100\,\frac{\text{V}}{\text{m}}$$

so that

$$E = 100\,\frac{\text{V}}{\text{m}} = \frac{\sigma}{\epsilon_0}$$

or

$$\sigma = \epsilon_0\left(100\,\frac{\text{V}}{\text{m}}\right) = (8.9 \times 10^{-12}\text{ C}^2/\text{N}\cdot\text{m}^2)\left(100\,\frac{\text{V}}{\text{m}}\right)$$

$$= 8.9 \times 10^{-10}\text{ C/m}^2$$

$$= 8.9 \times 10^{-4}\ \mu\text{C/m}^2$$

22.5 EQUIPOTENTIAL SURFACES

Just as electric field lines help us visualize electric fields, so *equipotential surfaces* help us visualize electric potential fields.

An equipotential surface is by definition a surface on which the electric potential has the same value at every point.

As an illustration, consider a single point charge. We have seen that the electric potential is given by (see Eq. 22.6)

$$V = \frac{kq}{r}$$

The surface on which $V = $ constant is a surface on which $r = $ constant, that is, a sphere. The equipotential surfaces for a single point charge are spheres that are concentric with the point charge, as shown in Fig. 22.8, where we illustrate the equipotential surfaces and the electric field lines of a single point charge. The electric potential has a different constant value on each equipotential surface.

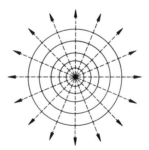

FIGURE 22.8 Equipotential surfaces (solid lines) and electric field lines (dashed lines) of a single point charge.

We notice in this simple example that the electric field lines intersect the equipotential surfaces perpendicularly. This result applies quite generally. Suppose that at some point the electric field did not intersect the equipotential

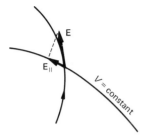

FIGURE 22.9 An electric field line that does not intersect an equipotential surface perpendicularly. The component of **E** is parallel to the surface; this is **E**$_{\parallel}$.

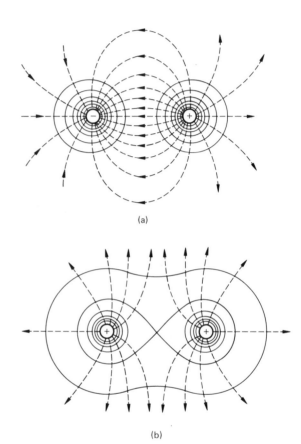

(a)

(b)

FIGURE 22.10 Electric field lines and equipotential surfaces of (a) two equal and opposite charges and (b) two equal positive charges. (Only the intersections of the equipotential surfaces with the plane of the page are shown.)

surface perpendicularly (Fig. 22.9). A component of the electric field would be tangent to the equipotential surface, and this would mean that the electric field would do work on a test charge being moved along the equipotential surface at this point. But if all points on the surface are at the same potential, no work is done by the electric field on a test charge moved along the surface. Therefore the electric field can have no component tangent to the equipotential surface.

The electric field lines and equipotential surfaces of two charges of equal magnitude and opposite sign and of two charges of equal magnitude and the same sign are illustrated in Fig. 22.10.

22.6 CONDUCTORS

A conductor is a substance or a medium, such as a metal or an ionic solution, in which there are charges (electrons or ions) that are free to move. In a nonconductor, such as glass or rubber, all the charges are bound in place and the material does not conduct electric charge. If an external electric field is applied to a conductor, the charges will move in response to the force created by that field.

In equilibrium, the state in which all charges are at rest, there can be no electric field inside a conductor. If there were a field in the conductor, the free charges would

move, and if they moved, the charge distribution would not be in equilibrium. Furthermore, there can be no net charge at any point in the interior of a conductor, for if there were, there would be an electric field in the interior. All charges on a conductor must reside on the surface in the steady state. It certainly is not obvious that there will be no field inside a conductor if the charges lie only on the surface. We will have to await Gauss's law to justify this assertion.

As an example of what happens when a conductor is placed in an external field, let us consider a neutral metal sphere next to a positive point charge (Fig. 22.11). If the conducting sphere remained everywhere neutral, then an electric field would be created inside the sphere by the external positive point charge. This electric field would cause charges to move within the conducting sphere: Positive charges would tend to move in the direction of the field and negative charges in the opposite direction of the field. Since in a metal only the electrons are free to move, they will be drawn toward the positive point charge, and since the sphere was initially neutral, the region vacated by the displaced electrons will be left positive. We have seen that there can be no charge in the interior of the conductor and that all charges, positive and negative, must reside on the surface of the sphere. Therefore the charge will continue to redistribute itself over the surface of the sphere until the electric field created by the surface charge exactly cancels the electric field of the point charge *at every point in the interior of the sphere.* When this is achieved the charges will remain at rest and we will have an equilibrium distribution.

We also observe in Fig. 22.11 that the electric field lines intersect the surface of the conductor perpendicularly. This occurs because all points within the conductor are at the same potential. In particular, all points on the surface are at the same potential, and we have seen that the electric field must always intersect an equipotential surface perpendicularly.

SUMMARY

We can visualize electric fields by means of electric field lines. An electric field line is a directed line whose tangent at a point gives the direction of the electric field at that point. The density of the electric field lines is a measure of

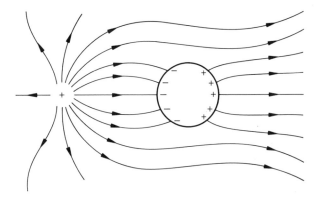

FIGURE 22.11   Electric field lines of a point charge and a neutral conducting sphere.

the strength of the electric field. Electric field lines begin and end only on electric charges; they do not intersect each other, and they intersect the surface of a conductor perpendicularly.

The electric potential V at a point in space is the potential energy of a unit charge at that point:

$$V = \frac{U}{q}$$

The electric potential at a point P is the work that the electric field would do on a unit test charge as it moved from P to infinity:

$$V_P = \int_P^\infty \mathbf{E} \cdot d\mathbf{s}$$

The electric field is related to the electric potential by the equations

$$E_x = -\frac{\partial V}{\partial x}$$

$$E_y = -\frac{\partial V}{\partial y}$$

$$E_z = -\frac{\partial V}{\partial z}$$

An equipotential surface is a surface on which the electric potential has the same value at every point. The electric field lines are everywhere perpendicular to the equipotential surfaces.

A conductor is a substance in which there are charges that are free to move. The electric field is zero within a conductor in equilibrium, and the electric potential is constant throughout a conductor in equilibrium. All electric charges must reside on the surface of a conductor.

PROBLEMS

22.A.1 Four charges of equal magnitude are arranged on the corners of a square as shown in the figure. Determine the potentials at A, B, and C in terms of the charge q and the side of the square a.

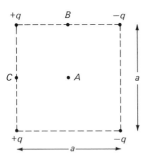

22.A.2 The electric potential along the x axis is indicated in the following figure. Determine the magnitude and direction of the x component of the electric field on the x axis between $x = 0$ and $x = 10$ cm.

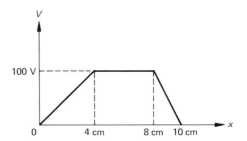

22.A.3 The potential difference between two large parallel conducting plates is 100 V, and the distance between the plates is 1 cm. What is the electric field between the plates?

22.A.4 The *electron volt* (abbreviated eV) is often used as a unit to measure energy. It is defined as the kinetic energy that an electron would acquire in moving through a potential difference of 1 V. What is an electron volt in joules?

22.A.5 An electron is emitted from the surface of the negative conductor in the following figure with negligible velocity. What is its velocity just before the electron strikes the positive conductor if the potential difference between the conductors is 1 V?

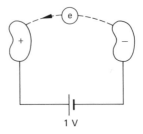

22.A.6 If $V = -e^z xy^2$, what are the three components of the electric field?

22.A.7 If $V = x^2 + 2xy + yz$, what are the three components of the electric field?

22.B.1 The electric field between the two charged plates in the accompanying figure is 100 N/C. Calculate the work done by the electric field on a positive unit charge along the paths (a) A to B, (b) A to C to B, and (c) A to D to B.

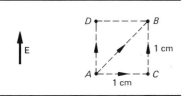

22.B.2

(a) Determine the electric potential on the axis of a uniformly charged disk of radius R.

(b) From (a) above, determine the electric field on the axis.

(c) Let R go to infinity and show that the electric field approaches that of an infinite plane.

22.B.3 A plot of the electric potential function on the x axis is a parabola and is described in the following figure. What is the x component of the electric field on the x axis?

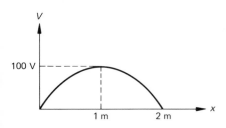

22.B.4 Determine (a) the electric potential and (b) the electric field of a uniform line charge of length L at a distance s from one end. The total charge on the line is q (see the figure).

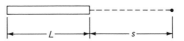

22.B.5 A 1-C charge is 1 m from a 2-C charge.

(a) Determine the electric potential at all points along the line joining the charges.

(b) From (a), determine the electric field at all points on the line joining the charges.

22.B.6 How much work must be done to bring a 3-μC charge from infinity to a point midway between a 1-μC charge and a 2-μC charge separated by a distance of 4 cm (see the figure).

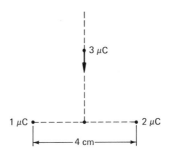

22.B.7 Four equal charges q are brought in one at a time from infinity to the four corners of a square of side a, as shown below. What is the work done in bringing in
   (a) the first charge?
   (b) the second charge?
   (c) the third charge?
   (d) the fourth charge?
   (e) What is the total potential energy of the system of four charges?

22.B.8 An electron passes through two small holes in two conducting plates connected to a battery, as shown in the following figure. Given that the potential across the plates is V and the velocity of the electron is v before it enters the first hole, determine the increase in velocity Δv assuming that $\Delta v \ll v$.

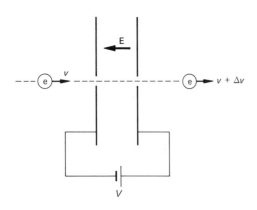

22.B.9 A positively charged body is attached to a cord. The other end of the cord is fixed, and the body rotates in a circle. Two charged parallel plates are introduced near the bottom of the trajectory in such a way that the fringe field from the plates accelerates the charged body on the cord. Every time the body passes the plates it is accelerated by the electric field (see the figure). This appears to be a violation of the conservation of energy theorem for an isolated system. Explain.

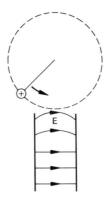

22.B.10 Two charges of 1×10^{-6} C and 2×10^{-6} C are at rest 1 cm apart. Each has a mass of 1 g. Because of mutual repulsion they move apart. What is the ultimate velocity of each charge?

22.C.1 A dipole consists of two equal and opposite charges separated by a distance a. The dipole moment p is defined as the product aq.
   (a) Show that the potential of a dipole oriented as shown in the figure is given by $V = kp \cos \theta / r^2$ if $r \gg a$.
   (b) Determine the electric field on the z axis.

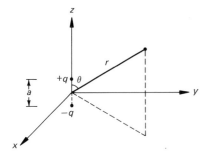

22.C.2 A 1-C charge and a −2-C charge are separated by a distance of 1 m.
  (a) Where is the electric field zero?
  (b) With the results of (a) as a clue, sketch the electric field lines.
  (c) Using (a) and (b), sketch the equipotential surfaces.

22.C.3 A 1-C charge and a 2-C charge are separated by a distance of 1 m.
  (a) Where is the electric field zero?
  (b) Using (a), sketch the electric field lines.
  (c) Using (a) and (b), sketch the equipotential surfaces.

22.C.4 A 2-g particle with a charge of 2×10^{-9} C is initially at rest at $x = -1$ cm, $y = 0$, and $z = 0$. At the same time a 1-g particle with a charge of 3×10^{-9} C is at rest at $x = 2$ cm, $y = 0$, and $z = 0$.
  (a) Where is the center of mass?
  (b) What is the total initial momentum?
  (c) What is the total initial energy?
  (d) The two particles are released and move apart. When the 2-g particle is at $x = -10$ cm, where is the 1-g particle?
  (e) What is the ultimate velocity of each particle?

23
GAUSS'S
LAW

WE HAVE SEEN that to compute electric fields by using Coulomb's law we must perform a vector summation over all the charges, and this may be a difficult calculation. The publication of Newton's theory of universal gravitation, for example, was delayed for some time while Newton attempted to determine the net gravitational force created by a spherical distribution of mass. We have assumed that a spherical distribution behaves like a point mass located at the center of the sphere, but this requires justification. We shall provide this in this chapter.

We may be able to simplify our task of determining fields from a given charge distribution by using the electric potential. This method requires only a scalar summation followed by a differentiation. But imagine the job of integrating dq/r over a spherical distribution of charge to obtain the potential outside the sphere. (Remember that r is the distance between the point at which the potential is being calculated and the element of charge dq. It is not the distance from dq to the center of the sphere.)

In this chapter we shall present a theorem that will allow us to calculate in a very simple manner the electric field associated with certain symmetrical charge distributions. This theorem, called *Gauss's law,* is a direct consequence of Coulomb's law and is not a new result.

Not only does Gauss's law have an advantage over Coulomb's law in the solution of certain problems, but it is a law that may be generalized to deal with problems in which the charges are not at rest. Gauss's law is preferred not simply because it gives the right answers for the fields of moving charges, but also because *there is no suitable way of modifying Coulomb's law to deal with moving charges*. We shall see, for example, that it is possible to have an electric field even when there is no net charge density anywhere.

A second law, similar in form to Gauss's law (they are both integral relations), may be deduced from Coulomb's law. This is simply a reformulation of our earlier observation that the electrostatic field is a conservative field, i.e., a field for which there exists an electric potential. This second law will require modification when we take up time-changing fields and will become Faraday's law of electromagnetic induction.

For static fields these two new laws, Gauss's law and what will become Faraday's law, are completely equivalent to Coulomb's law. Both may be derived from Coulomb's law and vice versa. For nonstatic fields we must abandon Coulomb's law, but our new integral relations may be suitably modified to deal with time-changing fields.

Let us begin by formulating Gauss's law. This law states that a certain integral of the electric field over a closed surface is proportional to the electric charge contained within that surface. An analog of this theorem for fluids offers the advantage of being more intuitive, and we shall discuss this first.

23.1 GAUSS'S LAW IN FLUID DYNAMICS

We have noted a certain similarity between the velocity field of fluids and the electric field of electric charges. Both are vector fields, and both may be visualized by means of field lines—streamlines in fluids and electric field lines in electrostatics. In dealing with fluids we never worried about the sources of the flow as we have about the sources of the electrostatic field. In electrostatics, fields are created by charges. What is the analog of electric charge in fluid dynamics? It must be something that creates fluid motion. A faucet creates fluid motion; so does a drain.

One can in principle imagine point sources and point sinks (or drains) in fluids that are analogous to positive and negative point charges in electrostatics. We may imagine water flowing radially away from a source and radially inward toward a sink. A measure of the strength of a source is the amount of water that flows from the source per second.

Let us consider a single source in an infinite incompressible fluid. What would the flow look like? Certainly the flow would be radial, but how would the velocity of the fluid vary with distance from the source? To determine this dependence, let us consider the volume of fluid that flows past a sphere of radius r, concentric with the source (Fig. 23.1). If $v(r)$ is the velocity at this radius and if we let Q be the volume of fluid that crosses the sphere per second, then[1]

$$Q = vA$$

where A is the area of the sphere and is equal to $4\pi r^2$.

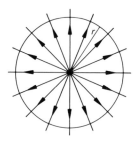

FIGURE 23.1   A point source in an infinite fluid.

Solving for v, we have

$$v = \frac{Q}{A} = \frac{1}{4\pi}\frac{Q}{r^2}$$

---

[1] This result follows from Eq. 11.5, $dm = \rho Av\, dt$. Here dm/dt is the mass crossing the area A per second and $(1/\rho)(dm/dt) = Av$ is the volume crossing the area A per second.

Now this expression for v is very similar to the expression for the electric field of a point charge. To make the analogy complete we need only show that Q is a constant and is characteristic of the strength of the source. But Q is the amount of fluid crossing a sphere per second, and for an incompressible fluid this must be the same for all spheres. If it were not the same for two spheres of radius r_1 and r_2, then there would have to be an increase in the amount of fluid between these two spheres. But it is not possible to increase the amount of an incompressible fluid in a fixed volume. Therefore Q is a constant and is equal to the output of the source.

We might express our result as a vector equation,

$$\mathbf{v} = \frac{1}{4\pi} \frac{Q}{r^2} \hat{\mathbf{r}} \qquad (23.1)$$

The velocity field of a sink is similar except that Q is negative. The combined velocity field of a source and an equal sink would be given by

$$\mathbf{v} = \frac{1}{4\pi} \frac{Q}{r_1{}^2} \hat{\mathbf{r}}_1 - \frac{1}{4\pi} \frac{Q}{r_1{}^2} \hat{\mathbf{r}}_2$$

where r_1 and r_2 are measured from the position of the source and sink, respectively. If we plotted the streamlines, they would be identical to the electric field lines of a dipole (Fig. 23.2). The fluid velocity fields can be added in exactly the same way that electric fields are added.

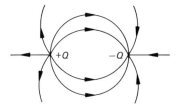

FIGURE 23.2 Streamlines of a source and a sink of equal strength.

Two important points about the fluid velocity field may be carried over to electric fields.

1. *The density of the field lines is a measure of the magnitude of the fluid velocity* We have already laid the groundwork for this result in Chap. 11. In Eq. 11.8 we showed that the conservation of fluid requires that

$$Av = \text{constant}$$

where A is the cross-sectional area of a bundle of streamlines. We derived this result by equating the mass of fluid that passes an area A ($\rho A v \, dt$) at one point in a bundle of streamlines to the mass that passes any other cross section. If ρ is a constant, we see that Av is a constant. The result we used earlier that Av is a constant for all spheres is a special case of this more general result.

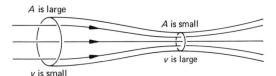

FIGURE 23.3 The cross-sectional area of a bundle of streamlines is inversely proportional to the velocity.

Now the cross-sectional area of a bundle of streamlines is a measure of how close the streamlines are to each other, i.e., of the density of the streamlines. Thus the density of the streamlines is a direct measure of the magnitude of the velocity.

2. *The net flow across any closed surface is a measure of the source strength inside that surface* (Gauss's law for fluids). Imagine a closed surface, and consider the net amount of fluid that crosses that surface per second. If

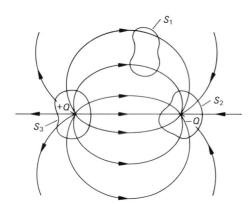

FIGURE 23.4   A source $+Q$ and a sink $-Q$ in an infinite fluid. The surface S_1 contains neither a source nor a sink, S_2 contains a sink, and S_3 contains a source.

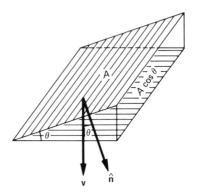

FIGURE 23.5   The flux through the area A and its projection $A \cos \theta$ must be the same. Here θ is the angle between the two surfaces or the angle between the normal $\hat{n}$ and the velocity vector **v**.

there are no sources within the volume bounded by the surface, then there will be no net flow out of the volume. If there are no sinks inside, then there will be no net flow into the volume. In Fig. 23.4, S_1 is a surface that encloses neither a source nor a sink. Fluid enters the volume bounded by S_1 and an equal volume of fluid must leave S_1; otherwise there would be an accumulation of fluid within S_1, and this cannot happen in an incompressible fluid. We can have a net accumulation of fluid within a surface only when there is a sink inside that surface, as in S_2.

Let us express this result quantitatively. We would like a general expression for the volume of fluid crossing a closed surface. We have seen that if the flow is perpendicular to the area, then the flow rate is given by

flow rate $= Av$

We would like to introduce a symbol to denote flow rate, which is the volume of fluid that passes an area per unit time. Let us call this rate the *fluid flux* and denote it by the symbol ϕ_v. (Later we will define an electric flux ϕ_E and a magnetic flux ϕ_B.) Thus

flow rate $= \phi_v$ $\qquad$ (23.2)

If the flow is not perpendicular to the area, then the flow rate or flux, as can be seen from Fig. 23.5, is given by

$\phi_v = Av \cos \theta$

where θ is the angle between the normal to the area and the velocity vector.

If we define a vector **A** by

$\mathbf{A} = A\hat{n}$

where $\hat{n}$ is a unit vector normal to the area, then

$\phi_v = \mathbf{A} \cdot \mathbf{v}$ $\qquad$ (23.3)

The flux across an arbitrary surface then is given by

$$\phi_v = \int \mathbf{v} \cdot d\mathbf{A}$$

where the integral is taken over the surface. If the surface is closed, then the net flux out of the surface is given by

$$\phi_v = \oint \mathbf{v} \cdot d\mathbf{A} \qquad (23.4)$$

where the integral is taken over the closed surface. (The symbol $\oint$ will be used to indicate an integral over a *closed* surface. The normal to the surface is always the outward normal.)

Let us compute the flux through a sphere of radius r with a source of strength Q at its center. The velocity is given by

$$\mathbf{v} = \frac{1}{4\pi} \frac{Q}{r^2} \hat{\mathbf{r}}$$

and is constant over the surface. Furthermore it is everywhere normal to the surface of the sphere. Therefore

$$\phi_v = \oint \mathbf{v} \cdot d\mathbf{A}$$

$$= \oint \frac{1}{4\pi} \frac{Q}{r^2} dA$$

$$= \frac{1}{4\pi} \frac{Q}{r^2} \oint dA = \frac{1}{4\pi} \frac{Q}{r^2} (4\pi r^2) = Q$$

since the area of a sphere is $4\pi r^2$. Hence the flux is equal to Q, which is not surprising since this is how Q was defined. But since the flux is independent of the shape and position of the surface, so long as it encloses the source, we know that

$$\phi_v = \oint \mathbf{v} \cdot d\mathbf{A} = Q$$

for *any* surface that encloses the source. If we have more than one source inside the surface, then

$$\phi_v = \oint \mathbf{v} \cdot d\mathbf{A} = Q_1 + Q_2 + \cdots$$

where $Q_1, Q_2, \ldots$ are the sources inside the surface of integration. Some of the sources may be drains, so some of the Q's may be negative; these of course are sinks. Sources or sinks outside the surface do not contribute to the integral since the amount of fluid that flows into the surface must also flow out. This is the mathematical statement of Gauss's law for fluids. We will often abbreviate this result and write

$$\phi_v = \oint \mathbf{v} \cdot d\mathbf{A} = Q \qquad (23.5)$$

where Q is the *net* source strength inside the closed surface of integration.

We might observe that this result is valid whether or not the sources or sinks are moving. We are still dealing with an incompressible fluid. We shall see that the electrodynamic counterpart of this law, Gauss's law, is of general validity. Unlike Coulomb's law, it applies to moving charges.

23.2 GAUSS'S LAW FOR ELECTRIC FIELDS

In Table 23.1 we compare our laws for fluids with those for electrostatics. The fields of point sources are the same except for a change of notation, namely $\mathbf{v}$ becomes $\mathbf{E}$ and Q becomes q/ϵ_0. The rule for adding the contribution from two or more sources is the same. Now these two rules are all that is needed to obtain the field created by any configuration of sources, and therefore Gauss's law, which is after all a mathematical operation on the fields, must be the same except for the change in notation. So if

$$\oint \mathbf{v} \cdot d\mathbf{A} = Q$$

then

$$\oint \mathbf{E} \cdot d\mathbf{A} = q/\epsilon_0 \qquad (23.6)$$

where q is the net electric charge within the closed surface of integration. Equation 23.6 is Gauss's law for electrostatics.

TABLE 23.1   Comparison of the velocity field of an incompressible fluid with the electrostatic field.

	Fluids	*Electrostatics*
Point source	$\mathbf{v} = \dfrac{1}{4\pi}\dfrac{Q}{r^2}\hat{\mathbf{r}}$	$\mathbf{E} = \dfrac{1}{4\pi\epsilon_0}\dfrac{q}{r^2}\hat{\mathbf{r}}$
Addition rule	$\mathbf{v} = \mathbf{v}_1 + \mathbf{v}_2$	$\mathbf{E} = \mathbf{E}_1 + \mathbf{E}_2$
Gauss's law	$\oint \mathbf{v} \cdot d\mathbf{A} = Q$	$\oint \mathbf{E} \cdot d\mathbf{A} = \dfrac{q}{\epsilon_0}$

We shall find it convenient to introduce an abbreviated notation for the integral in Eq. 23.6. We shall define the flux of the electric field in exactly the same way that we defined the fluid flux. The electric flux through an area is given by

$$\phi_E = \int \mathbf{E} \cdot d\mathbf{A} \qquad (23.7)$$

If the area is a closed area, then the electric flux through the closed surface is given by

$$\phi_E = \oint \mathbf{E} \cdot d\mathbf{A} \qquad (23.8)$$

We may then write Gauss's law in the form

$$\phi_E = \frac{q}{\epsilon_0}$$

where ϕ_E is the flux through a closed surface and q is the net charge within the surface. In integrals over closed surfaces, the element of area is always directed away from the interior.

23.3 APPLICATIONS OF GAUSS'S LAW

As we have said, the value of Gauss's law to us is twofold. First it can be a useful calculational tool, and second it may be generalized to nonsteady states, whereas Coulomb's law cannot. Let us now consider it as a tool to solve some complicated electrostatics problems.

We have mentioned that Newton was hampered in his development of the law of universal gravitation by the problem of determining the gravitational field outside a sphere. Let us investigate the electrostatic analog of this problem, the electrostatic field outside a spherical distribution of charge.

We let q be the net charge within a sphere of radius R. We need not assume that the distribution of charge is uniform, but it must be spherically symmetric. There may be a greater density of charge near the center. In any event, the net charge within the sphere is q. As a geometrical construction, draw a sphere of radius r concentric with the sphere of charge. This is called a *Gaussian surface,* and it is indicated by the dashed line in Fig. 23.6. Consider the electric flux through this surface,

$$\phi_E = \oint \mathbf{E} \cdot d\mathbf{A}$$

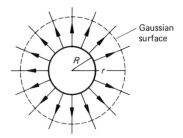

FIGURE 23.6 A spherically symmetrical distribution of charge inside a sphere of radius R. The spherical Gaussian surface (dashed line) of radius r is concentric with the sphere.

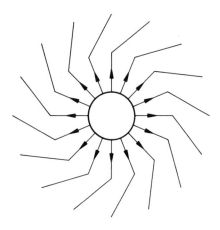

FIGURE 23.7 Electric field lines with kinks.

Now to evaluate this integral we must know something about **E** on the Gaussian surface. Since the charge distribution within the sphere is spherically symmetric, the electric field must also be spherically symmetric; i.e., if we rotate the sphere of charge through some angle, or equivalently rotate our head in the opposite direction through the same angle, the charge distribution will look the same and so the field must also look the same. We expect therefore that the electric field will be radial and of constant magnitude on the Gaussian sphere (Fig. 23.6). Any other configuration of field lines would not have spherical symmetry. To see this, suppose there are kinks in the field lines, as illustrated in Fig. 23.7. It might appear that this configuration looks the same regardless of the orientation of the viewer; and indeed no matter how you tilt your head the lines do look the same. But suppose you were on the other side of the page. The charge configuration would look the same, so the field should look the same. But it does not: The kink is in the opposite direction. The only possible configuration of field lines that is spherically

symmetric is that of radial field lines. The electric field **E** therefore is everywhere parallel to $d\mathbf{A}$ and constant in magnitude, so that

$$\phi_E = \oint \mathbf{E} \cdot d\mathbf{A} = \oint E \, dA = E \oint dA$$

$$= E(4\pi r^2) \tag{23.9}$$

But by Gauss's law

$$\phi_E = \frac{q}{\epsilon_0}$$

Therefore

$$E = \frac{1}{4\pi\epsilon_0} \frac{q}{r^2}$$

and so the electric field *exterior* to any spherically symmetric charge distribution is the same as the field produced by a point charge located at the center of symmetry whose magnitude is equal to the net charge within the sphere.

QUESTION

What is the electric field inside a spherically symmetric shell of charge?

ANSWER

Suppose we have a spherically symmetric distribution of charge between two spheres (Fig. 23.8) of radius R_1 and R_2 ($R_1 < R_2$).

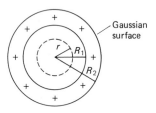

FIGURE 23.8 A spherically symmetrical charge distribution in a shell.

Consider a Gaussian surface of radius $r < R_1$ and concentric with the center of symmetry. From Gauss's law

$$\oint \mathbf{E} \cdot d\mathbf{A} = \frac{q}{\epsilon_0} = 0$$

since there is no charge within the surface of integration. From symmetry,

$$\oint \mathbf{E} \cdot d\mathbf{A} = E(4\pi r^2)$$

and so

$$E = 0$$

everywhere within the shell ($r < R_1$).

QUESTION

What is the electric field on either side of an infinite plate carrying a charge σ per unit area?

ANSWER

Let us choose as a Gaussian surface a pillbox, as shown in Fig. 23.9. Using symmetry arguments similar to those of the preceding question, we find that the electric field lines must be perpendicular to the sheet. If the sheet is positive, the field will be directed away from the sheet on both sides. There is no electric flux through the cylindrical sides of the pillbox. The flux through each end is EA, where E is the electric field at the ends and A is the area of the ends. Therefore the net flux out of the pillbox is

$$\oint \mathbf{E} \cdot d\mathbf{A} = 2EA$$

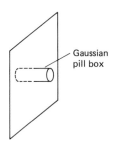

FIGURE 23.9 A Gaussian, cylindrical pillbox through an infinite sheet of charge. The end faces are equidistant from the surface.

The net charge within the box is σA, so by Gauss's law,

$$2EA = \sigma A / \epsilon_0$$

or

$$E = \sigma / 2\epsilon_0$$

We obtained this result earlier by integration of Coulomb's law. Note that the distance between the faces of the pillbox and the sheet does not enter into the answer. Therefore we have shown that the field is the same at all points.

How is electric charge distributed over a conductor in a steady state?

ANSWER

By definition a conductor is a body throughout which there are charged particles that are free to move. When there is any electric field in a conductor, these charges will move. In a steady state, therefore, there can be no electric field within the conductor. Thus the electric flux through any Gaussian surface within a conductor must be zero, and by Gauss's law there can be no electric charge within any Gaussian surface in the conductor. If there is any charge on the conductor, it must reside on the surface.

QUESTION

What is the electric field at a point within a spherical hole concentric with a charged spherical conductor?

ANSWER

We have shown earlier that the electric field is zero inside a spherical shell for any spherically symmetric charge distribution. The charge distribution on a conducting spherical shell is a special case of this, so the electric field must be zero in the cavity (Fig. 23.10).

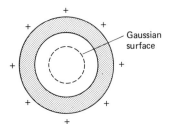

FIGURE 23.10   A spherical cavity in a charged spherical conductor.

In fact the electric field in any shape cavity within any shape conductor is zero. We cannot prove this result using Gauss's law alone, but must also use the fact that the electric field is conservative. (We will consider this point in Sec. 23.4; see also problem 23.B.8.) We can, however, make a rather convincing argument as follows. Consider a charged conductor of any shape. If there is no cavity within the conductor, then we know that the electric field is zero everywhere inside and the charges are distributed over the surface. Suppose we somehow cut out a portion of the interior to create a cavity (Fig. 23.11). If none of the charges on the outer surface move, then the electric field is still zero everywhere inside. In particular,

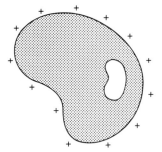

FIGURE 23.11   A cavity inside a conductor.

the electric field is zero within the conductor. It is also zero in the cavity even though it does not have to be. Certainly the original charge distribution is *a* distribution that satisfies the requirement that the electric field be zero within the conductor, but perhaps there is another solution. Perhaps some of the charge could distribute itself over the surface of the cavity, thereby creating an electric field within the cavity but no electric field within the conductor (Fig. 23.12). The total charge on the cavity surface must of course be zero.

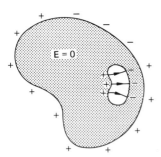

FIGURE 23.12 It is consistent with Gauss's law that there are charges on the inner surface of the conductor giving rise to an electric field within the cavity.

It is possible to show that the solutions to electrostatics problems are *unique,* so that if we find one solution, then that is *the* solution. Since *a* solution is that in which there is no electric field in the cavity, then this is *the* solution. Therefore there will be no charge on the surface of the cavity and no electric field within the cavity.

The field is zero within the cavity if and only if the electric field of a point charge satisfies Coulomb's law. If instead the field satisfied the equation

$$E = k \frac{q}{r^\alpha}$$

where α is some constant (but not equal to 2), then the electric field would not be zero within the cavity. Very precise measurements have been made to detect an electric field in a cavity. The experiments of Williams, Faller, and Hill have shown that if there is a field, it is so small that

$$1.9999999999999997 < \alpha < 2.0000000000000003$$

We shall assume that $\alpha = 2$.

23.4 A SECOND INTEGRAL EQUATION FOR THE ELECTRIC FIELD

We have seen that one consequence of Coulomb's law for a static distribution of charge is Gauss's law,

$$\oint \mathbf{E} \cdot d\mathbf{A} = \frac{q}{\epsilon_0}$$

Another integral relation follows from the fact that the electric force field derived from Coulomb's law is a conservative force field. If the electric field is conservative, then a unique potential exists. The potential at A is given by

$$V_A = \int_A^\infty \mathbf{E} \cdot d\mathbf{s}$$

The path from A to ∞ is immaterial since the integral is path independent. The potential difference between two points A and B is given by

$$V_{AB} = V_A - V_B = \int_A^B \mathbf{E} \cdot d\mathbf{s}$$

and is equal to the work done by the electric field on a positive unit charge as it moves from A to B (Fig. 23.13). Again the integral is path independent. Suppose A and B are coincident. Then since the work is independent of the path, the work must be zero, since one path is a path of zero length. Therefore the integral around any closed path of $\mathbf{E} \cdot d\mathbf{s}$ must be zero, that is,

$$\oint \mathbf{E} \cdot d\mathbf{s} = 0 \tag{23.10}$$

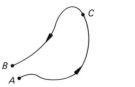

FIGURE 23.13 The work done by the electric field as a positive unit charge moves from A to B is the potential difference between A and B and is independent of the path.

where the symbol $\oint$ now denotes an integral around a closed line rather than a closed surface, as before. (One can readily see whether the integral denotes a line or a surface integral by looking at the differential.)

Another way of viewing this requirement is to observe that the integral around any closed path can be broken up into two parts: a part going out (path 1, Fig. 23.14) and a part coming back (path 2). Now the return integral from C to A (path 2) can be reversed, so that when we say that the round-trip integral is zero we are saying that the integral out from A to C is the same along either path, path 1 or path 2.

FIGURE 23.14 The integral $\oint \mathbf{E} \cdot \mathbf{ds}$ around the closed path from A to C and back again to A.

In summary, then, two integral requirements on the electrostatic field follow from the nature of Coulomb's law. These are Gauss's law,

$$\oint \mathbf{E} \cdot d\mathbf{A} = q/\epsilon_0 \tag{23.11}$$

where q is the net charge within the closed surface of integration, and the fact that

$$\oint \mathbf{E} \cdot d\mathbf{s} = 0 \tag{23.12}$$

for any closed path.

In electrostatics we can show conversely that Coulomb's law may be derived from Eqs. 23.11 and 23.12, and so either we may say that the field of a point charge is given by

$$\mathbf{E} = k\frac{q}{r^2}\hat{\mathbf{r}}$$

and the electric field of a collection of particles is the vector sum of the individual charges, or we may take the integral approach and say that

$$\oint \mathbf{E} \cdot d\mathbf{A} = q/\epsilon_0$$

and

$$\oint \mathbf{E} \cdot d\mathbf{s} = 0$$

These two formulations are equivalent in electrostatics. In nonsteady-state fields the integral equations may be modified to include the time-dependent effects but Coulomb's law must be abandoned.

23.5 SOLVING ELECTROSTATICS PROBLEMS

It may not be clear from our discussion so far how we might solve other problems in electrostatics which do not fit the pattern of the examples considered. We have seen that in some problems we can simply add the electric fields of the individual charges using Coulomb's law. This is possible when there is only a discrete set of point charges or a relatively simple continuous distribution, such as the ring charge or the uniformly charged plane. In other more complicated distributions, using Coulomb's law is too difficult, but there may be sufficient symmetry for us to use Gauss's law, as in the case of the spherical distribution of charge. But suppose we had to determine the electric field created by a uniformly charged ellipse. It would indeed be a formidable task to determine the electric field from Coulomb's law or our integral laws. Consider also the problem of determining the electric field of a point charge outside a conducting sphere. We know that a surface charge will develop on the sphere in order to cancel the electric field that the point charge would create within the conductor, but what is the surface distribution that does this? This is part of the problem! The surface distribution must be such that there is no net electric field within the sphere, and this distribution must be calculated.

These and similar problems are quite difficult and require methods other than those discussed here. We mention them so that the student will be aware that other methods besides Coulomb's law and our two integral conditions are available. We do not need new physical principles; rather we need new mathematical techniques.

SUMMARY

Gauss's law states that the electric flux through any closed surface is equal to the total electric charge within that surface divided by ϵ_0, so that

$$\oint \mathbf{E} \cdot d\mathbf{A} = q/\epsilon_0$$

A second integral requirement for the electric field is that

$$\oint \mathbf{E} \cdot d\mathbf{s} = 0$$

These two integral equations are equivalent to Coulomb's law for a static charge distribution.

PROBLEMS

23.A.1 A conducting spherical shell carries a net charge of $-1\ \mu C$. A 2-μC point charge is at the center of the shell. What are the net charges on the inner and outer surfaces of the shell?

23.A.2 If $E = 2$ V/m near the surface of a conducting sphere, what is the surface charge density?

23.A.3 Show that $\oint \mathbf{E} \cdot d\mathbf{s} = 0$ around a rectangle if $\mathbf{E}$ is the electric field of a point charge located at the center of the rectangle. (Do not evaluate the integral. Use symmetry arguments.)

23.A.4 Let $\mathbf{g}$ be the gravitational field, i.e., the force per unit mass. What is Gauss's law for the gravitational field?

23.A.5 A conducting shell (not necessarily spherical) carries a net charge of 4 μC.
  (a) What fraction of this charge resides on the inner surface and what fraction resides on the outer surface?
  (b) If a 2-μC point charge is introduced into the cavity of the shell, how will the 4-μC charge on the shell be redistributed?

23.B.1 Determine the electric field inside a uniformly charged cylinder of radius R. The charge density is ρ.

23.B.2 A charge q is distributed uniformly over the volume of a sphere of radius R. Determine the electric field
  (a) inside the sphere,
  (b) outside the sphere.

23.B.3 A charge q is distributed uniformly over the interior of a sphere or radius R. Determine the electric potential inside and outside the sphere.

23.B.4 A conducting sphere carrying a charge of 1 μC is concentric with two conducting shells. The inner shell carries a net charge of 2 μC and the outer shell a net charge of 3 μC. Determine the net charge on the inner and outer surfaces of each of the conducting shells.

23.B.5 A neutral spherical conducting shell has an inner radius of 10 cm and an outer radius of 20 cm. A point charge of 5 μC is located at the center of the cavity.
 (a) Determine the electric field at 5 cm, 15 cm, and 25 cm from the point charge.
 (b) What is the surface charge density on the inner surface of the shell?
 (c) What is the surface charge density on the outer surface of the shell?

23.B.6 Does there exist a charge distribution for which **E** is given by $E_x = y^2$, $E_y = 0$, and $E_z = 0$?

23.B.7 Show that if $E_x = x^2$, and $E_y = y^3$, and $E_z = 0$, then $\oint \mathbf{E} \cdot d\mathbf{s} = 0$ around any rectangle whose sides are parallel to the x and y axes.

23.B.8 Show that there exists a path for which $\oint \mathbf{E} \cdot d\mathbf{s}$ is not zero for the field illustrated in Fig. 23.12.

23.C.1 What is the electric flux through an infinite plane 1 m from a 2-μC point charge? (*Hint:* It is not necessary to evaluate an integral.)

23.C.2 If the electric field inside a sphere is radial and of constant magnitude, what is the volume charge density $\rho(r)$?

23.C.3 If $E_x = x^2$, $E_y = 0$, and $E_z = 0$, what is the charge per unit volume ($\rho(x)$) as a function of x? (*Hint:* Try a thin Gaussian pillbox of width dx.)

23.C.4 Three large parallel conducting plates carry net charges of 1 μC, 2 μC, and 3 μC from left to right in the figure below. Determine the net charge on the six surfaces of the three plates. (*Hint:* Gauss's law alone is not sufficient to solve this problem.)

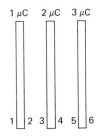

23.C.5 How much work must be done to charge a conducting sphere of radius R with a total charge q?

23.C.6 A thin hemispherical shell carries a uniform surface charge density. What is the direction of the electric field at any point on the equatorial plane bounded by the perimeter of the hemisphere?

24
CAPACITANCE

CERTAIN PROPERTIES OF objects are referred to as intrinsic properties. Examples of intrinsic properties are mass, electric charge, size, and shape. Examples of properties that are not intrinsic are position, velocity, acceleration, momentum, and energy. Intrinsic properties never change as long as the object maintains its identity.

Some properties of a system may be constant and still not be intrinsic properties. For instance, the energy and momentum of an isolated system are constant, but are not intrinsic properties; that is, the system may have any one of a number of different energies or momenta. The electron, however, cannot have any one of a number of different masses; its mass is an intrinsic property.

We would now like to demontrate that a pair of conductors has an intrinsic property called its capacitance, and we will use the symbol C to denote the value of the capacitance. The capacitance of a pair of conductors is defined in the following way. If a charge q is placed on one conductor and a charge $-q$ on the other conductor, and if the resulting potential difference between the two conductors caused by the charge distribution over the conductors if V, then the *capacitance* is the ratio of the magnitude of the charge on either body to the potential difference:

$$C = \frac{q}{V} \tag{24.1}$$

Now we find from experiment (and we shall prove from Coulomb's law) that the ratio q/V is a constant; i.e., if we place charges q and $-q$ on the conductors and observe a potential difference V, and then place charges q' and $-q'$ on the conductors and observe a potential difference V', then $q/V = q'/V'$. If this ratio is a constant, then it must depend on the position and shape of the two conductors and not on the charge q. It must be an intrinsic property of the conductor pair.

As an analogy, we observed that the ratio of the momentum to the velocity of a body was a constant. The momentum and the velocity may vary, but the ratio is always the same, namely the mass. (Momentum and velocity may be measured independently. The measurement process for the velocity is obvious: We might measure the momentum by shooting sticky balls of unit mass and unit velocity at the given object along a line of the velocity (see Fig. 3.6). The momentum of the object is numerically equal to the number of sticky balls necessary to bring the object to rest.)

Similarly, we might say that the spring constant

$$k = \frac{F}{x}$$

is an intrinsic property of a spring. The force and resulting displacement may vary, but the ratio is a constant.

Clearly a knowledge of such intrinsic properties is very useful, for they tell us that some relation among certain variables (certain nonintrinsic properties) is a constant. We can purchase a 0.2-kg potato or a quart of milk or a 3-μF (microfarad) capacitor and be sure of what we will have when we get home. On the other hand, if we purchase a zero-velocity potato at the store, we will not have a zero-velocity potato when we go out the door.

24.1 CAPACITANCE: AN INTRINSIC PROPERTY

We should be able to prove that the ratio of charge to voltage is a constant by applying the laws of electrostatics developed in the previous chapters. Suppose we place a total charge of $+q$ on one conductor and a charge of $-q$

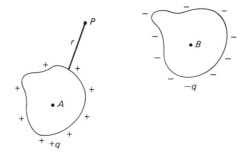

FIGURE 24.1 Charge distribution on two conductors.

on the other (see Fig. 24.1). Since these two bodies are conductors, the charge will distribute itself over the surfaces in such a way that the electric field is zero everywhere inside both conductors or, equivalently, the potential is a constant throughout each conductor. We have seen that the potential created by a point charge q at a distance r is given by

$$V = k\frac{q}{r}$$

The potential at some point P (see Fig. 24.1) is just the sum of the potentials at P created by all the charges on the surfaces of the two conductors, or

$$V_P = \int \frac{k\,dq}{r} \tag{24.2}$$

where dq is the quantity of charge on some small portion of the surface and r is the distance from that charge to the point P. Some of the charges dq will be positive and some negative. The sum is performed over all the charges on the surfaces of both conductors.

As we have said, the potential must be constant throughout each conductor (not the same constant for each conductor). At every point A inside the positively charged conductor, the potential will be, say, V_A. Inside the negatively charged conductor, the potential will be, say, V_B. The potential difference is given by

$$V = V_A - V_B$$

What happens when all the charges dq over both surfaces are doubled? From Eq. 24.2 it is clear that the potential at *every* point will also double. In particular, the potential V_A will double (while remaining uniform throughout the conductor) and V_B will double, and therefore the potential difference between the conductors will double. Clearly the total charge on each conductor will also double, so the ratio of the charge on either conductor to the potential difference will be unchanged. Similar changes will occur when the charges are tripled or quadrupled or increased by any constant factor. The ratio of the charge to voltage difference is a constant that we call the capacitance C. If C is independent of q and V, it must depend only on the geometry of the conductors.

This ratio is called the "capacitance" because it measures the capacity of the conductors to carry charge for a given potential difference. The simplest way to charge two conductors with equal and opposite charges is to connect a battery to the neutral conductors. This establishes a potential difference between the conductors that is equal to the potential of the battery, and whatever charge flows from one conductor through the battery must appear on the other conductor. Thus the two conductors carry equal and opposite charges. For a given battery of potential V, the charge on either conductor will be

$$q = CV$$

The larger C is, the larger is q; in other words, the greater the capacitance, the greater the capacity for carrying charge.

From the definition of capacitance, the unit of measurement is the coulomb per volt, which is called the *farad* (F).

$$1 \text{ farad} = \frac{1 \text{ coulomb}}{\text{volt}}$$

As we shall see, the farad is a very large unit, and capacitance is more often measured in microfarads (μF), where

$$1 \ \mu\text{F} = 10^{-6} \text{ F}$$

24.2 EXAMPLE OF THE INTRINSIC NATURE OF CAPACITANCE

Let us illustrate some of this theory with a couple of examples.

PARALLEL-PLATE CAPACITOR

Imagine two large rectangular plates of area A separated by a small distance s (see Fig. 24.2). We would like to show that the ratio of the charge to the voltage depends only on the geometry. Furthermore, we would like to obtain a formula, so that given the geometrical properties of the parallel-plate system, we may calculate the charge-to-voltage ratio or, equivalently, the capacitance.

FIGURE 24.2 Two large rectangular plates (in cross section) separated by a distance s. The charges of the respective plates are $+q$ and $-q$.

The task therefore is to find the voltage difference between the plates if the charges on them are q and $-q$. We have seen (Eq. 21.8) that the electric field between the plates is given by

$$E = \frac{\sigma}{\epsilon_0} = \frac{q}{\epsilon_0 A}$$

Now the electric field is the force on a unit charge,

$$E = \frac{F}{q'}$$

The potential difference between the plates is the work that the field would do on a positive unit charge as it is moved from one plate to the other. This work is the force E on the positive unit charge times the distance s it moves between the plates, that is,

$$V = Es$$

so

$$V = \frac{qs}{\epsilon_0 A}$$

Therefore

$$\frac{q}{V} = \frac{\epsilon_0 A}{s}$$

or

$$C = \frac{\epsilon_0 A}{s} \qquad (24.3)$$

We see explicitly in Eq. 24.3 that the charge-to-voltage ratio is a constant that depends only on the geometry, namely the area of the plates and the distance between them. (This ratio also depends on the nature of the medium between the plates, and we shall discuss this question later.)

QUESTION

What is the capacitance of a parallel-plate capacitor of area $A = 1 \text{ m}^2$ and separation $s = 0.1 \text{ cm}$?

ANSWER

From Eq. 24.3,

$$C = \frac{\epsilon_0 A}{s} = \frac{(8.9 \times 10^{-12} \text{C}^2/\text{N} \cdot \text{m}^2)(1 \text{ m}^2)}{(0.001 \text{ m})}$$

$$= 8.9 \times 10^{-9} \text{ F}$$

$$= 8.9 \times 10^{-3} \text{ } \mu\text{F}$$

QUESTION

If a 1-V battery were placed across the capacitor described in the preceding question, how much charge would pass through the battery?

ANSWER

The charge on either plate is

$$q = CV = (8.9 \times 10^{-9} \text{ F})(1 \text{ V})$$

$$= 8.9 \times 10^{-9} \text{ C}$$

If 8.9×10^{-9} C of charge were to flow through the battery, then the place that the charge came from would have a charge of 8.9×10^{-9} C less than before, and the place that the charge went to would have 8.9×10^{-9} C more than before. Thus 8.9×10^{-9} C would pass from one plate (leaving it with a charge of -8.9×10^{-9} C) through the battery to the other plate (giving it a charge of $+8.9 \times 10^{-9}$ C).

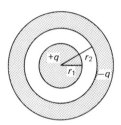

FIGURE 24.3 A spherical capacitor. The outer radius of the inner sphere (or shell) is r_1 and the inner radius of the outer shell is r_2.

SPHERICAL CAPACITOR

A spherical capacitor consists of two concentric conducting spheres separated by a nonconducting region (Fig. 24.3). If a total charge of $+q$ is placed on the inner sphere and a total charge of $-q$ on the outer sphere, the charges will distribute themselves over the outer surface of the inner sphere and the inner surface of the outer sphere. With this distribution of charge we may easily show, by applying the rules for fields generated by spherical charge distributions, discussed in Section 23.3, that the net electric field will be zero within both conducting bodies, as it must be. The electric field is not zero between the two conductors.

In Sec. 23.3 we observed that outside a spherical charge distribution the effect is the same as that in which the total charge is located at the center of the sphere. Therefore the field within the inner sphere is zero and the field between the spheres is that of a single charge $+q$ at the center. The field outside both spheres is again zero, since it is the field that is generated by a charge $+q$ at the center and an equal and opposite charge $-q$ also at the center; the net effect is zero.

The potential between the two spheres is that of a single charge $+q$ at the center. Thus the potential difference between the two surfaces of radii r_1 and r_2 is the potential difference between two points that are distances r_1 and r_2 from a single point charge; i.e., the potential difference between the spheres is

$$V = \frac{kq}{r_1} - \frac{kq}{r_2} = kq\left(\frac{1}{r_1} - \frac{1}{r_2}\right)$$

The capacitance of the concentric spheres is therefore given by

$$C = \frac{q}{V} = \frac{1}{k(1/r_1 - 1/r_2)} = \frac{r_1 r_2}{k(r_2 - r_1)}$$

As an important special case, let r_1 be approximately equal to r_2. Let $\Delta r = r_2 - r_1$. We may then write for the capacitance of the spherical capacitor

$$C = \frac{r_2 r_1}{k\,\Delta r} \approx \frac{r_1^2}{k\,\Delta r} = \frac{4\pi r_1^2}{4\pi k\,\Delta r}$$

$$= \frac{1}{4\pi k}\frac{A}{\Delta r} = \frac{\epsilon_0 A}{\Delta r}$$

Comparing this result with the capacitance of a parallel-plate capacitor (Eq. 24.3), we see that the result is nearly the same. The area of the plate is replaced by the area of the inner sphere, and the separation s between the plates is replaced by Δr, the separation between the spheres. In fact, we may think of the spherical capacitor as a parallel-plate capacitor bent over to form two concentric surfaces.

24.3 CAPACITORS CONNECTED IN SERIES AND IN PARALLEL

Let us imagine two sets of charged capacitors with one member of each set connected to one member of the other set by conducting wires (see Fig. 24.4). Let us suppose that the two capacitors are far enough apart, or that the

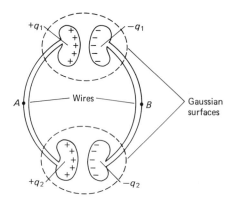

FIGURE 24.4 Two sets of capacitors connected by conducting wires.

electric field is localized for the most part to the region between opposing conductors, so that neither pair creates a significant electric field in the neighborhood of the other. If this condition is satisfied, then the electric charge will distribute itself so that there are equal and opposite charges on the opposing conductors and negligible charge on the connecting wires.

It is tempting to take this for granted, but it is not obvious. Suppose we put 8 C of charge at A and -8 C of charge at B (see Fig. 24.4). This charge must somehow distribute itself over the conducting surfaces, and it will do so in such a way that there is no net electric field within any conductor (this includes the wires). It is consistent with charge conservation that there would be a charge of $+5$ C on the upper left conductor, -4 C on the upper right conductor, $+3$ C on the lower left conductor, and -4 C on the lower right conductor.

To show that this does not happen, let us consider Gauss's law applied to the two Gaussian surfaces shown in Fig. 24.4. If there is no significant electric field on the Gaussian surfaces, then

$$\oint \mathbf{E} \cdot d\mathbf{A} = 0$$

over both surfaces. But this means that the net charge inside these surfaces must be zero, and therefore the charges on opposing conductors must be equal and opposite.

We still do not know what the equal and opposite charges are. Again, if we put 8 C of charge at A and -8 C at B in Fig. 24.4, then in the steady state we might have ± 4 C on the upper conductors and ± 4 C on the lower conductors or ± 5 C on the upper conductors and ± 3 C on the lower conductors. The additional condition needed to determine the division of the charge between the two conductors is that the electric potential must be a constant throughout any conducting body in the steady state. Since the upper and lower conductors are connected by a conducting wire, and the system is therefore one large conductor, both conductors on the left must be at the same potential. Similarly both conductors on the right must be at the same potential. From this it follows that *the potential differences between both capacitors must be the same.*

Suppose the capacitance of the upper pair of conductors is C_1 and the capacitance of the lower pair C_2. The charge on each would be given by

$$q_1 = C_1 V \tag{24.4}$$

and

$$q_2 = C_2 V \tag{24.5}$$

where V is the common voltage difference. If we know the total charge (say 8 C), then we know the sum $q_1 + q_2$.

Taking the ratio of Eqs. 24.4 and 24.5, we obtain the ratio of the charges

$$\frac{q_1}{q_2} = \frac{C_1}{C_2}$$

and from these two equations for the sum and the ratio, we can determine both q_1 and q_2.

Capacitors can be connected in various ways, and a convenient convention is used to represent the circuit. A single capacitor is represented by the symbol —||—. The connection indicated in Fig. 24.4 is represented in Fig. 24.5. A battery that is used in the circuit to charge the capacitors is represented by the symbol —|⊢. In this symbol the shorter line is the negative terminal and the longer line the positive terminal. In this convention positive charge flows from the positive terminal to the negative terminal in the external circuit. (It is of course the electrons that move, but it is simpler to think of a flow of positive charge and the effect is the same.) If positive charge flows from the positive terminal to the negative terminal, it follows that the positive terminal is at the higher potential.

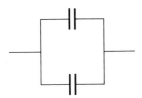

FIGURE 24.5 Symbolic representation of the conductors in Fig. 24.4.

QUESTION

A 6-V battery is connected to two capacitors $C_1 = 2 \ \mu F$ and $C_2 = 4 \ \mu F$ as shown in Fig. 24.6. What is the charge on each capacitor? What charge flows from the battery when the switch is closed?

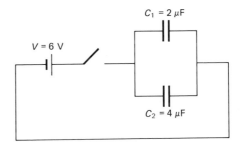

FIGURE 24.6 Two capacitors connected to a 6-V battery.

ANSWER

All points on a given conductor are at the same potential. Therefore, all points from the positive terminal of the battery to both left faces of the capacitors are at the same potential, and all points from the negative terminal to both right faces of the capacitors are at the same potential, which is 6 V less than the potential of the left faces. The potential across each capacitor is 6 V, so the charges on the capacitors are

$$q_1 = C_1 V = (2 \ \mu F)(6 \ V) = 12 \ \mu C$$

and

$$q_2 = C_2 V = (4 \ \mu F)(6 \ V) = 24 \ \mu C$$

The net charge is

$$q_1 + q_2 = 36 \ \mu C$$

and this is the charge that was delivered by the battery.

If we wanted to replace the two capacitors C_1 and C_2 in the preceding question by a single capacitor C, what should it be (see Fig. 24.7)?

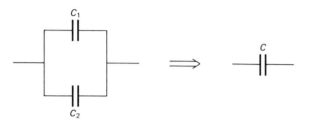

FIGURE 24.7   Two capacitors connected in parallel replaced by a single capacitor.

ANSWER

Imagine that the two capacitors C_1 and C_2 are inside a black box. In this case all we know when we close the switch is that a net charge of 36 μC flows into one side of the box and out the other side. What single capacitor would produce this same effect? Let us call the net charge q and the equivalent capacitor C. Then

$$q = CV$$

But

$$q = q_1 + q_2 = C_1 V + C_2 V = (C_1 + C_2)V$$

Therefore

$$CV = (C_1 + C_2)V$$

or

$$C = C_1 + C_2$$

so the equivalent capacitance is 2 μF + 4 μF = 6 μF.

   This manner of connecting the capacitors is called a *parallel* connection. When two or more capacitors are connected in parallel i.e., in such a way that the voltage is the same across each, the equivalent capacitance is given by

$$C = C_1 + C_2 + \cdots \tag{24.6}$$

Another way in which the capacitors may be connected is the so-called *series* connection, shown in Fig. 24.8. What is the single equivalent capacitor?

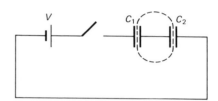

FIGURE 24.8   Two capacitors connected in series. The net charge within the dashed line is zero before the switch is closed and must remain so.

ANSWER

Once again we seek a single capacitor that will draw the same net charge from the battery. It is not true now that the potential is the same across each capacitor. The *sum* of the drops in potential across the capacitors must equal the battery potential:

$$V = V_1 + V_2$$

(The electrostatic field is conservative, so a potential field exists. The sum of the drops in potential around any closed loop must be zero. Let us start from the positive terminal of the battery, where the potential is V. There is a drop in potential when we cross the first capacitor, and a second drop in potential when we cross the second capacitor. There is no further change in potential until the battery is reached. Here there is a rise in potential of V when we cross to the positive terminal of the battery. Setting the sum of the changes in potential equal to zero, we have $-V_1 - V_2 + V = 0$.)

   Observing the portion of the conductor within the dashed line in Fig. 24.8, we see that if the charge is zero before the switch is closed, it must remain zero; there is no way for electric charge to reach this conductor. Therefore, the positive charge on the left face of C_2 must equal the negative charge on the right face of C_1. Since the charges are equal and opposite on opposing faces of each conductor, there must be the same equal and opposite charges on each capacitor. Therefore when the capacitors are connected in series, instead of the potentials being equal, as was the case for the capacitors connected in parallel, the charges are equal; so

$$q_1 = q_2$$

Now the equivalent capacitor must be such that it draws the same charge $q = q_1 = q_2$ from the battery when the switch is closed. Therefore

$$\frac{1}{C} = \frac{V}{q} = \frac{V_1}{q} + \frac{V_2}{q} = \frac{V_1}{q_1} + \frac{V_2}{q_2} = \frac{1}{C_1} + \frac{1}{C_2}$$

or

$$\frac{1}{C} = \frac{1}{C_1} + \frac{1}{C_2}$$

If more than two capacitors were connected in series, the equivalent capacitance would be given by

$$\frac{1}{C} = \frac{1}{C_1} + \frac{1}{C_2} + \frac{1}{C_3} + \cdots \qquad (24.7)$$

ANSWER

Let us reduce the system of three capacitors to a single equivalent capacitor. First the capacitors C_2 and C_3 are equivalent to a single 6-μF capacitor ($C = C_2 + C_3 = 2\ \mu F + 4\ \mu F = 6\ \mu F$). Next the 3-$\mu$F capacitor and the 6-$\mu$F capacitor connected in series are equivalent to a 2-μF capacitor ($1/2\ \mu F = 1/3\ \mu F + 1/6\ \mu F$), so in effect the 6-V battery is connected to a single 2-μF capacitor. The charge that passes through the battery to the 2-μF capacitor is given by

$$q = CV = (2\ \mu F)(6\ V) = 12\ \mu C$$

When 12 μC flow from the battery, then the same charge must accumulate on the left plate of the 3-μF capacitor (and of course there must be a charge of $-12\ \mu C$ on the right plate). The voltage across this capacitor is

$$V = \frac{q}{C} = \frac{12\ \mu C}{3\ \mu F} = 4\ V$$

Therefore the voltage across the 2-μF capacitor (and also across the 4-μF capacitor) must be 2 V.

QUESTION

What are the charge on the left face of the 3-μF capacitor and the voltage across the 2-μF capacitor in Fig. 24.9?

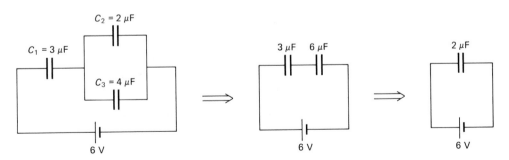

FIGURE 24.9 A capacitor circuit.

QUESTION

Why do the capacitances of capacitors connected in parallel add like scalars and the capacitances of capacitors connected in series add like reciprocal scalars?

ANSWER

We have seen the mathematics of this problem, but it would be nice to see it on a more intuitive level if possible. We have shown that the capacitance of a parallel-plate capacitor is given by

$$C = \frac{\epsilon_0 A}{s}$$

Suppose we have two capacitors with equal separation s but different areas A_1 and A_2. Let the capacitances be C_1 and C_2 and connect them in parallel, as shown in Fig. 24.10(a).

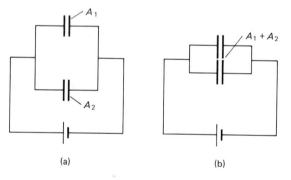

(a)   (b)

FIGURE 24.10 Two capacitors connected in parallel joined to form a single capacitor.

Neglecting fringe effects around the edges of the capacitors, we see that the capacitance will not change when we bring the two capacitors together as in Fig. 24.10(b). But now we

have a single capacitor of area $A_1 + A_2$ and separation s. The capacitance is

$$C = \frac{\epsilon_0(A_1 + A_2)}{s} = \frac{\epsilon_0 A_1}{s} + \frac{\epsilon_0 A_2}{s} = C_1 + C_2$$

Suppose, on the other hand, that we have two capacitors connected in series and of equal area but different separations s_1 and s_2, as shown in Fig. 24.11(a). Now the plates within the dashed line are isolated, so their net charge is zero, and therefore there will be equal and opposite charges on these plates.

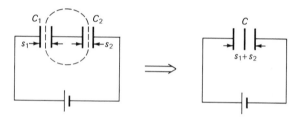

FIGURE 24.11 Two capacitors connected in series joined to form a single capacitor.

Suppose we shorten the wire joining these two plates so that the two capacitors are brought together (Fig. 24.11b). When we join the two inner plates, the equal and opposite charges will cancel, and we can remove this plate of no charge without changing the charges on the outside plates. We now have a single capacitor whose reciprocal capacitance is given by

$$\frac{1}{C} = \frac{s_1 + s_2}{\epsilon_0 A} = \frac{s_1}{\epsilon_0 A} + \frac{s_2}{\epsilon_0 A} = \frac{1}{C_1} + \frac{1}{C_2}$$

Of course the capacitors may not be parallel-plate capacitors, but it is always possible to replace a given capacitor by a parallel-plate capacitor of a given area (or separation) and then adjust the separation (or area) to yield the given capacitance.

QUESTION

How much work must be done to charge a parallel-plate capacitor?

ANSWER

Since the electrostatic field is a conservative field, we know that the work that must be done to create the final state is independent of the path or means by which the final state is reached. Let us imagine that initially we have two plates that are coincident and immediately next to each other, and that a charge q is frozen to one plate and a charge $-q$ frozen to the other. Since the plates are touching, the net charge is zero. Let us calculate how much work must be performed to pull these two plates apart to a separation s, thereby creating a charged parallel-plate capacitor with charges q and $-q$ on the two plates.

The force that the negatively charged plate exerts on the positively charged plate is the product of the electric field created by the negative plate, $\sigma/2\epsilon_0 = q/2\epsilon_0 A$ (see Eq. 21.6), multiplied by the charge on the positive plate q, or

$$F = \frac{q}{2\epsilon_0 A}q = \frac{q^2}{2\epsilon_0 A}$$

The work is the product of the force times the distance s, or

$$W = Fs = q^2\frac{s}{2\epsilon_0 A}$$

The capacitance of the parallel-plate capacitor is given by

$$C = \frac{\epsilon_0 A}{s}$$

so

$$W = \frac{\frac{1}{2}q^2}{C}$$

Since

$$q = CV$$

we may also write

$$W = \frac{\frac{1}{2}q^2}{C} = \frac{1}{2}qV = \frac{1}{2}CV^2$$

This calculation was performed for a parallel-plate capacitor. We shall leave it as an exercise (problem 24.C.2) for you to show that

$$W = \frac{\frac{1}{2}q^2}{C} = \frac{1}{2}qV = \frac{1}{2}CV^2 \tag{24.8}$$

for any capacitor regardless of its shape.

24.4 DIELECTRICS

We would like to consider the effect of the intervening medium on the interaction between charged bodies. Suppose, for example, that the medium between the capacitor plates were air or water or glass. What effect would this have on the capacitance?

One might ask how an electrically neutral medium could have any effect at all. We have seen earlier that electrically neutral bodies such as electric dipoles do produce electric fields and are themselves affected by electric fields. A dipole in an electric field will experience a torque that tends to align the dipole with the field (Fig. 24.12), so we might expect that a medium made up of electric dipoles—a dielectric—would have important electrical properties. Such a medium is affected by external electric fields and at the same time contributes to the electric field.

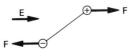

FIGURE 24.12 An external electric field creates a torque on a dipole that tends to align it with the electric field.

In the absence of an external electric field, the dipoles in a dielectric are randomly oriented and tend to cancel each other. When there is an external field, the dipoles line up and their combined effect can be considerable. The medium is said to be *polarized* when its dipoles line up.

Even a medium whose molecules have no permanent dipole moment can acquire polarization in the presence of an external electric field. The external field distorts the molecules by forcing positive charges in one direction and negative charges in the other, and thereby creates an electric dipole, called an *induced dipole* (see Fig. 24.13).

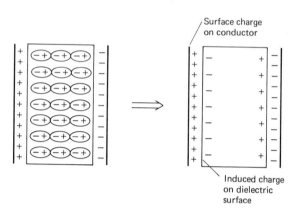

FIGURE 24.14 An exaggerated picture of the dipoles in a dielectric all aligned with the external field created by the electric field of the capacitor.

FIGURE 24.13 By deforming a molecule, an external electric field may induce a dipole moment.

To see what effect a dielectric can have on the electric field generated by free charges, let us consider a dielectric slab between the plates of a parallel-plate capacitor (see Fig. 24.14). For simplicity we have arranged the dipoles in an orderly fashion. The net effect is a neutral system within the dielectric, but surface charges on both faces. These surface charges will themselves create an electric field which, within the dielectric, will oppose the electric field created by the charge on the plates. In effect one might think of the net surface charge on the plates as being reduced by an amount equal to the induced charge on the dielectric surfaces.

When the electric field between the capacitor plates is reduced, the potential difference ($V = Es$) is reduced. The capacitance will therefore increase, since

$$C = \frac{q}{V}$$

and the free charge on the plates is the same but the potential difference is smaller.

Suppose we have two identical capacitors, one in a vacuum and with a capacitance C_0 and the other in a dielectric and with a capacitance C. The opposing plates of each capacitor carry equal charges q and $-q$. The capacitance of each is given by

$$C = \frac{q}{V}$$

and

$$C_0 = \frac{q}{V_0}$$

The ratio of the capacitances is

$$\frac{C}{C_0} = \frac{V_0}{V} \qquad (24.9)$$

Now if we assume that the degree to which the dielectric is polarized is directly proportional to the applied electric field, then the induced charge is directly proportional to the applied field. If the induced charge on the dielectric faces is directly proportional to the applied field, then the induced field is directly proportional to the applied field. But the fields are directly proportional to the potential differences, so we expect that

$$V \propto V_0$$

or

$$V_0 = \kappa V \qquad (24.10)$$

where κ is a proportionality constant characteristic of the dielectric and is called the *dielectric constant*. Since V is always less than V_0, it follows that

$$\kappa > 1$$

From Eqs. 24.9 and 24.10 we see that

$$\frac{C}{C_0} = \kappa$$

For a parallel-plate capacitor in a vacuum,

$$C_0 = \frac{\epsilon_0 A}{s}$$

so for a parallel-plate capacitor with a dielectric between the plates,

$$C = \frac{\kappa \epsilon_0 A}{s} \qquad (24.11)$$

It is customary to define the *electric permittivity* of a dielectric by the equation

$$\epsilon = \kappa \epsilon_0$$

Since κ is always greater than 1, ϵ is always greater than ϵ_0.

One finds in general that the ratio C/C_0 is the same for all capacitors regardless of their shape. This result is not restricted to parallel-plate capacitors.

If the potential difference between any two conductors in a dielectric medium is reduced by a factor $1/\kappa$, then

$$V = \frac{1}{\kappa} V_0$$

which suggests that the potential at any point within the dielectric medium is reduced by the same factor. (We may imagine that the surface charge σ on the conducting surfaces of the capacitor induces a local polarization charge that diminishes the net charge (see Fig. 24.14). It is this net effective charge that creates the electric potential at any point.) Since the electric field at a point is the gradient of the electric potential at that point, the electric field must be reduced by the same factor:

$$\mathbf{E} = \frac{1}{\kappa} \mathbf{E}_0$$

Since the electric field of a charge configuration is the vector sum of the electric fields created by the individual charges, we see that the electric field of a point charge must also be reduced by $1/\kappa$. Therefore the electric field of a point charge in a dielectric medium is given by

$$\mathbf{E} = \frac{1}{\kappa} \mathbf{E}_0 = \frac{1}{\kappa} \frac{1}{4\pi\epsilon_0} \frac{q}{r^2} \hat{\mathbf{r}} = \frac{1}{4\pi\epsilon} \frac{q}{r^2} \hat{\mathbf{r}}$$

In general we may apply all our previous results for charges in a vacuum to charges embedded in an infinite dielectric medium by replacing ϵ_0 by ϵ.

The dielectric constants for some materials are given in Table 24.1. Note from the table that the dielectric constant of water is very large. The water molecule has a very large dipole moment, and will therefore have a strong electrical interaction with other molecules. This makes it a good solvent. Furthermore, the large dielectric constant will greatly reduce (by a factor of 80) the force between electric charges (ions for example) dissolved in water. The large dielectric constant of water plays a crucial role in the chemistry of living organisms.

TABLE 24.1 Dielectric constants for various materials.

Material	κ
Vacuum	1
Air	1.0005
Paraffin	2.1
Carbon tetrachloride	2.2
Castor oil	4.7
Glass	4–8
Methyl alcohol	35
Water	80

QUESTION

The voltage difference across a capacitor in a vacuum is 6 V. Suppose the capacitor is dipped in oil and the new voltage is 1 V. What is the dielectric constant of the oil?

ANSWER

The voltage is reduced by a factor $1/\kappa$,

$$V = \frac{1}{\kappa} V_0$$

so

$$\kappa = \frac{V_0}{V} = \frac{6 \text{ V}}{1 \text{ V}} = 6$$

SUMMARY

The capacitance of a pair of conductors is a characteristic property of the pair and is defined by the equation

$$C = \frac{q}{V}$$

where q is the magnitude of the charge on either conductor and V is the resulting potential difference.

The capacitance of a parallel-plate capacitor is

$$C = \frac{\epsilon_0 A}{s}$$

Capacitors connected in series add according to the equation

$$\frac{1}{C} = \frac{1}{C_1} + \frac{1}{C_2} + \cdots$$

while capacitors connected in parallel add according to the equation

$$C = C_1 + C_2 + \cdots$$

A dielectric reduces the electric field at a point due to the orientation of permanent or induced electric dipoles. The reduction in the field is determined by the dielectric constant κ,

$$\mathbf{E} = \frac{1}{\kappa} \mathbf{E}_0$$

where $\mathbf{E}$ is the electric field in the dielectric and $\mathbf{E}_0$ is the field that would exist were it not for the dielectric. The dielectric constant $\kappa \geq 1$.

In a similar way

$$V = \frac{1}{\kappa} V_0$$

and

$$C = \kappa C_0$$

The dielectric permittivity of a medium is defined by the equation

$$\epsilon = \kappa \epsilon_0$$

PROBLEMS

24.A.1 A 6-V battery is connected across a 0.5-μF capacitor and a 1-μF capacitor connected in series.
  (a) What is the voltage across each capacitor?
  (b) What is the charge on each capacitor?

24.A.2 A 6-V battery is connected across a 3-μC capacitor and a 5-μF capacitor connected in parallel. What is the charge on each capacitor?

24.A.3 In the figure $C_1 = 3$ μF and $C_2 = 4$ μF and the plates carry the charges shown. What is the potential across each capacitor when the switches are closed?

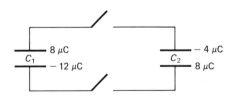

24.A.4 A parallel-plate capacitor with area 100 m² and plate separation 0.01 cm has a capacitance of 15 μF.
  (a) What is the dielectric constant of the material between the plates?
  (b) If the potential across the plates is 10 V, what is the net charge on the plates?
  (c) What is the induced charge on the dielectric?

24.A.5 Find the equivalent capacitance of the combination shown in the following figure.

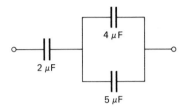

24.B.1 Determine the capacitance of the three plates shown in the figure; they are of area A and separation s. The upper and lower plates are connected by a thin wire.

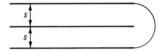

24.B.2 Determine the capacitance of two long coaxial cylinders of radii r_1 and r_2 and length L.

24.B.3 A 3-μF capacitor and a 6-μF capacitor are connected in series with a 10-V battery. The capacitors are then disconnected from each other and from the battery and reconnected positive plate to negative plate. What is the final potential across each capacitor?

24.B.4 What is the equivalent capacitance of the combination shown in the figure?

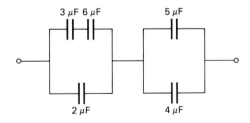

24.B.5 What is the equivalent capacitance of the combination shown in the figure?

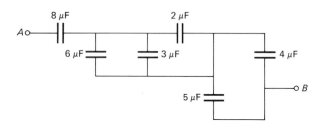

24.B.6 If the potential between A and B in the figure in problem 24.B.5 is 12 V, what is the charge on the 6-μF capacitor?

24.B.7 The potential across the battery in the following figure is 6 V. What is the potential across the 1-μF capacitor?

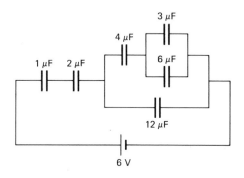

24.B.8 Two capacitors $C_1 = 3\ \mu$F and $C_2 = 4\ \mu$F have potential differences of 8 V and 5 V, respectively.
  (a) If the capacitors are connected positive plate to positive plate and negative plate to negative plate, what is the final potential difference?
  (b) If the capacitors are connected positive plates to negative plates, what is the final potential difference?

24.B.9 What is the force between two capacitor plates of area A and charge q? (The force is not qE, where E is the total electric field between the plates.)

24.B.10 How much work must be done to separate to a distance s two plates of charges q and $-q$ that are initially touching? If this is the work necessary to create a charged capacitor, how much energy is wasted when a battery is used to charge the capacitor? (How much work is done by the battery?)

24.B.11 A slab of glass 0.1 cm thick is inserted between the plates of a parallel-plate capacitor. The separation between the plates is 0.2 cm. If the dielectric constant of the glass is 5, what is the fractional change in the capacitance?

24.C.1 Determine the capacitance of the four plates shown in the figure of area A and separation s when
  (a) the two outer plates and the two inner plates are connected by thin wires, and
  (b) alternate plates are connected by thin wires.

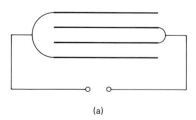

(a)

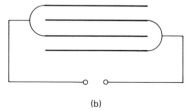

(b)

24.C.2 The work required to transfer a small quantity of charge dq from one plate to another of a parallel-plate capacitor is dqV, where V is the existing potential. If this process is continued until the final voltage is V_f, how much work is done starting from $V = 0$? Compare your answer with problem 21.B.10.

24.C.3 Starting with a box of 1-μF capacitors, how would you make a capacitor with a capacitance of $(n_1/n_2)\mu$F where n_1 and n_2 are any two integers?

24.C.4 Two large parallel conducting plates of area A are separated by a distance s, and between the plates are two dielectric layers of equal thickness (see the following figure). The dielectric constants are κ_1 and κ_2. What is the capacitance?

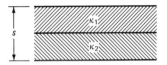

24.C.5 A parallel-plate capacitor is filled with a dielectric whose dielectric constant varies with position between the plates. The area of the plates is 1 m² and the separation, 0.1 cm. The dielectric constant $\kappa(x) = e^x$, where x ranges from 0 at one plate to 0.1 cm at the other. What is the capacitance?

25
ELECTRIC CURRENT AND RESISTANCE

UP TO THIS point we have considered only static distributions of electric charge. We have studied what happens when we close the switch in a circuit containing capacitors and a battery and a steady state is reached. But what of the time between the closing of the switch and the eventual static distribution of charge? There must be a flow of charge in the wires between the battery and the capacitors. What are the laws governing the flow of charge? To study this question, we must first introduce a quantitative measure for the flow of electric charge.

25.1 ELECTRIC CURRENT

When a switch is closed in a circuit containing a battery, an electric field is established in the wires. The wires are conductors and as such possess charges that are free to move, and this electric field will drive the charges through the wires. If during a short time dt an amount of charge dq passes a cross secion of wire (Fig. 25.1), then the electric current i at this place in the wire is defined as the ratio dq/dt, so

$$i = \frac{dq}{dt} \tag{25.1}$$

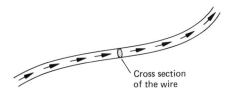

Cross section
of the wire

FIGURE 25.1 The current is the quantity of charge that passes a cross section per unit time.

We also assign a direction to the current, and this is the direction in which positive charge would move. In the case of a metallic conductor, only the electrons move, so the current is in the opposite direction of the electron flow. In an electrolytic solution both positive and negative ions move. If equal quantities of positive and negative charge cross a given cross section in opposite directions, they contribute equally to the current. A flow of positive charge in one direction is equivalent to an equal flow of negative charge in the opposite direction.

We treat a flow of positive charge in one direction in the same way as an opposite flow of negative charge because in a neutral conductor the effect is the same. In Fig. 25.2(a) regions A and B are electrically neutral. If 1 μC of electrons moves from A to B, region A will be left with 1 μC of positive charge and region B will acquire 1 μC of negative charge (Fig. 25.2(b)). If, on the other hand, 1 μC of positive charge moves from region B to region A (Fig. 25.2(c)), then region B will be left with 1 μC of negative charge and region A will acquire 1 μC of positive charge. The net effect is the same.

The unit for current is the coulomb per second, as is apparent from the definition. This ratio of units is abbreviated and called an *ampere* (A):

1 A = 1 C/s

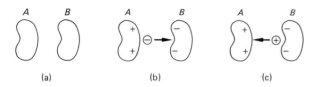

(a)   (b)   (c)

FIGURE 25.2 If A and B are originally neutral (a), then a flow of negative charge from A to B (b) is equivalent to a flow of positive charge from B to A (c).

25.2 ELECTRIC RESISTANCE

If electrons moved through conductors without impediment, then electric currents would be very large indeed. A metal has many free electrons, and the flow of charge created by the electric field in the metal would be enormous. The motion of the electrons is, however, greatly impeded by repeated collisions with other bound electrons in the conductor. The motion of the electrons through the conductor is a tortured path; they collide repeatedly with other electrons but are constantly driven by the electric field.

The thermal velocity of electrons in a metal is very large and the distance between collisions very small. If it were not for the electric field, the electrons in the metal would execute a random motion and there would be no net displacement one way or the other. However, because of the electric field, there is on the average a small displacement of the electrons in the direction of the electric force. The electrons tend to *drift* in the direction of the electric force (in the opposite direction of the electric field). The mean velocity of this drift is called the *drift velocity,* and it is many times smaller ($\sim 10^{-8}$) than the thermal velocity. The greater the electric field, the greater the drift velocity and hence the greater the current. We expect, therefore, that the current should be proportional to the electric field. The electric field in a wire is proportional to the gradient of the potential ($E_x = -\partial V / \partial x$), so the greater the potential difference between two points in a wire, the greater the current between these two points. If the current and the potential difference are directly proportional, we can write

$i \propto \Delta V$

If we let K be the proportionality factor, then

$$i = K \, \Delta V$$

We might call K the conductance of the portion of wire for which ΔV is the potential difference. The greater the conductance, the greater the current, and the greater the electric potential difference between two points, the greater the current. Now it is not customary to speak of the conductance of a piece of wire. Instead we shall define the *resistance R* of a piece of wire as the reciprocal of the conductance,

$$R = \frac{1}{K}$$

so

$$i = \frac{\Delta V}{R}$$

We can further simplify the notation by replacing ΔV by V and remembering that V is not the absolute electric potential at a point but the potential difference between two points for which the electric resistance is R. We may then write

$$i = \frac{V}{R}$$

or

$$V = iR \qquad (25.2)$$

Equation 25.2 is called *Ohm's law.*

Perhaps this equation should not be called a law. In establishing this relation we made a number of unjustifiable assumptions, the most important of which was that the mean drift velocity of the electrons is directly proportional to the electric field. It could have been proportional to the square root of the field strength, as it is in free fall in the air. We should regard this law of *electric* conductivity in much the same light as the law of *thermal* conductivity: It is an idealization that serves as a useful approximation.

We should also observe that the resistance of a conductor depends on its temperature. The resistance of a light bulb filament, for example, is substantially higher at its operating temperature than at room temperature.

This proportionality between the current and the voltage across a resistor is an *ad hoc* assumption, but the proportionality between the charge and the voltage across a capacitor is rigorously justifiable. The capacitance is a strict constant and has been proved to be so. If there is a dielectric medium, then once again we must resort to some "coefficient physics," and the linearity of the relation between the charge and the voltage depends on the constancy of the dielectric "constant."

25.3 RESISTANCE AND RESISTIVITY

We have seen that the current in a wire is proportional to the voltage drop along the length of the wire. We would expect the proportionality factor to depend on the material that the wire is made of and the geometry of the wire, in particular, its length and cross-sectional area. Using arguments similar to those we used to relate heat flow to temperature gradient, we may show that

$$R = \frac{\rho \ell}{A} \qquad (25.3)$$

where A is the cross-sectional area, ℓ the length, and ρ the electric resistivity, which depends on the nature of the conducting material. In these terms we may then write

$$i = \frac{A}{\rho} \frac{V}{\ell}$$

so that for a given V and l we will double the current if we double the area (in effect put two wires side by side). Furthermore the current is proportional to the gradient of the voltage (V/l) and not to the voltage drop alone. Equivalently the current is porportional to the electric field.

The resistivities of various substances at room temperature are given in Table 25.1. The unit for resistance is the volt per ampere, and this ratio is called an *ohm* and denoted by the symbol Ω.

TABLE 25.1 Electric resistivities of some substances at room temperature.

Material	$\rho(\Omega \cdot m)$
Aluminum	2.8×10^{-8}
Copper	1.7×10^{-8}
Carbon	3.5×10^{-5}
Graphite	8.0×10^{-6}
Iron	1.0×10^{-7}
Manganin	4.4×10^{-7}
Mica	9.0×10^{13}
Nickel	6.8×10^{-8}
Quartz	5.0×10^{-6}
Silver	1.6×10^{-7}
Steel	1.8×10^{-7}
Tungsten	5.5×10^{-8}

25.4 ENERGY DISSIPATED IN A RESISTOR

We have seen that an electron has to fight its way through a resistor. On collision with other electrons it loses energy. The electric field exerts a continual force on the electron, but nevertheless the electron continues to move with a constant average velocity. The energy lost by the electron is gained by the conductor in the form of heat.

To obtain a quantitative measure of the rate at which energy is dissipated in a resistor, we note that every charge that passes through the resistor undergoes a decrease in electric potential of V (that is, V is the electric potential drop across the resistor). Since V is potential energy per unit charge, it follows that the drop in electric potential energy of an element of charge dq that passes through the resistor is given by

loss in electric potential energy $= V \, dq$

The loss in electric potential energy per unit time is the power loss P and is given by

$$P = V\frac{dq}{dt} \tag{25.4}$$

Now it is not immediately obvious that dq/dt is the current i. It certainly looks like it, but we must remember that dq in Eq. 25.4 is the charge that passes through the conductor during the time dt, while dq in the equation $i = dq/dt$ is the charge that passes a cross section of the wire in the resistor in a time dt. That dq/dt and i are indeed the same follows from the fact that whatever charge flows into a resistor also flows out of the resistor, so that if a charge dq moves past a cross section at one end of the resistor, then an equal charge dq moves through a cross section at the other end in the same time. Thus the charge that moves through the resistor per unit time is the current i. Therefore the power dissipated is given by

$$P = iV \tag{25.5}$$

We can express this result in a number of ways. Since $V = iR$, we may write

$$P = iV = i^2 R = V^2/R \tag{25.6}$$

QUESTION

A water heater that will continuously deliver 1 kg of water per minute is required. The water is supplied by the city at 20°C, and an output temperature of 80°C is desired. What should be the resistance of the heating filament in the water if the line voltage is 220 V?

ANSWER

The power required to heat the water is given by

$$P = \frac{dQ}{dt} = C\frac{dm}{dt}\,\Delta T = \left(\frac{1\text{ cal}}{g}\right)\left(\frac{1000\text{ g}}{60\text{ s}}\right)(60°C)\left(4.2\frac{J}{cal}\right)$$

$$= 4200 \text{ W}$$

If this power is to be supplied by the heat dissipated in a heating filament, then

$$P = \frac{V^2}{R} = 4200 \text{ W}$$

or

$$R = \frac{V^2}{4200 \text{ W}} = \frac{(220 \text{ V})^2}{4200 \text{ W}} = 11.5 \text{ }\Omega$$

QUESTION

If you wanted to make the resistor required in the preceding question ($R = 11.5$ Ω) out of 10^{-3} m³ of copper, what length would you need? What if you used 10^{-3} m³ of manganin instead?

ANSWER

The resistance of a wire is given by

$$R = \frac{\rho\ell}{A}.$$

Multiplying the numerator and denominator by ℓ and observing that ℓA is the volume of the wire (10^{-3} m³), we have

$$R = \frac{\rho\ell^2}{A\ell}$$

and solving for ℓ, we obtain

$$\ell = \sqrt{\frac{RA\ell}{\rho}} = \sqrt{\frac{(11.5\text{ }\Omega)(10^{-3}\text{ m}^3)}{1.7\times 10^{-8}\text{ }\Omega\cdot\text{m}}} = 820 \text{ m}$$

This is a long wire.

If manganin were used,

$$\ell = \sqrt{\frac{RA\ell}{\rho}} = \sqrt{\frac{(11.5\text{ }\Omega)(10^{-3}\text{ m}^3)}{4.4\times 10^{-7}\text{ }\Omega\cdot\text{m}}} = 160 \text{ m}$$

SUMMARY

The electric current i in a wire is defined as the amount of charge that passes a cross section of the wire per unit time. Therefore

$$i = \frac{dq}{dt}$$

The electric current between two points is proportional to the potential difference between the two points. From this we obtain Ohm's law,

$$V = iR$$

where R is a proportionality constant and is called the resistance of a wire segment.

The resistance of a uniform wire is related to the cross-sectional area of the wire and the length according to the relation

$$R = \frac{\rho\ell}{A}$$

where ρ is the resistivity and depends on the material that the wire is made of.

The power loss in a resistor is given by

$$P = i^2R = iV = V^2/R$$

PROBLEMS

25.A.1 The power dissipated in a wire resistor 10 m long that is carrying a current of 8 A is 2 W. What is the cross-sectional area of the wire if the resistivity is $1.7 \times 10^{-8} \; \Omega \cdot m$?

25.A.2 A wire carrying a current of 10 A is dissipating 5 W of energy. If the resistivity is $2.3 \times 10^{-8} \; \Omega \cdot m$ and the cross-sectional area is 0.01 cm², what is the length of the wire?

25.A.3 Batteries are rated according to the amount of electric charge that may be drawn from them. For example, a 100 ampere-hour (A · h) battery will deliver 100 amperes for 1 hour, or $100 \, A \times 3600 \, s = 3.6 \times 10^5 \, C$. How long will a 6-V 100 A · h battery last if it is attached to a 100-W light. (Assume that the voltage remains constant.)

25.A.4 One meter of coiled heating filament generates 1000 W of heat when connected to a 200 V source. What length would generate 2000 W of heat when attached to the same source?

25.A.5 Power is being dissipated in a resistor at the rate of 500 W when the current is 2 A.
  (a) What is the resistance?
  (b) What is the voltage across the resistor?

25.B.1 In a certain metal there are 5×10^{23} free electrons per cubic centimeter. A wire made of this metal carries a current of 1 A, and its diameter is 0.2 cm.
  (a) What is the drift velocity of the electrons?
  (b) Compare this drift velocity with the thermal velocity of electrons at 300 K.

25.B.2 An electric wall heater draws 20 A. Compare the power loss in a line 10 m long of No. 12 gauge wire (diameter 2.5 mm) with that in a line of the same length of No. 14 gauge wire (diameter 1.6 mm). The wire is copper (resistivity $\rho = 1.7 \times 10^{-8} \; \Omega \cdot m$). Remember that there are two wires in the line.

25.B.3 An electric burner on a stove will bring 1 liter of water to a boil from 20°C in 4 min.
  (a) What is the maximum resistance of the filament if the line voltage is 220 V?
  (b) How long would it take to bring the same water to a boil on the same burner if the line voltage were 110 V?

25.C.1 The radius of a wire 1 m long varies with the distance along the wire as $x^{1/4}$ (0 m $\leq x \leq$ 1 m). Determine the resistance of 1 m of the wire as a function of ρ.

25.C.2 A copper sphere of radius 1 cm is immersed in a liquid whose resistivity is $0.02 \; \Omega \cdot m$. A second copper shell, concentric with the sphere, has an inner radius of 100 cm.
  (a) What is the resistance of the medium between the two copper spherical electrodes?
  (b) If the outer spherical shell has an inner radius of 10,000 cm, what is the resistance of the medium between the two copper electrodes?
  (c) In light of the answers to parts (a) and (b), can you explain why the resistance measured between two small electrodes placed on the human body is more-or-less independent of the distance between the electrodes so long as this distance is much greater than the radius of either electrode?

25.C.3 A refrigerator has an outside surface area of 4 m². The walls are 8 cm thick and have a thermal conductivity of 0.02 cal/s · m. The temperature inside the refrigerator is maintained at 5°C and the temperature outside is 45°C. What is the minimum current drawn by the refrigerator motor if the house voltage is 110 V? (*Hint:* It is necessary to consider the limitations imposed by the second law of thermodynamics.)

26
DC
CIRCUITS

WE HAVE SEEN that capacitors may be connected in many ways, and we have studied how the charge distributes itself over the different capacitors and the rules for calculating the voltage across each capacitor. We shall now do the same for resistors. Resistors may be connected in as many ways as capacitors may be connected. The magnitude and placement of resistors in a circuit can be used to control the magnitude and direction of the electric current.

We must also reexamine how a battery functions in driving a current through the circuit. When we employed batteries in Chap. 24 to charge capacitors, we were dealing with a static charge distribution. After the capacitors have become charged, there is no electric current. As soon as there is a current through the battery, the potential across the battery changes. There is a very simple model that accurately predicts this change in potential.

26.1 BATTERIES: EMF AND INTERNAL RESISTANCE

A battery is capable of maintaining steady currents through an external resistor. We know that every charge that flows through the resistor loses energy, and this energy must be supplied by the battery.

Let us consider a charge that begins a trip around the circuit in Fig. 26.1 at *b*, the positive terminal of the battery.

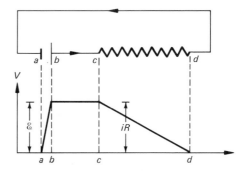

FIGURE 26.1 Electric potential as a function of position in a simple circuit.

411

The charge has just come from the battery and is in a high potential energy state. It moves down the wire to *c*, where it encounters a resistor. As it moves through the resistor, it continuously makes collisions and loses energy in the process. This energy loss per unit charge is the drop in potential across the resistor ($V = iR$) and appears in the form of heat within the resistor. After passing through the resistor, the charge is in a lower potential energy state and remains in this state until it reaches the negative terminal of the battery. As the charge moves through the battery, the battery does work on it to raise it back to the energy it had when it began the trip at *b*. The work done by the battery on a unit charge is called the *electromotive force*, or briefly the emf, of the battery, and is usually denoted by $\mathscr{E}$. This work is performed at the expense of chemical energy within the battery. In summary,

$\mathscr{E}$ = work done on a positive unit charge as it moves through the battery at the expense of chemical energy

Since the charge must return to its initial energy on completing the circuit, the *iR* drop plus the emf boost must be zero, so

$$-iR + \mathscr{E} = 0$$

or

$$\mathscr{E} = iR$$

When the external resistance is small or the battery is run down, we must also take into account the internal resistance *r* of the battery. Some internal resistance is inevitable in any battery, and its magnitude increases as the battery ages. The terminal voltage of a battery is the emf minus the internal *ir* drop. We may represent a battery, then, as an emf plus an internal resistance (see Fig. 26.2). The circuit equation for a single external resistance *R* becomes

$$\mathscr{E} - ir = iR$$

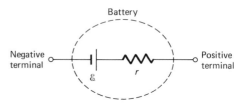

FIGURE 26.2 Schematic representation of a battery, including the internal resistance *r*.

In the following discussion we shall often neglect *r* in comparison with *R*.

QUESTION

A flashlight battery has an emf of 6 V. If the internal resistance is negligible, what is the resistance of a 10-W bulb? If the battery develops an internal resistance of 1 Ω as it ages, what is the effective power of the bulb?

ANSWER

The power delivered by the battery is given by

$$P = \frac{\mathscr{E}^2}{R}$$

If $P = 10$ W and $\mathscr{E} = 6$ V, then

$$R = \frac{\mathscr{E}^2}{P} = \frac{(6\text{ V})^2}{10\text{ W}} = 3.6\ \Omega$$

If the internal resistance in the battery increases to 1 Ω, then the current through the bulb is given by

$$\mathscr{E} - ir = iR$$

or

$$i = \frac{\mathscr{E}}{r + R} = \frac{6\text{ V}}{1\ \Omega + 3.6\ \Omega} = 1.3\text{ A}$$

The power dissipated in the bulb is

$$P_{\text{bulb}} = i^2R = (1.3\text{ A})^2(3.6\ \Omega) = 6.1\text{ W}$$

so the effective power is reduced from 10 W to 6.1 W.

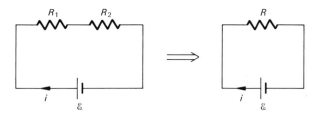

FIGURE 26.3 (a) Two resistors connected in series and (b) the equivalent circuit.

26.2 RESISTORS CONNECTED IN SERIES

We would like to lump together groups of resistors and replace them by a single resistor, much as we did for connected capacitors. Let us consider the circuit in Fig. 26.3. The two resistors R_1 and R_2 are connected in series, and we would like to replace them by a single equivalent resistor R. This resistor must have the property that the current in the circuit will be unchanged if the resistors R_1 and R_2 are replaced by the equivalent resistor R.

Now the current in the circuit in Fig. 26.3(a) is given by the circuit equation,

$$\mathscr{E} = iR_1 + iR_2$$

which states that the sum of the iR drops is equal to the emf boost. Solving for i, we have

$$i = \frac{\mathscr{E}}{R_1 + R_2}$$

The current through the equivalent resistor in Fig. 26.3(b) is given by

$$\mathscr{E} = iR$$

or

$$i = \frac{\mathscr{E}}{R}$$

These currents will be equal if

$$\frac{\mathscr{E}}{R} = \frac{\mathscr{E}}{R_1 + R_2}$$

or

$$R = R_1 + R_2$$

so that resistors in series add like scalars. When a third resistor is connected in series, the equivalent resistance is given by

$$R = R_1 + R_2 + R_3 \tag{26.1}$$

26.3 RESISTORS CONNECTED IN PARALLEL

The two resistors in Fig. 26.4 are connected in parallel. For the two resistors connected in series (Sec. 26.2), the current through each is the same and the sum of the potential drops across the resistors is equal to the applied emf.

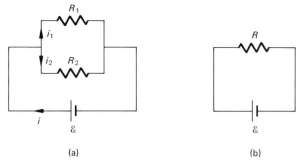

(a)   (b)

FIGURE 26.4 (a) Two resistors connected in parallel and (b) the equivalent circuit.

For two resistors connected in parallel, however, the sum of the currents through the two resistors is equal to the current through the battery, and the potential drop across each is equal to the emf. Therefore

$$i = i_1 + i_2$$

and

$$\mathscr{E} = i_1 R_1 = i_2 R_2$$

In the equivalent circuit

$$\mathscr{E} = iR$$

or

$$i = \frac{\mathscr{E}}{R}$$

but $i = i_1 + i_2$, so

$$\frac{\mathscr{E}}{R} = i_1 + i_2 = \frac{\mathscr{E}}{R_1} + \frac{\mathscr{E}}{R_2}$$

or

$$\frac{1}{R} = \frac{1}{R_1} + \frac{1}{R_2}$$

When three resistors are connected in parallel (see Fig. 26.5), the equivalent resistance is given by

$$\frac{1}{R} = \frac{1}{R_1} + \frac{1}{R_2} + \frac{1}{R_3} \tag{26.2}$$

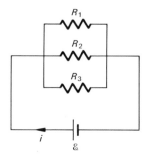

FIGURE 26.5   Three resistors connected in parallel.

QUESTION

What is the equivalent resistance for the resistor group in Fig. 26.6?

ANSWER

First we observe that the 3-Ω resistor is connected in parallel with the 6-Ω resistor. The pair is equivalent to a single resistor whose resistance is given by

$$\frac{1}{R} = \frac{1}{3\ \Omega} + \frac{1}{6\ \Omega} = \frac{1}{2\ \Omega}$$

or $R = 2\ \Omega$. This 2-Ω resistor is connected in series with the 4-Ω resistor, and the pair is equivalent to a single resistor whose resistance is given by

$$R = 2\ \Omega + 4\ \Omega = 6\ \Omega$$

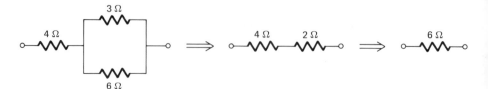

FIGURE 26.6   Three resistors reduced to a single equivalent resistor.

26.4 KIRCHHOFF'S RULES

There are many ways of connecting resistors so that they cannot be broken down into groups in series and groups in parallel. An example is illustrated in Fig. 26.7. If any one of the resitances is zero (or infinite), we can reduce the network as before. For example, if $R_3 = 0$, then by bending and stretching the network in Fig. 26.7, we can deform it into the network in Fig. 26.8. As another example in which the reduction method is not useful, consider the circuit in Fig. 26.9.

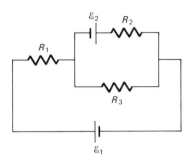

FIGURE 26.9 A circuit that cannot be reduced to groups of resistors connected in series and resistors connected in parallel.

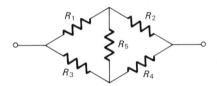

FIGURE 26.7 A group of resistors that cannot be reduced to resistors connected in series and resistors connected in parallel.

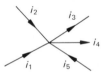

FIGURE 26.10 The net current entering the junction must be zero.

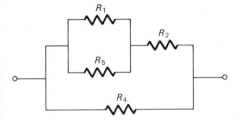

FIGURE 26.8 This network is equivalent to that of Fig. 26.7 if $R_3 = 0$.

To treat such circuits we may apply *Kirchhoff's rules,* which we may state as follows:

1. *Junction rule*   The sum of all the currents coming into a junction must be zero.

2. *Loop rule*   The total change in potential around any closed loop must be zero.

The first rule is a statement of charge conservation and the fact that charge cannot build up at a point if the system is in a steady state. If we have three or more current-carrying wires that meet at a junction (see Fig. 26.10), the amount of charge coming in must equal the amount of charge going out. If incoming currents are positive and outgoing currents are negative, then in Fig. 26.10 we must have

$$i_1 + i_2 - i_3 - i_4 + i_5 = 0$$

The second rule is a statement of the existence of unique electric potential in an electrostatic electric field, that is,

$$\oint \mathbf{E} \cdot d\mathbf{s} = 0$$

but

$$E_s = -\frac{\partial V}{\partial s}$$

so that

$$\oint \frac{\partial V}{\partial s} ds = \oint dV = 0$$

QUESTION

What is the current in the resistor R_2 in the circuit in Fig. 26.11?

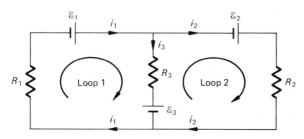

FIGURE 26.11 A multiloop circuit.

ANSWER

First the junction rule gives

$$i_1 - i_2 - i_3 = 0 \tag{26.3}$$

The loop rule around loop 1 gives

$$\mathscr{E}_1 - i_3 R_3 - \mathscr{E}_2 - i_1 R_1 = 0 \tag{26.4}$$

since $\mathscr{E}_1$ supports a current in the direction we take to proceed around the loop, while $\mathscr{E}_2$ opposes a current in this direction.

The choice for the direction of the currents is arbitrary, and the solution will tell us whether the assumptions are correct. If i_1 turns out to be negative, then the current is in the opposite direction. The direction we take around the loop is also arbitrary. If we had gone the other way, we would have obtained

$$-\mathscr{E}_1 + i_1 R_1 + \mathscr{E}_2 + i_3 R_3 = 0$$

Note that there is an increase in potential as we pass a resistor if we are proceeding in a direction opposite to the assumed direction of the current.

The loop rule around loop 2 gives

$$\mathscr{E}_2 - i_2 R_2 + \mathscr{E}_3 + i_3 R_3 = 0 \tag{26.5}$$

We know have three equations (Eqs. 26.3, 26.4, and 26.5) for three unknowns, and we can solve for all three currents, in particular for the current i_2 in resistor R_2.

Another loop is possible, and that is the loop around the entire perimeter through $\mathscr{E}_1$, $\mathscr{E}_2$, R_2, and R_1. This would give us no new information, and would in fact just give the sum of Eqs. 26.4 and 26.5.

SUMMARY

The electromotive force or emf of a battery is defined as the work that the battery does per unit charge that passes through it.

Resistors connected in series add according to the equation

$$R = R_1 + R_2 + \cdots$$

while resistors connected in parallel add according to the equation

$$\frac{1}{R} = \frac{1}{R_1} + \frac{1}{R_2} + \cdots$$

Kirchhoff's rules require that charge be conserved and that the sum of the electric potential differences around a closed path be zero. The two laws are:

1. *Junction rule*   The sum of all the currents coming into a junction must be zero.

2. *Loop rule*   The total change in electric potential around any closed loop must be zero.

PROBLEMS

26.A.1 Determine the equivalent resistance of the resistor group in the figure.

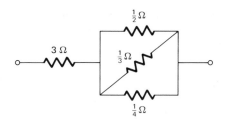

26.A.2 Suppose you have a 60-W bulb and a 100-W bulb. Which burns brighter if the bulbs are connected
   (a) in parallel?
   (b) in series?

26.A.3 Why do power companies transmit power at high voltage?
   (a) Suppose 10^6 W are being transmitted at 10,000 V. If the resistance in the line is r, what is the power loss in the line?
   (b) If 10^6 W are transmitted at 100 V and the resistance in the line is r, what is the power loss in the line?

26.A.4 A 5-W 12-V bulb is connected to a 12-V battery with an internal resistance of 2 Ω. What is the power delivered to the bulb? (A 5-W 12-V bulb is a bulb that will dissipate 5 W when a voltage of 12 V is applied.)

26.A.5 If a 5-W 12-V bulb (see problem 26.A.4) is connected to a 120-V source, what power is dissipated in the bulb?

26.B.1 How may a collection of 1-Ω resistors be connected to achieve an equivalent resistance of n_1/n_2 Ω, where n_1 and n_2 are any two integers?

26.B.2 Determine the resistor equivalent of the resistor groups in each of the following figures.

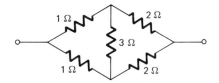

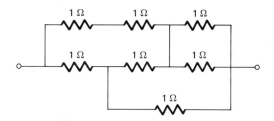

26.B.3 What is the current in the 5-Ω resistor in the figure?

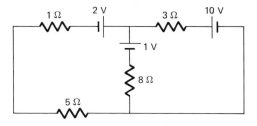

26.B.4 What is the current in the 5-Ω resistor in the figure?

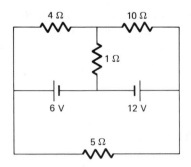

26.B.5 What is the current in the 1-Ω resistor in the figure?

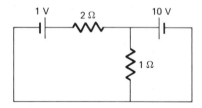

26.B.6 A 4000-W electric wall heater is wired with 20 m of No. 14 gauge copper wire (diameter, 1.6 mm; resistivity, 1.7×10^{-8} Ω · m). The line voltage is 220 V.
  (a) What is the power dissipated in the wires?
  (b) What is the power output of the heater? (Remember that there are two leads in the wire.)

26.B.7 A 4-Ω resistor is connected across the terminal of a battery, and the terminal voltage of the battery is 4.8 V. A 5-Ω resistor is connected across the same battery, and the terminal voltage is now 5 V.
  (a) What is the internal resistance of the battery?
  (b) What is the emf of the battery?

26.B.8 A circuit consists of a source of emf with an internal resistance r and an external resistance or load R. Show that the power delivered to the load is a maximum when $R = r$, that is, when the load matches the internal resistance. (This is known as "impedance matching." We shall discuss this point again in Chap. 32 when we take up the problem of how acoustic energy is transmitted from the air to the fluid in the inner ear.)

26.C.1 Determine the equivalent resistance of the resistor group in the figure, in which 12 1-Ω resistors are on the side of a cube. (Imagine that a current of 1 A enters the network at A.)

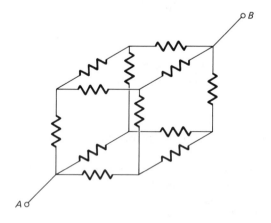

26.C.2 A device that measures current in a circuit is called an *ammeter*. Suppose a certain ammeter gives a full-scale deflection when the current through the ammeter is 1 mA. If one wanted an ammeter that gave a full-scale deflection when the current was 1 A, one could "shunt" the ammeter, i.e., place a resistance in parallel with the ammeter to divert some of the current (see the figure). If the resistance of the original ammeter was 10^{-3} Ω, what must be the resistance of the shunt for the ammeter to give a full-scale deflection when the current is 1 A?

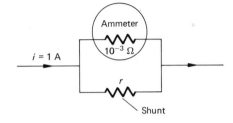

27
THE
MAGNETIC
FORCE

Up to this point we have considered only one-half of electromagnetic theory, namely the electric field, and only electrostatic fields at that. We shall now consider the sources and effects of magnetic fields.

It is difficult to develop this subject by beginning with an action-at-a-distance theory, as we did in electrostatics, and then proceeding to a field theory description. It is much simpler to begin immediately with a field theory. Let us define the magnetic field.

27.1 THE MAGNETIC FIELD

We shall consider the following hypothetical experiment: An electrically charged particle is moving in the neighborhood of a loop of wire carrying an electric current (Fig. 27.1). Some very unusual things happen that cannot be accounted for by either gravitational forces or electric forces. When the charge is at rest, it remains at rest (if we neglect the gravitational force); and when the charge is in motion, it does not move with a constant speed in a straight line. Therefore some *velocity-dependent force* must be

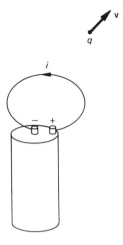

FIGURE 27.1 A charge in motion near a current loop.

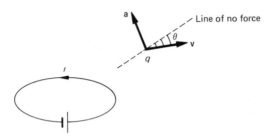

FIGURE 27.2 The acceleration, and hence the force, are zero when the velocity is directed along the line of no force; and as the angle θ between the velocity and the line of no force varies, the force varies with sin θ. The acceleration is directed perpendicularly to the velocity and the line of no force. The acceleration is proportional to the magnitude of the velocity and the quantity of electric charge q.

acting on the charge. The fact that this force depends on velocity differentiates it from either gravitational or electric forces, which are independent of velocity. The way in which this force depends on the velocity is also unusual. There is *one* direction in which a moving charge experiences no force, and we shall call the line in this direction the *line of no force.* The force is a maximum for velocities perpendicular to this line. (We must always remember that we measure force by observing the acceleration and multiplying by the mass.) In fact the magnitude of the force increases with sin θ where θ is the angle between the velocity and the line of no force (Fig. 27.2). Note that for $\theta = 0$ the sine is zero. We also find that the force is directly proportional to the magnitude of the velocity and the quantity of electric charge. The direction of the force is perpendicular to both the velocity and the line of no force.

Collecting this information, we have

$$F \propto qv \sin \theta$$

To say that F is proportional to $qv \sin \theta$ is to say that

$$F = Bqv \sin \theta \qquad (27.1)$$

where B is some constant depending on the current loop (its current, size, position, orientation, and shape) but independent of the parameters associated with the moving charge (its charge q, mass m, or velocity **v**). Since we can measure everything in Eq. 27.1 except B, we may use this equation to define B. The quantity B is called the *magnetic induction,* or the *magnetic field strength,* or simply the *magnetic field.* It has the unit newton-second per coulomb-meter, which is called the *tesla* (*T*). The unit *weber per square meter* (Wb/m²) is also used. It is equivalent to the tesla and, as we shall see, gives magnetic flux the unit of the weber. Another very common unit for measuring magnetic field is the *gauss* (G), where

$$1 \text{ T} = 10^4 \text{ G}$$

The gauss has the advantage of being of more reasonable size for most magnetic fields. For example, the magnetic field of the earth at the equator is about 0.5 G.

With the aid of Eq. 27.1 we can determine the magnetic field at any point. Thus the magnetic field is a field, i.e., it has a numerical value at every point in a domain, and we shall see shortly that it is a vector field.

If Eq. 27.1 were only a definition of B, then it would have no significance at all. We can define many things. Let Z be defined by the relation

$$Z = \frac{q^2 \tan \theta}{v^3}$$

We could measure Z at all points in space and we would have a new field. But what good is it?

Equation 27.1 has meaning because of the fact that B depends only on the current loop. For a given current loop, B is fixed at any given point. Having used our test charge to measure B, we know that any *new* charge q, with any velocity **v** making an angle θ with the line of no force, will experience an acceleration that can be *predicted by Eq. 27.1*. As we have said often, the test of a physical law is its ability to make predictions on untried experiments.

It is important to understand the development of this field theory for the magnetic force. We have discovered the law that allows us to determine the force given the field that exists at the point where the charge is located, and we have seen how we may measure the field. But something is missing. We would like to be able to *predict the field* from the structure of the current loop. *This is the other half of the field theory*. We would like to know how electric charges (or currents created by moving charges) create fields and then how these fields affect other charges. The field is the *link* in the interaction between moving charges. We will save this task of obtaining the magnetic field from the currents until Chap. 28. Let us assume for the time being that we know how to calculate magnetic fields or, failing this, how we may measure the field experimentally. We will concentrate for the moment on how charges behave in magnetic fields.

We have yet to incorporate into our force equation the information we gained from our experiments concerning the *direction* of **F**, namely that **F** is perpendicular to **v** and the line of no force. We can do this most expeditiously by using the vector cross product. We first define a vector **B** whose magnitude is as defined above and whose direction lies along the line of no force. The direction in which it *points* on this line is determined by the requirement that **v** × **B** be parallel to **F**. This is purely a convention. The vector **B** could just as well be defined so that **v** × **B** is antiparallel to **F**, and we would have to modify our rules for calculating **B** from a knowledge of the currents creating the **B** field.

We may finally put together in the following equation all the information gained from our experiments:

$$\mathbf{F} = q\mathbf{v} \times \mathbf{B} \tag{27.2}$$

This is both a definition of **B** and a physical law governing the force on particles moving in the **B** field.

27.2 THE TRAJECTORY OF A CHARGED PARTICLE IN A UNIFORM MAGNETIC FIELD

Let us consider first the simplest case of the motion of a charged particle in a magnetic field: a uniform field. It is customary to depict a magnetic field perpendicular to the plane of the page by a dot ($\cdot$) if the field is pointing out of the page and by a cross ($\times$) if the field is pointing into the page. As a mnemonic, think of an arrow with a point at one end and crossed feathers at the other. If the arrow is coming out of the page, we see the dot, and if the arrow has penetrated into the page, we see the cross.

In Fig. 27.3 we consider a particle of charge q and velocity **v** in the plane of the page. The magnetic field is

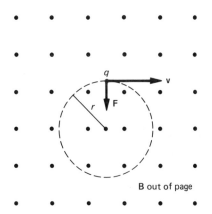

FIGURE 27.3   A charged particle moving in a magnetic field with **v** perpendicular to **B**.

directed out of the page. From Eq. 27.2 the force is perpendicular to the velocity. This of course is true in general and is not limited to this special example. Now we have seen that for any force in a plane (see Eqs. 2.15 and 2.16)

$$F_T = ma_T = m\frac{dv}{dt} \qquad (27.3)$$

and

$$F_R = ma_R = \frac{mv^2}{R} \qquad (27.4)$$

where F_T and F_R are the components of the force tangent and perpendicular to the path. We see that for any magnetic field F_T is zero and therefore $dv/dt = 0$. Now dv/dt is the time rate of change of the magnitude of the velocity, so the magnitude of the velocity cannot change in a magnetic field alone. The speed of the particle must remain constant.

Another way to state this is to say that F_T is zero and so the work ($\int F_T \, ds$) done on the particle by the field must be zero. If the work is zero, the change in kinetic energy must be zero. This is an important general result:

A particle moving in a magnetic field only cannot change its speed.

But the magnetic field does exert a force perpendicular to the path, so the direction of the velocity will change. In fact we have from Eqs. 27.2 and 27.4

$$\frac{mv^2}{R} = qvB$$

where R is the radius of curvature of the path and $\sin \theta = 1$ since v is perpendicular to B. Solving for R, we find

$$R = \frac{mv}{qB} \qquad (27.5)$$

But v and B are constants, and so therefore is the radius of curvature. Now the only path that has a constant radius

of curvature is the circle. *Therefore a particle travels in a circle of radius $R = mv/qB$ in a constant magnetic field if the velocity is perpendicular to the field.* Other parameters of interest are the angular velocity, the period, and the frequency of rotation. Now

$$\omega = \frac{v}{R} = \frac{qB}{m}$$

and

$$\omega = \frac{2\pi}{T}$$

so

$$T = \frac{2\pi}{\omega} = \frac{2\pi m}{qB}$$

and the frequency is given by

$$f = \frac{1}{T} = \frac{qB}{2\pi m}$$

Note that all three of these quantities (ω, T, and f) are independent of the velocity and the radius of the orbit. All particles of the same charge-to-mass ratio travel with the same angular velocity, period, and frequency in a uniform magnetic field.

The orbit of a charged particle in a constant magnetic field is a circle only if no component of velocity is parallel to the field. A component of velocity parallel to a magnetic field is unaffected by the field. The general motion of a charge in a uniform magnetic field is therefore a helix, which is a superposition of a circular motion perpendicular to the field and a constant velocity parallel to the field (Fig. 27.4).

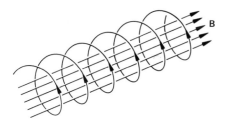

FIGURE 27.4 The path of a particle in a constant magnetic field is a helix.

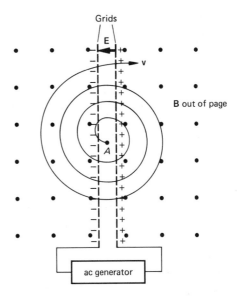

FIGURE 27.5 A positive charge rotating with ever-increasing speed in a constant magnetic field and an alternating electric field.

QUESTION

Suppose you want to accelerate a positively charged particle by applying an alternating electric field in such a way that the particle is accelerated by the electric field on every revolution in a transverse magnetic field. Such a device is illustrated in Fig. 27.5. The magnetic field is directed out of the page, and the particles are injected into the field at A, where their velocity is small. The radius of their orbit is

$$R = \frac{mv}{eB}$$

where e is the electronic charge. As the particle rotates, it eventually enters the electric field created by the grids. If the electric field at that moment is directed in the direction of the motion of the particle, the particle will accelerate, and after emerging from the area between the grids it will move in a slightly larger circle because of the increase in speed. (The electric field is confined more or less to the region between the grids much as it is in a parallel-plate capacitor.) After the particle completes another half revolution, it enters the area between the grids again, but the alternating generator across the grids has reversed the electric field in this half cycle, so the electric field again accelerates the charge. The particle moves in a still larger circle appropriate to the larger velocity until the next boost, and the cycle is repeated.

What is the frequency of the alternating generator? Does this frequency depend on the radius of the orbit?

ANSWER

The frequency of the electric field should equal the frequency of rotation of the charge, given by

$$f = \frac{eB}{2\pi m}$$

which is independent of the velocity or the radius R of the orbit! It is not necessary to adjust the frequency of the field to match the frequency of rotation. This is indeed fortunate, since the accelerating particles are of atomic size and it would be difficult to follow their motion and adjust the electric field according to their positions.

You may be wondering why the grids extend the full diameter of the particle orbit and are not just in the upper or lower half, where we might use a constant field and all particles, not just those in phase with the electric field, would be accelerated. It must be remembered that $\oint \mathbf{E} \cdot d\mathbf{s}$ is zero around any closed path in an electrostatic field. The total work done by an electrostatic field around a closed path is zero. A fringe electric field that extends beyond the grids opposes the motion of the charge after it leaves the region between the grids. The net effect of an electrostatic field would be zero.

The accelerator described here is similar to the operation of the cyclotron, although some liberties have been taken in the design.

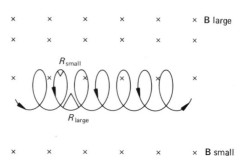

FIGURE 27.6   The magnetic field is directed into the page and increases in the vertical direction.

27.3 MOTION OF AN ELECTRIC CHARGE IN NONUNIFORM MAGNETIC FIELDS

If the field is nonuniform, the motion is not simple, but in many cases we can obtain a qualitative picture of the orbit. Suppose the magnetic field is directed into the page, but the strength increases in the vertical direction (see Fig. 27.6). The radius of curvature of a particle in a magnetic field (uniform or not) is given by

$$R = \frac{mv}{qB}$$

Now v is constant, so R is large where B is small and R is small where B is large. We can see from Fig. 27.6 that this causes a drift of the particle from left to right. Such a drift occurs for charged particles moving in the earth's magnetic field in the equatorial plane. The magnetic field decreases with distance from the earth, so positive particles drift

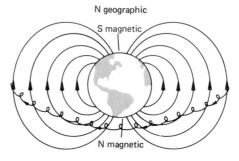

FIGURE 27.7   Positive particles drift from east to west in the earth's magnetic field in the equatorial plane.

from east to west and negative particles from west to east (see Fig. 27.7).

Drift may also be caused by a gravitational or electric field transverse to the magnetic field. If the magnetic field is into the page, and $\mathbf{g}$ or $\mathbf{E}$ is directed down, then a positive charge speeds up as it moves down (see Fig. 27.8). As the speed increases, the radius of curvature increases. As the particle moves up, it slows down and the radius of curvature decreases. The net effect is a drift to the right.

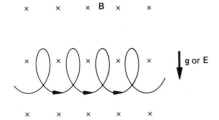

FIGURE 27.8 A positive charge will accelerate as it moves in the direction of **g** or **E**. With the increase in velocity the radius of curvature increases, and this produces a drift.

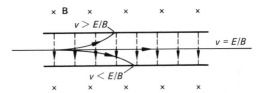

FIGURE 27.9 A parallel-plate capacitor creates a downward electric field. The magnetic field is directed into the page. If $v = E/B$, a particle experiences no net force and passes through the plates in a straight line. Particles with other velocities strike the plates. The crossed **E** and **B** fields act as a velocity selector.

If the **g** or **E** field is just right, it is possible for the particle to move in a straight line with constant speed. The net force is

$$\mathbf{F} = m\mathbf{g} + q\mathbf{v} \times \mathbf{B}$$

or

$$\mathbf{F} = q\mathbf{E} + q\mathbf{v} \times \mathbf{B}$$

If **g**, **v**, and **B** are mutually perpendicular and satisfy the equation

$$m\mathbf{g} + q\mathbf{v} \times \mathbf{B} = 0$$

or if **E**, **v**, and **B** are mutually perpendicular and

$$\mathbf{E} + \mathbf{v} \times \mathbf{B} = 0$$

then the net force is zero in each case.

We can use this fact to construct a velocity selector. If we have two long parallel plates with a narrow separation that produce a downward electric field, and if **B** is into the page (see Fig. 27.9), then a positive particle moving between the plates with a velocity $v = E/B$ will pass straight through. If v is greater or less than E/B, the particle will be deflected up or down and will strike the plates. Thus the device allows only particles of a select velocity to pass through.

If a charged particle is moving in a convergent magnetic field, an ever-increasing force opposes the motion into the convergent region. Suppose a positive particle is moving in a circle in a convergent magnetic field, as in Fig. 27.10. We see that the force has a component directed away from the region of stronger field, so the particle would experience a force directed toward the region of weaker field. (If the field were reversed, the velocity would also be reversed, so $\mathbf{F} = q\mathbf{v} \times \mathbf{B}$ would remain unchanged.)

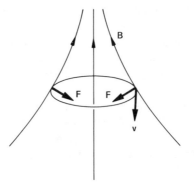

FIGURE 27.10 A positive particle rotates in a circular orbit in a convergent magnetic field. A component of the force is directed away from the region of stronger field.

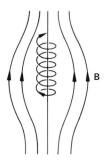

FIGURE 27.11 Charged particles are trapped in this magnetic bottle because of the convergence of the magnetic field lines. Particles move roughly in a helical path but are reflected in the region of convergent field.

This principle is useful in understanding the confinement of charged particles in the earth's Van Allen belts and also in the construction of magnetic bottles used to contain very hot gases, gases so hot that most of the atoms are ionized. Such gases are called plasmas. The sun's corona is a plasma; particles in a thermonuclear reactor are a plasma. A major effort is currently underway to create a *magnetic bottle* to contain such plasmas, so that the energy from fusion reactions can be controlled. The operation of these bottles is illustrated in Fig. 27.11. As particles spiral close to the convergent region, they are repelled by the opposing force.

A naturally occurring magnetic bottle is the magnetic field of the earth (see Fig. 27.12). At both poles the field lines converge, and here the particles are reflected. (At the same time there is the east-to-west drift that we spoke of earlier.)

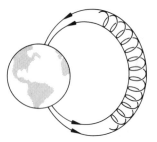

FIGURE 27.12 In the earth's magnetic field charged particles are reflected in the regions of converging magnetic field.

27.4 MAGNETIC FORCE ON AN ELECTRIC CURRENT

We have seen that a magnetic field exerts a force on a moving charge. An electric current is a lot of moving charges, so we must expect a magnetic force on electric currents. In fact, we should be able to determine the force from our basic force law on single charges. The force on a group of charges dq traveling in a wire at a mean velocity **v** is given by

$$d\mathbf{F} = dq\mathbf{v} \times \mathbf{B} \qquad (27.6)$$

Let dq be those charges that pass a cross section of the wire during a time dt. Then

$$dq = i\,dt \qquad (27.7)$$

Furthermore **v** dt will be the distance $d\ell$ traveled down the wire by the charges during the time dt, so

$$\mathbf{v}\,dt = d\ell \qquad (27.8)$$

Combining Eqs. 27.6, 27.7, and 27.8, we have

$$d\mathbf{F} = i\,d\ell \times \mathbf{B} \qquad (27.9)$$

Now $d\mathbf{F}$ was to be the force on a group of charges passing a certain cross section of the wire during a time dt. But these charges are the moving charges in a segment of wire of length dl, where $dl = \mathbf{v}\, dt$, so $d\mathbf{F}$ is the force on a segment of wire of length dl carrying a current i in a magnetic field $\mathbf{B}$.

QUESTION

What is the force on a straight wire of length L in a uniform magnetic field?

ANSWER

We obtain the force by integrating Eq. 27.9:

$$\mathbf{F} = \int d\mathbf{F} = \int i\, dl \times \mathbf{B}$$

But i and $\mathbf{B}$ are constant and therefore may be factored from the integral to give

$$\mathbf{F} = i\left(\int dl\right) \times \mathbf{B}$$

But $\int dl = \mathbf{L}$, so

$$\mathbf{F} = i\mathbf{L} \times \mathbf{B}$$

QUESTION

The earth's magnetic field is parallel to the earth's surface at the equator and directed from south to north. Its magnitude is 0.5 G. A 1-m length of straight wire with a mass of 50 g is carrying a current from west to east. What current will support the wire?

ANSWER

The gravitational force must balance the magnetic force (Fig. 27.13), so

FIGURE 27.13   Forces acting on a current-carrying wire.

$$iLB = mg$$

or

$$i = \frac{mg}{LB} = \frac{(0.05 \text{ kg})(9.8 \text{ m/s}^2)}{(1 \text{ m})(0.5 \times 10^{-4} \text{ T})} = 9800 \text{ A}$$

which is an impossibly large current.

QUESTION

What is the net force on a closed current loop of arbitrary shape in a uniform magnetic field?

ANSWER

The net force is given by

$$\mathbf{F} = \oint i\, dl \times \mathbf{B}$$

$$= i\left(\oint dl\right) \times \mathbf{B} = 0$$

since the integral of dl around a closed current loop is zero.

What is the torque on a plane loop of current if the normal to the plane makes an angle θ with the magnetic field?

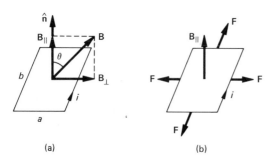

(a)   (b)

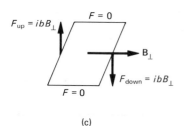

(c)

FIGURE 27.14 (a) A magnetic field **B** makes an angle θ with respect to the normal $\hat{n}$. (Of the two normals, we choose the one in the direction of the thumb of the right hand when the fingers are curled around the loop in the direction of the current.) The magnetic field is decomposed into its components (b) parallel and (c) perpendicular to the normal.

ANSWER

For simplicity we shall assume that the loop is rectangular, as shown in Fig. 27.14. The forces resulting from the parallel component of **B** add up to zero. They would tend to open up the loop, while the forces resulting from the perpendicular component of **B** would tend to rotate the loop to align the

normal with **B**. The forces on the two segments of length a are zero, and the two forces on the two segments of length b are equal and opposite and are given by

$$F_{up} = F_{down} = ibB_{\perp}$$

The moment of these forces about an axis through the center of the loop is

$$\tau = F_{down}\frac{a}{2} + F_{up}\frac{a}{2} = aF_{up}$$

$$= iabB_{\perp} = iabB \sin \theta$$

But $ab = A$, the area of the loop, so we may write

$$\tau = iAB \sin \theta$$

We may express the result as a vector equation:

$$\tau = i\mathbf{A} \times \mathbf{B}$$

We may generalize the result to a loop of arbitrary shape in a plane by dividing the shape into a large number of rectangular loops, all carrying a current i (Fig. 27.15). Opposing currents inside the loop cancel and give a net current of zero. The effective current is a current around the perimeter of the loop. Since the net force on each loop is zero, the net moment is the sum of the moments of each rectangle. But this is just $i\mathbf{A} \times \mathbf{B}$.

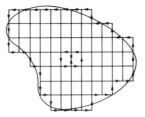

FIGURE 27.15 A current loop of arbitrary shape divided into a large number of rectangular loops.

The product $i\mathbf{A}$ is called the *magnetic dipole moment* μ of the current loop. The torque then may be expressed as

$$\tau = \mu \times \mathbf{B} \qquad (27.10)$$

This is to be compared with the torque on an electric dipole of moment $\mathbf{p}$, which is given by Eq. 21.7, that is,

$$\tau = \mathbf{p} \times \mathbf{E}$$

We have also seen that the potential energy of an electric dipole in a uniform electric field is given by $U = -\mathbf{p} \cdot \mathbf{E}$. The potential energy of a magnetic dipole in a uniform magnetic field is given by a similar expression,

$$U = -\mu \cdot \mathbf{B} \qquad (27.11)$$

We may obtain the torque on the dipole by differentiating the potential energy with respect to the angle of rotation.

SUMMARY

The force on a charge q moving in a magnetic field $\mathbf{B}$ with a velocity $\mathbf{v}$ is given by

$$\mathbf{F} = q\mathbf{v} \times \mathbf{B}$$

Since the force is perpendicular to the velocity, it can do no work on the charge, so the speed cannot change in a magnetic field alone.

The trajectory of a charge in a uniform magnetic field is a circle (or helix) whose radius is given by

$$R = \frac{mv}{qB}$$

A particle moving in crossed uniform electric and magnetic fields with a velocity perpendicular to both $\mathbf{E}$ and $\mathbf{B}$ will move in a straight line with constant speed if

$$v = \frac{E}{B}$$

The force on a current element in a magnetic field is

$$\mathbf{F} = i\, d\boldsymbol{\ell} \times \mathbf{B}$$

The torque on a plane current loop of area A in a uniform magnetic field is given by

$$\tau = i\mathbf{A} \times \mathbf{B}$$

PROBLEMS

27.A.1 The magnetic field in the figure is directed into the paper. Its strength increases uniformly in the x direction.
  (a) A positively charged particle is moving parallel to the positive y axis. Describe the motion qualitatively.
  (b) What would the path look like if the particle were negative?

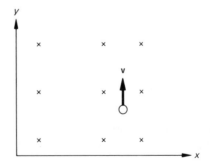

27.A.2 The magnetic field in the following figure is uniform and is directed into the page. A uniform electric field is directed along the x axis.
  (a) If positive particles were moving in the y direction, what would be the direction of the drift velocity?
  (b) What would it be if the particles were negative?

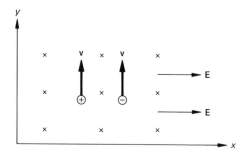

27.A.3 What are the magnitude and direction of the force on a 10-cm segment of straight wire carrying a current of 2 A at an angle of 30° with a uniform magnetic field of 3 T? (See the figure.)

27.A.4 If an electron rotates in a circle of radius 1 cm, what is the radius of the circular orbit of a proton traveling at the same speed in the same magnetic field?

27.A.5 What is the radius of curvature of a thermal electron ($T = 300$ K) moving transverse to the earth's magnetic field ($B = 0.5$ G)?

27.A.6 An electron moves through a velocity selector consisting of crossed electric and magnetic fields. If the electric field is 10,000 V/m, the magnetic field 0.1 T, and the path of the electron a straight line, what is the velocity of the electron?

27.A.7 Which way will the rod shown in the figure move when the switch is closed?

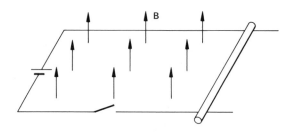

27.B.1 An electron has a velocity as shown in the figure and is moving in a magnetic field perpendicular to the page. The strength of the field decreases in the x direction and is constant in the y and z directions. Describe the motion.

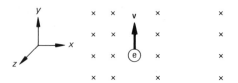

27.B.2 A particle moves in a helical path in a uniform magnetic field. If the pitch of the helix is equal to the radius, and the component of the velocity perpendicular to the magnetic field is 10 m/s, what is the component of the velocity parallel to the field?

27.B.3 The 5-cm rod in the following figure is sliding down the 37° incline with a constant velocity in a uniform magnetic field. The magnetic field strength is 0.1 T and is directed vertically. If the mass of the rod is 10 g, what is the current in the wire? (Neglect friction.)

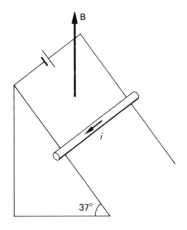

27.B.4 A charged particle moves through the crossed electric ($E = 20,000$ V/m) and magnetic ($B = 10$ T) fields of a velocity selector in a straight line. It then enters a uniform magnetic field ($B = 40$ T) and moves in a circular path with a diameter ($D = 10$ cm) determined by a darkening of the photographic plate (see the figure below).

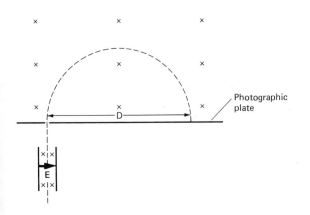

What is the mass of the particle? (This device is called a mass spectrometer.)

27.B.5 The following figure illustrates an airplane wing moving through the earth's magnetic field at a place where the field is vertical and has a magnitude of 0.5 G. The potential difference across the 10-m wingspan is 0.1 V. What is the speed of the plane?

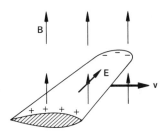

27.B.6 The accompanying figure illustrates a crude ammeter: a wire loop of 100 turns and an area of 40 cm² mounted on a board that balances when there is no current through the loop. A uniform magnetic field of 0.01 T is perpendicular to the normal to the loop and to the axis of rotation of the board. A current is passed through the loop, and a 5-g mass 10 cm from the axis of the board is required to maintain the balance. What is the current?

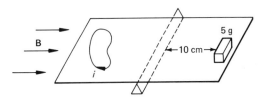

27.B.7 The current in the loop shown in the figure is 2 A, and the area of the loop is 100 cm². If the magnetic field has components $B_x = 0$, $B_y = 0.1$ T, and $B_z = 0.3$ T, what are the components of the torque on the loop?

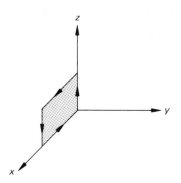

27.C.1 A disk of radius R carries a uniform surface charge density σ and rotates about its axis with an angular velocity ω. A magnetic field **B** is applied perpendicularly to the axis. What is the torque on the disk?

27.C.2 The measurement of blood flow rates in blood vessels is of considerable importance in medicine. One method of doing this without opening the blood vessel is to employ the "Hall effect." If a conductor is moving in a magnetic field, the free electrons will move in response to the magnetic force. In part (a) of the figure a conducting fluid (blood) is moving transverse to a magnetic field directed into the page. Positive ions will move up and negative ions down, and the ions will continue to move until the force created by the electric field generated by the charge separation exactly balances the force created by the magnetic field. (It makes its own velocity selector.) The voltage measured between two points A and B on the wall of the blood vessel separated by a distance s measured along a line perpendicular to the magnetic field and the flow velocity is V. If the magnetic field is **B**, what is the velocity of the blood?

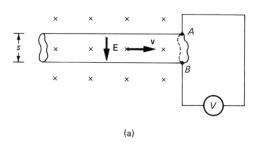

(a)

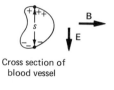

Cross section of
blood vessel

(b)

28
THE
MAGNETIC
FIELD

As we have said, a field theory has two aspects: How does the field affect a particle, and how does the particle create a field? We have dealt with the first question, and we shall now consider the second.

28.1 THE MAGNETIC FIELD OF A POINT CHARGE

The magnetic field of a moving point charge must be determined from experiment. By observing the magnetic interaction between charged particles, we find that the field of a single charge is given by

$$\mathbf{B} = \frac{\mu_0}{4\pi} \frac{q\mathbf{v} \times \mathbf{r}}{r^3} = \frac{\mu_0}{4\pi} \frac{q\mathbf{v} \times \hat{r}}{r^2} \qquad (28.1)$$

where μ_0 is the magnetic permeability of free space, determined from the relation

$$\frac{\mu_0}{4\pi} = 10^{-7} \text{ T} \cdot \text{m} \cdot \text{s/C} \qquad (28.2)$$

The 4π is introduced into Eq. 28.1 in order to avoid a 4π later in the integral formulation of magnetism.

As in electric fields, the vector $\mathbf{r}$ is directed *from* the source q of the field *to* the point of observation, i.e., the point at which the field is being calculated.

If we have a distribution of electric charge moving with a mean velocity $\mathbf{v}$, then the magnetic field created by an element of charge dq is given by

$$d\mathbf{B} = \frac{\mu_0}{4\pi} \frac{dq\, \mathbf{v} \times \mathbf{r}}{r^3}$$

We have seen (Sec. 27.4) that when the flow of charge takes place in a conducting wire,

$$dq\, \mathbf{v} = i\, d\ell$$

where i is the current in the wire and $d\ell$ is a differential segment of the wire. The magnetic field then is given by the law of *Biot and Savart*, namely

$$d\mathbf{B} = \frac{\mu_0}{4\pi} \frac{i\, d\ell \times \mathbf{r}}{r^3}$$

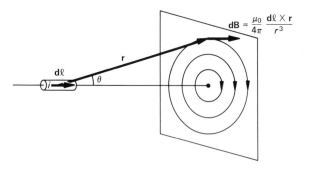

FIGURE 28.1   Magnetic field in an arbitrary plane perpendicular to the wire segment.

The magnetic field associated with a short segment of wire is illustrated in Fig. 28.1. The magnetic field lines are circles that lie in a plane perpendicular to the extension of the line segment, and the circles are concentric with the extension of the line segment. We can of course determine the direction of the magnetic field by applying the right-hand rule to the cross product $d\ell \times \mathbf{r}$ or by placing the thumb of the right hand in the direction of current and letting the coiled fingers indicate the direction of the magnetic field (see Fig. 28.2). In Fig. 28.1 only the magnetic field in one plane is shown. To obtain the direction of the magnetic field at any point, construct a plane that includes the point and is perpendicular to the extension of $d\ell$. The magnetic field is tangent to a circle in this plane through the point and concentric with the extension of $d\ell$.

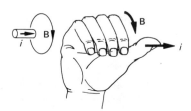

FIGURE 28.2   When the thumb of the right hand is pointed in the direction of the current, the fingers coil in the direction of the magnetic field of the current segment.

28.2 QUALITATIVE PICTURE OF THE MAGNETIC FIELD

Before we calculate some magnetic fields, we shall first give a qualitative picture of the field lines of some simple current configurations.

AN INFINITE STRAIGHT WIRE

Since the magnetic field lines of a short segment are concentric circles, it follows that the magnetic field lines of an infinite straight wire must also be circles. All circles will be concentric with the straight wire, and we obtain the sense of the field on the circle by the right-hand rule (see Fig. 28.3).

An analog with fluid dynamics is useful in picturing magnetic fields. We see that the magnetic field lines of a straight wire circulate about the wire instead of emanating from a source, as was the case for electric fields. It is possible to model the magnetic field on a fluid in which the flow is generated by a vortex rather than by sources and sinks, where by a vortex we mean a rotation. This rotation might be generated, for instance, by a rotating paddle

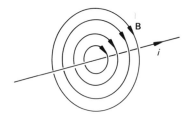

FIGURE 28.3   Magnetic field lines of an infinite straight wire.

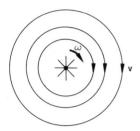

FIGURE 28.4   Rotating paddle wheel on a long shaft.

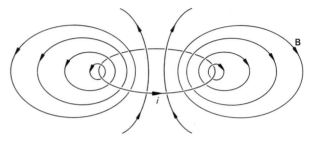

FIGURE 28.5   Magnetic field lines of a circular current loop. Only the field lines in the plane of the page are shown.

wheel with a very long shaft and very small blades. If we have a set of long thin blades attached to a long shaft immersed in an incompressible liquid, we may obtain a fluid flow very similar to that shown in Fig. 28.4. If the wire is bent, we may still imagine a series of blades distributed along the length of the wire. The rate of rotation of the blades is proportional to the strength of the electric current.

Since there are no sources or sinks, the field lines (of the fluid or the magnetic field) (a) never begin or end, and (b) always close on themselves.

We shall see an additional analog when we come to the integral formulation of the magnetic field equation.

A CIRCULAR CURRENT LOOP

The magnetic field lines of a circular current loop are illustrated in Fig. 28.5. The field lines are almost circles near the wire. Further from the wire they are no longer circular, but retain the property of closing on themselves. It is not difficult to imagine this field pattern as the flow pattern in an incompressible fluid generated by a circulation about the wire in a direction determined by the right-hand rule.

Remember that this field may be obtained quantitatively from the Biot-Savart law. We shall later calculate the field on the axis of the circle.

A SOLENOID

A solenoid is a right circular cylinder with an electric current flowing around the perimeter, perpendicular to the axis. It might be constructed in several ways: as a helical winding of wire on a long tube, or as a set of current rings such as that in Fig. 28.5, or as a current sheet such as that illustrated in Fig. 28.6(b). The resulting magnetic field is directed along the length of the cylinder. The field is strong inside the cylinder and much weaker outside (except at the ends). The field lines close on themselves (now shown in Fig. 28.6). If we tightly wind the helix or tightly pack the current rings, the field lines will become virtually straight lines of uniform density inside the solenoid, and the magnetic field will approach zero outside the solenoid.

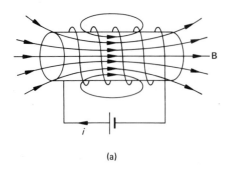

(a)

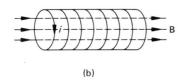

(b)

FIGURE 28.6   (a) Magnetic field lines of a solenoid of finite length. (b) Magnetic field lines of an infinite solenoid with a dense surface current.

The reason for this is that outside the solenoid the field lines have an infinite region in which to return in order to close on themselves. If the number of field lines is finite, we might expect the density of field lines outside to approach zero as the cylinder becomes longer and longer and the surface current more dense. (You may not find this argument entirely satisfying, but this result can be demonstrated rigorously.)

A CURRENT SHEET

An idealization that we shall find useful later is an infinite sheet of current, illustrated in Fig. 28.7. The magnetic fluid lines will be straight lines, parallel to the sheet and perpendicular to the current. They will be in opposite

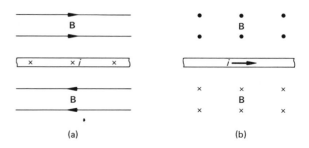

(a)   (b)

FIGURE 28.7   Magnetic field of an infinite current sheet perpendicular to the plane of the page. In (a) the current is directed into the page and the field lines lie in the plane of the page. In (b) the current lies in the plane of the page and the magnetic field is perpendicular to the plane of the page.

directions on either side of the sheet. The magnetic field will be uniform everywhere and will be directed according to the right-hand rule.

We obtain the magnetic field of two oppositely directed current sheets by summing the fields of the individual sheets. The field is illustrated in Fig. 28.8. The magnetic fields add between the sheets and cancel outside.

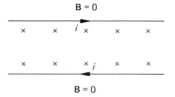

FIGURE 28.8   Magnetic field of two infinite current sheets (in cross section). The field is uniform between the sheets and zero outside.

28.3 CALCULATING THE MAGNETIC FIELD

Let us now determine the magnetic field of some simple current distributions from the Biot-Savart law.

AN INFINITE LINE CURRENT

Consider first an infinite straight wire carrying a current i. The magnetic filed $d\mathbf{B}$ created by a small current element dx (see Fig. 28.9) is directed out of the page. The magnitude of the field is given by

$$dB = \frac{\mu_0}{4\pi} \frac{i\, dx \sin\theta}{r^2}$$

Now

$$\sin\theta = \frac{R}{r}$$

and

$$r = \sqrt{x^2 + R^2}$$

so

$$B = \frac{\mu_0 iR}{4\pi} \int_{-\infty}^{\infty} \frac{dx}{(R^2 + x^2)^{3/2}}$$

$$= \frac{\mu_0 i}{2\pi R} \tag{28.3}$$

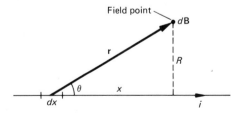

FIGURE 28.9 Magnetic field $d\mathbf{B}$ created by a small current element dx. The field $d\mathbf{B}$ is directed out of the page.

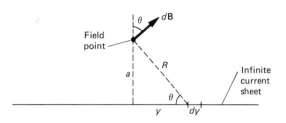

FIGURE 28.10 A strip dy of an infinite current sheet contributes a field $d\mathbf{B}$ at the field point. The current is directed into the page.

A CURRENT SHEET

We can derive the magnetic field of a current sheet from the field of a line current. The sheet of current can be thought of as a large number of strips of infinite length and width dy (see Fig. 28.10). The current in each strip is $K\, dy$, where K is the *surface current density*, i.e., the current per unit length of line perpendicular to the current. From Eq. 28.3 the magnetic field $d\mathbf{B}$ is given by

$$dB = \frac{\mu_0\, di}{2\pi R} = \frac{\mu_0 K\, dy}{2\pi R}$$

From the symmetry of the problem the only component of $d\mathbf{B}$ that will survive after the integration is the horizontal component $dB \sin\theta$. The total field is therefore

$$B = \frac{\mu_0 K}{2\pi} \int \frac{\sin\theta\, dy}{R}$$

$$= \frac{\mu_0 K}{2\pi} \int_{-\infty}^{\infty} \frac{a\, dy}{y^2 + a^2} = \frac{\mu_0 K}{2} \tag{28.4}$$

We see that the magnetic field has a constant magnitude that is independent of the distance from the sheet. This result should be compared with the electric field of an infinite sheet of charge density σ, $E = \sigma/2\epsilon_0$.

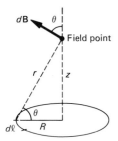

FIGURE 28.11 Magnetic field $d\mathbf{B}$ on the axis created by a current element $d\ell$ of a circular ring current i.

A RING CURRENT

The magnetic field of a circular ring current is in general difficult to calculate. However, we may determine the field on the axis of the ring very simply. The magnitude of the field created by a small segment of the ring of length $d\ell$ (see Fig. 28.11) is given by the Biot-Savart law,

$$dB = \frac{\mu_0}{4\pi} \frac{i \, d\ell}{r^2}$$

When the vector field $d\mathbf{B}$ is summed over all current elements on the ring, only the vertical component survives. Therefore

$$B = \frac{\mu_0 i}{4\pi} \int \frac{\cos\theta \, d\ell}{r^2}$$

But $\cos\theta$ and r are constant, so

$$B = \frac{\mu_0 i \cos\theta}{4\pi r^2} \int d\ell = \frac{\mu_0 i \cos\theta}{4\pi r^2} 2\pi R \qquad (28.5)$$

Now

$$\cos\theta = \frac{R}{r}$$

and

$$r = \sqrt{R^2 + Z^2}$$

so

$$B = \frac{\mu_0 i R^2}{2(R^2 + Z^2)^{3/2}} \qquad (28.6)$$

28.4 TWO PARALLEL CURRENTS

We have now looked at both aspects of the magnetostatic field theory. We have seen how moving charges or currents create magnetic fields (the Biot-Savart law) and how these fields exert forces on moving charges or currents. Let us put these two aspects together to solve a simple problem from first principles.

Let us imagine two infinite, parallel straight wires separated by a distance R and carrying currents i_1 and i_2, which we assume to be in the same direction. First let us determine the magnetic field at the second wire created by the current in the first wire (see Fig. 28.12). This field is given by Eq. 28.3:

$$B_1 = \frac{\mu_0 i_1}{2\pi R}$$

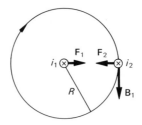

FIGURE 28.12 Two infinite parallel straight wires carrying currents i_1 and i_2 into the page.

The force on a length $d\ell_2$ of the second wire is

$$dF_2 = i_2 B_1 \, d\ell_2$$

or the force per unit length is

$$\frac{dF_2}{d\ell_2} = i_2 B_1 = \frac{\mu_0 i_1 i_2}{2\pi R} \qquad \text{(28.7)}$$

In a similar way we can determine the force on a unit length of the first wire created by the field of the second wire. We simply interchange the indices 1 and 2 in Eq. 28.7 to obtain

$$\frac{dF_1}{d\ell_1} = \frac{\mu_0 i_2 i_1}{2\pi R}$$

and so we see that the forces are of equal magnitude, as they must be by the law of action and reaction. It is not difficult to see that the forces are attractive. If the currents were in opposite directions, the forces would be repulsive.

QUESTION

Two large square flat plates are resting on top of one another, and they are connected as shown in Fig. 28.13 to a battery that produces equal and opposite currents in the plates. (There is a thin layer of electrical insulation between the plates.)

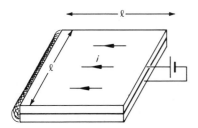

FIGURE 28.13 Two large square plates connected to a battery.

Assume that the current is uniform over the plates. If the upper plate has a mass of 10 g, what current is required to cause the upper plate to begin to rise?

ANSWER

The magnetic field of either plate is given by (see Eq. 28.4)

$$B = \frac{\mu_0 K}{2}$$

where K is the surface current density, and

$$K = \frac{i}{\ell}$$

where ℓ is the width of the plates. The magnetic force on the upper plate is

$$F = i\ell B = \frac{i\ell \mu_0 K}{2} = \frac{i^2 \mu_0}{2}$$

This force will balance the weight when

$$\frac{i^2 \mu_0}{2} = mg$$

or

$$i = \sqrt{\frac{2mg}{\mu_0}} = \sqrt{\frac{2(0.01 \text{ kg})(9.8 \text{ m/s}^2)}{(4\pi \times 10^{-7} \text{ T} \cdot \text{m/A})}}$$

$$= 395 \text{ A}$$

which is an extremely large current.

28.5 THE INTEGRAL FORM OF THE MAGNETIC FIELD LAWS

We have seen earlier that Coulomb's law for the electric field of a point charge,

$$\mathbf{E} = \frac{1}{4\pi\epsilon_0} \frac{q}{r^3} \mathbf{r}$$

and the law of vector addition of electric fields can be replaced by an equivalent pair of integral equations,

$$\oint \mathbf{E} \cdot d\mathbf{A} = \frac{q}{\epsilon_0}$$

and

$$\oint \mathbf{E} \cdot d\mathbf{s} = 0$$

We would now like to present the integral forms equivalent to the Biot-Savart law,

$$d\mathbf{B} = \frac{\mu_0}{4\pi} \frac{i \, d\boldsymbol{l} \times \mathbf{r}}{r^3}$$

The Biot-Savart law is more complicated than Coulomb's law, and it is more difficult to derive the integral equations that follow from it. We shall simply state the integral laws without proof and shall demonstrate that they are consistent with what we have learned from the Biot-Savart law.

MAGNETIC FLUX LAW

We have observed that the magnetic field lines are always closed, and this means that the net number of field lines leaving any closed surface must be zero. Therefore we must have

$$\oint \mathbf{B} \cdot d\mathbf{A} = 0 \qquad \text{(28.8)}$$

for any closed surface. It is convenient to express this result in terms of the *magnetic flux* ϕ_B through a surface, which is defined by the relation

$$\phi_B = \int \mathbf{B} \cdot d\mathbf{A}$$

The magnetic flux ϕ_B is the flux through the area over which the integral is performed. The area need not be closed, but if it is closed, then the flux is zero, from Eq. 28.8.

This flux condition for closed surfaces can be understood in terms of our fluid dynamics model for the magnetic field. In the fluid analog there are no sources or sinks, so $\oint \mathbf{v} \cdot d\mathbf{A}$ should be zero.

MAGNETIC CIRCULATION LAW

Let us return to our fluid analogy. We have seen earlier that the fluid flow originates in the vortexes. The rate of rotation of the fluid about a vortex is a measure of the strength of the vortex. A measure of the rate of rotation of the fluid about a vortex might be taken to be $\oint \mathbf{v} \cdot d\mathbf{s}$, where the integral is taken along some closed line that encloses the vortex. We might write

strength of vorticity = rate of rotation about the vortex

and

$$\text{rate of rotation about the vortex} = \oint \mathbf{v} \cdot d\mathbf{s}$$

so

$$\text{strength of vorticity} = \oint \mathbf{v} \cdot d\mathbf{s}$$

Imagine a long straight vortex, for example, a rod with blades attached along its length (see Fig. 28.14). The rate of rotation of the fluid about the vortex depends on where it is measured. (This is where our fluid model becomes dangerous. The fluid must not rotate with a uniform

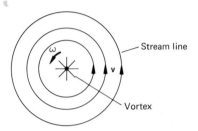

FIGURE 28.14   A line vortex.

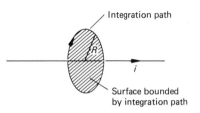

FIGURE 28.15   A circular path of integration and a plane area bounded by this circle.

angular velocity about the vortex. If it did, the velocity would increase with distance, not decrease as it must if it is to represent the magnetic field of a long line current. *The magnetic field decreases with distance from the line.*) The vortex strength is measured not by the velocity alone, but by the velocity (or rather its component along the path) and the length of the path taken around the vortex. This is a definition, and we have not proved anything.

The analog of this in magnetostatics is *Ampere's law,* which states that the vorticity of a magnetic field is proportional to the current, or

$$\oint \mathbf{B} \cdot d\mathbf{s} = \mu_0 i \tag{28.9}$$

where the integral is performed around any closed path. The current i is the net current that crosses any surface bounded by the closed path along which the integral is taken. The current is taken as positive or negative, depending on whether it is parallel or antiparallel to the thumb of the right hand when the fingers are curled in the direction of the path of integration.

As a simple example, consider a straight wire of infinite length, and take as a path of integration a circle of radius R concentric with the wire. Now the path of integration (see Fig. 28.15) is along a magnetic field line. We know then that $\mathbf{B} \cdot d\mathbf{s} = B\,ds$ since $\mathbf{B}$ and $d\mathbf{s}$ are parallel. Now B is constant on this line, so

$$\oint \mathbf{B} \cdot d\mathbf{s} = \oint B\,ds = B \oint ds = B2\pi R$$

From Ampere's law

$$B2\pi R = \mu_0 i \tag{28.10}$$

since a current i crosses the plane area enclosed by the circle. Furthermore, if the fingers of the right hand are coiled in the direction of integration around the loop, the thumb points in the direction of the current, so the current is considered positive. From Eq. 28.10 we have

$$B = \frac{\mu_0 i}{2\pi R}$$

which is just what we obtained from the Biot-Savart law (Eq. 28.3).

We did not have to take a circular path for the integration, since Ampere's law is valid for any path. Nor was it necessary for us to choose a plane area; we might have chosen a path such as that in Fig. 28.16. The integral is

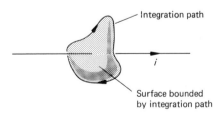

FIGURE 28.16   A path of integration that does not lie in a plane.

FIGURE 28.17 Around this contour $\oint \mathbf{B} \cdot d\mathbf{s} = 0$.

more difficult now, but the result must be the same, since the same current crosses the area bounded by the contour.

If the contour is such that the current does not cross the area bounded by the contour, then

$$\oint \mathbf{B} \cdot d\mathbf{s} = 0$$

Such a contour is illustrated in Fig. 28.17.

Not only are there several choices for the integration path, but many different surfaces are bounded by a given integration path. In Fig. 28.18 two different areas are bounded by the same contour. The current does not cross A_2 at all, but it crosses A_1 twice, once with a positive sense and once with a negative sense. Whatever area we choose, the current will cross it an even number of times, 0, 2, 4, ... The number of positive crossings will be the same as the number of negative crossings.

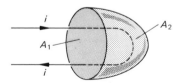

FIGURE 28.18 Two different areas A_1 and A_2 bounded by the same integration path.

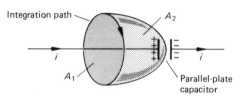

FIGURE 28.19 The current crosses A_1 but does not cross A_2.

Let us consider an example in which we do get different results depending on our choice of area, namely a line current charging a pair of capacitor plates. The current crosses the area A_1 in Fig. 28.19, but it does not cross the area A_2. This of course is not a static or steady-state situation. The charge on the capacitor plates is continuously increasing, and there is a time-dependent electric field. This is therefore not a problem we have to deal with in magnetostatics, but we shall have to face it later when we consider time-changing fields. Then we shall have to modify Ampere's law to deal with the inconsistency we have just observed.

28.6 CALCULATING MAGNETIC FIELDS USING AMPERE'S LAW

Just as we were able to calculate the electrostatic field of certain symmetrical electric charge distributions by using Gauss's law, so we may calculate the magnetic field of certain current configurations by using Ampere's law, provided there is sufficient symmetry. We have already seen an example of this procedure in Sec. 28.5, where we determined the magnetic field of a straight line current by integrating around a circle concentric with the current.

QUESTION

What is the magnetic field inside a long solenoid that is wrapped with a wire that makes N turns per unit length? The current is i.

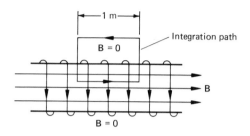

FIGURE 28.20   A rectangular path of integration. The length of the rectangle is 1 m and the width is arbitrary.

ANSWER

Take as an integration path the rectangle illustrated in Fig. 28.20. We have seen that the magnetic field lines are straight within the solenoid and that the field is zero outside. Therefore the contribution to the integral $\oint \mathbf{B} \cdot d\mathbf{s}$ is zero on the vertical sides and on the horizontal segment outside the solenoid. The only contribution is on the horizontal segment of 1-m length inside. Therefore

$$\oint \mathbf{B} \cdot d\mathbf{s} = B(1 \text{ m})$$

The net current crossing the area is just Ni, since there are N windings in a unit length and the current in each is i. From Ampere's law we have therefore

$$B = \mu_0 Ni \tag{28.11}$$

We see that the magnetic field is constant throughout the solenoid.

We shall have occasion to treat the current on the surface of the solenoid as a *surface current density K*. The surface current density is defined in general as the amount of charge that passes a unit length of line on the surface perpendicular to the direction of the current per unit time. If such a line is drawn on the solenoid parallel to the axis (perpendicular to the current in the wires), then the charge that crosses this line per unit time is the charge i that passes through each wire per unit time multiplied by the number N of wires per unit length, so

$$K = Ni$$

We may therefore express the magnetic field within the solenoid as

$$B = \mu_0 K \tag{28.12}$$

QUESTION

What is the magnetic field of an infinite current sheet with a surface current density K?

ANSWER

We have seen that the field reverses direction across the sheet. Integrating $\mathbf{B} \cdot d\mathbf{s}$ around the loop in Fig. 28.21, we have

$$\oint \mathbf{B} \cdot d\mathbf{s} = B(2 \text{ m})$$

since the horizontal segments contribute equally and the vertical segments make no contribution. The net current through the loop is the current K per unit length of line perpendicular to the current times the length of the current sheet intercepted by the rectangular integration loop, namely 1 m.

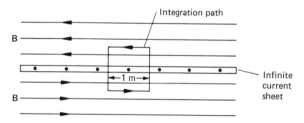

FIGURE 28.21   The current in the infinite sheet is directed out of the page. The integration loop is 1 m long and of arbitrary width.

Thus

$$B(2 \text{ m}) = \mu_0 K(1 \text{ m})$$

or

$$B = \frac{\mu_0 K}{2} \tag{28.13}$$

and since the result is independent of the height of the loop, the magnetic field does not depend on the distance from the sheet. This is the same result that we obtained earlier (Eq. 28.4) using the Biot-Savart law.

28.7 MAGNETIC MATERIALS

A detailed study of magnetic materials would carry us far beyond the level of this course, but it is important to recognize that all magnetic fields are created by currents. The earth's magnetic field is due to currents deep in the core of the earth, and the magnetic fields of bar magnets are due to atomic currents—motion of electrons about the atomic nucleus or the spin of charged particles about their axes.

Although the following discussion does not present an accurate picture (particles on an atomic scale must be treated by quantum mechanics), it nevertheless reconciles the existence of such things as bar magnets with our statement that all magnetic fields arise from currents. Let us consider the field of a long solenoid, illustrated in Fig. 28.22. The field outside the solenoid is typical of a bar magnet (see Fig. 28.23). The north pole of the magnet is by definition the end from which the field lines emanate,

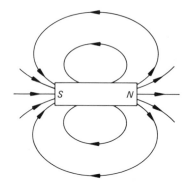

FIGURE 28.23   Magnetic field of a bar magnet.

and they converge on the south pole. Let us take as a model of the bar magnet a great many small current loops, all lined up as shown in Fig. 28.24. At any interior point in the bar the net current is zero, since opposing currents cancel. However, there are no opposing currents on the outside surface, so we have as an equivalent current system that of Fig. 28.25. But this is just the current of a solenoid, so we may achieve a magnetic field equivalent to that of a bar magnet by means of aligned current loops.

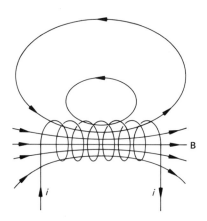

FIGURE 28.22   Magnetic field of a solenoid.

FIGURE 28.24   Rings of equal current aligned within a cylinder.

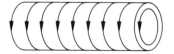

FIGURE 28.25   Current distribution equivalent to that of Fig. 28.24.

By the same arguments used to incorporate the properties of a dielectric into electrostatics, we can show that we can incorporate the magnetic effects associated with the medium into our theory by replacing μ_0 in all formulas by μ, where μ is the magnetic permeability appropriate to the medium. The one notable exception is ferromagnetic materials. For such materials we would have to permit μ to depend on the strength of the magnetic field and on the past history of the sample. In this case μ is no longer a constant associated with the material.

SUMMARY

The magnetic field generated by a charge q traveling with a velocity $\mathbf{v}$ is given by the Biot-Savart law,

$$\mathbf{B} = \frac{\mu_0}{4\pi} \frac{q\mathbf{v} \times \mathbf{r}}{r^3}$$

where

$$\frac{\mu_0}{4\pi} = 10^{-7} \text{ T} \cdot \text{m} \cdot \text{s/C}$$

and $\mathbf{r}$ is a vector from the charge to the field point.

The magnetic field of a current element is given by

$$d\mathbf{B} = \frac{\mu_0}{4\pi} \frac{i \, d\boldsymbol{\ell} \times \mathbf{r}}{r^3}$$

The magnetic field of an infinite line current is given by

$$B = \frac{\mu_0 i}{2\pi R}$$

and the magnetic field on either side of an infinite current sheet is given by

$$B = \frac{\mu_0 K}{2}$$

The force per unit length between two infinite parallel line currents is given by

$$\frac{F}{\ell} = \frac{\mu_0 i_1 i_2}{2\pi R}$$

The two integral equations that are the equivalent of the Biot-Savart law are

$$\oint \mathbf{B} \cdot d\mathbf{A} = 0$$

and Ampere's law,

$$\oint \mathbf{B} \cdot d\mathbf{s} = \mu_0 i$$

where i is the net current that crosses any surface bounded by the closed contour of integration.

PROBLEMS

28.A.1 Will the two current loops shown in the figure attract or repel each other?

28.A.2 When the switch in the following figure is closed, will the two sides A and B of the flexible wire loop come together or move apart?

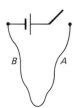

28.A.3 The two infinite parallel wires in the figure carry currents of 1 A in opposite directions. The distance between the wires is 1 m.
(a) What is the magnetic field at a point midway between the wires?
(b) What is the magnetic field at a point 1 m from each wire?

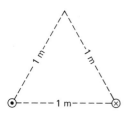

28.A.4 Two infinite parallel wires carry parallel currents of 1 A and 2 A. If the distance between the wires is 1 m, what is the line along which the magnetic field is zero?

28.A.5 The two infinite parallel wires in the figure carry equal currents of 10 A in the same direction. If the lower wire has a mass density of 1 g/m, how far apart must the wires be if the upper wire is to support the lower wire?

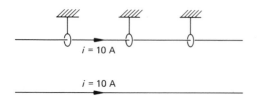

28.A.6 The three infinite sheets shown in cross section in the figure carry surface currents K, $K/2$, and $K/2$ in the indicated directions. Determine the magnitude and direction of the magnetic field at (a), (b), (c), and (d).

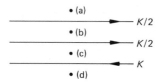

28.A.7 Determine the integral of **B** · d**s** over the contours of integration (a), (b), (c), and (d) in the following figure.

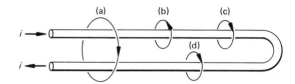

28.B.1 Two infinite wires carry equal currents of 1 A in the positive sense along the x and y axes. Determine the magnetic field at
(a) $x = 1$ m, $y = 1$ m, $z = 0$;
(b) $x = 0$, $y = 0$, $z = 1$ m.

28.B.2 Determine the magnetic field at point P in the figure.

28.B.3 A current i is distributed uniformly throughout an infinite solid cylinder of radius R. What is the magnetic field
(a) for $r < R$?
(b) for $r > R$?

28.B.4 Determine the magnetic field at the center of a square loop of side a and carrying a current i.

28.B.5 Determine the net force that the rectangular current loop in the figure exerts on the infinite straight wire. (Use the law of action and reaction.)

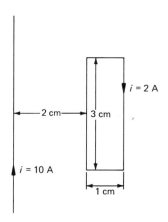

28.B.6 Determine the relation between the magnetic field at the end of a semi-infinite solenoid (A) and the magnetic field far from the end (B) (see the figure). What is the direction of the magnetic field at C?

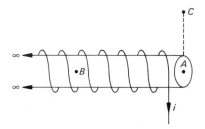

28.B.7 Using Ampere's law, determine the magnetic field within the toroid shown here as a function of distance r from the axis of symmetry. The current in the wire is i and there are N loops in all.

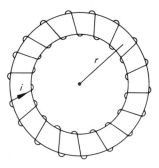

28.C.1 A charge q is uniformly distributed over a disk of radius R. If the disk is rotating with an angular velocity ω, what is the magnetic field on the axis of the disk?

28.C.2 A charge q is uniformly distributed over the surface of a sphere of radius R. If the sphere is rotating with an angular velocity ω, what is the magnetic field at the center of the sphere?

ELECTRODYNAMICS: DISPLACEMENT CURRENT AND ELECTROMAGNETIC INDUCTION

UP TO THIS point we have dealt only with static electric and magnetic fields. We have seen that the fields satisfy certain integral equations:

$$\oint \mathbf{E} \cdot d\mathbf{A} = q/\epsilon_0$$

$$\oint \mathbf{E} \cdot d\mathbf{s} = 0$$

$$\oint \mathbf{B} \cdot d\mathbf{A} = 0$$

$$\oint \mathbf{B} \cdot d\mathbf{s} = \mu_0 i$$

The first two equations are equivalent to Coulomb's law, and the second two to the Biot-Savart law. These equations tell us what kinds of fields are created by electric charges and currents. The force law is the other half of the field theory; it tells us how charges respond to the fields. When there are both electric and magnetic fields, the force law is

$$\mathbf{F} = q\mathbf{E} + q\mathbf{v} \times \mathbf{B}$$

(If there is a current, then $q\mathbf{v}$ is equivalent to $i \, d\ell$ and the magnetic force becomes $i \, d\ell \times \mathbf{B}$.)

We now have a complete picture of charges and currents in static fields. We must consider what happens when the fields are not constant in time. We shall show that two of our integral equations must be altered, while the other two integral equations and the force law remain unchanged.

QUESTION

If a quantity of electric charge is placed at the center of a conducting sphere and then released, it will move out to the surface of the sphere (Fig. 29.1). Assuming that the current associated with this charge flow is spherically symmetrical, determine the magnetic field associated with the current.

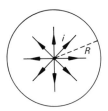

FIGURE 29.1 A charge flows radially from the center of a conducting sphere to the surface.

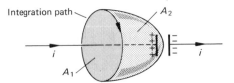

FIGURE 29.2 A capacitor that is being charged at a uniform rate.

ANSWER

Whatever the magnetic field, it must have spherical symmetry. The current flow looks the same regardless of the orientation of the sphere, and therefore the magnetic field must be independent of the orientation. Is there any way to construct magnetic field lines that close on themselves and yet have spherical symmetry? The answer is no, and therefore there can be no magnetic field. Since $\oint \mathbf{B} \cdot d\mathbf{s} = \mu_0 i$ and i is not zero, this equation must be incorrect. We shall see that a new term must be added to Ampere's law when the fields are not constant in time. This new term exactly cancels the current i, and so there is no magnetic field.

29.1 THE DISPLACEMENT CURRENT

Earlier we observed a problem with Ampere's law when there is a time-changing electric field (see Sec. 28.5). Let us consider Ampere's law applied to the charging of a capacitor by a constant current i. Specifically, let us consider the integral $\oint \mathbf{B} \cdot d\mathbf{s}$ around the closed path indicated in Fig. 29.2. The area A_1 is bounded by the integration path, and since the current crosses this area, we have from Ampere's law

$$\oint \mathbf{B} \cdot d\mathbf{s} = \mu_0 i$$

If we choose the area A_2, the current does not cross this area, so

$$\oint \mathbf{B} \cdot d\mathbf{s} = 0$$

The integral cannot have two different values.

Alternatively, we could have chosen the path of integration to be that of Fig. 29.3, so that the current crossing the area A_1 that contains the plane of the current is zero, but the current does cross A_2.

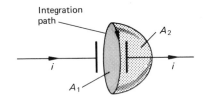

(a)

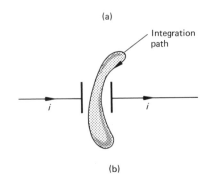

(b)

FIGURE 29.3 (a) The area A_1 is not cut by the current, but A_2 is. (b) The contour of integration does not lie in a plane.

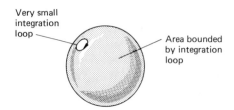

FIGURE 29.4 A very small contour of integration and an arbitrary area bounded by this loop.

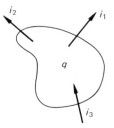

FIGURE 29.5 If a net charge flows from a closed surface ($i_1 + i_2 - i_3$), then the charge inside the surface must decrease.

If the path of integration does not lie in a plane, then of course we cannot choose one area over another in order to decide whether or not a current crosses the area bounded by the contour on which the integral of **B** · d**s** is evaluated.

Let us consider another integration path, a very small circle far from the wire and the capacitor, and let us consider an arbitrary surface bounded by this loop (see Fig. 29.4). As the size of the loop becomes smaller and smaller, the surface becomes an almost closed surface, a surface with a very small hole in it. If we shrink the radius of the loop to zero, the surface becomes closed. Since the integral of **B** · d**s** around this vanishingly small loop is zero, Ampere's law tells us that

The net current through a closed surface is zero.

But this is a conservation law for electric charge in a steady state. This should give us a pretty good clue of what modification is needed in Ampere's law for the nonsteady state. If a net current passes out of a closed surface, the amount of charge inside the surface must decrease, that is,

$$i = -\frac{dq}{dt}$$

where i is the net current passing out of a closed surface and q is the net charge inside the same closed surface (Fig. 29.5). But from Gauss's law (if we assume that it is still true for time-changing fields),

$$q = \epsilon_0 \oint \mathbf{E} \cdot d\mathbf{A}$$

so that

$$i = -\frac{dq}{dt} = -\epsilon_0 \frac{d}{dt} \oint \mathbf{E} \cdot d\mathbf{A}$$

or

$$i + \epsilon_0 \frac{d}{dt} \oint \mathbf{E} \cdot d\mathbf{A} = 0 \tag{29.1}$$

where i is the current through the closed surface of integration. This is the statement of conservation of electric charge for nonsteady-state systems.

Let us return to the example of the charging capacitor. If we choose a closed surface that cuts the wire and passes between the capacitor plates (see Fig. 29.6), then Eq. 29.1 states that the current flowing out of the surface plus the rate at which charge is being stored within the surface is zero.

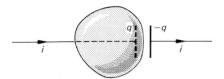

FIGURE 29.6 The rate of accumulation of charge within the closed surface is equal to the current entering (or equivalently the negative of the current leaving).

Let us now consider the following modification of Ampere's law:

$$\oint \mathbf{B} \cdot d\mathbf{s} = \mu_0 \left(i + \epsilon_0 \frac{d}{dt} \int \mathbf{E} \cdot d\mathbf{A} \right) \qquad \textbf{(29.2)}$$

where the direction of the surface element $d\mathbf{A}$ is determined by the right-hand rule, i.e., by the direction of the thumb when the fingers of the right hand are curled along the path of the $\mathbf{B} \cdot d\mathbf{s}$ integral. (Note that the integral of $\mathbf{B} \cdot d\mathbf{s}$ is around a closed line and the integral of $\mathbf{E} \cdot d\mathbf{A}$ is over any open surface bounded by the closed line.)

When Ampere's law is stated in this form, the value of the right-hand side no longer depends on our choice of area. If we pick two different areas A_1 and A_2 (see Fig. 29.7), then the law of conservation of charge (Eq. 29.1) assures us that the result will be the same. To see this, we need only reverse the normal on the area A_1, so that the areas A_1 and A_2 together make a closed surface for which the charge conservation equation is equivalent to the equality of the right-hand side of Ampere's law applied to each surface.

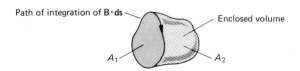

FIGURE 29.7 Two surfaces bounded by the same path.

This additional term in Ampere's law $[\epsilon_0 (d/dt) \int \mathbf{E} \cdot d\mathbf{A}]$ was first proposed by James C. Maxwell and is called the *displacement current*. It is called a *current* because it is added to the real current i to give the net vorticity ($\oint \mathbf{B} \cdot d\mathbf{s}$) of the magnetic field. It is called a *displacement* current for rather complex reasons of historical interest only.

QUESTION

A long straight circular wire carries a constant current i (Fig. 29.8) that is uniformly distributed throughout the wire. A small gap in the wire acts as a parallel-plate capacitor. Positive charge accumulates on the left face of gap and an equal negative charge accumulates on the right face. What are the electric and magnetic fields everywhere?

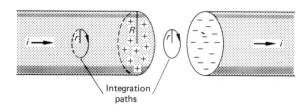

FIGURE 29.8 A long straight wire of radius R with a small gap.

ANSWER

First let us assume that the gap is small in comparison with the diameter of the wire, so that the electric field is confined primarily to the region of the gap. (A small electric field is necessary to drive the current in the wire. This field is negligibly small if the wire is a good conductor.)

(a) *Magnetic field within the wire.* Consider as an integration path a circle of radius r ($r < R$) that is concentric with the axis of the wire. From Ampere's law (Eq. 29.2),

$$\oint \mathbf{B} \cdot d\mathbf{s} = \mu_0 \left(i_r + \epsilon_0 \frac{d}{dt} \int \mathbf{E} \cdot d\mathbf{A} \right)$$

where i_r is the current that crosses the area bounded by the circular integration path. (We have assumed that the dielectric and magnetic permeabilities are those of a vacuum. We might have used the ϵ and μ appropriate to the medium.) If we choose as the area bounded by the integration contour a plane circular area of radius r, then since E is zero within the wire, the displacement current through this area is zero. (We might have chosen the area to be a cylinder open at the end of the contour and closed by a plane circle within the gap. In this case i_r would be zero and the displacement current would be nonzero.) The current i_r through the circular area of radius r is given by

$$i_r = \frac{\pi r^2}{\pi R^2} i = \frac{r^2}{R^2} i$$

since the current i is distributed uniformly over the wire and $\pi r^2 / \pi R^2$ is the fraction of the total cross section intercepted by the integration loop. We have then

$$\oint \mathbf{B} \cdot d\mathbf{s} = \frac{\mu_0 i r^2}{R^2}$$

From symmetry the magnetic field is uniform over the integration path and is everywhere parallel to $d\mathbf{s}$, so

$$\oint \mathbf{B} \cdot d\mathbf{s} = B 2\pi r$$

Therefore

$$B = \frac{\mu_0 i}{2\pi R^2} r \qquad (29.3)$$

so the magnetic field lines within the wire are circles concentric with the axis of the wire and the strength of the field increases linearly with the distance r from the axis.

(b) *Magnetic field outside the wire.* For $r > R$ we may apply Ampere's law as we have before and find (see Eq. 28.10),

$$\oint \mathbf{B} \cdot d\mathbf{s} = B 2\pi r = \mu_0 i$$

or

$$B = \frac{\mu_0 i}{2\pi r} \qquad (29.4)$$

Note that the magnetic field is continuous at $r = R$ (compare Eqs. 29.3 and 29.4 at $r = R$).

(c) *Electric field within the gap.* We may determine the electric field within the gap from Gauss's law. We did this earlier (see Eq. 21.7) and found that

$$E = \frac{\sigma}{\epsilon_0}$$

where σ is the surface charge density on the opposing faces of the gap. However σ is increasing uniformly with time because of the constant current i. The rate at which the charge q is accumulating on the faces is given by

$$\frac{dq}{dt} = i$$

or

$$q = it$$

Now

$$\sigma = \frac{q}{\pi R^2}$$

so

$$\sigma = \frac{it}{\pi R^2}$$

Therefore

$$E = \frac{it}{\pi \epsilon_0 R^2} \qquad (29.5)$$

(d) *Magnetic field within the gap.* Let us apply Ampere's law to a circular path coaxial with the axis of the gap (see Fig. 29.8) and of radius r:

$$\oint \mathbf{B} \cdot d\mathbf{s} = \mu_0 \left(i_r + \epsilon_0 \frac{d}{dt} \int \mathbf{E} \cdot d\mathbf{A} \right) \qquad (29.6)$$

Now there is no current i_r through the plane circular area bounded by the circular path of integration, but there is a displacement current and it is given by

$$\text{Displacement current} = \epsilon_0 \frac{d}{dt} \int \mathbf{E} \cdot d\mathbf{A} = \epsilon_0 \frac{d}{dt}(E\pi r^2)$$

$$= \epsilon_0 \pi r^2 \frac{dE}{dt}$$

But from Eq. 29.5

$$\frac{dE}{dt} = \frac{i}{\pi \epsilon_0 R^2}$$

so

$$\text{Displacement current} = \frac{r^2}{R^2} i$$

which, not surprisingly, is the same as i_r within the wire. We might have seen this by integrating the right-hand side of Ampere's law over a closed cylinder of radius r with one face in the wire and the other in the gap.

If we substitute this displacement current into Eq. 29.6, we find within the gap

$$\oint \mathbf{B} \cdot d\mathbf{s} = \frac{\mu_0 i r^2}{R^2}$$

But

$$\int \mathbf{B} \cdot d\mathbf{s} = B2\pi r$$

so

$$B = \frac{\mu_0 i}{2\pi R^2} r \qquad (29.7)$$

The magnetic field in the gap is therefore the same as it is in the wire. The displacement current created by the time-changing electric field creates a magnetic field in exactly the same way as the current i created by the charge flow does.

QUESTION

An alternating electric field with a frequency of 60 hertz (Hz) drives a current through a conducting wire of resistivity $\rho = 1.7 \times 10^{-8}\ \Omega \cdot$ m. What are the relative magnitudes of the displacement current and the electron current i?

ANSWER

The electron current is given by

$$i = \frac{V}{R} = \frac{E\ell}{R}$$

where ℓ is the length of the wire and V is the potential difference across the length of the wire.

The electric field E in the wire has a time dependence given by

$$E = E_0 \sin \omega t$$

The displacement current is therefore

$$\epsilon_0 \frac{d}{dt} \int \mathbf{E} \cdot d\mathbf{A} = \epsilon_0 \frac{dE}{dt} A$$

$$= \epsilon_0 \omega E_0 (\cos \omega t) A$$

The ratio of the maximum value of the displacement current to the maximum value of the electron current is

$$\frac{\epsilon_0 \dfrac{d}{dt} \displaystyle\int \mathbf{E} \cdot d\mathbf{A}}{i}$$

$$= \frac{\epsilon_0 \omega E_0 A}{E_0(\ell/R)} = \rho \epsilon_0 \omega$$

$$= (1.7 \times 10^{-8}\ \Omega \cdot \text{m})(8.9 \times 10^{-12}\ \text{C}^2/\text{N} \cdot \text{m}^2)(2\pi 60\ \text{Hz})$$

$$= 5.7 \times 10^{-17}$$

where we have made use of the fact that

$$R = \frac{\rho \ell}{A}$$

The displacement current is therefore 5.7×10^{-17} times the electron current. Whenever we have an electron current and a displacement current, we can usually neglect the displacement current. The displacement current is important in a vacuum where there is no electron current.

29.2 ELECTROMAGNETIC INDUCTION

We have just seen that Ampere's law requires modification for time-changing electric fields. We would now like to show that the integral equation

$$\oint \mathbf{E} \cdot d\mathbf{s} = 0$$

requires modification when there is a time-changing magnetic field.

To understand the need for a revision of the equation, consider a loop of conducting wire moving into a magnetic field, as in Fig. 29.9. The electrons in the segment of wire between a and b will experience a force ($q\mathbf{v} \times \mathbf{B}$)

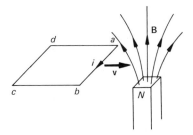

FIGURE 29.9   A rectangular conducting wire loop is moving into a magnetic field.

that will produce a current from a to b. The electrons in the segment between c and d will experience a force in the same direction, but of a lesser magnitude, since the magnetic field is weaker further from the magnet. We would expect then that there would be a current around the loop from a to b to c to d.

Now, rather than move the loop toward the magnet, suppose we move the magnet toward the loop, or, to put it another way, suppose the original experiment is viewed by someone sitting on the loop. This observer sees the magnet moving toward the loop. On the other hand, a person sitting on the magnet sees the loop moving toward the magnet. This observer sees a current in the loop, as mentioned in the preceding paragraph and understands its origin: $q\mathbf{v} \times \mathbf{B}$ forces. Now the other observer, the observer sitting on the loop, must also see a current, since clearly two people looking at the same thing cannot see different results. But how does the observer on the loop explain the origin of the current? This person sees a magnet coming toward the loop, and if this observer applies the same electromagnetic field equations as the other observer, no explanation is obtained! In this observer's frame of reference the wire and all the electrons in it are at rest. They will begin to move and create a current only when acted on by an electric field. For the observer on the loop the field equations would predict a time-changing magnetic field, but no electric field. The observer has two choices: Either there is a preferred frame of reference in which the electromagnetic equations work, or something is wrong with the equations. Einstein's theory of relativity tells us that there is no such thing as a preferred frame of reference. We therefore must assume that there exists a system of electromagnetic equations that are valid in all inertial frames of reference and we simply have to find out what they are.

We would like to find a modification of the equations

$$\oint \mathbf{E} \cdot d\mathbf{A} = q/\epsilon_0$$

$$\oint \mathbf{E} \cdot d\mathbf{s} = 0$$

$$\oint \mathbf{B} \cdot d\mathbf{A} = 0$$

and

$$\oint \mathbf{B} \cdot d\mathbf{s} = \mu_0\left(i + \epsilon_0 \frac{d}{dt}\int \mathbf{E} \cdot d\mathbf{A}\right)$$

which permits the observer sitting on the loop to understand (or, better, to predict) the current that arises in the wire. We have seen that this person could explain such a current only if there were an electric field. The charges are all at rest, and since $\mathbf{F} = q\mathbf{E} + q\mathbf{v} \times \mathbf{B}$, with $\mathbf{v} = 0$, only an electric field would generate the force to cause the motion and hence the current. But the four integral equations above would not predict an electric field. In particular, the equation

$$\oint \mathbf{E} \cdot d\mathbf{s} = 0 \qquad (29.8)$$

must clearly be wrong if there is to be an electric field parallel to the wire driving the current.

We shall postulate a new term on the right-hand side of Eq. 29.8 and shall show that it produces an electric field that drives the same current as that predicted by the observer at rest with respect to the magnet, i.e., a force on each electron equal to qvB.

This new term in the integral equation is very similar to the displacement current introduced into Ampere's law. We propose as the appropriate modification of Eq. 29.8

$$\oint \mathbf{E} \cdot d\mathbf{s} = -\frac{d}{dt}\int \mathbf{B} \cdot d\mathbf{A} = -\frac{d\phi_B}{dt} \qquad (29.9)$$

where $\mathbf{E} \cdot d\mathbf{s}$ is integrated over a closed contour and $\mathbf{B} \cdot d\mathbf{A}$ is integrated over any surface bounded by this contour. (The normal to the surface is determined by our previous right-hand rule: Coil the finger of the right hand in the direction of the contour integral and the thumb points in the direction of the normal.)

Let us see whether we can use Eq. 29.9 to determine the electric field as seen by an observer at rest on the wire loop. The magnetic field of a bar magnet (Fig. 29.9) is rather complicated, so we shall simplify the problem by looking at a magnetic field that is straight and uniform over half of space and zero over the other half. The magnetic field is constant to the right of the dashed line in Fig. 29.10 and zero to the left of the dashed line. From the point of view of the observer sitting on the loop, the magnetic field is advancing toward the loop with a velocity $\mathbf{v}$. (From the point of view of the observer sitting on the source of the magnetic field, the loop is advancing into a stationary magnetic field.)

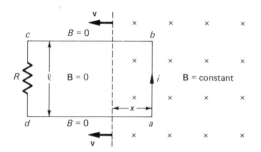

FIGURE 29.10 The magnetic field is uniform to the right of the dashed line and is directed into the page. The field is zero to the left of the dashed line. The field is advancing toward a wire loop with a velocity $\mathbf{v}$. The distance the field has advanced into the loop is the distance x.

We have seen that an electric field must be driving the current. Let us assume that this field is directed vertically. (This is the direction of the current in the wire when viewed by an observer at rest with respect to the source of the magnetic field. This person sees the wire advancing into the magnetic field with a velocity **v** parallel to the x axis. The $q\mathbf{v} \times \mathbf{B}$ force, where **v** is the velocity of the wire, would be directed vertically.) Evaluating the left-hand side of Eq. 29.9, we have

$$\oint \mathbf{E} \cdot d\mathbf{s} = \int_a^b \mathbf{E} \cdot d\mathbf{s} = E\ell$$

since we assume that the field is zero to the left of the dashed line and perpendicular to the segments bc and ad.

Next let us compute the right-hand side of Eq. 29.9. Now

$$\int \mathbf{B} \cdot d\mathbf{A} = -B\ell x$$

The reason for the minus sign is that **B** is antiparallel to $d\mathbf{A}$. Differentiating with respect to time, we obtain

$$\frac{d}{dt} \int \mathbf{B} \cdot d\mathbf{A} = -\frac{d}{dt}(B\ell x) = -B\ell\frac{dx}{dt} = -B\ell v$$

Equation 29.9 then becomes

$$E\ell = B\ell v$$

or

$$E = Bv \tag{29.10}$$

where E is the strength of the electric field, B the strength of the magnetic field, and v the velocity with which the magnetic field is advancing toward the loop. The force that this field would exert on a charge q would be

$$F = qE = qvB$$

since $E = vB$. Remember that v is not the velocity of the charge q. The charge is at rest from the point of view of the observer sitting on the wire loop.

But this is just the force that would act on a charge q moving with a velocity v into a stationary magnetic field B. We see then that from either point of view the force driving the charges that produce the current will be the same, so both observers will predict the same current if we modify the field equation $\oint \mathbf{E} \cdot d\mathbf{s} = 0$ to

$$\oint \mathbf{E} \cdot d\mathbf{s} = -\frac{d}{dt} \int \mathbf{B} \cdot d\mathbf{A} = -\frac{d\phi_B}{dt} \tag{29.11}$$

Note that the value of E obtained in Eq. 29.10 is independent of x. Thus to the observer moving into the magnetic field in Fig. 29.10 there would be a uniform electric field directed vertically. The electric field is zero where the magnetic field is zero.

When the entire loop has entered the magnetic field, the current will decay to zero, for now there is an opposing electric field in the wire segment cd. We see that $d\phi_B/dt$ is zero and of course $\oint \mathbf{E} \cdot d\mathbf{s}$ is also zero.

Let us summarize the findings of both observers.

1. *Observer at rest with respect to the source of the magnetic field.* The wire is moving into the static magnetic field with a velocity **v**, and the conduction electrons in the wire experience a force equal to $q\mathbf{v} \times \mathbf{B}$. This force produces a movement of the electrons along the wire and hence a current.

2. *Observer at rest with respect to the loop.* The loop is at rest in a time-changing magnetic field. The time-changing magnetic flux through the loop generates an electric field determined by the equation $\oint \mathbf{E} \cdot d\mathbf{s} = -d\phi_B/dt$. This electric field exerts a force on the stationary conduction electrons and they begin to move, thus creating a current in the wire.

QUESTION

The current in a long solenoid (Fig. 29.11) is increasing at a constant rate ($i = \alpha t$). The magnetic field inside the solenoid is given by Eq. 28.11,

$$B = \mu_0 Ni = \mu_0 N\alpha t$$

What is the electric field inside the solenoid?

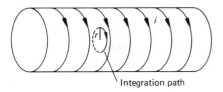

FIGURE 29.11   The solenoid current increases at a rate α.

ANSWER

Consider a circular path of integration of radius r about the axis of the cylinder. On this path

$$\oint \mathbf{E} \cdot d\mathbf{s} = -\frac{d}{dt} \int \mathbf{B} \cdot d\mathbf{A}$$

Now

$$\int \mathbf{B} \cdot d\mathbf{A} = \mu_0 N\alpha t \pi r^2$$

The electric field must have axial symmetry. If we assume that the electric field lines are circles about the axis of the cylinder and parallel to the integration path (see Fig. 29.11), then

$$\oint \mathbf{E} \cdot d\mathbf{s} = 2\pi r E$$

so

$$E = -\tfrac{1}{2}\mu_0 N\alpha r$$

or, since $i = \alpha t$,

$$E = -\tfrac{1}{2}\mu_0 N(di/dt)r \qquad\qquad (29.12)$$

We see that E is negative, so it points in the opposite direction from what we assumed; it points in the opposite direction from the current in the solenoid. We shall see in Sec. 29.3 that this is a general property of induced electric fields. The direction of the field will be the opposite of the direction of the field's cause, in this case the increasing current.

We should compare Eq. 29.12 with Eq. 29.7, where we derived the magnetic field in a parallel-plate capacitor (a gap in a wire) on which the charge was increasing at a uniform rate ($i = dq/dt$). There a continuously increasing, uniform electric field generated a circular magnetic field that was proportional to the rate of increase of the electric field and that increased linearly with the distance r from the axis of the capacitor. In Eq. 29.12 we have shown that a continuously increasing, uniform magnetic field generates a circular electric field that is proportional to the rate of increase in the magnetic field and that increases linearly with the distance from the axis of the solenoid. The reason for this duality is of course the similarity of the fundamental electromagnetic equations, in particular, Eqs. 29.2 and 29.11.

The left-hand side of Eq. 29.11 is called the induced electromotive force, or briefly the induced emf, and the symbol $\mathscr{E}$ is used. By definition

$$\mathscr{E} = \oint \mathbf{E} \cdot d\mathbf{s} \qquad\qquad (29.13)$$

The induced emf is defined only for a closed loop, which may coincide with a wire carrying a current in a circuit or may be simply a closed path in space.

From Eqs. 29.11 and 29.13 we see that

$$\mathscr{E} = -\frac{d}{dt} \int \mathbf{B} \cdot d\mathbf{A} = -\frac{d\phi_B}{dt} \qquad\qquad (29.14)$$

This relation was first proposed by Michael Faraday and is called *Faraday's law*.

We see from the definition that the emf is the work done by the electric field on a unit charge if it is taken around the integration loop.

We must distinguish between the induced emf defined by Eq. 29.13 and the emf of a battery, which we defined as the work done *by the battery* as it expends chemical energy to carry a unit charge through the battery *against* the electric field in the battery. We shall contrast these two different emf's after we have taken up Lenz's law and the concept of circuit inductance.

29.3 LENZ'S LAW

We would like to illustrate some of the many ways in which electric fields may be generated by time-changing magnetic flux and also to state a general rule (Lenz's law) that allows us to determine rather simply the direction of the induced emf.

We saw in Sec. 29.2 how an electric field is induced within a solenoid whose current is continuously increasing and that the respective fields are as illustrated in Fig. 29.12. We can always determine the direction of the

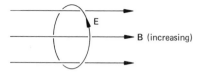

FIGURE 29.12 An increasing magnetic field generates an electric field that would drive a current whose field would oppose the increase in the magnetic field.

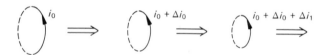

FIGURE 29.13 A small increase in current producing an emf that causes the current to increase without limit.

induced emf by the sign conventions to be used with Eq. 29.14; however, we can usually obtain the direction of the emf by examining the stability of the system. Consider a simple example: a perfectly conducting circular wire carrying a current i (Fig. 29.13). Suppose the current increases slightly by an amount Δi_0 because of the effect of some outside agent or a slight thermodynamic perturbation. This increase in current will increase the magnetic flux through the ring. Let us see what would happen if the emf generated by this time-changing magnetic flux created a further increase Δi_1 in the loop. In this case the initial effect would be repeated; i.e., the induced increase in current would induce a further increase in current, and so on. The current would increase indefinitely. We shall see in Chap. 30 that magnetic fields, or equivalently current rings, have energy, so that a small perturbation of the system would cause an indefinitely large increase in the energy of an isolated system—a violation of the energy principle.[1] We must assume therefore that the induced current will be in the opposite direction of the initial current. Let us propose a general law governing the stability of electromagnetic systems:

The emf induced by a time-changing magnetic flux will always act in a direction that opposes its cause.

This is called *Lenz's law*.

---

[1] If this energy were supplied by the thermal energy of the wire, we would instead have a violation of the second law of thermodynamics.

FIGURE 29.14 An increasing current in the first ring will induce an opposing current in the second ring.

Let us give some examples. In the case cited above of the single current ring, Lenz's law tells us that the induced emf generated by an increase in current will oppose its cause, i.e., the induced emf will generate a current opposite to the initial current. If the current were to decrease, the induced emf would support the current.

As a second example, suppose we have two wire rings side by side (see Fig. 29.14). If the current in the first ring increases, then magnetic flux through the second ring will increase. The cause of the induced current is the increase in magnetic field through the second ring. The induced emf and the current that it drives must oppose this increase in magnetic field, and therefore must be in the opposite direction of the current in the first ring.

If there were no second ring, there would still be an induced electric field. The integral of **E** · d**s** around the circular path previously occupied by the second ring would be the same. We may therefore use Lenz's law to determine the direction of an induced emf even when no currents are generated by the emf. We might amplify our statement of Lenz's law:

The emf generated by a time-changing magnetic flux will always be directed so that it cannot be used to support its cause.

In other words, if a wire *were* superimposed on a closed integration path, a current that opposes its cause *would* be induced.

QUESTION

What are the directions of the induced currents in the situations shown in Fig. 29.15?

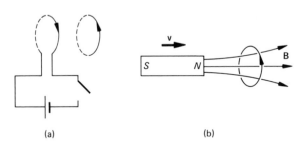

(a)   (b)

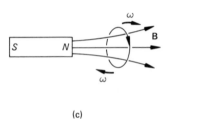

(c)

FIGURE 29.15 In (a) the switch is closed, in (b) the magnet is moved, and in (c) the wire ring is rotated.

ANSWER

(a) When the switch is closed the flux through the second loop increases. The current in this second loop will be in the opposite direction of the current in the first loop in order to decrease the flux.

(b) The flux through the loop is increasing, so the current will flow as in part (a) above.

(c) The flux through the loop is decreasing, so the current will flow in the opposite direction of that in parts (a) and (b) above in order to increase the flux. (This is the basis of an ac generator.)

29.4 INDUCTANCE

We have seen that a time-changing current in one loop can induce an emf in a second loop. Furthermore a time-changing current in a loop can induce an emf in the loop itself. These interactions are called *mutual induction* and *self-induction,* respectively. We shall first consider the latter.

Let us consider an arbitrary loop with a current i (Fig. 29.16), and let us determine the relation between the current and the magnetic flux through the loop. From the Biot-Savart law the magnetic field at any point in space is given by

$$\mathbf{B} = \frac{\mu_0 i}{4\pi} \int \frac{d\boldsymbol{\ell} \times \mathbf{r}}{r^3}$$

so that the *magnetic field is directly proportional to the current in the wire.*

FIGURE 29.16   A single current loop carrying a current i.

Let us consider next the net magnetic flux through the loop,

$$\phi_B = \int \mathbf{B} \cdot d\mathbf{A}$$

But **B** is directly proportional to the current. Since the current is constant, it may be factored from the integral, and so ϕ_B is directly proportional to the current,

$$\phi_B \propto i$$

Let the proportionality factor called the self-inductance be L, so that

$$\phi_B = Li \qquad\qquad\qquad (29.15)$$

This relationship between the flux through a loop and the current in the loop is similar in form and in manner of derivation to the relationship between the potential difference between two equal and oppositely charged conductors and the amount of charge, namely

$$V = \frac{1}{C} Q$$

The self-inductance L depends only on the geometry of the circuit, just as C depends only on the geometry of the conductors. The self-inductance L is a characteristic property of the loop.

The unit of inductance is the tesla—square meter per ampere or the weber per ampere, and this ratio is called a *henry* (H):

$$1 \text{ H} = 1 \text{ T} \cdot \text{m}^2/\text{A}$$

QUESTION

What is the self-inductance of a solenoid of length ℓ, cross-sectional area A, and N turns?

ANSWER

We calculate the inductance by determining the ratio of the flux to the current,

$$L = \frac{\phi_B}{i}$$

Now the magnetic field in the solenoid is given by Eq. 28.11 (N/ℓ is the number of turns per unit length),

$$B = \frac{\mu_0 N i}{\ell}$$

The flux in each turn of the coil is

$$\text{Magnetic flux per turn} = BA = \frac{\mu_0 NA}{l}i$$

The flux through all N turns of the solenoid is

$$\phi_B = \frac{\mu_0 N^2 A}{l}i$$

so

$$L = \frac{\phi_B}{i} = \frac{\mu_0 N^2 A}{l} \qquad \textbf{(29.16)}$$

This relationship between the flux and the current allows us to determine the emf induced in a closed loop as a result of a time-changing current. From Eq. 29.15

$$\mathscr{E} = -\frac{d\phi_B}{dt} = -L\frac{di}{dt}$$

Sometimes this is referred to as a *back emf,* for reasons that we will see in the following question.

QUESTION

In Fig. 29.17 the wire is connected to the battery at $t = 0$. If the emf of the battery is $\mathscr{E}_b$ and the total resistance of the wire is R, how does the current build up in the circuit? What is i as a function of time?

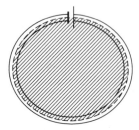

FIGURE 29.17   The wire is connected to the battery at $t = 0$ and charge begins to flow. The dashed line is an integration path.

ANSWER

Let us return to first principles. We do not want to give the impression that there is one set of rules for circuits and another for fields and particles. A circuit is a collection of particles that is acted on by electromagnetic fields and must obey the same laws.

We know that around any closed integration path

$$\oint \mathbf{E} \cdot d\mathbf{s} = -\frac{d\phi_B}{dt} \qquad \textbf{(29.17)}$$

Let us choose as an integration loop the dashed line in Fig. 29.17. This line runs through the center of the wire and through the battery. Let us consider first the left hand side of Eq. 29.17. The contribution to the integral through the wire is given by (see Eqs. 22.4 and 25.2)

$$\int \mathbf{E} \cdot d\mathbf{s} = iR \qquad \textbf{(29.18)}$$

where R is the resistance of the wire and $\mathbf{E}$ is the electric field in the wire. (This electric field propagates with a very large velocity, as we shall see in Chap. 30. For practical purposes we may assume that the electric field is established throughout the wire at the moment the connection is made with the battery.)

The contribution to the integral through the battery is

$$\int \mathbf{E} \cdot d\mathbf{s} = -\mathscr{E}_b \qquad \textbf{(29.19)}$$

since $\mathscr{E}_b$ is the work done by the battery against the electric field. Within the battery the electric field is in the opposite direction of the current, and hence the direction of integration around the circuit. Combining Eqs. 29.18 and 29.19, we have

$$\oint \mathbf{E} \cdot d\mathbf{s} = iR - \mathscr{E}_b \qquad \textbf{(29.20)}$$

Now this difference is not zero, as it was when we considered the steady state. Instead

$$\oint \mathbf{E} \cdot d\mathbf{s} = -\frac{d\phi_B}{dt} = -L\frac{di}{dt}$$

so

$$iR - \mathscr{E}_b = -L\frac{di}{dt} \qquad (29.21)$$

where L is the self-inductance of the circuit. In principle we could calculate L by evaluating in the steady state,

$$L = \frac{\phi_B}{i} = \frac{\displaystyle\int \mathbf{B} \cdot d\mathbf{A}}{i}$$

where $\int \mathbf{B} \cdot d\mathbf{A}$ is the magnetic flux through the area bounded by the circuit (the shaded region in Fig. 29.17) and i is the current that generates the field $\mathbf{B}$.

Before we proceed to solve Eq. 29.21 for i as a function of time, we should consider some important things. We see that it is no longer true that $\oint \mathbf{E} \cdot d\mathbf{s} = 0$ and therefore *there is no potential field,* as in electrostatics. We may not say, as we said before, that the sum of the potential changes around the circuit is zero.

We still have the principle of work and energy, however. We may always say that the work done on the system is equal to the change in energy, but we have lost the ability to determine the work by calculating a change in potential. We may, however, say that the work done on a unit charge by the electric field ($\oint \mathbf{E} \cdot d\mathbf{s}$) plus the work done by the battery on a unit charge ($\mathscr{E}_b$) is equal to the energy dissipated in the resistor by a unit charge[1] (iR), so

$$\oint \mathbf{E} \cdot d\mathbf{s} + \mathscr{E}_b = iR$$

or

$$-L\frac{di}{dt} + \mathscr{E}_b = iR \qquad (29.22)$$

In this form (Eq. 29.22), we may interpret $-L(di/dt)$ as the *back emf,* that is, the work done per unit charge by the electric field induced by the time-changing current.

---

[1] $\dfrac{dW}{dq} = \dfrac{dW}{dt}\dfrac{dt}{dq} = \dfrac{P}{i} = \dfrac{i^2 R}{i} = iR.$

Now the inductance of the circuit in Fig. 29.17 is small. The inductance would be greatly increased if the wire were coiled, for then the magnetic flux through the integration path would be the sum of the flux through each turn in the coil. (The integration path coincides with the wire.) Such an element in the circuit is generally indicated by the symbol ⦷⦷⦷. For example, Fig. 29.18 shows a circuit with a solenoid and a resistor in series. Although the inductance of the solenoid L is not the inductance of the circuit, since it represents only $\int \mathbf{B} \cdot d\mathbf{A}$ through the solenoid, often $\int \mathbf{B} \cdot d\mathbf{A}$ over the rest of the circuit area is negligible.

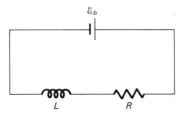

FIGURE 29.18   A resistor, an inductor, and a battery connected in series.

Returning to the differential equation for the current (Eq. 29.22), we find that the solution is given by

$$i = \frac{\mathscr{E}_b}{R}(1 - e^{-t/(L/R)}) \qquad (29.23)$$

as one may verify by substitution. At $t = 0$ the current is zero (the wire has just been connected), and as t becomes large compared to L/R the current approaches a constant value of $\mathscr{E}_b/R$, namely the current in the absence of the inductance. The inductor becomes unimportant after the current becomes constant.

We see that the smaller L is, the shorter the time it takes the current to reach its final value. The inductance serves as a kind of inertia in the circuit; it impedes the increase in the current or, equivalently, the acceleration of the electric charge. The induced electric field opposes its cause, the increase in the current.

QUESTION

We have just seen that there is no longer a potential field when the fields are time dependent. What would happen, however, if we placed voltmeters across each element of the circuit in Fig. 29.18? The fact that a potential does not exist does not mean that we cannot perform the experiment.

ANSWER

Let us apply Faraday's law to each of the three current loops of Fig. 29.19. Through the battery $\int \mathbf{E} \cdot d\mathbf{s}$ is just $-\mathscr{E}_b$. Through the voltmeter $\int \mathbf{E} \cdot d\mathbf{s}$ is $i_1'r$, where i_1' is the current shunted into the voltmeter and r is the resistance of the voltmeter. If we assume that the inductance of this circuit is negligible, then $d/dt \int \mathbf{B} \cdot d\mathbf{A} = 0$. Therefore

$$-\mathscr{E}_b - i_1'r = 0$$

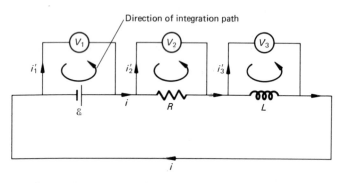

FIGURE 29.19   A circuit containing a battery, a resistor, and an inductor connected in series. Voltmeters are placed across each element in the circuit, and i' is the current that is shunted into the voltmeters.

Faraday's law applied to the second loop gives

$$-i_2'r + iR = 0$$

where we have assumed that $i_2' \ll i$.

Finally $\oint \mathbf{E} \cdot d\mathbf{s}$ around the third loop is just $-i_3'r$, since the electric field within the wire of the inductor is negligible if we assume that the resistance of the inductor is negligible. In the limit, where the wire becomes a perfect conductor, the resistance approaches zero. If there were any electric field in a wire of zero resistance, the current would be infinite.

The inductance of this third loop is not negligible, so

$$\int \mathbf{B} \cdot d\mathbf{A} = -L\frac{di}{dt}$$

Therefore

$$-i_3'r = -L\frac{di}{dt}$$

or

$$i_3'r = L\frac{di}{dt}$$

Let us assume that the reading on each voltmeter is instantaneously proportional to the current through that meter, that is, $V_1 \propto i_1'$, $V_2 \propto i_2'$, and $V_3 \propto i_3'$. Now

$$i_1' = \frac{-\mathscr{E}_b}{r} \tag{29.24a}$$

$$i_2' = \frac{iR}{r} \tag{29.24b}$$

and

$$i_3' = \frac{L(di/dt)}{r} \tag{29.24c}$$

We have also seen, applying Faraday's law to the circuit of Fig. 29.18, that

$$-L\frac{di}{dt} + \mathscr{E}_b = iR \qquad (29.25)$$

(see Eq. 29.22). It follows from Eqs. 29.24 and 29.25 that the sum of the voltmeter readings around the loop will be zero.

This is a general result and is often useful in analyzing complex circuits. Even though there is no potential field and $\int \mathbf{E} \cdot d\mathbf{s}$ is zero through an inductor (if we neglect its resistance), we may nevertheless say that a voltmeter across the inductor will read the value of $L(di/dt)$. The instantaneous sum of the voltmeter readings around any closed loop will be zero.

QUESTION

The switch is closed in the circuit of Fig. 29.20 at $t = 0$. At what rate does charge build up on the capacitor?

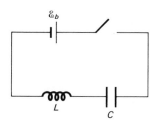

FIGURE 29.20   A capacitor, an inductor, and a battery connected in series.

ANSWER

Let us apply Faraday's law,

$$\oint \mathbf{E} \cdot d\mathbf{s} = -d\phi_B/dt = -L\frac{di}{dt}$$

where the path of integration is taken through the wire, the center of the capacitor, the inductor, and the battery in a clockwise direction. Neglecting the resistance of the wire we find that $\int \mathbf{E} \cdot d\mathbf{s}$ through the wire is zero. Through the capacitor $\int \mathbf{E} \cdot d\mathbf{s} = q/C$, where q is the charge on the right plate. Through the inductor $\int \mathbf{E} \cdot d\mathbf{s} = 0$. And through the battery $\int \mathbf{E} \cdot d\mathbf{s} = -\mathscr{E}_b$. Therefore

$$\frac{q}{C} - \mathscr{E}_b = -L\frac{di}{dt}$$

But $i = dq/dt$, so

$$L\frac{d^2q}{dt^2} + \frac{q}{C} = \mathscr{E}_b$$

The solution of this differential equation, which satisfies the initial conditions ($q = 0$ and $i = 0$ at $t = 0$), is given by

$$q = \mathscr{E}_b C\left(1 - \cos\frac{t}{\sqrt{LC}}\right)$$

as one may verify by substitution. The charge on the capacitor oscillates indefinitely between 0 and $2\mathscr{E}_b C$. If we had included the inevitable resistance in the circuit, we would have found that the charge on the capacitor eventually settles down to $\mathscr{E}_b C$.

SUMMARY

The electromagnetic equations must be modified for nonsteady states. The integral equations become

$$\oint \mathbf{E} \cdot d\mathbf{A} = q/\epsilon_0$$

$$\oint \mathbf{E} \cdot d\mathbf{s} = -\frac{d}{dt}\int \mathbf{B} \cdot d\mathbf{A}$$

$$\oint \mathbf{B} \cdot d\mathbf{A} = 0$$

and

$$\oint \mathbf{B} \cdot d\mathbf{s} = \mu_0\left(i + \epsilon_0\frac{d}{dt}\int \mathbf{E} \cdot d\mathbf{A}\right)$$

The two new terms $-(d/dt) \int \mathbf{B} \cdot d\mathbf{A}$ and $\mu_0\epsilon_0(d/dt) \int \mathbf{E} \cdot d\mathbf{A}$ are the time rate of change of the magnetic flux and the displacement current, respectively. The first is responsible for electromagnetic induction effects, and the second is necessary to assure charge conservation and is critical for the propagation of electromagnetic waves, as we shall see in Chap. 30.

The electromagnetic induction equation (the second equation in the group of four above) can also be written

$$\mathcal{E} = -\frac{d\phi_B}{dt}$$

where $\mathcal{E}$ is the electromotive force around a closed loop and ϕ_B is the magnetic flux through the loop.

Lenz's law states that the induced emf will always act in the opposite direction of its cause.

The electric current in a loop generates a magnetic flux through the loop that is proportional to the current. The proportionality factor depends only on the geometry of the loop and is called the self-inductance; hence

$$\phi_B = Li$$

The back emf caused by the self-inductance is given by

$$\mathcal{E} = -L\frac{di}{dt}$$

PROBLEMS

29.A.1 The two circular conducting rings in the following figure have a common perpendicular axis. The lower ring is fixed and carries a current i, and the upper ring has no current. The upper ring begins to fall.

(a) What is the direction of the induced current in the falling ring?
(b) The current induced in the falling ring will in turn induce an emf in the stationary ring. What is its direction?

29.A.2 The magnetic flux through a loop is given by $\phi_B = \frac{1}{2}e^{2t}$. What is the emf induced in the loop?

29.A.3 What is the self-inductance of a cylinder 10 cm long, 0.1 cm in radius, and wrapped with 1000 turns of wire?

29.A.4 The switch in the figure below is closed at $t = 0$. The current after 0.1 s is 2 A. What is the inductance L if the resistance is 2 Ω and the emf of the battery is 10 V?

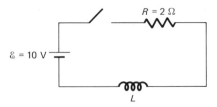

29.A.5 Determine the current in the circuit shown in the figure 1 s after the switch is closed.

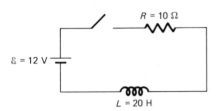

$R = 10\ \Omega$

$\mathcal{E} = 12\ V$

$L = 20\ H$

29.B.1 The magnetic flux through a loop caused by an *external* source is given by $\phi_B = 3t^2 + t$. Write down, but do not solve, a differential equation for the current induced in the loop. The self-inductance of the loop is L and the resistance is R.

29.B.2 A wire is rotating on a circular track of radius r (see the figure) with an angular velocity ω. A uniform magnetic field B is perpendicular to the plane of the circular track. What is the induced emf in the loop? What is the direction of the induced current in R?

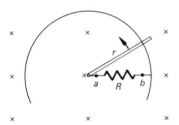

29.B.3 Determine the self-inductance of a toroid of cross-sectional area A with a total of N conducting loops (see the following figure). The mean radius of the toroid is R.

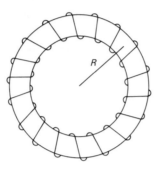

R

(Neglect the variation of the magnetic field over the cross section of the toroid.) (See problem 28.B.7.)

29.B.4 A rectangular loop with sides a and b has a resistance R. In the plane of the loop and parallel to the side of length a is an infinite straight wire carrying a current i that is increasing at a constant rate α ($i = \alpha t$). (See the following figure.) What is the steady state current in the rectangular loop? (If the current in the straight wire is not increasing at a constant rate, there is not a steady state. Although we may still calculate the instantaneous emf in the loop caused by the straight wire, we may not determine the instantaneous current in the loop from the equation $\mathcal{E} = iR$ unless we may neglect the self-inductance of the loop. The total emf in the loop is due to both the changing current i in the straight wire and the self-induced emf in the loop resulting from the changing current within the loop itself.)

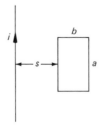

i

b

s

a

29.B.5 Determine the emf generated in a rotating loop of area A in a uniform magnetic field B as a function of time. The angular velocity of the loop is ω and the axis of rotation is in the plane of the loop (see the figure).

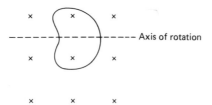

29.B.6 A rectangular conducting bar of length l slides down the incline formed by two parallel conducting rails, as shown in the figure. The conducting loop is completed by a resistance R that is much greater than the resistance of the rails or the bar. The incline makes an angle θ with the horizontal. A vertical magnetic field B is uniform throughout the loop. Show that the terminal velocity of the bar is given by

$$v = \frac{mgR \tan \theta}{B^2 l^2 \cos \theta}$$

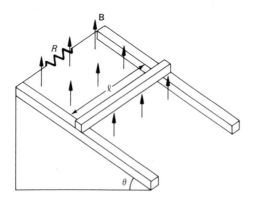

29.B.7 Two long straight concentric solenoids of equal length l have radii r_1 and r_2 ($r_1 < r_2$). They have N_1 and N_2 turns, respectively. If the current ($i = \alpha t$) in the outer solenoid increases at a constant rate, what emf is induced in the inner solenoid?

29.B.8 A circular parallel-plate capacitor of area A and separation s is charging at a constant rate α ($q = \alpha t$). Determine the magnetic field between the plates as a function of r, the distance from the axis of symmetry.

29.B.9 The potential across a parallel-plate 2-μF capacitor is increasing at the rate of 100 V/s. What is the net displacement current between the plates?

29.B.10 An external force F is required to push a rectangular loop of wire from a field-free region into a region of uniform magnetic field (see the following figure). If the force is constant, the velocity is constant. Determine the induced current in the loop from the following considerations: (1) The net force on the loop must be zero. (2) If the resistance of the loop is R, the induced current will dissipate power in the loop, and this power must equal the rate at which work is being done. Determine the current in terms of the magnetic field, the width of the loop, the velocity of the loop, and the resistance.

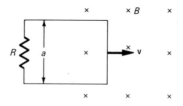

29.B.11 We observed at the beginning of this chapter that the magnetic field associated with a radial flow of charge in a conducting sphere must be zero from symmetry arguments.

(a) Show directly that

$$i + \epsilon_0 \frac{d}{dt} \oint \mathbf{E} \cdot d\mathbf{A} = 0$$

where i is the net current through a sphere of radius $r < R$ (Fig. 29.1) and $\oint \mathbf{E} \cdot d\mathbf{A}$ is the flux of the electric field through this same sphere.

29.C.1 Determine the instantaneous emf induced in the two independent loops (1) and (2) in the following figure.

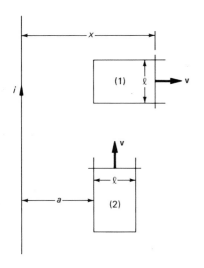

29.C.2 Two infinite coaxial cylinders of radii r_1 and r_2 carry equal and opposite surface currents parallel to their axis. What is the self-inductance per unit length?

29.C.3 Determine the self-inductance of a rectangular loop of wire 100 cm long and 1 cm wide. The radius of the wire is 0.1 cm. Neglect the end effects and the magnetic flux through the core of the wire.

29.C.4 Every circuit has some inductance; there will always be some magnetic flux through a current loop. Often it is possible to neglect the inductance, as in the circuit in the figure below. The flux through the circuit is significant only when the wire is wrapped around an axis many times. To see this more clearly, estimate the time it takes the current in the figure to reach a steady state. Let r be the radius of the wire, l the length of the rectangle, and s its width.

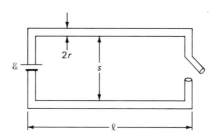

(a) Neglecting end effects ($s \ll l$) and the flux through the core of the wire, show that the self-inductance of the rectangular current loop is given by

$$L = \frac{\mu_0 l}{\pi} \ln \frac{s}{r}$$

(b) Neglecting the resistance of the ends, show that the resistance of the wire loop is $R = 2\rho l / \pi r^2$, where ρ is the resistivity of the wire.

(c) We saw in Eq. 29.17 and the discussion that followed that L/R is a measure of the time it takes the current to reach its steady-state value after the switch is closed. Suppose the wire is copper ($\rho = 1.7 \times 10^{-8} \ \Omega \cdot$ m) of radius $r = 0.1$ cm and the width of the rectangle is $s = 1$ cm. What time does it take the current to reach a steady value, i.e., what is L/R?

30
ENERGY
AND
ELECTROMAGNETIC
WAVES

LET US RESTATE our electromagnetic equations. The four integral relations are

$$\oint \mathbf{E} \cdot d\mathbf{A} = \frac{q}{\epsilon_0}$$

$$\oint \mathbf{E} \cdot d\mathbf{s} = -\frac{d}{dt} \int \mathbf{B} \cdot d\mathbf{A}$$

$$\oint \mathbf{B} \cdot d\mathbf{A} = 0$$

and

$$\oint \mathbf{B} \cdot d\mathbf{s} = \mu_0 \left(i + \epsilon_0 \frac{d}{dt} \int \mathbf{E} \cdot d\mathbf{A} \right)$$

These equations are called *Maxwell's equations*. Together with the force law, they constitute a complete theory for the interaction of electrically charged bodies. In the above integral form they are difficult if not impossible to use in calculations in which there is insufficient symmetry to permit integration. However, with a little vector calculus these equations can be converted to differential equations that may be more useful. The physical content is unchanged.

Let us go back and look at some things we discussed earlier to see whether they require modification in light of the addition of the new time-dependent terms to these equations. In particular, let us reexamine the question of the energy in a configuration of charges.

30.1 ENERGY IN AN ELECTROMAGNETIC FIELD

We have seen that the energy of an electric charge q at a point where the electric potential is V is given by qV. The potential difference between two points is given by

$$\Delta V = - \int \mathbf{E} \cdot d\mathbf{s}$$

where the integral is taken over any path joining the two points.

Suppose the path is a closed path. Then $\Delta V = 0$, so

$$\oint \mathbf{E} \cdot d\mathbf{s} = 0$$

around any closed path. But we now know that

$$\oint \mathbf{E} \cdot d\mathbf{s} = -\frac{d}{dt} \int \mathbf{B} \cdot d\mathbf{A}$$

so if there is a time-changing magnetic flux, then

$$\oint \mathbf{E} \cdot d\mathbf{s} \neq 0$$

in which case an electric potential field does not exist! How then do we determine the energy of a moving configuration of charges, a system in which there are time-changing magnetic fields?

We can no longer associate the energy with the charge configuration. Just as we are forced to give up the action-at-a-distance approach to the dynamics of moving charges and adopt a field theory, so we must look to this field to describe the energy of electromagnetic systems.

Perhaps a more striking illustration of why we must give up the idea of a potential energy associated with a given charge configuration is the existence of electromagnetic waves. Electromagnetic waves can exist independently of the charges that created them. For example, many of the stars we see in the sky have burned out long ago. The electromagnetic fields radiated by these stars carry energy, and this energy cannot be associated with the present positions and velocities of electric charges.

In *electrostatics* and *magnetostatics* we can associate the energy with either the charge configuration *or* the fields they create. Let us look at these static problems and see whether we can discover what energy we must associate with the fields to obtain agreement with our previous results using the charge configuration.

30.2 ELECTRIC FIELD ENERGY

The simplest system for studying the energy in an electric field is the parallel-plate capacitor. Let us see how much work must be done to establish an electric field between the plates. A charged parallel-plate capacitor may be created in many ways, and we shall adopt a very primitive method. Imagine two plates of area A carrying charges $+q$ and $-q$ that are initially superimposed. There is no electric field because the positive and negative charges cancel each other. Now suppose the two plates are pulled apart by an outside force to a separation s (Fig. 30.1). How much work must be done to effect the separation? This work will be the product of the force F and the distance s through which the force acts. The electric field created by the negative plate is given by (see Eq. 21.6)

$$E = \frac{\sigma}{2\epsilon_0} = \frac{q}{2\epsilon_0 A} \tag{30.1}$$

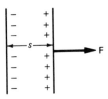

FIGURE 30.1 The positive plate is pulled a distance s by the external force $\mathbf{F}$. The negative plate is fixed.

The force that this field exerts on the positive plate is the product of the charge on the positive plate and the electric field at that plate created by the negative plate, that is,

$$F = qE = \frac{q^2}{2\epsilon_0 A}$$

This is the constant force that must be balanced by the outside agent separating the plates. The work performed by this force to separate two plates of charge $+q$ and $-q$ and of area A to a distance s is given by

$$W = Fs = \frac{q^2 s}{2\epsilon_0 A} \tag{30.2}$$

This work must equal the increase in the electrostatic potential energy. Thus Eq. 30.2 tells us the potential energy in terms of the charges and their distribution (represented by the area A of the plates and the separation s).

We would now like to express the energy in terms of the electric field created by the separation of charge. The net electric field between the plates is given by

$$E = \frac{\sigma}{\epsilon_0} = \frac{q}{\epsilon_0 A} \tag{30.3}$$

(Note that this is the field created by both plates.) Eliminating q between Eqs. 30.2 and 30.3, we find

$$W = U = \tfrac{1}{2}\epsilon_0 E^2 As \tag{30.4}$$

We shall interpret this as the energy U contained in the electric field between the plates.

For a general configuration of charge the electric field is not constant, as it is in the parallel-plate capacitor. We must therefore determine the energy associated with a nonuniform electric field.

In Eq. 30.4 we note that As is the volume between the parallel plates, so the energy per unit volume (u) between the plates is

$$u = \frac{\text{energy}}{\text{volume}} = \frac{U}{As} = \frac{1}{2}\epsilon_0 E^2 \tag{30.5}$$

We shall assume that we may obtain the energy in an arbitrary electric field by integrating the energy density u over the entire volume, that is,

$$U = \int u \, dx \, dy \, dz = \frac{1}{2}\epsilon_0 \int E^2 \, dx \, dy \, dz \tag{30.6}$$

QUESTION

How much work must be done to give a spherical conductor of radius R a total charge of q?

ANSWER

The work that must be done to charge the spherical conductor must equal the energy that is stored in the spherical conductor after it is charged. The energy stored in the charged spherical conductor is just the energy stored in the electric field associated with the charged sphere.

The electric field external to the spherical conductor is given by

$$E = \frac{kq}{r^2}$$

(The electric field inside the conductor is zero.) The energy per unit volume in the electric field outside the sphere is therefore

$$u = \frac{1}{2}\epsilon_0 E^2 = \frac{1}{2}\epsilon_0 k^2 q^2 \frac{1}{r^4}$$

The energy dU in a spherical shell of radius r and thickness dr is the energy density multiplied by the volume of the shell $(4\pi r^2 \, dr)$, or

$$dU = \frac{1}{2}\epsilon_0 k^2 q^2 \frac{1}{r^4} 4\pi r^2 \, dr$$

Integrating, we have

$$U = \frac{1}{2}\epsilon_0 k^2 q^2 4\pi \int_R^\infty \frac{dr}{r^2}$$

$$= \frac{\tfrac{1}{2}kq^2}{R}$$

(This is the same result we obtained earlier in problem 23.C.5.)

QUESTION

We shall find that the energy released in the fission (the breakup) of large nuclei is due to the Coulomb energy in the charged nucleus. We may approximate the charge density of the nucleus as that of a uniformly charged sphere. What is the energy in a uniformly charged sphere of radius R and total charge q?

ANSWER

To determine the energy of such a charge configuration, we compute the energy in the electric field. The electric field inside the sphere can be determined from Gauss's law to be (problem 23.B.2)

$$E = \frac{kq}{R^3}r$$

The energy in the field inside the sphere is

$$U_{\text{inside}} = \frac{1}{2}\epsilon_0 \int E^2 \, dV$$

$$= \frac{\epsilon_0 k^2 q^2}{2R^6} \int_0^R r^2 4\pi r^2 \, dr$$

$$= \frac{kq^2}{10R}$$

The electric field outside the sphere is the same as that outside the conducting sphere of the preceding question, so the energy in the field outside must be the same as that in the preceding question, namely

$$U_{\text{outside}} = \frac{kq^2}{2R}$$

The total energy in the field is then

$$\text{total energy} = \frac{kq^2}{10R} + \frac{kq^2}{2R}$$

$$= \frac{3}{5}\frac{kq^2}{R}$$

We shall use this result later when we study the energy of the nucleus, and we shall see that an enormous Coulomb energy is packed in the nucleus.

QUESTION

We have seen that the work required to charge a parallel-plate capacitor is given by (see Eq. 30.2)

$$W = \frac{q^2 s}{2\epsilon_0 A}$$

Now the capacitance of a parallel-plate capacitor is

$$C = \frac{\epsilon_0 A}{s}$$

Therefore

$$W = \frac{1}{2}\frac{q^2}{C}$$

or, since $C = q/v$,

$$W = \tfrac{1}{2}qV = \tfrac{1}{2}CV^2$$

How much work is required to charge an arbitrary capacitor?

ANSWER

As an alternative method of charging the capacitor, let us consider a process in which a small amount of charge dq is transferred from one conductor to the other. After this has been repeated many times, the net charge on each conductor is $+Q$ and $-Q$. With each transfer, the work that must be done against the existing potential difference is

$$dW = V \, dq$$

But the potential difference is related to the charge q already transferred by the relation

$$V = \frac{q}{C}$$

Therefore

$$W = \int dW = \frac{1}{C}\int_0^Q q \, dq = \frac{1}{2}\frac{Q^2}{C}$$

$$= \frac{1}{2}QV = \frac{1}{2}CV^2$$

These results apply quite generally to any capacitor, not just to the parallel-plate capacitor.

30.3 MAGNETIC FIELD ENERGY

We shall deal with the problem of determining the energy in a magnetic field in much the same way as we dealt with the problem of determining the energy in an electric field. We shall calculate the work that must be done to generate a known magnetic field and equate the work done to the magnetic field energy.

The simplest magnetic field is that of a long solenoid; in this case the magnetic field is uniform inside the solenoid and zero outside. (We shall assume either that the cylinder is long enough so that we may neglect the field energy associated with the field outside the cylinder at the ends, or that the cylinder is of infinite length and we are calculating the energy per unit length of cylinder.) How much work must be done to create this magnetic field? Let us imagine a very primitive experiment: A charge q is spread uniformly over a cylinder of length ℓ and radius R (Fig. 30.2). A crank on one end may be used to rotate the cylinder about its axis. Initially the cylinder is at rest and the magnetic field is zero. When the crank is turned, the cylinder begins to rotate, and the moving surface charge becomes a surface current. This current creates a magnetic field inside the cylinder. As the crank is turned faster and faster, the magnetic field increases, and this increasing magnetic field creates a time-changing magnetic flux ϕ_B through a cross section of the cylinder. From Faraday's law this time-changing flux will induce an electric field that will oppose its cause, the increasing velocity of the surface charges. The induced electric field will be circular, concentric with the axis of the cylinder, and in the opposite direction of the rotation. The electric field will exert a force F on the surface charge, given by

$$F = qE$$

If the surface has an instantaneous velocity v, then the power being supplied to move the surface charge through the opposing field is

$$P = Fv = qEv \tag{30.7}$$

This power is supplied by the person turning the crank. (We neglect the inertial effects of the cylinder.)

Now the magnetic field within the cylinder is given by Eq. 28.12,

$$B = \mu_0 K \tag{30.8}$$

where the surface current is the charge that passes a unit length of the cylinder per unit time. If we wait for one revolution, then the charge is q/ℓ, and the period of revolution is the time, so

$$K = \frac{q/\ell}{T} \tag{30.9}$$

The induced electric field is determined by the relation

$$\oint \mathbf{E} \cdot d\mathbf{s} = -\frac{d}{dt} \int \mathbf{B} \cdot d\mathbf{A}$$

Therefore

$$E2\pi R = \frac{dB}{dt}\pi R^2$$

or

$$E = \frac{1}{2}R\frac{dB}{dt} \tag{30.10}$$

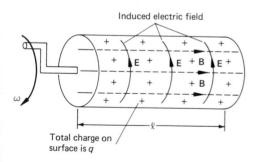

Induced electric field

Total charge on surface is q

FIGURE 30.2   A magnetic field is created in the uniformly charged cylinder when the crank is turned. The induced electric field opposes the acceleration of the cylinder.

Finally the velocity of rotation v is related to the period by the equation

$$v = \frac{2\pi R}{T} \qquad (30.11)$$

From Eqs. 30.7 through 30.11, we have

$$P = \frac{1}{\mu_0} B \frac{dB}{dt} \pi R^2 \ell$$

But $\pi R^2 \ell$ is the volume of the cylinder and P is the rate at which work is being done (dW/dt), so

$$\frac{dW}{dt} = \frac{1}{\mu_0} B \frac{dB}{dt} \text{(volume)}$$

Integrating with respect to time, we see that the total work done to create the magnetic field B is given by

$$W = \frac{1}{2\mu_0} B^2 \text{(volume)} \qquad (30.12)$$

or, equating the work done to the increase in energy, we have

$$u = \frac{\text{energy}}{\text{volume}} = \frac{1}{2\mu_0} B^2 \qquad (30.13)$$

which is the energy density in the magnetic field.

Once again this result was derived for a very special case, but it is of general validity.

QUESTION

We have just determined how much work must be performed to establish a current in a solenoid. How much work must be performed to create a current i in an arbitrary circuit?

ANSWER

As the current is increased, a back emf given by Faraday's law,

$$\mathscr{E} = -\frac{d\phi_B}{dt}$$

is generated. But

$$\phi_B = Li$$

where L is the inductance of the circuit, so

$$\mathscr{E} = -L \frac{di}{dt}$$

Work must be done at a rate of $-\mathscr{E}i$ to drive the charge against this back emf. Therefore

$$\frac{dW}{dt} = -\mathscr{E}i = Li \frac{di}{dt}$$

or the work required to establish a current i is given by

$$W = \tfrac{1}{2}Li^2 \qquad (30.14)$$

We shall leave it as an exercise for you to show that Eq. 30.12 is a special case of Eq. 30.14.

30.4 ELECTROMAGNETIC WAVES

Our original purpose in introducing electric and magnetic fields was to furnish a framework in which to describe the interaction between charged particles. We claimed that the field is a necessary intermediary because of the finite velocity of the propagation of electromagnetic signals. We have yet to discuss this propagation of electromagnetic signals. It is not a simple matter to show that the phenomenon of wave propagation is a consequence of Maxwell's equations. We shall find it necessary to make some assumptions and show that they are at least consistent with our field equations.

PRINCIPLE OF DELAYED EFFECTS

The most useful principle in understanding wave propagation is the *principle of delayed effects,* which may be stated as follows:

The electromagnetic field position P at time $t + r/c$ caused by a single particle depends only on the state (position P', *velocity, and acceleration) of the particle at time t. The quantity r is the distance between P and P' and c is a fundamental constant called the velocity of light.*

QUESTION

A single positive charge is at rest until time $t = 0$. Between time $t = 0$ and $t = 1$ s, the charge is moved about in some way and returned to its original position after 1 s. From 1 s on it remains at rest. What can we say about the electromagnetic field at $t = 3$ s?

ANSWER

Let us draw two spheres about the rest position of the charge. The outer sphere is of radius $3c$, and the inner sphere of radius $2c$. The principle of delayed effects tells us that the electric field outside the outer sphere and inside the inner sphere is that of a point charge at rest (see Fig. 30.3). In the shell between the two spheres the electric field varies in a manner that depends on the details of the motion of the charge between the times $t = 0$ and $t = 1$ s. The magnetic field will be zero outside the shell but must be different from zero within the shell, since the electric field is changing in time in this region. All we may

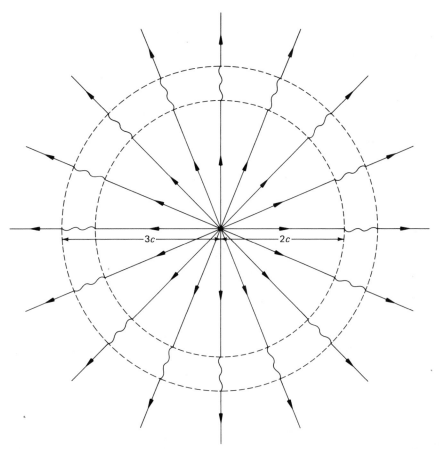

FIGURE 30.3 The electric field is that of a point charge at rest everywhere except within the shell between the two spheres.

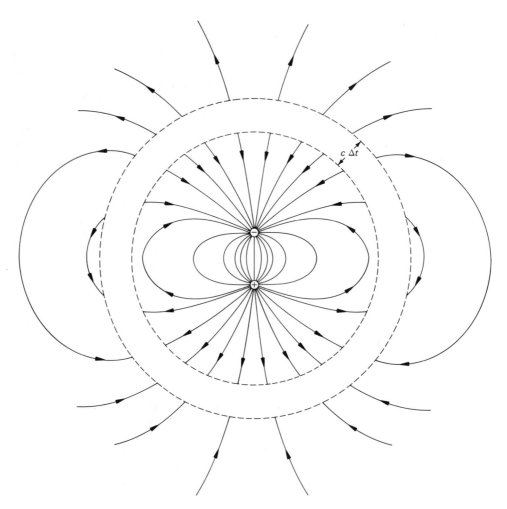

FIGURE 30.4   Static electric field at $t > \Delta t$ of a dipole that reversed its polarity at $t = 0$. The two dashed circles are of radius ct and $c(t + \Delta t)$.

say about the fields within the shell is that the electric field lines inside must connect with the electric field lines outside (electric field lines cannot begin or end in a vacuum) and the magnetic field lines must close on themselves without crossing.

Let us consider an example that allows us to determine some of the features of the time-changing electric field. Out ultimate objective is an oscillating dipole, i.e., two equal and oppositely charged particles oscillating on a straight line. Let us begin, however, with two equal and opposite charges at a fixed separation on a vertical line. At $t = 0$ the charges are accelerated toward each other for a short time Δt until they acquire a velocity v_0. They then continue to move with constant velocity v_0 for time t until they have interchanged positions. The polarity of the dipole has in effect been reversed. Let us attempt to construct the electric field at time $t + \Delta t$ with the information given.

In Fig. 30.4 the two circles (dashed lines) are of radii ct and $c(t + \Delta t)$. Outside the larger circle the field is that of a static dipole, and inside the smaller circle the field is that of two charges moving apart with a velocity v_0. If $v_0 \ll c$, we may approximate the field at time t as that of a dipole. How are these field lines to be connected so that they satisfy the requirements that the field lines may begin or end only on electric charges and that they may not cross?

We see in Fig. 30.4 that there is no way to join the field lines outside the larger circle with those inside the smaller circle without violating the requirements above. The only connection possible is that shown in Fig. 30.5.

As time goes on the two spheres will expand, but *the separation between them, namely c Δt, will remain the same.* We shall refer to the region between the two spheres as the *radiation zone* and the fields in the radiation zone as the *radiation fields,* in constrast to the static or quasistatic fields outside the radiation zone. Note again that the radiation field is associated with the *acceleration* of the charges.

A number of properties of the radiation field are obvious from Fig. 30.5:

1. The field lines are much denser in the radiation field than in the neighboring static field. Thus the electric field of the radiation field is much greater than the neighboring static field.

2. The radiation field is most intense in the direction perpendicular to the axis of the dipole.

3. Since the shell expands at a velocity c, the radiation field propagates with a velocity c.

4. When the spherical shell is far from the dipole, the radiation field is very nearly perpendicular to a line from the dipole to the field point.

5. A magnetic field will be induced within the shell as a result of the time-changing electric field or, equivalently, of the current associated with the charge motion. Note that $\oint \mathbf{E} \cdot d\mathbf{s}$ taken around the closed electric field lines that enter the shell from the outside cannot be zero since $\mathbf{E}$ and $d\mathbf{s}$ are everywhere parallel (or everwhere antiparallel) around the loop. Therefore there must be a time-changing magnetic flux through these loops. The magnetic field must be zero outside the shell because of the principle of delayed effects.

We may determine the direction of the magnetic field within the shell by examining the current associated with the motion of the charges. The current is vertically down, so the magnetic field lines must be circles whose centers lie on the vertical axis. By applying the right-hand rule, we see that the sense of the magnetic field on these circles is clockwise if we are looking down on the dipole from above.

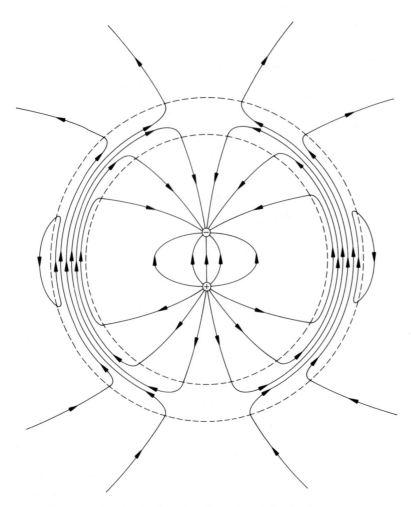

FIGURE 30.5   The electric field lines have been connected without crossing.

Within the shell the directions of **E** and **B** × **c** are the same. The velocity **c** is a vector in the direction of propagation of the wave (i.e., in the radial direction). We shall show later that this is a general property of electromagnetic waves.

If after the initial period of acceleration the acceleration of each particle were reversed, so that each returned to the position it occupied at $t = 0$, the field would be as shown in Fig. 30.6.

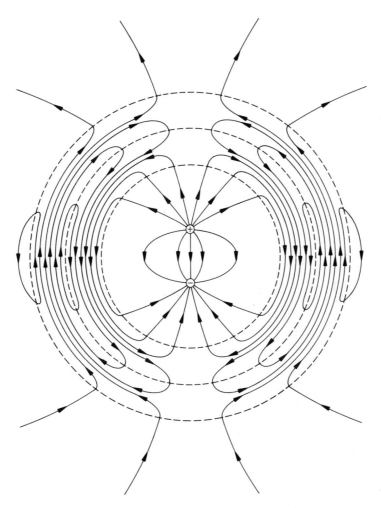

FIGURE 30.6 Electric field of a dipole whose charges have completed one full oscillation.

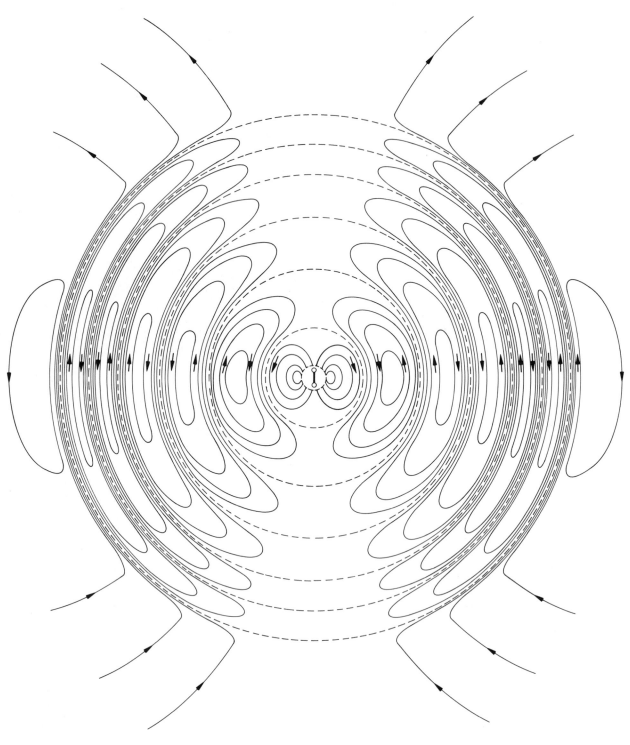

FIGURE 30.7 Radiation field of an oscillating dipole. This pattern propagates away from the source with a velocity c.

The electric field of a dipole in a state of continuous oscillation is illustrated in Fig. 30.7. This pattern is simply a superposition of the radiation field pattern of Fig. 30.6, because, by the principle of delayed effects, the field in any shell depends only on the motion of the charges during the time interval in which the charges underwent one complete oscillation. Far from the radiating dipole the radiation pattern is that of an electric field that is normal to the line from the field point to the dipole. An observer at rest would see a field that builds to a maximum, decreases to zero, reverses direction, etc. The electric field vector would oscillate with a frequency equal to that of the oscillating dipole. In a region of space far from the dipole the electric field would appear as shown in Fig. 30.8.

FREQUENCY SPECTRUM

We see that one of the important characteristics of the electromagnetic radiation from an oscillating dipole is the frequency of the wave, which is the same as the frequency of oscillation of the dipole. The wave frequency remains unchanged as the wave propagates, even if the wave is reflected by a mirror or diffracted by a lens.

The frequency spectrum of electromagnetic waves is often characterized by the source of the radiation (see Fig. 30.9). Radio waves are of course generated by circuits that produce an oscillating current in a broadcast antenna.

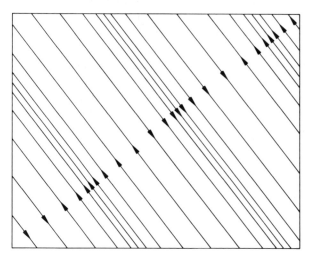

FIGURE 30.8 Radiation electric field far from the oscillating dipole.

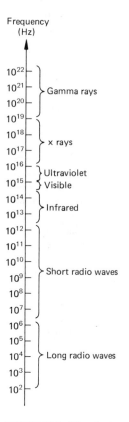

FIGURE 30.9 The electromagnetic spectrum.

The upper limit on the radiation frequency is determined by limits on the design of circuit elements. We have seen in the şimple *LC* circuit (Sec. 29.4) that an oscillating current is established in that circuit when the switch is closed. An upper limit on radiation frequencies is determined by such circuits.

Frequencies in the visible and near-visible range are produced by electrons in the outer shells of atoms. X rays are generated by electrons in the inner shells, and γ rays are generated by atomic particles within the nucleus.

Of course there are other means of generating any of these frequencies. For example, a lightning bolt generates a broad band of frequencies, and an electron striking a fixed target will give off electromagnetic radiation with a frequency range that depends on the kinetic energy of the electron.

QUESTION

Which has the higher frequency, red light or blue light?

ANSWER

As we add heat to a bar of metal, we find that it first becomes *red hot* and finally *white hot.* Now we may think of the radiation that is emitted by any body at a temperature above absolute zero as radiation from charged particles moving about in the body. (This is a classical description of a quantum mechanical phenomenon and is therefore somewhat dangerous. Nevertheless, it serves a useful conceptual purpose.) The hotter the body, the faster the charges move about, and as a consequence the higher the frequency of the emitted radiation. Not all the particles are moving at the same speed or oscillating about fixed equilibrium positions with the same frequency. But at higher and higher temperatures there will be more and more particles with higher and higher energies and higher and higher frequencies. If at a certain temperature a body glows red hot, this means that there is some radiation in the red frequencies and little radiation in the rest of the visible spectrum. If a body glows white hot, there must be significant radiation over the entire visible spectrum. (White light is a mixture of all frequencies in the visible spectrum.) Since red light is seen at lower temepratures and white light (including blue light) is seen at higher temperatures, and since an increase in temperature implies an increase in the number of charges oscillating at higher frequencies, it follows that red light is low frequency and blue light is high frequency.

QUANTITATIVE PICTURE OF THE RADIATION ELECTRIC FIELD

We may determine the manner in which the electric field varies with distance from the dipole by applying the principle of energy conservation. We have seen that the energy density in the electric field is proportional to the square of the electric field ($u = \frac{1}{2}\epsilon_0 E^2$). Let us consider the energy flux through a cone whose apex is at the dipole (see Fig. 30.10). In the steady state the rate at which energy crosses the area A_1 is $cu_1 A_1$, and the rate at which energy crosses the area A_2 is $cu_2 A_2$. Since these rates must be equal,

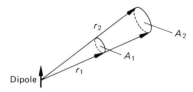

FIGURE 30.10   The energy flux through A_1 must equal the energy flux through A_2.

$cu_1 A_1 = cu_2 A_2$

but $A_1 \propto r_1{}^2$ and $A_2 \propto r_2{}^2$, so

$u_1 r_1{}^2 = u_2 r_2{}^2$

or

$ur^2 =$ constant

or

$u \propto \dfrac{1}{r^2}$

But $u \propto E^2$, so

$E = \dfrac{1}{r}$

or the radiation field decreases with the reciprocal of the distance from the source. This is a general property of all radiation fields, not just the dipole radiation field. Note that the radiation field decreases less rapidly than either the Coulomb field ($E \propto 1/r^2$) or the static dipole field ($E \propto 1/r^3$).

By a careful analysis of Maxwell's equations, one may show that the radiation electric field of a dipole that is oscillating harmonically with a frequency f has a maximum value given by

$$E = k\left(\frac{2\pi f}{c}\right)^2 \frac{qa}{r}\sin\theta \tag{30.15}$$

where θ is the angle between the axis of the dipole and the line joining the field point and the dipole (see Fig. 30.11). This formula reflects all the properties that our previous analysis has led us to expect.

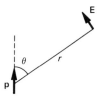

FIGURE 30.11   Radiation field of an oscillating dipole.

QUESTION

The maximum electrostatic field of a dipole is on the axis of the dipole and is given by $E_{\text{static}} = kqa/r^3$. Compare the electrostatic field of a static dipole with the radiation field of an oscillating dipole if the dipoles are on the sun and the observer is on the earth. Let the frequency of the radiation be that of red light ($f = 7.5 \times 10^{12}$ Hz).

ANSWER

$$\frac{E_{\text{rad}}}{E_{\text{static}}} = \frac{k(2\pi f/c)^2(qa/r)}{kqa/r^3} = \left(\frac{2\pi f}{c}\right)^2 r^2$$

$$= \left(\frac{2\pi \times 7.5 \times 10^{12}\ \text{Hz}}{3 \times 10^8\ \text{m/s}}\right)^2 (1.5 \times 10^{11}\ \text{m})^2$$

$$= 5.6 \times 10^{32}$$

so the electrostatic field is totally negligible in comparison to the radiation field.

POLARIZATION

We have seen that a vertical dipole will radiate an electromagnetic wave in which the electric field is perpendicular to the radius vector from the dipole to the field point and lies in the plane defined by the dipole and the radius vector, i.e., in a vertical plane. If the dipole is rotated 90° so

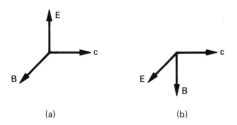

FIGURE 30.12 A wave polarized in (a) the vertical plane and (b) the horizontal plane. Note that in each case **E** and **B** × **c** are parallel.

that it is directed into or out of the page, then the electric field is also rotated by 90°. The orientation of the electric field determines the *polarization* of the wave. In Fig. 30.12(a) the polarization of the wave is vertical, while in Fig. 30.12(b) it is horizontal.

If the radiation field is generated by thermal radiation, we would expect to find a mixture of all possible polarizations. Different charges in the body will radiate waves of different polarizations. A wave of mixed polarization is called an *unpolarized wave* because no direction of the electric field is preferred. We should observe that in either case, polarized or unpolarized waves, the electric and magnetic fields are always perpendicular to the direction of propagation.

QUESTION

If the broadcast antenna of a radio station you wish to receive is vertical, how should you orient the antenna on your automobile?

ANSWER

Electrons that are oscillating up and down a vertical broadcast antenna are in effect radiating like a vertical dipole. The radiation field will also be vertical at the same level as the antenna. For this radiation field to effectively drive electrons

along the length of the receiving antenna, the antenna must also be vertical.

Note that with a vertical broadcast antenna most of the radiation is concentrated in a horizontal plane (see Eq. 30.15); this is fortunate since that is where the customers are.

30.5 THE VELOCITY OF LIGHT

We have seen how one may obtain some of the properties of electromagnetic waves simply by requiring that electric field lines satisfy certain rules: They must begin and end only on electric charges and they may not cross. These constraints were sufficient to give a rough picture of the electric field lines of an oscillating dipole.

We have said that we can deduce the nature of the propagation of electromagnetic waves from Maxwell's equations. In a region in which there are no charges and currents, these equations become

$$\oint \mathbf{E} \cdot d\mathbf{A} = 0$$

$$\oint \mathbf{B} \cdot d\mathbf{A} = 0$$

$$\oint \mathbf{E} \cdot d\mathbf{s} = -\frac{d}{dt} \int \mathbf{B} \cdot d\mathbf{A}$$

$$\oint \mathbf{B} \cdot d\mathbf{s} = -\epsilon_0 \mu_0 \frac{d}{dt} \int \mathbf{E} \cdot d\mathbf{A}$$

One of the properties of electromagnetic waves was that they propagate in a vacuum (a region with no charges and no currents) with a velocity c. The velocity c is a universal constant; it is the same for all electromagnetic waves and it has the same value regardless of the strength of the fields.

Now c does not appear as a parameter in Maxwell's equations. Since c is independent of the strength of the fields, it must depend only on quantities in Maxwell's equations other than **E** and **B**, namely ϵ_0 and μ_0. In particular, it must depend only on the product $\epsilon_0\mu_0$, since this is how ϵ_0 and μ_0 appear in Maxwell's equations. By looking at the dimensions of the last two integral equations, we can easily see that $\epsilon_0\mu_0$ has the dimensions of a time squared divided by a length squared. Thus from dimensional considerations alone we know that

$$c \propto \frac{1}{\sqrt{\epsilon_0\mu_0}}$$

We cannot determine the proportionality constant from dimensional considerations.

Let us examine a very special case of electromagnetic wave propagation, one with sufficient symmetry that we may employ Maxwell's equations in their integral form and determine the velocity of propagation, i.e., determine the constant of proportionality between c and $(\epsilon_0\mu_0)^{-1/2}$.

The example we shall consider is the propagation of the electromagnetic wave that charges a parallel-plate capacitor. When a battery is placed at the edge of a capacitor, it takes time for the entire capacitor to become charged. Electromagnetic signals do not propagate with an infinite velocity. A current passes down the upper plate (see Fig. 30.13) and back along the lower plate. This flow of charge i charges the upper plate positively and the lower plate negatively. We may assume that far from the battery the charge and current are uniformly distributed over the plate. The *wave front* divides the charged portion of the capacitor from the uncharged portion. (A more realistic example that is similar to the above is the propagation of an electromagnetic wave down a pair of telephone wires when a *connection* is made.) We have already determined the electric and magnetic fields associated with uniformly charged plates carrying uniform currents in our study of

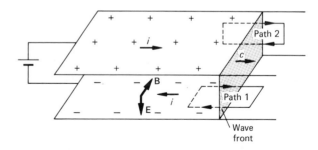

FIGURE 30.13 An electromagnetic wave is propagating down the parallel-plate capacitor.

electrostatics and magnetostatics. The electric field is directed downward from the positive upper plate to the negative lower plate. The magnetic field associated with the given current is perpendicular to the current and parallel to the plates, as shown in Fig. 30.13. (At the wave front the field lines will not be straight lines as we have pictured them because of edge effects, but far from the wave front they will be straight lines. We shall assume that our integration paths penetrate deeply into the region of uniform field, so we need not worry about the precise nature of the fields at the wave front.)

Let us apply two of Maxwell's equations to the two paths of integration shown in Fig. 30.13. On path 1 we consider

$$\oint \mathbf{B} \cdot d\mathbf{s} = \mu_0\epsilon_0 \frac{d}{dt} \int \mathbf{E} \cdot d\mathbf{A} \qquad (30.16)$$

For mathematical simplicity we shall assume that all sides of the integration path are of unit length. We have then

$$\oint \mathbf{B} \cdot d\mathbf{s} = B \qquad (30.17)$$

and

$$\frac{d}{dt}\int \mathbf{E} \cdot d\mathbf{A} = Ec \tag{30.18}$$

where c is the velocity with which the wave front is advancing. From Eqs. 30.16, 30.17, and 30.18, we have

$$B = \epsilon_0 \mu_0 Ec \tag{30.19}$$

On path 2 we consider

$$\oint \mathbf{E} \cdot d\mathbf{s} = -\frac{d}{dt}\int \mathbf{B} \cdot d\mathbf{A} \tag{30.20}$$

Now

$$\oint \mathbf{E} \cdot d\mathbf{s} = -E \tag{30.21}$$

and

$$\frac{d}{dt}\int \mathbf{B} \cdot d\mathbf{A} = cB \tag{30.22}$$

From Eqs. 30.20, 30.21, and 30.22, we have

$$E = cB \tag{30.23}$$

Comparing Eqs. 30.19 and 30.23, we see that the velocity of the wave front must be given by

$$c = \frac{1}{\sqrt{\epsilon_0 \mu_0}}$$

Substituting the values of ϵ_0 and μ_0, we have

$$c = \sqrt{(8.85 \times 10^{-12} \ C^2/N \cdot m^2)(4\pi \times 10^{-7} \ T \cdot m/A)}$$

$$= 3.0 \times 10^8 \ m/s$$

as the velocity of propagation of the wave front.

QUESTION

What is the relative proportion of the energy carried by the electric and magnetic fields?

ANSWER

The ratio of the energies is given by

$$\frac{\frac{1}{2}\epsilon_0 E^2}{(1/2\mu_0)B^2}$$

But

$$E = cB = \frac{1}{\sqrt{\epsilon_0 \mu_0}}B$$

so

$$\frac{\frac{1}{2}\epsilon_0 E^2}{(1/2\mu_0)B^2} = \frac{\frac{1}{2}\epsilon_0(1/\epsilon_0\mu_0)B^2}{(1/2\mu_0)B^2} = 1$$

In other words, the energy is carried equally by the electric and magnetic fields.

Let us summarize our findings. We have shown that

$$c = \frac{1}{\sqrt{\epsilon_0 \mu_0}} \tag{30.24}$$

and that the electric and magnetic fields are proportional, namely

$$E = cB \tag{30.25}$$

Furthermore, observing the directions of $\mathbf{E}$, $\mathbf{B}$, and the moving wave front in Fig. 30.12, we may write

$$\mathbf{E} = \mathbf{B} \times \mathbf{c} \tag{30.26}$$

Although these results have been derived for a very special wave, they can be shown to hold quite generally. The velocity is the same for all waves, and **E**, **B**, and **c** are mutually perpendicular and satisfy Eq. 30.26.

We determined the numerical value of the velocity of light by applying Maxwell's equations to an electromagnetic wave. Suppose another observer moving with respect to us carried out the same calculation also using Maxwell's equations. This other person must of course obtain the same value for the velocity of light. Let us call these two observers A and B, and let us denote the velocity of a light wave relative to observer A by $\mathbf{v}_{\ell/A}$ and the velocity of the same light wave relative to observer B by $\mathbf{v}_{\ell/B}$. We let $\mathbf{v}_{B/A}$ denote the velocity of observer B relative to observer A. We would expect that the relation between the velocity of light relative to the two observers would satisfy the relation

$$\mathbf{v}_{\ell/A} = \mathbf{v}_{\ell/B} + \mathbf{v}_{B/A} \tag{30.27}$$

If $\mathbf{v}_{B/A}$ is not zero, then how can $\mathbf{v}_{\ell/A} = \mathbf{v}_{\ell/B}$? How can both observers apply Maxwell's equations and obtain the same value for the velocity of light? This is one of the most profound paradoxes that has plagued science.

Scientists first assumed that Maxwell's equations applied only in a very special frame of reference, a frame of reference fixed with respect to the "ether." It was not exactly clear what the ether was, but it was some sort of medium for the propagation of light waves, just as air is the medium for the propagation of sound waves. Researchers assumed that Maxwell's equations required modification when applied to a frame of reference moving with respect to the ether. In this frame, when these modified Maxwell's equations were used, Eq. 30.27 would be satisfied.

Numerous experiments were performed to detect differences in the value of the velocity of light. The velocity of light was measured at various times during the day. If the

earth is moving through the ether, then different values for the velocity of light should have been obtained because of the rotation of the earth. The results, however, were always the same, $c = 3 \times 10^8$ m/s. We are forced therefore to assume that Maxwell's equations are valid in all inertial frames of reference, just as the laws of motion are valid in all inertial frames of reference. The velocity of light is a universal constant that is independent of the velocity of the observer (so long as this observer is not accelerating). Einstein accepted this fact and it became a fundamental principle of the special theory of relativity. By carefully examining the manner in which positions and velocities are measured Einstein was able to show that the formula for the addition of velocities (Eq. 30.27) is not correct. He derived an expression for the addition of velocities which reduces to our customary rule for adding velocities when all the velocities are small compared to the velocity of light but which is compatible with the uniformity of the velocity of light for all observers.

SUMMARY

The energy densities in the electric and magnetic fields respectively are given by

$$\tfrac{1}{2}\epsilon_0 E^2$$

and

$$(1/2\mu_0)B^2$$

Electromagnetic waves propagate with a velocity given by

$$c = \frac{1}{\sqrt{\epsilon_0 \mu_0}} = 3 \times 10^8 \text{ m/s}$$

The electric field, magnetic field, and direction of propagation are all mutually perpendicular. The ratio of the strengths of the electric and magnetic fields is equal to the propagation velocity,

$E/B = c$

It follows that the energy is carried equally by the electric and magnetic fields.

PROBLEMS

30.A.1 We have shown (Eq. 30.11) that the work required to establish a current i in an inductor is $\frac{1}{2}Li^2$. Show that for a solenoid $\frac{1}{2}Li^2 = (B^2/2\mu_0)A\ell$, where A is the cross-sectional area of the solenoid, ℓ is its length, and B is the uniform field within the solenoid.

30.A.2 Show that for a parallel-plate capacitor

$$\frac{1}{2}\frac{q^2}{C} = \frac{\epsilon_0 E^2}{2}As$$

where A is the area of the plates and s the plate separation.

30.A.3 Show that $\epsilon_0(d\Phi_E/dt)$ has the dimension of a current.

30.A.4 Given that $\mathbf{E} = \mathbf{B} \times \mathbf{c}$, determine the direction of the magnetic field throughout the shell in Fig. 30.5.

30.A.5 Given that $\mathbf{E} = \mathbf{B} \times \mathbf{c}$, determine the direction of the magnetic field everywhere in Fig. 30.8 if (a) the wave is propagating toward the upper right-hand corner of the box, and (b) the wave is propagating toward the lower left-hand corner of the box.

30.B.1 Show that the energy in the electric field of a parallel-plate capacitor is $\frac{1}{2}q^2/C$.

30.B.2 The work required to bring a charge dq from infinity to the surface of a spherical conductor is $dq\, V$, where V is the existing potential of the surface of the sphere. Calculate the net work that must be done to bring a total charge q to the surface of a neutral conducting sphere of radius R. Show that the result is the same as the total energy in the field of the sphere as calculated in Sec. 30.2.

30.B.3 Determine the capacitance per unit length of two coaxial cylinders of radii r_1 and r_2 by equating the energy in the field to $\frac{1}{2}q^2/C$.

30.B.4
  (a) Determine the energy in the electric field between two parallel-plate capacitors with charge q, area A, and separation x.
  (b) If x is increased by a small amount dx, what is the increase in field energy?
  (c) Equate the increase in field energy to the work that is done in increasing x by dx, and determine the force between the plates (see problem 21.C.1).

30.B.5 Determine the self-inductance per unit length of two coaxial cylinders of radii r_1 and r_2 by equating the energy in the magnetic field of a unit length to $\frac{1}{2}Li^2$, where L is the self-inductance of a unit length and i is the net current in the cylinders (see problem 29.C.2).

30.B.6 Energy from the sun is incident on the surface of the earth at the rate of 1400 W/m^2.
  (a) Given that this energy travels at the rate of 3×10^8 m/s, what is the energy density in the electromagnetic field at the surface of the earth?
  (b) From the result of part (a) above, what is the strength of the average electric and magnetic fields? (The average electric field is the square root of the time average of E^2.)

30.B.7 Determine the average value of the electric and magnetic fields 1 m away from a 100-W light bulb. (The average electric field is the square root of the time average of E^2.)

30.C.1 Show that the energy in the electric field between two concentric conducting spheres with equal and opposite charges is $\frac{1}{2}q^2/C$, where C is the capacitance.

30.C.2

(a) Determine the energy in the magnetic field between two parallel sheets of width a, length l, and separation x, each carrying a uniform current i (see the figure).

(b) Determine the self-inductance by equating the field energy to $\frac{1}{2}Li^2$.

(c) Determine the force between the plates by equating the work $F\,dx$ required to increase the separation by dx to the increase in field energy.

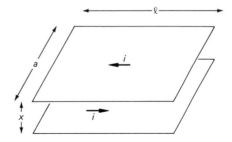

31
WAVE
MOTION

CERTAINLY EVERYONE HAS an idea of what the phrase *wave motion* means. We have seen water waves ranging from ocean breakers to ripples on a pond. It is common knowledge that light and sound are also propagated by means of wave motion. While our intuitive picture of wave motion, based on countless observations of the phenomenon, will be of great value in understanding the material of the following sections, it is much too imprecise and subjective to suffice as a foundation on which to build an exact physical theory of wave propagation. Our first task will be to devise a means of describing wave motion in a precise and unambiguous way. We shall begin by considering wave motion in a very simple system, in particular in a long uniform stretched string in which we can create waves simply by shaking one end. Fortunately, virtually all types of waves satisfy the same basic system of laws, so once we have finished analyzing the relatively simple but somewhat uninteresting case of waves traveling on a stretched string, we shall be able to apply all our results to the more important cases of sound waves, light waves, and even the matter waves of quantum theory, which are so important to any understanding of atomic structure.

31.1 STRING WAVES

First let us state that the system we shall be describing is actually an idealized model of a real string. We neglect the effects of air friction, the sag of the string resulting from gravitational force, and other factors that have only a small effect on the motion of a real string but would immensely complicate the theory. First we shall describe the observed phenomena in words, and then we shall present a method of representing wave motion in an exact mathematical fashion.

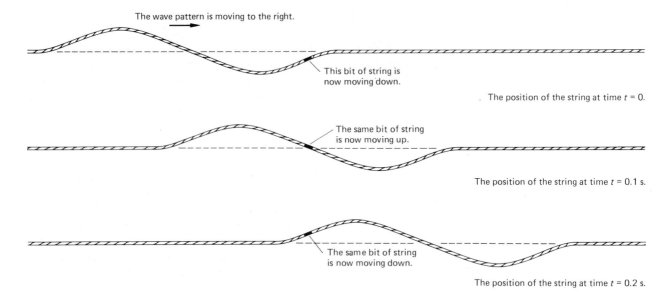

The wave pattern is moving to the right.

This bit of string is now moving down.

The position of the string at time $t = 0$.

The same bit of string is now moving up.

The position of the string at time $t = 0.1$ s.

The same bit of string is now moving down.

The position of the string at time $t = 0.2$ s.

FIGURE 31.1   Motion of a string with a traveling wave on it.

If one end of an initially stationary stretched string is moved quickly up and down, the motion will give rise to a *traveling wave.* A traveling wave is a pattern that moves along the string at constant speed with no change in shape. The wave may be moving in either direction. Figure 31.1 presents three consecutive "snapshots" of a string in which a traveling wave is moving to the right. The illustrations show the shape of the string at three different instants of time. Although the wave will move the full length of the string, it is obvious that any individual particle of the string moves very little. In Fig. 31.1 a point in the center of the string would first be at rest, then move downward a bit, then move upward a bit, and then return to its rest position. The actual motion of any particle of the string is in fact perpendicular to the direction of propagation of

the wave. For this reason string waves are classified as *transverse waves.* In a sound wave the actual motion of the particles is parallel to the direction of propagation of the wave. Waves of that type are classified as *longitudinal waves.*

THE ADDITION PRINCIPLE[1]

When both ends of a string are jiggled simultaneously, traveling waves from the two ends converge toward the center. We may calculate this motion of the string by using the following principle:

In the presence of simultaneous right-going and left-going waves, the displacement from the rest position of

---

[1] The principle presented here is more commonly called the *superposition principle,* but the authors feel that the name used here is more descriptive of the content of the principle.

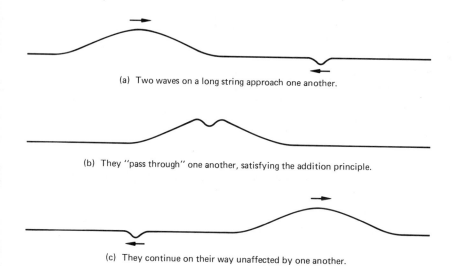

(a) Two waves on a long string approach one another.

(b) They "pass through" one another, satisfying the addition principle.

(c) They continue on their way unaffected by one another.

FIGURE 31.2   An illustration of the addition principle. As the two waves "pass through" one another, the vertical displacement of any point on the string is the algebraic sum of the vertical displacements of the point that would have been caused by each wave alone.

any point on the string at any instant of time is equal to the sum of the displacements that each wave would cause alone.

To illustrate the content of the addition principle, we show in Fig. 31.2 the position of a string at three instants of time.

31.2 WAVE FUNCTIONS

We have been describing the motion of a string by a combination of words and pictures. We shall now present a mathematical method of describing the motion of a string. We introduce x and y coordinate axes that are positioned so that the rest position of the string is along the x axis.

Consider that point on the string whose x coordinate is the number x, and call that point P_x. When a wave passes by the point, the y coordinate of P_x will fluctuate for awhile and then return to zero after the wave has passed. We introduce a function $u(x, t)$ that depends on the values of x and t and is defined as follows:

The function $u(x, t)$ is the vertical displacement of the point P_x from its equilibrium position at time t.

With the coordinate axes oriented as described above, $u(x, t)$ is simply the y coordinate of P_x at time t. For instance, $u(3.5, 1.2)$ is equal to the y coordinate that the point $P_{3.5}$

had at the time 1.2 s. The point $P_{3.5}$ is the point on the string whose x coordinate is 3.5. We call $u(x, t)$ the *wave function* of the string. Later we shall consider how to calculate the wave function for certain important types of wave motion. Right now we shall see, by means of an example, how we can use the wave function to describe the motion of the string if we know exactly what the wave function is. Suppose we know that the wave function for a string wave is

$$u(x, t) = \frac{1}{1 + (x - t)^2} \qquad (31.1)$$

where u is measured in centimeters, x in meters, and t in seconds. What does this mean in terms of how the string moves? If we set t equal to zero in Eq. 31.1, we get

$$u(x, 0) = \frac{1}{1 + x^2}$$

Remembering that $u(x, 0)$ is the y coordinate of the point P_x at time zero, we see that the above formula tells us where all the points on the string are at time zero. If we graph $u(x, 0)$ we will get a picture of the string at time zero (see Fig. 31.3). If we set t equal to 1 in Eq. 31.1, we

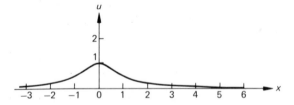

FIGURE 31.3   A graph of the function $u(x, 0) = 1/(1 + x^2)$.

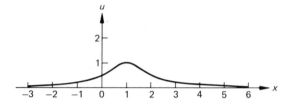

FIGURE 31.4   A graph of the function $u(x, 1) = 1/[1 + (x - 1)^2]$.

get a formula for the position of all points on the string at time $t = 1$ s (see Fig. 31.4):

$$u(x, 1) = \frac{1}{1 + (x - 1)^2}$$

Doing the same with $t = 2$ s, we obtain

$$u(x, 2) = \frac{1}{1 + (x - 2)^2}$$

The graph of this function is shown in Fig. 31.5. It is now easy to see what type of motion is described by the wave function we are considering. A smooth wave is moving, without changing its shape, down the string to the right. Since the x coordinate is being measured in meters, the speed of the wave is 1 m/s. (Of course our pictures have been greatly reduced in size so that 1 m along the x axis is represented by 0.7 cm on our graphs.)

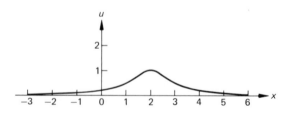

FIGURE 31.5   A graph of the function $u(x, 2) = 1/[1 + (x - 2)^2]$.

QUESTION

A traveling wave on a string is described by the wave function

$$u(x, t) = \frac{1}{1 + (x - 2t)^2}$$

How does this wave differ from the one we have just considered?

ANSWER

At times $t = 0$, $t = 1$ s, and $t = 2$ s, the wave function is

$$u(x, 0) = \frac{1}{1 + x^2}$$

$$u(x, 1) = \frac{1}{1 + (x - 2)^2}$$

$$u(x, 2) = \frac{1}{1 + (x - 4)^2}$$

Graphs of these three functions are shown in Fig. 31.6. It is clear that this wave has the same shape as the one discussed earlier, but this wave moves to the right with a speed of 2 m/s.

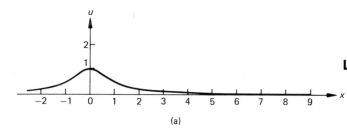

(a)

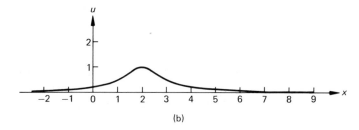

(b)

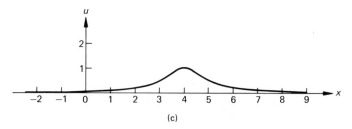

(c)

FIGURE 31.6   (a) A graph of $u(x, 0)$. (b) A graph of $u(x, 1)$. (c) A graph of $u(x, 2)$.

QUESTION

In a later section we show that the speed with which a string wave propagates depends on the tension in the string and the mass density of the string, not on the shape of the wave. On a given string all wave shapes propagate with the same speed. Suppose a right-going wave is propagating down a string whose wave speed is v meters per second. If the position of the string at time $t = 0$ is given by the graph of some function $f(x)$, what is the wave function describing the wave?

ANSWER

Let us choose some particular value of x and some particular instant t. We now ask: What is the y coordinate of the point P_x at time t? Certainly we know that it is equal to $u(x, t)$ because of the definition of the wave function. By considering Fig. 31.7 we can also see that the y coordinate of P_x at time t is the same as the y coordinate of P_{x_0} at time zero, where $x_0 = x - vt$. But, according to the statement of the original question, the y coordinate at time zero of P_{x_0} was $f(x_0)$. Therefore

$$u(x, t) = f(x - vt)$$

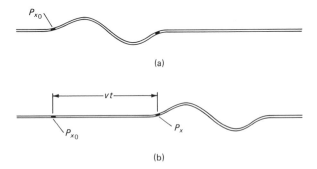

(a)

(b)

FIGURE 31.7   (a) Position of the string at time zero. (b) Position of the string at time t. If $x_0 = x - vt$, then the y coordinate of P_x at time t is the same as the y coordinate of P_{x_0} at time zero.

This gives the wave function of any right-going traveling wave in terms of the shape of the string at time zero. Note that this general formula is valid for the two previous cases we have considered. In both cases the function $f(x)$ was equal to $1/(1 + x^2)$. In the first case $v = 1$ m/s and in the second case $v = 2$ m/s.

In the preceding question we found that a general right-going wave on a string with wave velocity v is described by a wave function of the form

$$u(x, t) = f(x - vt) \quad \text{(right-going wave)} \tag{31.2}$$

A left-going wave on a string with wave velocity v has a wave function that differs from that of a right-going wave only by the fact that v is replaced by $-v$ (since the wave is traveling in the negative direction). Thus if we change the sign of v in the above formula, we obtain the wave function for a left-going wave:

$$u(x, t) = f(x + vt) \quad \text{(left-going wave)} \tag{31.3}$$

where $f(x)$ describes the shape of the string at time $t = 0$. We can now use the addition principle to obtain the wave function for a string of wave velocity v in the general case of simultaneous right-going and left-going waves. The wave function is

$$u(x, t) = f(x - vt) + g(x + vt)$$

where $f(x - vt)$ is the wave function of the right-going wave alone and $g(x + vt)$ is the wave function of the left-going wave alone. We can now restate the addition principle in a more general mathematical form. In this form it is also applicable to other types of waves (e.g., sound waves and light waves).

The Addition Principle   *If two or more waves are traveling in a system simultaneously, the resultant wave function is the sum of the wave functions of the individual waves.*

Nerve impulses are electrochemical waves that propagate along very long tubular extensions (axons) of nerve cells. The physical interpretation of the wave function $u(x, t)$ that we shall use to describe nerve impulse propagation is that $u(x, t)$ is the difference in electric potential between the inside and outside of the nerve axon at location x and time t. Suppose we measure, by means of two electrodes (one at point A within the cell and the other at point B just outside the cell membrane), the electric potential difference V_{AB} as a function of time as a left-going nerve impulse passes by (see Fig. 31.8). Let us assume that the values obtained are those plotted in Fig. 31.9. Suppose we have determined by other measurements that the propagation speed of nerve impulses in this particular axon is 10 m/s. What is the potential difference across the axon membrane as a function of position along the axon (i.e., as a function of x) at time $t = 0$?

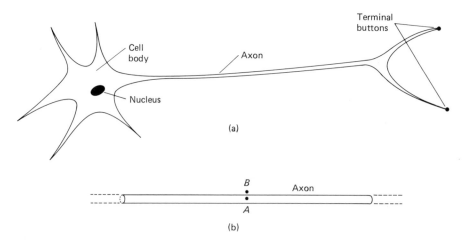

FIGURE 31.8   (a) A typical shape of a nerve cell involved in muscle control. (b) A section of the axon in which the electric potential difference across the axon membrane is being measured.

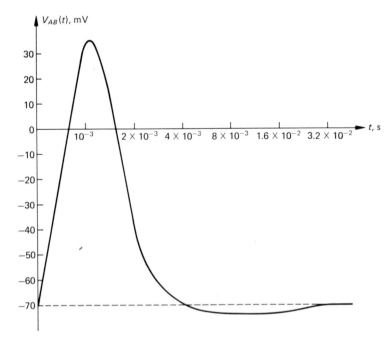

FIGURE 31.9   A graph of the potential difference across a location on the axon membrane as a nerve impulse passes. The potential difference is plotted in millivolts as a function of time. Note the nonlinear time scale.

TABLE 31.1   Values of $f(x)$ constructed by using Eq. 31.5.

t (s)	$V_{AB}(t)$ (mV)	$f(x)$ (mV)
0	-70	$f(0) = -70$
0.5×10^{-3}	-20	$f(0.005) = -20$
1×10^{-3}	35	$f(0.01) = 35$
1.5×10^{-3}	15	$f(0.015) = 15$
2×10^{-3}	-40	$f(0.02) = -40$
3×10^{-3}	-60	$f(0.03) = -60$
4×10^{-3}	-70	$f(0.04) = -70$
8×10^{-3}	-73	$f(0.08) = -73$
1.6×10^{-2}	-73	$f(0.16) = -73$
3.2×10^{-2}	-70	$f(0.32) = -70$

ANSWER

Let the x coordinates of the points A and B both be zero. Then Fig. 31.9 is a graph of the function $u(0, t) = V_{AB}(t)$. But we know from Eq. 31.3 that

$$u(x, t) = f(x + 10t) \qquad (31.4)$$

where $f(x)$ is the wave function at time zero, which is exactly what we have been asked to find. Setting x equal to zero in Eq. 31.4, we see that

$$f(10t) = u(0, t) = V_{AB}(t) \qquad (31.5)$$

By reading the values of $V_{AB}(t)$ from Fig. 31.9 we can construct a table of values of $f(x)$. (See Table 31.1) For instance, taking $t = 10^{-3}$ in Eq. 31.5, we see that $f(10^{-2}) = V_{AB}(10^{-3}) = 35$ mV.

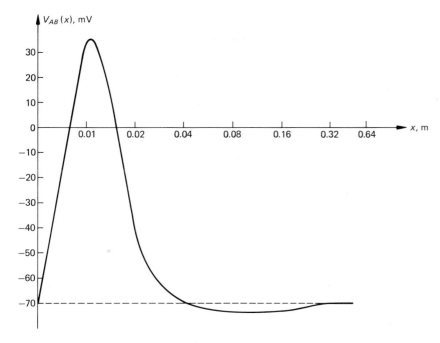

FIGURE 31.10 A graph of the potential difference across the membrane of the axon (as a function of location along the axon) at time $t = 0$.

From Fig. 31.10 we can see that for this somewhat typical nerve impulse the length of the wave that travels down the axon is about 20 cm. Thus it is only in fairly long nerve fibers that the impulse leaves its point of origin and travels freely down the axon for a time before reaching its destination.

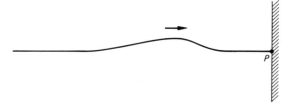

FIGURE 31.11   A wave propagating toward a fixed end of the string.

31.3 REFLECTION OF WAVES

Figure 31.11 shows a string in which a wave is propagating toward an end that is fixed at point P (i.e., at point P the string is attached to an immovable object). We want to

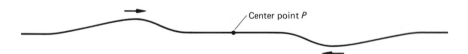

FIGURE 31.12   The imaginary string in which mirror-image waves propagate in both directions.

determine what happens when this wave reaches the fixed end of the string. One way of doing this is to imagine a string, of twice the length, in which our original wave plus an inverted mirror image wave are both propagating toward the center of the string (see Fig. 31.12).

A little thought will convince you that if we use the addition principle to calculate the motion of this imaginary double string, we shall find that the center point P never moves as the two waves pass one another. But the stationary condition of the point P was precisely the constraint imposed upon the motion of the actual string. In fact, the real string moves in exactly the same way as the left-hand side of the imaginary string. Upon reflection at a fixed boundary, a wave is changed into its negative mirror image (see Fig. 31.13).

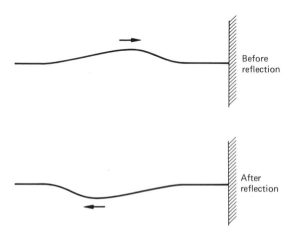

FIGURE 31.13   The actual string before and after reflection of the wave from the fixed end.

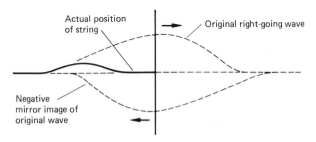

FIGURE 31.14   Position of the string while the wave is in the process of reflection.

As shown in Fig. 31.14, we can calculate the position of the string while the wave is being reflected by applying the addition principle to the imaginary double string and taking only the left-hand side.

QUESTION

At $t = 0$ a wave of the form show in Fig. 31.15 is propagating to the right at a speed of 2 m/s. What are the configurations of the string at $t = 1$ s and at $t = 2$ s?

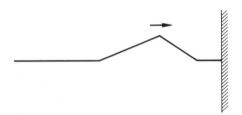

FIGURE 31.15   The wave at time $t = 0$.

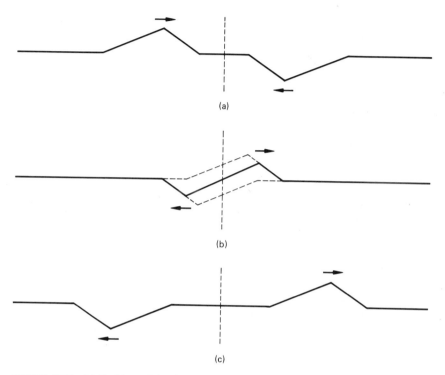

FIGURE 31.16 (a) Position of the double string at time $t = 0$. (b) Position of the double string at time $t = 1$ s. This may be calculated by use of the addition principle. (c) Position of the double string at time $t = 2$ s.

ANSWER

The configurations of the imaginary double string at times $t = 0$, $t = 1$ s, and $t = 2$ s are shown in Fig. 31.16. Taking the left half of the imaginary double string, we obtain the position of the actual string at the times $t = 1$ s and $t = 2$ s (see Fig. 31.17).

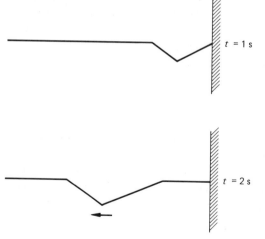

FIGURE 31.17 The positions of the actual string at times $t = 1$ s and $t = 2$ s are the same as the left-hand side of the double string in Fig. 31.16(b) and (c).

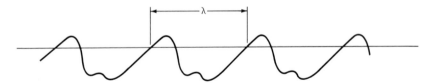

FIGURE 31.18 An example of a periodic wave form, with one wavelength indicated.

31.4 PERIODIC WAVES

The commonest and most important examples of waves are not short wave pulses but periodic waves in which the wave pattern repeats itself many times without change. An example of a periodic wave is shown in Fig. 31.18. For a periodic wave the *wavelength* λ is defined as the distance between equivalent points on the repeating pattern. The *period T* is defined as the time required for one wavelength to pass a given point on the string. If the wave is traveling at velocity v, then

$$\lambda = vT \tag{31.6}$$

The *frequency f* associated with a periodic wave is the number of wavelengths that pass a given point per second. Thus if the period is 0.1 s, the frequency is 10 per second or 10 hertz (Hz). [The unit 1 Hz is defined as $(1 \text{ s})^{-1}$.] In general the frequency is equal to the reciprocal of the period:

$$f = \frac{1}{T}$$

QUESTION

Equation 31.6 is valid not only for string waves but for any type of periodic wave. The speed of sound waves in air is about 330 m/s. The human ear is sensitive to frequencies in the range of about 20–20,000 Hz. What range of wavelengths does this correspond to?

ANSWER

The frequency of a periodic wave is equal to the reciprocal of its period. Thus the longest and shortest periods, are, respectively,

$$T_1 = \frac{1}{20 \text{ Hz}} = 0.05 \text{ s}$$

and

$$T_2 = \frac{1}{20{,}000 \text{ Hz}} = 0.5 \times 10^{-4} \text{ s}$$

The longest and shortest wavelengths are therefore

$$\lambda_1 = (330 \text{ m/s})(0.05 \text{ s}) = 16.5 \text{ m}$$

$$\lambda_2 = (330 \text{ m/s})(0.5 \times 10^{-4} \text{ s}) = 1.65 \text{ cm}$$

31.5 HARMONIC WAVES

The y coordinate of any point on a string in which a periodic wave is moving oscillates periodically with period T. If that vertical motion is simple harmonic motion, then the wave is called a *harmonic wave*. The wave function of a right-going harmonic wave is

$$u(x,\ t) = A \cos 2\pi \left(\frac{x}{\lambda} - \frac{t}{T} \right) \tag{31.7}$$

To confirm that this wave function actually describes a right-going harmonic wave, we must first show that it satisfies Eq. 31.2, which gives the general form for all right-going waves, and then show that a point on the string really does execute simple harmonic motion.

The first part is easy. Using the fact that $1/T = v/\lambda$, we can write the wave function as

$$u(x, t) = A \cos 2\pi\left(\frac{x - vt}{\lambda}\right)$$

which shows that the wave function is a function of $x - vt$ and therefore some type of right-going wave. To show that any point on the string moves up and down with simple harmonic motion, let us set x equal to some fixed position x_0 in Eq. 31.7. We obtain a formula for the y coordinate at time t of that point on the string whose x coordinate is x_0. The formula is

$$u(x_0, t) = A \cos\left(\frac{2\pi t}{T} + \text{const}\right)$$

But this is exactly the formula we derived in Chap. 9 for simple harmonic motion (see Eq. 9.19). Note that the amplitude of the simple harmonic vibration of any point on the string is equal to A. We can obtain the position of the string at $t = 0$ by plotting $u(x, 0)$ in terms of x using Eq. 31.7, with t set equal to zero. This gives

$$u(x, 0) = A \cos\left(\frac{2\pi x}{\lambda}\right)$$

It is clear from the graph shown in Fig. 31.19 that λ is the wavelength of the harmonic wave. The wave function of a left-going wave of wavelength λ and amplitude A is

$$u(x, t) = A \cos 2\pi\left(\frac{x}{\lambda} + \frac{t}{T}\right)$$

In Chap. 9 we introduced a quantity ω, called the *angular frequency* of a simple harmonic motion. This quantity is proportional to the frequency

$$\omega = 2\pi f = \frac{2\pi}{T}$$

We now introduce another quantity k, called the *wave number* associated with a harmonic wave. The wave number k is related to the wavelength λ in the same way that ω is related to the period T, that is,

$$k = \frac{2\pi}{\lambda}$$

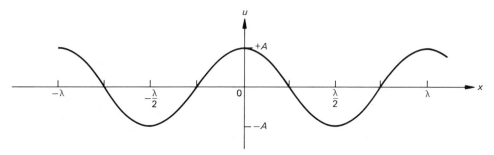

FIGURE 31.19   The graph of $u(x, 0) = A \cos(2\pi x/\lambda)$.

The unit of k in the mks system is the meter^{-1}. In terms of ω and k, the wave function of a right-going harmonic wave takes the simple form

$$u(x, t) = A \cos(kx - \omega t) \tag{31.8}$$

QUESTION

A traveling wave has the wave function (in mks units)

$$u(x, t) = 0.02 \cos(10x - 85t)$$

What are the amplitude, wavelength, frequency, and wave speed?

ANSWER

Comparing the wave function with the standard form given in Eq. 31.8, we see that the amplitude A is 0.02 m. Also

$$k = \frac{2\pi}{\lambda} = 10 \text{ m}^{-1}$$

which gives the wavelength

$$\lambda = \frac{2\pi}{10} = 0.63 \text{ m}$$

The frequency is given by

$$\omega = 2\pi f = 85 \text{ Hz}$$

or

$$f = \frac{85}{2\pi} = 13.53 \text{ Hz}$$

We can now determine the wave speed from the relation $\lambda = vT$, and we have

$$v = \lambda f = (0.63 \text{ m})(13.53 \text{ s}^{-1}) = 8.5 \text{ m/s}$$

31.6 STANDING WAVES

Consider an extremely long string that is fixed at $x = 0$ and extends in the negative x direction. By oscillating the free end of the string we can generate a continuous right-going harmonic wave in the string. The wave function of the right-going harmonic wave is $A \cos(kx - \omega t)$. When the right-going wave reaches the point at which the string is fixed, a negative mirror image reflected wave is generated. The wave function of the negative mirror image wave is $-A \cos(kx + \omega t)$, and this wave is a left-going wave. We obtain the actual wave function of the string by adding the wave functions of the incoming wave and the reflected wave. Thus

$$u(x, t) = A \cos(kx - \omega t) - A \cos(kx + \omega t)$$

[Note that $u(0, t) = 0$ for all values of t, as it should since the string is fixed at $x = 0$.] We can simplify this expression by using the trigonometric identity

$$\cos(a - b) - \cos(a + b) = 2 \sin a \sin b$$

Letting $a = kx$ and $b = \omega t$, we obtain

$$u(x, t) = 2 A \sin kx \sin \omega t$$

or, equivalently,

$$u(x, t) = B \sin\left(\frac{2\pi x}{\lambda}\right) \sin \omega t \tag{31.9}$$

where $B = 2A$ is the *standing wave amplitude*.

To help you visualize the motion that this wave function describes, we have drawn a series of pictures of the string at different instants of time (Fig. 31.20). From Eq. 31.9 we can see that the point on the string that has coordinate x undergoes simple harmonic motion (in the y direction) of angular frequency ω and of amplitude $B \sin(2\pi x/\lambda)$. Thus, because the value of $\sin \pi$ is zero, the point on the string whose x coordinate is $-\lambda/2$ does not vibrate at all.

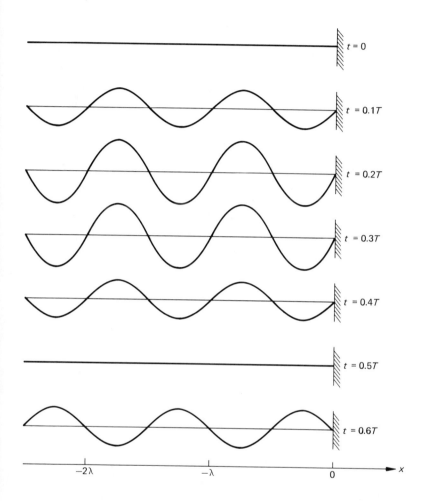

FIGURE 31.20 Pictures of the string at various times. The points on the string at $x = 0$, $-\frac{1}{2}\lambda$, $-\lambda$, $-\frac{3}{2}\lambda$, etc., remain stationary during the motion.

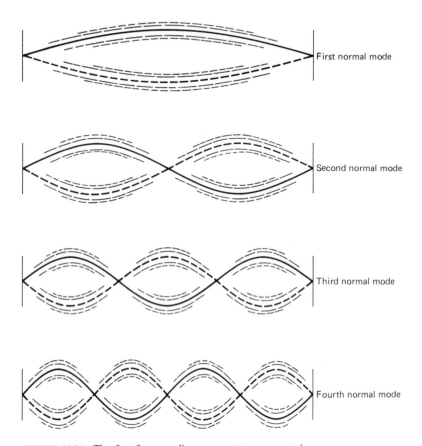

First normal mode

Second normal mode

Third normal mode

Fourth normal mode

FIGURE 31.21   The first four standing wave patterns on a string with fixed ends. The string oscillates between the extremes shown by the dashed and solid lines.

We may say the same for the point whose x coordinate is $-n\lambda/2$, where n is any integer. We refer to those stationary points as the *nodes* of the vibrational motion. The distance between adjacent nodes is $\lambda/2$. Since the nodes are stationary, we could fasten the string at any two of the nodes and remove the outside sections. We would then be left with a finite segment of string that was vibrating

harmonically. Such a motion of the string (and in general such a harmonic motion of any system) is called a *normal mode vibration* or a *standing wave*. A given string of length L that is fixed at the ends has a sequence of different normal modes of vibration. The sequence is illustrated in Fig. 31.2. For the first normal mode

$$\lambda = 2L$$

For the second normal mode

$$\lambda = L$$

For the third normal mode

$$\lambda = \frac{2L}{3}$$

In general, for the nth normal mode

$$\lambda = \frac{2L}{n}$$

Using Eq. 31.6, which relates the wavelength to the period of vibration, we find that the period of the nth normal mode is

$$T_n = \frac{2L}{nv}$$

The frequency of the nth normal mode is the reciprocal of the period:

$$f_n = \frac{nv}{2L}$$

31.7 WAVE SPEED OF STRING WAVES

At the end of this chapter we shall prove that traveling waves on a string move with a speed

$$v = \sqrt{\frac{\tau}{\mu}} \tag{31.10}$$

where τ is the tension in the string and μ is the mass per unit length of the string. This formula is reasonable in that it says that if one string is more massive (has a larger value of μ) than another, but both strings have the same tension, then the more massive string will react more slowly to the tensile forces in it, and its motion will therefore be a "slow-motion" version of the motion of the less massive string. On the other hand, if two strings have the same linear mass density (the same value of μ) but unequal tensions, then the string with greater tension will propagate waves faster.

QUESTION

What are the frequencies of the first two normal modes of a violin string of length 40 cm and mass 0.5 g if it is stretched with a tension of 80 N?

ANSWER

The mass density of the string is

$$\mu = \frac{5 \times 10^{-4} \text{ kg}}{0.4 \text{ m}} = 1.25 \times 10^{-3} \text{ kg/m}$$

The speed of waves on the string is

$$v = \sqrt{\frac{\tau}{\mu}} = \left(\frac{80 \text{ N}}{1.25 \times 10^{-3} \text{ kg/m}} \right)^{1/2} = 253 \text{ m/s}$$

The frequencies of the first two normal moles are

$$f_1 = \frac{v}{2L} = \frac{253 \text{ m/s}}{(2)(0.4 \text{ m})} = 316 \text{ Hz}$$

and

$$f_2 = \frac{2v}{2L} = 632 \text{ Hz}$$

FIGURE 31.22   The two sections S_L and S_R of the string in relation to the point P.

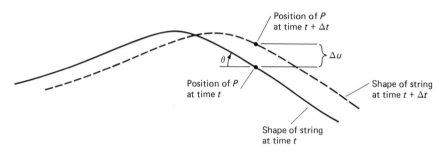

FIGURE 31.23   Motion of the wave and the point P during the time interval Δt.

31.8 WAVE ENERGETICS

As a wave propagates along a string, it carries energy with it. Some of this energy is the kinetic energy of motion of the string, and some is potential energy caused by the stretching of the string against tension.

We shall now calculate the rate at which this energy passes by a certain point P on the string, where we imagine P to be a little dot of ink we have put on the string. At the instant we look at the string, a wave is in the process of passing P from left to right. The point P itself just moves up and down (see Fig. 31.22).

That section of string to the right of P we call S_R. At this instant S_R has some energy E_R. At a later time Δt, the energy of S_R has changed to $E_R + \Delta E_R$ because more of the wave has propagated into that section of the string. That section of string to the left of P we call S_L. The section S_L exerts a force on S_R at only one point, namely the point P. Since this is a purely mechanical system, the only way

S_L can transmit energy to S_R is by doing mechanical work on S_R. The amount of work that S_L does on S_R in the time interval Δt is given by the distance P travels during the time interval Δt multiplied by the vertical component of the force that S_L exerts on S_R at point P. We take the vertical component because the point P moves only in the vertical direction, and the definition of work requires that the displacement be multiplied by the component of the force in the direction of the motion. Let us look at this section of the string in greater detail. The solid line in Fig. 31.23 shows the position of the string at the start of the time interval Δt. The dashed line shows the string at the end of the time interval Δt. The vertical displacement

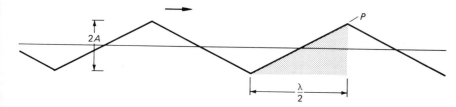

FIGURE 31.24   A traveling triangular wave.

of the point P during the time interval Δt is Δu. The magnitude of the force exerted by S_L at P is just τ, the constant tension in the string. The direction of the force is along the string, upward and to the left. The vertical component of that force is $\tau \sin \theta$. Thus the work done by S_L on S_R, which is equal to the change in energy of S_R, is given by

$$\Delta E_R = \tau \sin \theta \, \Delta u$$

We obtain the rate at which energy passes from left to right through the point P by dividing ΔE_R by the time interval Δt and taking the limit $\Delta t \to 0$:

$$\text{Rate of energy passage} = \frac{dE_R}{dt} = \tau \sin \theta \frac{\partial u}{\partial t}$$

Since we are interested only in small amplitude waves, we shall assume that $\sin \theta \simeq \tan \theta = -\partial u / \partial x$. This relation allows us to write the above formula in a form that will be more convenient for later applications:

$$\frac{dE_R}{dt} = -\tau \frac{\partial u}{\partial x} \frac{\partial u}{\partial t} \qquad \textbf{(31.11)}$$

As a first application of Eq. 31.11, we calculate the rate of energy transfer associated with the traveling wave pattern shown in Fig. 31.24. The wave is a periodic wave of wavelength λ. By considering the shaded triangle shown in the figure, we can see that

$$\frac{\partial u}{\partial x} = \frac{2A}{\lambda/2} = \frac{4A}{\lambda}$$

Also, during the half period associated with the shaded triangle, the point P undergoes a vertical displacement of $-2A$. Thus the velocity of the point is

$$\frac{\partial u}{\partial t} = \frac{-2A}{T/2} = -\frac{4A}{T}$$

Therefore the rate of energy transfer associated with this wave is

$$\frac{dE}{dt} = -\tau \frac{\partial u}{\partial x} \frac{\partial u}{\partial t} = 16\tau \frac{A^2}{\lambda T}$$

Using the facts that $\lambda = vT$ and $v = \sqrt{\tau/\mu}$, we can write the above equation in the form

$$\frac{dE}{dt} = 16 \sqrt{\tau\mu} \frac{A^2}{T^2}$$

The quantity

$$Z = \sqrt{\tau\mu}$$

is called the *wave impedance*. We may then write

$$\frac{dE}{dt} = 16Z \frac{A^2}{T^2}$$

Although the triangular wave shape considered above is easiest to analyze, it is obviously highly artificial. A much more important quantity is the rate at which energy is transmitted by a harmonic traveling wave of amplitude A and period T. A calculation similar in its principles to the one we have just made (see problem 31.C.7) shows that the rate of energy transfer associated with a harmonic wave is given by

$$\frac{dE}{dt} = 2\pi^2 Z \frac{A^2}{T^2}$$

$$= \tfrac{1}{2} Z \omega^2 A^2 \qquad \text{(harmonic wave)} \qquad (31.12)$$

QUESTION

Would it be possible for a person of normal strength to keep a reasonable size light bulb operating by transmitting energy to an electric generator by means of waves on a clothesline?

ANSWER

The mass of 10 m ($\simeq$30 ft) of typical clothesline is about $\tfrac{1}{2}$ kg ($\simeq$1 lb). Thus $\mu = (0.5 \text{ kg})/(10 \text{ m}) = 0.05$ kg/m. A person of reasonable strength might be expected to exert a tension of about 100 N ($\simeq$20 lb) while vibrating the rope with an ampli-

tude of 0.1 m and a period of 0.2 s. Thus a reasonable value for the power transmitted is

$$\frac{dE}{dt} = 2\pi^2 \sqrt{\tau\mu}\, \frac{A^2}{T^2} = 2\pi^2 \sqrt{(100 \text{ N})(0.05 \text{ kg/m})}\, \frac{(0.1 \text{ m})^2}{(0.2 \text{ s})^2}$$

$$= 11 \text{ W}$$

An 11-W light bulb is fairly small, so our estimate shows that it might be possible, but it would not be easy.

31.9 PARTIAL REFLECTION OF WAVES

When a sound wave traveling in water strikes the surface of the water, it is partially transmitted into the air and partially reflected back into the water. The same thing happens when light waves strike the air-glass surfaces of lens systems. In general, when any wave encounters an abrupt change in the medium in which it is traveling, it is partially reflected and partially transmitted past the discontinuity in the medium. Since so far we have discussed only string waves, we shall analyze the partial reflection of waves at the junction between two strings of different

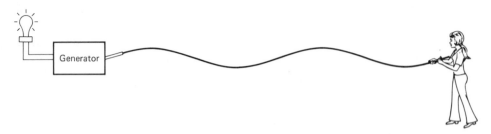

FIGURE 31.25   The energy carried by the waves in the clothesline is being used to operate an electric generator that lights the bulb.

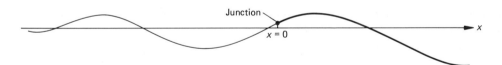

FIGURE 31.26 Two strings of different mass densities connected at a junction.

characteristics. We are actually interested in the much more important problems of sound and light wave transmission from one medium to another. For example, an understanding of some of the mathematical theory of sound wave reflection and transmission is essential to an appreciation of the necessity for the complex mechanical structure of the human ear. Air-breathing animals could only develop an effective hearing mechanism by evolving a solution to the following difficult problem: A nerve system that is sensitive enough to detect the exceedingly minute vibrations of sound waves of ordinary intensity would surely be much too delicate to be exposed directly to the air. The actual auditory nerve system is immersed in a protective fluid within the inner ear. However, when sound waves traveling in air strike an interface between air and water, only about one-thousandth of the acoustic energy is transmitted into the water; the rest is reflected from the interface back into the air. Thus the ear must include some mechanism for allowing reasonably efficient transmission of acoustic energy from air to liquid, and this mechanism is a system of bony levers within the middle ear. Before discussing the operation of the ear, we shall derive the necessary mathematical theory of wave transmission within the simpler context of string waves.

We consider two very long strings that are connected, as shown in Fig. 31.26, at the position $x = 0$. The strings are of different densities, and thus their wave velocities and wave impedances are different. A steady right-going harmonic wave of amplitude A_i and period T approaches the junction from the left. We call this the incoming wave (hence the subscript "i"). At the junction some of the energy being carried by the incoming wave will be transmitted to the second string and some of it will be reflected back toward the left in the form of a reflected wave. We want to determine exactly what fraction is transmitted and what fraction is reflected. Since not all the energy is reflected, the amplitude of the reflected wave will be smaller than A_i, and we call this amplitude A_r. The reflected wave is a left-going wave. Thus the wave function for the left-hand string is

$$u(x, t) = A_i \cos(kx - \omega t) + A_r \cos(kx + \omega t)$$

We obtain the vertical displacement of the junction at time t by setting x equal to zero in the above equation. We then have

$$u(0, t) = (A_i + A_r)\cos \omega t \qquad (31.13)$$

The junction oscillates up and down with period T, so the right-hand string will have a right-going harmonic wave on it, which is also of period T. This wave will not be of wavelength λ because the wave velocity of the right-hand string is different from that of the left-hand string. Its wavelength will be $\lambda' = v'T$, where $v' = \sqrt{\tau/\mu'}$ and μ' is the density of the right-hand string. We shall call the amplitude of the transmitted wave A_t. The wave function of the right-hand string is

$$u(x, t) = A_t \cos(k'x - \omega t)$$

with $k' = 2\pi/\lambda'$. We can again obtain an expression for the vertical coordinate of the junction by setting x equal to zero. Then

$$u(0, t) = A_t \cos \omega t \qquad\qquad \textbf{(31.14)}$$

In order for Eq. 31.14 to agree with Eq. 31.13, we must have

$$A_t = A_i + A_r \qquad\qquad \textbf{(31.15)}$$

If the amplitude A_i of the incoming wave is known, Eq. 31.15 gives us one equation for the two unknown amplitudes A_r and A_t, and to determine them we need one more relation. To obtain that relation we use the principle of conservation of energy. The rate at which energy is being delivered to the junction by the incoming wave is equal to the rate at which energy is being carried away by the transmitted wave and the reflected wave. To use the principle of conservation of energy, we recall that the rate at which energy is transmitted by a harmonic wave of period T and amplitude A is (see Eq. 31.12)

$$\frac{dE}{dt} = \tfrac{1}{2}Z\omega^2 A^2$$

The rate at which energy is being delivered to the junction by the incoming wave is given by the above equation with A equal to A_i. The rate at which energy is being carried away from the junction by the reflected wave is given by the same equation with A_i replaced by A_r:

$$\left(\frac{dE}{dt}\right)_r = \tfrac{1}{2}Z\omega^2 A_r^2$$

The rate at which energy is being carried away from the junction by the transmitted wave is

$$\left(\frac{dE}{dt}\right)_t = \tfrac{1}{2}Z'\omega^2 A_t^2$$

Setting

$$\left(\frac{dE}{dt}\right)_i = \left(\frac{dE}{dt}\right)_r + \left(\frac{dE}{dt}\right)_t$$

we see that

$$ZA_i^2 = ZA_r^2 + Z'A_t^2 \qquad\qquad \textbf{(31.16)}$$

We define a number R, called the *reflection coefficient*, as the ratio of the reflected wave amplitude to the incoming wave amplitude. Then

$$A_r = RA_i$$

Equation 31.15 then states that

$$A_t = (1 + R)A_i$$

Substituting these expressions for A_r and A_t into Eq. 31.16, we obtain a quadratic equation for the reflection coefficient:

$$Z = ZR^2 + Z'(1 + R)^2$$

This equation has two solutions. One solution is $R = -1$, which corresponds to complete reflection and would be appropriate if the point $x = 0$ were held fixed. The other solution, which is relevant to our problem, is

$$R = \frac{Z - Z'}{Z + Z'} \qquad\qquad \textbf{(31.17)}$$

We see that if Z' is equal to Z, then R is zero and there is no reflection. If the right-hand string is very light so that Z'

is very small, then R is very close to one. In that case the reflected wave will be a positive mirror image of the incoming wave. In the other extreme, if Z' is very much larger than Z, then R is very close to minus one. This case corresponds to having a fixed end on the left-hand string. (The right-hand string is too heavy to move.) As expected, we obtain a negative mirror image. For any other value of Z' the reflection coefficient lies somewhere between those two limits. The fraction of the incident energy that is reflected is

$$\frac{(dE/dt)_r}{(dE/dt)_i} = R^2 \qquad\qquad (31.18)$$

Thus we see that the incident energy is completely reflected in the two limiting cases, $R = +1$ and $R = -1$.

QUESTION

In the system shown in Fig. 31.27, the mass density of string 1 is twice that of string 2. What fraction of the energy in the traveling wave will be reflected at the junction?

ANSWER

The fraction of energy that is reflected is

fraction reflected $= R^2$

But R is given by Eq. 31.17, so

$$R = \frac{Z - Z'}{Z + Z'} = \frac{\sqrt{\tau\mu} - \sqrt{\tau'\mu'}}{\sqrt{\tau\mu} + \sqrt{\tau'\mu'}}$$

In this case $\tau = \tau'$ and $\mu = 2\mu'$. Therefore $Z = \sqrt{2}Z'$ and

$$R = \frac{\sqrt{2}Z' - Z'}{\sqrt{2}Z' + Z'} = \frac{\sqrt{2} - 1}{\sqrt{2} + 1} = \frac{0.414}{2.414} = 0.171$$

Calculating R^2 we find that the

fraction reflected $= R^2 = 0.03$ or 3%

31.10 THE WAVE EQUATION

Throughout this chapter we have considered many different forms of wave functions for string waves. All these forms are solutions of a single differential equation called the *wave equation*. The wave equation completely describes the dynamics of string wave motion; that is, it is the mathematical statement of the laws of dynamics as applied to a string. Exactly the same equation describes sound waves, waves on membranes, and electromagnetic waves. This equation is therefore the central element in more advanced mathematical treatments of wave motion. In this section

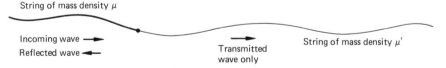

String of mass density μ

Incoming wave →
Reflected wave ◄

Transmitted wave only →

String of mass density μ'

FIGURE 31.27   A wave is being partially transmitted and partially reflected at the junction between a string of mass density μ and one of mass density μ', where $\mu = 2\mu'$.

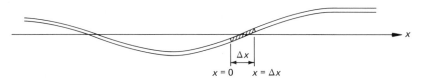

FIGURE 31.28   The mass of the system (i.e., the small segment of string) is $\mu \, \Delta x$. The velocity of the system is $\partial u / \partial t$. Therefore the momentum of the system is $P_y = \mu (\partial u / \partial t) \Delta x$.

we shall derive the wave equation from the laws of dynamics.

We use the following statement of the dynamical law:

The rate at which the momentum of any system changes is equal to the sum of the external forces on that system.

We consider a string of mass density μ and tension τ on which small-amplitude but otherwise arbitrary waves are traveling. Since the waves are of small amplitude, any point on the string moves only in the y direction. We take as our system a very short segment of the string, say the piece from $x = 0$ to $x = \Delta x$. The momentum of that piece of string can have only a y component, since no part of the string moves in the x direction. We call the momentum of the system $P_y(t)$. As the waves move along the string the momentum of the system changes. (Keep in mind the fact that *the system* we are considering is just that short *piece* of the string.) By the dynamical law we know that

$$\frac{dP_y(t)}{dt} = \begin{array}{l} \text{(the } y \text{ component of the force exerted} \\ \text{on the system at } x = 0) \\ + \text{ (the } y \text{ component of the force} \\ \text{exerted on the system at } x = \Delta x) \end{array} \qquad \textbf{(31.19)}$$

As shown in Fig. 31.28, the momentum of that small segment of string is $\mu(\partial u / \partial t)\Delta x$. (Since the segment is assumed to be extremely small, the derivative $\partial u / \partial t$ may be evaluated anywhere within the interval from 0 to Δx with negligible error. We shall in fact finally take the limit $\Delta x \to 0$.) Thus

$$P_y(t) = \mu \frac{\partial u}{\partial t} \Delta x$$

From this we see that the left-hand side of Eq. 31.19 is equal to

$$\frac{dP_y(t)}{dt} = \mu \frac{\partial^2 u}{\partial t^2} \Delta x$$

We must now evaluate the two terms on the right-hand side of that equation. Referring to Fig. 31.29, we see that, for small amplitude waves,

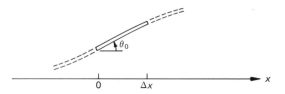

FIGURE 31.29   The magnitude of the force exerted on the system at $x = 0$ is equal to the tension τ. The y component of that force is $-\tau \sin \theta_0$, but for small values of θ_0 one can use the approximation $\sin \theta_0 \simeq \tan \theta_0 = (\partial u / \partial x)_0$. Thus the y component of the force exerted at $x = 0$ is $-\tau(\partial u / \partial x)_0$. [Note: $(\partial u / \partial x)_0$ means the slope of the string at $x = 0$.]

the y component of the force exerted on the system at $x = 0$ is equal to $-\tau(\partial u/\partial x)_0$.

A similar calculation of the force exerted on the system at $x = \Delta x$ gives

the y component of the force exerted on the system at $x = \Delta x$ is equal to $\tau(\partial u/\partial x)_{\Delta x}$

Thus Eq. 31.19 may be written as

$$\mu \frac{\partial^2 u}{\partial t^2} \Delta x = \tau \left(\frac{\partial u}{\partial x}\right)_{\Delta x} - \tau\left(\frac{\partial u}{\partial x}\right)_0$$

If we now divide the equation by Δx and make use of the fact that

$$\lim_{\Delta x \to 0} \frac{(\partial u/\partial x)_{\Delta x} - (\partial u/\partial x)_0}{\Delta x} = \frac{\partial^2 u}{\partial x^2}$$

we obtain what is known as the *wave equation* for string waves:

$$\mu \frac{\partial^2 u}{\partial t^2} = \tau \frac{\partial^2 u}{\partial x^2} \qquad (31.20)$$

To verify now that the types of string wave motion we have previously described do satisfy the laws of dynamics, we need only show that the corresponding wave functions are solutions of the wave equation given above. As an example of such a verification, let us consider the wave function for standing waves given in Eq. 31.9:

$$u(x, t) = B \sin kx \sin \omega t$$

The required partial derivatives are as follows:

$$\frac{\partial^2 u}{\partial x^2} = -k^2 B \sin kx \sin \omega t$$

and

$$\frac{\partial^2 u}{\partial t^2} = -\omega^2 B \sin kx \sin \omega t$$

Putting these into Eq. 31.20, we can see that the wave equation will be satisfied if

$$\mu \omega^2 = \tau k^2$$

or

$$\frac{\omega}{k} = \sqrt{\frac{\tau}{\mu}} \qquad (31.21)$$

Using the facts that $\omega = 2\pi/T$, $k = 2\pi/\lambda$, and $\lambda/T = v$, one can easily see that Eq. 31.21 is equivalent to the formula we have previously used (see Eq. 31.10) for the velocity of string waves:

$$v = \sqrt{\frac{\tau}{\mu}}$$

In a similar fashion we could confirm that all other wave functions we have used are also solutions of the wave equation (and thus the motions of the string described by them satisfy the laws of dynamics).

SUMMARY

A traveling wave on a string is a pattern that moves down the string with no change in shape.

As two waves pass through one another, the wave function is the sum of the wave functions of the individual waves.

For a string wave, $u(x, t)$ is the displacement from equilibrium, at time t, of the point on the string that has coordinate x.

Right-going and left-going waves have wave functions of the form $f(x - vt)$ and $f(x + vt)$, respectively. The curve $f(x)$ gives the shape of the string at $t = 0$ in both cases.

By reflection at a fixed end a wave is changed into its negative mirror image.

The wavelength λ is the distance between equivalent points on a periodic wave pattern.

The frequency f is the number of waves that pass a given point in a unit of time. It is also the frequency of oscillation of any point on the string.

The period T is the time it takes one wavelength to pass a given point:

$$f = \frac{1}{T}, \quad v = \frac{\lambda}{T}$$

The wave function of a right-going harmonic wave is

$$u(x, t) = A \cos 2\pi\left(\frac{x}{\lambda} - \frac{t}{T}\right)$$

This equation can also be written as

$$u(x, t) = A \cos(kx - \omega t)$$

where A is the amplitude of the wave, $k = 2\pi/\lambda$ is the wave number, and $\omega = 2\pi/T$ is the angular frequency. The angular frequency is 2π times the frequency.

The wave function of a standing wave is

$$u(x, t) = B \sin kx \sin \omega t$$

For a string of length L with both ends fixed, standing harmonic waves can occur only with one of the wavelengths

$$\lambda_n = \frac{2L}{n}, \quad n = 1, 2, \ldots$$

and frequencies $f_n = nv/2L$.

The rate at which energy passes a point x on the string, from left to right, is

$$\frac{dE}{dt} = -\tau \frac{\partial u}{\partial x}\frac{\partial u}{\partial t}$$

For a traveling harmonic wave the rate of energy transfer is

$$\frac{dE}{dt} = \tfrac{1}{2}Z\omega^2 A^2$$

where $Z = \sqrt{\tau\mu}$ is the wave impedance.

Waves on a string of mass density μ and tension τ propagate with speed

$$v = \sqrt{\frac{\tau}{\mu}}$$

At a junction between two strings a wave is partially reflected and partially transmitted. If the incoming wave (on string 1) has amplitude A, then the reflected wave will have amplitude RA and the transmitted wave will have amplitude $(1 + R)A$, where the reflection coefficient R is given by

$$R = \frac{Z_1 - Z_2}{Z_1 + Z_2}$$

The ratio of the energy in the reflected wave to the energy in the incoming wave is equal to R^2.

The differential equation satisfied by all string wave functions is

$$\mu\frac{\partial^2 u}{\partial t^2} = \tau\frac{\partial^2 u}{\partial x^2}$$

PROBLEMS

31.A.1 The figure shows a traveling wave on a string at two different instants of time. Draw a sketch of the wave at $t = 0.6$ s.

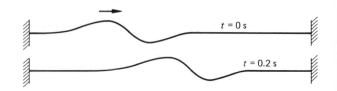

31.A.2 Draw a picture of the string in the following figure at the instant at which the center points of both waves pass the center point of the string.

31.A.3 The position of a string at $t = 0$ is shown in the accompanying figure. The wave speed in the string is 40 m/s. Draw sketches of the string at times $t = 1$ s, 2 s, 3 s, and 4 s.

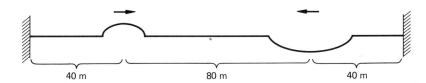

31.A.4 Draw a sketch of the wave shown in the figure and indicate on the sketch one wavelength.

31.A.5 What is the diameter of a steel wire that has a linear mass density of 1.4 g/m? (The density of steel is 7.8 g/cm³.)

31.A.6 If one end of a string of mass density 0.6 g/m is held fixed while the other end is subjected to a force of 8 N, what will be the velocity of the waves on the string?

31.A.7 Waves on a string have a velocity of 20 m/s. If the tension in the string is 20 N, what is the mass per unit length of the string?

31.A.8 A wave is partially reflected at the junction between two strings of different mass densities. If the mass density of the string carrying the transmitted wave (string 1) is three times that of the string carrying the incoming wave (string 2), what fraction of the energy is transmitted? What if the mass density of string 1 is one-third that of string 2?

31.B.1 For the waveform shown in the figure in problem 31.A.1, draw a graph of the vertical displacement of the midpoint of the string as a function of time for the time interval $0 < t < 1$. Assume that the maximum displacement is 1 cm.

31.B.2 On a *yt* graph plot the vertical displacement of point *P* as a function of *t* for the wave shown in the following figure. (Let the time axis run from 0 to 3 s.) The speed of the wave is 2 m/s.

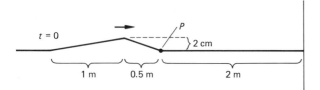

31.B.3 A wave of the shape shown in the figure travels toward a fixed end of the string. The wave speed is 10 m/s. Draw pictures of the string at times $t = 0.15$ s and $t = 0.20$ s.

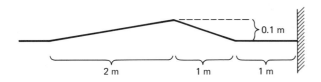

31.B.4 A wave on a 10-m string is described by the wave function

$$u(x, t) = \frac{10}{(x - 2t)^4 + 1}$$

Plot the shape of the string at times $t = 0$ and $t = 2$ s.

31.B.5 A right-going wave is traveling in a string with a wave velocity of 3 m/s. At time $t = 0$ the shape of the string is given by the function

$$u(x, 0) = \frac{\sin x}{x}$$

What is the wave function for all values of *x* and *t*?

31.B.6 The wave function of a string whose wave speed is 1 m/s and in which simultaneous right-going and left-going waves are traveling is

$$u(x, t) = f(x - t) + g(x + t)$$

Suppose the wave function of the right-going wave is $f(x - t) = \frac{1}{2}\cos(x - t)$ and that of the left-going wave is $g(x + t) = \frac{1}{2}\cos(x + t)$. Draw a sketch of the string at times $t = 0$, $\pi/6$ s, $\pi/3$ s, and $\pi/2$ s. [*Hint:* Use $\cos(a + b) = \cos a \cos b - \sin a \sin b$.]

31.B.7 A typical rope has a mass density of 0.1 kg/m. Suppose such a rope is held taut with a tension of 100 N (about 20 lb), and waves of period 0.5 s, amplitude 10 cm, and of the form shown in Fig. 31.25 are created. How many watts of power are carried down the rope by the waves?

31.B.8 As the tension in a wire increases, the wave speed increases. However, there is a limit to the wave speed we can obtain, because at too high a value of the tension the wire will snap. It is reasonable to assume that the maximum tension a wire made of a particular material can withstand is proportional to the cross-sectional area of the wire, $\tau_{max} = KA$. Show that the maximum wave speed is $v_{max} = \sqrt{K/\rho}$, where ρ is the volume mass density of the material that the wire is made of.

31.B.9 The first normal mode of a string is called the fundamental mode. A certain string is tuned to give middle C (256 Hz) when under 200 N tension. How much should the tension be increased in order to produce high C (512 Hz) as a fundamental mode?

31.B.10 A steel wire has a mass density of 5 g/m. The wave speed on the wire is 10 m/s. If the angle θ shown in the figure is $2°$, how much energy is transferred from S_L to S_R in a time interval of 10^{-4} s? What is the rate of energy transfer?

S_L S_R

31.B.11 A string is connected at one end to a small massless ring around a vertical frictionless rod (see the following figure). Using the fact that no energy can be transmitted to the frictionless rod, show that the string will move in such a way that its end always remains horizontal. Such an end is called a *free end* in contrast to a fixed end. The reflected wave from a free end is a *positive* mirror image of the incoming wave.

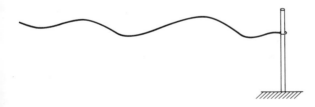

31.B.12 Use the result of problem 31.B.11 to derive the following formula for the wavelengths and frequencies of normal modes of a string of length L with one fixed end and one free end:

$$f_n = \left(\frac{2n-1}{4}\right)\left(\frac{v}{L}\right)$$

31.B.13 By vibrating one end at a frequency of 60 Hz, a harmonic wave of amplitude 2 cm and wavelength 30 cm is created in a long string of mass density 2 g/m. How many watts of electric power would be required to run the vibrator (see the figure)?

31.C.1 A monophonic phonograph record represents sound by horizontal displacements of the record groove. Consider a $33\frac{1}{3}$ r/min record that has a 12-in diameter and a 4-in label in the center. If the music recorded on it has a frequency range of 20–20,000 Hz, what is the range of wavelengths of the record groove (see the figure)?

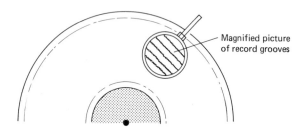

Magnified picture of record grooves

31.C.2 Suppose we simply guessed that the formula for the speed of traveling waves on a string was
$v = C\tau^K\mu^L$
where C was a dimensionless constant and K and L were unknown exponents. Using the facts that the dimensions of μ are kilograms per meter and those of τ are newtons, show that the only possible dimensionally correct formula is $K = \frac{1}{2}$ and $L = -\frac{1}{2}$.

31.C.3 A string with mass density 0.002 kg/m has a tension of 5 N.
(a) What is the speed of traveling waves on the string?
(b) If the string has a right-going wave on it and its wave function at time $t = 0$ is $u(x, 0) = 0.03 \cos(4x)$ (where x is in meters and u is in centimeters), what is its wave function at time t?
(c) At time $t = 0.01$ s, what is the y coordinate of the point on the string whose x coordinate is 0.75 m?
(d) What are the wavelength and period of the wave?

31.C.4 What is the energy per unit length in a traveling harmonic wave of amplitude A and period T?

31.C.5 A standing wave on a string is composed of two traveling waves. By using the fact that the energy in the standing wave is the sum of the energies in its constituent traveling waves, derive a formula for the total wave energy in a string of length L that is vibrating in its nth normal mode with amplitude B. (*Hint:* Use the results of Problem 31.C.4.)

31.C.6 Two strings of mass densities μ_1 and μ_2 and tensions τ_1 and τ_2 are connected through a massless lever, as shown in the following figure. What must be the ratio of the lever arms l_1/l_2 for perfect transmission to occur? (Assume that the waves are of small amplitude.)

31.C.7 A right-going harmonic wave has a wave function $u(x, t) = A \cos(kx - \omega t)$.
(a) Using Eq. 31.11, show that the rate at which energy is flowing past the point with coordinate $x = 0$ is

$$\frac{dE}{dt} = \omega^2 Z \sin^2 \omega t$$

(b) By integrating the rate obtained in part (a) over one complete period and then dividing by T, show that the average rate of energy transfer is that given in Eq. 31.12.

31.C.8 An electric motor turns a disk with a period of T seconds (see the figure). The disk has a radius A. String of mass density μ is attached to the edge of the disk and stretched with tension τ. Thus a traveling spiral wave of amplitude A is created in the string. The wavelength of the spiral is $\lambda = vT$, where $v = \sqrt{\tau/\mu}$. At the point of contact the string makes an angle θ with a line drawn perpendicular to the face of the disk. Show that $\tan \theta = 2\pi A/\lambda$.

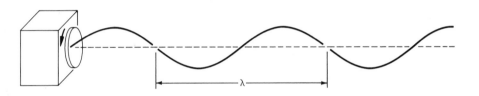

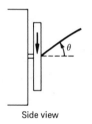

Side view

The point at which the string meets the disk moves in a circle. What are the direction and magnitude of the velocity of that point? The string exerts a force on the disk at that point. What is the component of the force in the direction of the velocity of the point? At what rate is the disk performing work on the string? If you have calculated everything correctly, your answer would be

$$\frac{dE}{dt} = \frac{2\pi A\tau}{T} \sin \theta$$

If A is much smaller than λ, then θ is very small and $\sin \theta \simeq \tan \theta = 2\pi A/\lambda$. In that case show that

$$\frac{dE}{dt} = 4\pi^2 \sqrt{\tau\mu} \left(\frac{A}{T}\right)^2$$

Show that one could obtain the spiral wave by adding two harmonic waves that vibrate in perpendicular planes. Using that fact, comment on the comparison of the above formula and Eq. 31.12 of the text.

31.C.9 The formula $f = v/2L$ for the lowest normal mode frequency of a string was derived with the assumption that the ends of the string were perfectly stationary. Sometimes the ends have a little "give," and therefore move a small amount as the string vibrates. The motion of the end supports will affect the normal mode frequencies. Consider the string shown in the figure. Assume that, as the string vibrates with amplitude A, the ends move harmonically (in phase with the string) with amplitude αA, where $\alpha \ll 1$. If the string is of length L and wave velocity v, what is the fundamental normal mode frequency. (*Hint:* Compare the shape of the string with that of a slightly longer string that is rigidly fixed at the ends.)

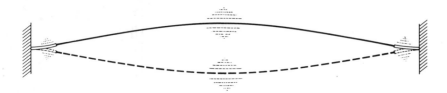

32 SOUND WAVES

SOUND WAVES ARE compressional waves in a gas, liquid, or solid. A sound wave in air might be produced by a device such as that shown in Fig. 32.1. The air in front of a vibrating piston is alternately compressed and expanded. As it is compressed, energy and momentum are transferred to the air directly in front of the piston. This layer of compressed air moves slightly forward, compressing the air in front of it and delivering its energy and momentum to that layer of air, which then moves forward, and so on. Regions of compression sandwiched between regions of expansion propagate away from the face of the piston with a speed called the speed of sound.

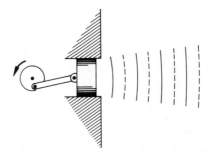

FIGURE 32.1 An oscillating piston producing sound waves.

32.1 WAVE FRONTS

Suppose the density of the air when the piston is idle is ρ kilograms per cubic meter. When the piston is running and sound waves are propagating through the air, the density at any point fluctuates. At a given instant the density is larger than ρ in those regions in which the air is compressed and smaller than ρ in those regions in which the air is expanded. If we connect those points at which the density has just attained a temporary maximum, we obtain a set of surfaces that we call *positive wave fronts*.

If we connect all those points at which the density has just achieved a temporary minimum, we obtain a set of surfaces that we call *negative wave fronts.* We shall always indicate the positive wave fronts by solid lines and the negative wave fronts by dashed lines. The positive and negative wave fronts alternate and move away from the source at the speed of sound. The adjectives "positive" and "negative" refer to the fact that at the location of a positive (or negative) wave front the density wave function is positive (or negative). The density wave function is defined as the deviation of the density from its average value.

32.2 PLANE WAVES

When the wave fronts form a collection of parallel planes, we call the sound wave pattern a *plane wave.* Within a fixed finite region of space at a large distance from the piston, the wave pattern is approximately a plane wave (see Fig. 32.2). Let us focus our attention on a sample of the air within that region whose dimensions are very small in comparison to the wavelength of the plane wave. We refer to that very small sample as an "element of air." As the wave passes, the element moves back and forth

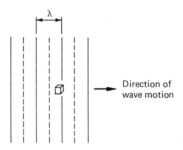

FIGURE 32.3 The element of air represented by the box vibrates back and forth as the wave passes.

periodically (see Fig. 32.3). Note that the motion of the air is not perpendicular to the direction of propagation of the wave but rather is parallel to the direction of propagation. Sound waves are longitudinal waves. The maximum displacement of the element of air from its rest position is called the *amplitude* of the waves.

32.3 SOUND WAVE VELOCITY

The velocity of the propagation of waves in a gas or liquid is given by a formula that is similar to that for the velocity

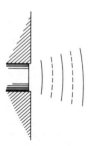

FIGURE 32.2 Far from the piston the wave fronts are approximately plane surfaces.

of string waves. The formula (whose derivation we shall omit) is

$$v = \sqrt{\frac{B}{\rho}} \qquad (32.1)$$

where ρ is the mass density of the fluid in kilograms per cubic meter, and B is a property of the fluid called the *adiabatic bulk modulus.* The adiabatic bulk modulus of a fluid is defined as follows.

Consider a volume V_0 of the fluid contained in a cylinder with a piston at one end. A force is maintained on the piston so that the fluid is under a pressure p_0 (see Fig. 32.4).

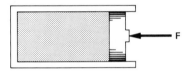

FIGURE 32.4 The adiabatic bulk modulus is measured by compressing the fluid within a thermally nonconducting cylinder.

Both the cylinder and the piston are thermally insulating; that is, they will not conduct heat. If the force is changed slightly, then the pressure will also change slightly to a new value $p + \Delta p$, and the fluid volume will change slightly to a new volume $V + \Delta V$. The adiabatic bulk modulus is defined in terms of the ratio of the change in pressure to the change in volume:

$$B = -\frac{\Delta p}{\Delta V} V$$

From this equation we see that B must have units of pressure, namely newtons per square meter.

32.4 ACOUSTIC ENERGY FLUX

We can convert Eq. 31.10 for the speed of string waves to Eq. 32.1 for the speed of sound waves by replacing the linear mass density μ and the tension τ by the volume mass density ρ and the bulk modulus B. We can also convert the other important equations that were derived for string waves to equations involving sound waves by the same substitution. For instance, suppose we consider a plane harmonic wave of frequency f and amplitude A in a fluid and ask at what rate energy is transported by the wave through a unit area (see Fig. 32.5). The answer is given by a formula that is identical to Eq. 31.12, but with the above-mentioned substitutions. The rate at which energy is transmitted per unit area is called the *energy flux,* and

$$\text{energy flux} = \tfrac{1}{2} Z \omega^2 A^2$$

where $\omega = 2\pi f$ and $Z = \sqrt{\rho B}$.

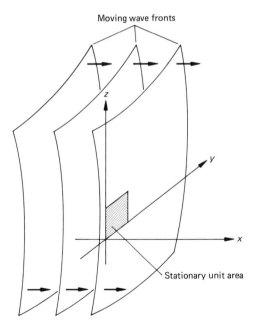

FIGURE 32.5 The rate at which energy passes through a unit area (placed parallel to the wave fronts) is called the energy flux.

QUESTION

The amplitude of vibration of a painfully loud sound at 3000 Hz is about 0.01 mm. Taking the effective area of the ear as 1 cm², determine how much power is delivered to the ear by such a sound. (For air Z is 414 kg/m² s.)

ANSWER

Let S by the area of the ear ($S = 10^{-4}$ m²). Then the power delivered to the ear is

power

$$= \text{(flux)(area)} = \tfrac{1}{2}Z\omega^2 A^2 S$$

$$= (0.5)(414 \text{ kg/m}^2 \text{ s})(2\pi \times 3000 \text{ Hz})^2(10^{-5} \text{ m})^2(10^{-4} \text{ m}^2)$$

$$= 7.35 \times 10^{-4} \text{ W}$$

When a sound wave propagating in air encounters an interface between the air and a region that is filled with some other fluid, some of the energy in the wave is reflected at the boundary and some is transmitted into the fluid (Fig. 32.6). If the amplitude of the incident sound wave is A and its wave fronts are parallel to the interface, then the amplitude of the reflected sound wave will be RA, where

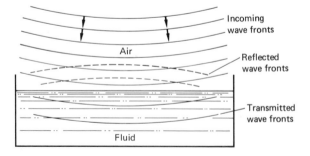

FIGURE 32.6 Sound waves are partially transmitted and partially reflected at an interface between two different materials.

the reflection coefficient R is given by a formula that is identical in form to that for string waves:

$$R = \frac{Z - Z'}{Z + Z'}$$

where $Z = \sqrt{B\rho}$ and $Z' = \sqrt{B'\rho'}$. The unprimed quantities refer to the air, and the primed quantities refer to the other fluid. Using the data in Table 32.1, we can determine the fraction of acoustic energy that is reflected at an air-water interface. The reflection coefficient is

$$R = \frac{414 - 1.5 \times 10^6}{414 + 1.5 \times 10^6} = -0.9995$$

It is clear that the reflection coefficient is so close to -1 that there is almost perfect reflection at the interface.

TABLE 32.1 Acoustic data for air and water

Property	Air	Water
ρ (kg/m³)	1.3	1000
B (N/m²)	1.4×10^5	2.3×10^9
Z (kg/m² s)	414	1.5×10^6

32.5 IMPEDANCE-MATCHING DEVICES

To illustrate how the problem of transmitting sound waves from air to water might be solved, we shall analyze a simple device that has some structural similarity to the human ear. We strongly emphasize, however, that the following calculation cannot be construed as a mathematical analysis of the operation of the actual ear. Such an analysis would be far beyond the level of this course. The intent of the calculation is simply to show that a mechanical device of this sort could solve the problem of getting the acoustic energy from air to water. The device is illustrated in Fig. 32.7, and its operation can be described as follows.

The sound waves in air move the piston on the left. The motion of the piston must have the same amplitude A as

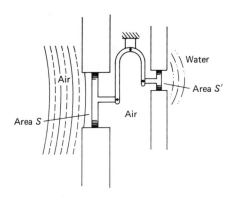
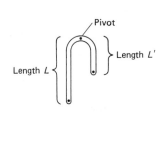

FIGURE 32.7 A mechanical device for transmitting sound waves from air to water without reflection.

the sound waves if there is to be no reflected wave. The piston and lever are assumed to have negligible mass. Through the lever, the motion of the piston on the left causes the piston on the right to oscillate with an amplitude $A' = (L'/L)A$. The motion of this piston creates sound waves of amplitude A' in the water. If the incoming sound wave is a harmonic wave, then the rate at which energy lands on the left-hand piston is

$$\left(\frac{dE}{dt}\right)_i = (\text{flux})(\text{area}) = \tfrac{1}{2}Z\omega^2 A^2 S$$

where Z is the impedance of the air, S is the area of the piston, and the subscript "i" indicates "incoming." If there is to be no reflected wave, then all this energy must be delivered to the sound wave in the water. The rate at which energy is being carried away by the sound wave in the water is

$$\left(\frac{dE}{dt}\right)_o = \tfrac{1}{2}Z'\omega^2 A'^2 S'$$

where Z' is the impedance of water and the subscript "o" indicates "outgoing." If we set

$$\left(\frac{dE}{dt}\right)_i = \left(\frac{dE}{dt}\right)_o$$

and use the fact that $A' = (L'/L)A$, we obtain a relationship involving the dimensions of the device that must be satisfied for perfect transmission:

$$ZSL^2 = Z'S'(L')^2$$

or, equivalently,

$$\frac{S}{S'}\left(\frac{L}{L'}\right)^2 = \frac{Z'}{Z}$$

From Table 32.1 we see that

$$\frac{Z'}{Z} = \frac{1.5 \times 10^6}{414} = 3.6 \times 10^3$$

If, for example, the ratio of the diameter of the left piston to that of the right piston were 6:1 and the ratio of the respective lever arms were 10:1, then one could achieve complete energy transmission.

QUESTION

One hears that television antenna lead-in wire has an impedance of 300 Ω and that certain loudspeaker systems have an impedance of 8 Ω. Does that impedance have any relation to the wave impedance introduced for string waves and sound waves?

ANSWER

Yes. For instance, if one were mistakenly to splice together 50 ft of 300-Ω line and 50 ft of 90-Ω line to make a 100-ft antenna lead-in, then most of the incoming signal would be reflected at the junction in the two lines and sent back to the antenna lead-in, then part of the incoming signal would be systems, the input impedance of the speaker must be matched to the output impedance of the amplifier by an impedance-matching device such as the audio output transformer. If this is not done, there will be poor energy transfer from the amplifier to the speaker system. In general, whenever a harmonic signal is being transferred from one device or medium to another, the rate at which energy is transferred is proportional to $\omega^2 A^2$, where A is the amplitude of the signal. The proportionality constant for a given device is defined to be one-half the impedance. Since both energy transmission and signal amplitude must match at the junction, we can only obtain perfect transmission without an impedance-matching device when the impedance of the input device is the same as the impedance of the output device.

32.6 WAVES ON MEMBRANES

A membrane is a thin sheet of elastic material. It can propagate waves only when it is under tension, that is, when it is being stretched. The two characteristics of a membrane that determine its wave-carrying properties are the mass density of the membrane and the stress in the membrane. The mass density of the membrane is defined as the mass per unit area of the membrane. The stress in the membrane is defined by considering an imaginary straight line segment of length l on the membrane. The portion of the membrane on one side of the line segment exerts a pulling force on that portion of the membrane on the other side of the line segment, and the force is proportional to l:

force $= sl$

The proportionality constant s is called the *stress* in the membrane (see Fig. 32.8). The propagation speed of waves on a membrane of stress s and mass density σ is given by a formula that is an obvious analog of the formulas for the speed of string waves and sound waves:

$$v = \sqrt{\frac{s}{\sigma}} \tag{32.2}$$

If s is given in newtons per meter and σ in kilograms per square meter, then v is obtained from Eq. 32.2 in meters per second.

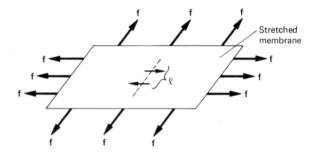

FIGURE 32.8 The force communicated across a line segment of length l on the particles on one side of the line *by* the particles on the other side of the line is equal to sl.

32.7 NORMAL MODE VIBRATIONS OF A MEMBRANE

A membrane that is held fixed at its periphery has certain natural frequencies of vibration. The characteristic vibrations of the membrane associated with those frequencies are called the *normal modes* of the membrane. We shall consider only two different shapes of membranes: a circular membrane and a membrane in the shape of a very long strip of uniform width.

THE CIRCULAR MEMBRANE

The lowest-frequency normal mode vibration of a circular membrane is a motion in which all parts of the membrane except the edge move together in simple harmonic motion (see Fig. 32.9). The amplitude of vibration is largest at the center of the membrane and goes to zero at the periphery.

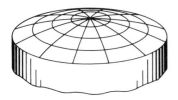

FIGURE 32.9 A circular membrane (a drumhead) vibrating in its lowest normal mode.

For each of the higher-frequency normal modes the amplitude of vibration is zero along certain lines on the membrane surface. These lines are called the *nodal lines* of the vibration. The portions of the membrane on opposite sides of any nodal line vibrate exactly out of phase: When one is up, the other is down, and vice versa (see Fig. 32.10).

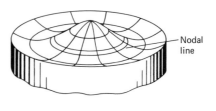

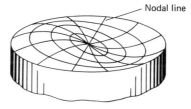

FIGURE 32.10 Two higher normal modes of a drumhead with their nodal lines indicated.

THE LONG-STRIP MEMBRANE

The lowest-frequency normal mode of a membrane in the form of a long strip with its edges held fixed is a vibration of the form shown in Fig. 32.11. This vibration has a wavelength that is equal to twice the width W of the membrane:

$$\lambda = 2W \tag{32.3}$$

The frequency of the vibration is therefore

$$f = \frac{v}{\lambda} = \frac{\sqrt{s/\sigma}}{2W} \tag{32.4}$$

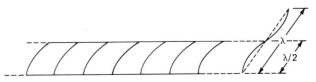

FIGURE 32.11 Fundamental mode of vibration of a long-strip membrane. The vibration has a wavelength equal to twice the width of the membrane.

32.8 FORCED OSCILLATIONS OF A SYSTEM

We have considered the free oscillations of a segment of string and of circular and strip membranes. The free harmonic oscillations of these systems occur only at certain frequencies associated with the normal modes of the systems. If one of these systems is acted on by a small external driving force whose strength varies periodically with frequency f, then the system can be made to oscillate with the frequency of the driving force. However, the amplitude of such a forced oscillation will remain relatively small unless the frequency of the driving force is equal to one of the normal mode frequencies of the system. If, for example, one end of a string segment is made to vibrate harmonically with small amplitude, then the amplitude of vibration will be small everywhere on the string unless the frequency at which the end is being driven is a normal mode frequency of the string, in which case the amplitude of vibration of most of the string will become much larger than the amplitude at the driven end (see Fig. 32.12). It is the forced

Vibrator

FIGURE 32.12   A segment of string set into forced vibration by a vibrator at one end. If the vibrator frequency is a normal-mode frequency, then the amplitude of vibration is much larger at the center of the string than at the end.

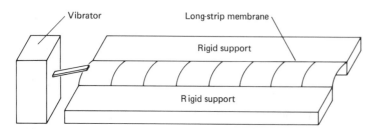

Vibrator   Long-strip membrane

Rigid support

Rigid support

FIGURE 32.13   A vibrator creating forced oscillations of a long-strip membrane.

oscillations of a long-strip membrane that will be important in the analysis of the physics of the ear. If one end of a long-strip membrane is moved up and down with a small amplitude and a frequency f, as shown in Fig. 32.13, the amplitude of vibration of all parts of the membrane will usually remain small. However, if f is the frequency of the fundamental normal mode (given by Eq. 32.4), then the membrane will be put into a large-amplitude normal mode vibration.

32.9 THE PHYSICS OF HEARING

A very simplified diagram of the portions of the ear that are involved in hearing is shown in Fig. 32.14. (Another major portion of the inner ear is involved in maintaining our balance or orientation with respect to the gravitational field.) The outer ear consists of a short canal that is open to the air at one end and closed by a membrane (the eardrum or tympanic membrane) at the other end. The middle ear is an air-filled cavity that is maintained at external air pressure by an opening through the auditory canal that opens to the outside when we swallow. Within the middle ear is a bone lever system (the ossicles) that transmit acoustic vibrations from the tympanic membrane to a much smaller membrane that separates the middle ear from the fluid-filled inner ear. By a combination of reduction in the areas of the two membranes and reduction in the distance of travel of the two ends of the lever, a highly

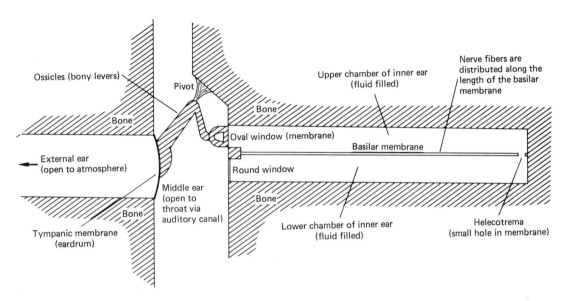

Ossicles (bony levers)

Pivot

Bone

Bone

External ear
(open to atmosphere)

Middle ear
(open to
throat via
auditory canal)

Bone

Tympanic membrane
(eardrum)

Oval window (membrane)

Round window

Bone

Upper chamber of inner ear
(fluid filled)

Nerve fibers are
distributed along the
length of the basilar
membrane

Bone

Basilar membrane

Lower chamber of inner ear
(fluid filled)

Helecotrema
(small hole in membrane)

FIGURE 32.14   Schematic diagram of the basic parts of the human ear.

efficient transmission of acoustic energy from air to fluid is accomplished.

We have taken great liberties in simplifying our diagram of the inner ear so that we could discuss the operation of the ear at a level that is reasonably commensurate with present understanding of the functioning of the ear. The true structure has a great many complex anatomical details whose exact functions are still not completely understood. The inner ear is a closed chamber that is almost separated into two parts by a membrane system that runs the length of the chamber but contains a small hole (called the helicotrema) at one end through which fluid can pass from one side of the chamber to the other. The membrane system has a number of layers between which reside the auditory nerve fibers. We shall refer to the membrane system as the basilar membrane, although that name usually denotes only one of the membrane layers. The nerve fibers are electromechanical transducers, that is, devices that convert the mechanical stimulation of being stretched to an electric impulse. The basilar membrane is essentially a long strip that varies in width and mass density from one end to the other. Stimulation by a harmonic sound wave causes a standing wave pattern to be set up on the membrane. The location of the maximum amplitude of the standing wave pattern is a function of the frequency of the sound wave.

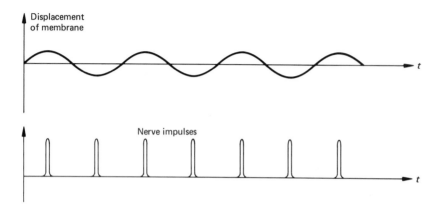

FIGURE 32.15 At low frequencies the nerve impulses originating from one location on the basilar membrane are in phase with the vibration there.

32.10 FREQUENCY DETERMINATION BY THE EAR

Since nerve fibers are distributed all along the basilar membrane, the nerve impulses originating in those fibers supply to the brain information on the details of the vibrational pattern set up on the basilar membrane by incoming sound. At frequencies of up to a few hundred hertz, nerve impulses originate from the basilar membrane at a fixed point in each cycle of the basilar membrane vibration (see Fig. 32.15). This train of timed impulses obviously supplies to the brain information that is related in a very simple way to the frequency of the incident sound. However, if the frequency of the incoming sound is increased beyond a few hundred hertz, the nerve fibers are incapable of generating nerve impulses at a rate comparable to the vibrational frequency. The brain then determines the frequency by analyzing the vibrational pattern set up on the basilar membrane. The simplest frequency-dependent characteristic of the vibrational pattern is the location on the membrane of the point of largest-amplitude vibration. As the frequency of the sound increases, that point shifts along the basilar membrane from the helicotrema end to the window end, and thus the point acts as a frequency indicator. According to Eq. 32.4, the natural frequency of

vibration of a strip membrane depends on its stress s, its mass density σ, and its width W. All the quantities vary gradually from one end of the basilar membrane to the other. They all change in such a way as to increase the normal mode frequency as one goes from the helicotrema end to the window end. The basilar membrane is narrowest near the windows and widest near the helicotrema, as shown in Fig. 32.16. Its length is about one hundred times its average width. The membrane is under greatest tension at the window end. The effective mass density of the membrane is much greater at the helicotrema end than at the window end. The effective mass density of a section of the basilar membrane has little to do with the actual mass density of the membrane itself, but is determined primarily by the mass of fluid (contained in the two chambers) that must be set into motion if that section of the membrane is to vibrate (see Fig. 32.17). By referring to Eq. 32.4 one can see that decreasing the stress, increasing the mass density, and increasing the width all cause the resonant frequency of a strip membrane to decrease. Thus the fundamental frequency of vibration of the section of the basilar membrane near the helicotrema is much lower than that of the section of the membrane near the windows (see Fig. 32.18). The purpose of the second

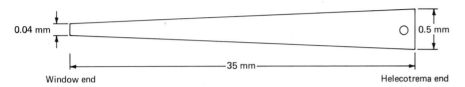

FIGURE 32.16 Top view of the basilar membrane with approximate dimensions (not drawn to scale).

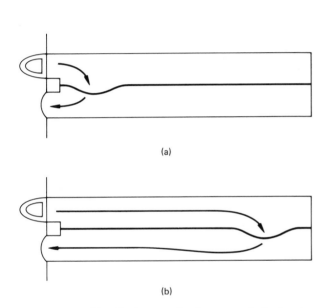

FIGURE 32.17 The effective mass density at a particular location depends on the flow pattern for vibration at that location. (a) For vibration at the window end, only a small amount of fluid is set into motion. The effective mass is therefore small. (b) For vibration at the helecotrema end, a large mass of fluid must be set into motion. The effective mass is therefore large.

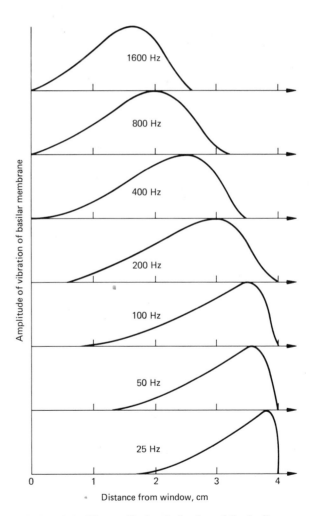

FIGURE 32.18 The amplitude of vibration of the basilar membrane as a function of position when the membrane is excited by sounds of various frequencies.

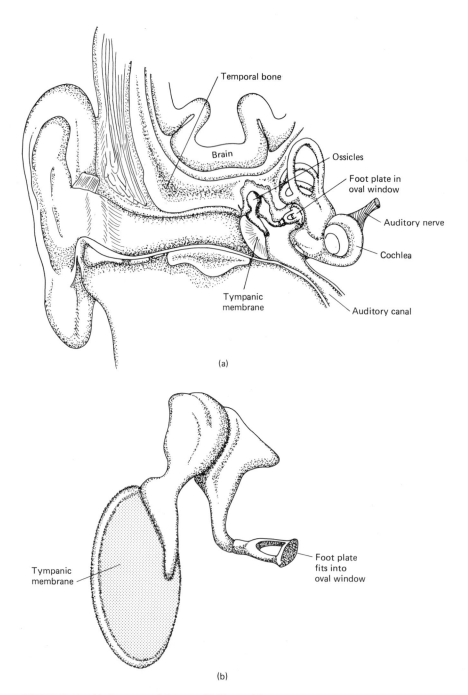

FIGURE 32.19   (a) Anatomy of the ear. (b) The ossicles.

membrane "window" between the inner and middle ear is to improve the impedance match at low frequencies where the lever system is not sufficient to accomplish an efficient transfer of energy from air to fluid.

The actual anatomy of the ear differs somewhat from what we have shown in our schematic diagrams. The cochlea, a structure that includes the upper and lower chambers, is curled up into a spiral or snail-shell curve. (In fact, the word "cochlea" is simply the Greek word for a snail shell.) Attached to the cochlea is a system of three semicircular canals that play no part in hearing but form a sense organ used to detect rotational accelerations of the head and its orientation with respect to the direction of the gravitational field. Figure 32.19 shows the actual shape and location of the elements of the ear.

32.11 ANALYSIS OF COMPLEX SOUNDS

A body vibrating with simple harmonic motion creates a harmonic sound wave in the surrounding air. If the vibration continues with constant frequency and amplitude, then the pressure variation (the instantaneous air pressure minus the average air pressure) at a fixed point in space, such as the location of the ear, is a simple cosine function of time:

$$p(t) = A \cos(\omega t + \phi)$$

(Note that in this formula A gives the maximum change in pressure, not the displacement of the element of air.) The pressure variation to which the ear is subjected by a harmonic sound wave is thus completely defined by three quantities: the amplitude A, the angular frequency ω, and the phase constant ϕ. To illustrate the meaning of

those three quantities, in Fig. 32.20 we have drawn a graph of $p(t)$ and then shown what happens to the graph when each one of the quantities is separately changed.

Since we may change the phase at will simply by changing the zero of our time scale, it is not surprising that our ears are entirely insensitive to the value of ϕ. Although we cannot detect a change in the overall phase of a sound, we can sense any difference in the relative phase of the sound waves reaching our two ears. When the pressure variations at the right ear occur slightly sooner than the pressure variations at the left ear, we can detect the time lag. We sense it not as a relative phase difference, but as a directionality of the sound; that is, the sound seems to be coming from the right (see Fig. 32.21).

When the amplitude is held fixed and the frequency is varied, we detect that change subjectively as a change in the *pitch* of the sound. When the frequency is kept constant and the amplitude is varied, we note a change in the *loudness* of the sound. Careful observation reveals that things are not quite as simple as the last two statements seem to indicate. The subjective pitch of a sound is not purely a function of its frequency, but depends slightly on its amplitude. A subject who is asked to determine whether two sounds of different amplitude have the same pitch will say that they do have the same pitch when in fact they have somewhat different frequencies. On the other hand, the subjective experience of loudness created by a sound is in part dependent on its frequency. Two sounds of different frequencies that have the same energy flux are not perceived as being equally loud. When the energy flux is kept constant and the frequency is varied, the loudness appears to be a maximum at about 3000 Hz and decreases for higher or lower frequencies. From the foregoing discussion it would appear that two sound sources of the same loudness and pitch would be indistinguishable, but such a conclusion is obviously false. We can easily

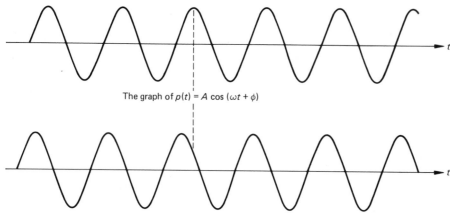

The graph of $p(t) = A \cos (\omega t + \phi)$

If ϕ is increased, the variations all take place at an earlier time. That is, the graph is shifted to the left but otherwise unchanged.

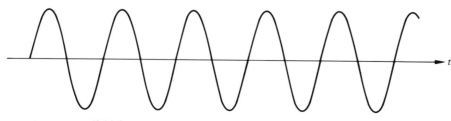

If A is increased the magnitude of the pressure variations increases.

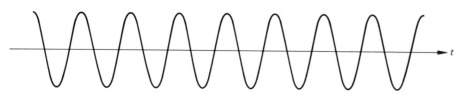

If ω is increased the time interval for each pressure variation decreases.

FIGURE 32.20   Effects of separately increasing each of the quantities ϕ, A, and ω.

FIGURE 32.21 When sound waves come from one side, the wave fronts strike one ear before the other, and the result is a relative phase difference.

differentiate between a voice and a violin both making sounds of the same loudness and pitch. The reason we seem to arrive at this false conclusion is that we have been discussing only harmonic waves. In acoustic theory harmonic waves are referred to as *pure tones.* Instruments such as pianos and violins do not produce pure tones. While the pressure variation they produce is a periodic (repeating) function of time, that function is not a simple cosine function. It is clear that an infinite number of different periodic functions with the same period and amplitude exist (see Fig. 32.22). Must we analyze each one independently in order to have an adequate theory of sound propagation? Fortunately we need not. A very important mathematical theorem attributable to Jean Baptiste Fourier states that we can obtain any periodic wave of frequency f by adding (by the addition principle) pure tones of frequencies f, $2f$, $3f$, etc. Let us describe Fourier's

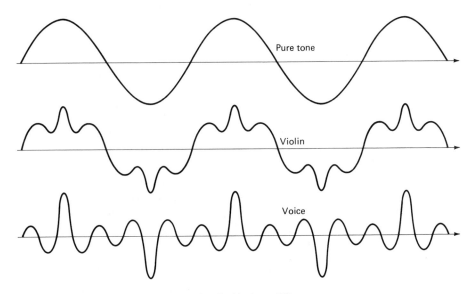

FIGURE 32.22 The wave functions associated with three different sounds of the same period and amplitude.

theorem more precisely. Suppose we add a pure tone of frequency f, amplitude A_1, and phase ϕ_1 to another pure tone of frequency $2f$, amplitude A_2, and phase ϕ_2. By the addition principle the resultant pressure variation will be the sum of the pressure variations caused by each pure tone separately:

$$p(t) = A_1 \cos(\omega t + \phi_1) + A_2 \cos(2\omega t + \phi_2)$$

The sound is no longer a simple pure tone, since the pressure variation cannot be written as single cosine, but it is still periodic with period $T = 1/f$. To see this, note that the second term in the sum has a period of $\frac{1}{2}T$; that is, it repeats itself after each time interval of length $\frac{1}{2}T$. But it then automatically also repeats itself after each time interval of length T. Since both terms repeat after each time interval T, the sum does also. The same argument would show that any sum of the form

$$p(t) = A_1 \cos(\omega t + \phi_1) + A_2 \cos(2\omega t + \phi_2)$$
$$+ A_3 \cos(3\omega t + \phi_3) + \cdots$$

gives a pressure variation function $p(t)$ that is periodic with frequency f and period $T = 1/f$. Such a sum is called a *Fourier series*. The frequency f is called the *fundamental frequency*; the frequency $f_2 = 2f$ is called the *second harmonic*; etc. That such a Fourier series yields a periodic function is fairly obvious. What is not at all obvious, but was proved by Fourier, is that *any periodic function can be written as a Fourier series*. (Of course Fourier had the modesty not to use the term Fourier series.) We shall not prove Fourier's theorem.

Thus any complex sound that is periodic is entirely equivalent to a pure tone of the same frequency plus another pure tone of twice the frequency plus another

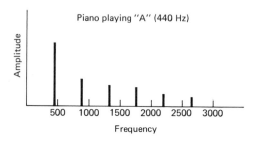

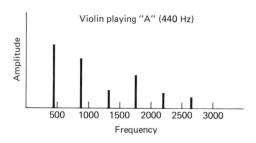

FIGURE 32.23 Fourier amplitudes of piano and violin tones.

pure tone of three times the frequency, etc. To reproduce the sound with perfect precision, one may need an infinite number of harmonics; but in fact to reproduce it to any preassigned accuracy, one needs only a finite number of terms in the Fourier series. (Just as one needs an infinite decimal 1.414 . . . to define $\sqrt{2}$, but one actually uses a finite decimal in any real calculation.) The numbers A_1, A_2, A_3, . . . are called the Fourier amplitudes associated with the periodic function $p(t)$, and the numbers ϕ_1, ϕ_2, ϕ_3, . . . are called the Fourier phases. The Fourier amplitudes of violin and piano tones are shown in Fig. 32.23. There are formulas that allow one to calculate the amplitudes and phases associated with any given function, but we shall not consider them.

By the use of electronic signal generators, one can produce complex sounds that have any desired set of Fourier amplitudes and phases. When this is done and highly trained listeners attempt to discriminate between different complex sounds of the same fundamental frequency, it is found that two complex sounds with the same set of amplitudes but entirely different sets of phases are almost completely indistinguishable by the human ear. This is in spite of the fact that the graphs of the pressure variations for the two sounds may look quite different. The subjective experience produced by a sound is determined almost entirely by the Fourier amplitudes of those harmonics that lie within the sensitivity range of the ear. The pitch we assign to a complex sound is basically a measure of the frequency of the fundamental. Strangely enough, the last statement is valid even when the amplitude of the fundamental (the number A_1) is zero. That is, a complex sound made up of pure tones with frequencies $2f$, $3f$, $4f$, $5f$, etc., will be assigned a pitch corresponding to the frequency f rather than the frequency $2f$. Of course, if the amplitudes of all the odd harmonics are also set equal to zero, then the sound is heard as one of frequency $2f$.

SUMMARY

Sound waves are compressional waves in a gas, liquid, or solid.

A positive (negative) wave front is a locus of points at which the density has just reached a maximum (minimum).

A plane wave is a wave with plane wave fronts.

The formulas for the velocity, wave impedance, and energy flux are the same as those for string waves with the replacements $\mu \rightarrow \rho$ and $\tau \rightarrow B$; that is,

$$v = \sqrt{\frac{B}{\rho}}, \quad Z = \sqrt{B\rho}, \text{ and } \quad \text{energy flux} = \tfrac{1}{2}Z\omega^2 A^2$$

where B is the adiabatic bulk modulus and is given by

$$B = -\frac{\Delta p}{\Delta V} V$$

where Δp and ΔV are the changes in pressure and volume of a sample of the fluid in a thermally insulating container.

The energy flux is the amount of energy transmitted through a unit area in a unit time by the sound wave.

The reflection coefficient for reflection at the boundary between two different materials is

$$R = \frac{Z - Z'}{Z + Z'}$$

The tension force exerted across any line segment l in a membrane is equal to sl, where s is the stress in the membrane.

The velocity of waves on a membrane is

$$v = \sqrt{\frac{s}{\sigma}}$$

where σ is the mass density per unit area of the membrane.

The lowest-frequency normal mode of vibration of a circular drumhead has no nodes. The higher-frequency normal modes have nodal lines that are either circles (concentric with the edge) or straight lines through the center.

The fundamental mode of a long-strip membrane of width W has a wavelength $\lambda = 2W$ and a frequency $f = v/2W$.

The amplitude of forced oscillations of a system is small unless the frequency of the driving force is one of the normal mode frequencies of the system.

In the ear the auditory nerves are contained in a long-strip membrane whose width, stress, and effective mass density vary from one end to the other. The frequency of the sound determines the position of maximum vibration of the membrane. The acoustic energy is transmitted from the outer ear to the inner ear by an impedance-matching device containing a bone lever system.

A periodic sound is one for which the pressure at any point varies periodically:

$$p(t + T) = p(t).$$

Any periodic function can be written as a Fourier series:

$$p(t) = A_1 \cos(\omega t + \phi_1) + A_2 \cos(2\omega t + \phi_2)$$
$$+ A_3 \cos(3\omega t + \phi_3) + \cdots$$

PROBLEMS

32.A.1 The equation $f_n = nv/2L$ for the normal mode frequencies of a string with fixed ends can also be used for the frequencies of an organ pipe with closed ends. (One must then use the speed of sound for the wave speed.) Calculate the lowest-frequency normal mode of a 2-m organ pipe.

32.A.2 Using the values of B and ρ given in Table 32.1, find the velocity of sound waves in air and in water.

32.A.3 Find the fraction of energy reflected at an interface between water and another liquid with $B = 1.8 \times 10^9$ N/m² and $\rho = 1500$ kg/m³.

32.A.4
(a) Find the energy flux of a sound wave in air with an amplitude $A = 10^{-6}$ m and a frequency $f = 100$ Hz.
(b) If a sound wave with the same frequency and energy flux is traveling in water, what is its amplitude?

32.B.1 The ear canal is about 3 cm long. Consider it as a pipe with one closed (fixed) end and one open (free) end. Calculate the lowest normal mode frequency and compare it with the frequency of maximum hearing sensitivity (approximately 3000 Hz). (*Hint:* See problem 31.B.12.)

32.B.2 Show that the right-hand side of Eq. 32.1 has units of meters per second.

32.B.3 A thermally insulated cylinder 30 cm long and 10 cm in diameter is filled with water and closed at the end by a movable piston. The water is under normal air pressure (approximately 10^5 N/m²). What distance will the piston move if an added force of 100 N is applied (see the figure)?

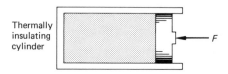

Thermally insulating cylinder — F

32.B.4 The sensitivity of the human ear is quite miraculous. Under the right conditions, a human can detect sounds at a frequency of approximately 3000 Hz whose amplitudes are approximately 10^{-11} m. This distance is about one-tenth the size of a hydrogen atom! Taking 1 cm² as the effective area of the ear, calculate the power (in watts) that arrives at the eardrum for barely detectable sounds.

32.B.5 A painfully loud sound has an energy flux about 10^{14} times that of a barely detectable sound (see problem 32.B.4). What is its amplitude?

32.B.6 An explosion takes place on the surface of the ocean, and the sound of the explosion travels through both the air and the water. How long does it take the sound to travel 1 km in air and how long to travel the same distance in water?

32.B.7 Experienced fishermen know that, when fishing by the shore of a lake, one must step very gently in order not to frighten away the fish. However, loud talking has no such frightening effect on the fish. Can you explain this apparent contradiction?

32.B.8 A plane sound wave bounces back and forth between the top and bottom of a pool that is 3 m deep. Assuming that there is perfect reflection at the bottom of the pool, determine how long it takes for the wave to lose 1% of its energy.

32.C.1 A violin string of length 30 cm and mass density 0.8 g/m is vibrating in its first normal mode with a frequency of 1200 Hz and an amplitude of 1 mm. If the string is left alone, its amplitude diminishes by a factor of 0.9 during 0.1 s as a result of interaction with the air. (The violin is creating sound waves that carry away energy.) Assuming that all the energy loss goes into producing sound waves, determine the power (in watts) of the sound waves being produced? (*Hint:* The solution to problem 31.C.5 is $\frac{1}{4}\mu\omega_n{}^2 B^2 L$.)

32.C.2 The long rubber band shown in the figure can be considered either a strip of membrane, in which case the wave speed in it would be calculated by Eq. 32.2, or else a string, in which case the wave speed would be calculated by Eq. 31.10. Show that both methods would give the same answer.

32.C.3 In Fig. 32.21, what is the phase difference between the signals being received by the two ears if the loudspeaker shown is at an angle of 45° with respect to the listener's nose and is producing sound of frequency 3000 Hz? Use an ''average'' head size.

32.C.4 Taking the range of human hearing as 20–20,000 Hz, determine the maximum number of pure tones that would be needed to match any periodic sound well enough so that the error would be undetectable by a human listener.

33
WAVES
IN
TWO
AND
THREE
DIMENSIONS

WAVES SUCH AS sound waves, water waves, or waves on a membrane that propagate in two-dimensional or three-dimensional media exhibit phenomena not seen in one-dimensional string waves. The comparative complexity of wave phenomena in multidimensional media is simply a reflection of the much greater complexity and richness of the geometry of two and three dimensions in comparison to the trivial geometry of one dimension.

Although it is possible to produce in three dimensions traveling waves whose wave fronts are geometrical surfaces of many different shapes, we shall consider only the three most important possibilities, namely plane waves, spherical waves, and cylindrical waves.

33.1 SPHERICAL WAVES

A spherical body that is pulsating periodically, so that its radius increases and decreases harmonically in time, produces a sound wave whose wave fronts are spheres centered about the body (see Fig. 33.1). Such waves are called *spherical waves* (or more exactly spherical harmonic waves). The wavelength and period of a spherical harmonic wave are defined in the same way as those of a plane harmonic wave. Provided the undisturbed fluid is uniform, the waves will propagate outward at constant velocity. The wavelength will therefore not depend on the distance, but the amplitude will diminish with increasing distance from the source. It is not difficult to derive the exact relation between distance and amplitude by using the energy conservation principle. The *acoustic power* of the source is the rate at which the source radiates acoustic energy; it is measured in watts (i.e., joules per second). If the energy is not dissipated in the intervening space, the total energy that passes through any spherical surface centered about the source during one second must be

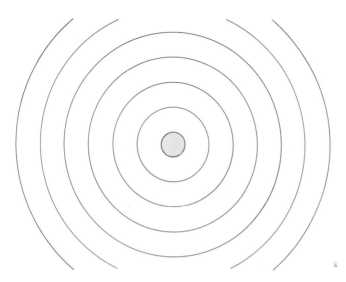

FIGURE 33.1 Spherical waves emanating from a pulsating body.

equal to the acoustic power. The rate of transport of energy through a spherical surface of radius r is equal to the energy flux at the location of the surface multiplied by the surface area. If $A(r)$ is the amplitude of the wave at radius r, then the acoustic power P is given by

$$P = \tfrac{1}{2}Z\omega^2 A^2(r) \times 4\pi r^2 \tag{33.1}$$

We could use this equation to determine the amplitude at radius r in terms of the source strength, the period, and the impedance. But we obtain a more generally useful relation by noting that Eq. 33.1 implies that $A(r)$ is proportional to $1/r$ if P is to be independent of r; that is,

$$A(r) = \frac{\text{const}}{r} \quad \text{(Spherical waves)} \tag{33.2}$$

QUESTION

What is the formula for $A(r)$ in terms of r, P, Z, and ω?

ANSWER

From Eq. 33.1 we get

$$A^2 = \frac{P}{2\pi Z\omega^2 r^2}$$

or

$$A = \frac{(P/2\pi Z)^{1/2}}{\omega r}$$

QUESTION

As mentioned in problem 32.B.4, under the right conditions it is possible to detect sounds at a frequency of approximately 3000 Hz whose amplitudes are approximately 10^{-11} m. Under ideal conditions, at what distance could the human ear just detect a loudspeaker with an acoustic power of 1 W and radiating spherical waves at 3000 Hz?

ANSWER

According to the results of the preceding question, the amplitude of the sound wave at a distance r is

$$A = \frac{(P/2\pi Z)^{1/2}}{\omega r}$$

In this case $A = 10^{-11}$ m, $P = 1$ W, $Z = 414$ kg/m² s, and $\omega = 6\pi \times 10^3$ Hz. Thus

$$r = \frac{(1/2\pi \times 414)^{1/2}}{(6\pi \times 10^3)(10^{-11})} = 1.04 \times 10^5 \text{ m}$$

If it were possible to set up ideal listening conditions, a person with sharp hearing could detect a 1-W sound source at a distance of 100 km!

CYLINDRICAL WAVES

A *cylindrical wave* is a wave whose wave fronts are concentric cylinders. If we assume that we have a source of harmonic cylindrical waves and we consider the rate at which energy flows through a cylindrical surface of radius r and unit length, then by an argument that is almost identical to that given for spherical waves we find that the amplitude of the wave at radius r is proportional to $1/\sqrt{r}$:

$$A(r) = \frac{\text{const}}{\sqrt{r}} \qquad \text{(Cylindrical waves)}$$

33.2 INTERFERENCE

When more than one source of waves exists, the wave amplitude at any point in space may be determined by the addition principle. One of the most important examples of this involves two sources of cylindrical waves of equal amplitude and frequency. We shall assume that the two sources are *in phase,* which means that they emit similar wave fronts simultaneously. Such a situation might occur if a plane sound wave were incident upon a rigid wall in which two narrow parallel slits were separated by a distance d. Each of the slits would act as a source of cylindrical waves (see Fig. 33.2). We shall want to determine the

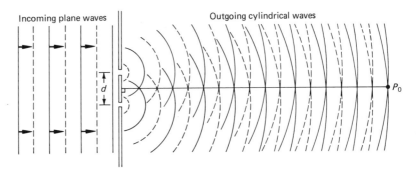

FIGURE 33.2 A cross section of a rigid wall in which two long slits have been cut. The pressure oscillation at the location of the slits causes cylindrical waves to emanate from each slit.

amplitude of the wave at points that are far from the two sources (that is, points at which the distance from the sources is much larger than the separation of the sources). At any point that is equally distant from both slits (such as the point P_0 shown in Fig. 33.2), the wave fronts from both slits will arrive simultaneously. For instance, a high-pressure wave front from the top slit will arrive at P_0 simultaneously with a high-pressure wave front from the bottom slit. The amplitude of the pressure variation at P_0 will therefore be twice as large as the amplitude that would have resulted from a single slit. The waves from the two slits are said to interfere *constructively* at point P_0.

If we now consider a different point that has the property that its distance from the upper slit is greater than its distance from the lower slit by exactly one-half a wavelength,

then the separate waves from the two slits arrive *out of phase*. A low-pressure wave front from one slit will arrive simultaneously with a high-pressure wave front from the other slit. The resultant pressure variation at that point is zero; that is, the pressure at that point does not fluctuate at all but remains equal to the value it would have in the absence of both waves. The two waves are said to interfere *destructively*. The condition that must be satisfied by those points at which destructive interference occurs is that $l_2 - l_1$ be equal to a half-integral number of wavelengths, where l_1 and l_2 are the distances from P_0 to the two slits (see Fig. 33.3). That is, we must have

$$d \sin \theta = (n + \tfrac{1}{2})\lambda \qquad \text{(Destructive interference)}$$

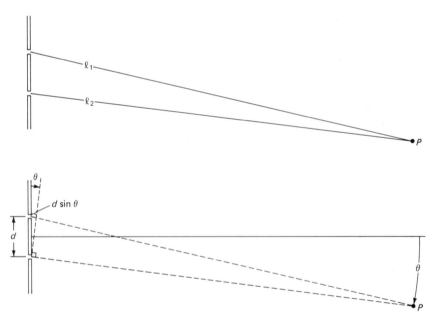

FIGURE 33.3 The distances between point P and the two slits differ by an amount $d \sin \theta$.

where n is any positive or negative integer. For constructive interference $l_2 - l_1$ must equal an integral number of wavelengths, or

$$d \sin \theta = n\lambda \qquad \text{(Constructive interference)}$$

The point P_0 in Fig. 33.2 satisfies the above equation with n equal to zero.

QUESTION

At an outdoor rock concert two loudspeakers are placed as shown in Fig. 33.4. What would the listener indicated hear when the speakers were emitting a complex periodic sound of fundamental frequency 3317 Hz?

0.5 m

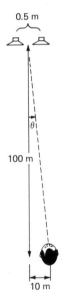

100 m

10 m

FIGURE 33.4   Position of loudspeakers relative to a listener.

ANSWER

The angle θ shown is 5.71°. (The angle is determined from the fact that $\tan \theta = \frac{1}{10}$.) At that angle destructive interference would occur at any wavelength given by

$$\lambda = \frac{(0.5 \text{ m})(\sin 5.71°)}{n + \frac{1}{2}}$$

Using the relation between wavelength and frequency ($f = v/\lambda$), we see that destructive interference would occur at frequencies satisfying the equation

$$f = \frac{(330 \text{ m/s})}{(0.5 \text{ m})(\sin 5.71°)}\left(n + \frac{1}{2}\right)$$

For $n = 0$, 1, and 2 those frequencies are

$$f = 3317 \text{ Hz, } 9950 \text{ Hz, and } 16,580 \text{ Hz}$$

Constructive inteference would occur at the frequencies given by

$$f = \frac{(330 \text{ m/s})}{(0.5 \text{ m})(\sin 5.71°)} n$$

which are

$$f = 6634 \text{ Hz, } 13,260 \text{ Hz, and } 19,900 \text{ Hz}$$

The complex sound is composed of pure tones of frequencies 3317, 6634, 9950, 13,270, 16,580, and 19,900 Hz (that is, f, $2f$, $3f$, etc.). Since the pure tones at frequencies 3317, 9950, and 16,580 Hz would be canceled by destructive interference, the listener would hear a complex sound made up of frequencies 6634, 13,270, and 19,900 Hz. Thus the sound would be periodic and of frequency 6634 Hz. (It would sound one octave higher in pitch.)

If light waves rather than sound waves are impinging upon an opaque screen with two narrow slits and we place a sheet of film parallel to the screen and at a large distance

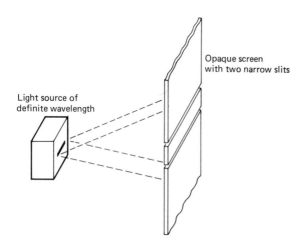

Light source of
definite wavelength

Opaque screen
with two narrow slits

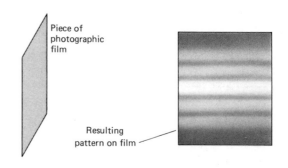

Piece of
photographic
film

Resulting
pattern on film

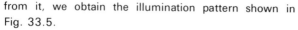

FIGURE 33.5   Production of a two-slit interference pattern.

from it, we obtain the illumination pattern shown in Fig. 33.5.

At this point we should take special note of the unique and peculiar feature of wave addition. If either one of the two slits is closed, the intensity (that is the energy flux) in the dark regions increases. It seems inconceivable that one could get such an effect if a beam of light consisted of a shower of particles. Rather it would seem that closing one slit would simply eliminate the particles emerging from that slit and would thereby reduce the intensity. Nevertheless, we shall contend later that in a certain clearly defined sense light does consist of particles.

DIFFRACTION GRATINGS

The interference effect we have noted is used in a device, called a diffraction grating, that is used to measure accurately the wavelength of light sources. The grating consists

of a thin glass plate, one side of which contains a very large number N (typically in the tens of thousands) of evenly spaced slits. A plane wave impinges upon the grating from one side, and the transmitted light is observed at a large distance from the grating on the other side. In certain directions the waves transmitted by all the transparent slits add in phase, so that the resultant wave amplitude and intensity are very large. From Fig. 33.6 we can see that the condition for such constructive interference is exactly what it was for the double slit, namely

$$d \sin \theta = n\lambda \tag{33.3}$$

The feature that distinguishes the many-slit interference grating from the two-slit interference grating is that in the

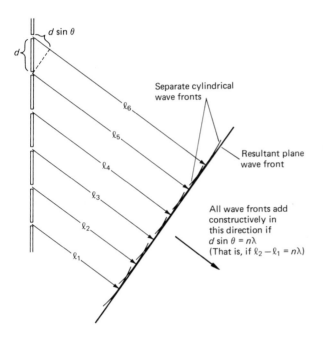

FIGURE 33.6   A diffraction grating.

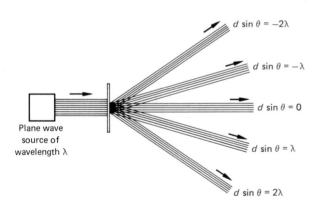

FIGURE 33.7   When a beam of light of wavelength λ strikes an interference grating, beams emerge from the other side in those directions θ satisfying the equation $d \sin \theta = n\lambda$.

former there is almost perfect destructive interference at all angles other than those given by Eq. 33.3. If a beam of light of a definite frequency strikes an interference grating, then a number of well-defined beams emerge from the other side of the grating, as shown in Fig. 33.7.

33.3 THE ELECTROMAGNETIC SPECTRUM

Electromagnetic waves, whose physical nature we described in Chap. 30, can have any wavelength between zero and infinity. Near one end of the spectrum of wavelengths are the waves produced by 60-Hz power lines; these have wavelengths of about 5×10^6 m. (The amplitude of the waves generated by a power line is extremely small, and thus the power lost by radiation is completely negligible.) Near the other end of the spectrum of observed electromagnetic radiation are the waves produced by astrophysical cosmic ray sources; these have wavelengths of less than 10^{-12} m (see Fig. 33.8). The source of these extremely short wavelength electromagnetic waves is still unknown.

Visible light consists of electromagnetic waves in the wavelength range from about 0.4×10^{-6} m to about 0.7×10^{-6} m. As the wavelength decreases (i.e., as the frequency increases), the observed color of the light changes from red to orange to yellow to green to blue to violet. Most light sources produce light that is a mixture of many different frequencies, just as most sound sources produce complex sounds made up of many different pure tones. A diffraction grating may be used to separate the light mixture into its different frequency components. If a thin beam of white light strikes a diffraction grating,

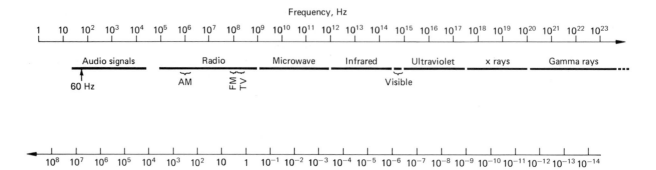

FIGURE 33.8 The electromagnetic spectrum.

then the undeviated beam will still contain all components mixed together (i.e., it will still be white). However, the first deviated beam for that component that has wavelength λ will occur at an angle θ, given by $\sin \theta = \lambda / d$. Since θ will be different for each wavelength component of the mixture, the deviated beam will not leave in a single direction but will fan out according to wavelength, and will produce the *first-order spectrum*. Thus the wavelength mix of the complex light beam may be determined (see Fig. 33.9). The second deviated beam for each wavelength component produces the *second-order spectrum,* and so on.

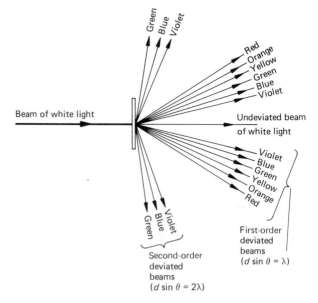

FIGURE 33.9 In the deviated beams the different wavelength components leave in different directions.

In our previous discussion of interference phenomena we used the fact that when a plane wave is incident from the left upon an opaque screen with a thin slit, the slit acts as a source of cylindrical waves in the region to the right of the screen. If the opaque screen contained a small hole instead of a slit, the hole would act as a source of spherical waves (see Fig. 33.11). We shall now introduce a method

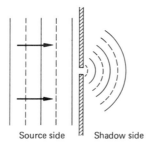

FIGURE 33.11 A small opening in an opaque screen acts as a source of spherical waves on the shadow side.

of calculating the radiation on the shadow side of an opaque screen that has an opening of finite size and general shape. The method is called *Huygens' principle*. It was first proposed by Christian Huygens and later proved by Gustav Kirchhoff, who used the mathematical theory of wave propagation. Huygens' principle states that one may treat a large aperture as a collection of very small openings each of which is a source of spherical waves (see Fig. 33.12). Thus we can calculate the wave amplitude at a large distance from the aperture by using the addition principle to sum up the contributions of the spherical waves from all the small areas into which the larger opening has been decomposed. The amplitude of each spherical wave is proportional to the amplitude of the plane wave that strikes the illuminated side of the screen. The proportionality constant may be calculated, but we shall not do the calculation here. Using Huygens' principle, one can calculate the wave pattern in the region to the right of an

---

QUESTION

A thin beam of light strikes an interference grating etched with 10,000 lines per centimeter. The light is composed of two components: one of wavelength 0.5 μm and one of wavelength 0.6 μm. What will be the angle between the outgoing deviated beams resulting from the two wavelength components (i.e., the angle α shown in Fig. 33.10)?

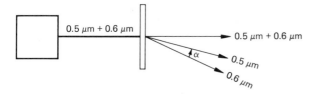

FIGURE 33.10 The wavelength components, 0.5 μm and 0.6 μm are separated by an angle α in the first-order spectrum.

ANSWER

The distance between adjacent lines in the grating is 0.0001 cm, so $d = 1$ μm. The angles that the 0.5 μm and 0.6 μm deviated beams make with the undeviated beam are θ and θ', where

$$\sin \theta = \frac{0.5 \ \mu m}{1 \ \mu m} \quad \text{or} \quad \theta = 30°$$

and

$$\sin \theta' = \frac{0.6 \ \mu m}{1 \ \mu m} \quad \text{or} \quad \theta' = 36.87°$$

Therefore

$$\alpha = 6.87°$$

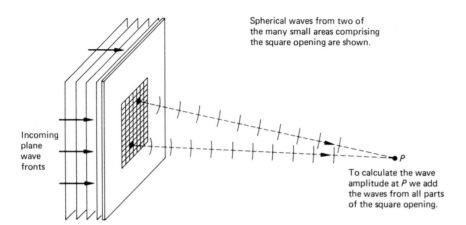

Spherical waves from two of
the many small areas comprising
the square opening are shown.

Incoming
plane
wave
fronts

To calculate the wave
amplitude at P we add
the waves from all parts
of the square opening.

FIGURE 33.12   A large opening in an opaque screen can be considered a collection of very small openings, each a source of spherical waves. The net wave amplitude at P is the sum of the contributions from all the small openings.

opaque screen in which an aperture of arbitrary shape is being illuminated from the left. The calculation, however, is quite complex, and thus we shall simply describe the results without a mathematical derivation.

If the hole in the screen is small in comparison to the wavelength of the illuminating wave, then the wave pattern on the shadow side of the screen is that of a spherical wave emanating from the hole, as shown in Fig. 33.13. If the hole is large in comparison to the wavelength, then most of the plane wave will pass through the hole undeviated. At the edge of the shadow region there will not be an entirely abrupt change in the intensity, but rather there will be a narrow interference pattern, as shown in Fig. 33.14. The bending of light (or any waves) around the edges of an obstacle is called *diffraction.* Although we cannot calculate the diffraction pattern in detail for any

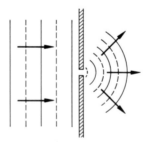

FIGURE 33.13   We get spherical waves on the right if the diameter of the hole is smaller than λ.

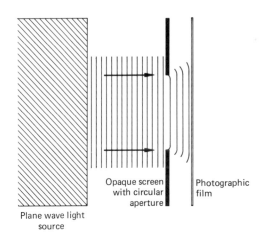

Plane wave light source

Opaque screen with circular aperture

Photographic film

Actual shape of aperture

Shadow of aperture on film

FIGURE 33.14 If the aperture is much larger than λ, then most of the light is undeviated. However, the edge of the shadow is not completely sharp but exhibits a narrow diffraction pattern.

aperture, we shall examine one important case in which we can use Huygens' principle to obtain most of the important details of the pattern. This is the case of a very long single slit.

33.5 SINGLE-SLIT DIFFRACTION

An opaque screen has in it a very long slit of width W that is being illuminated by a plane wave of wavelength λ. We want to calculate the wave amplitude at a point P very far from the slit in the direction θ (see Fig. 33.15). The first thing we shall show is that the intensity at point P is zero whenever θ is equal to one of the angles θ_1, θ_2, θ_3, etc., where

$$\sin \theta_n = \frac{n\lambda}{W} \tag{33.4}$$

Thus, if a sheet of photographic film were placed a large distance from the slit, then at the angle θ_n one would obtain a dark line.

Even before we prove the above statement, let us note that for the case in which the slit is of width less than λ (that is, if $w < \lambda$) there are no solutions to Eq. 33.4 since $\sin \theta$ cannot be greater than 1; that is, there are no dark

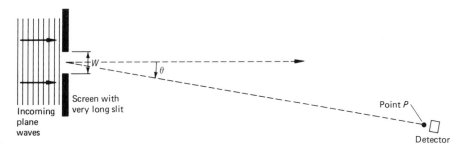

Incoming plane waves

Screen with very long slit

Point P

Detector

FIGURE 33.15 Single-slit diffraction setup. The distance from the aperture to point P is much larger than the width of the aperture.

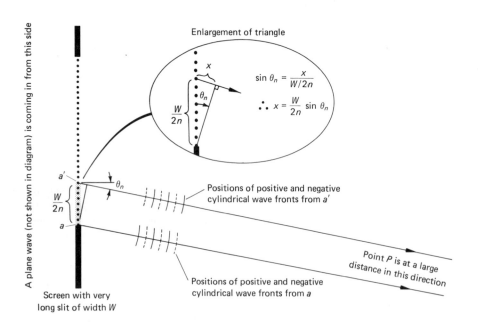

Enlargement of triangle

$$\sin \theta_n = \frac{x}{W/2n}$$

$$\therefore x = \frac{W}{2n} \sin \theta_n$$

A plane wave (not shown in diagram) is coming in from this side

Positions of positive and negative cylindrical wave fronts from a'

Point P is at a large distance in this direction

Positions of positive and negative cylindrical wave fronts from a

Screen with very long slit of width W

Note: In this diagram $n = 2$

FIGURE 33.16 The slit of width w has been decomposed into a large number of very narrow slits that are represented by the spaces between the dots. The waves from the narrow slit a' arrive at the detector exactly one-half wavelength behind those from the narrow slit a.

fringes. Thus, if the slit width is less than the wavelength, every place on the shadow side of the screen is illuminated to some extent.

In Fig. 33.16 we have assumed that the detector is at point P, which is a very large distance in the direction θ_n for the case $n = 3$. Using Huygens' principle, we have considered the slit, which is of finite width W, as being composed of a large number of extremely narrow slits, represented by the spaces between the dots in the figure. Each narrow slit is a source of cylindrical waves on the shadow side. We have selected two of the narrow slits and labeled them a and a'. They are separated by a distance $W/2n$. As we can see from the figure, the distance from a' to the point P is larger than the distance from a to P by

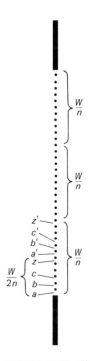

FIGURE 33.17   The waves from a' cancel those from a at the detector; the waves from b' cancel those from b; and so on. Thus all the waves from the section of width w/n add up to zero at the detector. This is true for each of the n sections of the slit, and therefore the total intensity at the detector is zero.

FIGURE 33.18   Single-slit diffraction pattern.

an amount $(W/2n)\sin \theta_n$. But by the definition of θ_n, we see that

$$\frac{W}{2n}\sin \theta_n = \frac{W}{2n}\frac{n\lambda}{W} = \frac{1}{2}\lambda$$

Thus the cylindrical waves arriving at P from a' exactly cancel the cylindrical waves arriving at P from a, because the two waves arrive at P one-half wavelength out of phase. Referring now to Fig. 33.17, we see that the waves arriving at P from b and b' also cancel, as do those from c and c', and so on to z and z'. Thus the net wave amplitude at P created by all the narrow slits in the section (of width W/n) shown in the figure is zero. The same is true for each of the n sections that make up the complete slit of width W. Therefore the net wave amplitude at P is zero, and this is what we set out to prove. Note that an essential ingredient in the argument was the assumption that the point P was at an angle θ_n, which was a solution of Eq. 33.11 for the integer n. At some arbitrary angle one could not carry through the above argument simply by dividing the slit width into n equal parts.

If the distance between the slit and the light detector is kept constant while the angle θ made by the normal to the screen and the line drawn to the detector is varied, the light intensity measured by the detector will vary. It will be a maximum when the detector is directly in front of the slit. The light intensity will go to zero at the angles $\theta = \theta_1$, θ_2, etc. Between the two angles θ_n and θ_{n+1}, it will rise and then fall. On a photographic film one would obtain a bright line between those two dark lines. The pattern obtained is somewhat reminiscent of the two-slit interference pattern, but differs from it in the following important respect. *The intensity of the nth bright line decreases rapidly with increasing* n. Most of the energy passing through the slit goes into the central bright line centered about $\theta = 0$. When the slit is wide enough so that secondary, tertiary, etc., bright lines exist, their intensities are much less than that of the central line. This can easily be seen in Figs. 33.18 and 33.19.

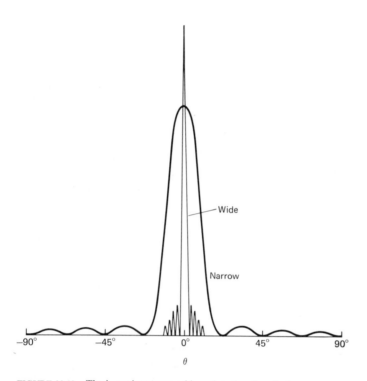

FIGURE 33.19 The intensity measured by a detector placed a large distance from a single slit given as a function of angle for the cases of narrow and wide slits.

DIFFRACTION BY APERTURES OF GENERAL SHAPE

Although the details of the diffraction pattern produced by an aperture depend on the exact shape of the aperture, the analysis we have just completed can give us a useful estimate of the range of angles into which a plane wave is diffracted by an aperture whose smallest dimension is W. The estimate we shall use is the angle between the central maximum and the first minimum at θ_1. Thus an aperture

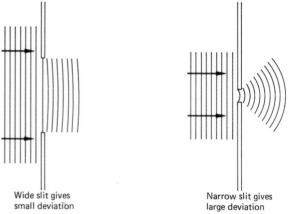

FIGURE 33.20 The angular spread of the waves on the shadow side depends on the size of the slit in comparison to the wavelength.

whose smallest dimension is W spreads an incoming plane wave front over a range of angles θ given by

$$\theta = \frac{\lambda}{W} \qquad (33.5)$$

where we have assumed that λ/W is small enough so that we can use the approximation $\sin \theta = \theta$.

QUESTION

As we shall show in detail later, it is the diffraction of light passing through the pupil of the eye that limits the resolution of the eye (i.e., the sharpness of the image on the retina). What is the angular spread of yellow light passing through a 2-mm pupil?

ANSWER

Yellow light has a wavelength of about 0.58 μm. Thus the diffraction angle θ is

$$\theta = \frac{0.58 \times 10^{-6} \text{ m}}{2 \times 10^{-3} \text{ m}} = 2.9 \times 10^{-4} \text{ rad}$$

Details that subtend an angle smaller than θ cannot be distinctly resolved. At a distance of 1 km the size of a detail that subtends an angle of 3×10^{-4} rad is 0.3 m (about 1 ft).

33.6 REFRACTION

When a plane wave strikes a plane interface between two media, it is partially reflected and partially transmitted. We shall now see that the transmitted wave fronts are not parallel to the incident wave fronts. We shall also determine the law of bending of the wave fronts. Figure 33.21 shows the incident and transmitted wave fronts at times

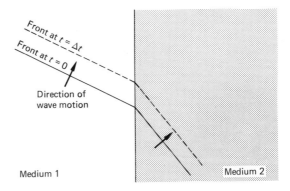

FIGURE 33.21 Propagation of a wave front from one medium into another.

$t = 0$ and $t = \Delta t$. We have not indicated the reflected wave, although one exists, because it is irrelevant to our analysis.

We assume that the velocities of wave propagation in the two media are different; in medium 1 the velocity is v_1 and in medium 2 it is v_2. The distances that the wave fronts would propagate during the time interval Δt are $v_1 \Delta t$ and $v_2 \Delta t$ in medium 1 and medium 2, respectively. From Fig. 33.22 we see that this implies that

$$\sin \theta_1 = \frac{v_1 \Delta t}{s}$$

and

$$\sin \theta_2 = \frac{v_2 \Delta t}{s}$$

where s is the distance between the wave fronts at the interface. Therefore

$$\frac{\sin \theta_1}{v_1} = \frac{\sin \theta_2}{v_2} \qquad (33.6)$$

Equation 33.6 is called *Snell's law*.

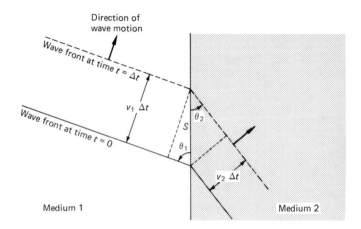

FIGURE 33.22   Geometrical construction for Snell's law.

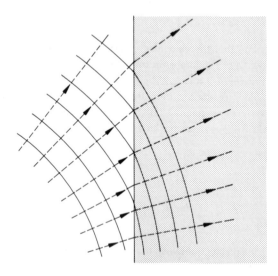

FIGURE 33.23   The system of rays associated with a set of curved wave fronts propagating from one medium into another.

It is often convenient to describe the progress of a wave in two or three dimensions by a system of lines called rays. A *ray* is a line that is drawn so that it is normal to the wave front at every point. An example of such a construction is shown in Fig. 33.23. The propagation speed in the right-hand medium is smaller than that in the left-hand medium. The rays are shown as dashed lines. As we can see from Fig. 33.24, the angle that a wave front makes with the interface is equal to the angle that the corresponding ray makes with the normal to that interface. Using this fact we can interpret Snell's law as a law for the bending of light rays at an interface (see Fig. 33.25). For light propagation through transparent materials, it is common to write Snell's law in terms of a property of the material called its index of refraction. The *index of refraction n* of a material is defined as the ratio of the speed of light propagation in a vacuum to the speed of light propagation in the material:

$$n = \frac{c}{v}$$

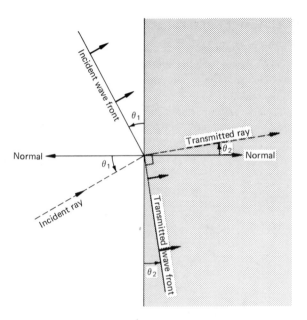

FIGURE 33.24   The light ray in each medium is normal to the wave front in that medium; therefore the angle that the light ray makes with the normal to the interface is the same as the angle the wave front makes with the plane of the interface.

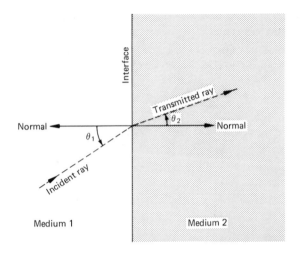

FIGURE 33.25   Snell's law (Eq. 33.6) gives the relation between the angles made by the incident and transmitted light rays with the normal to the interface. Note that both rays lie in a single plane.

TABLE 33.1   Indices of refraction for a number of transparent substances

Substance	Index of refraction
Water	1.33
Glass (plate)	1.52
Glass (flint)	1.66
Quartz crystal	1.54
Diamond	2.42
Air	1.00

where c is the speed of light in a vacuum. Careful measurements have shown that $c = 2.997925 \times 10^8$ m/s. In all our calculations we shall use the simpler approximate value $c = 3 \times 10^8$ m/s. In terms of the indices of refraction of the two media, Snell's law takes the form

$$n_1 \sin \theta_1 = n_2 \sin \theta_2$$

Table 33.1 gives the indices of refraction of various transparent substances.

QUESTION

In going from a vacuum into a piece of glass, a light ray is bent as shown in Fig. 33.26. What is the speed of light propagation in the glass?

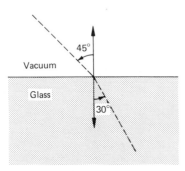

FIGURE 33.26   Bending of a light ray as it passes from a vacuum into a piece of glass.

ANSWER

The index of refraction of a vacuum is 1, and we call the index of refraction in the glass n_g. Thus, according to Snell's law, $\sin 45° = n_g \sin 30°$, so the index of refraction of the glass is

$$n_g = \frac{\sin 45°}{\sin 30°} = 1.415$$

The speed of light in the glass is therefore

$$v = \frac{c}{n_g} = 2.12 \times 10^8 \text{ m/s}$$

33.7 THE LAW OF REFLECTION

Figure 33.27 shows a wave front being (completely or partially) reflected at a flat interface between two materials. The wave front is shown at times $t = 0$ and $t = \Delta t$. By an argument similar to the one given in our derivation of Snell's law, we can show that the reflected wave front makes the same angle with the interface as did the incident wave front:

$$\theta_r = \theta_i \tag{33.7}$$

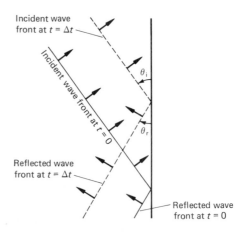

FIGURE 33.27   Incident and reflected wave fronts at times $t = 0$ and $t = \Delta t$. The angles θ_i and θ_r are equal.

We can also use Eq. 33.7 as a law of reflection of light rays (see Fig. 33.28).

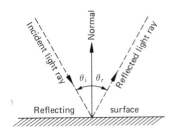

FIGURE 33.28 The reflected light ray makes the same angle with the normal to the surface as does the incident light ray.

33.8 TOTAL INTERNAL REFLECTION

When two transparent materials have different indices of refraction, the material with the larger index of refraction (and hence the smaller wave propagation speed) is called the *optically denser* material. Snell's law predicts that a light ray passing into a medium of greater optical density will bend toward the normal; that is, θ_2 will be smaller than θ_1. On the other hand, when a light ray enters a material of lower optical density, it bends away from the normal to the interface. A peculiar problem seems to arise in the latter case when one attempts to use Snell's law to calculate the angle θ_2 taken by the ray in the less optically dense medium. According to Snell's law,

$$\sin \theta_2 = \frac{n_1}{n_2} \sin \theta_1$$

where, by assumption, n_1/n_2 is greater than 1. If the angle of the incident ray is sufficiently large, then the right-hand side of this equation will yield a number larger than 1. But $\sin \theta_2$ cannot be larger than 1, and thus there would be no solution to the equation for θ_2! What actually happens to the incident light ray in this paradoxical situation in which Snell's law can make no prediction?

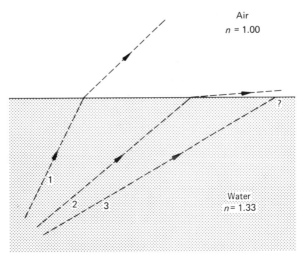

FIGURE 33.29 Snell's law can be used to calculate the directions of light rays 1 and 2, but for ray 3 it gives absurd results.

We illustrate the problem in Fig. 33.29 by showing three different incident rays in water and two of the corresponding transmitted rays. We resolve the paradox by remembering that there are reflected rays that are not indicated in the figure. The fraction of the incident light that is reflected depends on the direction of the incident ray. If we gradually increase the angle of the incident ray, we find that the reflected ray gradually increases in amplitude until all the light is reflected when θ_1 is equal to or larger than the angle θ_0, which is given by

$$\sin \theta_0 = \frac{n_2}{n_1}$$

The angle θ_0 is called the *critical angle,* and has the following property. If a ray is incident upon an interface at the angle θ_0 from the optically denser material, then Snell's law will predict that the transmitted ray will leave the interface at an angle of 90° (that is, it will just graze the interface). In actuality there will be no transmitted ray if the incident ray makes an angle of θ_0 or more with the normal to the interface. Thus when Snell's law gives absurd results for the angle of the transmitted ray, there is no transmitted ray. In this case all the light is reflected at the interface back into the optically denser medium.

QUESTION

What is the smallest value of the index of refraction for which a 90° prism with no mirror coating on its back surface will give complete reflection of the light ray (see Fig. 33.30)? The index of refraction of air is 1.00.

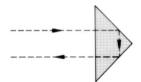

FIGURE 33.30   A light ray striking a 90° prism.

ANSWER

The angle at which the light ray strikes the back surface of the prism is 45°. If the index of refraction is barely sufficient to cause complete reflection, then the critical angle must be 45°. Therefore

$$\sin 45° = \frac{n_1}{n_2} = \frac{1}{n_2}$$

or

$$n_2 = \sqrt{2}$$

SUMMARY

The amplitude of a spherical wave diminishes with distance from the source by the relation

$$A = \frac{\text{const}}{r}$$

where r is the distance from the source. For a cylindrical wave the corresponding relation is

$$A = \frac{\text{const}}{\sqrt{r}}$$

If the distances from two different sources of waves to a point P differ by an integral number of wavelengths, the waves from the two sources add constructively at P. If the distances differ by $(n + \frac{1}{2})\lambda$, the waves from the two sources interfere destructively at P.

For an opaque screen with two slits the condition for constructive interference is

$$d \sin \theta = n\lambda, \qquad n = 0, \pm 1, \pm 2, \ldots$$

where d is the distance between the slits, λ is the wavelength, and θ is the angle between a line from the slits to point P and the normal to the opaque screen containing the two slits. The condition for destructive interference (zero intensity) is

$$d \sin \theta = (n + \frac{1}{2})\lambda, \qquad n = 0, \pm 1, \pm 2, \ldots$$

For a diffraction grating the condition for constructive interference is the same as for the two-slit screen, but very little light is transmitted in any direction that does not satisfy the condition for constructive interference.

Radiowaves, infrared light, visible light, ultraviolet light, x rays, and γ rays are all the same type of waves. They differ only in wavelength and frequency.

The wave pattern that passes through an aperture of arbitrary shape in an opaque screen may be calculated using Huygens' principle, which states that each tiny element of area within the aperture acts as a source of spherical waves. The wave amplitude at any point in space (on the shadow side of the opaque screen) may be calculated by adding the contributions of all the spherical waves and taking account of their relative phases. Huygens' principle correctly predicts that the shadow of a sharp edge is not perfectly sharp.

A plane wave, in passing through an aperture whose smallest dimension is W, is spread out into a diverging beam of angle

$$\theta \simeq \frac{\lambda}{W}$$

When a wave passes from one medium to another, the rays (normals to the wave fronts) are bent in accordance with Snell's law:

$$\frac{\sin \theta_1}{v_1} = \frac{\sin \theta_2}{v_2}$$

where v_1 and v_2 are the wave speeds in the two media and θ_1 and θ_2 are the angles made by a normal to the interface and the ray in the corresponding media. Snell's law may also be written in terms of the indices of refraction of the media:

$$n_1 \sin \theta_1 = n_2 \sin \theta_2$$

where $n = c/v$ (c = speed of light in a vacuum and v = speed of light in the material).

When a light ray is reflected at a surface, the angle that the reflected ray makes with the normal to the surface is equal to the angle that the incident ray makes with the normal.

If a light ray approaches a surface and the material on the other side of the surface has a lower index of refraction than the material that the ray is passing through, then the ray will be completely reflected at the surface if the angle it makes with the normal is larger than θ_0, where

$$\sin \theta_0 = \frac{n_{lower}}{n_{higher}}$$

PROBLEMS

33.A.1 At 40 m from a source of 4000-Hz spherical sound waves, the amplitude is 2×10^{-6} m. What is the amplitude at 120 m? What is the power of the source?

33.A.2 A two-slit interference device is illuminated by light of wavelength 0.8 μm. The first dark lines occur at angles of $\pm 10^{-4}$ rad. What is the distance between the slits?

33.A.3 In Fig. 33.7, if the angle between the undeviated beam and the beam marked $d \sin \theta = \lambda$ is 20°, what is the angle between the undeviated beam and the beam marked $d \sin \theta = 2\lambda$?

33.A.4 What is the range of frequencies of visible light? The range of wavelengths is 0.4–0.7 μm.

33.A.5 A beam of light of wavelength 0.5 μm is diffracted in passing through a narrow slit of width 0.05 mm. At what angle is the first dark line in the diffraction pattern?

33.A.6 A light ray is bent in going from air to glass. If the incident ray makes an angle of 45° with the normal to the interface and the refracted ray makes an angle of 20°, what is the index of refraction of the glass?

33.A.7 What is the velocity of light propagation in a material of index of refraction $n = 2.2$?

33.B.1 A loudspeaker radiates acoustic energy in the form of sound waves at 2000 Hz. Assuming that the energy is radiated uniformly over a hemisphere and that the total power is 1 W, determine the wave amplitude at 3 m from the speaker. At what distance from the speaker will the wave amplitude by 10^{-8} cm?

33.B.2 The intensity of sound is commonly measured in *decibels* (abbreviated dB). If I is the intensity of a sound in watts per square meter, then its intensity in decibels is given by

$$I_{dB} = 10 \log_{10} \frac{I}{I_0}$$

where $I_0 = 10^{-12}$ W/m^2.
(a) If we double I, what happens to I_{dB}?
(b) What is I_{dB} at 3 m from a 10-W hi-fi that radiates equally in all directions?

33.B.3 Light waves pass through two parallel slits a distance of 0.1 mm apart and then impinge upon a wall 2 m away. If the distance between adjacent maxima in the center of the interference pattern is 1.2 cm, what is the wavelength of the waves?

33.B.4 The index of refraction of pure water is 1.33. Suppose the index of refraction of a particular sugar solution is 1.42. If we construct a cubical glass container that is partitioned diagonally and fill one half with water and the other half with the sugar solution, by how many radians will the light ray shown in the figure be deviated?

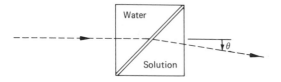

33.B.5 Two fish that are equally far below the surface of a quiet lake are 4 ft apart. What is the maximum depth at which the fish will see each other perfectly reflected on the surface of the lake (see the following figure)?

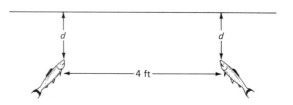

33.C.1 The speed of a water wave depends on the depth of the water, and water waves travel more slowly in shallower water. Use these facts to explain why the waves at the beach are always coming directly at the shoreline.

33.C.2 As viewed from a depth d below the surface of a quiet lake, the image of all the "outside world" forms a circle of radius R. What is the value of R?

33.C.3 A ray of light passes through a uniform sheet of glass whose index of refraction is n and whose thickness is d (see the figure). Show that the transmitted ray is parallel to the incident ray but shifted by a distance s given by

$$s = d \sin \theta \left(1 - \frac{\cos \theta}{\sqrt{n^2 - \sin^2 \theta}}\right)$$

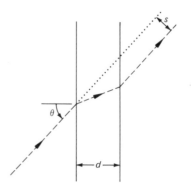

33.C.4 An ambitious junior detective constructs a "snoopermike" by nailing together two large boards at a 90° angle and placing a microphone along the line of symmetry, as shown. For maximum sensitivity for 3000 Hz, what should the distance d be? Estimate the increase in the amplitude of the pressure variation caused by the reflector.

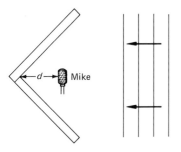

34
OPTICS

IN THIS CHAPTER we shall see how we may apply the basic laws governing the refraction, reflection, and diffraction of light to obtain an understanding of the operation of a wide variety of optical devices. The first problem we shall consider is how we can use Snell's law of refraction to calculate the focusing power of a lens.

34.1 SINGLE-SURFACE LENSES

If the interface between two media with different indices of refraction is curved, then the shape of a wave front will be changed as it passes from one medium into the other. A plane wave in one medium may become a spherical wave in the other, or it may become a wave front with no simple geometrical shape. A single-surface converging lens is an interface between two media that is so shaped that it can convert that portion of an expanding spherical wave front that strikes the interface into a contracting spherical wave front (see Fig. 34.1). We shall now show that under certain conditions a spherically shaped interface will act as a converging lens. In the analysis we will make certain approximations, so that the predictions we make will not agree exactly with the observed phenomena.

Let us consider an interface between two media with indices of refraction n_1 and n_2. We assume that the interface is a spherical surface of radius R (see Fig. 34.2) and that the point C is the center of the spherical interface (see Fig. 34.1). In medium 1 a point source of light is at position A. We want to prove that all those rays that leave A at small angles with respect to the line between A and C will converge to a focus at some point B behind C. We have drawn one of those rays in the figure. By Snell's law

$$n_1 \sin \theta_1 = n_2 \sin \theta_2$$

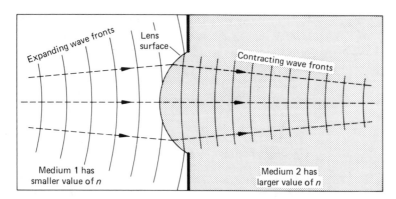

FIGURE 34.1   A curved surface can convert expanding wave fronts into contracting wave fronts.

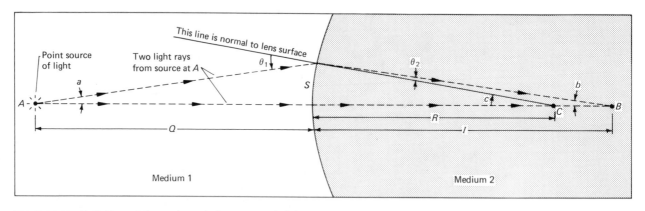

FIGURE 34.2   Definition of the angles and distances needed for the analysis of the single-surface lens. Two rays emanating from the object at A intersect at the image point B.

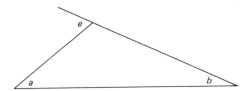

FIGURE 34.3   The exterior angle e is equal to the sum of the two opposite interior angles a and b.

If θ_1 and θ_2 are small, we can use the approximation $\sin \theta \simeq \theta$ to write Snell's law in the form

$$n_1\theta_1 = n_2\theta_2 \qquad (34.1)$$

We need the following simple theorem from geometry: In a triangle such as that shown in Fig. 34.3, any exterior angle e is equal to the sum of the two internal angles a and b. Using this theorem, we see that

$$\theta_1 = a + c \qquad (34.2)$$

and

$$c = \theta_2 + b \qquad (34.3)$$

In Fig. 34.2 we have defined an *object distance* Q and an *image distance* I. From the definition of radian measure we know that the angle c is the ratio of the arc length S to the radius R:

$$c = \frac{S}{R}$$

If the angles a and b are sufficiently small, we can use the approximations

$$a = \frac{S}{Q}$$

and

$$b = \frac{S}{I}$$

Substituting these expressions for a, b, and c into Eqs. 34.2 and 34.3, we have

$$\theta_1 = \frac{S}{R} + \frac{S}{Q}$$

and

$$\theta_2 = \frac{S}{R} - \frac{S}{I}$$

Finally, putting these expressions for θ_1 and θ_2 into Eq. 34.1 and canceling the common factor S, we see that

$$\frac{n_1}{R} + \frac{n_1}{Q} = \frac{n_2}{R} - \frac{n_2}{I}$$

or

$$\frac{n_1}{Q} + \frac{n_2}{I} = \frac{n_2 - n_1}{R} \qquad (34.4)$$

This relation tells us the distance at which the two rays in the figure cross. Since the crossing distance I does not depend on the angle a, we can see that *all* rays from A come together at a common point. This is exactly what we set out to prove. In order for all the rays in medium 2 to converge to a common point, the wave fronts in medium 2, which are perpendicular to those rays, must be converging spherical wave fronts.

Before we proceed any further in the mathematical development, let us consider what the physical or observational

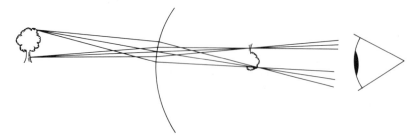

FIGURE 34.4 All rays that leave a single point on the object come together at a single point on the image.

consequences are of the imaging property we have just proved. Suppose an illuminated object is in medium 1 (e.g., the tree shown in Fig. 34.4). Each point on the surface of the object will be a source of diverging light rays. The light rays that diverge from any one point on the object will all intersect at a point in medium 2 called the *image* of that point. The rays will pass through the image and again diverge. If viewed from the far right (see Fig. 34.4), the set of diverging rays will make it appear as if an object is at the image location. A white screen placed at the image location will show a picture of the object. If we know the object distance Q, the indices of refraction n_1 and n_2, and the radius R of the spherical surface, then Eq. 34.4 allows us to predict the position of the image.

QUESTION

A candle is placed 3 m away from a spherical fishbowl of diameter 20 cm (see Fig. 34.5). Will an image be produced inside the fishbowl? (The index of refraction of water is 1.33.)

FIGURE 34.5 Will an image be produced within the fishbowl?

ANSWER

The object distance Q is 3 m and the radius R of the spherical surface is 0.1 m. The indices of refraction n_1 and n_2 are equal to 1.00 and 1.33, respectively. Substituting these values into Eq. 34.4, we obtain an equation for the image distance I:

$$\frac{1.00}{3\text{ m}} + \frac{1.33}{I} = \frac{0.33}{0.1\text{ m}}$$

Solving for I, we see that

$$I = 0.45\text{ m}$$

Since 0.45 m is larger than the diameter of the fishbowl, no image is formed inside the bowl. (However, an image is formed outside the bowl to the right.)

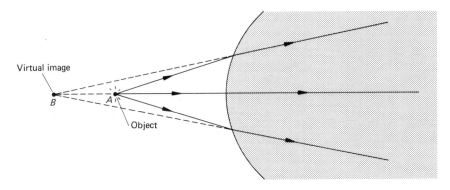

FIGURE 34.6 When I is negative, the rays on the right side are still diverging.

In our derivation of Eq. 34.4 we have assumed that the quantities R, I, and Q are all positive. Actually Eq. 34.4 is correct and has a physical interpretation when any of these quantities is negative. The meanings of negative R and I are given by the following rules. (The meaning of negative Q will be discussed later.)

1. When the lens surface, as viewed from the object side, is concave, the value of R to be used in Eq. 34.4 is minus the radius of curvature.

2. When the image distance I is negative, the rays within medium 2 diverge as if they were coming from a point B (called the *virtual image* of A) that is a distance $-I$ to the left of the lens surface (see Fig. 34.6).

QUESTION

Why does an object submerged in water appear (from outside the water) to be closer to the surface than it actually is?

ANSWER

We can understand this phenomenon and calculate the apparent position of a submerged object by considering the flat surface of the water as the surface of a sphere of infinite radius. We can then use Eq. 34.4 with $R = \infty$. As an example, consider a point source of light on the bottom of a pool that is 2 m deep. From outside the pool the rays will appear to be diverging from a point a distance d below the surface (see Fig. 34.7). To calculate d we note that in this case $n_1 = 1.33$,

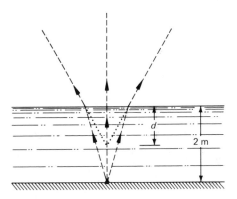

FIGURE 34.7 The rays above the water seem to be coming from a point d meters below the surface.

$n_2 = 1.00$, $R = \infty$, and $Q = 2$ m. Therefore Eq. 34.4 gives

$$\frac{1.33}{2\ \text{m}} + \frac{1.00}{I} = \frac{-0.33}{\infty} = 0$$

or

$$I = -\frac{2\ \text{m}}{1.33} = -1.5\ \text{m}$$

According to our stated interpretation of negative image distances, the rays outside the water are diverging from a point 1.5 m below the surface. Thus the source would appear to be at that depth.

34.2 THE DOUBLE-SURFACE LENS

Although the single-surface lens is very important in biological systems (the major lens of the mammalian eye is the curved front surface of the cornea, a single-surface lens), the lenses used in optical equipment almost always have two surfaces. It is easy to use our analysis of the single-surface lens to obtain an equation describing the imaging properties of double-surface lens. We shall make the assumption that the lens is a thin lens so that we may measure the image and object distances from the center of the lens rather than from its surface.

Let us consider the double-surface lens shown in Fig. 34.8. We assume that the point object is so close to the lens that the left surface of the lens alone does not form a real image. Inside the lens the rays still diverge from a point a distance $-I'$ to the left, where I' is given by the single-surface lens equation

$$\frac{n_a}{Q} + \frac{n_g}{I'} = \frac{n_g - n_a}{R_1} \tag{34.5}$$

The final image, produced by the second surface, would be the same as the image produced by the single-surface lens shown in Fig. 34.9. Applying the single-surface lens

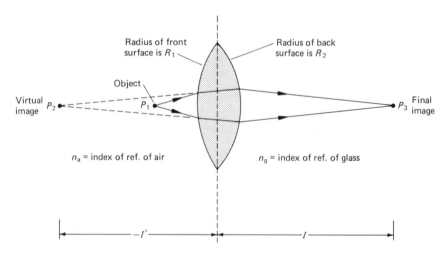

FIGURE 34.8 The double-surface lens. The virtual image of the first surface acts as the object for the second surface.

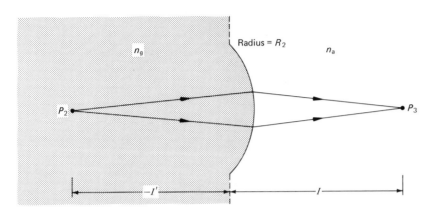

FIGURE 34.9 The rays that strike this single-surface lens are the same as those that strike the second surface of the double-surface lens shown in Fig. 34.8.

formula again, with $n_1 = n_g$, $n_2 = n_a$, $Q = -I'$, and $R = R_2$ (note that I' is negative, so that $Q = -I'$ is positive), we get

$$\frac{n_g}{-I'} + \frac{n_a}{I} = \frac{n_a - n_g}{R_2} \qquad (34.6)$$

If we add corresponding sides of Eqs. 34.5 and 34.6, the I' term drops out and we find that

$$\frac{n_a}{Q} + \frac{n_a}{I} = (n_g - n_a)\left(\frac{1}{R_1} + \frac{1}{R_2}\right) \qquad (34.7)$$

Equation 34.7 allows us to predict where an image would be formed by a thin lens. We can simplify the equation by introducing a distance F, called the *focal length* of the lens and defined as follows:

$$\frac{n_a}{F} = (n_g - n_a)\left(\frac{1}{R_1} + \frac{1}{R_2}\right) \qquad (34.8)$$

With this definition of the focal length, Eq. 34.7 can be written in the simple form

$$\frac{1}{Q} + \frac{1}{I} = \frac{1}{F} \qquad (34.9)$$

Rays that are parallel as they approach the lens can be imagined as coming from an object at an infinite distance from the lens. Using Eq. 34.9 with $Q = \infty$ (and thus $1/Q = 0$), we see that such parallel rays are brought to a focus at a distance F behind the lens. Thus $I = F$ when $Q = \infty$ (see Fig. 34.10). We have already seen that a

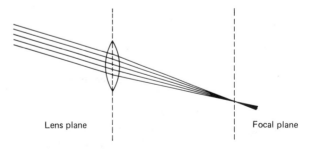

Lens plane   Focal plane

FIGURE 34.10 Distant objects are brought to a focus at the focal plane of a lens. One can use this fact to determine the focal length of a lens without measuring the curvatures of the surfaces of the lens.

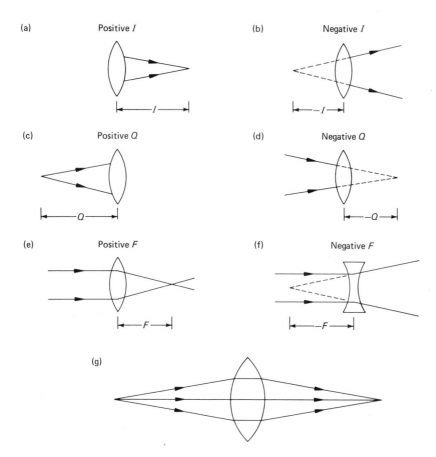

FIGURE 34.11   The optical sign rules.

negative value of the image distance could be given a meaningful interpretation. In fact, all variables (I, Q, and F) in Eq. 34.9 may be either positive or negative. The meanings assigned to positive and negative values of the variables are indicated in Fig. 34.11 and by the following corresponding rules.

SIGN RULES

(a) If after passing through the lens the rays are *converging*, then the *image* is said to be *real*, and the positive number I is equal to the distance from the lens plane to the point of convergence of the rays.

(b) If after passing through the lens the rays are *diverging*, then the *image* is said to be *virtual*, the positive number $-I$ is equal to the distance from the lens plane to the point of divergence of the rays.

(c) If as they enter the lens the rays are *diverging,* then the *object* is said to be *real,* and the positive number Q is equal to the distance between the lens plane and the point of divergence of the rays.

(d) If as they enter the lens the rays are *converging,* then the *object* is said to be *virtual,* and the positive number $-Q$ is equal to the distance between the lens plane and the point of convergence of the rays.

(e) If parallel rays that enter the lens *converge* to a point after leaving the lens, then the positive number F is equal to the distance between the lens plane and the point of convergence of the rays.

(f) If parallel rays that enter the lens *diverge* after leaving the lens, then the positive number $-F$ is equal to the distance between the lens plane and the point of divergence of the rays.

Rather than trying to memorize these sign rules, the reader can simply remember that everything (R_1, R_2, F, I, and Q) in Fig. 34.11(g) is positive. Anything that deviates qualitatively from this picture is negative.

QUESTION

At what point would the optical system shown in Fig. 34.12 form an image of the point object shown? The focal lengths of the lenses (in centimeters) are as indicated.

ANSWER

We first ignore the second lens and consider what happens to the light as it passes through the first lens. The rays are diverging when they meet the first lens; therefore, the object is real (which was obvious anyway) and $Q = 8$ cm. We know that $F = 4$ cm, so using Eq. 34.9 we see that

$$\frac{1}{8 \text{ cm}} + \frac{1}{I} = \frac{1}{4 \text{ cm}}$$

or

$$I = 8 \text{ cm}$$

Since I is positive, we know that when the rays leave the first lens they are converging toward a point 8 cm to the right of that lens. Of course, the converging rays strike the second lens before they come to a focus. Since the rays entering the second lens are converging, the object for that lens is virtual. The object distance is $Q = -6$ cm, since the converging rays, as they hit the lens, are aimed at a point 6 cm to the right of the second lens. The focal length of the second lens is -12 cm so, again applying Eq. 34.9, we obtain the formula for the image distance,

$$\frac{1}{-6 \text{ cm}} + \frac{1}{I} = \frac{1}{-12 \text{ cm}}$$

which gives

$$I = 12 \text{ cm}$$

We find again that the image distance is positive, and this indicates that a real image is formed 12 cm to the right of the second lens.

FIGURE 34.12 Optical system consisting of two double-surface lenses.

QUESTION

A converging lens is made with one side flat and the other side of radius R. The focal length of the lens in air is 10 cm, and the focal length of the lens when it is submerged in water is 20 cm

FIGURE 34.13   A converging lens (a) in air and (b) in water.

(see Fig. 34.13). What is the index of refraction of the glass and what is the value of R?

ANSWER

Applying Eq. 34.8 to the case of the lens in air, we obtain (using $n_a = 1.00$)

$$\frac{1}{10 \text{ cm}} = (n_g - 1)\left(\frac{1}{R}\right)$$

When we apply Eq. 34.8 to the case of the lens in water, we must replace n_a by the index of refraction of water, namely $n_w = 1.33$, and we obtain

$$\frac{1.33}{20 \text{ cm}} = (n_g - 1.33)\left(\frac{1}{R}\right)$$

Multiplying the first equation by $10 R$ and the second equation by $(20/1.33)R$ and noting that $(20/1.33) \simeq 15$, we have the following two linear equations:

$$R - 10n_g + 10 = 0$$

$$R - 15n_g + 20 = 0$$

The solution of this pair of linear equations for R and n_g is

$$R = 10 \text{ cm}$$

$$n_g = 2$$

34.3 THE HUMAN EYE

The most familiar optical system by far is the human eye (see Fig. 34.14). As an optical system it is relatively simple. It has two lenses (a single-surface lens and a double-surface lens) for focusing the light rays, a variable opening (the pupil) to control the total amount of light admitted to the lens, a system of light-sensitive receptors (the retina) at the back surface. The lens forms a real image on the retinal surface of the object being viewed, and the photosensitive receptors on the retinal surface convert the image to a pattern of nerve impulses.

The front surface of the eye is covered by a tough transparent membrane called the cornea. Between the cornea and the crystalline lens is a region filled with a fluid (the aqueous humor) whose index of refraction is about the same as that of water. The corneal surface therefore acts as a single-surface converging lens. The crystalline lens is a two-surface convex lens made of somewhat flexible material. The space behind the crystalline lens is filled with jellylike transparent material called the vitreous humor. The crystalline lens is attached to a system of muscles at its periphery, and these are capable of varying its convexity

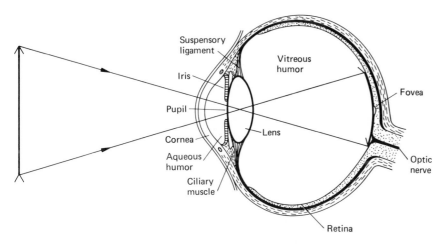

FIGURE 34.14 Principal components of the right eye as viewed from above.

and therefore of controlling its focal length. When the muscles are relaxed, an object at infinity is brought to a focus on the retinal surface. As the muscles flex, the focal length of the crystalline lens decreases and nearer and nearer objects are brought to a focus on the retina. For an average person of twenty years of age with no eye defects, the closest distance at which objects can be brought to a sharp focus (the *near point* of the eye) is about 10 cm. As a person ages, the near point distance usually increases. A typical value of the near point of a mature person is 20 cm. The radius of curvature of the corneal surface is about 0.8 cm. The index of refraction of the aqueous humor directly behind the corneal sheath is about 1.33. Thus in the absence of the crystalline lens, the image of a distance object ($Q = \infty$) would be produced at a distance I behind the cornea, where

$$\frac{1.33}{I} = \frac{0.33}{0.8 \text{ cm}}$$

or

$$I = 3.2 \text{ cm}$$

Since the distance between the cornea and the retina of an adult human eye is about 2.2 cm, the image would fall well behind the retina if it were not for the added converging power of the crystalline lens. The index of refraction of the substance of the crystalline lens is not constant throughout the lens. Its average value is about 1.41, which is larger than that of the aqueous and vitreous humors that surround it. Thus because of its convex shape it acts as a converging lens. However, its ability to converge the light rays passing through it is augmented by the fact that its index of refraction is larger toward the center of the lens

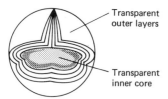

FIGURE 34.15   Anatomical structure of the crystalline lens.

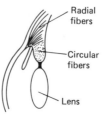

FIGURE 34.16   The two types of ciliary muscle fibers.

than it is near its periphery. Anatomically the lens is constructed like a transparent onion (see Fig. 34.15), the substance of the inner layers, and particularly the nucleus, being of higher index of refraction than that of the outer layers. The position of the lens is maintained by the suspensory ligaments that radiate outward from its periphery in all directions. The suspensory ligaments end on the ciliary muscle, which circles the lens and makes contact with the outer capsule of the eye. When the ciliary muscle contracts, the tension in the suspensory ligaments is reduced. This is exactly the opposite of what one would expect by looking at Fig. 34.14. We can resolve the apparent paradox by noting that the ciliary muscle fibers do not run radially outward from the suspensory ligaments to the periphery. In fact, the ciliary muscle fibers are of two types. One type runs in a circular direction around the circumference of the circle defined by the lens plus suspensory ligaments. It is easy to see how contraction of those fibers would lead to a reduction in the tension of the suspensory ligaments. The second set of fibers runs from the inner surface of the cornea backward and peripherally, as shown in Fig. 34.16. Their contraction also serves to reduce the suspensory ligament tension. When the tension on the periphery of the lens is reduced, the lens

bulges out more in the center, thus increasing the curvature of its surfaces and decreasing its focal length. Objects at a closer distance can then be brought into focus. In its relaxed state the eye focuses at infinity. Activities such as reading that require constant contraction of the ciliary muscles may tire the eye.

The retina, on which the optical image is brought to a focus, is a complex multilayered sheet covering most of the inner surface of the eye. It is more-or-less transparent, but is backed (in the human eye) by a black light-absorbing layer. In certain animals, including cats, the retinal layer is backed by a reflective layer rather than an absorbing layer. This increases the sensitivity of the eye since the retinal cells get an opportunity to interact with the light twice, once as it passes through the retina from front to back and again as it is reflected from back to front. (In such eyes there is, however, a corresponding decrease in resolution as a result of the random scattering of the light.) This reflective layer causes cat's eyes to shine when they are illuminated by a strong narrow beam of light such as a flashlight (see Fig. 34.17). Contrary to popular myth, cat's eyes do not shine when there is a general low level of illumination such as moonlight, and they do not shine at all in complete darkness.

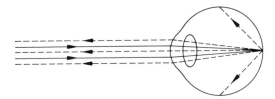

FIGURE 34.17 Representation of the shining of a cat's eye. When the eye is focused on the light source, the parallel incoming rays (solid lines) are brought to a focus on the retina. The randomly back-scattered rays (dashed lines) are then converged into a parallel beam by the lens. The outgoing beam is aimed back at the source of light, and thus the eye appears to be shining.

STRUCTURE OF THE RETINA

In the human eye the darkly pigmented layer forms the outer boundary of the retina. Just in front of it is the layer of rods and cones, the two basic types of light-sensitive cells. In these cells the pattern of illumination produced by the optical system of the eye is converted to a complex of nerve impulses that carry the informational content of the optical image to the visual cortex of the brain. The details of the physical mechanism by which light energy initiates a nerve impulse are still unclear. The nerve impulses are not communicated without change directly from the photosensitive cells to the visual cortex. Instead a complex nerve network forms the innermost layers of the retina, and the nerve impulses produced by the rods and cones are transmitted to these retinal nerve cells, where the first stages of the synthesis and analysis of the visual information are accomplished. The outgoing nerve impulses from this nerve network are carried to the visual cortex by a layer of nerve fibers lying on the inner surface of the retina. The fibers converge toward an area on the retina called the optic disk, at which point they leave the eye through the optic nerve. The optic disk is devoid of rods and cones, and thus the portion of the visual image falling on that region is not registered. For this reason the field of vision of each eye contains a *blind spot*. The blind spots of the fields of vision of the two eyes do not overlap. Hence these are spots usually not noticed, since one eye is able to supply the information lacking in the image being transmitted by the other eye. Figure 34.18 shows how one may easily detect the blind spot. At that place on the retina directly opposite the cornea, there is a region of specialized structure called the fovea centralis. In this region the inner nerve layers are very much reduced in thickness, and the photosensitive cell layer consists almost entirely of tightly packed cone cells. Here vision is most acute (i.e., the resolution is best). Thus, to resolve small details in an image, one must look directly at the details so that their image will fall within the fovea.

□   ●

FIGURE 34.18 A test for the blind spot. Close the left eye and with the right eye continue to look directly at the square as you vary the distance between the paper and your eye. At a distance of about 8 in the image of the round spot will fall on the optic disk and the square will disappear. Note that the blind spot is "filled in" by the background color (it is not dark).

34.4 MAGNIFYING INSTRUMENTS

The amount of detail we distinguish when viewing an object obviously depends on the size of the image produced on the retina, and this is determined by the angle that the object subtends as measured from the surface of the eye. For a given object, we may increase that angle by bring-

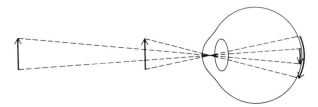

FIGURE 34.19 The same object will produce a larger image on the retina if it is brought closer to the eye. However, if it is brought too close, a sharp image cannot be formed by the cornea and lens.

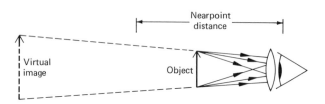

FIGURE 34.20 The additional converging power of a magnifying glass allows one to view an object closer than the near point, and thus a larger image is produced on the retina.

ing the object nearer, but the near point of the eye places a limit on how closely we may view an object (see Fig. 34.19). A simple magnifying glass is a converging lens that allows us to view an object at a closer distance than our near point by producing a virtual image that lies outside our near point. Our eye is capable of bringing the rays from the virtual image to a focus on the retina (see Fig. 34.20).

The *magnifying power* of any lens or other optical instrument is defined as the ratio of the size of the image produced on the retina by an object being viewed through the lens (or instrument) to the size of the image produced on the retina by the same object being viewed by the naked eye. To make this definition precise, one must specify the exact distance from the eye at which one views the object both with the naked eye and with the lens or optical instrument. For instruments that are intended for viewing specimens at close distances, such as microscopes and magnifying glasses, the distance taken for naked-eye viewing is the *standard near point,* which is 25 cm. For magnifying lenses the specimen is assumed to be viewed through the lens at a distance that would put the virtual image at the standard

near point. For microscopes the location of the object when being viewed through the instrument is fixed by the construction of the instrument. For instruments that are intended for viewing distant objects, such as telescopes and opera glasses, the viewing distances with and without the instrument are taken to be equal and large. (The value obtained for the magnifying power does not depend on the viewing distance as long as it is much larger than the focal lengths of the lenses in the instrument.)

QUESTION

What is the magnifying power of a lens of focal length $F = 10$ cm?

ANSWER

The size of the image produced on the retina is proportional to the angle subtended at the eye by the object (see Fig. 34.21). If the object is of size d and is viewed at 25 cm by the naked eye, that angle is (for $d \ll 25$ cm)

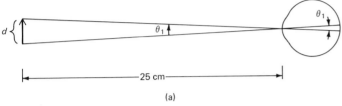
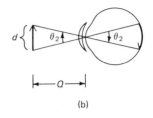

(a)

(b)

FIGURE 34.21 (a) When viewed by the naked eye at 25 cm, the object subtends an angle $\theta_1 = d/(25\text{ cm})$. (b) When viewed through the magnifying lens with the virtual image at 25 cm, the object subtends an angle $\theta_2 = d/Q$.

$$\theta_1 = \frac{d}{25\text{ cm}}$$

When the object is viewed through the lens at a distance such that its virtual image is at 25 cm, the angle subtended at the eye is

$$\theta_2 = \frac{d}{Q}$$

where

$$\frac{1}{-25\text{ cm}} + \frac{1}{Q} = \frac{1}{10\text{ cm}}$$

or

$$Q = \frac{25}{3.5}\text{cm}$$

The magnifying power (m.p.) is the ratio of θ_2 to θ_1:

$$\text{m.p.} = \frac{\theta_2}{\theta_1} = \frac{d/(25/3.5)}{d/25}$$

$$= 3.5$$

QUESTION

A typical person can comfortably distinguish details that subtend an angle as small as 10^{-3} rad, corresponding to an object of size 0.25 mm at a distance of 25 cm. What magnifying power is required for viewing objects of size d that are smaller than 0.25 mm?

ANSWER

Referring to Fig. 34.21, we can see that $\theta_1 = d/25$ cm, while $\theta_2 = D/25$ cm, where D is the size of the virtual image. If the object is to be viewed comfortably, the virtual image must be of size 0.25 mm. Since the magnifying power is the ratio of θ_2 to θ_1, the magnifying power required is

$$\text{m.p.} = \frac{\theta_2}{\theta_1} = \frac{D}{d} = \frac{0.25\text{ mm}}{d}$$

THE MICROSCOPE

A simple microscope is a two-stage device. First a converging lens (the objective) is used to produce a greatly magnified real image of the object. Then another converging lens (the ocular) acts as a simple magnifying glass to allow us to view the real image at a close distance (see

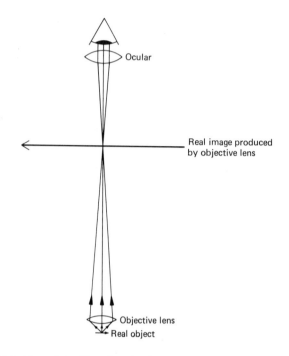

FIGURE 34.22 Representation of the working of a simple microscope.

Fig. 34.22). The *linear magnification M* of the objective is defined as the ratio of the size of the real image S_i to the size of the object S_o. From Fig. 34.23 we see that

$$M = \frac{I}{Q}$$

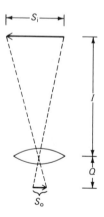

FIGURE 34.23 Because the two triangles shown are similar, the ratio S_i/S_o is equal to the ratio I/Q.

QUESTION

The microscope represented by Fig. 34.24 contains two lenses with the focal lengths shown. The distance $L = 20$ cm is fixed, but the distance l may be varied at will. What is the minimum size object that could be resolved? Assume that the near point of the eye is 25 cm.

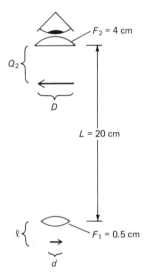

FIGURE 34.24 Representation of a simple microscope, with the focal lengths of the two lenses as shown.

ANSWER

The objective lens produces a real image below the ocular. The closer that real image is to the ocular, the larger an angle it subtends at the eye. The rays from that real image pass through the ocular and form a virtual image at a distance $-I_2$, where

$$\frac{1}{I_2} + \frac{1}{Q_2} = \frac{1}{F_2}$$

Q_2 is the distance between the ocular and the real image produced by the objective lens and $F_2 = 4$ cm. In order to be resolvable by the eye, the virtual image must not be closer than 25 cm, which is the near point. If it is exactly 25 cm, then $I_2 = -25$ cm and

$$\frac{1}{Q_2} = \frac{1}{F_2} - \frac{1}{I_2} = \frac{1}{4 \text{ cm}} + \frac{1}{25 \text{ cm}}$$

or $Q_2 = 3.45$ cm. The real image is a distance of 3.45 cm from the ocular, and in order to be resolvable it must subtend an angle of 10^{-3} rad. Thus $D/Q_2 = 10^{-3}$ or

$$D = 3.45 \times 10^{-3} \text{ cm}$$

The real image is a distance $I_1 = 16.55$ cm from the objective lens, which has a focal length of 0.5 cm. Thus

$$\frac{1}{l} + \frac{1}{16.55 \text{ cm}} = \frac{1}{0.5 \text{ cm}}$$

or $l = 0.515$ cm. The ratio of D to d is the same as the ratio of I_1 to l, and therefore

$$d = \left(\frac{0.515}{16.55}\right)(3.45 \times 10^{-3} \text{ cm})$$

$$= 1.1 \times 10^{-6} \text{ m}$$

34.5 ULTIMATE RESOLUTION OF AN OPTICAL SYSTEM

According to the ideas we have been using, it seems that we could investigate structures of indefinitely small size with a microscope simply by making the magnification sufficiently large. Actually there is a fundamental limit to the amount of detail we can resolve with a light microscope. No matter how ingeniously a microscope is designed or how carefully it is made, it cannot resolve details whose

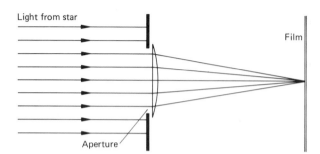

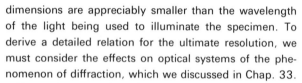

FIGURE 34.25   A single-lens camera photographing a star.

FIGURE 34.26   Diffraction pattern produced by a circular aperture.

dimensions are appreciably smaller than the wavelength of the light being used to illuminate the specimen. To derive a detailed relation for the ultimate resolution, we must consider the effects on optical systems of the phenomenon of diffraction, which we discussed in Chap. 33.

To convince ourselves that the effects of diffraction will require us to alter somewhat the conclusions of the optical theory we used in the preceding section, let us consider the simple optical system shown in Fig. 34.25. Parallel light rays from a distant star pass through an aperture of width w and are focused by a converging lens on a film. Using the ray-tracing method that we have been employing throughout this chapter, we would conclude that the light converges to a geometrical point at the location of the film. Let us recall, however, that according to Eq. 33.5, when the originally parallel light rays pass through the aperture, they are dispersed into a range of angles whose maximum, θ, is given by

$$\theta = \frac{\lambda}{W}$$

where λ is the wavelength of the light. In other words, the light wave to the right of the aperture is not simply a sharply cut-off section of a plane wave, but rather a wave whose wave fronts have been distorted, very slightly in the center but quite a bit at the edges. The lens will not focus the wave front to a point; instead a small diffraction pattern will be produced on the film (see Fig. 34.26). If light is entering the aperture from two stars whose angular separation is less than the diffraction angle θ, then the diffraction pattern produced by one star will resemble the image produced by a single star of twice the brightness. The diffraction angle θ thus represents a lower limit to the detail that can be resolved with an optical instrument of aperture diameter W using light of wavelength λ (see Fig. 34.27).

ANGULAR RESOLUTION OF THE EYE

The *angular resolution* of the human eye is defined as the minimum angular distance at which two point sources of light can be distinguished from a single point source.

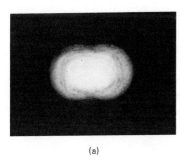

(a)

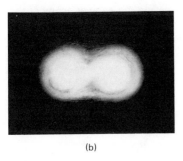

(b)

FIGURE 34.27 (a) Diffraction pattern produced by two point sources that are barely resolvable. (b) Diffraction pattern produced by two point sources that are well resolved.

From Fig. 34.28 we see that the analysis that we just carried out to determine the angular resolution of a camera can be applied without change to the human eye.

The diameter of the input port is simply the diameter of the pupil, which varies from 3 to 6.5 mm. The wavelength of visible light is in the range 4×10^{-7} to 7×10^{-7} m. To estimate the angular resolution, we should choose a large pupil opening and a short wavelength. Let us take the pupil opening as 5 mm and the wavelength as 5×10^{-7} m. Then the smallest angle that can be resolved by the eye is

$$\theta = \frac{\lambda}{a} = \frac{5 \times 10^{-7}}{5 \times 10^{-3}} = 10^{-4} \text{ rad}$$

Note that we have previously used a value of 10^{-3} rad as the required angular separation for the comfortable resolution of ordinary details (not pairs of point sources). If two point sources are separated by an angle θ, then on the retina they produce images that have a linear separation of θd, where d is the diameter of the eye ($d \simeq 2.3$ cm). For $\theta = 10^{-4}$ that separation is

$$\theta d = (10^{-4})(2.3 \times 10^{-2} \text{ m}) = 2.3 \times 10^{-6} \text{ m}$$

This distance is the typical diameter of a single cone cell in that region of the retina where our vision is most acute

Two point sources

Overlapping diffraction patterns

FIGURE 34.28   Two point sources closer than the minimum resolvable angle produce overlapping images on the retina.

(the fovea centralis). We see that we could not improve our visual acuity significantly by evolving smaller receptor cells in the retina. On the other hand, if the cone cells were larger than 2.3×10^{-6} m in diameter, then our visual acuity would be reduced. Here is another example of evolutionary forces solving a physics problem by trial and error.

THE INSECT EYE

The insect eye is a simple structure. It is essentially a hemisphere composed of conical subunits called om-

matidia. Each ommatidium is transparent at the large end and contains photosensitive nerve fibers that exit at the small end (see Fig. 34.29). An ommatidium is simply an open tube that samples the light coming from a narrow solid angle. Since there is no imaging system within the ommatidium, the minimum resolvable angle must be greater than $\alpha = a/R$, where R is the radius of the eye and a is the size of the open end of an ommatidium; that is,

$$\theta > \alpha = \frac{a}{R}$$

But another limitation on the angular resolvability prevents natural selection from increasing the angular acuity of the

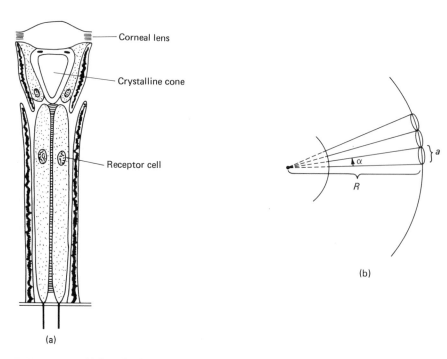

(a)

(b)

FIGURE 34.29 (a) Longitudinal section of an ommatidium from the compound eye of an insect. (b) Schematic model of the insect eye with the quantities R, a, and α indicated.

FIGURE 34.30 Because of bending resulting from diffraction, some light reaches the nerve fibers that would not if light traveled only in straight lines.

eye by simply making a smaller. Because of diffraction effects, the minimum resolvable angle cannot be less than λ/a, where λ is the wavelength of the light to which the eye is sensitive (see Fig. 34.30). Therefore we also know that $\theta > \lambda/a$, which means that the minimum resolvable angle, θ, must lie above both curves in the graph in Fig. 34.31. It is clear that the value of a that permits the greatest visual acuity is that value at which the curves cross; that is,

$$\frac{\lambda}{a} = \frac{a}{R}$$

or

$$a = \sqrt{\lambda R} \qquad\qquad\qquad \textbf{(34.10)}$$

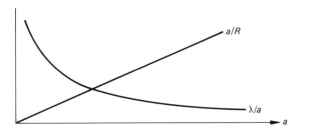

FIGURE 34.31 The angle θ must be above both curves.

In the bee the diameter of the eye is about 4 mm ($R = 2 \times 10^{-3}$ m), and the bee's eye has maximum sensitivity for light of wavelength near $\lambda = 4 \times 10^{-7}$ m. Substituting those numbers into Eq. 34.10, we get $a = 2.8 \times 10^{-5}$ m. This is remarkably close to the observed value of 3×10^{-5} m.

These examples again make it clear that the laws of physics form the "legal" framework within which evolution proceeds. A fair degree of understanding of those laws is an essential prerequisite to any understanding of the evolutionary pressures on organisms and the limited freedom available to an organism in responding to those pressures. Many aspects of biological organisms are determined by the maximization of certain useful properties (such as visual acuity) subject to the constraints imposed by basic physical laws.

34.6 RESOLUTION OF A MICROSCOPE

We shall now make a similar analysis in order to determine the limit of resolution of a microscope. The method of analysis we shall use is designed to emphasize the fact that one cannot resolve details smaller than the limit of resolution by making improvements in the internal design or construction of the microscope. (Of course, one can do much worse than the limit of resolution by designing or constructing the microscope poorly.) In our analysis we shall not need any information regarding the design of the microscope other than the size of the input aperture and its distance from the specimen. We shall consider the case of a specimen being viewed by transmitted light, although almost identical considerations would apply to the case of viewing by reflected light. Thus Fig. 34.32 specifies the instrument in sufficient detail for our purposes. We call the angle ϕ the *aperture angle.*

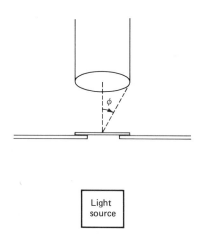

FIGURE 34.32 The aperture angle θ is the angle subtended by one-half of the objective lens as measured from the specimen plane.

If the microscope is capable of resolving details of size s, then it would be capable of resolving any distinct pattern, such as a pattern of circular black dots or a pattern of lines, whose details were all of size s or larger. Let us consider the case of a specimen consisting of a glass plate on which is printed a pattern of parallel straight black lines of width s and separation s. By a happy coincidence we have already considered in detail what happens when light of wavelength λ passes through such a specimen. This pattern of opaque and transparent parallel lines is just what we called a diffraction grating. The light that leaves the specimen and travels toward the microscope objective is a collection of plane wave "beams" at angles 0, θ_1, θ_2, etc., where

$$\sin \theta_n = n\frac{\lambda}{2s} \qquad n = 0, \pm 1, \pm 2, \dots \qquad (34.11)$$

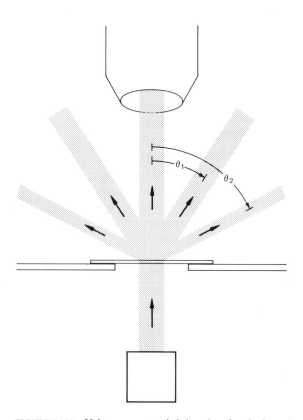

FIGURE 34.33 If the aperture angle is less than θ, only the undeviated plane wave enters the aperture.

Note that we obtained Eq. 34.11 from Eq. 33.3 by letting $d = 2s$. In Eq. 33.3, d is the distance between the centers of the transparent slits. If the aperture angle ϕ is less than θ_1, then the only light that enters the objective is the undeviated beam at $\theta = 0$ (see Fig. 34.33). In that case the light that enters the aperture is exactly the same as the light that would enter if the object being viewed were a uniform, partially absorbing sheet of glass. Since none of the information needed to reconstruct the pattern at the image plane has actually entered the optical system, there will be no trace of the striped pattern in the final image. (If the microscope is being used for direct viewing by the eye, that final image is produced on the retina.) It would be quite reasonable for the reader to make the objection that this analysis is only applicable to uniformly striped patterns, but this objection is not really valid. By a much more complicated mathematical analysis, one can show that the basic effect persists for all patterns. That is to say, the light waves that are needed to reconstruct the fine detail in the image all leave the specimen plate at large angles with respect to the undeviated beam. Thus for a microscope of small aperture angle, they are permanently lost, and no amount of magnification can produce those small details in the image. For a microscope of aperture angle ϕ in which light of wavelength λ is being used, the size s of the smallest resolvable detail is given by setting θ_1 of Eq. 34.11 equal to ϕ. That would mean that at least the first deviated beam just got into the edge of the aperture. The resolvability criterion we obtain is

$$s = \frac{\lambda}{2 \sin \phi} \tag{34.12}$$

Note that Eq. 34.12 shows that any microscope using light of wavelength λ cannot resolve details smaller than $\lambda/2$, because $\sin \phi$ cannot be greater than 1 for any aperture angle.

QUESTION

If we are using light of wavelength 5×10^{-7} m and an objective with an aperture angle of 70°, what is the size of the smallest resolvable detail?

ANSWER

Using Eq. 34.12 with $\lambda = 5 \times 10^{-7}$ m and $\sin \phi = \sin 70° = 0.94$, we obtain

$$s = \frac{5 \times 10^{-7} \text{ m}}{2 \times 0.94} = 2.7 \times 10^{-7} \text{ m}$$

OIL-IMMERSION OBJECTIVES

It is obvious from Eq. 34.12 that one way of decreasing the size of the minimum resolvable details is to decrease the wavelength of the light. If we try to decrease the wavelength by increasing the frequency, we quickly run into the problem that the eye is only sensitive to light of a restricted frequency range. (However, see the discussion of the ultraviolet microscope below.) Since the wavelength is related to the frequency by the equation $\lambda = v/f$, it is possible to decrease λ at a fixed frequency by using a medium in which the velocity of light propagation is less than it is in air. The common practice is to fill the space between the specimen plate and the objective lens with an oil that has a high index of refraction (see Fig. 34.34). If λ is the wavelength in air of the light being used for viewing, then its wavelength in the oil is λ/n. The size of the minimum resolvable details is then given by

$$s = \frac{\lambda}{2n \sin \phi}$$

which can easily be a factor of 2 smaller than the size we would obtain with an equivalent objective without oil immersion.

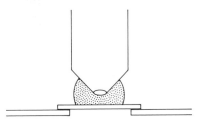

FIGURE 34.34   The oil-immersion objective.

THE DARK-FIELD MICROSCOPE

In a microscope the light source is behind the specimen, which is viewed by light that has been scattered during transmission. The specimen being viewed often consists of small objects surrounded by a medium that does not scatter the light but transmits it undeviated. The dark-field microscope is a device that removes the undeviated light that has simply passed through the medium, so that the scattered light from the specimen may be viewed against a dark background. This procedure often improves the contrast of the image. The principle of the device is very simple. The specimen is illuminated by a light source that produces a hollow cone of converging light rays, with the specimen near the apex of the cone. The unscattered light above the specimen forms a hollow diverging cone. The objective is placed within the hollow portion of the cone so that the unscattered rays do not enter it (see Fig. 34.35).

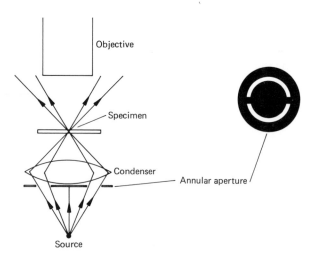

FIGURE 34.35 The light source for a dark-field microscope.

THE ULTRAVIOLET MICROSCOPE

Since the minimum resolvable size of a specimen is proportional to the wavelength of the light being used to view the specimen, an obvious way to decrease the limit of resolution is to use light of shorter wavelength. In microscopes designed for direct viewing by the eye, one is obviously limited by the fact that our eyes are fairly insensitive to light of wavelength shorter than half a micrometer $(0.5 \times 10^{-6}$ m). However, photographic film is sensitive to much shorter wavelengths, and therefore the sensitivity range of the human eye is no serious obstacle. There are two barriers to the use of light of very short wavelength, such as x rays. First, no materials exist that refract light of that frequency to an appreciable extent. Thus no x-ray lenses are available. Second, many substances transmit x rays without noticeable scattering; that is, many of the specimens of interest are quite invisible to x rays. Neither of these difficulties occurs with ultraviolet light. For microscopes using ultraviolet light, lenses must be made of quartz because glass absorbs ultraviolet light, but that creates no serious complications. The greatest advantage afforded by ultraviolet microscopy is not actually the increased resolution (the limit of resolvability is increased by only a factor of 2), but the fact that nucleic acids absorb strongly at those wavelengths, and thus their distribution in the cell may be easily determined.

THE PHASE-CONTRAST MICROSCOPE

Ordinary microscopes are capable of producing images of biological specimens only because different portions of the specimen absorb light in different amounts. Actually most biological specimens in their natural state absorb little or no light and are therefore quite indistinct, if not invisible, in an ordinary microscope. They are rendered observable only by staining processes that always kill and sometimes distort the biological sample of interest. The phase-contrast microscope is a device that allows us to view unstained live specimens. Although most unstained cells absorb very little light, they do scatter light because the index of refraction of the cell varies from point to point. Thus the live cell could be seen in a dark-field microscope, but its structural details could not be ascertained because all portions of the cell would scatter light with approximately the same strength, and thus there would be very little contrast within the image of the cell. With a dark-field microscope the image of a transparent object such as that shown in Fig. 34.36 would be an undifferentiated luminous area. However, because of variations in thickness and in the index of refraction, the light rays emerging from various parts of the specimen have different phases; that is, the wave fronts of the spherical waves leaving one point on the specimen

are ahead of or behind the corresponding wave fronts emerging from another part. The phase lag is typically small; that is, the waves emerging from different parts of the specimen differ in phase by a small fraction of a wavelength. Since photographic plates and the eye are both sensitive only to the light intensity of the image and not to the phase, this variation in phase, which could be used to distinguish different portions of the image, is undetected by an ordinary microscope. The phase-contrast microscope converts this phase variation into an intensity variation. Unfortunately the theory of the phase-contrast microscope is subtle.

We begin our analysis by considering what happens to a plane light wave when it passes through a specimen that does absorb light and therefore could be seen in an ordinary microscope (see Fig. 34.37). We shall then compare this with the propagation of a plane wave through a

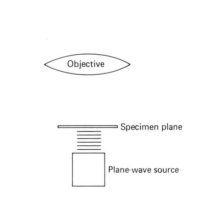

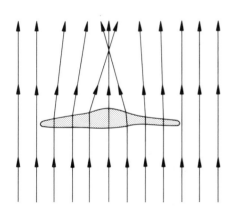

FIGURE 34.36 Light rays passing through a transparent object.

FIGURE 34.37 The simplified model of a microscope that we shall use in this section. There is no ocular, and the image produced by the objective lens is recorded directly on photographic film.

transparent specimen of varying index of refraction. We take a particularly simple model of such a specimen, namely a thin sheet of glass whose index of refraction is completely uniform. The sheet of glass has been nonuniformly stained with a light-absorbing dye, so that different portions of the glass absorb different fractions of the light passing through it. The specimen is being illuminated by a plane light wave from below. If the glass were not stained, then the wave function of the light wave at any point on the top surface of the specimen would be a harmonic wave of the form

$$u = A \cos \omega t$$

The amplitude A would be the same at all points on the surface since, in the absence of the dye, there would be no absorption. Actually, because of the staining the wave function on the top surface of the glass is

$$u = (1 - \alpha)A \cos \omega t$$

where the absorption coefficient α is some number between zero and one that measures the fractional reduction in the amplitude of the wave as a result of absorption by the dye. The number α will be different at different points on the glass; it will be zero at those points where there is no dye and will be larger at those points where the dye is denser. We could write the above wave function as the sum of two wave functions:

$$u = A \cos \omega t - \alpha A \cos \omega t \equiv u_0 + u_1 \qquad (34.13)$$

where u_0 is a wave function of a constant amplitude plane wave and u_1 is the wave function of a wave that is in phase with u_0 but has an amplitude that varies across the surface of the specimen. The addition principle tells us that we could calculate the total wave function at any point on the image plane by first calculating how the uniform plane wave u_0 would propagate to that point on the image plane, then calculating how the complicated wave u_1 would propagate to that point on the image plane, and then adding the two wave functions.

We are now ready to consider what happens to a wave when it passes through a specimen that is not absorbing

light but whose index of refraction varies from point to point. Again we take a very simple form of such a specimen, namely a sheet of clear glass of uniform thickness but of nonuniform index of refraction. Because of the nonuniformity of the index of refraction, the portions of the plane wave fronts that are all in phase at the bottom surface of the glass will no longer be in phase at the top surface of the glass, since it will take more time for the wave to propagate through some parts of the glass than through others. Since no absorption takes place in the glass, the wave amplitude will be the same at every point on the top surface. Thus the wave function on the top surface of the specimen will be of the form

$$u = A \cos (\omega t + \beta)$$

where the phase difference β varies from point to point. We shall make the assumption, which is valid in cases of biological specimens, that the phase shift β is small. We then apply the trigonometric formula for the cosine of $x + y$, where x is ωt and y is β:

$$\cos (\omega t + \beta) = \cos \omega t \cos \beta - \sin \omega t \sin \beta$$
$$\simeq \cos \omega t - \beta \sin \omega t$$

In the second line we have used the approximations for $\cos \beta$ and $\sin \beta$ that are valid for small β. Thus the wave function u has the form

$$u = A \cos \omega t - \beta A \sin \omega t$$

This is very similar to the wave function in the case of an absorbing specimen (Eq. 34.13). It is the sum of a uniform plane wave and a wave whose amplitude varies from place to place on the surface. The only real difference is that in this case the two waves are out of phase by an amount $\pi/2$. If there were some way of shifting only the second term in the above wave function by an amount $\pi/2$, then [since $\sin(\omega t + \pi/2) = \cos \omega t$] the above wave would have exactly the same form as in the case of absorption, and thus the image produced would be the same as that obtained with a specimen that had a varying absorption proportional to β. However, since both waves are in the same region (on top of the specimen), it seems impossible to shift one and leave the other unchanged. We now note that at a certain stage in their propagation from the specimen plane to the image plane these two waves (actually one should say these two parts of the single wave u) separate themselves in space and can be separately modified. To see where this occurs and how the modification is accomplished, let us again recall that according to the addition principle the two parts of the wave propagate as if the other were not there. The first term is a uniform plane wave and will therefore come to a focus on the optical axis at the focal plane of the objective, after which it will spread out before hitting the image plane (see Fig. 34.38). The second term, being a complicated wave of varying amplitude, will not be brought to a focus at the focal plane, but will pass through the focal plane spread out over a large area. In the phase-contrast microscope a sheet of glass is placed at the focal plane. The sheet has a small hole in it on the optical axis so that the uniform plane wave that has come to a focus there can get through the hole without being slowed down. The other part of the wave must pass through the sheet of glass, whose thickness has been chosen so as to retard the wave by exactly the amount needed to make the combined wave function identical with the wave function of an absorbing specimen. The image obtained on the image plane thus reveals variations in the index of refraction as if they were variations in absorption. A transparent specimen of variable index of refraction produces an image of varying intensity that can be photographed or seen by the eye.

SUMMARY

A spherical surface of radius R will produce an image at a distance I of an object at a distance Q, where

$$\frac{n_1}{Q} + \frac{n_2}{I} = \frac{n_2 - n_1}{R}$$

A double-surface lens with radii of curvature R_1 and R_2 has a focal length F, where

$$\frac{1}{F} = \frac{1}{Q} + \frac{1}{I}$$

and

$$\frac{n_a}{F} = (n_g - n_a)\left(\frac{1}{R_1} + \frac{1}{R_2}\right)$$

An object at infinity (whose rays are parallel when they enter the lens) will be brought to a focus at a distance F behind the lens.

All the quantities I, Q, F, and R can be either positive or negative, with the interpretation of the two signs given in Fig. 34.39.

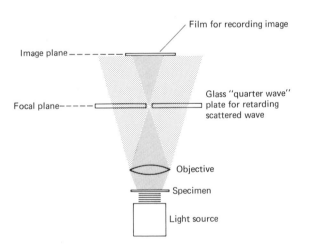

FIGURE 34.38 The elements of a phase-contrast microcope. The uniform plane wave comes to a focus at the focal plane of the objective, and thus gets through the hole without being slowed down by traveling through the glass "quarter-wave plate."

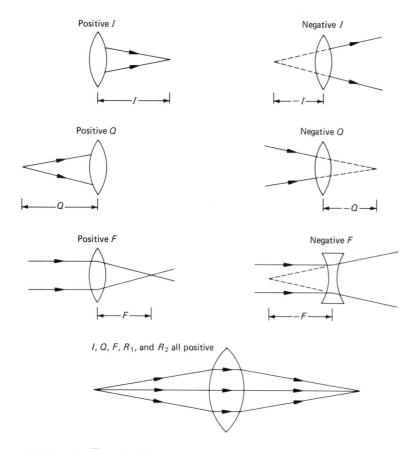

FIGURE 34.39   The optical sign rules.

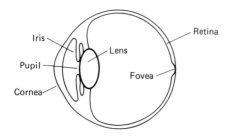

FIGURE 34.40   Principal elements of the eye.

The principal parts of the human eye are as follows (see Fig. 34.40):

Cornea   A transparent spherical front surface that acts as a single-surface converging lens.

Iris   An opaque membrane in front of the lens with an adjustable aperture to control the intensity of light on the retina.

Pupil   An aperture in the iris.

Lens   A double-surface lens of variable focal length that keeps objects in focus on the retina.

Retina   A light-sensitive nerve coating on the inner back surface of the eye.

Fovea centralis   The region on the retina of maximum acuity.

A simple magnifying glass produces a virtual image, beyond the near point, of an object held closer to the eye than the near point. The virtual image can be focused on the retina.

In a microscope the objective lens produces a large real image of the specimen. The ocular acts as a simple magnifying glass to allow us to view that real image at a close distance.

The linear magnification of a lens is the ratio of the image size to the object size:

$$M = \frac{S_i}{S_o} = \frac{I}{O}$$

Because of diffraction, the resolution of any optical system is limited. For a system with an input aperture of diameter a, two distant point sources cannot be distinguished from a single point source unless they are separated by an angle larger than $\theta = \lambda/a$, where λ is the wavelength of the light being used. The angle between the sources is measured from the input aperture.

A microscope cannot resolve details of the specimen smaller than

$$S = \frac{\lambda}{2 \sin \phi}$$

where ϕ is the aperture angle (see Fig. 34.41).

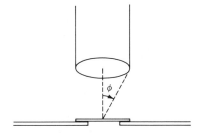

FIGURE 34.41   The aperture angle ϕ.

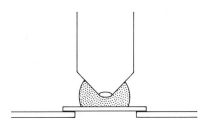

FIGURE 34.42   The oil-immersion objective.

For an oil-immersion objective the minimum resolvable size is reduced by a factor of $1/n_{oil}$ (see Fig. 34.42).

In a dark-field microscope the light that passes through the specimen undeviated does not enter the objective.

The ultraviolet microscope achieves better resolution by using shorter wavelengths.

The phase-contrast microscope can be used for viewing a specimen that has a varying index of refraction but negligible light absorption. By means of a "quarter" wave plate with a hole in it at the focal plane, the scattered light can be phase shifted while the uniform plane wave component is unchanged. This causes variations in phase to appear on the image plane as variations in intensity.

PROBLEMS

34.A.1 In each of the parts of the figure, calculate the position and type (real or virtual) of the final image produced by each of the lens systems shown.

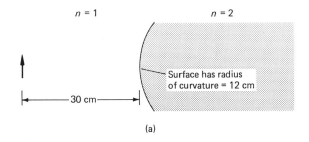

(a)

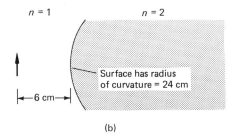

(b)

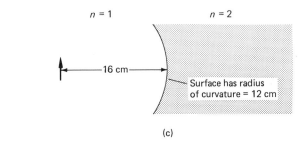

(c)

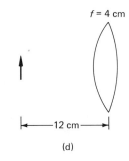

(d)

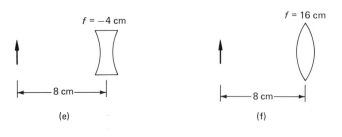

(e)   (f)   (g)

34.A.2 What is the focal length of the lens shown in the figure? The front surface has a radius of curvature of 20 cm and the back surface has a radius of curvature of 40 cm. The index of refraction of the glass is 2.5.

34.B.2 As shown in the figure, an object is 20 cm from the curved face of a hemispherical lens. (The lens is *not* a thin lens.) The index of refraction of the lens is 1.5 and its radius is 1 cm. What types of images are formed and where are they formed?

34.B.3 Two lenses are made with glass of index of refraction 1.6. Each lens has one curved surface and one flat surface. The radius of curvature of the curved surface of one lens is 15 cm, and that of the other lens is 40 cm. What are the focal lengths of the two lenses? If the lenses are combined by having their flat sides placed together, what will be the focal length of the composite lens?

34.B.1 Calculate the position and type (real or virtual) of the final image produced by each of the lens systems shown in the following figure. The focal lengths of the lenses (in centimeters) are given in the figure, and all distances are in centimeters.

34.B.4 A microscope has an aperture angle of 40°. Using light of wavelength 0.5 μm, could one resolve cell components of diameter 0.8 μm with this microscope?

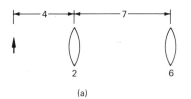

(a)

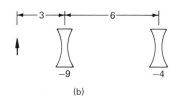

(b)

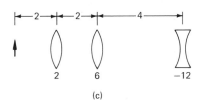

(c)

34.B.5 What is the limit of resolution of an ultraviolet microscope that has an aperture angle of 80°, uses light of wavelength 0.2 μm, and has an oil-immersion objective with oil of index of refraction 1.6?

34.B.6 A glass sphere of diameter 1 cm has an index of refraction of 1.5. As viewed through the glass, how far from the front surface will a point on the back surface appear (see the accompanying figure)?

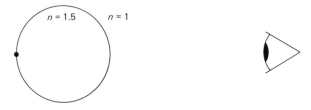

34.C.1 Myopia, or nearsightedness, is an abnormality in which the eye, in its resting state, focuses objects at some distance D (not infinity) on the retina. The distance D is called the *far point* of the eye. Since contracting the eye muscles serves to decrease the focal length of the eye, there is no way a myopic person can bring into focus objects that are further away than the far point. Suppose a nearsighted person with a far point of 20 cm obtains a corrective lens to bring the far point to infinity.
   (a) What is the focal length of the lens?
   (b) Where would the virtual image be of an object that was 100 cm from the corrective lens?
   (c) If the angle subtended by this object is 10^{-2} rad as viewed by a person with 20 $\times$ 20 vision, what will the angle be for the nearsighted person? (The object is 100 cm from the eye in both cases.)

34.C.2 Use the analysis leading to Eq. 34.10 to estimate the angular resolution of a bee's eye. What would be its angular resolution if the bee's eye had its maximum sensitivity at a wavelength of 4.7×10^{-7} m, at which frequency the sun's radiation is most intense?

34.C.3 An object of size 2×10^{-3} cm is illuminated with light of wavelength 5×10^{-5} cm. The eye of the viewer has a near point of 25 cm and a minimum angle of resolution of 10^{-3} rad.
   (a) If this object is to be viewed through a microscope, what magnifying power is required?
   (b) What will be the size of the image on the retina?

34.C.4 Using a phase-contrast microscope of the design shown in Fig. 34.38, a bacterium immersed in an aqueous medium ($n = 1.33$) is viewed. The index of refraction of the bacterium is 1.32 and its thickness is 2 μm.
   (a) If light of frequency 6×10^{14} Hz is used, what is the phase shift β of the light that passes through the center of the bacterium?
   (b) What is the ratio of the light intensity in the center of the image of the bacterium to the intensity in the image of the medium surrounding the bacterium?

35
THE
THEORY
OF
RELATIVITY

WITH THE THEORY of relativity we begin for the first time to study the physics of the twentieth century. Twentieth-century physics seems very different in style from any of the scientific developments, that preceded it. In fact, modern physics has successfully pushed the boundaries of human knowledge well beyond the range of direct ordinary human experience. These explorations have shown that certain assumptions about the structure of the world—which seemed so obvious and fundamental that they could not possibly be false—were in fact quite false when applied to events far outside the range of our everyday experience. For instance, to a nineteenth-century scientist the statement ''Two explosions, one on the earth and one on Mars, occurred simultaneously'' seemed perfectly clear and precise. The statement might have been false, but its meaning was certainly clear. To a twentieth-century physicist this statement seems as meaningful as the sentence ''Mary is taller,'' which clearly is incomplete. Mary is taller than what? The word ''taller'' indicates a comparison between two things or persons; it does not describe a property of one thing. On the other hand, it probably seems perfectly clear to the reader that the simultaneity of two events is an absolute property: Either the two events are simultaneous or they are not. It appears that in discussing the simultaneity of two events we need not refer to any other events. But we shall see that this is not so, that two events can be simultaneous with respect to one person and not simultaneous with respect to another person. We will have to revise our concept of simultaneity, in fact all our ideas of space and time, to achieve a real understanding of the subject of relativity. And later we will have to radically revise our fundamental ideas of the structure of the world to understand the quantum theory of microscopic phenomena, in which point particles can exist without being in any particular place. In sum, twentieth-century thinkers have found that in totally unfamiliar realms the scientific method, which combines precise reasoning with experimental verification, is a very effective guide, while many notions that we have learned from birth, or that have been genetically programmed into our brains by millennia of evolution, must be questioned and explicitly verified before they can be trusted.

35.1 THE NEWTONIAN PRINCIPLE OF RELATIVITY

We can illustrate what a principle of relativity is by using as an example the Newtonian principle of relativity. It is common knowledge that one experiences no difficulty in walking in a train, bus, or airplane that is moving at constant velocity. Sitting in a train that is traveling at 50 mi/h, one can toss up a ball and confidently expect it to behave the same way it would for a person sitting on a bench in the station. In fact, in a smoothly moving train with the window shades down, there is no way to detect the motion of the train by means of mechanical experiments. This situation illustrates the Newtonian principle of relativity, which states that Newton's laws of motion are valid without modification if all position measurements are made relative to an object (i.e., the train) that is moving at constant velocity without rotation. Since a ball tossed in the air by someone on a railroad station platform moves on a trajectory that satisfies Newton's equations of motion, the principle of relativity states that a ball tossed into the air by a person on a moving train should move on the same trajectory *relative to the person on the train*. This principle is not a new law of mechanics to be added to the three laws introduced earlier; rather it is a direct consequence of the mathematical form of those laws and the law of addition of velocities.

To see that this is so, let us consider the motion of a ball tossed into the air by the passenger on the train. We assume that the passenger is an experimental physicist with a meager travel allowance who must therefore do her traveling on flatcars (see Fig. 35.1). Using a large carpenter's square, she has set up a coordinate system on the flatcar with respect to which she measures the velocity of the moving ball. The velocity she observes at time t we

shall call $\mathbf{v}(t)$. Her more sedentary counterpart on the station platform has also set up a coordinate system in which he measures the velocity of the same ball. He naturally observes a different velocity which we shall call $\mathbf{v}'(t)$. The relationship between $\mathbf{v}'(t)$ and $\mathbf{v}(t)$ is given by the velocity addition formula

$$\mathbf{v}'(t) = \mathbf{v}(t) + \mathbf{V} \tag{35.1}$$

where $\mathbf{V}$ is the constant velocity of the train with respect to the platform. The gravitational force on the ball is $m\mathbf{g}$, where $\mathbf{g}$ is the gravitational field vector, which points downward. Newton's law of motion tells us that

$$\frac{d\mathbf{v}'}{dt} = \mathbf{g}$$

Differentiating both sides of Eq. 35.1 and using the fact that $d\mathbf{V}/dt = 0$ (i.e., the velocity of the train is constant), we get

$$\frac{d\mathbf{v}'}{dt} = \frac{d\mathbf{v}}{dt}$$

Thus

$$\frac{d\mathbf{v}}{dt} = \mathbf{g}$$

which shows that the ball has exactly the same acceleration when its position is measured by the uniformly moving observer as it has when its position is measured by the observer at rest on the platform.

Newton himself was quite familiar with the fact that his laws of mechanics satisfied a relativity principle. Clearly, if the laws of mechanics did not satisfy such a relativity principle, then their application to motion on the surface of the earth would be extremely complicated. If we assumed that the laws of mechanics were valid in the frame of reference of the sun but not in any system of coordinates moving at constant velocity with respect to the sun, then, since a piece of the earth's surface is in high-speed motion, we would have to know our velocity through space at a given instant in order to predict the motion of any mechanical object near us. The local rules of mechanics that we would have to use in such simple operations as walking would change from hour to hour as the combination of our rotational velocity around the earth's axis and the velocity of the earth in its orbit around the sun added in different ways to give different total velocities with respect to the sun. Actually, the laws of Newtonian mechanics are not exactly valid if position and velocity measurements are made with respect to the surface of the earth. The surface of the earth does not move at constant velocity, but rather accelerates, and, unlike constant-velocity motion, accelerated motion does have

observable effects. Fortunately, although the velocity of the earth's surface is very large by ordinary standards, its acceleration is very small and can therefore usually be ignored. When it cannot be ignored, as in calculating satellite orbits, things really do get very complicated.

35.2 THE EFFECT OF ELECTROMAGNETIC THEORY ON THE NEWTONIAN RELATIVITY PRINCIPLE

When Maxwell developed the mathematical theory of electromagnetic wave propagation, it seemed as though the principle of relativity would have to be abandoned. To Maxwell and all his contemporaries it seemed perfectly

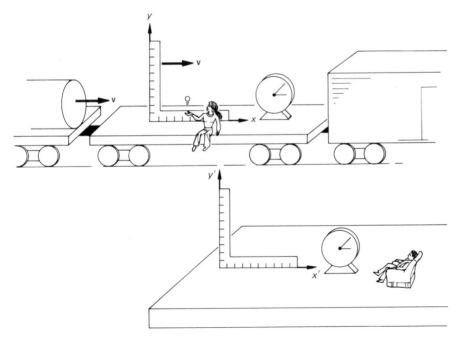

FIGURE 35.1   Observers in two different reference frames recording the motion of a ball.

obvious that a wave was some kind of disturbance that traveled through a medium. Sound waves were pressure variations that traveled through air or other materials, water waves were height variations that traveled through water, and electromagnetic waves must be variations in some other quantity associated with the electric and magnetic fields that traveled through some yet undescribed medium. The medium or material that carried electromagnetic waves came to be called the *ether*. Maxwell and others tried without much success to construct theoretical models of a mechanical medium that would propagate waves in accordance with Maxwell's equations for electromagnetic wave propagation. We need not discuss these attempts in any detail; we mention them only to point out that nineteenth-century physicists universally expected that one would see light waves propagate at the same speed in all directions only if one were at rest with respect to the ether. A coordinate system at rest with respect to the ether was called an *absolute rest frame*. By referring to Fig. 35.2 one can see that it seems quite obvious that if the two light signals are traveling in opposite directions with speed c with respect to the man, then they must be traveling at speeds $c + v$ and $c - v$ with respect to the

woman. A relativity principle for electromagnetic theory would state that motion of a closed railroad car at constant velocity could not be detected by means of electromagnetic experiments within the railroad car. No such principle seemed valid because, by measuring the velocity of light propagation, one could determine the speed of the railroad car with respect to the ether.

Considering the fact that electromagnetic theory seemed not to satisfy any principle of relativity, scientists found certain aspects of the theory very puzzling or strangely coincidental. These puzzling aspects of the theory are exemplified by the situation shown in Fig. 35.3. Two people approach one another with constant relative speed v. The woman holds a bar magnet in her hand, with the north pole pointed at the man. The man holds a large loop of wire in his hand, and in the loop of wire is an ammeter. Both persons see the ammeter register a pulse of

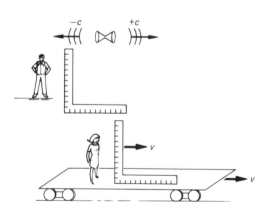

FIGURE 35.2 As observed by the woman, the two light signals would propagate with speeds $c + v$ and $c - v$.

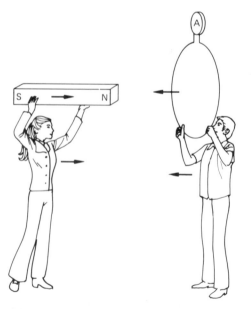

FIGURE 35.3 Both observers can explain the current generated in the wire loop, but the explanations differ.

current in the loop as the magnet approaches and enters it. From the point of view of the man the generation of the current would be explained as follows.

Man's Explanation   The loop is stationary. The electrons in the loop are also initially stationary and therefore experience no force caused by the magnetic field of the magnet. Since the magnet is moving, the magnetic field in the vicinity of the loop is changing. Changing magnetic fields create electric fields. The electric field created is in such a direction as to drive a current through the loop of just the observed value.

From the point of view of the woman the observed current would be explained as follows.

Woman's Explanation   The magnet is stationary, and therefore no electric field exists anywhere. The electrons in the loop are moving, along with the loop, toward the magnet. A charged particle moving in a magnetic field experiences a force given by

F $= q$**v** $\times$ **B**

That force is in the direction of the loop and therefore causes the electrons to move around the loop. Again a detailed calculation of the current that should be caused by the magnetic force gives a result that agrees with the ammeter reading.

Each person has no difficulty in predicting the current in the loop by applying Maxwell's equations in his or her own frame of reference. This seemed to be a very odd coincidence for a theory that was supposed to be applicable only in the ether rest frame.

35.3 ATTEMPTS TO MEASURE THE VELOCITY OF THE ETHER

In order to clarify the details of the electromagnetic theory of light, investigators devised a number of experiments whose aim was to detect the effects of ether motion on light propagation. We shall describe only one of them.

THE MICHELSON-MORLEY EXPERIMENT

The most sensitive of all the experiments performed to detect the motion of the earth through the ether was that carried out by the American physicists A. A. Michelson and E. W. Morley in 1887. Essentially the same experiment, but with modifications to improve the sensitivity, was repeated a number of times in subsequent years. We shall consider first a drastically simplified model of the experimental apparatus and then a more realistic model.

The aim of the experiment is to compare the times it takes wave fronts to travel from the source to the two mirrors and back to the source (see Fig. 35.4). We shall assume

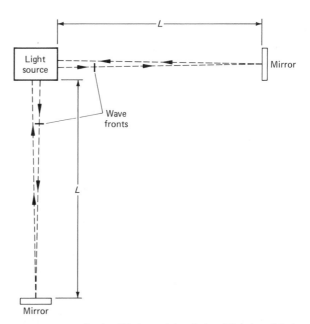

FIGURE 35.4 A simplified model of the Michelson-Morley experiment.

that the whole apparatus is rigidly constructed and is moving, through the stationary ether, to the right with a velocity v that is less than the velocity of light c.

With respect to the apparatus, the light wave travels at a speed $c - v$ when moving toward the right-hand mirror and with a speed $c + v$ when returning to the source (Fig. 35.5). The total time taken for the trip back and forth in the direction parallel to the velocity of the apparatus is

$$T_{\parallel} = \frac{L}{c - v} + \frac{L}{c + v} = \frac{2cL}{c^2 - v^2}$$

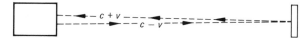

FIGURE 35.5 The light wave moving back and forth in the parallel arm of the Michelson-Morley experiment.

If we now look at the wave front that travels toward the mirror that is in a direction perpendicular to the motion of the apparatus through the ether, we see that we must determine the velocity of the wave front with respect to the apparatus by using the vector form of the velocity addition law. We know that the velocity of the ether with respect to the apparatus is to the left with a magnitude v. We know that the magnitude of the velocity of the wave front with respect to the ether is just the velocity of light c. We also know that the resultant velocity of the wave front with respect to the apparatus must be in the direction toward the mirror. This tells us that the magnitude of the velocity of the wave front with respect to the apparatus is $\sqrt{c^2 - v^2}$. (Apply the Pythagorean theorem to the vector diagram in Fig. 35.6.) The magnitude of the velocity on the return path is the same, as we can see by drawing a similar

FIGURE 35.6 The light wave moving back and forth in the perpendicular arm of the Michelson-Morley experiment.

vector diagram at the mirror. Thus the total time taken for the round trip by the wave front is

$$T_{\perp} = \frac{2L}{\sqrt{c^2 - v^2}}$$

Unless $v = 0$, the times $T_{\parallel}$ and $T_{\perp}$ are different. Two wave fronts that start from the source simultaneously will not return to the center of the apparatus at the same time. Given L and c, the difference in arrival times for the two wave fronts should allow one to calculate v. Thus one could determine the velocity of the apparatus through the ether.

The experiment we have described is not a practical one because of the extremely high speed of light waves. However, by an interference method, Michelson and Morley were able to measure the tiny time differences involved. In the actual Michelson-Morely experiment a half-silvered mirror[1] is placed at the intersection of a T attached to a heavy marble slab. The mirror is set at a 45° angle, as shown in Fig. 35.7. The marble slab floats in a pool of mercury so that the slab may be slowly turned by the application of extremely weak forces. Fully coated mirrors are set up at two ends of the T, and a screen on which to view an interference pattern is set up at the other end (Fig. 35.7). A monochromatic light beam from an external source strikes the half-silvered mirror, where it is broken into a transmitted beam and a reflected beam. These beams travel in perpendicular directions to two ends of the T, where they are reflected back toward the half-silvered mirror. Half of each beam is then sent to the screen; the other half is sent back toward the source and absorbed. On the screen the two beams create an interference pattern, and the smallest change in the relative phase of the two beams causes noticeable shifts in the interference pattern. Such a change in relative phase could be caused by a slight change in the time it takes one of the beams to travel to the fully coated mirror and back. The interference pattern was observed with different orientations of the T. This was done at various times of day and at different times of the year to eliminate any possibility that the apparatus was accidentally at rest in the ether at the time of observation. No significant shift in the interference pattern was detected. The Michelson-Morley experiment and all similar experiments failed to detect any

motion of the earth relative to the ether. A number of attempts were made to explain the null result of the experiments. Some of the important explanations and the reasons for their failure are as follows.

1. The earth and other bodies drag the ether with them as they move.

This explanation is untenable because such a flow of the light-conducting medium would cause a deflection of light traveling to the earth from outer space. This light would be detectable as a constant wandering of the apparent positions of the stars.

2. The propagation velocity of light depends on the motion of the source in the same way that the velocity of a bullet depends on the motion of the gun. That is, light travels at constant velocity with respect to its source only, not with respect to the ether.

This explanation is untenable on two counts. First, Maxwell's equations, which have been verified in great detail, predict that the velocity of light propagation is independent of the motion of the source. Second, if light from a receding source traveled to the earth more slowly than light from an approaching source, there would be a detectable change in the apparent motion of astronomical objects such as the satellites of Venus and Jupiter, which are constantly changing their velocities relative to the earth.

3. Solid bodies are distorted by motion through the ether. In particular, they shrink in the direction of the motion by a factor $\sqrt{1 - (v/c)^2}$. That is, a square of side L would become a rectangle of sides L and $L\sqrt{1 - (v/c)^2}$ (see Fig. 35.8).

This would mean that the arm of the Michelson-Morley apparatus that was in the direction of the motion would shrink by just the right amount to cancel the effect that had

---

[1] An ordinary mirror is made by coating the back surface of a sheet of glass with an uncolored metal such as silver or aluminum. A "half-silvered" mirror is made in the same way except for the fact that the metallic coating is so light (it is only a fine spray of metal) that half the light is transmitted and only half reflected.

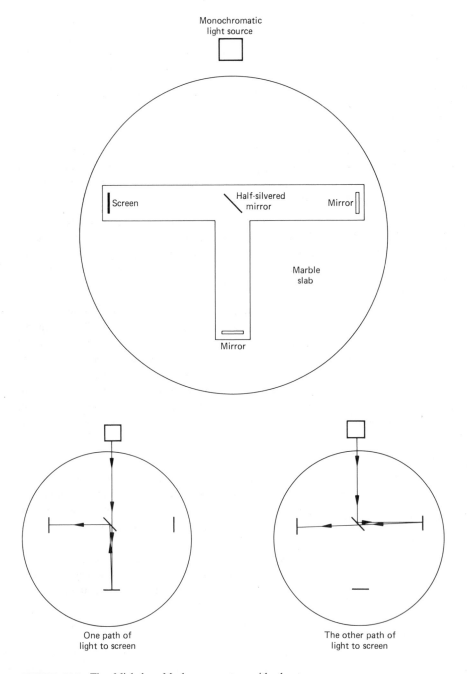

FIGURE 35.7 The Michelson-Morley apparatus, with the two possible paths of light from the source to the screen shown.

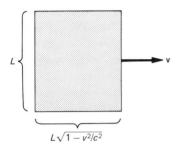

FIGURE 35.8   The Lorentz-Fitzgerald contraction.

been expected. This contraction, called the Lorentz-Fitzgerald contraction, was analyzed and explained in great detail by Lorentz. If one assumed that electromagnetic forces between the particles of a body were what made a rigid body rigid, then one could show that just such a contraction could be expected, so that the explanation was not quite as unconvincing as it appeared at first sight. One of the difficulties with the Lorentz-Fitzgerald ''explanation'' was that it could not explain the null results of certain other experiments that would have detected the motion of the earth through the ether in spite of the contraction. In actuality, Lorentz' explanation was simply superseded by the publication, a short time later, of Einstein's first paper on relativity, which explained the null result of all such experiments and at the same time predicted a number of startling new effects that were subsequently confirmed. Incidentally, a length contraction equal to the Lorentz-Fitzgerald contraction is a consequence of the relativity theory.

35.4 EINSTEIN'S THEORY OF RELATIVITY

The experiments seem to show that the relativity principle is valid not only for mechanics but for electromagnetic theory as well. One might reasonably ask why people who were familiar with those experiments did not immediately recognize the fact that the equations of electromagnetic theory were valid in any constant-velocity frame of reference. (A constant-velocity frame is usually called an *inertial frame*. More exactly, an inertial frame is any system of coordinates in which isolated particles travel at constant velocities.) The reason physicists were reluctant to accept the principle of relativity for electromagnetic theory is that it seemed to lead to clear logical contradictions. To illustrate some of these contradictions, let us consider the following imaginary experiment.

A man stands in the center of a long dark train. At each end of the train are light-sensitive devices that will cause a firecracker to pop as soon as they detect any light. We imagine ourselves riding on the train with the man. He lights a match. Now if the train were stationary the light would travel in both directions with speed c and would cause the firecrackers to pop simultaneously. But if unaccelerated motion is undetectable, then to the observers on the train (even if it is moving with respect to the surface of the earth) the light will still travel at the same speed in both directions and will cause the firecrackers to pop simultaneously. Now let us suppose that we are on the ground as the train moves by and we view the same sequence of events. We see the man light the match, and we also see[2] the light move in both directions at constant velocity *with respect to us on the ground.* But we see the back end of the train move toward the light while the front end of the train moves away from the light. Therefore we would see that the firecracker in the rear of the train

---

[2] The word ''see'' here should not be taken literally. One cannot see light propagate; nor could the human eye resolve a sequence of events that took place with such rapidity as the events described here.

popped before the firecracker in the front of the train! This experiment is illustrated from both points of view in Fig. 35.9 and 35.10.

We seem to have shown that any attempt to combine Maxwell's electromagnetic theory with a principle of relativity leads to internal logical inconsistencies. The resulting theory predicts that the firecrackers pop simultaneously and that the firecrackers do not pop simultaneously. Before we conclude that this experiment proves that the theory is absurd, we must recall that there is no contradiction at all in saying that an object is both moving and not moving. An object on the train may be stationary according to all observers on the train and moving according to all observers on the ground. By making a careful analysis of the experimental procedures we use to verify simultaneity, Einstein realized that two events happening at different places in space might be simultaneous and not simultaneous in the same sense that an object might be moving and not moving. The events would be simultaneous to one set of observers and not simultaneous to another set of observers. This idea is not as outrageous as it sounds at first. Suppose we look up into the night sky and see explosions occur simultaneously on two distant stars. Were those two explosions actually simultaneous? If we

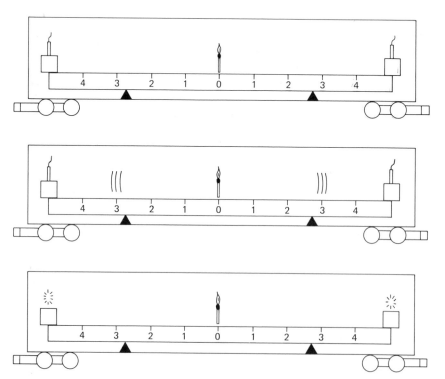

FIGURE 35.9   The experiment as viewed by a person on the train. If light travels at speed c in both directions within this frame of reference, then the firecrackers pop simultaneously.

were completely naive, we would simply say that if we saw them occur at the same time, then they must have happened simultaneously. Being more sophisticated than that, we realize that it may take many years for the light from a star to reach our eyes. Thus, if one star is twice as far from us as the other, events on the two stars that appear simultaneous on the earth may have been separated by years of time. In order to determine simultaneity, we must first know the distances to the stars and then take into account the time required for light signals to travel from the stars to our eyes, using the fact that light travels at velocity c. Being still more sophisticated, we might realize that if exactly the same thing is done by another intelligent being on another planet that is moving with respect to ours, then the conclusion this person comes to regarding the simultaneity of the two explosions might be different from ours. In other words, the inhabitant of the other planet might assign time values to events in the universe that are

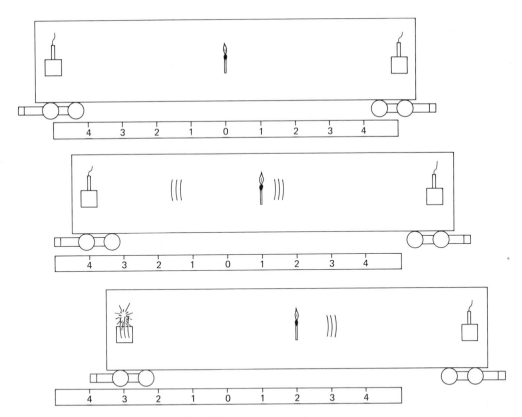

FIGURE 35.10 The experiment as viewed by a person on the ground. If light travels at speed c in both directions within this frame of reference, then the firecracker in back pops before the one in front.

different from the time values we assign to them. (We mean more here than the trivial difference of using a different zero of the time scale.) This may be true even though the other person uses the same basic rules for assigning time values to events that we do. Furthermore, the laws of physics may have exactly the same form when expressed in terms of his space and time values as they do when expressed in terms of ours.

35.5 THE LORENTZ TRANSFORMATION

It is clear that the only way we shall be able to settle the question of whether the principle of relativity is self-contradictory or not is by a very careful logical analysis. While we make the analysis we shall have to be on our guard against making assumptions based on ''common sense,'' since our common sense notions have grown out of our everyday experiences and none of us has any experience with velocities that are comparable to the velocity of light. We have suggested that events that are simultaneous for one observer may not be so for another. We have also mentioned the possibility that rigid bodies shrink when they move. Certainly a good place to start is by establishing the relationship between the position and time of a particular event as measured by one observer and the position and time of the same event as measured by another observer who is moving with respect to the first. In order to do this we shall make certain definite assumptions. We shall try to be very explicit in our reasoning.

We picture a Chief Observer (call him Mr. Unprime) sitting on an infinitely long meter stick (Fig. 35.11). He is situated at the zero of the meter stick and he holds a clock. By choosing a peculiar unit of time, namely $\frac{1}{3} \times 10^{-8}$ s, he has arranged that the velocity of light is equal to 1. We call that extremely short time unit a subsecond (abbreviated ss). We temporarily introduce this artificial time unit only to prevent the reader from being distracted from following the central logic of the analysis by the necessity of having to work through the more complicated algebra that occurs with ordinary units. However, at the end of this section it will be quite easy to convert our results to the form they take in ordinary units. Mr. Unprime has employed an army of Junior Observers who are distributed at small intervals along the stick to take measurements of the time and place of events that do not occur close to the zero of his meter stick. Each of the Junior Observers also has a clock. In order to synchronize the clocks, Mr. Unprime sends a light pulse in both directions each unit of time. The nameless Junior Observers are told that these light pulses are separated by exactly 1 ss in time. When the first pulse arrives at the Junior Observer who is at the 17-m mark, he sets his clock to 17 ss (since he knows that it took the light that long to reach him). He also checks that his clock is running at the same rate as the Chief's.

Mr. Unprime and his army of Junior Observers are aware of the fact that close by another endless meter stick is passing them at the speed of v meters per subsecond. Sitting at the zero of the other meter stick is another Chief Observer, Ms. Prime, who has also synchronized the clock

FIGURE 35.11 Mr. Unprime and his subordinates. Collectively they make up one inertial frame.

she carries with her own army of Junior Observers on her meter stick. When the two Chief Observers pass, they both notice that their clocks read exactly zero. Whenever an event occurs (e.g., a firecracker pops, an object passes), the nearest of Mr. Unprime's observers records the time and place of the event: We shall call his measurements t and x. The nearest of Ms. Prime's observers also records her measurements of the time and place of the event; we shall call these t' and x'. Our problem is now to calculate the relationship between the measurements taken by Mr. Unprime's observers and those taken by Ms. Prime's observers. We make the following assumptions:

1. Any isolated object (an object with no force applied to it) moves without acceleration according to both observers; that is, both frames are inertial frames.

2. Light pulses propagate at the velocity of ± 1 m/ss according to both observers.

3. The origins of the meter sticks coincide when both clocks read zero.

4. The equation for the position of Ms. Prime, namely $x' = 0$, is transformed into $x = vt$, while the equation for the position of Mr. Unprime, namely $x = 0$, is transformed into $x' = -vt'$. That is, Ms. Prime travels with velocity v through the unprimed frame, while Mr. Unprime travels with velocity $-v$ through the primed frame.

Assumptions 1 and 2 express the principle of relativity as it applies to two of the simplest laws of mechanics and electromagnetic theory, respectively. Assumptions 3 and 4 are of a much more trivial nature. Assumption 3 simply guarantees that both observers use a mutually convenient zero for their time scales. Assumption 4 simply expresses the fact that their relative speed is v meters per subsecond.

Before we proceed with the formal derivation, let us reiterate exactly what we aim to do. We consider an event that happens at position x' and time t' according to Ms. Prime's observers. This same event can also be observed by Mr. Unprime's observers, and they determine that it happened at position x and time t. We shall see that, with the four assumptions given, we can determine the relationship between the pair of numbers (x', t') and the pair of numbers (x, t). That is, if we know the space and time coordinates that Ms. Prime assigns to an event, we shall have formulas that allow us to calculate the space and time coordinates that Mr. Unprime assigns to the same event. They will be of the form

$$x = f(x', t') \quad \text{and} \quad t = g(x', t') \tag{35.2}$$

where $f(x', t')$ and $g(x', t')$ are two functions that we shall determine.

First we make use of assumption 1. The equation that describes the motion of any unaccelerated object in Ms. Prime's frame is of the form

$$x' = u't' + x'_0 \tag{35.3}$$

where u' is the velocity of the object and x'_0 is the location of the object at time $t' = 0$. Thus any event that occurs at the location of the object must have space and time coordinates that satisfy Eq. 35.3. We can picture the object as a brick moving freely down the x' axis with no forces on it. In order to keep our attention focused on events rather than objects, we imagine that, on the brick, is an ant with hiccups. Each hiccup is an event. If we plot the (x', t') coordinates of these events, we get a series of points that all lie on the straight line given by Eq. 35.3 (Fig. 35.12).

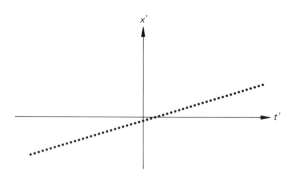

FIGURE 35.12   The times and places of the ant's hiccups according to Ms. Prime.

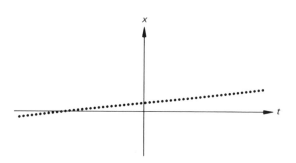

FIGURE 35.13   The times and places of the ant's hiccups according to Mr. Unprime.

Now the unprimed observers are also observing the ant's distress. If they plot the space and time coordinates that they assign to the hiccups, they must also obtain points on a straight line (Fig. 35.13). This is so because they also see the ant moving along on an isolated object, and isolated objects move without acceleration in both frames. (In general, the speed of the brick and its initial position are not the same in the two frames.) Thus the transformation given by Eq. 35.2 must have the property that when the points on any straight line in the (x', t') variables are plotted in the (x, t) plane, they must still lie on a straight line. A transformation with this property is called a *linear transformation*. One can show that a linear transformation is always of the form

$$x = Ax' + Bt' + C \qquad (35.4)$$

and

$$t = Dt' + Ex' + F \qquad (35.5)$$

We must still determine the constants A, B, C, D, E, and F. An event that happens right at the place and time at which

the two Chief Observers pass one another has space and time coordinates $x' = 0$ and $t' = 0$ according to Ms. Prime and $x = 0$ and $t = 0$ according to Mr. Unprime. (See assumption 3.) Putting the space and time coordinates of such an event into Eqs. 35.4 and 35.5, we see that $C = 0$ and $F = 0$. Thus

$$x = Ax' + Bt' \qquad (35.6)$$

and

$$t = Dt' + Ex' \qquad (35.7)$$

We have still made no use of Assumptions 2 or 4. The equation for a light pulse sent out from the origin to the right at the instant both observers pass is, in the primed coordinates, $x' = t'$. This equation must transform into $x = t$. Putting $x' = t'$ into Eqs. 35.6 and 35.7, we get

$$x = (A + B)t'$$

and

$$t = (D + E)t'$$

This gives $x = t$ only if

$$A + B = D + E \qquad (35.8)$$

The equation of motion of a left-going light pulse sent out from the origin just as the Chief Observers pass is $x' = -t'$. This must transform into $x = -t$. From this we obtain another equation relating the unknown constants,

$$-A + B = -D + E \qquad (35.9)$$

Equations 35.8 and 35.9 imply that $A = D$ and $B = E$. We have so far reduced our transformation to the form

$$x = Ax' + Bt'$$

and

$$t = At' + Bx'$$

Already we can see a rather strange symmetry between time and space variables. Also we can see that if Ms. Prime determines that an event took place at $t' = 0$ at any place other than the origin, Mr. Unprime will not feel that that event took place at $t = 0$ even though they have synchronized their clocks on passing. Separated events that are simultaneous to one observer are not simultaneous to another. We now want to use Assumption 4. We set $x' = 0$. Our transformations should then give us $x = vt$. What we get is

$$x = Bt'$$

and

$$t = At'$$

or

$$\frac{x}{t} = \frac{B}{A}$$

This is the equation $x = vt$ only if $B = vA$. We can now eliminate B and write the transformation as

$$x = A(x' + vt') \qquad (35.10)$$

and

$$t = A(t' + vx')$$

Had we done the same analysis to calculate the transformation from x and t to x' and t', the only difference would have been that the velocity of Mr. Unprime in Ms. Prime's system would be $-v$ rather than v. We would thus have obtained the result

$$x' = A(x - vt)$$

$$t' = A(t - vx)$$

Substituting these relations for x' and t' into Eq. 35.10, we obtain

$$x = A[A(x - vt) + vA(t - vx)]$$

$$= A^2(1 - v^2)x$$

which determines the final constant A:

$$A = \frac{1}{\sqrt{1 - v^2}}$$

Thus the relationship between the space and time co-ordinates assigned to the same event by observers in the two frames is

$$x = \frac{1}{\sqrt{1 - v^2}}(x' + vt')$$

and

$$t = \frac{1}{\sqrt{1 - v^2}}(t' + vx')$$

We must now convert these formulas from the sub-second time unit to the usual time unit. We do this by noting that

t (in subseconds) $= ct$ (in seconds)

and

v (in meters/subsecond) $= v/c$ (in meters/second)

Thus simply replacing t, t', and v by ct, ct', and v/c does the trick. We then get

$$x = \frac{1}{\sqrt{1 - \beta^2}}(x' + \beta ct') \tag{35.11}$$

and

$$ct = \frac{1}{\sqrt{1 - \beta^2}}(ct' + \beta x') \tag{35.12}$$

where

$$\beta = v/c$$

As we mentioned before, we obtain the transformation in the other direction by replacing v by $-v$ in these formulas:

$$x' = \frac{1}{\sqrt{1 - \beta^2}}(x - \beta ct) \tag{35.13}$$

and

$$ct' = \frac{1}{\sqrt{1 - \beta^2}}(ct - \beta x) \tag{35.14}$$

By methods similar to the ones we have used, one can show that, if both primed and unprimed observers make measurements in all three space dimensions, the transformations involving their y and z coordinates are trivial. That is,

$$y = y'$$

and

$$z = z'$$

This transformation was first derived by H. A. Lorentz, and is therefore called the *Lorentz transformation*, although Lorentz did not have a complete appreciation for its relationship to the principle of relativity.

In order to become more familiar with these somewhat peculiar transformation equations, let us apply them to a few important cases.

35.6 TIME DILATION

We shall henceforth assume that all observers have ordinary clocks that tick once per second. The ticks of Ms. Prime's clock are a sequence of events that happen at $x' = 0$ and $t' = 1, 2, 3$, etc. According to Mr. Unprime, the nth tick of Ms. Prime's clock takes place at the following place and time (put $x' = 0$ and $t' = n$ into Eqs. 35.11 and 35.12):

$$x_n = \frac{vn}{\sqrt{1 - \beta^2}} \text{ meters}$$

and

$$t_n = \frac{n}{\sqrt{1 - \beta^2}} \text{ seconds}$$

Therefore, according to Mr. Unprime, the time interval between two successive ticks of Ms. Prime's clock is

$$\Delta t = \frac{1}{\sqrt{1 - \beta^2}} \text{ seconds}$$

This number is larger than 1. In other words, as measured by Mr. Unprime, Ms. Prime's clock is running slow. What is most surprising is that, if we perform the same analysis using the inverse transformation (Eqs. 35.13 and 35.14), we find that, according to Ms. Prime, Mr. Unprime's clock is running slow!

One's first reaction might be to say that "in reality" both clocks run at the same rate but, because of some sort of illusion created by the velocities, each observer's clock "appears" to be running slow to the other observer. The time-dilation effect cannot be explained away as an illusion because it can have real physical effects. For instance, suppose v, the relative speed between the two observers, is nine-tenths the speed of light (that is, $\beta = 0.9$); then, according to Mr. Unprime, Ms. Prime's clock is running slow by a factor of about 2.2. Thus, if Ms. Prime lived to be 100 years old by her own clock, she would be more than 220 years old according to Mr. Unprime and she could therefore travel (at the speed she is going according to Mr. Unprime) a distance of about 200 light years. Since we shall find that nothing can travel at a speed larger than the speed of light, a distance of 200 light years would be out of the range of a human being if it were not for the time-dilation effect. There is obviously no way of brushing aside as somehow unreal the fact of a person arriving someplace while still alive rather than 100 years dead.

Although technology has not succeeded in moving human beings at $0.9c$, many instances exist of elementary particles moving at velocities very close to c. Of particular interest in terms of the time-dilation effect are the μ mesons produced at the top of the atmosphere by high-energy cosmic-ray particles from space. When they are at rest μ mesons last only about two-millionths of a second before spontaneously disintegrating into an electron and a neutrino. Thus, during this time, traveling at the speed of light, they could move a distance

$$(3 \times 10^8 \text{ m/s})(2 \times 10^{-6} \text{ s}) = 600 \text{ m}$$

Since the atmosphere is much thicker than 600 m, almost none of these particles should reach the earth's surface.

But in fact many of them do. The decay rate of the μ mesons can be explained perfectly if one takes into account the fact that the lifetime of a moving particle is increased by a factor of $1/\sqrt{1 - (v/c)^2}$, where v is the particle's velocity.

In brief, if $\Delta t'$ is the time interval between two events that occur *at the same place in the primed frame,* then the time interval between the same two events (that will occur at different places) in the unprimed frame is (see Fig. 35.14).

$$\Delta t = \frac{\Delta t}{\sqrt{1 - \beta^2}} \qquad (35.15)$$

35.7 LENGTH CONTRACTION

Ms. Prime sits at the origin of her meter stick. As we have seen, the motion of Ms. Prime through the unprimed co-ordinate frame is described by the equation

$$x = vt$$

Ms. Prime's Junior Observer at the 1-m mark has a trajectory in the unprimed frame that is found by setting x' equal to 1 in Eq. 35.13. That is,

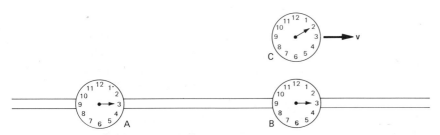

FIGURE 35.14   The time-dilation effect. A clock moving through an inertial frame runs slow in comparison to the clocks in that frame.

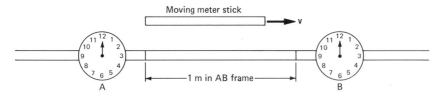

FIGURE 35.15   The length-contraction effect. An object moving through an inertial frame is shortened (in the direction of its motion) in comparison to similarly constructed objects at rest in that frame.

$$1 = \frac{1}{\sqrt{1-\beta^2}}(x - vt)$$

or

$$x = vt + \sqrt{1-\beta^2}$$

Thus in the unprimed frame the distance between Ms. Prime and her first observer is only $\sqrt{1-\beta^2}$ meters. Hence the length of meter rod between them, which is 1 m in the frame in which the rod is at rest, appears to be shortened by a factor of $\sqrt{1-\beta^2}$ in the frame in which the rod moves at velocity v. In general, an object that has a length L_0 in its rest frame will appear shortened by a factor of $\sqrt{1-\beta^2}$ (in the direction of its motion) in a frame in which it moves at velocity $v = \beta c$. Thus in the new frame it will have a length L given by

$$L = \sqrt{1-\beta^2}L_0 \tag{35.16}$$

35.8 THE GEOMETRY OF SPACETIME

One can easily imagine the reader at this point muttering in frustration, "I can follow the equations. If I'm careful I can even follow the logic; but I find it just impossible to form any clear picture of what this is all about." What the reader desires is a visualization of these space-time measurement processes which will agree with his or her fundamental intuitive notions of the geometry of space and

the flow of time and which will reveal "in pictures" the geometric meaning of the equations we have derived. That desire will never be completely satisfied because those fundamental intuitive ideas of space and time are incorrect. Our powers of geometric visualization have been determined by a combination of a lifetime of experiences and a genetically determined intellectual apparatus formed in the crucible of evolutionary history. Since our usual concepts of space and time are fully adequate for describing ordinary experiences, and brain cells were not wasted in producing conceptual powers of no survival value, we have no ability to form pictures of geometries other than three-dimensional Euclidean space. It is possible to mathematically define a four-dimensional Euclidean space and to prove theorems about spheres and cubes and the like in that four-dimensional space, but nobody (not even the people who prove the theorems) can form a picture in his or her mind of any four-dimensional object. All our daily experiences can most efficiently be organized in terms of a picture of space and time equivalent to that described by Newton in the *Principia:* "Absolute space, in its own nature, without relation to anything external, remains always similar and immovable. Absolute, true, and mathematical time, of itself, and from its own nature, flows equable without relation to anything external." Our world is one of objects located in three-dimensional Euclidean space. Their locations and their shapes vary as a function of a universal, absolute time.

The basic geometric element in this world is a *point in space,* and the point in space may be occupied by a material body or it may be unoccupied. It exists throughout all time. This Euclidean-Newtonian picture of the world is different in certain fundamental ways from the geometric picture of the world that physicists have been forced to adopt in order to construct a theoretical framework that agrees with experimental observations. The basic element in the new geometry is the *event.* Geometrically an event is *a certain place at a certain time,* that is, a point in *spacetime.* Physically an event is anything that happens at that particular place and time: the lighting of a match, two electrons bumping, or any other imaginable thing that can happen. An object is simply the persistent sequence of events that happen at the location of that object: the reflection of light off the surface of the object, the whirling around of the electrons in the object, etc. Something is always happening at the location of an object. The collection of all events in spacetime forms a four-dimensional set.

Since we have no powers to picture four-dimensional sets, we can best get a picture of the new geometry by the following device. We discard one space dimension; that is, we imagine that the spatial world is a plane rather than a three-dimensional space. For convenience we assume that all the objects in our world are little particles whose positions can be represented by single dots. We can then represent the Newtonian view of such a world by having this collection of little dots swarm around on the two-dimensional spatial plane as time progresses. We construct a model of the relativistic view of the world as follows. On a clear plastic sheet we print dots representing the positions of all the particles at one instant of time. Using equally and closely spaced instants of time, we do the same on a sequence of plastic sheets. If we now stack the sheets together in sequence, we can easily see what appear to be the continuous trajectories of each particle. This is still a Newtonian model since we could easily pull out one sheet

to determine exactly where all the particles were at a particular instant of universal time. To produce a relativistic model of the world, we stack all the sheets together carefully and heat them just enough so that they fuse and the sequences of dots become continuous curves. We now have our model of the relativistic spacetime: a solid piece of plastic with some curves embedded in it to represent the trajectories of material particles (see Fig. 35.16).

Since spacetime is infinite in all dimensions, it is a somewhat unwieldy model. Let us take out just a big sphere of it and throw away the rest. We can appreciate the important difference between our spacetime and the Newtonian model we started with by considering what happens if we attempt to reverse the process of fusion. We bring our plastic sphere to a man with a band saw and we ask him to cut our model up into space slices. ''Certainly,'' he says. ''Which way do you want me to slice it?'' ''Perpendicular to the time direction,'' we tell him. ''What time direction?'' he asks. ''All I see is a big ball of plastic with some wiggly lines in it.'' The realization suddenly dawns on us that we have lost track of which way the space sheets ran. It seems that we must be very careful in deciding which way to cut because, once we have chosen a particular slice, we are making a statement that all events that happen in that slice are happening at the same time. We do not want to make such a decision without good grounds.[3] We finally get a good idea. We shall investigate a particular method of slicing our spacetime into uniform sheets and see whether, when we use that slicing, the spatial positions of the particles as a function of time satisfy the laws of physics. We choose a direction and try it (without actually feeding it into the band saw). To our delight we find that we have

---

[3] One might object on the grounds that our trouble stems only from our having cut a spherical sample, and if we had the whole model, we could see the direction of the surface. But remember, the original model was infinite in all directions, as is real spacetime.

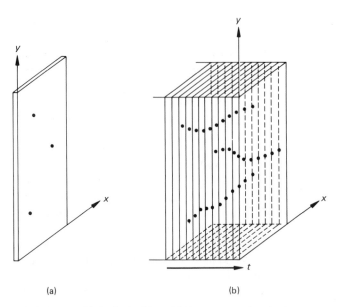

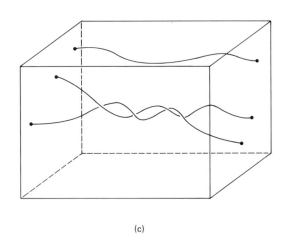

(a)   (b)   (c)

FIGURE 35.16   (a) A single sheet with dots representing the locations of particles at one instant. The full sheet is infinite in both directions. (b) A stack of individual sheets making a Newtonian model of the world. The full stack is infinite in all three directions. (c) The relativistic model of spacetime. When the model is extended to infinity there is no unique way of cutting it up into space slices and thus defining the collection of simultaneous events.

luckily hit upon just the right direction, because the laws of physics are satisfied to a tee. But our delight fades when we discover that by choosing a different orientation for the time slices, we get a different description of the motion of the particles but one that still satisfies all the laws of physics. In fact, we discover that the laws of physics are satisfied by any slicing within certain limits. The laws of physics are of no use in trying to discover the absolute time direction in spacetime. Since the motions of objects and the laws of physics are the only things we have to work with, we are forced to accept the conclusion that no absolute time direction exists. An infinite number of spacetime coordinate systems can be used to describe the motion of things, and every one is as good as every other one.

35.9 THE RELATIVISTIC ADDITION OF VELOCITY LAW

At the beginning of this chapter we saw that the law for addition of velocities combined with the law of constant light speed led to a contradiction of the principle of relativity. We have since then assumed that the law of constant light speed and the principle of relativity are both correct. The law of velocity addition, which we had used before, must therefore be false. We shall now use the Lorentz transformation formulas to derive the correct law for the addition of velocities.

Ms. Prime is traveling through the unprimed coordinate system at velocity v. Suppose an object is traveling at a velocity u' down Ms. Prime's meter stick (see Fig. 35.17). How fast would it be traveling as measured by Mr. Unprime? Let us assume that the object just crosses the origins of both coordinate systems as the Chief Observers pass one another (at $t = t' = 0$). Then its initial coordinate in both frames is zero.

Consider any event that occurs at the location of the object. According to Ms. Prime the event will have space-time coordinates (x', t') that satisfy the relation

$$x' = u't' \tag{35.17}$$

The same event will have unprimed coordinates (x, t) given by

$$x = \frac{x' + vt'}{\sqrt{1 - \beta^2}} \tag{35.18}$$

and

$$ct = \frac{ct' + \beta x'}{\sqrt{1 - \beta^2}} \tag{35.19}$$

If we use Eq. 35.17 to write x' in terms of t' in Eqs. 35.18 and 35.19, we see that

$$\frac{x}{t} = \frac{u' + v}{1 + u'v/c^2}$$

or

$$x = ut$$

with

$$u = \frac{u' + v}{1 + u'v/c^2} \tag{35.20}$$

QUESTION

Suppose Ms. Prime is moving at a velocity of $0.9c$ with respect to Mr. Unprime. A moondog is moving at a velocity of $0.9c$ down the x' axis as measured by Ms. Prime (Fig. 35.18). What is the velocity of the moondog with respect to Mr. Unprime? Is it larger than c?

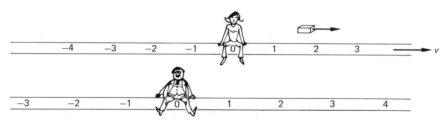

FIGURE 35.17 The addition of velocities. The brick travels at velocity u' with respect to Ms. Prime. Ms. Prime travels at velocity v with respect to Mr. Unprime. At what speed does the brick travel with respect to Mr. Unprime?

FIGURE 35.18   What is the speed of the moondog?

ANSWER

We use the formula for velocity addition, with $v = 0.9c$ and $u' = 0.9c$. Thus the velocity of the moondog with respect to Mr. Unprime is

$$u = \frac{u' + v}{1 + u'v/c^2} = \frac{1.8c}{1 + (0.9)^2} = 0.994c$$

This velocity is not larger than c. The addition of any two velocities less than c gives a velocity less than c. In fact, if object B is moving at $0.9c$ with respect to object A, and object C is moving at $0.9c$ with respect to object B, and object D is moving at $0.9c$ with respect to object C, etc., then, with respect to object A, the objects B, C, D, etc., move at velocities that get closer and closer to c as we go further out in the sequence, but nothing moves at a velocity larger than c with respect to anything else.

35.10 THREE-DIMENSIONAL VELOCITY ADDITION

In the case considered above the moving object has only an x component of velocity; that is, as measured in the un-primed frame it is moving in the same direction as Ms. Prime, although at a different speed. If in the primed frame the object is moving in some arbitrary direction with velocity components u'_x, u'_y, and u'_z, then the velocity transformation laws for the y' and z' components are different than that for the x' component. It is left as an exercise (problem 35.C.3) for you to show that, according to Mr. Unprime, the object will have velocity components u_x u_y and u_z given by

$$u_x = \frac{u'_x + v}{1 + u'_x v/c^2} \tag{35.21}$$

$$u_y = \frac{\sqrt{1 - \beta^2}\,u'_y}{1 + u'_x v/c^2} \tag{35.22}$$

$$u_z = \frac{\sqrt{1 - \beta^2}\,u'_z}{1 + u'_x v/c^2} \tag{35.23}$$

QUESTION

Does any simple device exist that would help us to visualize these Lorentz transformations?

ANSWER

Yes, with a piece of white paper and a piece of clear paper we can make the following *Lorentz transformer.* Mark axes on both papers and label them as shown in Fig. 35.19. It is best to draw the primed axes in one color and the unprimed axes in another. Now pin the pieces of paper together through their origins.

Each point on the plane now represents an event. To find the x and t coordinates of any particular event (that is, any particular point), draw perpendicular lines to the x and ct axes

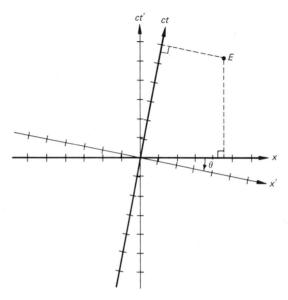

FIGURE 35.20 Measuring the spacetime coordinates of event E in the unprimed system.

and read off the values of x and ct (see Fig. 35.20). Doing the same thing with the x' and ct' axes gives the primed coordinates of the same event. The two sets of coordinates are related by a Lorentz transformation. The value of $\beta = v/c$ for the Lorentz transformation is related to the angle between the two sheets by the formula

$$\beta = \sin \theta$$

Figure 35.21 shows two events that occur at the same place in the unprimed frame; that is, the events have the same x coordinate. Clearly the time interval between the two events assigned in the unprimed frame is larger than the time interval

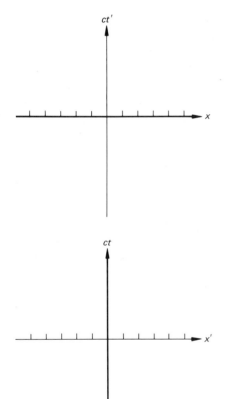

FIGURE 35.19 Making a Lorentz transformer.

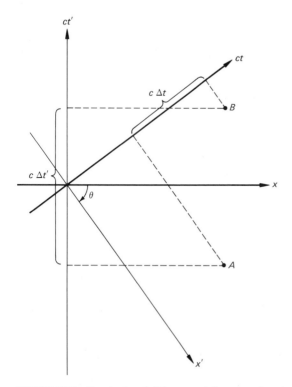

FIGURE 35.21 Events A and B happen at the same place in the (x, t) frame. The time interval between them is smaller in the (x, t) frame than it is in the (x', t') frame.

in the primed frame. This example illustrates the time-dilation effect. The Lorentz contraction effect can be illustrated in a similar way (see problem 35.C.5).

35.11 MOMENTUM CONSERVATION

We now want to investigate whether the relativistic transformation laws are consistent with Newton's laws of

dynamics. One of the simplest and most fundamental principles of Newtonian dynamics is the law of momentum conservation. Let us analyze a simple two-particle scattering experiment by using the momentum-conservation law and the relativistic velocity addition law (see Eq. 35.21 through 35.23).

Suppose we are traveling with Ms. Prime, and two identical particles are moving toward one another along the y' axis, with velocities $+u'$ and $-u'$. The particles have an elastic (energy-conserving) collision and move off along the positive and negative x' directions, again with velocities $+u'$ and $-u'$ (see Fig. 35.22). Using the Newtonian formula for momentum, we see that momentum

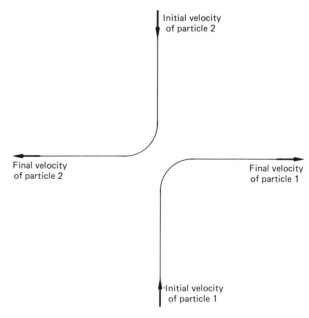

FIGURE 35.22 A simple two-particle scattering process, viewed in the primed frame.

is obviously conserved in this collision. For instance,

$$(P_{x'})_{\text{final}} = mu' + m(-u') = 0 = (P_{x'})_{\text{initial}}$$

Now let us look at the same collision as observed by Mr. Unprime (see Fig. 35.23). In particular, we want to see whether the x component of momentum is conserved. We obtain the x components of velocity of the two particles before the collision by using the velocity addition formulas, with $u'_x = 0$ and $u'_y = \pm u'$. The initial values of u_x for the two particles are

Particle 1: $(u_x)_{\text{initial}} = v$

and

Particle 2: $(u_x)_{\text{initial}} = v$

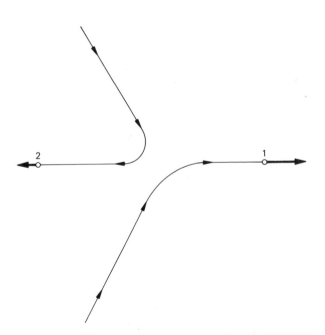

FIGURE 35.23 The same scattering process as shown in Fig. 35.22, as viewed by observers in the unprimed frame.

The final values of u_x are given by the velocity addition formulas, with $u'_x = \pm u'$ and $u'_y = 0$. They are

Particle 1: $(u_x)_{\text{final}} = \dfrac{u' + v}{1 + u'v/c^2}$

and

Particle 2: $(u_x)_{\text{final}} = \dfrac{-u' + v}{1 - u'v/c^2}$

The x component of the total momentum before the collision was

$$(P_x)_{\text{initial}} = mv + mv = 2mv$$

The x component of the total momentum after the collision is

$$(P_x)_{\text{final}} = m\frac{u' + v}{1 + u'v/c^2} + m\frac{-u' + v}{1 - u'v/c^2}$$

Putting the fractions over a common denominator, we obtain

$$(P_x)_{\text{final}} = 2mv\frac{1 - (u'/c)^2}{1 - (\beta u'/c)^2}$$

$$= \frac{1 - (u'/c)^2}{1 - (\beta u'/c)^2}(P_x)_{\text{initial}}$$

where $\beta = v/c$. Since $1 - (\beta u'/c)^2 > 1 - (u'/c)^2$, the expression multiplying $(P_x)_{\text{initial}}$ in the above equation must be less than 1. Therefore, according to Mr. Unprime,

$$(P_x)_{\text{final}} < (P_x)_{\text{initial}}$$

We see that if momentum (from the Newtonian formula) is conserved in Ms. Prime's coordinate frame, it is not conserved in Mr. Unprime's coordinate frame. Thus either the laws of Newtonian dynamics are incorrect in all coordinate frames, or the laws of dynamics are different in different coordinate frames. The second possibility would violate the principle of relativity, which states that the laws of physics are the same for all unaccelerated observers. Repeated and varied experiments have demonstrated that the relativity principle is true and that Newtonian mechanics must be modified. Newtonian dynamics cannot be made consistent with relativity because it treats the time co-ordinate t completely differently from the space co-ordinates x, y, and z. But we have already seen that there is a great deal of symmetry between space and time vari-ables; one person's time interval is another person's space and time intervals. The cause of the lack of symmetry in Newtonian dynamics is that the space coordinates are expressed as functions of the time coordinates $x(t)$, $y(t)$, and $z(t)$. When one transforms the variables, the time and space coordinates get jumbled and the equations look completely different.

35.12 RELATIVISTIC VELOCITY AND MOMENTUM

What we must do is to write all four coordinates in terms of a parameter τ that would not change from one observer to another. We can see how to define such a parameter by considering the following experiment. Mr. Unprime takes some object and attaches to it a small standard clock (one that ticks, when stationary with respect to his clock, once per second). He then applies forces to the object so that it moves and accelerates. As the object moves through his coordinate system, all his Junior Observers mark down, when it passes them, its space coordinates x, y, z, the time t, and the reading on the clock attached to the object, τ. Because of the time-dilation effect, the reading on the moving clock, τ, will not be the same as the time in the unprimed frame, t. After all this information is sent to the Chief, he compiles plots of the functions $x(\tau)$, $y(\tau)$, $z(\tau)$, and $t(\tau)$, that is, the position and time coordinates of the object when its own clock read τ. We call the parameter τ the *proper time* of the object. We now defined the *relativistic*

velocity of the object by taking the rate of change of the coordinate with respect to the proper time. We shall write relativistic velocities with capital letters.

$$U_x = \frac{dx(\tau)}{d\tau}$$

$$U_y = \frac{dy(\tau)}{d\tau}$$

$$U_z = \frac{dz(\tau)}{d\tau}$$

We can also define a *time component* of relativistic velocity

$$U_t = c\frac{dt(\tau)}{d\tau}$$

where the factor c has been added so that U_t will have the units of a velocity. The four components of relativistic velocity and the three components of "ordinary velocity" of an object are not unrelated. If we know the ordinary velocity components u_x, u_y, and u_z, we can calculate the relativistic velocity easily enough. Consider a time interval dt. During that time interval a clock moving with the object (at speed u) would, since it runs slow, register a time interval (see Eq. 35.15).

$$d\tau = dt\sqrt{1 - u^2/c^2}$$

Thus we see that

$$U_t = c\frac{dt}{d\tau} = \frac{c}{\sqrt{1 - u^2/c^2}}$$

If for any velocity u we introduce the definition

$$\gamma_u \equiv \frac{1}{\sqrt{1 - u^2/c^2}}$$

then U_t may be written in the form

$$U_t = \gamma_u c$$

We can also relate U_x, U_y, and U_z to u_x, u_y, and u_z. For instance,

$$U_x = \frac{dx}{d\tau} = \frac{dt}{d\tau}\frac{dx}{dt} = \gamma_u u_x$$

Similarly $U_y = \gamma_u u_y$ and $U_z = \gamma_u u_z$. We can express all this by the two equations

$$U_t = \gamma_u c \quad \text{and} \quad \mathbf{U} = \gamma_u \mathbf{u} \qquad (35.24)$$

We are now going to define four components of *relativistic momentum* in terms of U_t and $\mathbf{U}$.

$$P_t = mU_t \quad \text{and} \quad \mathbf{P} = m\mathbf{U} \qquad (35.25)$$

35.13 TRANSFORMATION LAWS OF RELATIVISTIC VELOCITY AND MOMENTUM

We shall assume that for any isolated system of particles (one without any external forces) each component of the total relativistic momentum remains constant. We can then show that if such a momentum-conservation law is true for one observer, it must be true for every observer. We will find that the fourth component of momentum is related to the energy, so the energy-conservation theorem becomes combined with the momentum-conservation theorem in relativity. First, we have to see how relativistic velocity is transformed from one coordinate system to another. Suppose we are situated in Ms. Prime's coordinate system and we see an object moving with a clock attached to it (the proper-time clock). At position x'_1, y'_1, z'_1 and time t'_1 the hand on the proper-time clock passes the number τ_1. According to Mr. Unprime this same event happens at

$$x_1 = \gamma_v(x'_1 + \beta ct'_1)$$

$$y_1 = y'_1$$

$$z_1 = z'_1$$

and t_1, where

$$ct_1 = \gamma_v(ct'_1 + \beta x'_1)$$

At position x'_2, y'_2, and z'_2 and at time t'_2 Ms. Prime sees the hand on the proper-time clock pass the τ_2 mark. Mr. Unprime sees this event at

$$x_2 = \gamma_v(x'_2 + \beta ct'_2)$$

$$y_2 = y'_2$$

$$z_2 = z'_2$$

and t_2, where

$$ct_2 = \gamma_v(ct'_2 + \beta cx'_2)$$

When Mr. Unprime calculates the x component of relativistic velocity, he obtains

$$U_x = \frac{x_2 - x_1}{\tau_2 - \tau_1} = \frac{\gamma_v(x'_2 + vt'_2) - \gamma_v(x'_1 + vt'_1)}{\tau_2 - \tau_1}$$

$$= \gamma_v \frac{x'_2 - x'_1}{\tau_2 - \tau_1} + \gamma_v v \frac{t'_2 - t'_1}{\tau_2 - \tau_1}$$

$$= \gamma_v(U'_x + \beta U'_t)$$

where U'_x and U'_t are the components of relativistic velocity as calculated by Ms. Prime. If we do the same for the other components of relativistic velocity, we find that the four components of relativistic velocity transform from one coordinate system to another in exactly the same way as the spacetime coordinates x, y, z, and ct.

$$U_x = \gamma_v(U'_x + \beta U'_t)$$

$$U_t = \gamma_v(U'_t + \beta U'_x)$$

$$U_y = U'_y$$

$$U_z = U'_z$$

$$(35.26)$$

Since the relativistic momentum is simply the mass times the relativistic velocity, it is obvious that the relativistic momentum components also transform according to the Lorentz transformation.

$$P_x = \gamma_v(P_x' + \beta P_t')$$

$$P_t = \gamma_v(P_t' + \beta P_x')$$

$$P_y = P_y'$$

$$P_z = P_z'$$

(35.27)

Any four physical observables A_1, A_2, A_3, and A_4 that have the property that their values, as measured in two inertial frames, are related by a Lorentz transformation are collectively called a *four-vector*. Thus the quantities x, y, z, and ct are the components of the position four-vector of an event in spacetime. The relativistic velocity and relativistic momentum are also four-vectors.

35.14 CONSERVATION OF FOUR-MOMENTUM

Suppose now we are in Ms. Prime's frame and we see a collision between two particles. The particles initially have the following components of relativistic momentum:

Particle 1: $(P_x')_{initial}$, $(P_y')_{initial}$, $(P_z')_{initial}$, and $(P_t')_{initial}$

Particle 2: $(Q_x')_{initial}$, $(Q_y')_{initial}$, $(Q_z')_{initial}$, and $(Q_t')_{initial}$

After the collision the particles have the following relativistic momenta:

Particle 1: $(P_x')_{final}$, $(P_y')_{final}$, $(P_z')_{final}$, and $(P_t')_{final}$

Particle 2: $(Q_x')_{final}$, $(Q_y')_{final}$, $(Q_z')_{final}$, and $(Q_t')_{final}$

If all four components of relativistic momentum are conserved, then

$$(P_x')_{final} + (Q_x')_{final} = (P_x')_{initial} + (Q_x')_{initial}$$

and similar equations hold for the y', z', and t' components. Now let us see that as Mr. Unprime views the same collision he will automatically find that the relativistic momentum is conserved if Ms. Prime did. In Mr. Unprime's coordinate frame the x component of the total initial relativistic momentum is

$$(P_x)_{initial} + (Q_x)_{initial} = \gamma_v[(P_x')_{initial} + \beta(P_t')_{initial}]$$
$$+ \gamma_v[(Q_x')_{initial} + \beta(Q_t')_{initial}]$$

The x component of the total final relativistic momentum is

$$(P_x)_{final} + (Q_x)_{final} = \gamma_v[(P_x')_{final} + \beta(P_t')_{final}]$$
$$+ \gamma_v[(Q_x')_{final} + \beta(Q_t')_{final}]$$

The difference between the two expressions gives the change in the x component of momentum as seen by Mr. Unprime.

$$(P_x)_{final} + (Q_x)_{final} - [(P_x)_{initial} + (Q_x)_{initial}]$$
$$= \gamma_v\{(P_x')_{final} + (Q_x')_{final} - [(P_x')_{initial} + (Q_x')_{initial}]\}$$
$$+ \gamma_v\beta\{(P_t')_{final} + (Q_t')_{final} - [(P_t')_{initial} + (Q_t')_{initial}]\}$$
$$= 0$$

since each expression enclosed in braces is zero if all components of relativistic momentum are conserved in the primed coordinate system. Now it is clear why we must assume that *all* components (including the time component) of relativistic momentum are conserved. If the time component were not conserved in the primed coordinate system, then the x component would not be conserved in the unprimed coordinate system. The question we must now ask is: What is the physical meaning of the conservation of the time component of the relativistic momentum? In terms of the ordinary velocity, P_t is given by

$$P_t = \frac{mc}{\sqrt{1 - (u/c)^2}} \tag{35.28}$$

For the typical velocities of material particles, which are much less than the velocity of light, $(u/c)^2$ is a very small number. We can then use the approximation

$$\frac{1}{\sqrt{1 - (u/c)^2}} = \left[1 - \left(\frac{u}{c}\right)^2\right]^{-1/2} \simeq 1 + \frac{1}{2}\left(\frac{u}{c}\right)^2$$

We then see that

$$P_t \simeq \frac{mc^2 + \frac{1}{2}mu^2}{c} \tag{35.29}$$

Except at velocities near the velocity of light, the time component of momentum of a particle is proportional to a constant plus the usual Newtonian kinetic energy of the particle. For the types of collision processes we have considered, the constant term mc^2 can be ignored since it simply adds a fixed number to the initial value of P_t and the same number to the final value, thus having no effect on the conservation law. For velocities much less than c, the law of conservation of P_t is completely equivalent to the law of conservation of Newtonian kinetic energy. Experiment has shown that, when u is not much less than c, the Newtonian kinetic energy is not conserved but the time component of the total momentum is conserved.

QUESTION

Suppose a constant force f is applied to a particle of mass m. If the particle is initially stationary, what will be its ordinary velocity u at time t?

ANSWER

We choose the x direction as the direction of the force and call P_x simply P. (P_y and P_z are zero and we shall never have to consider P_t.) Then, according to Eq. 35.24 and 35.25,

$$P = mU_x = \frac{mu}{\sqrt{1 - u^2/c^2}}$$

Squaring both sides of this equation, we get

$$P^2 = \frac{m^2 u^2}{1 - u^2/c^2}$$

which, when solved for u, gives the formula

$$u = \frac{P/m}{\sqrt{1 + P^2/m^2 c^2}} \tag{35.30}$$

The force on a particle in relativistic mechanics has the same meaning that it does in Newtonian mechanics, namely the rate at which momentum is transferred to the particle. Since the momentum begins at zero and increases with the constant rate f, the momentum at time t must be

$$P(t) = ft$$

Therefore $u(t)$ is given by

$$u(t) = \frac{ft/m}{\sqrt{1 + (ft/mc)^2}} \tag{35.31}$$

Let us consider what this formula means. The quantity ft is the total momentum that has been delivered to the particle by time t. If ft is much smaller than mc, the particle undergoes constant acceleration, as predicted by the nonrelativistic formula

$$u(t) = \frac{f}{m}t$$

As ft continues to increase, the acceleration diminishes and the particle velocity approaches closer and closer to the velocity c.

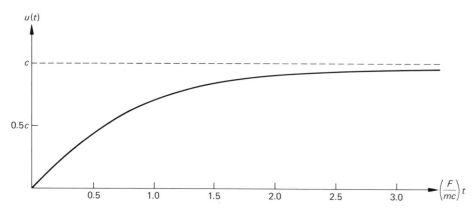

FIGURE 35.24   The ordinary velocity $u(t)$ for a particle subjected to a constant force.

Thus, no matter how long we push on a particle, we can never get it to the velocity $u = c$ although its momentum increases without bound (see Fig. 35.24).

QUESTION

Could the time-dilation effect possibly be used to keep a person young?

ANSWER

It depends very much on what is meant by "keeping a person young." As we shall see, it is possible for you, by keeping yourself moving at high velocity, to still be young and vigorous when everyone else is old and gray.[4] Then everybody else would congratulate you on how wonderfully young you have remained. However, in terms of your personal experience you would not have lived an exceptionally long time at all. Traveling around in

_____
[4] This assumes that you are presently young and vigorous.

your spaceship you would only have had time to have the normal number of heartbeats, to think the normal number of thoughts, and to feel the normal number of emotions for a person of your apparent physical age. From your own point of view, all you would have accomplished by your incessant flitting about would be to make everyone else old.

Let us now consider the question in a more detailed, mathematical fashion. It has been well confirmed by experiments involving unstable particles moving in accelerators that the proper time τ measures the rate at which the internal disintegration mechanism of the particle proceeds. This is true even when the particle is undergoing acceleration caused by external forces, so long as the forces are not so violent as to disrupt the internal structure of the particle. We know from everyday experience that a force of mg newtons applied to a person of mass m has no detrimental physical effects. Let us see what the effects would be of applying that gentle force over long periods of time. We picture a brother and sister, each ten years old, and at rest relative to one another. They are in separate airtight compartments in space, and each compartment contains a clock. We assume that the compartments have negligible mass. A steady external force of mg newtons in the positive x direction is applied to the sister's compartment, and this causes it to accelerate in a way we have analyzed in a previous question. The force is maintained for twenty years as measured by the brother's clock. The force is then reversed in direction and applied for another twenty years, and this causes the sister to gradually slow down and come to a stop on the brother's fiftieth birthday. She is, however, a very great distance from her brother. The entire process is then reversed, and this brings the sister back to the side of her aged brother as he celebrates his ninetieth birthday. We now calculate the time elapsed on her clock.

During the first twenty years (his time) her velocity as a function of his time is given by Eq. 35.31, with $F = mg$:

$$u(t) = gt/\sqrt{1 + g^2 t^2/c^2} \tag{35.32}$$

During his time interval dt her clock changes by an amount $d\tau$, where

$$d\tau = dt\sqrt{1 - u^2/c^2}$$

If we substitute the expression for $u(t)$ from the first equation into the second, we obtain

$$d\tau = dt/\sqrt{1 + g^2 t^2/c^2} \tag{35.33}$$

Letting t range from 0 to $t = T = 20$ yr $= 6.3 \times 10^8$ s, we integrate to get the range of τ:

$$\int_0^\tau d\tau = \int_0^T \frac{dt}{\sqrt{1 + g^2 t^2/c^2}}$$

This gives

$$\tau = \frac{c}{g}\log\left(\frac{gT}{c} + \sqrt{1 + \frac{g^2 T^2}{c^2}}\right) \tag{35.34}$$

Evaluating the right-hand side for $T = 6.3 \times 10^8$ s, $g = 9.8$ m/s^2, and $c = 3 \times 10^8$ m/s, we get

$$\tau = 1.14 \times 10^8 \text{ s} = 3.6 \text{ yr}$$

This is just one-quarter of the sister's journey. Each of the other three-quarters makes an equal contribution to her age, and this brings her, upon her return to her brother's side, to the age of

$$(10 + 4 \times 3.6) \text{ yr} = 24.4 \text{ yr}$$

About the age of one of his granddaughters!

35.15 "DOES THE INERTIA OF A BODY DEPEND UPON ITS ENERGY CONTENT?"[5]

What about the constant term mc^2? Does its inclusion in the time component of momentum have any significance? We shall see that the constant term is exactly equal to the energy content of the object when it is at rest. To indicate the meaning of this peculiar statement, let us consider the following simple experiment, viewed in Mr. Unprime's frame.

We have two identical blocks, each of mass m, and a spring of mass m_s. We compress the spring between the blocks and connect them with a thin thread (see Fig. 35.25). We measure the mass of the entire system and obtain M. The system is stationary, and therefore it has a time component of momentum

$$P_t = Mc$$

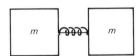

FIGURE 35.25 Two masses with a spring compressed between them are held together by a thin thread.

Now we carefully cut the thread in the center, and the two blocks move in opposite directions with velocities $+u$ and $-u$. The spring is left stationary (see Fig. 35.26). The time component of the total momentum is now

$$P_t = m_s c + 2\frac{mc}{\sqrt{1 - (u/c)^2}}$$

[5] This is a translation of the title of a two-page paper published by a junior patent clerk named Albert Einstein in 1905 [*Annalen der Physik* **17** (1905)].

FIGURE 35.26   The thread has been cut and the two masses move away with speed u, leaving the spring stationary.

Since P_t is conserved,

$$M = m_s + 2\frac{m}{\sqrt{1 - (u/c)^2}} > m_s + 2m$$

Thus before we broke the thread the mass of the system must have been more than the sum of the masses of its constituent parts! If the velocity u is much smaller than c, we can use the approximation

$$2\frac{m}{\sqrt{1 - (u/c)^2}} \simeq 2\left[m + \frac{1}{2}m\left(\frac{u}{c}\right)^2\right] = 2m + \frac{E_K}{c^2}$$

where $E_K = mu^2$ is the combined kinetic energy of the two blocks. But the kinetic energy of the blocks after the spring is released is equal to the potential energy that was stored in the spring while it was compressed. Thus

$$M = m_s + 2m + \frac{E_p}{c^2}$$

The mass of the system before the thread was broken was larger than the sum of the masses of the two blocks plus the extended spring by an amount equal to the energy stored in the compressed spring divided by c^2. In other words, if the internal energy of an object is increased by an amount ΔE, its inertial mass also increases by an amount Δm, where

$$\Delta E = \Delta mc^2 \tag{35.35}$$

Thus if we take a block of metal and increase its internal energy by heating it, its mass (and weight) will increase in proportion to the heat energy added. Before the reader tries to confirm the effect with a pot of hot water on a bathroom scale, it must be pointed out that the change in mass is very small. If a million calories are added to a body, the extra mass comes to

$$\Delta m = 4.18 \times 10^6/(3 \times 10^8)^2$$

$$= 4.64 \times 10^{-11} \text{ kg}$$

This is much too small a change to be detected on a scale.

Although the mass changes associated with ordinary energy-transfer processes, such as heating a body or stretching a spring, are too small to be detected, the mass changes involved in energy release by nuclear-transformation processes are significant and easily measured. As an example, in the deuterium fusion process[6] two atoms of deuterium combine to form one atom of helium, with the release of a large amount of radiant energy. The deuterium atom is made up of one electron, one proton, and one neutron. It has the same chemical characteristics as ordinary hydrogen. A helium atom contains two electrons, two protons, and two neutrons. Since the helium atom has exactly the constituents of two deuterium atoms, one might expect it to have twice the mass of a deuterium atom. However, the constituents in a helium atom are more tightly bound together, and this is why energy is released in the transformation of deuterium to helium. The measured mass of a deuterium atom is 3.3442×10^{-27} kg, and that of a helium atom is 6.6459×10^{-27} kg. Thus, according to Eq. 35.35, the energy released when two deuterium atoms become one helium atom is

$$E = (2m_D - m_{He})c^2$$

$$= 3.825 \times 10^{-12} \text{ J}$$

To get a better idea of what this energy value means, let us calculate the energy released when Avogadro's number of

---

[6] The fusion process is discussed in much more detail in Chapter 39.

deuterium atoms (about two grams of deuterium) are converted into half that number of helium atoms.

$$E = \left(\frac{6.022 \times 10^{23}}{2}\right)(3.825 \times 10^{-12} \text{ J})$$

$$= 1.152 \times 10^{12} \text{ J}$$

$$= 320,000 \text{ kWh}$$

This relation between energy release and mass change has been accurately verified by experiment.

QUESTION

The quantity $E = cP_t$ is called the relativistic energy of a body. At low velocities it is equal to the Newtonian kinetic energy plus the constant mc^2. Using the Newtonian formulas for momentum ($\mathbf{p} = m\mathbf{v}$) and kinetic energy ($E_K = \frac{1}{2}mv^2$), one can easily derive the following relation between the momentum and kinetic energy of a particle moving at low speed:

$$E_K = \tfrac{1}{2}p^2/m \quad \text{(for } v \ll c\text{)} \tag{35.36}$$

What is the corresponding relation between the relativistic energy E and the relativistic momentum $\mathbf{P}$ of a moving particle?

ANSWER

Using Eq. 35.24 and 35.25, we can express the relativistic energy and momentum in terms of the ordinary velocity $\mathbf{u}$ in the following way:

$$E = cP_t = \frac{mc^2}{\sqrt{1 - (u/c)^2}} \tag{35.37}$$

and

$$\mathbf{P} = \frac{m\mathbf{u}}{\sqrt{1 - (u/c)^2}} \tag{35.38}$$

We want to combine these two equations so as to eliminate the velocity $\mathbf{u}$. To do this we first square both equations to get rid of the radical sign:

$$E^2 = \frac{m^2c^4}{1 - (u/c)^2}$$

$$P^2 = \frac{m^2u^2}{1 - (u/c)^2}$$

We now write E^2 in terms of P^2 by using the relation

$$\frac{1}{1 - u^2/c^2} = \frac{1 - u^2/c^2}{1 - u^2/c^2} + \frac{u^2/c^2}{1 - u^2/c^2} = 1 + \frac{u^2/c^2}{1 - u^2/c^2}$$

This gives

$$E^2 = m^2c^4 + c^2P^2$$

or

$$E = c\sqrt{m^2c^2 + P^2} \tag{35.39}$$

which is the relativistic equivalent of Eq. 35.36.

QUESTION

Is the change in the relativistic energy of a body equal to the work done on the body by external forces?

ANSWER

Yes. Consider a body of mass m being accelerated by a force $\mathbf{f}$. The rate at which work is being done on the body is equal to $\mathbf{u} \cdot \mathbf{f}$, where we define $\mathbf{f}$ by

$$\mathbf{f} = \frac{d\mathbf{P}}{dt}$$

We must therefore show that

$$\mathbf{u} \cdot \frac{d\mathbf{P}}{dt} = \frac{dE}{dt}$$

The relativistic energy of a body is related to its momentum by Eq. 35.39. If we differentiate that equation with respect to t [using the rule for differentiation of the square of a vector, namely $dP^2/dt = 2\mathbf{P} \cdot (d\mathbf{P}/dt)$], we obtain

$$\frac{dE}{dt} = \frac{c\mathbf{P} \cdot (d\mathbf{P}/dt)}{\sqrt{m^2c^2 + P^2}} = c^2\frac{\mathbf{P} \cdot (d\mathbf{P}/dt)}{E}$$

By dividing Eq. 35.38 by Eq. 35.37 we can see that

$$c^2\frac{\mathbf{P}}{E} = \mathbf{u} \tag{35.40}$$

Therefore

$$\frac{dE}{dt} = \mathbf{u} \cdot \frac{d\mathbf{P}}{dt} = \mathbf{u} \cdot \mathbf{f} \tag{35.41}$$

Thus, if a particle moves along a known trajectory under the influence of a known force, we may calculate its change in energy in the same way as we would in Newtonian dynamics:

$$E_{\text{final}} - E_{\text{initial}} = \int \mathbf{f} \cdot d\boldsymbol{l} = \text{work done by force along path} \tag{35.42}$$

35.16 PARTICLES OF ZERO MASS

The relationship between the energy and momentum of a moving particle (Eq. 35.39) gives nonzero values of E even for the case of zero mass. That is, if we set m equal to zero in that equation, we get the following relation between E and P:

$$E = cP \tag{35.43}$$

It is an experimental fact that physical particles that satisfy this relationship between E and P do exist. The most important class of such zero mass particles is the set of particles called *photons*. In a manner that we describe in chapter 37 on quantum theory, photons are the carriers of the energy and momentum associated with electro-magnetic waves.

Taking the magnitude of the vectors in Eq. 35.40, we see that the velocity of any particle is related to its energy and momentum by

$$u = c^2 P/E \tag{35.44}$$

For a particle of zero mass (one that satisfies Eq. 35.43), this relation gives

$$u = c^2 P/cP = c$$

Zero-mass particles always travel at the velocity of light! If sufficient energy and momentum are available, a zero-mass particle can be created, but when it comes into existence, it is already moving with velocity c. Zero-mass particles can be absorbed and have their energy and momentum transferred to the absorbing body; however, they can never be slowed down and collected. Zero-mass particles can never be the permanent constituents of physical objects in the way that electrons, protons, and other particles of finite mass are. They are certainly rather bizarre but physically important solutions of the equations of relativistic dynamics. Some of the more common types of elementary particles of both zero and finite mass are listed in Table 35.1.

TABLE 35.1   Masses and charges of some elementary particles

Particle name	Symbol	Mass (u) ($1\,u = 1.66043 \times 10^{-27}$ kg)	Electric charge ($e = 1.6 \times 10^{-19}$ c)
Photon	γ	0	0
Electron	e^-	5.4860×10^{-4}	$-e$
Positron	e^+	5.4860×10^{-4}	$+e$
Proton	p	1.00728	$+e$
Neutron	n	1.00867	0
Pi plus	π^+	0.149848	$+e$
Pi minus	π^-	0.149848	$-e$
Pi zero	π^0	0.144910	0
Mu minus	μ^-	0.113432	$-e$
Mu plus	μ^+	0.113432	$+e$
e neutrino	ν	0	0
μ neutrino	ν_μ	0	0

SUMMARY

At the end of the nineteenth century it was believed that the laws of mechanics were valid in all inertial frames (i.e., satisfied a relativity principle) but that Maxwell's equations were valid only in the ether frame. All experiments to detect the motion of the earth through the ether gave null results.

The theory of relativity assumes that all laws of physics have the same form in every inertial frame. With that assumption one can derive the *Lorentz transformation* between the spacetime coordinates of an event in one inertial frame and the spacetime coordinates of the same event in another inertial frame. If the origins of the two frames coincide at $t = 0$ and $t' = 0$, and if the primed frame moves through the unprimed frame with velocity v in the x direction, the transformations are

$$x = \frac{x' + \beta ct'}{\sqrt{1 - \beta^2}} \qquad (\beta = v/c)$$

$$ct = \frac{ct' + \beta x'}{\sqrt{1 - \beta^2}}$$

$$y = y'$$

$$z = z'$$

Time dilation is defined as follows: Two events that occur at the same place in the primed frame and are separated by a time interval $\Delta t'$ will, in the unprimed frame, be separated by a longer time interval, namely

$$\Delta t = \frac{\Delta t'}{\sqrt{1 - \beta^2}}$$

Thus to the unprimed observers the primed clock will appear to run slow. The effect is symmetric: to the primed observers the unprimed clock appears to run slow.

In a Lorentz contraction an object of length L_0 in a frame of reference in which it is at rest will have a length

$$L = L_0\sqrt{1 - v^2/c^2}$$

in a frame in which it moves at velocity v in a direction parallel to its length.

If an object moves through the primed frame with velocity components $u'_x = dx'/dt'$, $u'_y = dy'/dt'$, and $u'_z = dz'/dt'$, it will be moving through the unprimed frame with velocity components $u_x = dx/dt$, $u_y = dy/dt$, and $u_z = dz/dt$, where

$$u_x = \frac{u'_x + v}{1 + u'_x v/c^2}$$

$$u_y = \frac{\sqrt{1 - \beta^2}\, u'_y}{1 + u'_x v/c^2}$$

$$u_z = \frac{\sqrt{1 - \beta^2}\, u'_z}{1 + u'_x v/c^2}$$

Newtonian dynamics must be modified in order to be made consistent with the principle of relativity. One can accomplish this modification by writing everything in terms of four-vectors.

A *four-vector* is a set of four physical observables whose measured values in one inertial frame are related to their measured values in another inertial frame by a Lorentz transformation.

The *relativistic velocity* of an object is a four-vector whose components are

$$U_x = \frac{dx}{d\tau}$$

$$U_y = \frac{dy}{d\tau}$$

$$U_z = \frac{dz}{d\tau}$$

and

$$U_t = c\frac{dt}{d\tau}$$

where τ is called the *proper time* and is the time that would be registered by a clock moving with the object.

The relativistic velocity is related to the ordinary velocity, $\mathbf{u} = d\mathbf{r}/dt$, by

$$\mathbf{U} = \gamma_u \mathbf{u} \quad \text{and} \quad U_t = \gamma_u c$$

where

$$\gamma_u = \frac{1}{\sqrt{1 - u^2/c^2}}$$

The *relativistic momentum* of an object is a four-vector that is equal to the mass of the object times its relativistic velocity:

$$\mathbf{P} = m\mathbf{U} = m\gamma_u \mathbf{u}$$

and

$$P_t = mU_t = mc\gamma_u$$

All four components of the total relativistic momentum are conserved for an isolated system.

The time component of momentum is related to the energy of the particle:

$$E = cP_t = \frac{mc^2}{\sqrt{1 - u^2/c^2}}$$

where $\mathbf{u}$ is the ordinary velocity of the particle.

At velocities much less than c, the relativistic energy is

$$E \simeq mc^2 + \tfrac{1}{2}mu^2$$

If the primed frame moves with velocity v in the x direction with respect to the unprimed frame, then the measurements of the relativistic momentum of any object in the two frames are related by

$$P_t = \gamma_v(P_t' + \beta P_x')$$

$$P_x = \gamma_v(P_x' + \beta P_t')$$

$$P_y = P_y'$$

$$P_z = P_z'$$

If the internal energy of an object changes by an amount ΔE, then its inertial mass will change by an amount Δm, where

$$\Delta E = \Delta mc^2$$

Particles of zero mass always travel at the velocity of light. Their energy and momentum are related by

$$E = cP$$

where $P = |\mathbf{P}|$.

The relationship between momentum $\mathbf{P}$ and energy E for particles of mass m is

$$E = c\sqrt{m^2c^2 + P^2}$$

PROBLEMS

35.A.1 As shown in Fig. 35.1, an observer with coordinate axes (x, y) is moving at a low velocity, $v = 2$ m/s, in the x direction with respect to another observer with coordinate axes (x', y'). The velocities are low enough to allow us to use purely Newtonian physics. The frames coincide at time $t = 0$. In the unprimed frame a pebble is tossed in the air and moves with a trajectory $y(t) = (8 - \tfrac{1}{2}gt^2)$ and $x(t) = 5t$, where t is measured in seconds and distances in meters. What is the trajectory in the primed frame?

35.A.2 With regard to the Michelson-Morley experiment, show that

$$T_{\parallel} - T_{\perp} = \frac{Lv^2}{c(c^2 - v^2)}$$

35.A.3 Show that a contraction of the parallel arm of the Michelson-Morley apparatus by a factor of $\sqrt{1 - v^2/c^2}$ would cause a null result for the experiment.

35.A.4 The primed frame is moving at speed $v = 0.4c$ with respect to the unprimed frame. An explosion occurs at the time and place $x' = 6 \times 10^3$ m and $t' = 2$ s. Where and when does the same event occur according to the unprimed observers?

35.A.5 Ms. Prime is moving at speed $\frac{1}{3}c$ with respect to Mr. Unprime. An event occurs in the primed frame at time $t' = 10$ s and position $x' = 4 \times 10^8$ m. What spacetime coordinates does Mr. Unprime assign to the event?

35.A.6 In the frame of reference in which they are at rest, π^+ mesons have an average lifetime of 2.6×10^{-8} s. Neglecting relativistic effects, calculate the distance a newly created π^+ would travel before it disintegrated if it were moving at $0.99c$. Do the same, taking the time-dilation effect into account.

35.A.7 Consider the collision shown in Fig. 35.23. If $u = v = c/2$ and $m = 10^{-27}$ kg, what are the values of the total momentum before and after the collision?

35.A.8 Describe the range of possible values that the following can take.
(a) u_x (b) U_x (c) $u = |\mathbf{u}|$, (d) $U = |\mathbf{U}|$, (e) U_t.

35.A.9 Show that $P_t^2 - |\mathbf{P}|^2 = (mc)^2$.

35.A.10 An atom of ^{8}Be is composed of four protons, four neutrons, and four electrons. It has a mass of 8.005308 u. (1 u = 1 atomic mass unit = 1.66043×10^{-27} kg.) An atom of ^{4}He is composed of two protons, two neutrons, and two electrons. It has a mass of 4.002605 u. Note that two ^{4}He atoms have less mass than one ^{8}Be atom even though they are composed of the same particles. Left to itself, an atom of ^{8}Be will spontaneously disintegrate into two ^{4}He atoms. The excess mass is converted into the kinetic energy of the helium atoms. Assuming that the ^{8}Be atom is initially at rest, calculate the final velocity of the ^{4}He atoms.

35.A.11 At equilibrium a typical spring weighs 50 g, is 10 cm long, and has a force constant of 1 N/cm. If the spring is stretched by 5 cm, how much does its mass increase? The mass increase corresponds to the mass of how many electrons?

35.B.1 The motion of the earth around the sun has a speed of about 30 km/s. If the arms of the Michelson-Morley apparatus were 1 m long and the light being used had a wavelength of 400 nm, by what fraction of a wavelength would Michelson and Morley have expected the waves from the two arms to be shifted with respect to one another? (*Hint:* See problem 35.A.2.)

35.B.2 Suppose you were a patent examiner and someone submitted a patent for a device that would measure the speed of a train (on a windless day) by timing the back-and-forth travel of a sound wave (see the following figure). Should you grant the patent?

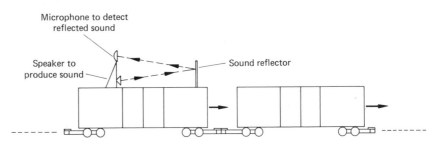

35.B.3 Show that, according to the Lorentz transformation equations, for any values of x, y, z, and t, the following identity holds:

$$(ct)^2 - (x^2 + y^2 + z^2) = (ct')^2 - (x'^2 + y'^2 + z'^2)$$

35.B.4 A rod of length $L_0 = 200$ m is at rest in the primed frame. Two explosions occur simultaneously at the ends of the rod. What is the distance between the explosions in the unprimed frame? Here $\beta = 0.6$. Explain why the space intervals between the two events in the two frames should or should not (take your choice) be related by a Lorentz contraction.

35.B.5 A spaceship travels at velocity v meters per second toward a distant star (as measured from the earth).

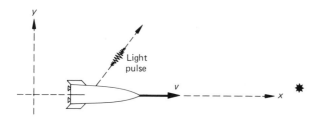

The occupants of the spaceship send off a pulse of light in a direction that they determine to be perpendicular to the line joining the spaceship and the star. What would be the components of velocity of the light pulse as determined by an observer on the earth (see the figure)? Show that the speed of the light pulse is c according to the observer on the earth.

35.B.6 For a certain particle $x(0) = 0$, $U_x(\tau) = \alpha$, and $U_y = U_z = 0$. Find $x(t)$.

35.B.7 In Ms. Prime's frame a particle is moving with relativistic velocity components $U'_x = 0.5c$, $U'_y = 0.6c$, and $U'_z = 0$.
  (a) Calculate U'_t.
  (b) Calculate U_x, U_y and U_t in Mr. Unprime's frame. The speed of Ms. Prime through the unprimed frame is $v = 0.7c$.

35.B.8 A particle of mass m_1 that is initially stationary is struck by another particle of mass m_2 that is traveling at speed v. Both particles combine to form a particle of mass M moving at speed V. No radiation is given off. Calculate M and V given $m_1 = m_2$ and $v = 0.8c$.

35.B.9 At the distance of the earth (1.5×10^{11} m) the density of radiant energy from the sun is about 1.5 kW/m^2. Calculate the rate at which mass is being converted to energy by the sun. Write your answer in metric tons per year.

35.B.10 A particle of mass m is initially stationary at $x = 0$. Starting at $t = 0$ it is subjected to a constant force $f = dP/dt$. Find x as a function of t.

35.B.11 If a constant force of 1000 N (about 200 lb) is applied to a person of 100-kg mass, how long would it take for the velocity of the person to reach $0.9c$? The time should be given in the frame in which the person is initially at rest.

35.B.12 How much work would be done by a force of mg newtons applied for twenty years to a person of mass $m = 75$ kg? The person is initially at rest. Determine the average rate at which work is being done in kilowatts.

35.B.13 A particle of mass m and charge e moves within a region of uniform magnetic field. The magnetic field has magnitude B and points in the z direction. The particle moves in a circle in the xy plane.

(a) Show that the relationship between the angular frequency of the circular motion ω and the radius of the circle R is

$$\omega = \frac{c\lambda}{\sqrt{c^2 + \lambda^2 R^2}}$$

where $\lambda = eB/m$.

(b) If the magnetic field is 2 T and a proton moves in a circle with radius $R = 40$ cm, what is the energy of the proton?

35.C.1 In Ms. Prime's frame a straight rod makes an angle θ' with respect to the x' axis. According to Mr. Unprime the rod is moving in the x direction at velocity v and is oriented at an angle θ with respect to the x axis. Determine the angle θ as a function of θ' and v.

35.C.2 A round disk is punched out of a very thin metal sheet and a hole is left in the sheet. The sheet is held stationary while the disk is moved through the hole as shown in the following figure. The disk is always parallel to the sheet and the angle of approach is small. The disk, being Lorentz contracted and therefore shorter than the hole, fits through easily. What does the motion look like when viewed by an observer moving with the disk? The hole will then be Lorentz contracted. Will the disk fit through?

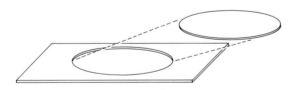

35.C.3 Using the Lorentz transformation laws, derive Eq. 35.21, 35.22, and 35.23.

35.C.4 Near the surface of the earth, in Ms. Prime's frame, a long straight iron bar is held up parallel to the x' axis until time $t' = 0$, when all parts of it are simultaneously

released. The bar remains parallel to the x' axis as it falls. Any point on the bar has a trajectory

$$y'(t') = y_0' \qquad \text{for} \quad t' < 0$$

and

$$y'(t') = y_0' - \tfrac{1}{2}g(t')^2 \qquad \text{for} \quad t' > 0$$

By looking at the motion of the two end points and the center point of the rod in Mr. Unprime's frame, show that the rod bends in that frame.

35.C.5 On a diagram similar to Fig. 35.21, use dashed lines to plot the trajectories of the two ends of a rigid rod of length L at rest in the unprimed frame. Take two events that occur simultaneously in the primed frame at the ends of the rod. By projection of those two events onto the x' axis, find the length L' of the rod in the primed frame. Indicate L and L' on your diagram.

35.C.6 The Stanford linear accelerator produces a beam of electrons that have 20 GeV of energy (1 giga-electron-volt $= 10^9$ eV $= 1.6 \times 10^{-10}$ J). What is the speed of the electrons as a fraction of c?

35.C.7 An instantaneous rest frame for a particle undergoing accelerated motion is any inertial frame in which the particle is instantaneously at rest. Suppose a particle begins at rest in Mr. Unprime's frame at $t = 0$. It then moves in such a way that, in its instantaneous rest frame, its acceleration is always g (and in a constant direction down the x axis).

(a) Calculate the object's ordinary acceleration, velocity, and position as determined by Mr. Unprime.

(b) How far would the object get during the time interval $t = 0$ to $t = 10^9$ s, which is about 32 yr of Mr. Unprime's time?

(c) How far would the object get during a proper time interval of 10^9 s?

(d) What would be the value of t when $\tau = 10^9$ s?

35.C.8 A K+ meson at rest decays into a π^+ and a π^0. What is the velocity of the π^0? What is the energy (including the rest energy, mc^2) of the π^+? (The mass of a K+ is 966.7 m_e.)

35.C.9 As viewed in a certain frame, which we call the center-of-mass frame, two protons approach one another with equal but opposite velocities. The speed of each is 0.999c. Calculate the relativistic energy of one of the protons in the center-of-mass frame. What we call the laboratory frame is a frame of reference in which one of the protons is at rest. Calculate the relativistic energy of the moving proton in the laboratory frame. What is the ratio of E_{lab} to $E_{c.m.}$? This analysis is applicable to high-energy proton-proton scattering experiments in which one proton is part of a stationary target while another proton, from a proton accelerator, approaches it with energy E_{lab}. The kinds of elementary particle reactions that can be obtained depends on the value of $E_{c.m.}$. One of the problems faced by elementary particle physicists is that $E_{c.m.}$ is much smaller than E_{lab} for high-energy particles.

36
THE
ORIGINS
OF
QUANTUM
PHYSICS

36.1 CLASSICAL PHYSICS

ACCORDING TO THE ideas generally accepted during the late nineteenth century, physical reality consists of two types of entities: particles and fields. A particle is something that has a mass, a charge, and a location in space. A field has none of those properties; it is everywhere at once. The momentum and energy carried by the field are distributed throughout the space in which the field exists. This picture is usually referred to as classical physics. The classical theory is able to describe and accurately predict an incredibly wide range of phenomena. It can, for example, be used to calculate the trajectories of the planets with great precision, and it can account for light, heat, sound, magnetism, and electricity. Of course, the theory of relativity produced significant changes in the classical picture, but it did not alter the fundamental separation of the elements of reality into particles and fields. Nor did relativity theory change the general concept of a particle and a field. Starting in the late nineteenth century and continuing through the first third of the twentieth century, another major line of experimental and theoretical work was being carried out, and this eventually led to a revision of our fundamental ideas of the structure of the world that was even more radical than the revision brought about by relativity theory. This line of work, which led to what is now called *quantum theory,* was, for the most part, independent of relativity theory. Relativistic physics must be used when one is considering systems containing objects moving at speeds that are comparable to c. In contrast, quantum physics must be used when one is considering systems of atomic dimensions or smaller. Both these statements are only rough "rules of thumb." On the one hand, situations exist in which one must use relativity for low-velocity phenomena or quantum theory for large-sized objects. On the other hand, cases exist in which one can get away with using nonrelativistic classical physics even though it seems that this theory should not be valid. When analyzing a small-sized system containing high-speed particles, one must use the principles of relativistic quantum physics. But not all such systems can be understood, because we are still struggling with the task of developing relativistic quantum mechanics.

36.2 CAVITY RADIATION

In the late nineteenth and early twentieth centuries, scientists realized that in spite of its general success the classical picture of the physical world contained certain flaws. One of the most troublesome problems arose when people attempted to combine Maxwell's electromagnetic theory with thermodynamics. Consider a completely evacuated container that contains an object at some finite temperature (see Fig. 36.1). The walls of the container are perfectly reflecting mirrors. The object, being composed of charged particles in thermal motion, will radiate energy into the container, and it will also absorb radiant energy that has been reflected from the walls. Since no energy leaves the container, the entire system will eventually reach equilibrium at some temperature T. The problem posed is to calculate the density of radiant energy within the container as a function of the equilibrium temperature. First let us show that the equilibrium radiant energy density cannot depend on the detailed structure of the body.

FIGURE 36.1 What is the radiant energy density when this system is in equilibrium?

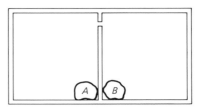

FIGURE 36.2 At equilibrium the radiant energy density in the two containers must be the same regardless of the nature of objects A and B.

Suppose two different bodies, A and B, are both at temperature T and in equilibrium with the radiation within their separate enclosures. Suppose further that the radiant energy density within enclosure A is greater than that within enclosure B. If we now open a small window between the two containers (see Fig. 36.2), a greater radiant energy flux will be incident on the window from container A than from container B because of the higher radiation density in A. Radiation will flow from A to B, and this will cause the radiant energy density of B to increase while that of A decreases. Bodies A and B will no longer be in equilibrium with the radiation within their containers. Body A will begin to radiate energy into its container and its temperature will decrease; body B will absorb some of the increased radiant energy and its temperature will increase. This process will continue until the radiation energy densities within the two containers are equal, at which time the net flux of energy through the window will become zero. The two bodies will be left at unequal temperatures T_A and T_B, with $T_B > T_A$. During the time interval after the window had been opened, energy would have been flowing from the cooler to the hotter body in obvious violation of thermodynamic laws. The only way to avoid this contradiction of the laws of thermodynamics is to assume that the density of radiant energy in a container in equilibrium depends only on the temperature. This conclusion is in agreement with experimental measurements of the radiant energy density within containers at various temperatures. By making a small window in the container and measuring the radiation passing through the window, one can determine the energy density inside. In fact, one can measure not only the radiant energy density but much more detailed characteristics of the radiation inside the cavity. Figure 36.3 shows a typical setup for making such measurements. An evacuated oven or cavity is maintained at some temperature T. The cavity contains a small window. The radiation inside the cavity can be considered to be composed of a large (or infinite) number of plane waves. The wavelength, frequency, and direction of propagation of an electromagnetic wave can all be specified by giving the wave vector $\mathbf{k}$. The direction of $\mathbf{k}$ gives the direction of propagation, while the wavelength λ and angular frequency ω are related to the magnitude of $\mathbf{k}$ by

$$\lambda = 2\pi/k$$

and

$$\omega = ck$$

Because of the opaque screen, only those plane waves leaving the window whose wave vectors lie within a narrow range of directions get into the detector. The prism then spreads out the radiation passing through the hole in the screen. The angle of deviation depends on the magnitude of $\mathbf{k}$, and thus the detector measures the intensity of radiation being emitted by the window within a narrow range of wave vectors. With this information one can easily calculate the energy density of waves of the same wave vector inside the cavity.

Results indicate that the energy density is independent of polarization and independent of the direction of $\mathbf{k}$, as would be expected. The energy per unit volume of those waves whose wave vectors lie within a region d^3k, centered at $\mathbf{k}$, is proportional to d^3k (see Fig. 36.4). The proportionality constant $E(k)$ is called the energy distribution function for cavity radiation. The measured energy distribution functions are shown in Fig. 36.5 for a number of temperatures. Max Planck noticed that these curves can

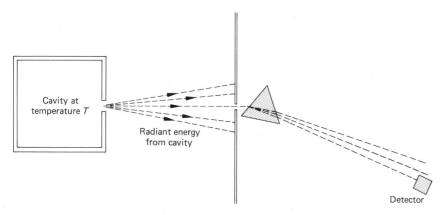

FIGURE 36.3   A setup for measuring the cavity-radiation distribution function.

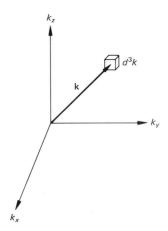

FIGURE 36.4   Each plane wave is represented by a point in k *space.* The energy per unit volume of those plane waves whose wave vectors fall within the element d^3k is $E(k)$.

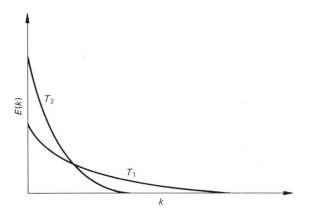

FIGURE 36.5   The Planck distribution function at two different temperatures, $T_2 > T_1$.

be matched perfectly by what is now called the Planck distribution formula:

$$E(k) = \frac{\hbar\omega}{4\pi^3 (e^{\beta\hbar\omega} - 1)} \qquad (36.1)$$

where $\omega = ck$ is the angular frequency associated with a plane wave of wave vector k, $\beta = 1/k_BT$ is the inverse of the temperature times Boltzmann's constant,[1] and $\hbar$ (pronounced "aitch bar") is a new constant with the units of joule-seconds. This constant, which was quite inexplicable at the time, is now called Planck's constant. Its value is

$$\hbar = 1.0545 \times 10^{-34} \text{ J} \cdot \text{s}$$

QUESTION

What is the total radiant energy inside a 1-m³ cavity at 27°C (300K)?

ANSWER

We obtain the total radiant energy per unit volume by integrating the energy distribution function over all values of the wave vector **k**. Using the fact that $\omega = ck$, we get

$$E = \frac{\hbar c}{4\pi^3} \int \frac{k \, d^3k}{(e^{\beta\hbar ck} - 1)}$$

If we define a three-dimensional variable **x** by $\mathbf{x} = \hbar c\mathbf{k}$, we get

$$E = \frac{1}{4\pi^3\hbar^3c^3\beta^4} \int \frac{x \, d^3x}{e^x - 1}$$

The integral can be evaluated with an integral table, and we obtain

$$E = \frac{\pi^2(k_B T)^4}{15(c\hbar)^3}$$

With $T = 300$ K, this gives an energy density of

$$E = 6.13 \times 10^{-6} \text{ J/m}^3$$

Thus not much energy is required to heat up a vacuum to room temperature.

---

[1] We have added the subscript B to Boltzmann's constant to avoid confusion with the wave vector.

Planck further showed that this distribution formula could be derived by thermodynamic arguments if one made the following strange assumption:

*The electromagnetic field energy carried by those waves of wave vector **k** can take only the values 0, ℏω, 2ℏω, 3ℏω, ...*

We say that the energy associated with these waves is quantized, with the value of the *energy quantum* being $\hbar\omega$. Planck understood perfectly well that this was no ''explanation'' of the cavity radiation formula at all, since the quantization assumption disagreed completely with the predictions of Maxwell's electromagnetic equations. One could not simply replace Maxwell's equations by the quantization hypothesis because the quantization rule by itself was not a complete theory. In fact, in carrying out the theoretical analysis of cavity radiation, Planck had to use Maxwell's equations throughout the calculation and had only to impose the strange quantization rule at the last stage of the calculation.

36.3 THE PHOTOELECTRIC EFFECT

The evidence for the quantization hypothesis was made stronger, but no less puzzling, by Einstein's theoretical analysis of the *photoelectric effect*. When light shines on a clean metal surface, electrons are ejected from the surface with some distribution of velocities that depends on the frequency and intensity of the light and on the nature of the metal (see Fig. 36.6). Using the following assumptions, Einstein made some definite predictions regarding the dependence of the electron velocities on the frequency and intensity of the light.

1. The energy in the light is carried and delivered to the electrons in discrete quanta of amount $\hbar\omega$, where ω is the angular frequency of the light and $\hbar$ is Planck's constant.

2. An electron within the metal is bound to the metal by an energy W, which depends on the nature of the metal. The energy W is called the *work function* of the metal.

3. The probability that one electron will absorb more than one quantum of energy is negligibly small at ordinary light intensities. (For very high-intensity light one can see the effects of multiple quantum absorption.)

An electron in the metal that absorbs one quantum of energy will have a kinetic energy of $\hbar\omega$. If the velocity of the electron is directed out of the metal, then it will fly out of the surface. If, on its way out, it does not lose any

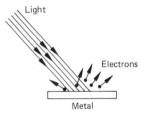

FIGURE 36.6 The photoelectric effect.

energy by collisions with other particles in the metal, then it will have an energy of

$$E = \hbar\omega - W \tag{36.2}$$

when it gets far from the metal surface. The work W is the work that the electron had to do against the attractive forces tending to pull it back into the metal. Many of the electrons, after absorbing a quantum of energy, will lose some of the energy because of collisions before they leave the surface. Thus the energy given by Eq. 36.2 is the maximum energy that ejected electrons will have.

Figure 36.7 shows an experimental setup for measuring the maximum kinetic energy of electrons emitted by the photoelectric effect. A small sphere of the test metal is placed at the center of a larger spherical anode that contains a hole through which a beam of light can be focused on the test metal. If no electric field opposes the motion of the ejected electrons, they will all eventually strike the anode. If the electrostatic potential difference V

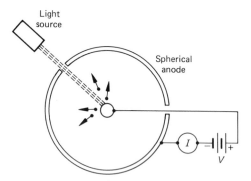

FIGURE 36.7 If eV is greater than the maximum energy of the photoelectrons being emitted, then none of them reach the spherical anode and the current is zero.

between the test metal and the anode satisfies the inequality

$$eV < \hbar\omega - W$$

then some of the electrons will reach the anode. One can then adjust V to find the cutoff voltage at which there is no photoelectric current. According to the quantization hypothesis, the cutoff voltage should be a linear function of the angular frequency of the light with a slope equal to $\hbar/e$, and it should go to zero when $\hbar\omega$ is equal to the work function of the metal. The cutoff voltage should be independent of the intensity of the light. According to the quantization hypothesis, the intensity of the light tells how many quanta of light energy strike the surface per second, and therefore is proportional to the magnitude of the photoelectric current whenever V is less than the cutoff voltage (Fig. 36.8). All these simple predictions have been confirmed in many experiments, yet they disagree completely with anything that might be expected from a classical picture of the photoelectric process. In the classical picture the intensity of the light is proportional to the square of the electric field strength associated with the light wave. Therefore, increasing the intensity should increase the force on the electrons, and this should increase the energy of the ejected electrons.

36.4 THE COMPTON EFFECT

That electromagnetic waves are capable of carrying energy and momentum was a prediction of the classical theory that had been experimentally confirmed. This fact was not disturbed by the quantization hypothesis, which stated only that the energy carried in the wave came in units or quanta equal to $\hbar\omega$. But the momentum carried by electromagnetic waves is related to the energy carried by the waves by the equation

$$p = \frac{E}{c} \tag{36.3}$$

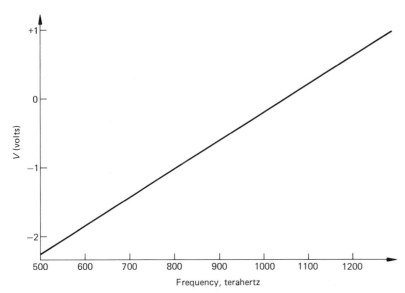

FIGURE 36.8 The cutoff voltage as a function of the light frequency. According to Einstein's theory $V = (2\pi\hbar\nu - W)/e$, which means that the cutoff voltage should be proportional to the light frequency.

If the energy is quantized in units of $\hbar\omega$, then it seems unavoidable that the momentum carried by electromagnetic waves must also be quantized. The quantum units of momentum would be

$$p = \frac{\hbar\omega}{c} = \hbar k$$

where k is the magnitude of the wave vector. Since the direction of wave propagation, **k**, is also the direction of the electromagnetic momentum, we can write this as a vector equation:

$$\mathbf{p} = \hbar\mathbf{k} \tag{36.4}$$

Assuming that these energy and momentum quanta exist, one might picture them as some sort of particles, com-monly called *photons,* which are associated with the electromagnetic wave. The energy and momentum of a photon satisfy the relation

$$E = cp$$

which is the relation derived in Chap. 35 for particles of zero mass. In 1922 A. H. Compton performed an experiment that clearly illustrated the quantization of electromagnetic momentum and energy. The experiment can be described as the scattering of a zero-mass photon by an electron. The electron is initially stationary (see Fig. 36.9). The incoming photon has a momentum **p** and an energy cp, and the scattered photon has a momentum **p**′ and an energy cp'. The relativistic energy of the electron is mc^2 before the collision and $E = c\sqrt{P^2 + m^2c^2}$ after the

FIGURE 36.9   Scattering of an electron by a photon.

collision. The momentum and energy conservation laws state that

$$\mathbf{p} - \mathbf{p}' = \mathbf{P} \tag{36.5}$$

and

$$cp + mc^2 - cp' = E \tag{36.6}$$

But, referring to the definition of E, we see that

$$\frac{E^2}{c^2} = P^2 + m^2 c^2 \tag{36.7}$$

Using the fact that $(\mathbf{p} - \mathbf{p}')^2 = p^2 + p'^2 - 2pp' \cos \theta$, we get, by substituting Eq. 36.5 and 36.6 into 36.7,

$$(p - p' + mc)^2 = p^2 + p'^2 - 2pp' \cos \theta + m^2 c^2$$

If we square out the left-hand side, cancel identical terms on the two sides of the equation, and then divide by $mcpp'$, we obtain the equation

$$\frac{1}{p'} - \frac{1}{p} = \frac{1 - \cos \theta}{mc} \tag{36.8}$$

Using Eq. 36.4 and the fact that $k = 2\pi/\lambda$, we can write Eq. 36.8 in the form

$$\lambda' - \lambda = (1 - \cos \theta)\lambda_c \tag{36.9}$$

where

$$\lambda_c = \frac{2\pi\hbar}{mc} \tag{36.10}$$

is called the *Compton wavelength*. The Compton wavelength has the dimension of length but is not actually the wavelength of anything.

Equation 36.9 predicts that the radiation scattered at an angle θ with respect to the incoming beam will have a wavelength larger than the incoming radiation by an amount $(1 - \cos \theta)\lambda_c$. In the experiment performed by Compton, a beam of x rays of wavelength about 0.7 Å was scattered by a sample of graphite. The wavelength of the scattered radiation was then measured and compared with the angle of scattering. The results indicated that the scattered radiation had two components (see Fig. 36.10). One component had a wavelength that satisfied Eq. 36.9 very well. This component was due to a scattering process in which an electron, weakly bound to one of the graphite carbon atoms, scattered a photon and was ejected from the carbon atom in the process. Since a photon of wavelength 0.7 Å has an energy of about 20,000 eV, which is much larger than the binding energy of the electron in the carbon atom, the electron could be considered as essentially free, and the analysis leading to Eq. 36.9 could be applied. The other component of the scattered radiation had the same wavelength as the incoming radiation. It was due to another kind of scattering process in which the energy and momentum are transferred to the entire carbon atom rather than being taken up by a single electron. One could still apply Eq. 36.9 and 36.10, but in this case the mass to be used in Eq. 36.10 is the mass of a carbon atom rather than the mass of an electron. The carbon atom has such a large mass that the wavelength shift predicted by Eq. 36.9 is undetectable.

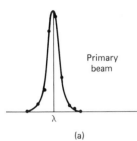
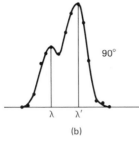
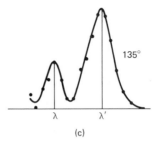
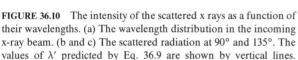

FIGURE 36.10 The intensity of the scattered x rays as a function of their wavelengths. (a) The wavelength distribution in the incoming x-ray beam. (b and c) The scattered radiation at 90° and 135°. The values of λ' predicted by Eq. 36.9 are shown by vertical lines.

The net effect of all the experiments that exhibited the particlelike or quantum properties of electromagnetic radiation was to produce an extremely bewildering situation in physics. On the one hand, the classical theory of electromagnetism was being used with more and more success to describe the propagation, reflection, and diffraction of electromagnetic waves. On the other hand, the photoelectric effect, the Compton effect, and other experiments we have not mentioned seemed clearly to show that the energy and momentum transported by an electromagnetic wave are somehow carried by zero-mass particles. No one had any idea how to make these conflicting pictures of electromagnetic radiation consistent.

During the same interval of time in which the above experiments involving light quanta were carried out (about 1890–1923), another line of experiments and theoretical work was producing a similar paradox regarding the structure of the atom. We shall now give a brief description of some of the more important of those experiments and theoretical investigations.

36.5 ALPHA-PARTICLE SCATTERING

Certain radioactive elements were known to emit a type of nonelectromagnetic radiation called α radiation. In the chapter on nuclear physics we describe in detail the physical process involved in α radiation. By means of experiments in which a thin pencil of "α rays" were observed passing through electric and magnetic fields (see Fig. 36.11), investigators determined that the radiation consisted of a shower of high-energy positively charged particles. The particles, called α particles, had a mass nearly equal to that of a helium atom and a positive charge equal to $2e$, where the charge of an electron is $-e$. In 1910 Ernest Rutherford analyzed an experiment in which an extremely thin gold foil was bombarded by a stream of α particles of known energy. The positively charged α particles were deflected because of the microscopic electric fields within the gold atoms and passed out of the gold foil, moving in all possible directions. The flux of α particles at a given angle with respect to the incident

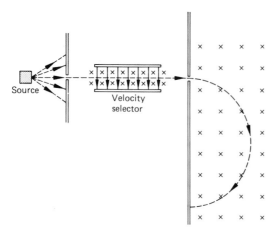

FIGURE 36.11 Identification of particles by means of their motion through electric and magnetic fields. In the velocity selector **E** points downward and **B** points into the paper. A particle will pass through without deflection only if $qE = qvB$ or $v = E/B$. In the uniform magnetic field to the right, the particle (of known velocity) moves in a circle of radius $R = mv/qB$. This allows a determination of m/q, from which one can identify the particle.

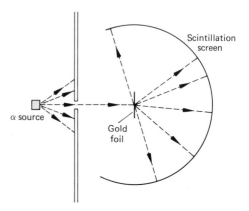

FIGURE 36.12 Scattering of α particles. The scintillation screen gives off a little flash of light when struck by an α particle.

beam was measured by means of a scintillation counter (see Fig. 36.12).

Rutherford showed, by means of an argument that used purely classical physics, that the measured α particle flux at all angles could be simply accounted for if one assumed that all the positive charge and almost all the mass of an atom were concentrated in a tiny region in the center of the atom. Rutherford called that region the *nucleus*. The negative charge that was needed to neutralize the atom consisted of a collection of electrons that were somehow distributed within the rest of the space occupied by the atom. The size of the nucleus was extremely small, even in comparison with the tiny size of the whole atom. A typical atom has a diameter of about 1 Å $= 10^{-10}$ m, and a typical nucleus has a diameter of 1 fermi (fm) $= 10^{-15}$ m.

Thus the volume of the nucleus is about 10^{-15} times the volume of the atom and, since most of the mass resides in the nucleus, the mass density within a nucleus is about 10^{15} times the mass density of the ordinary material we are used to.

According to classical physics, a planetary atom such as that shown in Fig. 36.13 should be completely unstable. The moving electrons would be expected to produce electromagnetic waves that would carry away energy, and as the electrons lost energy they would spiral ever closer to the attracting nucleus. The whole assemblage would contract to a point, with a burst of radiation, in a short time. Even if one ignored the problem of stability, there was no way of understanding, within the planetary atom model, the most fundamental property of atoms, namely the fact that every atom of an element is like every other atom of the same element. Using the planetary picture of, say, a helium atom, one would expect the size of the atom to depend on the radii of the electron orbits around the nucleus. But there is nothing in classical physics that would prevent different atoms from having different electron orbits. It is not necessary to belabor the obvious point

FIGURE 36.13 The planetary atom.

that an entity as changeable as a collection of electrons orbiting a nucleus bears little resemblance to the picture of rigid, identical, indestructible atoms that is necessary to explain the chemical and physical properties of matter. In spite of the difficulty physicists experienced in trying to harmonize Rutherford's planetary model of the atom with the known properties of atoms, later repeats of his experiment and a number of different experiments clearly confirmed that the basic structure of an atom consists of a heavy positive nucleus surrounded in some way by light negative electrons.

36.6 ENERGY QUANTIZATION IN ATOMS

At the time that the ideas of electromagnetic energy quanta developed, scientists had already known for a long time that isolated atoms, when energized by thermal collisions, radiated their excess energy in the form of electromagnetic waves of definite discrete frequencies. The collection of possible radiation frequencies, called the *emission spectrum* of the atom, is different for different elements (see Fig. 36.14). Before the introduction of the Rutherford planetary model of the atom, the existence of discrete emission frequencies did not seem difficult to understand in principle. It was assumed that the atom was a fairly rigid but slightly deformable structure made up of electrically charged material. As such it would be expected to have certain normal modes of vibration, with associated discrete normal-mode frequencies. It seemed reasonable that these vibrational frequencies of the semirigid atomic structure were observed as the emission frequencies of electromagnetic waves.

After Planck's postulate of electromagnetic energy quantization, Einstein suggested a complete reinterpretation of the emission spectrum of an atom. He reasoned that if the electromagnetic field energy could have only quantized values, then when an atom emitted electromagnetic radiation it could do so only in discrete, whole quanta. Einstein viewed the process of radiation by an atom as one in which the atom emitted a complete quantum or photon of radiation. The photon carried away an amount of energy $\hbar\omega$, where ω is the angular frequency of the emitted radiation. The fact that he could not really explain the relationship between the particlelike photons and the continuous electromagnetic wave naturally troubled Einstein and everyone else. The detailed logical connection between the particles and the wave functions was presented only with the later development of a complete quantum theory. [We shall introduce that theory in the next chapter.] Since the photon carries away an amount of energy $\hbar\omega$, the energy state of the atom must change by a corresponding amount:

$$E_{\text{before}} - E_{\text{after}} = \hbar\omega$$

where E_{before} and E_{after} are the energies of the atom before and after the photon is emitted. The fact that a particular kind of atom only radiates photons of those discrete frequencies that are in its spectrum means that the energy changes that such atoms undergo are always of certain amounts. An obvious way to explain such discrete energy changes by an atom is to assume that the energy of the atom itself can have only certain particular discrete values. Let us temporarily assume that this is so and see what consequences follow from the assumption. We assume that the energy of a particular atom can, for some unknown reason, have only the values $E_1, E_2, E_3, \ldots$ and no value in between these discrete values. We call the set of numbers $E_1, E_2, E_3, \ldots$ the *energy spectrum* of the

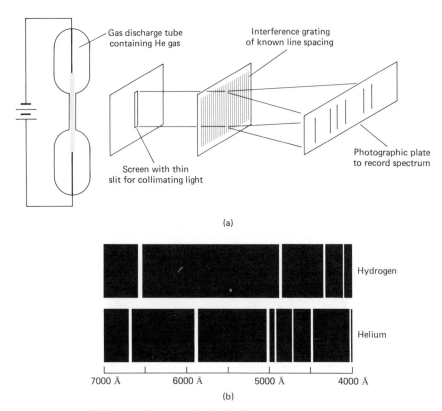

(a)

(b)

FIGURE 36.14   (a) A setup for measuring the helium spectrum. (b) The visible portions of the emission spectra of hydrogen and helium.

atom, and we call any one of the numbers in it, E_n, an *energy level.* We call the lowest energy level, the *ground-state energy.* If the atom is in a state with energy E_m and then emits a photon of energy $\hbar\omega$, in the process changing to the energy state E_n, then by energy conservation the photon energy must be equal to the difference in the two atomic energy levels:

$$\hbar\omega = E_m - E_n$$

If we call the angular frequency of the photon emitted in this process ω_{mn}, then

$$\omega_{mn} = \frac{E_m - E_n}{\hbar} \tag{36.11}$$

To see that this assumption leads to a simple verifiable prediction, let us look at all the possible energy transitions among the three levels E_1, E_2, and E_3. (We always order the levels so that $E_1 \leq E_2 \leq E_3$, etc.) There are three possible transitions, $E_2 \rightarrow E_1$, $E_3 \rightarrow E_2$, and $E_3 \rightarrow E_1$, and they lead to the following values for the radiation frequencies:

$$E_2 \rightarrow E_1: \quad \omega_{21} = \frac{E_2 - E_1}{\hbar}$$

$$E_3 \rightarrow E_2: \quad \omega_{32} = \frac{E_3 - E_2}{\hbar}$$

$$E_3 \rightarrow E_1: \quad \omega_{31} = \frac{E_3 - E_1}{\hbar}$$

From these equations it is a trivial matter to see that

$$\omega_{31} = \omega_{32} + \omega_{21}$$

In general, for any three levels $E_\ell \leq E_m \leq E_n$, we get

$$\omega_{n\ell} = \omega_{nm} + \omega_{m\ell} \tag{36.12}$$

We are thus led to expect, in the emission spectrum of every element, many instances of two emission frequencies which, when added, yield another frequency in the spectrum. Because these emission frequencies can usually be measured with great precision, this is a very strong prediction. That many combinations of spectral lines satisfy Eq. 36.12 was already known at the time of Einstein's analysis. This fact was called the *Ritz combination principle*.

36.7 THE FRANCK-HERTZ EXPERIMENT

In 1914 J. Franck and G. Hertz reported on an experiment that strongly corroborated Einstein's idea that the discrete emission spectrum of an atom was caused by the existence of discrete atomic energy levels. Figure 36.15 shows Franck and Hertz's apparatus. A metal, when heated sufficiently, will emit electrons through its surface. This effect, called *thermionic emission,* was of great importance when electronic devices were made with vacuum tubes

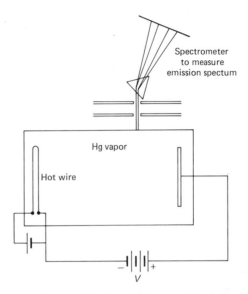

FIGURE 36.15 The Franck-Hertz experiment. The hot wire emits electrons that are accelerated toward the cathode. By collisions they excite mercury atoms, which then emit radiation.

rather than transistors. The basic mechanism is quite understandable. Because of thermal agitation, some of the electrons in the metal gain insufficient energy to overcome the work function at the surface and fly off. A typical thermionic electron has an energy of about $E \simeq k_B T$, where T is the absolute temperature of the metal. For typical hot-wire temperatures ($\sim 10^3$ K), the energy of thermionic electrons is about 0.1 eV.

In the Franck-Hertz apparatus, electrons are emitted by a hot wire into a very dilute mercury vapor. An accelerating potential V draws the electrons toward the cathode. If an electron does not collide with a mercury atom on the way, it will have an energy of eV electron volts by the time it reaches the cathode and is absorbed. Thus it is possible for the electrons, by colliding with the mercury atoms, to impart to the atoms any energy up to, but not exceeding, eV. If the mercury atoms are excited by this electron bombardment, they will give off their extra energy as radiation. Franck and Hertz observed the emitted spectrum as the accelerating potential V was varied, and they found that any frequency in the mercury emission spectrum remained unobserved until eV became larger than $\hbar\omega$, where ω was the angular frequency of the spectral line. The interpretation of this observation within the Einstein picture of radiative emission is very simple. Unless the electrons have sufficient energy to excite the atom from its ground state to some state with energy larger than $\hbar\omega$, there is no way for the atom to emit a photon of that frequency. If instead one tries to relate the emission spectrum to some sort of normal modes of vibration of the atom, there is no obvious way to understand the Franck-Hertz effect. Thus the Franck-Hertz experiment very much strengthened the belief in the existence of discrete energy levels in atoms. Of course, it did not explain why discrete energy levels existed; it only convinced people that they did.

36.8 THE HYDROGEN ATOM

As one can easily understand, active research workers became very excited, but also very puzzled, by this strange quantization phenomenon. As physicists always do, they looked for the simplest system in nature that exhibited this effect, and the obvious candidate was the hydrogen atom. Experiments had already confirmed that a hydrogen atom was composed of a single proton and a single electron. If physicists could not understand hydrogen, there was little chance that they would understand more complex atoms with many interacting electrons. Not only was the picture of a hydrogen atom simple, but it was known that, according to Einstein's interpretation of the emission spectrum,

the energy levels of hydrogen satisfied a tantalizingly simple formula. The nth energy level has the value

$$E_n = \frac{-A}{n^2} \tag{36.13}$$

where the constant

$$A = 2.179 \times 10^{-18} \text{ J} = 13.60 \text{ eV}$$

The challenge to theorists could be presented in a very direct form: Construct a theory that will yield the constant A in terms of the basic parameters of the system, such as the electron and proton masses, their charges, the velocity of light, and Planck's constant. Niels Bohr was the first to meet that challenge.

The Bohr theory begins with a completely classical picture of the hydrogen atom, namely that the atom consists of an electron orbiting, according to Newton's laws of dynamics, around a proton. Since the proton is so much heavier than the electron, we shall neglect its "wobble" and consider it as fixed in space at the origin of our coordinate system. For simplicity, Bohr considered only circular orbits of the electron (see Fig. 36.16). The theory was later extended by Arnold Sommerfeld to include elliptical orbits as well. An electron moving at speed v in a circle of radius r has a centripetal acceleration equal to v^2/r. The force attracting the electron to the proton is, according to Coulomb's law, equal to ke^2/r^2, where k is the Coulomb force constant and e is the charge of both the

FIGURE 36.16 The Bohr model of the hydrogen atom.

electron and the proton. By Newton's second law, we obtain

$$\frac{mv^2}{r} = \frac{ke^2}{r^2} \tag{36.14}$$

If we multiply this equation by $\frac{1}{2}r$, we obtain a simple relation between the kinetic energy ($E_K = \frac{1}{2}mv^2$) and the potential energy ($E_p = -ke^2/r$):

$$E_K = \frac{ke^2}{2r} = -\tfrac{1}{2}E_p$$

The total energy of the atom can therefore be expressed in terms of the orbital radius:

$$E = E_K + E_p = \tfrac{1}{2}E_p = -\frac{ke^2}{2r}$$

Since, from a classical point of view, any value of r is possible, we get no sign of any discrete energy levels from this analysis. Bohr then asked himself how he could introduce Planck's constant $\hbar$ into the theory in such a way as to obtain the observed energy levels. The units of $\hbar$ are joule-seconds or (energy $\times$ time). Writing this in terms of fundamental units (mass, length, and time), we get

$$\text{units of } \hbar = (\text{mass})\left(\frac{\text{distance}}{\text{time}}\right)^2(\text{time})$$

$$= (\text{mass})\left(\frac{\text{distance}}{\text{time}}\right)(\text{distance})$$

$$= \text{units of } mvr$$

But mvr is just the angular momentum of the atom. Bohr proposed the following rather arbitrary quantization postulate:

Only those orbits are possible for which the angular momentum mvr is equal to $\hbar$, $2\hbar$, $3\hbar$, ...

Postponing a discussion of the legitimacy of this strange assumption for a moment, let us see how it leads directly to the correct energy levels. If we assume that

$$mvr = n\hbar$$

where n is a positive integer, we get, by squaring the equation and dividing by mr^3, the following relation:

$$\frac{mv^2}{r} = \frac{n^2\hbar^2}{mr^3} \tag{36.15}$$

If we now substitute this into Eq. 36.14, we can solve for $ke^2/2r$ and obtain

$$E_n = -\frac{mk^2e^4}{2\hbar^2n^2} \tag{36.16}$$

Comparing this with Eq. 36.13, we get a formula for A:

$$A = \frac{mk^2e^4}{2\hbar^2}$$

If we use the known values of all the fundamental constants involved to evaluate the right-hand side of this equation, we obtain

$$\frac{mk^2e^4}{2\hbar^2} = 2.180 \times 10^{-18} \text{ J}$$

The small discrepancy between this value and the value for A given before is due to the approximation we made during the calculation of treating the proton as a completely stationary object. If we had taken into account the finite mass of the proton, we would have obtained a result that was smaller by a factor of $M/(M + m)$, where M is the proton mass. This would have eliminated the discrepancy completely (see problem 36.A.7).

The essential elements of the Bohr theory of the hydrogen atom are the following:

1. Only certain discrete allowed orbits exist for the electron. These orbits are determined by the quantization condition that the angular momentum is equal to $n\hbar$, where n is any positive integer.

2. While the electron moves in one of the allowed orbits it does not radiate. This, of course, violates Maxwell's equations.

3. A hydrogen atom that is not in its ground state might, at any time, jump from the state it is in into some lower state. In doing so it would emit the excess energy as a single photon. The theory did not predict when this would happen; nor did it explain how the electron moved in going from one allowed orbit to another.

Bohr never considered this assemblage of classical concepts and arbitrary quantization conditions as a real theory of the hydrogen atom. He knew that it was only a useful step toward a complete self-consistent physical theory of atomic structure. Shortly after Bohr presented his first work, Arnold Sommerfeld was able to modify the theory slightly to include elliptical orbits and relativistic effects. All attempts to extend the theory to atoms containing more than a single electron failed.

As we shall see later, there is no such thing as an orbit of an electron in an atom according to the correct quantum theory of the atom. In spite of the fact that an erroneous concept (the electron orbit) plays such a fundamental role in the theory, the Bohr theory does correctly predict many properties of single-electron atoms. The ions, He^+, Li^{2+}, etc., are all one-electron systems to which the Bohr theory can be applied. The only modification that one would have to make in the analysis in order to apply it to an atom consisting of one electron moving around a nucleus of charge Ze would be to replace one of the factors of e in each equation by Ze. Thus the energy levels of such a system would be

$$E_n = -\frac{mk^2Z^2e^4}{2\hbar^2n^2}$$

We can easily calculate the radius of the circular orbit that in the Bohr theory gives rise to the nth energy level by combining Eqs. 36.14 and 36.15. This radius is $r_n = n^2\hbar^2/mke^2$. For a one-electron ion of nuclear charge Z, the corresponding formula is

$$r_n = \frac{n^2\hbar^2}{mkZe^2}$$

Figure 36.17 shows the first Bohr orbits, with correct relative sizes, for H, He^+, and Li^{2+}. The radius of the ground-state orbit in hydrogen is called a *Bohr radius* and is written a_0:

$$a_0 = \frac{\hbar^2}{mke^2} = 0.53 \ \text{Å}$$

As a matter of fact, a_0 does give a good estimate of the size of a hydrogen atom in its ground state.

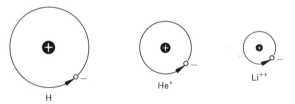

FIGURE 36.17   Relative sizes of H, He^+, and Li^{2+} according to the Bohr theory.

36.9 DE BROGLIE WAVES

In 1924 Louis de Broglie, a doctoral candidate at the Faculty of Sciences of the University of Paris, presented a thesis in which some very bold, but at the time unsubstantiated, conjectures were made. His studies concerned the relationship between waves and particles in the quantum theory. He described the initial stages of his work as follows:

The necessity of assuming for light two contradictory theories—that of waves and that of corpuscles—and the inability to understand why, among the infinity of motions which an electron ought to be able to have in the atom according to classical concepts, only certain ones were possible: such were the enigmas confronting physicists at the time I resumed my studies of theoretical physics.

When I started to ponder these difficulties two things struck me in the main. Firstly the light-quantum theory cannot be regarded as satisfactory since it defines the energy of a light corpuscle by the relation $W = \hbar\omega$ which contains a frequency ω. Now a purely corpuscular theory does not contain any element permitting the definition of a frequency. This reason alone renders it necessary in the case of light to introduce simultaneously the corpuscle concept and the concept of periodicity.

On the other hand the determination of the stable motions of the electrons in the atom involves whole numbers, and so far the only phenomena in which whole numbers were involved in physics were those of interference and of normal mode vibrations. That suggested the idea to me that electrons themselves could not be represented as simple corpuscles either, but that a periodicity had also to be assigned to them too.

I thus arrived at the following overall concept which guided my studies: for both matter and radiations, light in particular, it is necessary to introduce the corpuscle concept and the wave concept at the same time. In other words the existence of corpuscles accompanied by waves has to be assumed in all cases. However, since corpuscles and waves cannot be independent because, according to Bohr's expression, they constitute two complementary forces of reality, it must be possible to establish a certain parallelism between the motion of a corpuscle and the propagation of the associated wave. The first objective to achieve had, therefore, to be to establish this correspondence.[2]

Thus de Broglie assumed that, just as there were particles associated with light waves, there were waves of some yet undetected variety associated with massive particles such as electrons. What he pictured was a particle moving within a wave packet. A *wave packet* is a small region of waves that move together through space. One can create a wave packet by taking a flashlight that has a very narrow beam (a laser would be much better) and turning it on and off within an extremely short time span (see Fig. 36.18). If a packet of light waves travels through a region in which the index of refraction varies smoothly from place to place, it will not propagate in a straight line. Nor

FIGURE 36.18 A well-focused flashlight turned on and off in a ridiculously short time interval would produce a small wave packet.

will it move at constant speed. The packet of light waves will follow some curved trajectory at nonuniform speed. De Broglie realized that if a particle were somehow confined to, or guided by, a packet of "matter waves," then the particle would execute all the motions of the wave packet. But it was already known that a particle accelerated along a curved path whenever it passed through a region of nonconstant potential, $V(x, y, z)$. Thus de Broglie was faced with the problem of finding some relationship between the characteristics of the wave packet, such as its wavelength and frequency, and the characteristics of the particle, such as its momentum and the potential energy at its location. This relationship had to have the property that the motion of the wave packet that one could calculate using the laws of geometric optics would agree exactly with the motion of the particle that one could calculate using Newton's laws of dynamics.

Although de Broglie considered the more general case of a smoothly varying potential and even used the correct relativistic equations of dynamics, we shall take only the simple case of a nonrelativistic particle moving from one two-dimensional region of constant potential into another region in which the potential is also constant but different from that of the first region. We can make a model of such a system by considering a particle that is sliding on one of two smooth horizontal surfaces that are at different levels and have a rounded-off step between them. The particle rapidly changes its direction and speed whenever it goes from one region to the other (see Fig. 36.19a). The change in the direction and magnitude of the particle's

[2] Niels H. Heathcote, *Nobel Prize Winners in Physics, 1901–1950*, Abelard-Schuman, New York, 1954, p. 291.

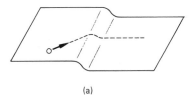

(a)

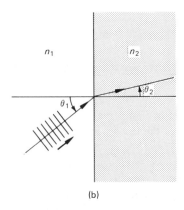

(b)

FIGURE 36.19   The wave packet in (b) changes its speed and direction of motion in a way that is analogous to the motion of the particle in (a).

momentum can be determined from the conservation laws for the energy and the y component of momentum:

$$\frac{p_1^{\,2}}{2m} + V_1 = \frac{p_2^{\,2}}{2m} + V_2 \tag{36.17}$$

$$p_{1y} = p_{2y} \tag{36.18}$$

Similarly, when a wave packet moves from a region in which the phase velocity (the velocity of the wave fronts) has one value, v_1, into a region in which it has another value, v_2, the wave packet is refracted according to Snell's law (see Fig. 36.19b). We can determine the change in wavelength and direction of propagation of the wave packet by two rules. One is that the frequency in the new medium is the same as the frequency in the old medium. This is an immediate consequence of the fact that the number of wave fronts that approach any point on the interface of the two regions in one second must equal the number of wave fronts that leave the same point. Wave fronts do not disappear at the interface. Thus

$$\omega_1 = \omega_2 \tag{36.19}$$

The other rule is Snell's law, which states that

$$\frac{\sin \theta_1}{v_1} = \frac{\sin \theta_2}{v_2} \tag{36.20}$$

If we multiply corresponding sides of Eqs. 36.19 and 36.20, and use the fact that the magnitude of the wave vector $k = \omega/v$, we get the relation

$$k_1 \sin \theta_1 = k_2 \sin \theta_2$$

or

$$k_{1y} = k_{2y} \tag{36.21}$$

Comparing Eqs. 36.17 and 36.18 with Eqs. 36.19 and 36.21 shows that the particle and the wave packet will have the same trajectory if we assume the following relations between the properties of the particle and the properties of the wave:

1. ω is proportional to $E = \dfrac{p^2}{2m} + V$

2. $\mathbf{k}$ is proportional to $\mathbf{p}$

Comparison with the case of photons certainly suggests that the constant of proportionality is Planck's constant:

$$\frac{p^2}{2m} + V = \hbar\omega \tag{36.22}$$

and

$$\mathbf{p} = \hbar\mathbf{k} \tag{36.23}$$

These relations, which were first suggested in de Broglie's thesis, are called the *de Broglie relations*. It can be shown that the de Broglie relations lead to agreement between the particle and wave packet motion for the case of a continuously varying potential.

De Broglie also pointed out that the concept of matter waves led to a simple interpretation of the Bohr quantization condition for hydrogenic orbits. The allowed Bohr orbits turn out to be exactly those orbits for which the de Broglie wave has an integral number of wavelengths along the orbital path (see Fig. 36.20). To see this, we multiply Eq. 36.14 by mr and get

$$p^2 = \frac{mke^2}{r}$$

But Eq. 36.23 states that

$$p^2 = \hbar^2 k^2 = \left(\frac{2\pi\hbar}{\lambda}\right)^2$$

which gives

$$\frac{mke^2}{r} = \left(\frac{2\pi\hbar}{\lambda}\right)^2 \qquad (36.24)$$

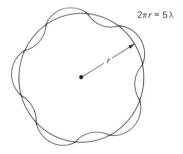

$2\pi r = 5\lambda$

FIGURE 36.20 De Broglie could obtain the Bohr quantization condition by demanding that the electron's orbit contain an integral number of matter wavelengths.

In order to fit an integral number of wavelengths exactly around the orbit, we would need to have

$$2\pi r = n\lambda \qquad (36.25)$$

Using Eq. 36.25 to eliminate λ in Eq. 36.24, we get the usual equation for the radius of a Bohr orbit:

$$r = \frac{\hbar^2 n^2}{mke^2}$$

36.10 THE DAVISSON-GERMER EXPERIMENT

At the time of publication of de Broglie's work, two physicists at the Bell Laboratories in New York, C. J. Davisson and L. H. Germer, were puzzling over some strange results they had obtained from an electron-scattering experiment. They had aimed a beam of electrons of known energy E at a single crystal of nickel and had observed the intensity of those scattered electrons whose energies were close to the energy of the incoming electrons. Scattering without energy loss is called *elastic scattering*. The two physicists had measured the intensity of elastically scattered electrons as a function of the angle of scattering for a number of different orientations of the electron detector with respect to the crystal planes of the nickel crystal (see Fig. 36.21a). They had found pronounced maxima for certain scattering angles and energies (see Fig. 36.21b). During a trip to England, Davisson discussed the peculiar results of his electron-scattering experiments with a group of physicists who suggested that they might be explained in terms of de Broglie's electron wave hypothesis. On returning to New York, Davisson discovered that his experiments did indeed supply a clear and simple confirmation of de Broglie's conjecture. The analysis needed is rather simple.

Let us assume that the hypothesized electron waves are in some way scattered by the nickel atoms in the crystal. We consider a perfect crystal on which a uniform plane electron wave (whatever that is) is incident at an angle θ

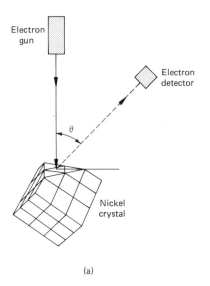

(a)

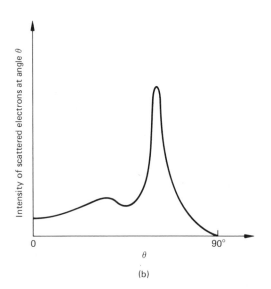

(b)

FIGURE 36.21 (a) The Davisson-Germer experiment. (b) The scattered electron intensity as a function of angle shows strong peaks at certain angles.

with respect to a certain set of crystal planes (see Fig. 36.22). If we were to add the amplitudes of the waves scattered, in some particular direction, θ', by all the atoms in the crystal, we would be very likely to obtain a small resultant wave amplitude because the waves coming from the different atoms would have different relative phases and would, for the most part, cancel one another. We now ask ourselves: Under what conditions will the waves from all atoms add in phase? Figure 36.23 shows that the waves from all the atoms of a single plane add in phase if $\theta' = \theta$. Even if all the waves originating from a single plane add in phase, we will still have cancellation unless the waves from consecutive planes also add in phase. As shown in Fig. 36.24, the condition for this is that

$$2d \cos \theta = n\lambda \qquad \text{(36.26)}$$

Since Davisson and Germer knew the spacing between adjacent crystal planes in nickel, they could use Eq. 36.26 to determine the wavelength of the scattered electron waves. The results of their experiments confirmed the de Broglie relations very well (see Fig. 36.25).

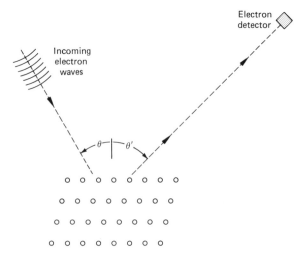

FIGURE 36.22 Microscopic view of electron wave scattering.

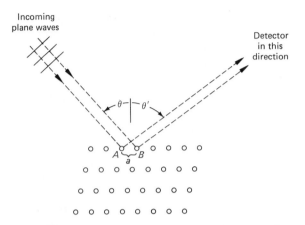

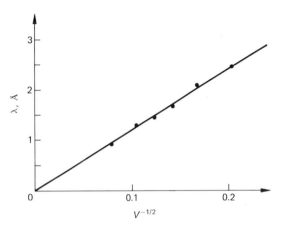

FIGURE 36.23 The wave fronts have to travel an extra distance $a \sin \theta$ to reach atom B in comparison with atom A. Thus the scattered waves leave B later than they leave A. But the waves from A must travel an extra distance $a \sin \theta'$ to get to the detector. If $\theta = \theta'$ the two effects cancel and the waves from A and B arrive in phase.

FIGURE 36.25 Relationship between the measured de Broglie wavelength and the potential through which the electrons have been accelerated. According to the de Broglie relation, $p = 2\pi\hbar/\lambda$. Combining this with $p^2/2m = eV$ gives $\lambda = 2\pi\hbar/\sqrt{2meV} = 1.23$ Å$/\sqrt{V}$.

SUMMARY

Within an evacuated cavity at temperature T, the electromagnetic density of those waves whose wave vectors lie within a region d^3k centered at $\mathbf{k}$ is given by the Planck formula

$$E(k) = \frac{\hbar\omega}{4\pi^3(e^{\beta\hbar\omega} - 1)}$$

where $\beta = 1/k_B T$, $\omega = ck$, and $\hbar$ is a new fundamental constant called *Planck's constant:*

$$\hbar = 1.0545 \times 10^{-34} \text{ J} \cdot \text{s}$$

The energy and momentum carried by an electromagnetic wave exist only in discrete units called quanta. Each

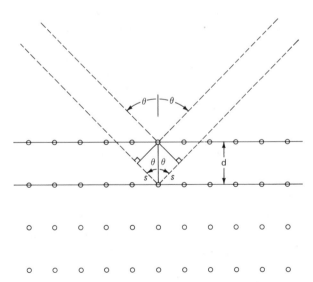

FIGURE 36.24 The waves scattered from the second plane have to travel an extra distance $2s = 2d \cos \theta$ in comparison with those scattered from the first plane. All waves will arrive in phase if $2d \cos \theta = n\lambda$ ($n = 1, 2, \ldots$).

quantum of energy and momentum can be pictured as a zero-mass particle (a *photon*), with

$$E = \hbar\omega$$

and

$$\mathbf{p} = \hbar\mathbf{k}$$

The *photoelectric effect* is the name given to the property of metals of emitting electrons when irradiated by light of sufficiently high frequency.

The maximum energy of the photoelectrons that are emitted is related to the frequency of the incident light ω and the work function of the metal W by

$$E_{max} = \hbar\omega - W$$

The number of photoelectrons is proportional to the light intensity, but E_{max} is not related to the light intensity.

One can understand the photoelectric effect by assuming that each photoelectron absorbs one photon, and thereby receives an amount of energy $\hbar\omega$. The photoelectron must do an amount of work W against attractive forces at the metal surface in order to escape from the metal.

The *Compton effect* is the scattering of photons by free electrons. The scattered photons have lower energy and therefore longer wavelength than the incident photons. Conservation of energy and momentum leads to the following formula for the change in wavelength of those photons that are scattered by an angle θ with respect to the direction of the incoming photon beam:

$$\Delta\lambda = (1 - \cos\theta)\lambda_c$$

where the *Compton wavelength* is given by

$$\lambda_c = \frac{2\pi\hbar}{mc}$$

Rutherford showed that one could understand the scattering of α particles by the atoms in a thin gold foil by assuming that the atoms were composed of an extremely small nucleus surrounded by light electrons.

An isolated atom can radiate electromagnetic waves only at certain discrete frequencies. The set of possible frequencies is called the *emission spectrum* of the atom.

The frequencies in the emission spectrum are given by

$$\hbar\omega_{nm} = E_n - E_m$$

where E_n and E_m are the discrete energies of the atomic states before and after the transition.

The set of possible energies of an atom is called the atom's *energy spectrum*. The lowest energy level is the *ground-state energy*.

Franck and Hertz excited the atoms in a mercury vapor by electron collision and found that no line in the mercury emission spectrum became visible until the electron energy exceeded $\hbar\omega$, where ω is the angular frequency of the line. That is, unless the electrons have sufficient energy to excite the atom from its ground state to some state with energy greater than $\hbar\omega$, there is no way for an atom to emit a photon of that energy.

Niels Bohr showed that one could derive the hydrogen energy levels by using Newtonian mechanics plus the strange assumption that the angular momentum of the atom could have only the values

$$mvr = n\hbar \qquad (n = 1, 2, 3, \ldots)$$

From this he obtained the experimentally observed energy levels

$$E_n = \frac{-A}{n^2} \qquad (n = 1, 2, 3, \ldots)$$

where $A = mk^2e^4/2\hbar^2 \simeq 2.18 \times 10^{-18}$ J. The radius of the nth Bohr orbit in a hydrogen atom is

$$r_n = n^2 a_0$$

where the *Bohr radius* is

$$a_0 = \frac{\hbar^2}{mke^2} = 0.53 \text{ Å}$$

De Broglie showed that if one assumed that some kind of matter wave was associated with a particle of mass m and that the frequency and wave vector of the matter waves were related to the energy and momentum of the particle by the *de Broglie relations*

$$\hbar\omega = \frac{p^2}{2m} + V$$

and

$$\hbar\mathbf{k} = \mathbf{p}$$

then, according to geometric optics, the motion of a matter wave packet would satisfy Newton's equations of motion for a particle of mass m. De Broglie also showed that one could derive the Bohr angular momentum quantization conditions by demanding that the electron orbit contain an integral number of matter wavelengths.

The existence of electron matter waves was confirmed by Davisson and Germer in experiments on electron scattering by crystals.

PROBLEMS

36.A.1 The temperature at the center of the sun is estimated to be about 3 million degrees kelvin. What would be the radiant energy density in a cavity at that temperature?

36.A.2 For a certain metal the photoelectric cutoff voltage for light of wavelength $\lambda = 0.3$ μm is 2 V. What is the work function of the metal?

36.A.3 Suppose you have a battery and wish to run a car. You are given the choice of using the battery to power a flashlight that you shine out the back window or using the battery to power an engine. Which would you choose? (Assume all processes are 100% efficient.) We know light has momentum, and just as the hydrogen atom recoils when it emits a photon, so will the car recoil from the photons emitted by the flashlight.

36.A.4 At what maximum velocity would photoelectrons be ejected from a metal with a work function of 1.4 eV by light of wavelength 0.5 μm?

36.A.5 A photon interacts with an electron that is initially at rest. The photon is scattered but with a new frequency. What is the energy given to the electron if the change in the photon frequency is 10^5 Hz?

36.A.6 Suppose the human eye could detect light of such an intensity that 10 photons enter the pupil per second. What intensity light source (i.e., what wattage) at the distance of the moon could be observed? Assume that 30% of the energy is given off in the form of visible light (average wavelength equal to 0.6 μm).

Diameter of pupil $= 0.5$ cm

Distance to moon $= 4 \times 10^8$ m

36.A.7 Show that by replacing the electron mass m in Equation 36.16 by the "reduced mass," $\mu = mM/(M + m)$, where M is the proton mass, we obtain a value of A (to be used in Equation 36.13) of 2.179×10^{-18} J. This replacement makes the correction for the finite mass of the proton.

36.A.8 Suppose a system had just three electronic energy levels:

$$E_1 = 2 \times 10^{-16} \text{ J},$$

$$E_2 = 3 \times 10^{-16} \text{ J},$$

$$E_3 = 6 \times 10^{-16} \text{ J}.$$

What would be the wavelengths of the emission spectrum? At what frequencies could the system absorb electromagnetic radiation?

36.A.9 Using the de Broglie relations, calculate and compare the frequencies associated with electrons, neutrons, and photons all of 1 Å wavelength. What are the corresponding energies?

36.B.1 The specific heat at constant volume of a system is defined as

$$C_v = \frac{\partial E(T, V)}{\partial T}$$

Calculate the specific heat of a cubic-meter cavity. At what temperature is its specific heat equal to the constant volume specific heat of one mole of helium (treated as an ideal gas)?

36.B.2 In an x-ray tube, electrons are accelerated between electrodes with a potential difference of 50 kV. Assume that, in the anode, each electron produces one photon of the same energy. Calculate the ratio of the photon wavelength to the electron wavelength.

36.B.3 A photon of initial wavelength $\lambda = 10$ Å is scattered through an angle of $45°$ by a free electron. What is the recoil velocity of the electron?

36.B.4 Derive a formula for the de Broglie wavelength of the electron in the nth Bohr orbit.

36.C.1 The energy of one photon with wave vector $\mathbf{k}$ is $\hbar\omega$. We can thus interpret $E(k)/\hbar\omega$ as the density of photons of wave vector k. Derive an expression for the total density of photons in a cubic meter cavity at temperature T in the form

$$\text{density of photons} = \int_0^\infty f(k)\, dk$$

36.C.2 For Compton scattering, derive a formula for the angle between $\mathbf{P}$ and $\mathbf{p}$ in terms of p, p', and θ.

37

THE SCHRÖDINGER WAVE EQUATION

IT IS NOT unlikely that the reader, at this point, is suffering from a bad case of conceptual indigestion brought about by too rapid ingestion of a bewildering sequence of experimental results and incompletely baked explanations of those experiments. Before we present the theory that finally brought order into the very puzzling intellectual situation described in the last chapter, it might help if we try to unify and review the results of most of the experiments we have discussed. We shall then have fixed clearly in our mind what it is that the theory has to explain. We do this by describing the following simple, idealized experiment, which combines most of the results of the experiments described in the last chapter.

37.1 DIFFRACTION OF LIGHT AND ELECTRONS

Let us consider the experimental apparatus diagramed in Fig. 37.1. It consists of a source, an opaque screen with a single slit, and a phosphorescent screen. By the flick of a switch the source can be made to emit either a beam of monochromatic ultraviolet light or monoenergetic electrons. The intensity of the beam (in either case) may be adjusted at will. The slit width can also be varied. When irradiated by electrons or light the screen emits light. We

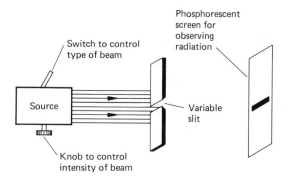

FIGURE 37.1 Experimental setup described in Sec. 35.2.

begin with a large slit width and an intense beam of electrons. A clearly defined glowing stripe appears on the phosphorescent screen where the beam of electrons strikes it. We next switch to an intense beam of ultra-violet light. Again we see a well-defined bright stripe on the phosphorescent screen. So far we cannot distinguish between the electrons and the light. In order to do so we switch the source to electrons at very low intensity. Sure enough, we now see little flashes on the phosphorescent screen as individual electrons strike it. We keep track of exactly where each of the flashes occurs and find that they are randomly distributed over that area of the phosphorescent screen that had been occupied by the bright stripe. We now switch the source back to an intense beam of light and then make the slit very narrow. As expected, we see a single-slit diffraction pattern on the screen. The pattern consists of a series of alternating bright and dark bands. The total width of the pattern is much larger than the slit width. All this is just what we would have predicted from classical theory; the electrons behave as particles and the light behaves as waves. Now we open the slit wide and switch to a low-intensity light source. We see little localized flashes as "light particles" hit the phosphorescent screen. The light seems to be a wave in one case and particles in another. To further compare light with electrons, we now switch to a high-intensity electron source and narrow the slit. In this case we see a single slit diffraction pattern on the phosphorescent screen. The electron beam is behaving like waves. Both the electron beam and the light beam behaved like waves for small slit width and like particles for low intensity. What will happen if we make the slit width small *and* the intensity low? We switch to a low-intensity electron beam and a small slit width. On the phosphorescent screen we see flashes as individual electrons strike the surface. At first it seems that the electrons are behaving like particles, for they seem to be striking the screen in random positions. But when we keep careful account of exactly where the electrons hit, we notice that no electrons ever hit the screen where the electron diffraction pattern we obtained before was completely dark. In fact, if we average over a very long time, we find that the rate at which electrons hit any small area on the screen is proportional to the intensity of the single-slit diffraction pattern at that point. If we replace the phosphorescent screen with a piece of photographic film and expose the film for a long enough time so that very many electrons have hit it, each one producing a microscopic dot, then on developing the film we get an excellent photograph of the single-slit diffraction pattern. The electrons have devised a way of being particlelike and wavelike simultaneously. Our classical dichotomy of particles and fields has been outflanked. As expected, we get exactly the same phenomenon when we use a low-intensity light source and a narrow slit width (see Fig. 37.2). Nature does not consist of certain elements that are particles and other elements that are fields; instead each entity is both particle and field. The theory that quantitatively describes these particle-field entities is called quantum theory.

37.2 QUANTUM THEORY

There are two fundamental elements in quantum theory.

1. A wave equation, which may have different forms for different kinds of particles. Shortly we shall introduce the wave equation for electrons, called *Schrödinger's equation* after the man who first suggested it. The wave equation for photons is simply the ordinary wave equation (see Equation 31.20), which can be derived from Maxwell's equations.

Intensity	Slit width	Type of beam	Pattern on screen
High	Wide	Light	
High	Wide	Electrons	
Low	Wide	Light	
Low	Wide	Electrons	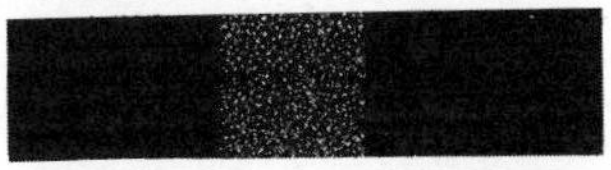
High	Narrow	Light	
High	Narrow	Electrons	
Low	Narrow	Light	
Low	Narrow	Electrons	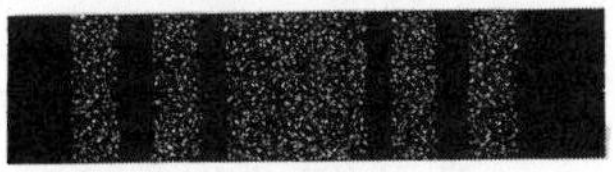

FIGURE 37.2   Summary of the results of the experiment described in Sec. 37.2.

2. We have a rule for interpreting the wave function in terms of the observed particle flux or the observed particle density. That is, if we have found the solution of the electron wave equation (the Schrödinger equation) that is appropriate for a particular physical setup, we can use the rule for predicting the electron flux at any point in space.

Before we develop the mathematical structure of the theory any further, we can help communicate the physical content of the quantum theory by describing first how a working physicist uses the theory to make calculations and second how the results of the calculations are interpreted in terms of experimental prediction.

In carrying out the calculations, we simply ignore the particle properties of the electrons or photons. Suppose we want to analyze the scattering of electrons of momentum p by stationary protons. We use the Schrödinger equation, which is appropriate for electron waves. We consider the case of an incoming plane electron wave with wave number given by the de Broglie relation $p = \hbar k$. By using the Schrödinger equation we can calculate the electron wave radiation given off by the proton (see Fig. 37.3). Just as we had a formula for the energy flux caused by a sound wave, a corresponding formula gives the electron flux caused by a traveling electron wave. We can use that formula to calculate the electron flux in any direction caused by the scattered electron wave (see Fig. 37.4). Once we have made that calculation, we are in a position to interpret the result in terms of experimental observation. The interpretation is as follows. If we place an electron detector at a particular position, then the rate at which electrons enter the detector will be given by the product of the area of the input port of the detector and the calculated electron flux at that

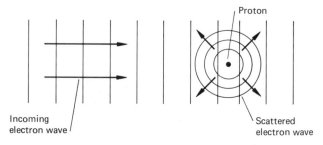

FIGURE 37.3 An incoming plane electron wave being scattered by a proton.

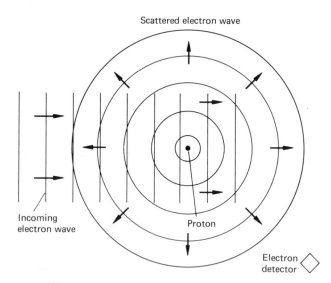

FIGURE 37.4 The average electron flux at any position can be computed using the Schrödinger equation.

point. The exact times at which electrons enter a given detector cannot be predicted by the quantum theory in any way. Only the average rate is calculable.

37.3 THE RELATIONSHIP BETWEEN QUANTUM AND CLASSICAL PHYSICS

Since the method of making and interpreting calculations in the quantum theory is so totally different from the method used in the classical theory, one is immediately faced with the question of why the classical theory ever gives the correct results, if in fact the quantum theory is the true description of nature. In preparation for answering that question, let us consider how one would analyze the scattering of electrons by a proton using the classical theory. One would consider a single electron of momentum p that had a definite position in the electron beam. (In the classical theory the electron beam is simply a stream of electrons, all of momentum p, but distributed at random positions within the beam.) For that particular electron one would use Newton's second law and Coulomb's law to calculate the trajectory and, in particular, to determine the direction of travel of the electron after it is scattered by the proton (see Fig. 37.5). One would carry out such a calculation for all possible initial positions of the electron

in the beam. Having done that, one would then assume that the electrons were randomly distributed throughout the beam and would calculate how many electrons per unit time would be scattered in a particular direction by the proton. That would allow one to predict the flux of electrons at the detector for any position of the detector. The result would agree precisely with the quantum theory calculation! Actually the case of electron-proton scattering is an exceptional case in that respect. Usually the quantum theory calculation of scattered electron flux agrees with the classical theory calculation only for large values of the electron's momentum, that is, for very small values of the electron wavelength. For small values of the momentum the two calculational methods give different predictions, and only the quantum theory agrees with the observations.

The wave aspects of electrons went unnoticed for so much longer than the wave aspects of photons because the wavelength of typical electrons is extremely small even in comparison with the very short wavelengths of visible photons. Consider the case of an electron that has been accelerated from rest by moving between two metal plates that have a potential difference of 100 V (Fig. 37.6). The energy of the electron when it strikes the positive plate is

$$E = eV = 1.6 \times 10^{-17} \text{ J}$$

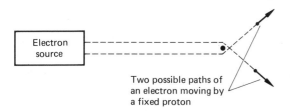

FIGURE 37.5 Scattering of electrons by a proton as pictured by the classical theory.

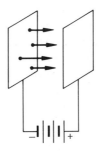

FIGURE 37.6 Electrons are accelerated in the electric field between the plates.

The momentum of the electron is related to its energy by $E = p^2/2m$, which gives

$$p = \sqrt{2mE} = \sqrt{(2)(9.1 \times 10^{-31} \text{ kg})(1.6 \times 10^{-17} \text{ J})}$$

$$= 5.4 \times 10^{-24} \text{ kg} \cdot \text{m/s}$$

The corresponding wavelength is

$$\lambda = \frac{2\pi\hbar}{p} = 1.2 \times 10^{-10} \text{ m} = 1.2 \text{ Å}$$

This distance is close to the diameter of a hydrogen atom. In comparison, the wavelength of yellow light is about 6000 Å. When electron waves are sent through devices such as single or double slits (which give rise to noticeable diffraction patterns for light waves), the wavelength is so much smaller than the slit width that it is extremely difficult to detect the diffraction pattern. However, one can clearly demonstrate interference effects involving electron waves by sending the electron beam through a crystal lattice, as is done in x-ray diffraction. The typical crystal lattice spacing is a few angstroms for simple crystals and is thus of a size comparable to the electron wavelength.

At this point we should note that a limit is imposed on the resolution of an electron microscope because of the wave aspects of electrons. For an electron microscope using a 15-kV potential difference to accelerate the electrons, the wavelength of the electrons is about 0.1 Å. This is so much smaller than the limits on resolvability resulting from lens imperfections that we can completely ignore the wave nature of the electrons in analyzing the electron microscope. That is, we can use the method of ray tracing, as we did in geometrical optics. The rays corresponding to the wavefronts of electron waves turn out to be exactly the trajectories of electrons calculated according to Newton's second law.

Having seen how Newtonian mechanics and quantum theory can make the same predictions in many situations, we shall turn our attention to the related question of how the quantum theory can agree with the field theory of electromagnetism. The two theories do not always agree. The quantum theory states that the electromagnetic field gives the photon flux at any point and that any charged particle near that point will absorb energy from the field by absorbing photons, each of energy $\hbar\omega$. The classical theory states that the electromagnetic field gives the force on a charged particle and that the charged particle absorbs energy from the field continuously as that force does work on the particle. The greatest disagreement occurs when the frequency of the electromagnetic wave is very large (so that $\hbar\omega$ is large) but the amplitude of the wave is very small. The classical theory states that if such an electromagnetic wave falls on a collection of charged particles, the particles will all absorb energy very slowly because of the small amplitude of the electric field. In contrast, the quantum theory states that whenever one of the charged particles absorbs a photon it will receive a large amount of energy because of the large value of $\hbar\omega$, but that such absorptions will be very infrequent, because the photon flux, which is proportional to the square of the amplitude, is very small. In a process such as photosynthesis, in which a large amount of energy is needed in a very small locality in order to drive a chemical reaction, the classical theory gives completely false predictions, while the quantum theory adequately explains the observed reaction rates. Although the two theories give completely different descriptions of the details of the energy absorption process, they agree in many cases on the average rate of energy absorption. Infrequent absorptions of large amounts of energy can give the same net absorption as continuous absorption of small amounts of energy. Thus the two theories would agree on how much heat would be generated by sunlight falling on a black surface.

The predictions of the two theories agree in detail concerning the motion of macroscopic charged bodies in slowly varying electric and magnetic fields. As an example, let us consider the effect of a passing electromagnetic wave of frequency 10^3 Hz on the motion of a small charged body of mass 10^{-3} g. If the body is initially at rest and absorbs one photon, its energy will be

$$E = 2\pi\hbar f = (6.6 \times 10^{-34} \text{ J})(10^3 \text{ Hz})$$

$$= 6.6 \times 10^{-31} \text{ J}$$

The velocity of the body is then

$$v = \sqrt{2E/m} = \sqrt{1.3 \times 10^{-30} \text{ J}/10^{-6} \text{ kg}}$$

$$= 1.14 \times 10^{-12} \text{ m/s}$$

Since this velocity is obviously undetectable, we would not notice the absorption of individual photons but would detect only the resultant effect of very large numbers of photons. It can be shown that, if we average over large numbers of photons, then the resultant momentum imparted to the body will agree with the predictions of the classical theory.

Before going deeper into the physical interpretation of the theory, we must introduce more of its mathematical structure. In particular, we shall introduce the fundamental equation satisfied by the electron-field wave function.

37.4 THE SCHRÖDINGER EQUATION

The electron wave effects we have discussed are roughly comparable to what was known to Erwin Schrödinger at the time he began his major work on quantum theory a half century ago. Electron wave interference effects had been experimentally verified. (The electron wave interference effects were created in crystal scattering experiments in which an electron beam was substituted for the usual x-ray beam.) The de Broglie relations had been verified for electrons and photons.

The question Schrödinger asked was: "What is the wave equation satisfied by the electron wave function?" We shall present an imaginary reconstruction of the train of thought that led Schrödinger to postulate the equation that bears his name. What is presented here should not be interpreted as a description of historical reality. The arguments that were actually presented by Schrödinger in support of his wave equation involve advanced mathematical formulations of mechanics that cannot be meaningfully repeated here.

We shall assume that the wave number and angular frequency of the plane-wave solutions of the wave equation are related to the observed momentum and energy by the de Broglie relations

$$p = \hbar k \tag{37.1}$$

and

$$E = \hbar \omega \tag{37.2}$$

The values of p and E must satisfy the well-established relation between momentum and energy for a free particle of mass m, namely

$$E = \frac{p^2}{2m}$$

Therefore we need a wave equation whose plane-wave solutions have

$$\omega = \frac{\hbar k^2}{2m} \tag{37.3}$$

Let us consider a uniform plane wave of unit amplitude ($A = 1$) traveling in the positive x direction. It is natural to assume that this wave has a wave function of the form

$$\psi(x, t) = \cos(kx - \omega t) \tag{37.4}$$

The wave equations for string waves, sound waves, and electromagnetic waves are all of the form

$$\frac{\partial^2 \psi}{\partial t^2} = \beta \frac{\partial^2 \psi}{\partial x^2} \tag{37.5}$$

where the constant β depends on the type of wave being considered. We shall presently see that we will have to modify both the plane-wave solution given by Eq. 37.4 and the wave equation given by Eq. 37.5 in order to obtain a wave equation whose solutions satisfy Eq. 37.3. Let us substitute our assumed solution (Eq. 37.4) into the wave equation; that is, let us assume that $\psi(x, t) = \cos(kx - \omega t)$. Then

$$\frac{\partial^2 \psi}{\partial t^2} = -\omega^2 \cos(kx - \omega t)$$

and

$$\frac{\partial^2 \psi}{\partial x^2} = -k^2 \cos(kx - \omega t)$$

Using these expressions in Eq. 37.5, we get

$$\omega^2 \cos(kx - \omega t) = \beta k^2 \cos(kx - \omega t)$$

which gives the following relation between ω and k:

$$\omega^2 = \beta k^2$$

This relation is obviously different from Eq. 37.3. The simplest way to obtain a first power of ω rather than a second power is to take only a first time derivative rather than a second. Therefore let us try a wave equation of the form

$$\frac{\partial \psi}{\partial t} = \beta \frac{\partial^2 \psi}{\partial x^2} \tag{37.6}$$

If we now try a solution of the form $\psi = \cos(kx - \omega t)$ in our new wave equation, we get

$$\omega \sin(kx - \omega t) = \beta k^2 \cos(kx - \omega t) \tag{37.7}$$

The constant terms, ω and βk^2, come out fine, but the first derivative has converted the cosine into a sine. The two sides of Eq. 37.7 cannot possibly be equal for all values of x and t. If we are going to maintain our new wave equation, we must modify our plane-wave solution. We note that the function $\sin(kx - \omega t)$ also describes a plane wave traveling in the positive x direction. Let us try to find a solution of our new equation that is a combination of cosine and sine terms. We try

$$\psi(x, t) = \cos(kx - \omega t) + \alpha \sin(kx - \omega t)$$

where α is as yet undetermined. The terms on the two sides of Eq. 37.6 will then be

$$\frac{\partial \psi}{\partial t} = \omega \sin(kx - \omega t) - \alpha\omega \cos(kx - \omega t)$$

and

$$\beta \frac{\partial^2 \psi}{\partial x^2} = -\beta k^2 \cos(kx - \omega t) - \beta\alpha k^2 \sin(kx - \omega t)$$

Equation 37.6 would therefore be satisfied if

$$\omega = \alpha\beta k^2 \tag{37.8}$$

and

$$\alpha\omega = \beta k^2$$

Dividing the first equation by the second, we get

$$\alpha^2 = -1$$

or

$$\alpha = \sqrt{-1} = i$$

This means that our plane-wave solution is of the form

$$\psi(x, t) = \cos(kx - \omega t) + i \sin(kx - \omega t)$$

The combination $\cos \theta + i \sin \theta$ occurs frequently in the study of complex numbers (see Appendix A.4), and is usually written as $e^{i\theta}$. Thus our plane-wave solution is

$$\psi(x, t) = e^{i(kx - \omega t)}$$

If we use $\alpha = i$ in Eq. 37.8, we get

$$\omega = -i\beta k^2$$

which is identical to the desired relation between ω and k (Eq. 37.3) if we choose

$$\beta = \frac{i\hbar}{2m}$$

We are thus forced to accept an imaginary number in our basic wave equation! Putting this value of β into Eq. 37.6 and multiplying the equation through by $i\hbar$, we get the form of the Schrödinger equation that is commonly used:

$$i\hbar \frac{\partial \psi}{\partial t} = -\frac{\hbar^2}{2m} \frac{\partial^2 \psi}{\partial x^2} \tag{37.9}$$

37.5 THE PHYSICAL MEANING OF WAVE FUNCTION

We are now in the situation that Schrödinger was in half a century ago. We have a wave equation but we do not know the physical meaning of the solutions of the equation. If we solve the equation, we shall end up with a complex electron-wave function $\psi(x, t)$. But what does an "electron wave function" mean? We noted before, in the description of the single-slit diffraction of electrons and light, that the detectable electron flux exhibited a pattern similar to the energy flux in a classical sound wave or light wave. For those classical waves the energy flux is proportional to the square of the wave function. In order to obtain agreement between the solutions of the Schrödinger equation and the observed electron flux, we must assume that the particle flux associated with a plane wave of wave vector k is given by

$$\text{electron flux} = \frac{\hbar k}{m} \psi^* \psi \tag{37.10}$$

where ψ^* is the complex conjugate of the complex number $\psi(x, t)$.

When Eq. 37.10 is combined with the de Broglie relation $p = \hbar k$, we are naturally led to an even simpler physical interpretation of the electron wave function ψ. The quantity

$$v = \frac{\hbar k}{m} = \frac{p}{m}$$

is the classical velocity of the electrons in the beam that is described by the plane-wave function. If the flux of electrons in the beam is $v\psi^*\psi$, then the density of electrons (i.e., the number of electrons per unit volume) must be given by

$$\text{electron density} = \psi^*\psi$$

We shall present the physical interpretation of the electron-wave function in one more equivalent form. Consider the case of an electron beam of very low intensity—of such low intensity that we see a flash on the phosphorescent screen on an average of once every few minutes. It would sound silly to talk about the electron density in such a beam; this would be something like saying that the number of pet cobras per person is 10^{-6}.

What we really mean in the latter example is that the probability that a person chosen at random will own a pet cobra is 10^{-6}. Similarly, what we really mean by the electron density in a beam of very low intensity is that, if we choose a volume ΔV in the beam and we simultaneously search everywhere within ΔV for an electron, then the probability of finding one in ΔV is equal to ΔV times the electron density. Thus we can say that the

probability of finding an electron in $\Delta V = \psi^*\psi\,\Delta V$ (37.11)

The advantages of this formula are twofold:

1. We can also use it for high-intensity beams simply by making ΔV extremely small.

2. We can also use it for other types of solutions of the Schrödinger equation. In particular, we can use it for standing-wave solutions of the equation for which the electron flux interpretation of the wave function has no meaning.

The factor $\psi^*\psi$ in Eq. 37.11 is called the probability density or, less precisely, the electron density.

QUESTION

Is there really any practical way of simultaneously searching all parts of a small volume for an electron?

ANSWER

Yes. By shooting a laser light pulse through the volume, one could detect any electron in it with near certainty. The photon density is so high in a laser pulse that any electron in the path of the pulse is sure to scatter a large number of photons that could then be detected.

QUESTION

Is it possible to measure the electron field ψ directly rather than detecting its existence by means of interactions with the associated electrons?

ANSWER

No. It is a basic tenet of the quantum theory that the only way that one can physically interact with the electron field is by interacting with whole electrons. For instance, if light is passed through a region in which an electron field exists, then either a number of photons are scattered or absorbed by electrons or the light is completely unaffected.

QUESTION

Since it is impossible to measure the electron field except by interactions with electrons, why not formulate the theory in terms of the motion of electrons and drop any reference to the seemingly fictitious electron field?

ANSWER

During the past 50 years no one has been able to formulate any theory (that agrees with experiment) in terms of a deterministic motion of electrons rather than an electron field with a probabilistic interpretation. But feel free to try yourself. In order to appreciate the difficulties you would face in formulating a theory without an electron field, consider the following experimental effect (see Fig. 37.7).

A low-intensity electron beam is incident upon a metal sheet in which there are two slits. At a distance from the slits a small electron detector is located at a position at which the Schrödinger equation predicts complete destructive interference. No electrons are ever detected in the detector no matter how long we wait. We now close one of the slits by means of a little door. Immediately we begin to detect electrons. Any deter-

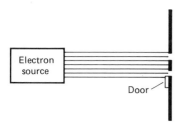

Door

Detector

FIGURE 37.7   Experimental setup in which a low-intensity electron beam is incident upon a metal sheet in which there are two slits.

ministic electron theory is going to have to explain how one gets more electrons to the detector by closing off one of the paths between the source and the detector.

37.6 STANDING-WAVE STATES AND DISCRETE ENERGY VALUES

In addition to the plane-wave solutions we have been considering, solutions for standing waves can be obtained from the Schrödinger equation. The standing-wave solutions are applicable to situations in which the electron-field oscillations are confined to a finite region of space. The simplest of those situations is a one-dimensional electron system confined to a bounded region of space. This is an unrealistic example in that real electron systems are three dimensional, but it shows all the same properties as the real system, and the mathematics is much easier. We look for wave functions $\psi(x, t)$ that are solutions of the Schrödinger equation in the presence of electron-repelling "walls" at $x = 0$ and $x = L$. Because of the walls, no electrons will be found outside the interval of length L. Since the electron density is zero outside that interval, the wave function must be zero there. Thus we must look for a

solution of the equation that is zero at $x = 0$ and $x = L$. This is reminiscent of the situation we had for a string of length L with both ends fixed. In that case the solutions of the wave equation were of the form (see Eq. 31.9)

$$u(x, t) = A \sin kx \sin \omega t$$

As it stands, this function does not satisfy the Schrödinger equation. It must be modified in the same way that the plane-wave solution was modified. In problem 37.B.3 you will be asked to verify that the following function is a solution of the Schrödinger equation:

$$\psi = A \sin(kx)e^{-i\omega t}$$

where

$$\omega = \frac{\hbar k^2}{2m}$$

Since the sine of zero is zero, the above solution is zero, as it should be, at $x = 0$. The solution must also be zero at $x = L$. This will be the case only if $kL = n\pi$, where n is any integer. Thus for each integer n there is a standing-wave solution of the form

$$\psi = A \sin(k_n x)e^{-i\omega_n t}$$

with

$$k_n = \frac{n\pi}{L} \quad \text{and} \quad \omega_n = \frac{\hbar k_n^2}{2m}$$

Using the fact that $(e^{-i\omega_n t})^*(e^{-i\omega_n t}) = 1$ (see the last equation in Appendix A.4), we can easily evaluate the probability density for these standing-wave solutions:

$$\psi^*\psi = A^2 \sin^2 k_n x$$

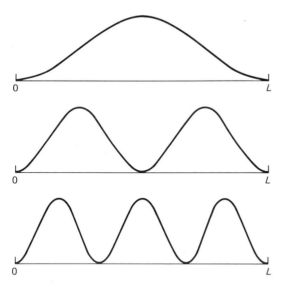

FIGURE 37.8   Plots of the probability density for the standing-wave solutions representing electrons confined to a box of length L.

The probability density for the cases $n = 1, 2,$ and 3 is plotted in Fig. 37.8. Although the wave function for the standing-wave state varies periodically with t because of the factor $e^{-i\omega_n t}$, the probability density remains constant. For that reason such solutions are called *stationary states*. They are also called *quantum states*. In general, any solution of the Schrödinger equation of the form $\psi(x, t) = u(x)e^{-i\omega t}$ gives a stationary state, as we can see by computing the probability density $\psi^*\psi$. As we stated before, the angular frequency of a quantum state is related to the observed energy of the system by the de Broglie relation. Thus the only possible energies of a particle in a one-dimensional box of length L are the discrete energies

$$E_n = \hbar\omega_n = \frac{\hbar^2 k_n^2}{2m}$$

Using $k_n = n\pi/L$, we get

$$E_n = \frac{n^2\pi^2\hbar^2}{2mL^2} \tag{37.12}$$

No solution to the wave equation has a frequency other than one of the normal mode frequencies, and thus no solution has a corresponding energy other than one of the normal mode energies. This effect occurs in any system in which the electrons remain localized in space. The energy of the electrons is related to the frequency of the electron wave. The only solutions for which the electron field (and thus the electron density) remains localized are standing-wave solutions that have discrete frequencies and hence discrete energies. For a given system, the set of allowed energy values $E_1, E_2, E_3, \ldots$ is called the *energy spectrum* of that system. The relationship between the energy spectrum of a system such as a hydrogen atom and the frequencies of electromagnetic radiation given off or absorbed by such a system was given in Sec. 36.6.

Let us briefly review the elements of the quantum theory presented so far.

1. There is an equation (the Schrödinger equation) for the electron field wave function. The equation has both traveling-wave solutions and standing-wave solutions.

2. The traveling-wave solutions are associated with electron beams or fluxes. The average rate at which electrons cross a unit area is given by a formula that is similar to the formulas that give the energy flux in string or sound waves. The experimentally observable energy and momentum of the electrons are related to the frequency and wavelength of the electron waves by the de Broglie relations.

3. The standing-wave solutions are associated with bound systems of electrons such as an electron in a closed box or the electron bound to the nucleus of a hydrogen atom. The square of the magnitude of the wave function at a particular location gives the probability density for finding an electron near that location. The energy of the electron is related to the frequency of the standing wave by the de Broglie relations.

QUESTION

As we shall see later, the hydrogen atom is a somewhat fuzzy entity according to the quantum theory. Its size is therefore not well defined. A reasonable estimate of the outer diameter of a hydrogen atom is about 1.5 Å. In its lowest energy state, the kinetic energy of the electron of a hydrogen atom is about 2.2×10^{-18} J. How well does this compare with the energy of an electron trapped inside a one-dimensional box of length $L = 1.5$ Å?

ANSWER

The lowest energy state of an electron in a one-dimensional box of length $L = 1.5 \times 10^{-10}$ m is given by Eq. 37.12, with $n = 1$:

$$E = \frac{(\pi \hbar / L)^2}{2m}$$

The value of $\hbar$ is about 1×10^{-34} J · s and, for an electron, $m = 9.1 \times 10^{-31}$ kg. These values give

$$E = 2.4 \times 10^{-18} \text{ J}$$

Just knowing the size of the region in which an electron is confined allows us to estimate its kinetic energy.

QUESTION

If a particle that is confined to a finite region of space can have only certain discrete energies, why was this discreteness of the energy not noticed in macroscopic phenomena long ago?

ANSWER

For macroscopic systems the spacing of the allowed energies is so close that the set of possible energies is effectively con-tinuous. For example, the allowed energies for a 1-g particle in a 1-cm box are

$$E_n = \frac{(\pi \hbar n / L)^2}{2m}$$

$$= \frac{n^2 [(3.14 \times 10^{-34})/10^{-2}]^2}{2 \times 10^{-3}}$$

$$\simeq (5 \times 10^{-61}) n^2 \text{ J}$$

In this form the equation is difficult to interpret in terms of classical ideas. Let us set E_n equal to $\frac{1}{2} m v_n^2$ and see what the change in velocity would be in going from the nth allowed energy to the $(n + 1)$st. If we set

$$E_n = \tfrac{1}{2} m v_n^2$$

then

$$v_n = \frac{\pi \hbar n}{mL}$$

and

$$v_{n+1} - v_n = \frac{\pi \hbar}{mL} = 3.14 \times 10^{-29} \text{ m/s}$$

Such minute changes in velocity could not be detected.

37.7 THE EMISSION AND ABSORPTION SPECTRUM

The discreteness of the possible energy levels of a bound electron system has a profound effect on the interaction of the system with the electromagnetic field. A description of the interaction between the electron field and the elec-tromagnetic field requires the following three elements:

1. The electron is a charged particle, and electric charge is exactly conserved (i.e., it never changes) during changes in the electron field. Thus if we search for electrons and

we find one electron in a closed system, we shall never find two electrons or no electron at a later time, unless there has been some electron flux into or out of the system. (That would mean, of course, that the system had not really been closed between the measurements.) Thus if a system with one electron is initially in a certain quantum state, it can change to another quantum state as a result of inter-action with the electromagnetic field but it cannot change to a state in which the electron wave function is zero, because that would mean that there was no longer any electron in the system.

2. The photon is not a charged particle. Nothing prevents the electromagnetic field from changing from a zero-photon state to a one-photon state.

3. The total energy is exactly conserved during any change in the electron and electromagnetic fields. Thus if the electron system changes from a quantum state of energy E_n to another quantum state of energy E_ℓ, where E_n is greater than E_ℓ, then the electromagnetic field must pick up the extra energy, usually in the form of a single photon of energy $E_n - E_\ell$. The frequency of the photon emitted during this transition is given by the de Broglie relation

$$\hbar \omega_{n\ell} = E_n - E_\ell \tag{37.13}$$

The frequencies of electromagnetic radiation given off by a bound system thus give slightly indirect information about the energy levels of the system. By measuring the electromagnetic radiation frequencies, we can determine the energy differences between pairs of levels. Since we usually do not know exactly which pair of quantum states is involved in a particular transition, measuring the radiation frequencies of a given system leaves us with the mathematical problem of finding a set of numbers E_1, E_2, . . . whose differences will give the observed fre-

quencies. The set of frequencies of electromagnetic radiation given off by any system is called the *emission spectrum* of that system. It is clear from Eq. 37.13 that the emission spectrum of a system could be completely unchanged if we added the same constant to all the energy levels. The value of one of the E_n's must be known (or in some cases chosen arbitrarily) and the emission spectrum used to determine the other levels with respect to the known one.

A typical experimental apparatus for measuring an atomic emission spectrum is shown in Fig. 37.9. Such a device is called a *spectrometer*. If electromagnetic radi-ation is passing through a region of space containing an electron system, then, because of the interaction between the electron field and the electromagnetic field, it is pos-sible for the electron field to shift from one quantum state to another. A photon, supplying the required difference in energy, is absorbed from the electromagnetic field in the process. Since the energy of the absorbed photon must equal the difference in energy of two levels in the energy spectrum, it is clear that any frequencies in the *absorption spectrum* will also be in the emission spectrum.

Absorption spectrum measurements are particularly important in astrophysics. When analyzed by a spectrom-eter, light coming from the sun or distant stars shows dark lines caused by absorption of particular frequencies by atoms in the cooler surface layers of the star. By analyzing the absorption lines, one can determine both the chemical composition and the temperature of the outer layers.

In the spectrum of light from distant stars and galaxies, dark lines are also created by absorption as the light passed through interstellar gas or dust clouds. The study

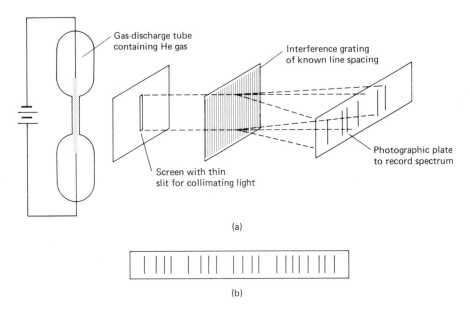

(a)

(b)

FIGURE 37.9   (a) A simplified diagram of the experimental setup for recording the helium emission spectrum. (b) The spectrum obtained with such an apparatus.

of such absorption lines is the only way available to obtain information on certain important components of the universe.

QUESTION

Is there any way, without going through the complete quantum theory of radiation, to get an idea of why an electron in an energy eigenstate does not radiate even though the wave function oscillates with angular frequency ω_n, and why, when the electron makes a transition, the angular frequency of the radiation is $\omega_{nm} = \omega_n - \omega_m$? In other words, is there any way of identifying anything that oscillates at that frequency and can be considered as the source of the electromagnetic radiation?

ANSWER

We have identified $\psi^*\psi \, \Delta V$ as the probability of finding an electron in the volume element ΔV. If we are a little bit careless about our definitions, we can interpret $\rho(x, t) = -e\psi^*(x, t) \psi(x, t)$ as the electric charge density associated with the electron density $\psi^*\psi$. If the electron is in an energy eigenstate, then $\psi(x, t)$ is of the form

$$\psi(x, t) = \psi_n(x, t) = e^{-\omega_n t} u_n(x)$$

where $u_n(x)$ is a real function. This gives an electric charge density of

$$\rho(x, t) = -e u_n^2(x)$$

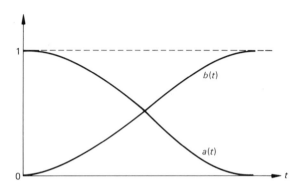

FIGURE 37.10   The functions $a(t)$ and $b(t)$.

which is constant in time and would therefore not be expected to create any electromagnetic radiation.

To answer the second part of the question, let us consider a wave function that describes an electron making a slow transition from the quantum state $\psi_n(x, t)$ to the quantum state $\psi_m(x, t)$. Such a wave function might be written in the form

$$\psi(x, t) = a(t)\psi_n(x, t) + b(t)\psi_m(x, t)$$

where a and b are real functions that gradually change from $a = 1, b = 0$ to $a = 0, b = 1$ (see Figure 37.10).

The electric charge density associated with this state is

$$\rho(x, t) = \psi^*\psi = -e[a^2\psi_n^*\psi_n + b^2\psi_m^*\psi_m + ab(\psi_n^*\psi_m + \psi_m^*\psi_n)]$$

If we now use the facts that $\psi_n = e^{-\omega_n t}u_n(x)$ and

$\psi_m = e^{-\omega_m t}u_m(x)$, we obtain

$$\rho(x, t) = -e[a^2u_n^2(x) + b^2u_m^2(x) + 2abu_nu_m \cos(\omega_n - \omega_m)t]$$

By assumption, $a(t)$ and $b(t)$ vary slowly in time. Therefore the major time dependence is in the term containing the factor $\cos(\omega_n - \omega_n)t$, which represents a charge density that is oscillating at the angular frequency ω_{nm} and would thus be expected to produce radiation of that frequency.

37.8 NORMALIZATION OF THE WAVE FUNCTION

The ground-state wave function for an electron in a one-dimensional box of length L was shown to be

$$\psi(x, t) = A \sin(kx)e^{-i\omega t}$$

where $k = \pi/L$ and $\omega = \hbar k^2/2m$. We can make the standing wave of any amplitude A without affecting the wavelength or the frequency. How are we to choose the appropriate amplitude? Because of the fact that the electron field is not directly measurable, there does not seem to be any physical quantity that corresponds to the amplitude. As we shall now see, the probabilistic interpretation of the wave function is enough to determine the amplitude. Suppose we know by prior measurement that there is one electron in the box. The probability that the electron would be found in an interval of small length dx centered at x is

$$(A^2 \sin^2 kx) \cdot dx$$

If we search for the electron within the finite interval $a \le x \le b$, the probability of detecting it there is

$$P_{ab} = \int_a^b A^2 \sin^2 kx \, dx$$

If we choose $a = 0$ and $b = L$, then the interval becomes the complete box within which we know the electron will be found somewhere. Thus for that case the probability P_{ab} must be equal to 1:

$$1 = \int_0^L A^2 \sin^2 kx \, dx$$

This equation determines the amplitude A (see problem 37.C.4):

$$A = \sqrt{\frac{2}{L}} \tag{37.14}$$

In general, the wave function $\psi(x, t)$ describing any system that contains just one electron must satisfy a *probability normalization condition:*

$$\int_a^b \psi^*\psi \, dx = 1 \qquad (37.15)$$

where the interval $a \leq x \leq b$ is the space available to the system. If the system has no definite bounds, then $a = -\infty$ and $b = +\infty$.

37.9 THE HEISENBERG UNCERTAINTY RELATION

When we try to apply the normalization condition (Eq. 37.15) to the wave function of an electron of momentum p, we encounter a paradox. Such a state of the electron field is described by a plane wave

$$\psi(x, t) = Ae^{i(kx-\omega t)}$$

There are no bounds on the system, and therefore the normalization condition is

$$\int_{-\infty}^{\infty} \psi^*\psi \, dx = 1$$

But $(e^{i(kx-\omega t)})^*(e^{i(kx-\omega t)}) = 1$, so

$$\int_{-\infty}^{\infty} \psi^*\psi \, dx = A^2 \int_{-\infty}^{\infty} 1 \cdot dx = A^2 \cdot \infty$$

An exact plane-wave state is spread out uniformly over an infinite interval and cannot be normalized. An exact plane-wave state is simply not a physically realizable state. A realizable state would have a wave function similar to the one depicted in Fig. 37.11. Such a wave function is spread out over a large but finite region of space. At any

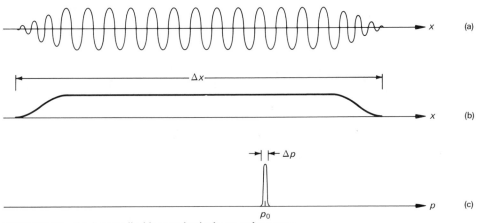

FIGURE 37.11   (a) A normalizable state that is almost a plane wave. The wave function is actually complex, so only the real part of it is plotted here. (b) The position probability density $\psi * \psi$ associated with the wave function shown in part (a). (c) The momentum probability density associated with the wave function shown in part (a).

instant of time there is some interval of length Δx within which the electron would almost surely be detected if it were searched for. We shall call Δx the *uncertainty in position* of the electron. An electron whose wave function was an exact plane wave of wave number k_0 would, if its momentum were measured, always be found to have momentum $p_0 = \hbar k_0$ exactly. In other words, the probability of finding the electron with momentum $p = p_0$ would be one and the probability of finding the electron with any momentum p other than p_0 would be zero. If the wave function is not an exact plane wave, this is no longer true. The probability distribution associated with the momentum is just like that associated with the position of the electron. As shown in Fig. 37.11, it is very narrow for a state that is almost a plane wave with large position uncertainty. A formula for computing the momentum probability density from the wave function exists. We shall not present it, but shall merely state without proof a very important theorem called the *Heisenberg uncertainty relation:*

For any wave function whatsoever the spread in the position probability density Δx and the spread in the momentum probability density Δp are related by the inequality

$$\Delta x \, \Delta p \geq \hbar$$

No state in quantum theory corresponds to the classical idea of a particle with a definite position x and a definite momentum p. This is true for electrons, photons, and all other particles.

For particles in three dimensions (i.e., real particles) the uncertainty principle relates the uncertainty in a given coordinate x, y, or z to the uncertainty in the corresponding component of momentum p_x, p_y, or p_z; that is,

$$\Delta x \, \Delta p_x \geq \hbar$$

$$\Delta y \, \Delta p_y \geq \hbar \tag{37.16}$$

$$\Delta z \, \Delta p_z \geq \hbar$$

We have approached the uncertainty principle from a mathematical point of view. To fully appreciate its meaning, we must now consider what happens if we experimentally attempt to thwart the principle by constructing a device that will produce a beam of photons whose y component of position and y component of momentum are more well defined than Eq. 37.16 says is possible. We begin with a device that produces an electromagnetic plane wave that is extended over a large region and has very well-defined wave fronts that propagate in the x direction. The photons associated with such a beam would have $p_y \simeq 0$ with a very small uncertainty (see Fig. 37.12). The photon beam is, however, extensively spread out in the y direction. The y coordinate uncertainty is correspondingly large. Since Δp_y is already acceptably small, we attempt to reduce Δy by placing in the path of the beam a screen that will only allow photons whose y coordinate lies in a narrow range to pass through. We shall then have produced a beam with simultaneously small values of Δp_y and Δy. What actually happens is that the electromagnetic wave, in passing through the narrow slit, is diffracted into a range of directions $\theta \simeq \lambda/\Delta y$ (see Eq. 33.6 and Fig. 37.13).

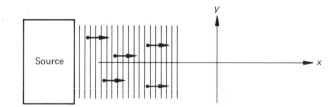

FIGURE 37.12   A source of photons of very small momentum uncertainty.

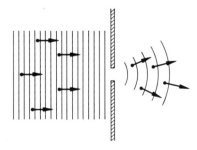

FIGURE 37.13 In passing through the narrow slit the electromagnetic wave is diffracted, and this causes the photons on the right to have a wider range of p_y's.

The associated photons will no longer be found to have very small values of p_y. In fact, if p is the magnitude of the momentum, then a typical value of p_y would be

$$p_y = p \sin \theta \simeq p\theta = \hbar k \frac{\lambda}{\Delta y}$$

$$= \frac{\hbar 2\pi}{\lambda} \frac{\lambda}{\Delta y} = 2\pi \frac{\hbar}{\Delta y} \qquad (37.17)$$

The value of p in Eq. 37.17 can be taken as an estimate of Δp_y for the beam after it passes through the slit. We see that

$$\Delta y \, \Delta p_y \simeq 2\pi\hbar > \hbar$$

Our attempt to reduce Δy has caused a corresponding increase in Δp_y. Our failure to violate the uncertainty principle could have been predicted from the outset. Since there is a mathematical theorem that states that no wave function exists that has position and momentum uncertainties whose product is smaller than $\hbar$, it is obvious that no device is going to produce an electromagnetic beam with a wave function that has such a property.

QUESTION

If there is no state in which a particle has definite momentum and position, then how can classical mechanics, which assumes that particles have definite momentum and position, ever give correct predictions?

ANSWER

The minimum uncertainty allowable by the uncertainty principle is too small to be observable in the motion of macroscopic objects. As an example, let us suppose that we have an object of mass 1 g moving with a velocity of 100 m/s. Suppose that the velocity is known to within one part in a trillion. The uncertainty in v is then

$$\Delta v = 10^{-10} \text{ m/s}$$

This means that

$$\Delta p = m \, \Delta v = 10^{-13} \text{ kg} \cdot \text{m/s}$$

With that uncertainty in momentum, the uncertainty principle states that the uncertainty in the position of the object cannot be less than

$$\Delta x = \frac{\hbar}{\Delta p} = \frac{10^{-34}}{10^{-13}} = 10^{-21} \text{ m}$$

This distance is about one hundred billionth the diameter of a hydrogen atom. The uncertainty principle is not a significant constraint on the accuracy of a classical position determination.

37.10 THE CLASSICAL LIMIT OF QUANTUM MECHANICS

A one-particle state in which the particle has fairly well-defined momentum and fairly well-defined position is described by a wave function of the form shown in Fig. 37.11. Such a wave function is referred to as a *wave packet*. As we saw in the preceding question, for a particle of 1-g mass, it is possible to have a quite well-defined momentum and yet have a wave packet of very small spatial extent. According to the quantum theory, this little wave packet moves through space in accordance with the Schrödinger equation. The particle will always be found within the wave packet. If we can show that the wave packet moves in such a way as to satisfy the laws of Newtonian mechanics, then we shall have shown that Newtonian mechanics is a logical consequence of the Schrödinger equation. In Sec. 37.11 we shall show that the Schrödinger equation leads to Newtonian mechanics in the limit in which the wavelength is very small in comparison to the wave-packet size. It must also be assumed that the classical potential energy function $U(x)$ is almost constant over a distance as small as the wave packet (Fig. 37.14).

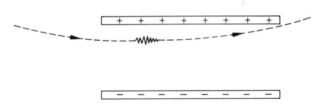

FIGURE 37.14   The motion of a small electron wave packet through space satisfies the laws of classical mechanics.

37.11 THE SCHRÖDINGER EQUATION WITH A POTENTIAL

The form of the Schrödinger equation we have been using was designed to ensure that the de Broglie relations led to the classical energy equation

$$E = \frac{p^2}{2m}$$

This classical equation is valid only if no external forces are applied to the particle. If the particle is subject to a conservative force, then the above energy equation must be replaced by the equation

$$E = \frac{p^2}{2m} + U(x) \tag{37.18}$$

where $U(x)$ is the potential function associated with the force.

Let us now consider what the appropriate wave equation for electrons subject to a force is. The force is assumed to be a conservative one with an associated potential function $U(x)$. Newtonian mechanics can be derived from the Schrödinger equation under appropriate conditions, but one cannot logically derive the Schrödinger equation from the laws of Newtonian mechanics. In spite of the fact that the more general theory (quantum theory) cannot be derived from the less general theory (Newtonian mechanics), it is still true that the demand that quantum theory should reduce to Newtonian mechanics in the appropriate limit so constrains our freedom to choose a wave equation for the electron field that there seems to be only one simple modification of Eq. 37.9 (the Schrödinger equation for electrons without forces) that will lead to Eq. 37.18 in the classical limit. Let us consider a case in which classical Newtonian mechanics should certainly be valid. We consider a particle whose mass m is of the order of 1 g moving in a potential $U(x)$ that varies appreciably over distances of the order of 1 m. The wave function for the particle is some moving wave packet. As we have seen above, the value of $\hbar$ is so small that we may assume that within the wave packet the wave function differs very little from a pure plane wave and at the same time assume that the total length of the wave packet is so small that the potential is essentially constant over the packet. Then Eq. 37.18 can be replaced by the equation

$$E = \frac{p^2}{2m} + U_0$$

where U_0 is a constant (over the wave packet). The value of the "constant" will actually vary gradually as the wave packet moves from one place to another, but at any one time the variation of $U(x)$ over the packet will be completely negligible. If we assume that the de Broglie relations are still valid, we must search for an equation that, for a constant potential, has a plane wave solution

$$\psi(x, t) = Ae^{i(kx - \omega t)} \qquad (37.19)$$

with

$$\hbar\omega = \frac{\hbar^2 k^2}{2m} + U_0 \qquad (37.20)$$

When applied to a plane wave solution such as Eq. 37.19, the terms we now have in the Schrödinger equation give

$$i\hbar\frac{\partial\psi(x, t)}{\partial t} = \hbar\omega\psi(x, t)$$

and

$$-\frac{\hbar^2}{2m}\frac{\partial^2\psi(x, t)}{\partial x^2} = \frac{\hbar^2 k^2}{2m}\psi(x, t)$$

It is clear that the simplest possible term we might add to the Schrödinger equation in order to obtain the relation given in Eq. 37.20 rather than the relation $\hbar\omega = \hbar^2 k^2/2m$ is a term of the form $U_0\psi(x, t)$. That is, if we assume that the equation is of the form

$$i\hbar\frac{\partial\psi}{\partial t} = -\frac{\hbar^2}{2m}\frac{\partial^2\psi}{\partial x^2} + U_0\psi$$

then it has plane-wave solutions that satisfy Eq. 37.20. In order for Eq. 37.20 to be satisfied wherever the wave packet happens to be, we must have U_0 be appropriate to that location. This can be done by replacing U_0 by $U(x)$. Thus, if we assume that the fundamental equation of quantum theory is

$$i\hbar\frac{\partial\psi(x, t)}{\partial t} = -\frac{\hbar^2}{2m}\frac{\partial^2\psi(x, t)}{\partial x^2} + U(x)\psi(x, t) \qquad (37.21)$$

we are assured of obtaining Newtonian mechanics in those situations in which we know that Newtonian mechanics is valid. Equation 37.21 was first proposed by E. Schrödinger. That the Schrödinger equation is also valid in those cases in which one cannot use Newtonian mechanics can only be confirmed by finding solutions to the equation in those cases and comparing the predictions of the theory with experimental observations. During the past 50 years this has been done in great detail for an extremely wide variety of systems with uniformly positive results. The quantum theory may, sometime in the future. become a limiting case of a more general physical theory just as Newtonian mechanics is the low-velocity limit of relativistic mechanics and the short-wavelength limit of quantum mechanics. However, within the wide realm of atomic, molecular, and solid-state phenomena, the quantum theory has been so thoroughly confirmed that its essential correctness is beyond reasonable doubt.

37.12 QUANTUM STATES OF A PARTICLE IN A POTENTIAL

The possible energy values of a particle of mass m in a one-dimensional potential $U(x)$ are $\hbar\omega_n$, where the numbers ω_1, ω_2, . . . are the angular frequencies of the standing-wave solutions of the Schrödinger equation 37.21. A standing wave has a wave function of the form

$$\psi(x, t) = u(x)e^{-i\omega t}$$

The function $u(x)$ describes the wave pattern of the normal mode. The function ψ oscillates with the angular frequency ω. If we put this form of solution into Eq. 37.21 and use the fact that

$$i\hbar \frac{\partial(ue^{-i\omega t})}{\partial t} = \hbar\omega u e^{-i\omega t} = Eue^{-i\omega t}$$

we see that $u(x)$ must be a solution of the equation

$$-\frac{\hbar^2}{2m}\frac{d^2u(x)}{dx^2} + U(x)u(x) = Eu(x) \tag{37.22}$$

The physical interpretation of the solution is that $\int_a^b \psi^*(x, t)\psi(x, t)\, dx$ is the probability of finding the particle within the interval $a \leq x \leq b$ at time t. Because of the special form of the standing-wave solution, that probability can be written more simply in terms of $u(x)$:

$$\begin{matrix}\text{probability of finding} \\ \text{the particle within the} \\ \text{interval } a \leq x \leq b\end{matrix} = \int_a^b |u(x)|^2\, dx$$

This interpretation of the normal mode wave function makes sense only if $u(x)$ satisfies the normalization condition

$$\int_{-\infty}^{\infty} |u(x)|^2\, dx = 1 \tag{37.23}$$

Thus the mathematical problem that one must solve in order to find the quantum states of a particle in a potential $U(x)$ is that one must find solutions to Eq. 37.22 that satisfy the normalization condition (Eq. 37.23). Usually

such solutions exist only if E has certain particular values $E_1, E_2, \ldots$. For all other values of E it is possible to find solutions of Eq. 37.22, but the solutions do not satisfy the normalization condition. For each value of E in the energy spectrum, $E_1, E_2, \ldots$, there is a corresponding solution of Eqs. 37.22 and 37.23, $u_1(x)$, $u_2(x), \ldots$. Thus the *Schrödinger energy equation* is

$$-\frac{\hbar^2}{2m}\frac{d^2u_n(x)}{dx^2} + U(x)u_n(x) = E_n u_n(x)$$

37.13 THE HARMONIC OSCILLATOR

We shall not consider in any detail the difficult mathematical problems encountered in finding solutions of the Schrödinger energy equation (Eq. 37.22). As an example of the results one obtains from such an analysis and their physical consequences, we shall describe qualitatively some of the solutions of the equation for the case of a harmonic oscillator potential. Recall that the potential energy function for a harmonic oscillator is of the form (see Eq. 5.31)

$$U(x) = \tfrac{1}{2}kx^2 \tag{37.24}$$

The quantum theory of the harmonic oscillator should not be viewed as being relevant to a macroscopic mass attached to a spring. Although the theory is technically applicable to such a system, the distinctly quantum mechanical aspects of the solutions, such as the discrete energy levels, would be completely undetectable for a system of macroscopic size. Rather, the theory is appropriate and leads to observable consequences for the many microscopic systems in which a particle of atomic dimensions is bound to other particles by forces that can be well approximated by harmonic oscillator forces over a finite range of distances. An excellent example of such a system is a diatomic molecule. In such a molecule the two atoms would ordinarily be found close to their equilibrium distance from one another. The equilibrium distance is that distance at which the potential energy resulting from the interaction between the atoms is a minimum. If the separation between the atoms is made larger or smaller than the equilibrium distance, a restoring force will be encountered which, over a distance that is different for different diatomic molecules, can be approximated by a "Hooke's law" force. Such a force leads to a potential energy function of the form given in Eq. 37.24. The Schrödinger energy equation (also called the *time-independent Schrödinger equation*) for a particle of mass m in a potential $U(x) = \frac{1}{2}kx^2$ is

$$-\frac{\hbar^2}{2m}\frac{d^2u(x)}{dx^2} + \frac{1}{2}kx^2u(x) = Eu(x) \qquad (37.25)$$

By mathematical methods that are too complex to be given here, this equation can be shown to have a solution that satisfies the normalization condition only if E has one of the values $E_0, E_1, E_2, \ldots$, where E_n is given by

$$E_n = (n + \tfrac{1}{2})\hbar\sqrt{k/m}$$

Note the appearance of the combination $\sqrt{k/m}$ in the energy spectrum of the harmonic oscillator. That is just the angular frequency of vibration predicted by the classical theory of the oscillator. We can therefore write the formula for the energy levels of the quantum oscillator in the form

$$E_n = (n + \tfrac{1}{2})\hbar\omega_0$$

where $\omega_0 = \sqrt{k/m}$ is the classical angular frequency of vibration. The spacing between adjacent energy levels is constant and equal to $\hbar\omega_0$ (Fig. 37.15). This fact has the interesting consequence that if such a quantum oscillator is

FIGURE 37.15 The energy spectrum of the quantum harmonic oscillator.

initially in a quantum state with energy E_n and makes a transition to the next lower energy state, emitting a photon with the excess energy in the process, then the frequency of the emitted photon would be given by

$$\hbar\omega_{\text{photon}} = E_n - E_{n-1} = \hbar\omega_0$$

or

$$\omega_{\text{photon}} = \omega_0$$

Thus a quantum harmonic oscillator would emit electromagnetic radiation whose frequency was equal to the expected classical frequency of vibration. Because of this fact many classical theories of the electromagnetic properties of solids that assumed that the particles were subject to Hooke's law forces were at least partially successful. However, their degree of success was quite limited, and they can hardly be compared with the very

Wave functions

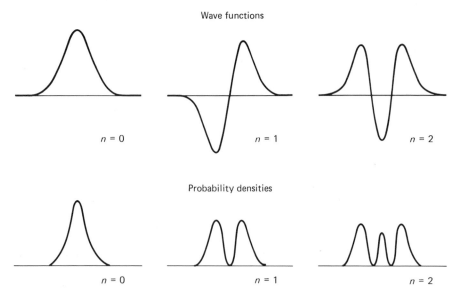

Probability densities

$n = 0$ $n = 1$ $n = 2$

FIGURE 37.16   Wave functions and probability densities for the first three quantum states of a particle in a harmonic oscillator potential.

detailed and accurate calculations presently made with the quantum theory of solids.

Associated with each energy value E_n is a wave function $u_n(x)$. The first three wave functions are illustrated in Fig. 37.16. Their mathematical forms are given in problem 37.C.3.

QUESTION

The plane-wave solutions of the Schrödinger equation for a free particle are $\psi = Ae^{i(kx-\omega t)}$, where $\omega = \hbar k^2/2m$. This wave could be written as

$$\psi = Ae^{ik(x-ut)} = f(x - ut)$$

where the wave velocity u is given by $u = \omega/k$. But $\omega/k = \hbar k/2m = p/2m$. This is not the correct velocity for free particles. Free particles should travel at a velocity $v = p/m$. Why does the wave velocity not come out equal to the velocity of the particles?

ANSWER

The wave function is not directly measurable, so there is no good reason to equate the wave velocity with the particle velocity. The directly measurable quantity, namely the particle density $\psi^*\psi$, is simply a constant, and thus one cannot tell by looking at it how fast the particles in the beam are traveling. If we take the ratio of the particle flux to the particle density (see Eqs. 37.10 and 37.11), we get the expected particle velocity $v = p/m$. But this is a very unconvincing argument, since we used the particle velocity in deriving Eq. 37.11. What we need to do to get a believable formula for the particle velocity from the Schrödinger equation is to choose a solution for which the electron density is not so smooth. If the electron density is lumpy, then we can watch how the lumps of electron density move and we can say with reasonable confidence that the only way a lump of electrons can move with velocity v is for the electrons in the lump to move with that velocity. This is not really as difficult as it sounds. If we add two plane waves of slightly different momenta, we obtain a lumpy electron density for which we can easily calculate the velocity of the lumps. We take one plane wave with wave number $k + \Delta k$ and angular frequency $\omega + \Delta\omega$ and another with wave number $k - \Delta k$ and angular frequency $\omega - \Delta\omega$, where Δk and $\Delta\omega$ are very small. In order for both wave functions to satisfy the Schrödinger equation, we need to have

$$\omega + \Delta\omega = \frac{\hbar(k + \Delta k)^2}{2m} \simeq \frac{\hbar k^2}{2m} + \frac{\hbar k\,\Delta k}{m}$$

and

$$\omega - \Delta\omega = \frac{\hbar(k - \Delta k)^2}{2m} \simeq \frac{\hbar k^2}{2m} - \frac{\hbar k\,\Delta k}{m}$$

Subtracting the bottom equation from the top, we get

$$\Delta\omega = \frac{\hbar k\,\Delta k}{m},$$

a fact we shall use later. If we assume that each plane wave has an amplitude $\frac{1}{2}A$, then the wave function for the addition of the two plane waves is

$$\psi = \tfrac{1}{2}Ae^{i(kx+\Delta kx-\omega t-\Delta\omega t)} + \tfrac{1}{2}Ae^{i(kx-\Delta kx-\omega t+\Delta\omega t)}$$

$$= \tfrac{1}{2}Ae^{i(kx-\omega t)}(e^{i(\Delta kx-\Delta\omega t)} + e^{-i(\Delta kx-\Delta\omega t)})$$

$$= Ae^{i(kx-\omega t)}\cos(\Delta kx - \Delta\omega t)$$

Looking at the electron density, we get

$$\psi^*\psi = A^2\cos^2(\Delta kx - \Delta\omega t)$$

The measurable electron density now comes in a series of clumps that move to the right with a velocity given by

$$v = \frac{\Delta\omega}{\Delta k} = \frac{\hbar k}{m} = \frac{p}{m} \tag{37.26}$$

When we make a series of wave packets by adding more than one plane wave, we find that the wave packets move with the classically expected velocity.

This analysis supplies the missing element in our discussion of the classical limit of quantum mechanics. We showed that the energy was related to the momentum in the proper way $[E = p^2/2m + U(x)]$. We did not show that the wave packet with momentum p actually moved at the classical velocity $v = p/m$. We can now do so simply by noting that Eq. 37.20,

$$\hbar\omega = \frac{\hbar^2 k^2}{2m} + U_0$$

also leads to the relation $\Delta\omega = \hbar k\,\Delta k/m$. Therefore the rest of our analysis can be used without change for the case of a particle in a potential, and thus Eq. 37.26, which states that $p = mv$, will be valid for a particle moving in a slowly varying potential.

SUMMARY

The classical picture of nature, in which certain things such as electrons and protons are purely particles while other things such as electromagnetism and gravitation are purely fields, does not agree with experiment. Instead there are electron particles associated with an electron field, proton particles associated with a proton field; photon particles associated with an electromagnetic field, etc.

A wave equation is satisfied by each one of the fields. For traveling waves a particle flux is associated with the waves.

The wavelength and frequency of the field are related to the momentum and energy of the associated particles by the de Broglie relations

$$p = \hbar \frac{2\pi}{\lambda} = \hbar k \quad \text{and} \quad E = \hbar \frac{2\pi}{T} = \hbar \omega$$

For electrons the known relation between energy and momentum is $E = p^2/2m$, which leads, via the de Broglie relations, to a relation between ω and k:

$$\omega = \frac{\hbar k^2}{2m}$$

The simplest wave equation that has solutions satisfying this relation is the Schrödinger equation

$$i\hbar \frac{\partial \psi}{\partial t} = -\frac{\hbar^2}{2m} \frac{\partial^2 \psi}{\partial x^2}$$

It has complex plane-wave solutions,

$$\psi = A e^{i(kx - \omega t)}$$

where $\omega = \hbar k^2/2m$

The physical interpretation of the wave function is that the positive real number $\psi^*(x, t)\psi(x, t) \cdot \Delta x$ gives the probability for finding an electron at time t in the interval Δx.

The Schrödinger equation also has standing-wave solutions, which are associated with particles that remain localized in space. The standing-wave solutions have certain discrete possible frequencies and therefore the bounded system can have only certain discrete energies $E_1, E_2, \ldots$. For a particle in a one-dimensional box of length L, the quantum states are

$$\psi_n = A \sin k_n x e^{-i\omega_n t}$$

where $k_n = n\pi/L$ and $\omega_n = \hbar k_n^2/2m$. The allowed energy levels are $E_n = n^2 \pi^2 \hbar^2 / 2mL^2$.

When an electron system goes from a quantum state of energy E_n to one of energy E_m, where $E_n > E_m$, the excess energy is usually given off as a single photon of angular frequency

$$\omega_{nm} = \frac{E_n - E_m}{\hbar}$$

If the system initially has energy E_m, it can absorb a photon of angular frequency ω_{nm} and change to a quantum state of energy E_n.

Because of its interpretation in terms of probability density, the wave function of a one-electron system must satisfy the probability normalization condition

$$\int_a^b \psi^* \psi \, dx = 1$$

where the interval $a < x < b$ is the space available to the system.

The Heisenberg uncertainty principle shows that it is impossible to have an electron in a quantum state in which its uncertainty in position Δx and its uncertainty in momentum Δp_x do not satisfy the inequality

$$\Delta x \, \Delta p_x \geq \hbar$$

For systems of macroscopic size there exist solutions of the Schrödinger equation in the form of moving wave packets that are well localized in space. The motion of the packet satisfies the equations of classical dynamics.

For a particle in a potential $U(x)$ the correct form of the wave equation is

$$i\hbar \frac{\partial \psi}{\partial t} = -\frac{\hbar^2}{2m} \frac{\partial^2 \psi}{\partial x^2} + U(x)\psi$$

The stationary states (or quantum states) are solutions of the Schrödinger equation of the form

$$\psi(x, t) = u_n(x)e^{-i\omega_n t}$$

The function $u_n(x)$ is a solution of the Schrödinger energy equation

$$E_n u_n = -\frac{\hbar^2}{2m}\frac{d^2 u_n}{dx^2} + U(x)u_n$$

A typical problem in which the Schrödinger energy equation is used is a particle in a harmonic oscillator potential $U = \frac{1}{2}kx^2$. The energy spectrum of the harmonic oscillator has energy levels E_n, where

$$E_n = (n + \tfrac{1}{2})\hbar\sqrt{k/m} \qquad (n = 0, 1, 2, \ldots)$$

PROBLEMS

37.A.1 If a photographer took a photograph of a still life by using very-low-intensity light and leaving the shutter open for a very long time, which one of the statements below would be true? (Explain your reasoning.)

(a) The film will remain unexposed because the photons of very-low-intensity light do not have sufficient energy to carry out the chemical reaction in the film.

(b) The image obtained will be simply a random scatter of dots where individual photons hit the film.

(c) The image obtained will be similar to the picture that would be obtained with normal intensity, but will be much more "grainy" because of the particlelike properties of low-intensity light.

(d) The picture obtained will be the same as the picture the would be obtained with normal-intensity light and a short shutter-opening time.

37.A.2 The figure shows a light source, a screen with two holes that can be separately opened or closed, and a photon detector at a large distance from the screen. When only the upper hole is open, the rate at which photons are detected is R_1. When only the lower hole is open, the rate is R_2. When both holes are open, the rate at which photons are detected is R_{12}. According to the quantum theory, which of the following is true? (Explain your reasoning.)

(a) $R_{12} = R_1 + R_2$

(b) R_{12} may be smaller than, the same as, or larger than $R_1 + R_2$, depending on the location of the detector.

(c) R_{12} is always less than $R_1 + R_2$ because of interference between the two holes.

(d) $R_{12} = (R_1 + R_2) \sin \theta$.

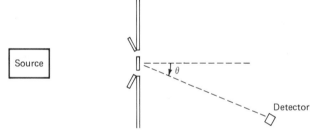

37.A.3 What are the wavelength and frequency associated with an electron that has been accelerated through a potential difference of 40,000 V?

37.A.4 Calculate the first two energy levels of an electron in a one-dimensional box of length 1 Å.

37.A.5 What are the first three energy levels of an electron in a one-dimensional harmonic oscillator potential $U(x) = \frac{1}{2}kx^2$ with $k = 8$ N/m?

37.A.6 What is the de Broglie wavelength of a 1-g ping-pong ball traveling at 3 m/s?

37.B.1 Find the value of c for which the wave function

$$u(x) = \frac{c}{\sqrt{x^2 + a^2}}$$

is normalized within the region $-\infty < x < \infty$.

37.B.2 A photon interacts with an electron that is initially at rest. The photon is scattered but with a new frequency. What is the energy given to the electron if the change in the photon frequency is 10^5 Hz?

37.B.3
(a) By taking the required partial derivatives, verify that the wave function

$$\psi(x, t) = A \sin(kx)e^{-i\omega t}$$

satisfies Eq. 35.9 if $\omega = \hbar k^2/2m$.
(b) Show that $\psi(L, t) = 0$ only if k has one of the values $k_n = n\pi/L$.

37.B.4 Show that one obtains a standing-wave solution of Eq. 37.9 that is equivalent to the one given in problem 37.B.3 by adding two plane-wave solutions of opposite momenta, p and $-p$. The solution obtained is not identical to that given in problem 37.B.3. Why is the difference of no physical importance?

37.B.5 The diameter of a hydrogen atom in its ground state is about 1.5 Å.
(a) Calculate the minimum uncertainty in Δp_x of an electron whose position uncertainty Δx is 1.5 Å.
(b) Calculate the kinetic energy of an electron with momentum equal to the value of Δp_x you obtained in part (a).

37.B.6 A certain system has only six lines in its emission spectrum. Their angular frequencies are 0.5×10^{16} Hz, 1.2×10^{16} Hz, 1.3×10^{16} Hz, 1.8×10^{16} Hz, 2.5×10^{16} Hz, and 3.0×10^{16} Hz. Taking the ground state as the zero of your energy scale, determine the energy spectrum of the system.

37.C.1 Consider a classical particle of mass m in a one-dimensional harmonic oscillator potential $U(x) = \frac{1}{2}kx^2$
(a) Assume that the particle has an energy $E = \frac{1}{2}\hbar\sqrt{k/m}$ and calculate its amplitude of vibration A.
(b) Calculate the ground-state energy of a quantum mechanical particle in a one-dimensional box of length $2A$. (The length $2A$ is the total distance that the classical particle in part (a) traveled.)

37.C.2 An incoming photon of energy $\hbar\omega$ is scattered directly backward by an electron (see the figure). The photon leaves with energy $\hbar\omega'$. The electron leaves with a velocity of 10^6 m/s. What are the values of ω and ω'?

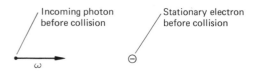

37.C.3 The first two quantum states of the harmonic oscillator have wave functions

$$u_0(x) = e^{-\alpha x^2} \quad \text{and} \quad u_1(x) = xe^{-\alpha x^2}$$

Show that these wave functions satisfy the Schrödinger energy equation only if $\alpha = \sqrt{km}/2\hbar$, $E_0 = \frac{1}{2}\hbar\sqrt{k/m}$, and $E_1 = \frac{3}{2}\hbar\sqrt{k/m}$.

37.C.4 By evaluating the integral $\int_0^L \sin^2 kx\, dx$, where $k = n\pi/L$, show that the amplitude of the normalized wave function must be $\sqrt{2/L}$, as stated in Eq. 37.14.

37.C.5 Consider a quantum system of a particle of mass m confined to a one-dimensional box of length L. The system is in its ground state. Assume that the right wall at $x = L$ is movable. If it is moved rapidly, the particle may make a transition to a different quantum state. However, it can be shown that if the wall is moved very slowly, then the system will always remain in its ground state. Suppose that the length of the box is decreased very slowly from L to $L - \Delta L$. The energy of the particle will then increase from $(\pi\hbar/L)^2/2m$ to $[\pi\hbar/(L - \Delta L)]^2/2m$. The increased energy of the particle must come from the work done by the outside force needed to move the wall.

(a) Use this fact to determine the force exerted by the particle on the wall.

(b) Compare the answer you obtained in part (a) with the average force that a classical particle of the same energy, bouncing back and forth between the two walls, would exert on the wall.

38
QUANTUM THEORY OF ATOMIC STRUCTURE

ONE OF THE most satisfactory aspects of the quantum theory has been its ability to explain, without additional assumptions, the physical and chemical properties of the elements. Beginning with nothing more than the Schrödinger equation and the Coulomb potential (to describe the force between electrically charged particles), one can predict the layout of the periodic table of the elements, including the lengths of the periods and the valences of the elements. One can also predict the geometrical structure of chemical compounds: the hexagonal benzene ring, the tetrahedral methane molecule, the double spiral of DNA. The mathematical problems encountered in trying to apply the quantum theory to systems as complex as atoms and molecules are truly formidable. We shall only be able to sketch the barest essentials, and even in doing that we shall present only an approximate version of the true quantum theory of many particle systems.

38.1 THE SCHRÖDINGER EQUATION IN TWO AND THREE DIMENSIONS

In discussing the Schrödinger equation for a one-dimensional particle with a potential function $U(x)$, we began with the classical energy equation

$$E = \frac{p^2}{2m} + U(x)$$

and saw that the demand that quantum theory agree with classical theory in those circumstances in which classical theory is known to be valid suggested that the appropriate wave equation was

$$i\hbar \frac{\partial \psi(x,\,t)}{\partial t} = -\frac{\hbar^2}{2m} \frac{\partial^2 \psi(x,\,t)}{\partial x^2} + U(x)\psi(x,\,t) \tag{38.1}$$

If we want to mimic as closely as possible the argument we used but apply it to a particle in a two-dimensional space (with coordinates x and y), we must begin with the two-dimensional form of the energy equation, namely

$$E = \frac{1}{2m}(p_x^2 + p_y^2) + U(x, y)$$

where p_x and p_y are the two components of the momentum vector, and $U(x, y)$ is the potential energy of the particle at location (x, y). The wave function for such a system would be of the form $\psi(x, y, t)$. An argument, similar in every respect to that given in Sec. 37.11, then leads to the two-dimensional wave equation:

$$i\hbar\frac{\partial\psi}{\partial t} = -\frac{\hbar^2}{2m}\left(\frac{\partial^2\psi}{\partial x^2} + \frac{\partial^2\psi}{\partial y^2}\right) + U(x, y)\psi \qquad (38.2)$$

This strongly (and correctly) suggests that for a three-dimensional system we need only add one more term of the form

$$-\frac{\hbar^2}{2m}\frac{\partial^2\psi}{\partial z^2}$$

to the right-hand side of Eq. 38.2. As before, we obtain the Schrödinger energy equation for the quantum states of a particle in two dimensions by assuming that the wave function varies harmonically in time with an angular frequency ω that is related to the energy E by the de Broglie relation. The two-dimensional form of the Schrödinger energy is

$$-\frac{\hbar^2}{2m}\left(\frac{\partial^2 u}{\partial x^2} + \frac{\partial^2 u}{\partial y^2}\right) + U(x, y)u = Eu \qquad (38.3)$$

where

$$\psi(x, y, t) = u(x, y)e^{-i\omega t}$$

and

$$E = \hbar\omega$$

38.2 TRANSLATIONAL AND ROTATIONAL INVARIANCE IN CLASSICAL AND QUANTUM THEORY

In a two-dimensional system, if the potential energy function U, which we ordinarily expect to be a function of both coordinates, is a function of only one coordinate, then we say that the system has a *translation invariance*. For instance, if $U(x, y)$ is actually $U(x)$, then the potential is invariant with respect to a translation of the particle in the y direction. That is, $U(x, y + a) = U(x, y)$ for any displacement a. According to Newtonian mechanics, if a particle moves in a potential that has such an invariance, then the y component of momentum of the particle remains constant. To prove this we need only recall the equation that gives the rate of change of p_y in terms of the y component of the force:

$$\frac{dp_y}{dt} = F_y = -\frac{\partial U}{\partial y} = 0$$

where the last step is a consequence of our assumption that U is actually a function of x only. Thus for a one-particle classical system with translation invariance in the y direction, the y component of the linear momentum is constant. We shall now see that for a one-particle quantum system with such an invariance, we can make certain definite statements about the energy quantum states of the system. For such a system the Schrödinger energy equation would be

$$Eu(x, y) = -\frac{\hbar^2}{2m}\left(\frac{\partial^2 u(x, y)}{\partial x^2} + \frac{\partial^2 u(x, y)}{\partial y^2}\right)$$

$$+ U(x)u(x, y) \tag{38.4}$$

Suppose $u_0(x)$ is a solution of the one-dimensional Schrödinger equation with an energy value E_0. That is, suppose

$$E_0 u_0(x) = -\frac{\hbar^2}{2m}\frac{d^2 u_0(x)}{dx^2} + U(x)u_0(x)$$

We can now construct a solution of the two-dimensional equation by multiplying $u_0(x)$ by a plane wave of momentum $p_y = \hbar k$ in the y direction. This would give a function $u(x, y)$ of the form

$$u(x, y) = u_0(x)e^{iky}$$

This function satisfies Eq. 38.4 with the energy E given by

$$E = E_0 + \frac{p_y^2}{2m}$$

Thus a two-dimensional system with translational invariance in the y direction has energy quantum states with definite values of p_y. The dependence of the wave function on the variable y is of the form of a plane wave.

A more common type of invariance for two-dimensional and three-dimensional systems is *rotational invariance*. We can best describe this invariance by introducing polar coordinates in the plane (see Fig. 38.1). In terms of polar coordinates the potential energy of a particle on the plane is a function $U(r, \phi)$. Often the potential energy depends only on the distance of the particle from the origin. This would be the case, for instance, if the force associated with the potential energy were the Coulomb force created by a fixed charge Q at the origin of the coordinate system.

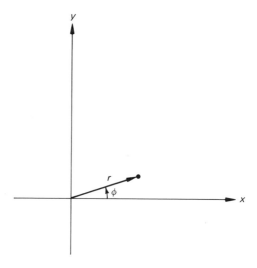

FIGURE 38.1 The two polar coordinates r and ϕ specify the location of a point in the plane.

In that case the potential energy of a charge q at location (r, ϕ) would be

$$U(r, \phi) = \frac{kQq}{r}$$

In general, if the potential energy is a function $U(r)$ of r only, we say that the system exhibits invariance with respect to rotations about the origin. For a classical system with such rotational invariance, one can immediately predict that the angular momentum of the particle, measured from the origin, remains constant during the motion of the particle. This is due to the fact that a potential $U(r)$ is associated with a force that points directly toward or away from the origin. Such a force applies no torque on the particle about the origin. If L is the angular momentum of the particle about the origin, then

$$\frac{dL}{dt} = \tau = 0$$

where τ is the torque about the origin. Thus L is a constant.

For a quantum system of a particle in a rotationally invariant potential, one can make a statement about the energy quantum states that is very similar to the statement made for the case of translational invariance. If the potential U does not depend on the angle ϕ, then the quantum states are of the form

$$u(r, \phi) = u_n(r)e^{im\phi} \tag{38.5}$$

where the constant m is associated with the angular momentum of the particle by the de Broglie relation

$$L = \hbar m \tag{38.6}$$

The constant m should not be confused with the mass of the particle, although traditionally the same symbol is used.

It is not difficult to see why a wave function of the form $e^{im\phi}$ should be associated with angular momentum about the origin. If we look at the wave function within a small area at a distance r from the origin (see Fig. 38.2), we see that within that region the wave function is approximately a plane wave with the direction of the electron flux pointing tangent to a circle about the origin. A circulating electron flux is associated with the wave function $e^{im\phi}$. It is that electron flux that gives rise to the angular momentum of the quantum state.

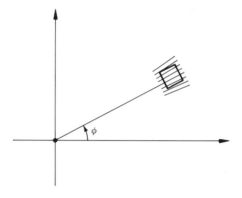

FIGURE 38.2   The lines along which the phase of the wave function $e^{im\phi}$ is constant are the lines $\phi = $ constant. Within any small area such as the one shown, the constant phase lines look like a set of parallel, equally spaced straight lines.

38.3 QUANTIZATION OF ANGULAR MOMENTUM

The fact that the angle ϕ ranges from 0 to 2π has an important effect on the possible values of the angular momentum of a quantum mechanical system. The point with coordinates $(r, \phi + 2\pi)$ is the same as the point with coordinates (r, ϕ). Thus one must demand that the wave function have the same value for $(r, \phi + 2\pi)$ that is has for (r, ϕ), that is,

$$u(r, \phi + 2\pi) = u(r, \phi)$$

Using the form of the wave function given in Eq. 38.5, we see that we must have

$$e^{im(\phi + 2\pi)} = e^{im\phi}$$

But

$$e^{im(\phi + 2\pi)} = e^{im\phi}e^{im2\pi}$$

Therefore we need to have

$$e^{i2\pi m} = 1$$

This relation is satisfied only if

$$m = 0, \pm 1, \pm 2, \ldots$$

By the de Broglie relation this implies that the angular momentum of the particle about the origin will always have one of the values

$$L = 0, \pm \hbar, \pm 2\hbar, \ldots$$

Thus the quantum states of a particle in a two-dimensional rotationally invariant potential can be labeled with an integer, m, which specifies their angular momentum in units of $\hbar$. For a given value of m there exist many different quantum states that have the angular momentum $m\hbar$. To uniquely specify the quantum state, we need another label, n, which distinguishes among the states with the same angular momentum. Thus the quantum states are labeled

$u_{nm}(r, \phi)$, where

$$u_{nm}(r, \phi) = u_n(r)e^{im\phi} \tag{38.7}$$

The quantum state $u_{nm}(r, \phi)$ has associated with it an angular frequency ω_{nm} and a discrete energy

$$E_{nm} = \hbar\omega_{nm}$$

The integer n that we have introduced simply as a label to distinguish the various quantum states with the same angular momentum actually has a simple interpretation in terms of the graph of the function $u_n(r)$, namely

$$n - 1 = \text{number of nodes in } u_n(r)$$

A node is a value of r at which the wave function $u_n(r)$ is equal to zero. In Fig. 38.3 functions $u_1(r)$, $u_2(r)$, and $u_3(r)$ are shown. They have zero, one, and two nodes, respectively.

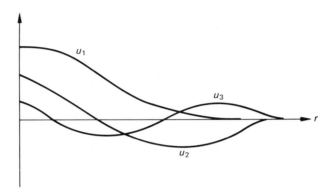

FIGURE 38.3 Graphs of the functions $u_1(r)$, $u_2(r)$, and $u_3(r)$, which have zero, one, and two nodes, respectively.

38.4 THREE-DIMENSIONAL QUANTUM STATES IN A SYMMETRICAL POTENTIAL

In three dimensions a rotationally invariant potential is a function $U(r)$ of the variable $r = \sqrt{x^2 + y^2 + z^2}$. In discussions of the wave functions of a particle in such a potential, it is best to introduce a coordinate system that includes the variable r as one of the coordinates. The coordinate system we shall use is shown in Fig. 38.4 and is called a *spherical coordinate system*. The simplest quantum states of a particle in a three-dimensional spherically symmetrical potential $U(r)$ are functions of the variable r only. For reasons that will soon be made clear, we shall call those quantum states $u_{n00}(r)$. (That is, u_{100}, u_{200}, etc.) They are states of zero angular momentum. As before, the integer n gives the number of nodes of the wave function and it can have the values 1, 2, ... For these

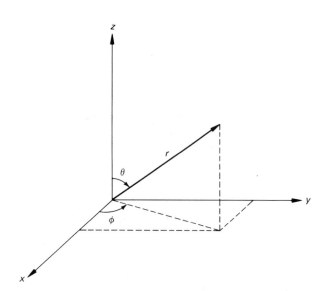

FIGURE 38.4 The location of a point in three-dimensional space is described by the three coordinates r, ϕ, and θ, where r is the distance of the point from the origin, the angle ϕ measures rotation about the z axis, and the angle θ is the angle between the radius vector and the z axis.

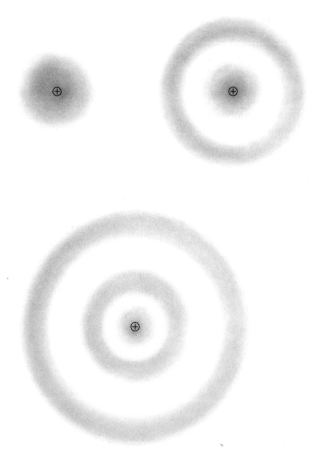

FIGURE 38.5 A picture of the electron probability density in a plane that includes the nucleus for the first three zero-angular-momentum states of a hydrogen atom. The third state is actually twice as large as shown here. The spherically symmetrical potential is $U(r) = -e^2 k/r$. The states shown have zero, one, and two radial nodes. The darkness of shading is proportional to $u_{n00}^2(r)$.

zero-angular-momentum states the electron probability density is spherically symmetrical, just as the potential is (see Fig. 38.5).

Other quantum states do not have zero angular momentum. The wave functions of those quantum states depend on the angular variables ϕ and θ as well as on the radial variable r. They are of the form

$$u_{n\ell m}(r, \phi, \theta) = u_n(r) f_\ell(\theta) e^{im\phi} \tag{38.8}$$

The three integers n, ℓ, and m are called the *quantum numbers* of the state. The quantum numbers n and m have the same meaning as before. The quantity $n - 1$ is the number of radial nodes in the function $u_n(r)$. The number m, which can be a positive or negative integer or zero, gives the angular momentum about the z axis (the axis around which ϕ is measured). If m is positive, then a right-handed electron flux around the z axis is associated with the state. We now have to give the physical interpretation of the new quantum number ℓ. Recall that the angular momentum of a system, according to classical mechanics, is a vector quantity with components L_x, L_y, and L_z. The quantum number m is associated with L_z. In fact,

$$L_z = m\hbar \tag{38.9}$$

The quantum number ℓ is associated not with another component of the angular momentum, but rather with the magnitude of the angular momentum vector. In particular, the square of the length of the angular momentum vector is given by

$$L_x^2 + L_y^2 + L_z^2 = \ell(\ell + 1)\hbar^2 \tag{38.10}$$

The total angular momentum quantum number ℓ must be a nonnegative integer:

$$\ell = 0, 1, 2, \ldots \tag{38.11}$$

The value of L_z^2 cannot be larger than the value of $L_x^2 + L_y^2 + L_z^2$. This means that the absolute value of m cannot be larger than ℓ. Thus there exists a quantum state with quantum numbers n, ℓ, m only if m has one of the values

$$m = -\ell, -\ell + 1, -\ell + 2, \ldots, \ell \tag{38.12}$$

QUESTION

It is possible to find quantum states for which L_x, L_y, and L_z all have definite values?

ANSWER

No. There is an uncertainty relation involving any two components of angular momentum that is similar to the uncertainty relation for linear momentum and position. The angular momentum uncertainty relation states that, except for the one case in which all components of angular momentum are zero, there exists no quantum state for which more than one component of momentum has a definite value. We can grasp the origin of this relation by noting that the quantum states that have a definite component of angular momentum about the z axis have a probability density that is symmetrical about the z axis (see Fig. 38.6). If we specify that the state be symmetrical about the z axis, we cannot demand that it be simultaneously symmetrical about the x axis or the y axis unless the state is completely rotationally symmetrical. The completely symmetrical states are just the zero-angular-momentum states. We can choose the orientation of the z axis arbitrarily and then find quantum states that are symmetrical about the

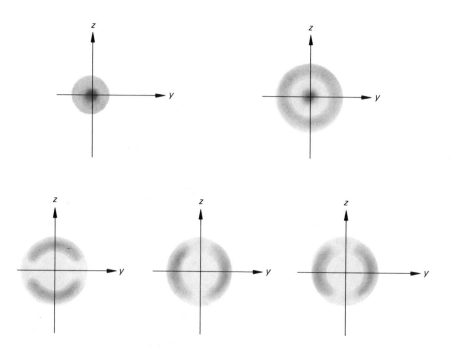

FIGURE 38.6 The probability density in the yz plane for the first five quantum states of the hydrogen atom. To obtain the probability density in three dimensions, one should rotate the pattern shown around the z axis. The first two states will then have a probability density that is completely symmetrical about the origin. The other three will have a probability density that is symmetrical about the z axis only.

axis we have chosen, but we cannot then demand that they also be symmetrical about some other axis.

38.5 THE HYDROGEN ATOM

The quantum theory of the hydrogen atom illustrates very well all the properties of a three-dimensional system with a spherically symmetrical potential. The potential energy of an electron at a distance r from a proton is

$$U(r) = -\frac{ke^2}{r}$$

The mass of the proton is about two thousand times that of the electron, so we can ignore the motion of the proton without great loss in accuracy. We treat the proton as being rigidly fixed in space and choose its location as the origin of our coordinate system. We can obtain the Schrödinger energy equation, whose solutions give the possible standing-wave patterns of the electron field around the proton, from Eq. 38.3 by adding one more term to the left-hand side. The equation is then

$$-\frac{\hbar^2}{2m}\left(\frac{\partial^2 u}{\partial x^2} + \frac{\partial^2 u}{\partial y^2} + \frac{\partial^2 u}{\partial z^2}\right) - \frac{ke^2}{r}u = Eu \qquad \text{(38.13)}$$

where $r = \sqrt{x^2 + y^2 + z^2}$ and $u = u(x, y, z)$. The quantity m is the mass of an electron.

Finding solutions of Eq. 38.13 is a rather complex matter that we shall not go into. However, the solutions, when written in terms of the spherical coordinates shown in Fig. 38.4, are fairly simple in form. We can write them most simply in terms of the *Bohr radius*,

$$a = \frac{\hbar^2}{mke^2}$$

The Bohr radius gives a measure of the radius of a hydrogen atom. The hydrogen atom has no well-defined size since the wave function is actually continuous and extends to infinite distance; however, if we choose as the radius of the hydrogen atom that distance from the proton at which the electron is most likely to be found, then for a hydrogen atom in its ground state the radius is equal to a. Written in terms of the spherical coordinates, r, ϕ, and θ, the Schrödinger equation has a solution $u_{n\ell m}(r, \phi, \theta)$ for each set of the quantum numbers (n, ℓ, m) for which

$$n = 1, 2, \ldots \infty$$

$$\ell = 0, 1, \ldots, n - 1$$

and

$$m = 0 \pm 1, \ldots, \pm\ell$$

The energy value associated with the quantum state $u_{n\ell m}$ is

$$E_n = -\frac{ke^2}{2an^2} \qquad \text{(38.14)}$$

The energy depends only on the quantum number n. Thus a number of different quantum states have the same energy. When this happens we say that the energy level is degenerate. The *degeneracy* of the energy level is defined as the number of quantum states with that energy. As n becomes larger and larger, the energy of the quantum state rises but always remains negative. The energy values become more and more closely spaced, approaching infinite density at zero energy (see Fig. 38.7). The first few wave functions are

$$u_{100} = Ae^{-(r/a)}$$

$$u_{200} = B\left(1 - \frac{r}{2a}\right)e^{-(r/2a)}$$

$$u_{210} = Cre^{-(r/2a)}\cos\theta \qquad \text{(38.15)}$$

$$u_{211} = Cre^{-(r/2a)}(\sin\theta)e^{i\phi}$$

$$u_{21-1} = Cre^{-(r/2a)}(\sin\theta)e^{-i\phi}$$

where A, B, and C are constants that are determined by the normalization condition and u_{100} and u_{200} are spherically symmetrical states with zero angular momentum. Further states in this series would be of the same general form, that is, a polynomial in r times an

FIGURE 38.7   The energy spectrum of the hydrogen atom.

exponential function of r. The angular momentum of the quantum states u_{210}, u_{211}, and u_{21-1} has magnitude

$$L = \sqrt{L_x^2 + L_y^2 + L_z^2} = \sqrt{\ell(\ell + 1)}\hbar = \sqrt{2}\hbar$$

The z components of angular momentum of these three states are 0, $+\hbar$, and $-\hbar$, respectively. The charge-density patterns associated with all five states are shown in Fig. 38.6.

38.6 THE ZEEMANN EFFECT

According to the physical interpretation we have given to the wave function, the state $u_{n\ell m}$ has associated with it an electron flux or current about the z axis if m is not zero. If m is positive, the flux about the z axis is "right-handed." (Put your right thumb in the direction of the z axis and your fingers will wrap around the axis in the direction of the flux.) If m is negative, the flux about the z axis is "left-handed." Since the electron has a negative electric charge, a persistent right-handed electron current about the z axis would constitute a left-handed electric current about that axis. This is just the current pattern one would need to produce a magnetic dipole. The electron flux associated with the quantum state acts essentially like a small current loop around the z axis. The current is in a direction opposite to the electron flux because the electron charge is negative. The electron circulation is responsible for both the angular momentum and the magnetic moment of the quantum state. It can be shown, in either the classical or the quantum theory, that the ratio of the magnetic moment of the magnetic dipole to the angular momentum is exactly one-half the ratio of the charge to the mass of the circulating particles. This is true of any system of orbiting particles as long as nonrelativistic mechanics can be used. If the circulation velocities are close to the velocity of light, the simple result is not valid.

To get some idea of the origin of the simple relation between angular momentum and magnetic dipole moment, we shall consider a simple classical system. The system is a collection of N similar charges, each of mass m and charge q, moving in a circle at velocity v around the origin in the xy plane (see Fig. 38.8). The angular momentum of the system is N times the angular momentum of one of the particles:

$$L = Nrmv \tag{38.16}$$

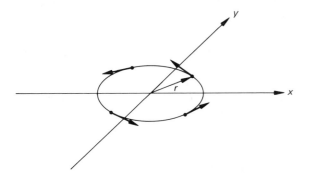

FIGURE 38.8 N charges moving in a circle give rise to an angular momentum L_z and a magnetic dipole.

The collection of moving particles constitutes a current loop whose area is

$$A = \pi r^2$$

The current in the loop is equal to the average rate at which charge passes any point in the loop. If T is the time it takes for one charge to complete a circular orbit, then during a time T an amount of charge Nq passes any point in the loop. The time T is given by

$$T = \frac{2\pi r}{v}$$

The current in the loop is

$$I = \frac{Nq}{T} = \frac{Nqv}{2\pi r}$$

The magnetic dipole moment of a loop current is equal to the product of the current and the area of the loop (see Sec. 27.4). Thus, if we call the magnetic dipole moment μ, we get

$$\mu = IA = \frac{Nqv}{2\pi r}\pi r^2$$

or

$$\mu = \tfrac{1}{2}Nqvr$$

Comparing this formula for μ with Eq. 38.16 for L, we see that

$$\mu = \frac{1}{2}\frac{q}{m}L \tag{38.17}$$

It is true, although not obvious, that the same relation holds between any component of the angular momentum and the corresponding component of the magnetic dipole moment for any quantum state of a charged particle. The above relation between magnetic moment and angular momentum determines the way in which an atom reacts to being placed in a magnetic field. A magnetic dipole that is placed in a uniform magnetic field of strength B has an energy that depends on its orientation with respect to the field (Fig. 38.9). In particular, if the field B is in the z direction, then the energy of the dipole is equal to (see Eq. 27.11)

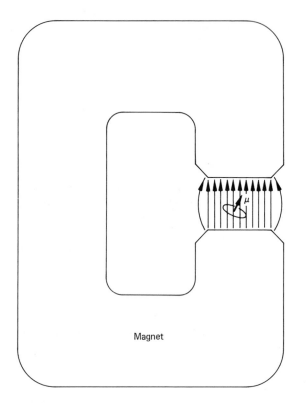

FIGURE 38.9   The energy of a magnetic dipole depends on its orientation with respect to the magnetic field.

$$E_{\text{dipole}} = B\mu_z \qquad (38.18)$$

where μ_z is the z component of the magnetic dipole vector. The magnetic dipole vector has a magnitude equal to the magnetic dipole moment and a direction perpendicular to the plane of the current loop. Using Eq. 38.17 (with $q = -e$), we can write the magnetic dipole energy of a hydrogen atom in a magnetic field in terms of the component of angular momentum in the direction of the field (which we take as the z direction):

$$E_{\text{dipole}} = \frac{1}{2}\frac{e}{m_e}BL_z \qquad (38.19)$$

Because of the interaction of the atom with the magnetic field, we should expect that the energy value for quantum states will depend on the quantum number m, since

$$L_z = m\hbar$$

The energy associated with the quantum state $u_{n\ell m}$ would no longer depend only on the quantum number n. It would be

$$E_{n\ell m} = E_n + \frac{1}{2}\frac{e}{m_e}B\hbar m \qquad (38.20)$$

The energy spectrum would therefore change in the manner shown in Fig. 38.10. This change in the spectrum is called the *normal Zeemann effect*.

38.7 ELECTRON SPIN

If a sample of hydrogen is actually placed in a magnetic field and its energy spectrum is determined, the result differs very much from the normal Zeemann effect shown in Fig. 38.10. With no field the energy levels are given by Eq. 38.14, as expected. As the field is turned on, the degenerate energy levels separate into a set of closely spaced levels called a *multiplet*. The separation is proportional to the magnitude of the magnetic field B, as we would expect from Eq. 38.20. The aspect of the spectrum that is very different from what we would expect is the

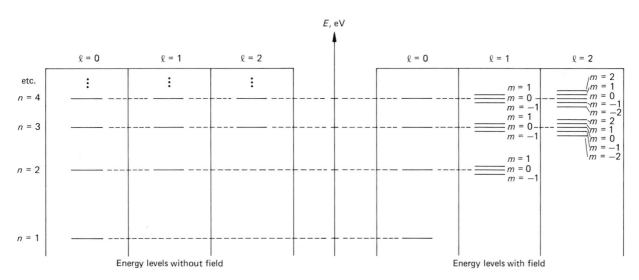

FIGURE 38.10   Energy level shifts in the normal Zeemann effect.

degeneracy of each level. There are twice as many levels in each multiplet as we expect from the known combinations of m and l. Even the $l = 0$ states separate into two levels, as if they had a magnetic moment that could have two different possible values.

There is another source of angular momentum in the hydrogen atom in addition to the angular momentum caused by the electron flux around the nucleus. That other source of angular momentum is the intrinsic *spin angular momentum* of the electron itself. One might picture the electron as spinning on its own axis so that the total angular momentum of the hydrogen atom is the vector sum of the spin angular momentum of the electron about an axis through its center plus the angular momentum of the electron caused by its motion about the proton. This essentially classical picture of the electron as a small charged spinning ball should only be considered an inadequate substi-

tute for an honest theory of electron spin angular momentum. A proper theory does exist, but it is rather complicated. It predicts all properties of electrons and photons with astonishing accuracy. The full theory is called *quantum electrodynamics.* To the present no experimental contradiction of the predictions of quantum electrodynamics has ever been detected, even though some experiments have been able to test predictions with a precision of one part in 10^7. This theory is a relativistic quantum theory of the electromagnetic and electron fields. It predicts that the particles associated with the electron field have an angular momentum whose magnitude s is always

$$s = \sqrt{\tfrac{3}{4}}\hbar = \sqrt{\tfrac{1}{2}(\tfrac{1}{2} + 1)}\,\hbar \qquad (38.21)$$

Equation 38.21 should be compared with Eq. 38.10, which gives the magnitude of the orbital angular momentum. The component of the electron spin angular momentum in any preassigned direction, which we shall call the z direction, can have either of the two values

$$s_z = \tfrac{1}{2}\hbar \quad \text{or} \quad s_z = -\tfrac{1}{2}\hbar$$

A magnetic dipole moment is associated with the electron spin, just as a magnetic moment is associated with the electron flux around the nucleus. However, the relativistic quantum theory predicts that the ratio of the electron spin magnetic moment to the electron spin angular momentum is twice the value given by Eq. 38.17. Thus

$$\mu_{z(\text{spin})} = -\frac{e}{m} s_z \qquad (36.22)$$

When the electron spin magnetic moment is taken into account, the observed energy spectrum of a hydrogen atom in a magnetic field is explained completely. Because of the existence of the electron spin, the quantum numbers n, l, and m are not adequate to completely specify the possible quantum states of an electron in a hydrogen atom. We must add one more quantum number, which we call σ, to define the z component of the spin angular momentum. The possible quantum states of the hydrogen atom are therefore labeled $u_{n\ell m\sigma}$, where σ can have the values of $+1$ or -1 and s_z is related to σ by

$$s_z = \sigma\frac{\hbar}{2}$$

In the absence of a magnetic field, the energy value associated with the quantum state does not depend on σ, and thus one does not detect the existence of the spin quantum number. When the hydrogen atom is placed in a magnetic field, the spin magnetic dipole moment interacts with the magnetic field and causes the otherwise degenerate energy levels to separate into the number of distinct levels observed.

38.8 THE EXCLUSION PRINCIPLE

We have restricted our discussion of atomic structure to the hydrogen atom because we must still introduce one more basic principle before we can present a quantum theory of many-electron atoms. That basic principle was first postulated by Wolfgang Pauli before the modern version of the quantum theory had been developed. It was later derived from the fundamental principles of the quantum theory. That derivation was also done by Pauli. This postulate, called the *Pauli exclusion principle,* states that:

Two electrons will never be found in the same quantum state.

An equivalent way of stating the exclusion principle is:

No two electrons in a many-electron system can have exactly the same set of quantum numbers.

38.9 THE PERIODIC TABLE OF THE ELEMENTS

With a combination of the exclusion principle and our previous analysis of the quantum states of an electron in a spherically symmetrical potential, we can explain, at least qualitatively, the striking relationship between the chemical characteristics of the elements and their atomic numbers. The *atomic number* of an element is simply the number of electrons in a neutral atom of that element.

The modern picture of one atom of an element of atomic number Z is the following. Most of the mass of the atom is concentrated in a *nucleus* of extremely small volume and high density. The densities of all atomic nuclei are about 10^{14} g/cm^3. This is about ten trillion times the density of lead. The nucleus is composed of two types of particles, neutrons and protons. The neutrons have no electric charge, but the protons each carry a positive electric charge of the same magnitude as the electronic charge. There are Z protons in the nucleus. Surrounding the nucleus is a distribution of Z electrons in Z different electronic quantum states. If the atom is in its lowest energy state (i.e., its ground state), then those electronic quantum states in which the Z electrons would be found are the states of lowest energy. The two lowest energy quantum states have no radial nodes ($n = 1$) and no angular momentum ($l = 0$). They differ from one another by the value of the z component of spin angular momentum ($\sigma = +1$ and $\sigma = -1$). In the terminology commonly used by chemists and atomic physicists, both these states are called 1s states. The number 1 refers to the principle quantum number n, and the letter s indicates the angular momentum quantum number l in a scheme in which further values of the angular momentum are assigned letter symbols according to the code given in Table 38.1. In the hydrogen atom a single 1s state is occupied by the one electron. In helium both 1s states are occupied. The *configuration* of an atomic ground state is a listing of the occupied states. The configuration of helium is written as 1s^2, where the superscript 2 indicates that both 1s states are occupied. The ground state of the lithium atom, which has three electrons, has a configuration 1$s^2$2s^1. The configurations of the first 18 elements are shown in Table 38.2.

TABLE 38.1 The letter symbols used for the first six angular momentum values in the atomic spectroscopist's notation. Further states are given by letters in simple alphabetical order after h. The first three entries in this seemingly arbitrary sequence of letters describe characteristics of the emission spectral lines associated with the states (s, sharp; p, principal; d, diffuse).

$l = 0 - s$
$l = 1 - p$
$l = 2 - d$
$l = 3 - f$
$l = 4 - g$
$l = 5 - h$

TABLE 38.2 The electron configuration of the elements in the first three periods of the periodic table

Name	Symbol	Atomic number	Configuration
Hydrogen	H	1	1s^1
Helium	He	2	1s^2
Lithium	Li	3	1s^2 2s^1
Beryllium	Be	4	1s^2 2s^2
Boron	B	5	1s^2 2s^2 2p^1
Carbon	C	6	1s^2 2s^2 2p^2
Nitrogen	N	7	1s^2 2s^2 2p^3
Oxygen	O	8	1s^2 2s^2 2p^4
Fluorine	F	9	1s^2 2s^2 2p^5
Neon	Ne	10	1s^2 2s^2 2p^6
Sodium	Na	11	1s^2 2s^2 2p^6 3s^1
Magnesium	Mg	12	1s^2 2s^2 2p^6 3s^2
Aluminum	Al	13	1s^2 2s^2 2p^6 3s^2 3p^1
Silicon	Si	14	1s^2 2s^2 2p^6 3s^2 3p^2
Phosphorus	P	15	1s^2 2s^2 2p^6 3s^2 3p^3
Sulfur	S	16	1s^2 2s^2 2p^6 3s^2 3p^4
Chlorine	Cl	17	1s^2 2s^2 2p^6 3s^2 3p^5
Argon	Ar	18	1s^2 2s^2 2p^6 3s^2 3p^6

From Fig. 38.12 we can immediately see that the noble gases, which are chemically unreactive, have what are called filled shells. All the electronic states of one type, such as the $1s$ or $2p$ states, are occupied by a maximum number of electrons, while all further states are completely empty. Thus the noble gas helium occurs at the filling of the $1s$ shell; the noble gas neon occurs at the filling of the $2p$ shell; and the noble gas argon occurs at the filling of the $3p$ shell.

38.10 THE CHEMICAL BOND

When two initially isolated atoms are brought close to one another, the electronic quantum states of each atom are affected by the presence of the other atom. Rather than work through the mathematics, we shall try to develop an intuitive understanding of the effect of one atom on another by using, as a pictorial analog, the vibration patterns of circular drumheads.

Let us first consider a system composed of two noninteracting circular drumheads. We could manufacture such a system by cutting two circular holes in a board and covering the holes with membranes (see Fig. 38.11). The two lowest-frequency normal modes of this system are very simple. They have identical frequencies. One of them consists of a vibration pattern in which only the left drumhead is vibrating in its fundamental mode while the right drumhead remains at rest. The other normal mode of the

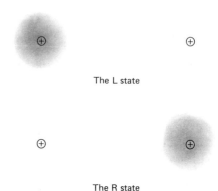

The L state

The R state

FIGURE 38.12   The lowest-frequency quantum states of noninteracting hydrogen atoms.

same frequency has the right drumhead vibrating while the left remains at rest. We call the two modes of vibration the L mode and the R mode, respectively. We compare this system to a pair of protons that are widely separated from one another (see Fig. 38.12). The two lowest-frequency (and thus lowest-energy) quantum states of the electron field surrounding the isolated protons are a $1s$ state centered on the left proton and a $1s$ state centered on the right proton. These states have exactly the same frequency, and we call these the L state and the R state, respectively. We have temporarily ignored the electron spin variable in counting the states but will reintroduce it shortly. The point of the analogy will become clear when we introduce interacting drumheads.

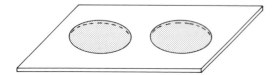

FIGURE 38.11   Two noninteracting circular drumheads.

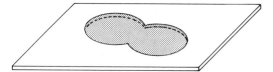

FIGURE 38.13   A drumhead composed of two intersecting circles.

We construct a system of interacting drumheads by cutting two slightly overlapping circular holes in a board and covering the opening with a membrane that is glued down at the edge (see Fig. 38.13). The system now has no normal mode in which only one-half of the membrane vibrates. The two lowest-frequency normal modes of the system now have different frequencies. The mode of lowest frequency is a vibration pattern in which both halves of the membrane vibrate in phase. We call this mode of vibration the *symmetrical mode.* Because the effective area of the combined membrane is larger than either single area of the noninteracting membrane, the frequency of the symmetrical mode is lower than the frequency of the fundamental modes of the noninteracting drumheads (see Fig. 38.14).

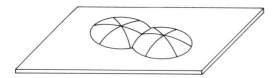

FIGURE 38.14   Symmetrical mode of vibration. Both halves move up and down in phase.

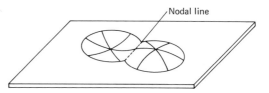

FIGURE 38.15   Antisymmetrical mode of vibration. The two halves move in opposite directions with a nodal line in between.

FIGURE 38.16   The antisymmetrical mode has the same frequency as a drumhead in the shape of an incomplete circle.

The second normal mode of the interacting drumheads is one in which the left and right halves vibrate with opposite phase. When the right half is up, the left half is down, and vice versa. We call this the *antisymmetrical mode* (see Fig. 38.15). The pattern of vibration of the right half is exactly the same as the fundamental mode of a drumhead that has the form of a circle with a slice taken off the left side (Fig. 38.16). Since this shape has a smaller area than the full circle, the fundamental normal mode frequency for the antisymmetrical mode is higher than the frequency of the noninteracting drumhead.

Thus the interaction between the drumheads causes the frequencies of the two lowest normal modes, which are equal for the noninteracting system, to split into two unequal frequencies, one of which is higher than the noninteracting normal-mode frequency and the other lower.

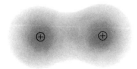

(a) The (L+R) state

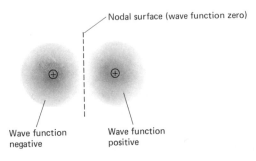

Nodal surface (wave function zero)

Wave function   Wave function
negative   positive

(b) The (L−R) state

FIGURE 38.17   Bonding and antibonding states of the hydrogen molecule.

For a system of an electron near two protons, the lowest-frequency electronic quantum states show the same behavior as the drumhead. As the protons are brought closer together, the two originally degenerate quantum states are replaced by a symmetrical quantum state of lower frequency that has no nodes and an antisymmetrical quantum state whose wave function is positive around one proton and negative around the other. The latter state has a higher frequency than the original degenerate state. It has a nodal surface that is the bisector of the line between the two protons (Fig. 38.17). The symmetrical state is called the *bonding state,* and the antisymmetrical state is called

the *antibonding state.* If two hydrogen atoms are gradually brought together, the two electrons, which are initially in 1s states around the separated protons, will both go into the lower-frequency bonding state with opposite spins. Since the energy of each electron in the bonding state is lower than it would be in the separated 1s state, a certain amount of energy would be required to separate the pair of hydrogen atoms. It is this energy that is seen as the energy of chemical combination. When initially isolated hydrogen atoms come together, the difference in energy between the separated states and the bound state is either radiated in the form of electromagnetic waves or transmitted to nearby atoms in the form of kinetic (i.e., heat) energy. The hydrogen-hydrogen bond is a perfect example of the *covalent bond.* With this type of bond the two atoms share equally the pair of electrons in the bonding state. For dissimilar atoms the bonding state may be concentrated mostly around one nucleus, in which case we have an *ionic bond.* No sharp distinction between ionic and covalent bonds can be made. Intermediate cases in which the bonding state is a little more on one atom than on the other are common.

The lack of chemical reaction between helium atoms can easily be understood from the same point of view. In the separated helium atoms the 1s states around each atom are doubly occupied by electrons with opposite spins. If the nuclei are brought closer, the quantum states are gradually changed from 1s states localized on the individual nuclei to a bonding and an antibonding state, just as they are for the hydrogen atom. The crucial difference between the two cases is that we must now distribute four electrons rather than two in the two lowest states. Thus both the bonding and the antibonding states must each be occupied by a pair of electrons. The increase in energy of the antibonding state more than compensates for the decrease in energy of the bonding state, so the net effect is an increase in total energy as the helium atoms are brought together. Since the energy of separated helium atoms is lower than the energy of the system when the nuclei are close, the atoms can only be brought together by some external force and will fly apart as soon as that external force is removed. For this reason there is no such thing as a stable diatomic helium molecule.

Let us now look at a few atoms of higher atomic number. The next atom in the sequence is lithium. A lithium atom consists of a nucleus with a triple positive charge surrounded by three occupied electronic quantum states. For a lithium atom in its ground state, the $1s$ state is doubly occupied while the $2s$ state is only singly occupied. As we can see in Fig. 38.5, the negative electronic charge density associated with the $1s$ state (there called u_{100}) is closely concentrated around the nucleus. The effect of the two negative electrons in the $1s$ state is to cancel two-thirds of the electric charge of the nucleus. The third electron, which is in the $2s$ state and therefore lies further from the nucleus, feels the attractive force of only a single positive charge. Since it is already at an appreciably larger distance from the nucleus and the electrostatic potential diminishes as the distance increases, this outer electron can be removed entirely from the atom with relative ease. In contrast, removing one of the $1s$ electrons from a lithium atom requires much more energy (see Table 38.3). The loosely bound outer electron can have its wave function easily distorted in the presence of other atoms. This causes lithium to be a highly reactive element that forms chemical bonds easily.

TABLE 38.3 Energy required to remove consecutive electrons from an atom, starting with the outermost electron (in electron volts).

Atomic number	Element	Electron being removed						
		$1s$	$1s$	$2s$	$2s$	$2p$	$2p$	$2p$
1	H	13.6						
2	He	54.6	24.6					
3	Li	122.3	75.9	5.7				
4	Be	217.7	153.5	18.2	9.3			
5	B	340.4	259.3	37.9	25.1	8.3		
6	C	490.0	392.3	64.6	47.7	24.4	11.3	
7	N	667.0	552.0	97.9	77.5	47.0	29.6	14.5

In general, the quantum states available for occupation by the electrons of all atoms occur in groups called *shells*. Two quantum states in the same shell have their charge density (i.e., probability density) concentrated at about the same distance from the nucleus. They also have comparable energy values. In contrast, quantum states belonging to different shells have very different energy values and are well separated with respect to their charge density. For this reason the chemical properties of a particular atom depend principally on how many quantum states are occupied or empty in the outermost shell. All elements that have only one occupied quantum state in the outermost shell have similar chemical characteristics. Lithium is one such element. Others are sodium, potassium, rubidium, cesium, and francium. These are all monovalent alkalis. They are all metals, which means that, when large numbers of the atoms are brought together to form a crystal of the element, the loosely bound outer electrons easily move from atom to atom, giving the material a large electric conductivity. When an electromagnetic wave is

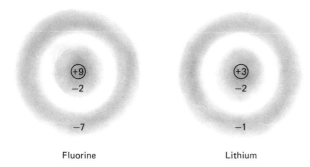

Fluorine   Lithium

FIGURE 38.18   The inner core of a fluorine atom has a charge of $+7e$ while that of a lithium atom has a charge of $+e$.

incident upon such a metallic crystal, the oscillating electric field associated with the wave easily causes the loosely bound outer electrons to oscillate and thus re-radiate the electromagnetic wave in the form of a re-flected wave. Thus, if the metals have a smooth surface so that the reflected wave comes off in a uniform direction, they are shiny.

Elements with two electrons in the outermost shell are divalent metals. Beryllium, magnesium, and calcium are examples of these. It is clear from what we have said that the chemical valence of an atom is determined by the number of occupied or empty quantum states in the outer-most shell. Monovalent metals have a valence of $+1$, and divalent metals have a valence of $+2$. Negative valences occur when the outermost shell is almost but not quite full. Fluorine, chlorine, bromine, and iodine all have outermost shells in which there is room for only one more electron. These elements have a valence of -1.

Consider what would happen if a fluorine atom and a lithium atom were brought close together. One of the quantum states in the fluorine atom is occupied by only one electron, although it has room for two electrons of opposite spin. The singly occupied state in the lithium outer shell and the singly occupied state in the fluorine outer shell can interact to form a bonding and an antibonding state. The two electrons available go into the bonding state with opposite spins and form a lithium fluoride molecule. The bonding state is heavily concentrated on the side of the fluorine atom because an electron in the outer shell of fluorine experiences a charge of $+7e$ while an electron in the outer shell of lithium experiences a charge of only $+e$ (see Fig. 38.18). Thus it is a fairly good approximation to say that the lithium atom simply loses an electron to the outer shell of the fluorine atom and becomes a positive ion that is electrostaticially attracted to the negative fluorine ion. The quantum mechanical description, however, has the advantage of also being able to explain what happens when two fluorine atoms come together to form the diatomic fluorine molecule. In this case the two electrons that began in the singly occupied states of the fluorine atoms go into the evenly shared bonding state of the fluorine molecule. The fluorine molecule thus has a covalent bond. When an atom has all the quantum states of its outer shell completely occupied, the element is a noble gas. Helium, neon, argon, krypton, and radon are noble gases.

The types of quantum states comprising the first five major shells are shown in Table 38.4. Note that the d states (the states with $l = 2$) are in a higher shell than would be expected from a glance at their principle quantum number n. For hydrogen the energy of any quantum state is determined completely by the quantum number n, but for a larger atom this is not so. The electronic charge has an effect on the electrostatic potential $U(r)$ that is used in the Schrödinger equation to determine the quantum state. When the potential is not a simple Coulomb potential, as it is in hydrogen, the energy level of the quantum state depends on both quantum numbers n and l. The fact that the $l = 2$ states are pushed up in energy and outward in position by just large enough amounts to put the states

TABLE 38.4 Types of quantum states in the first five major shells.

Shell	Maximum number of electrons	Types of states in shell
1st	2	$1s$
2nd	8	$2s$, $2p$
3rd	8	$3s$, $3p$
4th	18	$3d$, $4s$, $4p$
5th	18	$4d$, $5s$, $5p$

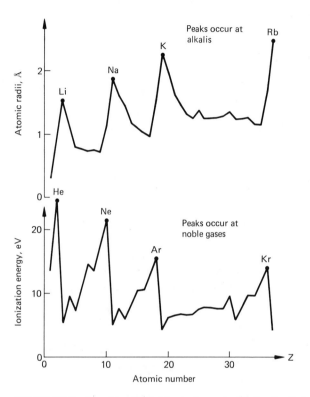

FIGURE 38.19 Atomic radii and ionization energies for the first 37 elements.

in the next higher shell can be predicted only by solving the Schrödinger equation in detail. That many significant properties of atoms depend primarily on the number of electrons in the outermost shell is convincingly illustrated by the data collected in Fig. 38.19.

SUMMARY

The Schrödinger equation for a two-dimensional system with potential $U(x, y)$ is

$$i\hbar \frac{\partial \psi}{\partial t} = -\frac{\hbar^2}{2m}\left(\frac{\partial^2 \psi}{\partial x^2} + \frac{\partial^2 \psi}{\partial y^2}\right) + U(x, y)\psi$$

The periodic solutions of the equation are of the form

$$\psi(x, y, t) = u(x, y)e^{-i\omega t}$$

where $u(x, y)$ is a solution of the Schrödinger energy equation

$$Eu(x, y) = -\frac{\hbar^2}{2m}\left(\frac{\partial^2 u}{\partial x^2} + \frac{\partial^2 u}{\partial y^2}\right) + U(x, y)u(x, y)$$

If $U(x, y + a) = U(x, y)$ for any values of x, y, and a, the system has translational invariance in the y direction. In this case the solutions of the Schrödinger energy equation are of the form

$$u(x, y) = u_n(x)e^{iky}$$

where $p_y = \hbar k$, $E = E_n + p_y^2/2m$, and $u_n(x)$ is a solution of the one-dimensional Schrödinger energy equation with energy E_n.

If, for a two- or three-dimensional system, the potential depends only on the distance from the origin [that is, $U = U(r)$], then the system has rotational invariance. In this case the solutions of the Schrödinger energy equation are of the form

$$u(r, \theta) = u_n(r)e^{im\theta} \qquad \text{(Two dimensions)}$$

$$u(r, \theta, \phi) = u_n(r)f_\ell(\theta)e^{im\phi} \qquad \text{(Three dimensions)}$$

The number of values of r for which $u_n(r) = 0$ (the number of nodes) is equal to $n - 1$.

The z component of angular momentum is $L_z = \hbar m$, where $m = 0 \pm 1, \pm 2, \ldots$.

The square of the angular momentum vector is $L^2 = \ell(\ell + 1)\hbar^2$, where $\ell = 0, 1, 2 \ldots$.

The magnitude of L_z cannot be larger than L; therefore $|m| \le \ell$.

The Schrödinger energy equation for the hydrogen atom has a solution for each set of quantum numbers (n, ℓ, m) that satisfies the restrictions

$$n = 1, 2, 3 \ldots$$

$$0 \le \ell \le n - 1$$

and

$$-\ell \le m \le \ell$$

For hydrogen (but not other rotationally invariant systems), the energy depends only on the quantum number n and is given by

$$E_n = -\frac{ke^2}{2an^2}$$

where $a = \hbar^2/mke^2$ is the Bohr radius.

For a system of electrons the magnetic moment associated with the circular electron flux is related to the angular momentum by

$$\mu_z = -\frac{e}{2m_e}L_z$$

An intrinsic spin angular momentum is associated with the electron. It has a magnitude s, where $s^2 = \frac{1}{2}(\frac{1}{2} + 1)\hbar^2$, and a z component of $\pm\frac{1}{2}\hbar$. Associated with the spin angular momentum is a magnetic moment whose z component is

$$\mu_{z(\text{spin})} = -\frac{e}{m_e}s_z$$

The Pauli exclusion principle states that no two electrons can have exactly the same set of quantum numbers.

In an atom of atomic number Z the first Z single particle quantum states are occupied.

The quantum states are arranged in shells of about the same energy. Atoms in which the outermost shell is completely filled are called noble gases and are chemically unreactive. Elements in which the outermost shell contains only one electron are monovalent alkalis.

When two atoms are brought close together, the quantum states that were initially localized on each atom are transformed into bonding and antibonding states on both atoms. If the bonding state is nearly evenly distributed between the two atoms, the associated chemical bond is a covalent bond. If the bonding state is concentrated mostly on one atom, the chemical bond is an ionic bond.

Many of the chemical and physical properties of atoms depend on the number of filled or empty quantum states in the outermost shell.

PROBLEMS

38.A.1
   (a) Write the Schrödinger energy equation for a particle in two dimensions that is subject to a force derivable from the potential $U(x, y) = \frac{1}{2}kx^2$.
   (b) Using the information given in the statement of problem 37.C.3, write one solution to the equation you obtained in part (a).

38.A.2 If a hydrogen atom undergoes a transition from a state with $n = 3$ to a state with $n = 2$, what will be the frequency and wavelength of the emitted photon? In what part of the electromagnetic spectrum (i.e., visible, ultraviolet, etc.) will it lie?

38.A.3 How much energy (in electron volts) is required to remove both $2s$ electrons from a beryllium atom? How much energy is required to remove both electrons from a Be^{2+} ion?

38.A.4 If there were three possible states of electron spin instead of two, what would be the nuclear charge of the first monovalent alkali?

38.A.5 If a hydrogen atom exhibited the normal Zeemann effect, what would be the frequency of the photon emitted in a transition $(n, \ell, m) = (2, 1, -1) \rightarrow (n, \ell, m) = (2, 0, 0)$ for a hydrogen atom in a magnetic field of 1.5 T?

38.B.1 What are the spherical coordinates r, θ, ϕ of a point that has rectangular coordinates $(x, y, z) = (1, 2, 1)$?

38.B.2 Prove that $L_z{}^2 < L^2$ if $L_z = m\hbar$, $L^2 = \ell(\ell + 1)\hbar^2$, and $|m| \leq \ell$ (see Eqs. 38.10, 38.11, and 38.12).

38.B.3
   (a) Write the probability density u^*u for the five quantum states given in Eq. 38.15.
   (b) For which states is the probability density symmetrical about the z axis?
   (c) For which states is it symmetrical about the x axis?

38.B.4
   (a) Calculate the ratio of the magnetic moment μ to the mass of the atom for a lithium atom. (Lithium has a net spin angular momentum of $\frac{1}{2}\hbar$.)
   (b) If all the spins in a 1-kg sample of lithium were lined up parallel to each other and the sample were placed in a 0.5 T magnetic field, how much energy would have to be supplied to rotate the sample a half turn so that the magnetic moments were all pointed against the field?

38.B.5 What are the mks units of magnetic dipole moment? In these units, what is the value of μ_z for an electron (see Eq. 38.22)?

38.B.6 The photon, like the electron, has intrinsic spin angular momentum. The spin angular momentum of the photon is $\hbar$. Because of its spin, the photon emitted when an atom makes a transition carries away an amount $\hbar$ of angular momentum. Thus the angular momentum of the atom must change by one unit during the transition. Taking this fact into account, draw arrows (from one level to another) to indicate all the possible one-photon transitions in a diagram just like the left side of Fig. 38.10. Are there any states that could not lose energy by means of a one-photon transition?

38.C.1 What are the rectangular coordinates (x, y, z) of a point that has spherical coordinates $(r, \theta, \phi) = (2, \pi/3, \pi/4)$?

38.C.2 According to the classical theory, an electron in a circular orbit around a hydrogen nucleus should emit electromagnetic radiation whose angular frequency is equal to the angular frequency of the orbital motion.

  (a) Calculate the classical orbital angular frequency of an electron with an energy E_n given by Eq. 38.14.

  (b) Calculate the angular frequency of the photon emitted in the transition $E_n \to E_{n-1}$ according to the quantum theory.

  (c) Compare the two angular frequencies for the cases $n = 2$ and $n = 100$.

38.C.3 According to the theory of the normal Zeemann effect (i.e., neglecting electron spin), in what strength magnetic field would the $(n, \ell, m) = (2, 1, 1)$ quantum state have the same energy as the $(n, \ell, m) = (3, 1, -1)$ quantum state for a hydrogen atom?

38.C.4 A two-dimensional harmonic oscillator has a potential function

$$U(x, y) = \tfrac{1}{2}k(x^2 + y^2) = \tfrac{1}{2}kr^2.$$

  (a) Write the Schrödinger energy equation (Eq. 38.4) for such a system.

  (b) Let $u_n(x)$ be the solution of the one-dimensional harmonic oscillator Schrödinger equation (Eq. 38.5) that has energy value $E_n = (n + \tfrac{1}{2})\hbar\omega_0$. Show that the function

$$u_{nm}(x, y) = u_n(x)\, u_m(y)$$

    is a solution of the two-dimensional equation with an energy value $E_{nm} = (n + m + 1)\hbar\omega_0$.

  (c) What are the degeneracies of the first three energy levels of the two-dimensional oscillator?

38.C.5 (Follows problem 38.C.4.) The two-dimensional harmonic oscillator potential has rotational invariance in the sense of Sec. 38.2. Therefore the solutions of the Schrödinger equation should be of the form shown in Eq. 38.5.

  (a) Using the explicit forms of $u_0(x)$ and $u_1(x)$ given in problem 37.C.3, write the functions $u_{00}(x, y)$, $u_{01}(x, y)$, and $u_{10}(x, y)$ explicitly.

  (b) Show that $u_{00}(x, y)$ is of the form given in Eq. 38.5, but that u_{01} and u_{10} are not.

  (c) Show that the functions

$$u_+(x, y) = u_{10} + iu_{01}$$

    and

$$u_-(x, y) = u_{10} - iu_{01}$$

    are of the form given in Eq. 38.5 and are also solutions of the two-dimensional Schrödinger equation.

39
NUCLEAR
PHYSICS

IN CHAP. 38 the nucleus of the atom was taken to be nothing more than a positively charged massive point object. From the point of view of the atomic electrons surrounding it, this is quite a good description of the atomic nucleus. On an atomic scale the size of the nucleus is extremely small. If we expanded a carbon atom until it was the size of a typical room, its nucleus would be the size of a grain of finely ground salt. In spite of its very small size, the nucleus accounts for 99.995% of the mass of the whole atom. The repulsive electrostatic forces between the positively charged protons in this tiny object are billions of times stronger than the electrostatic forces involved in atomic structure. Surely there must be a powerful new attractive force to overcome the disruptive force of like charges brought so close to one another. There is. It is a totally different kind of force, called the *strong force,* which attracts protons to protons in spite of electrostatic repulsion. It also attracts protons to neutrons and neutrons to neutrons. The nuclear force is not only adequate to overcome the electrostatic repulsive forces. At typical nuclear distances it is so strong that, in analyzing the internal structure of the nuclei of lighter elements such as carbon, one can neglect the comparatively weak electrostatic forces.

39.1 THE NUCLEAR FORCE

The most direct way to study the forces between nuclear particles is by means of two-nucleon scattering experiments. The term *nucleon* indicates either of the two major constituents of nuclei, namely protons and neutrons. In two-nucleon scattering a substance rich in hydrogen nuclei (i.e., protons) such as paraffin is used as a target.

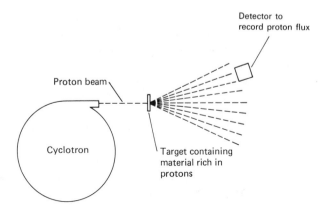

FIGURE 39.1   A proton-proton scattering experiment.

The target is placed directly in a beam of high-velocity protons or neutrons. The flux of particles of each type leaving the location of the target is measured carefully with an appropriate type of particle detector (see Fig. 39.1). One then compares the observed distribution of particle flux with the flux that is predicted by the Schrödinger equation for various conjectured nucleon-nucleon forces. Actually the potential function associated with the nucleon-nucleon force is directly used in the calculation. When one can theoretically reproduce the observed results of all two-nucleon scattering experiments by using the Schrödinger equation with a single potential function, one then says that an acceptable nucleon-nucleon potential has been obtained. Of course, as scattering experiments are improved, the nucleon-nucleon potential may have to be refined. A broad outline of the results of these investigations is as follows.

1. The nuclear force, in contrast to the electromagnetic force, is the same between two protons, two neutrons, or a proton and a neutron. Thus it is said to be *charge independent.*

2. The nuclear force is very strong. Because of this, typical binding energies of nucleons in a nucleus are in millions of electron volts (MeV).

3. The nuclear force has a range of only about 10^{-15} m. The distance 10^{-15} m is called one *fermi* (abbreviated fm). At a distance of a few fermis the nucleon-nucleon force is complete negligible.

4. The nuclear force is strongly repulsive at very short distances (less than about 0.5 fm) but attractive at larger distances (see Fig. 39.2).

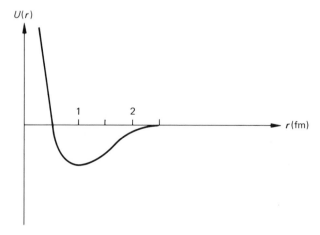

FIGURE 39.2   An approximate graph of the potential energy of two nucleons separated by a distance *r*.

5. The nuclear force is really much more complicated than Fig. 39.2 indicates. The proton and neutron are both particles with spin angular momentum $\frac{1}{2}\hbar$. The force between two nucleons depends not only on their separation r but also on the relative orientation of their spin axes and their relative velocities.

39.2 NUCLEAR FLUID

As Fig. 38.21 shows, the sizes of atoms fluctuate greatly as the number of electrons is increased. There is only a slight trend for the size to increase as Z increases. If we compare atoms of similar type such as helium and krypton, which are both noble gases, we see that krypton has less than twice the radius of helium even though it contains 18 times as many electrons. The average electron density increases appreciably as we go to atoms of higher and higher Z. This is quite understandable. The larger attractive electric charge of the higher Z nucleus tends to pull in all the surrounding quantum states. This effect very much reduces the rate of growth of atoms in comparison to what one would expect if a fixed set of quantum states were filled with more and more electrons as Z increased.

Because the nuclear force cuts off sharply at a few fermis and has a strongly repulsive core, there is no corresponding effect of continuously increasing density within the nucleus as the number of nucleons is increased. Because of the repulsive core, the nucleons in a larger nucleus must maintain a certain distance from one another. A particular nucleon only experiences the force of the other nucleons within its immediate neighborhood. Thus, once the nucleus we are considering reaches a certain size, the potential experienced by a given nucleon is independent of the total number of nucleons in the nucleus (see Fig. 39.3). The basic structure of the nucleus bears a

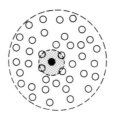

FIGURE 39.3 The force on any nucleon in the interior of the nucleus is due to its immediate neighbors and does not change when more nucleons are added to the system.

strong qualitative resemblance to the structure of a liquid drop, but the similarity is only qualitative. In a nucleus the size of the repulsive core is much smaller in comparison to the average interparticle distance than it is in a liquid. The number of particles in a nucleus is vastly smaller than in a liquid, and the exclusion principle plays an important role in the structure of nuclei but not of liquids. The particle density of the nuclear fluid is about 1.5×10^{44} particles per cubic meter. Since the mass of each nucleon is about 1.67×10^{-27} kg, this gives a nuclear fluid mass density of

$$\rho_{nuc} = (1.5 \times 10^{44} \text{ nucleons/m}^3)$$
$$\times (1.67 \times 10^{-27} \text{ kg/nucleon})$$
$$= 2.5 \times 10^{17} \text{ kg/m}^3$$

A cubic centimeter of nuclear fluid would weigh a quarter of a billion metric tons!

QUESTION

What is a neutron star?

ANSWER

A neutron star is a sphere of pure nuclear fluid with a typical diameter of 20 km. This gives it a total mass of a few times the mass of the sun. Neutron stars are created when the compressive forces caused by gravitational attraction within a star are so great that they overcome the forces that usually hold atoms apart from one another. The electronic structures surrounding the initially separate atomic nuclei collapse under the gravitational pressure and the nuclei finally fuse, forming a giant nucleus of kilometer dimensions.

The liquid-drop analogy can be carried a little further in the following way. It is clear from Fig. 39.3 that a nucleon near the surface of a nucleus would very likely feel a different potential than a nucleon well inside the nucleus. This is exactly the effect that was shown in Sec. 12.1 to lead to the phenomenon of surface tension. The nucleus also exhibits surface tension, and many dynamical effects that one would have difficulty analyzing by considering the mechanics (or quantum mechanics) of individual particles in the nucleus can be treated more simply and with acceptable accuracy if one considers the nucleus a liquid drop of electrically charged fluid.

39.3 BINDING ENERGIES OF NUCLEI

An excellent example of the way in which the simple liquid-drop model of the nucleus can be used to understand gross features of nuclei is the use of the model to explain the main effects determining nuclear binding energies. A nucleus is identified by two integers:

$N \equiv$ number of neutrons in the nucleus

$Z \equiv$ number of protons in the nucleus

In terms of N and Z, the total number of nucleons A, which is also called the *mass number* of the nucleus, is given by

$$A = N + Z$$

The value of Z determines the chemical characteristics of an atom, and therefore tells us to what element a given nucleus belongs. For a given Z the possible N values that can be coupled with it to form reasonably stable nuclei determine the different *isotopes* of that element. For instance, hydrogen can have N equal to 0, 1, or 2. Those three different isotopes are denoted ^{1}H, ^{2}H, and ^{3}H, where the superscript on the left is the mass number of the nucleus.

The masses of separated protons and neutrons are

$$m_p = 1.6727 \times 10^{-27} \text{ kg}$$

and

$$m_n = 1.6750 \times 10^{-27} \text{ kg}$$

However, the mass of a ^{12}C nucleus, which is composed of six protons and six neutrons, is (see Table 39.1)

$$M(^{12}\text{C}) = 19.921 \times 10^{-27} \text{ kg}$$

which is not equal to the combined masses of six protons plus six neutrons:

$$6m_p + 6m_n = 20.085 \times 10^{-27} \text{ kg.}$$

The mass of the combined system of 12 particles is less than the sum of the separated masses by an amount

$$\Delta m = 1.7 \times 10^{-28} \text{ kg}$$

TABLE 39.1   Masses of selected nuclei

Isotope	N	Z	Mass of nucleus (without electrons) in atomic mass units	in units of 10^{-27} kg
^{1}H	0	1	1.00728	1.67265
^{2}H	1	1	2.01356	3.34364
^{3}H	2	1	3.01551	5.00743
^{4}He	2	2	4.00152	6.64476
^{12}C	6	6	11.9967	19.9213
^{16}O	8	8	15.9906	26.5533
^{30}Si	16	14	29.9662	49.7606
^{38}Ar	20	18	37.9529	63.0231
^{38}Ca	18	20	37.9654	63.0439
^{48}Ca	28	20	47.9417	79.6100
^{54}Cr	30	24	53.9258	89.5471
^{56}Fe	30	26	55.9207	92.8507
^{60}Ni	32	28	59.9156	99.4934
^{108}Cd	60	48	107.878	179.134
^{124}Sn	74	50	123.878	205.707
^{232}Th	142	90	231.989	385.232
^{232}U	140	92	231.987	385.229
^{235}U	143	92	234.994	390.222
^{236}U	144	92	235.996	391.885
^{238}U	146	92	238.001	395.215

This is exactly the mass equivalent of the energy that would have to be supplied in order to separate the 12 particles in the nucleus against the action of the nuclear force ($E = \Delta mc^2$). This same energy would be released as heat and radiant energy if the separated nucleons were allowed to combine to form a nucleus of ^{12}C. This energy of combination is called the total *binding energy* of ^{12}C and is given by

$$B(^{12}C) = (\Delta m)c^2 = 1.54 \times 10^{-11} \text{ J}$$

All known nuclei have masses that are smaller than the sum of the masses of the particles that compose them. For the nucleus of neutron number N and proton number Z, the binding energy is given in terms of the mass of the nucleus, $M(N, Z)$, by

$$B(N, Z) = [Nm_n + Zm_p - M(N, Z)]c^2 \qquad (39.1)$$

A formula called the *empirical mass formula* predicts the binding energies of most nuclei remarkably well (see Fig. 39.4). We shall first present the formula and then explain the origin of each term in the formula by means of the liquid-drop model.

$$B(N, Z) = Ab_{volume} - 4\pi R^2 b_{surface}$$
$$- \frac{3}{5} \frac{kZ^2 e^2}{R} - \frac{(N - Z)^2}{A} b_{symmetry} \qquad (39.2)$$

where $A = N + Z$ and R is the radius of a sphere containing A particles at the density of nuclear fluid. That is,

$$A = \tfrac{4}{3}\pi R^3 (1.5 \times 10^{44} \text{ nucleons/m}^3)$$

or

$$R = (1.2 \times 10^{-15} \text{ m}) \times A^{1/3} = 1.2 A^{1/3} \text{ fm} \qquad (39.3)$$

The constants b_{volume}, $b_{surface}$, and $b_{symmetry}$ must be determined experimentally. Actually one can calculate them from the Schrödinger equation by using the nuclear potential, but the calculation is extremely complex. The numerical values of the constants are

$$b_{volume} = 2.49 \times 10^{-12} \text{ J}$$
$$b_{surface} = 1.43 \times 10^{17} \text{ J/m}^2 \qquad (39.4)$$
$$b_{symmetry} = 3.73 \times 10^{-12} \text{ J}$$

Before we make any use of the formula, let us consider the physical meaning of each term.

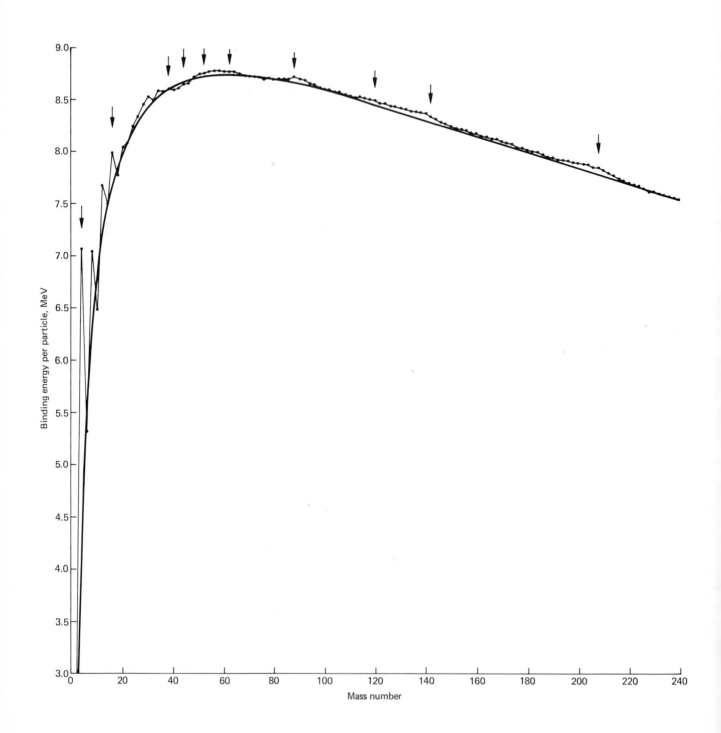

The constant b_{volume} is the binding energy per particle that one would obtain for a very large piece of nuclear fluid with equal numbers of neutrons and protons if one could switch off the electrostatic interaction. All the other terms in the formula should be viewed as corrections to the term Ab_{volume} which tend to reduce the binding energy.

The term $-4\pi R^2 b_{surface}$ is the correction to take account of the fact that the particles near the surface do not form as many potential energy bonds as do those in the interior. The constant $b_{surface}$ has exactly the same interpretation as the surface tension T introduced in Sec. 12.1.

The term

$$-\frac{3}{5}\frac{kZ^2 e^2}{R}$$

is the electrostatic energy of an amount of charge Ze distributed uniformly throughout a volume of radius R (see Sec. 30.2). The fact that it is negative indicates that the repulsive electrostatic forces act to reduce the cohesion of a nucleus. They are disruptive forces.

The term

$$-\frac{(N-Z)^2}{A}b_{symmetry}$$

is known as the symmetry energy. It can be easily understood from the following quantum theoretical argument. We know that the nucleus has a fairly definite surface; that is, it is unlikely that the nucleons will be found at appreciable distances outside the surface. Let us replace the spherical nucleus by a spherical volume of the same radius with hard walls (see Fig. 39.5). We approximate the interactions among the nucleons by a constant potential

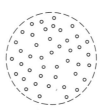 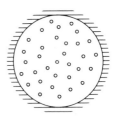

FIGURE 39.5 In our analysis the spherical surface of the nucleus is replaced by a spherical container.

throughout the volume that is equal to the average nuclear potential that a nucleon in the actual nucleus experiences. This is the same for protons and neutrons because of the charge independence of the nuclear potential. We ignore the electrostatic potential because its effect has already been taken into account in the electrostatic energy term. To calculate the energy of A nucleons in our simplified model of the nucleus, we first solve the Schrödinger energy equation (which has the same form for the proton or neutron fields as for the electron field) in three dimensions. From that equation we obtain a sequence of quantum states $u_1, u_2, \ldots$ and corresponding energies $E_1, E_2, \ldots$. To obtain the lowest energy for a system of A nucleons in this spherical box, we fill up A of the quantum states in such a way as to obtain the lowest total energy. Both protons and neutrons must satisfy the Pauli exclusion principle. We obtain the lowest energy when we put two neutrons (with opposite spins) and two protons (also with opposite spins) in the lowest $A/4$ quantum states. (If A is not evenly divisible by 4, then the last quantum state will not be completely filled.) This gives us a nucleus with equal numbers of protons and neutrons. For that state $N-Z=0$, and thus the expressions for the symmetry energy would

FIGURE 39.4 A comparison of the actual binding energies of nuclei with the predictions of the empirical mass formula. For each value of A the values of N and Z are those that give the most stable nucleus. The smooth curve is the prediction of the empirical mass formula. The arrows are explained in Sec. 37.7.

give zero. If we now want to modify the state so that we get a nucleus with the same number of nucleons A but unequal numbers of protons and neutrons (let us say $N > Z$), then we must remove a number of protons and replace them by neutrons. But when we do so we cannot put the neutrons into the same quantum states that the protons were in because those quantum states already are occupied by two neutrons each. We must therefore put the new neutrons in higher energy states than those occupied by the protons we removed (see Fig. 39.6). Thus the energy (not including the electrostatic energy) is greater for a

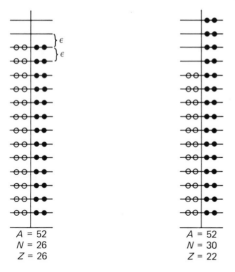

$$A = 52 \qquad A = 52$$
$$N = 26 \qquad N = 30$$
$$Z = 26 \qquad Z = 22$$

FIGURE 39.6 (a) Occupation of the levels for the nucleus $(N, Z) = (26, 26)$. Neutrons are represented by solid circles and protons by empty circles. (b) Corresponding diagram for the nucleus $(30, 22)$. The energy of the nucleus $(30, 22)$ is greater than the energy of the nucleus $(26, 26)$ by the amount $\epsilon(N - Z)^2/8$, where $(N - Z) = 8$ in this case. It can be shown that ϵ is proportional to A^{-1}. Thus the energy difference is proportional to $(N - Z)^2/A$.

nucleus with unequal numbers of protons and neutrons than for a nucleus of the same size with equal numbers of protons and neutrons. The symmetry energy term describes that increase in energy. (An increase in the total energy is a decrease in the binding energy.)

39.4 NUCLEAR DECAY MODES

A nucleus may spontaneously transform itself to reach a lower energy state in a number of ways. The three most important processes of spontaneous transformation are known as α decay, β decay, and γ decay.

In γ *decay* a photon is emitted. A nucleus that is in a state other than its ground state may make a transition to a lower energy state by emitting a photon that carries off the difference in energy between the two states. Since typical nuclear energy states are in the megaelectron volt range, the photons emitted are in the energy range referred to as γ rays (see Fig. 39.7).

In the process of α *decay* a nucleus emits an α particle, which is just a helium nucleus. It is a bound system composed of two protons and two neutrons. In α decay both the proton number and the neutron number of the nucleus are reduced by two. Thus the nucleus (N, Z) is transformed by α decay into the nucleus $(N - 2, Z - 2)$. Because of the electrostatic energy term in the binding energy, all the large nuclei $(Z > 82)$ are unstable with respect to α decay. Left to themselves, they will eventually transform by emitting an α particle.

The process of β *decay* is somewhat more exotic. In the other two decay processes the particle ejected is either a massless uncharged particle or a preexisting part of the nucleus. In β decay a charged massive particle is created at the instant of decay. That particle is either an electron or a positron. The positron is a particle that is identical to an electron in every respect except its charge: It has a charge

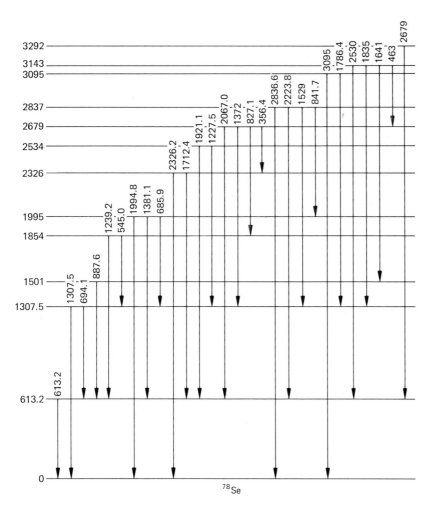

FIGURE 39.7 Some of the energy levels of the ^{78}Se nucleus, with the energies of observed emitted γ rays shown. All energies are in kiloelectron volts.

of $+e$ rather than a charge of $-e$. We have previously stated that the number of electrons in a system could not change because the total charge must remain constant. In β decay the requirement that electric charge be conserved is satisfied by having a neutron change into a proton at the instant at which the electron is created. If the particle ejected is a positron, then a proton must be transformed into a neutron at the instant of decay. A β decay in which the ejected particle is an electron is called a β^- decay; one in which the ejected particle is a positron is called a β^+ decay. One of the simplest and most important examples is the β^- decay of free neutrons. An individual neutron has a mass that is larger than the combined masses of a proton and an electron:

$$m_n = m_p + m_e + 1.4 \times 10^{-30} \text{ kg}$$

Thus if the neutron underwent β^- decay, the resulting proton and electron would be expected to have a total kinetic energy of

$$(1.4 \times 10^{-30} \text{ kg})c^2 = 1.26 \times 10^{-13} \text{ J}$$

In actuality, when β^- decay of neutrons is observed, the resulting particles have a wide range of kinetic energies but always less than the expected amount. Even more surprising is the fact that a neutron that is at rest may give rise to an electron and proton whose total momentum is not zero. Thus the decay process seems to violate both energy and momentum conservation (see Fig. 39.8). The missing energy and momentum are related by $E = cp$. The paradox has been resolved by the discovery that another particle is involved in β^- decay. This is a zero-mass particle of spin $\frac{1}{2}\hbar$ called an *antineutrino* (written $\bar{\nu}$). The β^- decay reaction is actually of the form

$$n \rightarrow p + e^- + \bar{\nu}$$

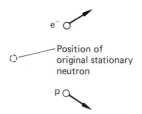

FIGURE 39.8 The β^- decay of a neutron does not seem to conserve momentum.

The antineutrino has no electric charge. It has the peculiar property that its spin angular momentum is always in the same direction as its linear momentum vector. There exists another particle, called the *neutrino,* that has exactly the same properties as the antineutrino but whose spin angular momentum vector is always in the opposite direction of its momentum vector (Fig. 39.9). The neutrino is denoted by ν in reaction equations. It is emitted along with the positron in β^+ decay. In that reaction the nucleus (N, Z) is changed to the nucleus $(N + 1, Z - 1)$.

$$(N, Z) \rightarrow (N + 1, Z - 1) + e^+ + \nu$$

In another type of β decay, called *electron capture,* a proton plus one of the atom's electrons combine and are transformed into a neutron, and a neutrino is emitted.

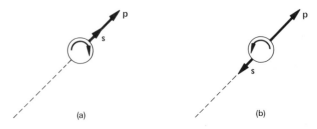

FIGURE 39.9 (a) Antineutrino and (b) neutrino.

39.5 THE LAW OF RADIOACTIVE DECAY

One characteristic that is common to all modes of nuclear decay is that *the probability that any given unstable nucleus will spontaneously decay during the short time interval dt is proportional to the time interval and independent of the past history of that nucleus and the state of any surrounding nuclei.* If at time $t = 0$ we begin with a sample containing N_0 unstable nuclei of one type, they will undergo decay in a way that allows us to predict the overall rate of disintegrations quite accurately, although we cannot predict the exact time when any particular nucleus will disintegrate. Let $N(t)$ be the number of unstable nuclei still left at time t. During the interval from t to $t + dt$ the probability that any one nucleus will decay is equal to $\lambda\,dt$, where λ is some constant that depends on the type of nucleus under consideration. The number of nuclei that will decay during the time interval dt is therefore $N(t)\lambda\,dt$. Thus the change in $N(t)$ during the time interval

will be $dN = -N(t)\lambda\,dt$. This gives us the basic equation of radioactive decay:

$$\frac{dN}{dt} = -\lambda N(t) \tag{39.5}$$

The solution of the equation that satisfies the requirement that $N(0)$ be equal to the given number N_0 is

$$N(t) = N_0 e^{-\lambda t} \tag{39.6}$$

The constant λ is called the *decay constant* for that particular decay process. It has the unit second^{-1}.

Another constant that is often used to describe the rate of a radioactive decay process is called the half-life of the nucleus. The *half-life* is defined as the time required for exactly half of the original nuclei to decay. Thus the half-life T is given by the equation $N(T) = \frac{1}{2}N_0$. Using Eq. 39.6, we can relate the half-life to the decay constant (see Fig. 39.10):

$$\tfrac{1}{2}N_0 = N_0 e^{-\lambda T}$$

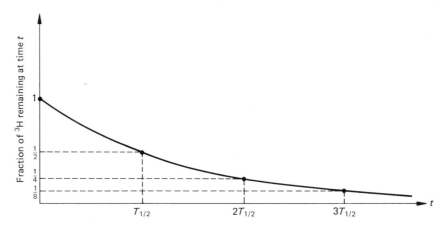

FIGURE 39.10 The β^- decay of ^{3}H to produce ^{3}He. In this case $T_{1/2} = 12.33y$.

which gives

$$T = \frac{\ln 2}{\lambda}$$

39.6 NUCLEAR STABILITY

By undergoing β^+ or β^- decay, a nucleus may adjust its electric charge Z without affecting its mass number A. In the two types of β decay the nuclear transformations are

$$\beta^-: \quad (N, Z) \rightarrow (N - 1, Z + 1)$$
$$\beta^+: \quad (N, Z) \rightarrow (N + 1, Z - 1)$$

(39.7)

Most values of N and Z would give nuclei that were unstable with respect to β decay. For given A there are typically very few values of Z that yield β-stable nuclei. They are shown in Fig. 39.11.

It is obvious that the β-stable nuclei form a definite curve in the NZ plane. We can understand the physical processes at work producing that curve, and even predict the detailed curve fairly well, by making use of the empirical mass formula:

The total energy of the nucleus (N, Z) is the mass energy of the nucleons minus the binding energy (see Eq. 39.1):

$$E(N, Z) = M(N, Z)c^2$$
$$= Nm_n c^2 + Zm_p c^2 - B(N, Z)$$

(39.8)

The nucleus (N, Z) will be stable with respect to β decay if either of the decays (Eq. 39.7) yields a nucleus of higher energy, that is, if

$$E(N - x, Z + x) > E(N, Z)$$

(39.9)

for x equal to $+1$ or -1. Certainly inequality 39.9 would be satisfied if $E(N - x, Z + x)$, considered as a function of x, had a minimum at $x = 0$. The condition for a minimum at $x = 0$ is that

$$\frac{dE(N - x, Z + x)}{dx}\bigg|_{x=0} = 0$$

(39.10)

If we use the empirical mass formula expression for $B(N, Z)$ in Eq. 39.8 and then take the derivative indicated in Eq. 39.10, we obtain an equation involving N, Z, and A. Putting in numerical values for all the constants appearing in the equation (m_n, m_p, $b_{symmetry}$), we obtain the following equation for β-stable nuclei:

$$N - Z = \frac{A^{5/3} - 1.8A}{A^{2/3} + 130}$$

(39.11)

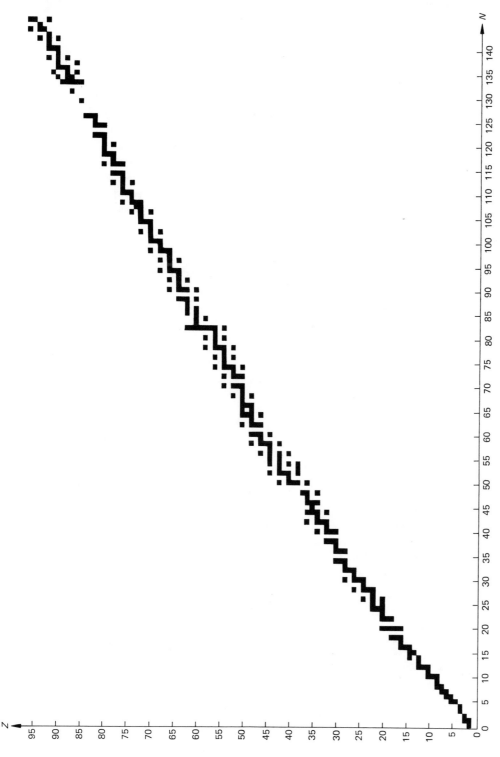

FIGURE 39.11 The values of N and Z that yield nuclei that are stable with respect to β decay.

QUESTION

For the nucleus with $A = 185$, what values of N and Z are predicted by Eq. 39.11? How does the prediction compare with reality?

ANSWER

From Eq. 39.11 we obtain

$$N - Z = \frac{(185)^{5/3} - 1.8(185)}{(185)^{2/3} + 130} = 35$$

Thus $N + Z = 185$ and $N - Z = 35$. This gives $N = 110$ and $Z = 75$. Experimentally it is found that the only β-stable nucleus of mass number 185 is (110, 75), which is ^{185}Re.

In Fig. 39.12 the predicted curve of β stability is superimposed on the chart of β-stable nuclei. For small values of A the β-stability curve follows fairly closely the line $N = Z$, and this indicates that the neutron and proton quantum states are occupied equally. This is clear evidence for the charge independence of the nuclear potential. The electrostatic potential that is certainly not charge independent is negligible for small Z, because the basic electrostatic interaction is much weaker than the nuclear interaction. However, as Z increases the electrostatic interaction becomes more and more important. Because of the long range of the Coulomb potential, a given proton interacts electrostatically with all other protons in the nucleus. Its nuclear interaction takes place only with its near neighbors. The large number of protons compensates for the intrinsic weakness of the electrostatic interaction, and the electrostatic term in the energy becomes more and more significant. The energy of the proton levels is thus shifted upward, and it is advantageous for the nucleus to take on more neutrons than protons in spite of the fact that the symmetry energy acts to equalize proton and neutron numbers. For nuclei with Z larger than 82, the electrostatic term dominates, and all such nuclei are unstable with respect to α decay, which is a method by which the nucleus can reduce the proton number and thus lower the electrostatic potential.

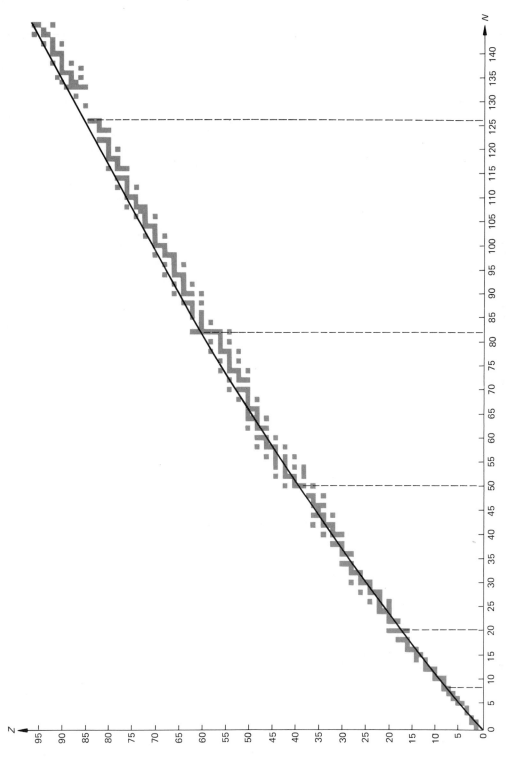

FIGURE 39.12 Comparison between the actual curve of β stability and that predicted using the empirical mass formula.

39.7 NUCLEAR SHELLS

From Fig. 39.13 we see that the discrepancies between the actual binding energies of nuclei and the predictions of the empirical mass formula are not at all random. Particularly beyond $A = 50$ the actual binding energies are not simply scattered on both sides of the smooth curve. Instead in certain regions of the curve the nuclei are consistently more strongly bound than the simple theory predicts. The underlying cause of these consistent deviations is the existence of a definite shell structure within the nuclear energy levels. Just as the energy levels of atoms are not distributed uniformly but are grouped into shells with significant gaps between the highest energy level in one shell and the lowest energy level in the next, the energy levels of nuclei come in distinct groups. The common practice of calling a group of such energy levels a nuclear shell can be misleading. There is no spatial separation in the shells with an associated variation in the particle density as a function of the distance from the center of the nucleus. The particle density within a nucleus is quite uniform. It is the energy levels that are grouped together into distinguishable shells. Both theoretical calculations and experimental observation indicate that the shells become filled at the particle numbers 2, 8, 20, 28, 50, 82, and 126. These "magic numbers" are the same for neutrons and protons. Thus, when N (or Z) approaches a magic number, the binding energy of each further neutron (or proton) is increased until the magic number is reached. Once the magic number is reached, the next neutron (or proton) is much more weakly bound.

In Fig. 39.12 the positions at which either N or Z is equal to a magic number are indicated by straight lines. We should expect the binding energy curve to deviate from the empirical mass formula curve whenever the line

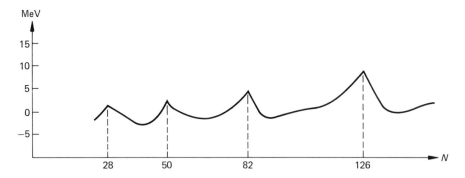

FIGURE 39.13   Actual binding energy of nuclei of neutron number N minus the empirical mass formula prediction of the binding energy. The curve shown gives the average for all nuclei of the given neutron number. At the "magic numbers" 28, 50, 82, and 126 the actual binding energies are larger than those predicted by the empirical mass formula.

of stable nuclei cuts a magic number line. This happens at the following values of (N, Z):

(N, Z)	A
(2, 2)	4
(8, 8)	16
(20, 18)	38
(24, 20)	44
(28, 24)	52
(34, 28)	62
(50, 38)	88
(70, 50)	120
(82, 60)	142
(126, 82)	208

These values of A are indicated by small arrows on Fig. 39.4. Except for the case of $A \leq 40$, the closing of nuclear shells by either neutrons or protons is associated with a significant increase in the binding energy (see Fig. 39.13.)

39.8 NUCLEAR FISSION AND FUSION

It is common knowledge that vast amounts of energy are released in nuclear reactions. Why this should be expected is clear from Fig. 39.4, which gives the binding energy per particle of the most stable nuclei as a function of the mass number of the nucleus. The greatest per-particle binding energies occur around the mass numbers between 50 and 60. This range includes the elements chromium, manganese, iron, cobalt, and nickel. The nucleus ^{30}Si contains 16 neutrons and 14 protons and has a mass of 29.966 atomic mass units (abbreviated u). The nucleus ^{60}Ni contains 32 neutrons and 28 protons and has a mass of 59.915 u. Because the binding energy per particle of ^{60}Ni is greater than that of ^{30}Si, the mass of the ^{60}Ni

nucleus is less than the mass of two ^{30}Si nuclei even though the particle composition is identical. If two ^{30}Si nuclei were to combine to form one ^{60}Ni nucleus, the mass difference would be given off in the form of radiant energy. The mass difference, in atomic mass units, is 2×29.966 u $- 59.915$ u $= 0.017$ u. In kilograms this is

$$0.017 \text{ u} = (0.017 \text{ u})(1.66 \times 10^{-27} \text{ kg/u})$$

$$= 2.8 \times 10^{-29} \text{ kg}$$

This mass difference, converted to energy, would give

$$E = (\Delta m)c^2 = (2.8 \times 10^{-29} \text{ kg})(9 \times 10^{16} \text{ m}^2/\text{s}^2)$$

$$= 2.5 \times 10^{-12} \text{ J}$$

This may not seem like a great deal of energy, but 1 kg of silicon contains 2×10^{25} nuclei. If that kilogram of silicon were converted to nickel, the energy released would be 25 trillion joules. One can obtain the same magnitude of energy release by breaking up a very heavy nucleus to form lighter, more tightly bound nuclei. Processes of the first type are called *fusion* reactions. Those of the second type are called *fission* reactions.

With the large differences that exist in the binding energies of nuclei, it is no surprise that highly exothermic (energy-releasing) nuclear reactions exist. The question that really needs to be answered is why these reactions are not taking place spontaneously all the time. Oxygen and hydrogen atoms interact chemically with a much smaller release of energy than is typical of nuclear reactions, and yet having large amounts of oxygen and hydrogen gas mixed together would present a serious health hazard. All that would be needed is a spark. In contrast ^{108}Cd is a completely stable nucleus even though it contains the same constituents as two ^{54}Cr nuclei and would release substantial energy if it ever underwent fission to form two of the lighter nuclei. The reason most nuclei are quite stable in spite of the fact that one can conceive of strongly exothermic reactions involving them is that all the imaginable reactions have the property that large amounts of energy would have to be temporarily supplied to the nucleus in order to get the reaction to occur. Of course, this initiation energy would then be returned together with the calculated reaction energy. However, if nothing is available to supply the initiation energy, then the reaction will never take place, and the nucleus will be effectively stable.

Two forces are at work in nuclei: the nuclear force and the electrostatic force. In both fission and fusion one of these forces is primarily responsible for the energy released (the *reaction energy*) while the other opposes the reaction and is thus responsible for the energy that must be supplied to initiate the reaction (the *initiation energy*). The roles played by the two forces are reversed in fission and fusion.

FUSION

In a fusion reaction the source of the energy released is, according to our liquid-drop model of the nucleus, the surface energy term. The combined nucleus formed by the fusion of the two smaller nuclei has a smaller surface area than the two separated nuclei. An equivalent way of stating this fact is to say that the nucleons in the product nucleus have, on an average, more nearest neighbors with which they can form strong nuclear force bonds. Therefore the nuclear force is the source of the energy released in the reaction. The electrostatic force opposes the fusion reaction. In order for fusion to occur, the two smaller nuclei must be brought close enough to each other for the attractive nuclear force to take effect. But both smaller nuclei are positively charged objects and thus repel one another. If the two nuclei have charges $Z_1 e$ and $Z_2 e$ and radii r_1 and r_2, respectively, then we may obtain a good estimate of the energy that must be supplied to initiate the fusion reaction by calculating the potential energy of two charged spheres of those radii just at the touching distance (see Fig. 39.14). This potential energy is

$$U = k \frac{Z_1 Z_2 e^2}{r_1 + r_2}$$

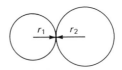

FIGURE 39.14   Because of the short range of the nuclear force, the two fusing nuclei must essentially come into contact in order to react.

QUESTION

What is the estimated initiation energy for the reaction

$$^{30}\text{Si} + {}^{30}\text{Si} \rightarrow {}^{60}\text{Ni?}$$

At what temperature would the average kinetic energy of a silicon atom be equal to that initiation energy?

ANSWER

Using Eq. 39.3, we estimate the radius of ^{30}Si as

$$R = (1.2 \times 10^{-15} \text{ m})(30)^{1/3} = 3.7 \times 10^{-15} \text{ m}$$

The estimated initiation energy is therefore

$$E_{\text{init}} = \left(9 \times 10^9 \frac{\text{J} \cdot \text{m}}{\text{C}^2}\right) \frac{(14)^2 (1.6 \times 10^{-19} \text{ C})^2}{(2)(3.7 \times 10^{-15} \text{ m})}$$

$$= 6 \times 10^{-12} \text{ J}$$

At absolute temperature T the average kinetic energy of an atom is $\frac{3}{2}kT$. Setting this equal to E_{init} we obtain the following formula for T:

$$T = \frac{2}{3}\frac{E_{\text{init}}}{k} = \frac{2}{3}\frac{6 \times 10^{-12} \text{ J}}{1.4 \times 10^{-23} \text{ J/K}}$$

$$= 3 \times 10^{11} \text{ K}$$

This is about 20,000 times the estimated temperature at the center of the sun. It is about a billion times room temperature, which is why this reaction is never seen on earth. Reactions like this do take place, however, in the late stages of the gravitational collapse of stars.

The most potentially useful fusion reaction is the fusion of two deuterium nuclei to form a nucleus of ^{4}He. Deuterium is the second commonest isotope of hydrogen. It is ^{2}H, which indicates that its nucleus is composed of one proton and one neutron. This reaction is particularly advantageous from the point of view of energy generation in that the nucleus of ^{4}He (that is, the α particle) is doubly magic ($N = Z = 2$) and therefore very tightly bound while the ^{2}H nucleus (the deuteron) is quite weakly bound. Thus the reaction energy is large. The supply of deuterium is effectively unlimited; one in every 6000 hydrogen atoms has a deuteron nucleus. Having twice the mass of a normal hydrogen atom the deuterium atoms can be easily separated. Since the reacting nuclei have only one unit of charge, the initiation energy is as small as possible. The nuclear force has a range of a couple of fermis. If we take 3 fm as the distance between each pair of particles, we should expect deuteron-deuteron reactions if we can bring the protons within 9 fm of one another (see Fig. 39.15). The temperature at which the average kinetic energy of the particles would be sufficient to bring the protons that close together is about one billion degrees

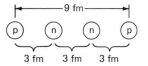

FIGURE 39.15 At a distance of 9 fm between protons, the two deuterons can undergo fusion.

Kelvin, which is far too high to be attained in any controlled situation. However, the particles in a gas do not all have the average kinetic energy. Instead there is a broad distribution of velocities among the particles, so that even at a much lower temperature, say a few million degrees, a small fraction of the particles will have enough energy to undergo fusion. The problem of creating and maintaining a controlled fusion reaction thus is reduced to the technical problem of maintaining a sufficiently dense sample of deuterium for a long enough time at a few million degrees to allow a reasonable fraction of the deuterium nuclei to fuse.

FISSION

The process of fission is the breaking up of a nucleus into a number of smaller constituents. Although the word "fission" is not commonly used in that context, it is clear that α decay is an example of a fission reaction. The term is more often used to describe a process in which a large nucleus breaks up into a few pieces of comparable size.

In order to determine the relative importance of surface energy, electrostatic energy, and symmetry energy in the fission of a large nucleus, we shall evaluate the change in each term separately for the case of an even binary fission of ^{232}U. We thus begin with the nucleus $(N, Z) = (140, 92)$ and end with two nuclei $(70, 46)$. The nucleus $(70, 46)$ falls below the curve of β-stable nuclei and would therefore be expected to transform by β decay to the nearest stable species, namely ^{116}Cd, which is $(68, 48)$. Since β-decay processes have half-lives that are much longer than the time involved in the actual fission process, we can ignore the final β decay since we are interested in the dynamics of fission. Actually this whole process in which the nucleus neatly separates into only two equal pieces is a quite artificial case whose only justification is its simplicity. Table 39.2 shows the value of each term in the empirical mass formula for the initial and final nuclei. Although the total reaction is strongly exothermic, releasing 208 MeV per nucleus of uranium, it is clear that the surface term and the electrostatic term are in strong opposition to one another.

TABLE 39.2   Values of terms in the empirical mass formula for the fission of ^{232}U into ^{116}Pd.

Nucleus	R(fm)	$4\pi R^2 b_{\text{surface}}$ (MeV)	$\dfrac{3}{5}\dfrac{kZ^2e^2}{R}$ (MeV)	$\dfrac{(N-Z)^2}{A}b_{\text{symmetry}}$ (MeV)
^{232}U, (140, 92)	7.37	610	992	231
^{116}Pd, (70, 46)	5.85	385	312	116
Change (2 × second row − first row)		+160	−368	0

Left to itself, the ^{232}U nucleus will undergo spontaneous fission. But it is a rather slow process, with a half-life of 72 yr. The origin of the relative stability of a nucleus such as this, which has available to it such a strongly exothermic decay reaction, is the following fact. For small shape deformations of the ^{232}U nucleus, the electrostatic energy decreases while the surface energy increases, but the surface energy increases more rapidly than the electrostatic energy decreases. The net effect is an energy increase, and thus the nucleus is pulled back to its equilibrium shape. A large shape deformation must occur before the electrostatic term will overcome the surface term and initiate fission (see Fig. 39.16).

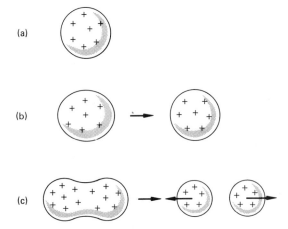

FIGURE 39.16   (a) Equilibrium shape of a large fissionable nucleus. (b) For small deformations the surface tension pulls the nucleus back to its equilibrium shape. (c) For large deformations the Coulomb force dominates and the nucleus breaks up.

We intended to explain why the nucleus is relatively stable and does not fly apart immediately, but we seem to have overshot our mark. We have apparently shown that unless some external forces create a sufficiently large deformation of the nucleus, it will never undergo fission. But in fact the ^{232}U nucleus does spontaneously decay by fission with the above-mentioned half-life. Where does

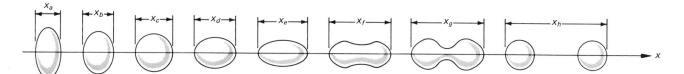

FIGURE 39.17 The three-dimensional shape of the nucleus is obtained by rotating the figure about the x axis.

the deformation energy come from? Our *deus ex machina* is the fact that the nucleus is a quantum mechanical system and cannot be described completely adequately with classical concepts. We shall present the quantum mechanical theory of nuclear fission in only enough detail to communicate the way in which the paradox we have encountered is resolved.

Figure 39.17 shows a sequence of possible shapes of the ^{232}U nucleus arranged in order of elongation in the x direction. The shapes are labeled by a parameter x, which gives the length of the nucleus in the x direction. The equilibrium value of x is x_c. If we plot the sum of the electrostatic and surface energies of the nucleus (which are the only terms relevant to the fission process) as a function of the parameter x, we get a graph of the form shown in Fig. 39.18. As a system satisfying classical dynamics, the nucleus can easily be pictured as undergoing a kind of oscillatory normal-mode vibration between configurations x_b and x_d. Ordinary liquid drops are known to carry out just such vibrations. For the nucleus this mode of vibration is *quantized;* that is, one can write a wave function $\psi(x)$ whose square gives the probability that the nucleus would actually be found with that value of the deformation parameter. Thus we must discuss the question of what the solutions would be of the Schrödinger equation for a system with a potential such as the one

shown in Fig. 39.18. We shall not attack that mathematical problem directly, but instead shall obtain the essential results by using an analogy with a simple string system.

Let us consider the system shown in Fig. 39.19. A string of infinite length is attached to a wall at one end, and some distance from the wall a very heavy mass M is attached to the string. If M were infinite, the string segment would be effectively fixed at both ends and could vibrate in a normal mode as shown. Because the mass M is only large but not infinite, the mass will actually vibrate with a small amplitude. This will create right-going traveling waves of small amplitude on the string beyond the mass. An energy flux is associated with these traveling waves. Since there is no source of energy in the system, the energy being carried away by the traveling wave must be coming from the energy in the normal-mode motion of the string segment. Thus the amplitude of the normal-mode motion will not remain constant, but will slowly diminish as the energy leaks out of the system to the right. The amplitude of vibration of the mass is proportional to the amplitude of vibration of the string segment, and the amplitude of the traveling wave is equal to the amplitude of vibration of the mass. Therefore the amplitude of the traveling

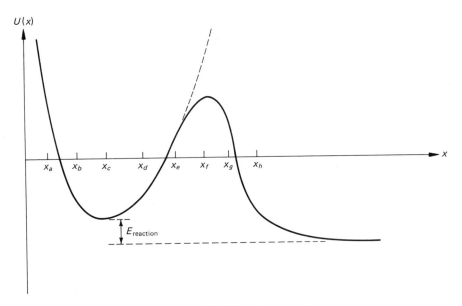

FIGURE 39.18 The "deformation potential" $U(x)$ is the sum of the electrostatic and surface energy terms for various values of the deformation parameter x. It has a local minimum for $x = x_c$ but a true minimum for the separated nucleus at $x = \infty$. The difference between the values of U at these points is the energy released in the fission reaction. The dashed line shows a harmonic oscillator potential that matches $U(x)$ near the local minimum.

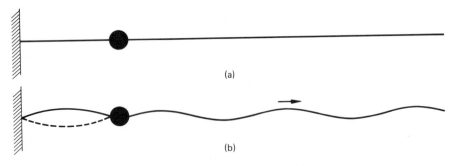

FIGURE 39.19 (a) A string of infinite length has a large mass M attached to it at some distance from the wall. (b) When the string segment vibrates in its fundamental normal mode, the mass moves slightly, thus causing small-amplitude right-going traveling waves to the right of the mass.

wave is proportional to the amplitude of vibration of the string segment:

$$A_{tr} = \alpha A_{vib} \tag{39.12}$$

Squaring Eq. 39.12, we obtain

$$A_{tr}{}^2 = \alpha^2 A_{vib}{}^2$$

But the total energy of vibration of the string segment is proportional to the square of the amplitude of vibration:

$$E \sim A_{vib}{}^2$$

and the rate at which energy is being carried away by the traveling wave is proportional to the square of the amplitude of the traveling wave:

$$\frac{dE}{dt} \sim -A_{tr}{}^2$$

Thus

$$\frac{dE}{dt} = -\lambda E \tag{39.13}$$

where λ is some constant of proportionality that we shall not bother to calculate. This equation gives an exponential decay in the energy of the string segment:

$$E(t) = E(0)e^{-\lambda t} \tag{39.14}$$

Equations 39.13 and 39.14 are similar to Eq. 39.5 and 39.6. This is no accident. As we shall see when we draw the analogy between this string system and the fission decay of a nucleus, this is exactly the same mathematical phenomenon that leads to the law of radioactive decay.

Let us now return to our consideration of the Schrödinger equation for a system with the potential shown in Fig. 39.18. Near the local minimum of the potential it is similar to a harmonic oscillator potential. In that region the lowest-energy solution of the Schrödinger equation is

very similar to the function shown in Fig. 37.16. The ground-state wave function is not exactly zero anywhere; it just approaches zero gradually as $x \to \infty$. But for large x the potential is nothing like a harmonic oscillator potential, and one would therefore expect the wave function to be very different from a harmonic oscillator wave function. And it is. The very-small-amplitude harmonic oscillator wave function gradually changes to a small-amplitude right-going traveling wave (see Fig. 39.20). Near the local minimum the wave function oscillates harmonically in time with a factor $e^{-i\omega t}$, where $\hbar\omega$ is very close to the ground-state energy of the harmonic oscillator. However, the amplitude gradually diminishes in such a way that the probability of finding the system with a deformation parameter near the local minimum decreases exponentially in time. The traveling wave is associated with a flux. When one refers to Fig. 39.17 for the interpretation of the deformation parameter (in particular the rightmost figure labeled x_h), one realizes that the flux is a flux of the fission-decay products. As time goes on it becomes less and less likely that the nucleus will be found intact and more and more likely that it will have undergone fission.

NEUTRON-INDUCED FISSION

In applications of the fission reaction to power generation and nuclear weapons, one does not take a sample of spontaneously radioactive nuclei and simply wait for them to disintegrate at their own pace. Rather one induces fission in the nuclei by supplying the needed deformation energy by means of a neutron-capture reaction. The basic process can be understood in terms of the ideas we have already developed. Suppose a low-energy neutron approaches a

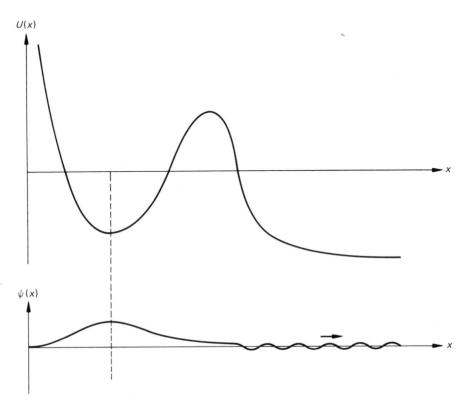

FIGURE 39.20   The wave function $\psi(x)$ looks like a harmonic oscillator wave function near the local minimum but connects to a small-amplitude traveling wave at large x.

large unstable nucleus. The neutron, being uncharged, will not be repelled by the electric charge of the nucleus. When it comes within a few fermis of the nuclear surface, it will come under the influence of the attractive nuclear potential and will be drawn into the nucleus with tremendous force. The energy of the incoming neutron will be quickly transmitted to the nucleus as a whole, and the new nucleus will be left in a highly excited state. Some of the excitation energy will go into the deformation vibrational modes, driving them beyond the range of stability and

causing fission of the nucleus. We have discussed only the simple case of binary fission in which the parent nucleus breaks up into two pieces. More commonly a large number of pieces result, including some single neutrons. These emitted neutrons are then free to move to other nuclei, causing them to undergo fission and to give rise to still more free neutrons. If this process continues undiminished or grows, we say that a *chain reaction* has been initiated. Whether a chain reaction is sustained and the rate at which it grows depends on the balance between the rate at which new neutrons are being created and the rate at which neutrons are being lost through the surface of the sample of fissionable material. This in turn depends on the surface-to-volume ratio of the sample. With increasing sample size the surface-to-volume ratio decreases. Thus there is a certain sample size, referred to as the *critical mass,* for which a chain reaction will sustain itself or grow.

THE CONTROLLED FISSION REACTOR

In a controlled fission reactor the nuclear reaction rate of a large sample of nuclear fuel containing fissionable material is controlled by the insertion of long rods of neutron-absorbing material (called *control rods*) into the fuel sample. If the control rods are inserted all the way, they absorb a large fraction of the emitted neutrons and the fission rate is much reduced. If the control rods are pulled out, those neutrons are free to cause fission reactions, and thus the fission rate increases. However, the concentration of fissionable material in the fuel elements is kept low enough so that even without control rods the reaction rate would not grow without any bound. For the types and amounts of fuel used in nuclear power—generating reactors, the reaction rates that would occur if all control were lost are sufficient to produce enormous amounts of heat energy but are not comparable to those occurring in nuclear explosions.

SUMMARY

The nucleus of an atom consists of two types of nucleons: protons, which are positively charged, and neutrons, which are uncharged. The nuclear force is the same for protons and neutrons and is therefore said to be charge independent. The nuclear force is very strong but very short ranged. Its range is about 10^{-15} m $= 1$ fm.

The structure of a nucleus is qualitatively similar to the structure of a liquid drop. The particle density in the nuclear fluid is about 1.5×10^{44} particles per cubic meter.

The mass number of a nucleus is $A = N + Z$, where N is the neutron number and Z is the proton number. Nuclei with the same number of protons (and thus the same chemical characteristics) but different mass numbers are called different isotopes of the same element.

The mass of the nucleus (N, Z) is related to its binding energy by

$$M(N, Z) = Nm_n + Zm_p - B(N, Z)/c^2$$

The binding energies of nuclei are fairly well predicted by the empirical mass formula:

$$B(N, Z) = Ab_{volume} - 4\pi R^2 b_{surface}$$
$$- \frac{3}{5} \frac{kZ^2 e^2}{R} - \frac{(N - Z)^2}{A} b_{symmetry}$$

where the values of b_{volume}, $b_{surface}$, and $b_{symmetry}$ are given in Eq. 39.4.

The three most important processes of nuclear transformation are known as α decay, β decay, and γ decay. The process of α decay is the emission of a ^{4}He nucleus, which is called an α particle. In β decay a neutron changes into a proton plus an electron and the electron is emitted along with a massless neutrino. In another type of β decay, called electron capture, a proton plus one of the atom's electrons combine and are transformed into a neutron and a neutrino is emitted. The process of γ decay is the emission of a high-energy photon as the nucleus makes a transition from one quantum state to another of lower energy.

The law of radioactive decay states that for all spontaneous decay processes, the rate at which decays occur in a large sample of nuclei is proportional to the number of undecayed nuclei still left. Therefore

$$\frac{dN}{dt} = -\lambda N(t)$$

The number of undecayed nuclei at time t is

$$N(t) = N_0 e^{-\lambda t}$$

where N_0 is the number at time zero. The time for half of the original nuclei to decay is called the half-life and is given, in terms of the decay constant λ, by

$$T = \frac{\ln 2}{\lambda}$$

Of all the possible combinations of integers N and Z, only a small fraction actually describe stable nuclei. The stable nuclei all lie close to the curve given by the equation

$$N - Z = \frac{A^{5/3} - 1.8A}{A^{2/3} - 130}$$

The energy levels of nuclei are grouped into shells similar to atomic shells. The shell closings occur when N or Z is equal to one of the magic numbers 2, 8, 20, 28, 50, 82, or 126.

Nuclear fission is a process in which a large nucleus breaks up into a number of smaller pieces with the release of nuclear energy. Nuclear fusion is a process in which two smaller nuclei fuse, with the release of energy, to form a larger, more tightly bound nucleus.

PROBLEMS

39.A.1 Assuming uniform density, calculate the diameter of the ^{238}U nucleus.

39.A.2 The ^{16}O nucleus has a mass of 15.9906 u. What is its binding energy?

39.A.3 Using the empirical mass formula, estimate the mass of the ^{232}Th nucleus. Compare your result with the actual mass.

39.A.4 Using the empirical mass formula, calculate the binding energy of ^{12}C.

39.A.5 What is the wavelength of the γ ray emitted when the ^{78}Se nucleus decays from its excited state at 613.2 keV to its ground state (see Fig. 39.7)?

39.A.6 ^{32}P decays by β^- decay with a half-life of 14.2 days. What is its decay constant? What is the resulting nucleus? How long would it take for 0.9 of the original ^{32}P to have undergone decay?

39.A.7 According to Eq. 39.11, what should be the β-stable isotope of mass number 80?

39.A.8 Estimate the initiation energy for the reaction ^{24}Mg + ^{4}He → ^{28}Si.

39.B.1 The nuclear force becomes strongly repulsive at an internucleon distance of about 0.5 fm. If we construct a sphere about the center of each nucleon of diameter 0.5 fm and call it the repulsive core of the nucleon, then, in a large nucleus, what fraction of the nuclear volume is occupied by the repulsive cores of the nucleons?

39.B.2 Estimate the number of nucleons that are within a distance of 2 fm from the surface of a nucleus of mass number A. Evaluate your estimate for the cases ^{12}C and ^{238}U.

39.B.3 Use the empirical mass formula to estimate the excitation energy (the energy above the ground-state energy) that the ^{235}U nucleus would be left with after the slow neutron-capture reaction ^{234}U $+ n \rightarrow {}^{235}$U.

39.B.4 Draw a rough graph of the ratio of electrostatic energy to surface energy of stable nuclei from $A = 4$ to $A = 238$.

39.B.5 Calculate separately the electrostatic, surface, and symmetry energies of each of the following three nuclei: ^{9}Be, ^{59}Co, and ^{209}Bi.

39.B.6 The nuclei ^{38}Ca and ^{38}Ar have $(N, Z) = (18, 20)$ and $(20, 18)$, respectively. They are called *mirror nuclei.* According to the empirical mass formula, the mass difference of two mirror nuclei is attributable entirely to the difference in electrostatic energy. All other terms in the formula have the same values for both nuclei. Use the empirical mass formula to estimate the difference in electrostatic energies for these two nuclei and compare your answer with the actual difference in their binding energies.

39.B.7 Suppose a particular nucleus had a probability of 0.001 of emitting an α particle during any second and also a probability of 0.002 of emitting a positron during any second. What would be the half-life of the nucleus (i.e., the time required for half the nuclei to decay by any means)?

39.B.8 Using the actual masses of the nuclei involved, calculate the total energy that would be released by the fusion of 1 kg of tritium (^{3}H) with an equal number of protons to form ^{4}He nuclei.

39.C.1 In the gravitational collapse of a white dwarf to form a neutron star, the nucleus (N, Z) swallows up Z electrons to become a pure neutron nucleus $(A, 0)$. These neutron nuclei then fuse to form the giant, almost pure-neutron, nucleus that is the neutron star. Considered as a nuclear reaction, the reaction $(N, Z) + Ze^- \rightarrow (A, 0)$ is highly endothermic (energy absorbing). The energy to drive the reaction comes from the gravitational pressure. The energy absorbed from the gravitational field when the atom collapses is $E = pV$, where p is the pressure and V is the original volume of the atom. Taking the radius of the Fe atom as 1 Å, estimate the pressure required to drive this reaction for the nucleus ^{56}Fe.

39.C.2 Using the analysis that goes with Fig. 39.6 and the numerical value of $b_{symmetry}$, estimate the spacing between nucleon energy levels in ^{78}Se. Compare this with the level spacing shown in Fig. 39.7.

39.C.3 Calculate the electrostatic force on a proton at the surface of a nucleus (N, Z).

39.C.4 In the β^- decay of a stationary neutron, the proton and electron come off with the relative angle shown in the figure. The electron has an energy of 3×10^{-15} J. What are the energy and direction of the antineutrino?

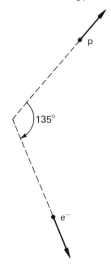

39.C.5 Using Eq. 39.11 calculate four well-spaced points on the curve of β-stable nuclei and draw the curve.

39.C.6 A ^{235}U nucleus breaks up into the following components:

$$^{235}U \rightarrow {}^{124}Sn + 2\ {}^{48}Ca + {}^4He + 11\ n$$

What is the energy released by the reaction?

39.C.7 Draw an approximate curve of the energy of an α particle as a function of x, where x is a coordinate along an axis through the center of a ^{232}U nucleus (see the following figure). At the surface of the nucleus you can only estimate the nuclear interaction energy, but in other regions try to get the most accurate value you can.

39.C.8 Carry out the detailed analysis leading from Eq. 39.10 to Eq. 39.11.

APPENDIXES   Appendix A

A.1 MATHEMATICAL FORMULAS

QUADRATIC FORMULA

If $ax^2 + bx + c = 0$, then

$$x = \frac{-b \pm \sqrt{b^2 - 4ac}}{2a}$$

TRIGONOMETRIC IDENTITIES

$$\sin(x \pm y) = \sin x \cos y \pm \cos x \sin y$$

$$\cos(x \pm y) = \cos x \cos y \mp \sin x \sin y$$

$$\sin 2x = 2 \sin x \cos x$$

$$\cos 2x = \cos^2 x - \sin^2 x$$

SERIES EXPANSIONS

$$(1 + x)^n = 1 + nx + \frac{n(n-1)}{2!} x^2$$

$$+ \frac{n(n-1)(n-2)}{3!} x^3 + \cdots$$

$$\sin x = x - \frac{x^3}{3!} + \frac{x^5}{5!} - \cdots$$

$$\cos x = 1 - \frac{x^2}{2!} + \frac{x^4}{4!} \cdots$$

$$e^x = 1 + x + \frac{x^2}{2!} + \frac{x^3}{3!} + \cdots$$

DERIVATIVES

$$\frac{dx^n}{dx} = nx^{n-1}$$

$$\frac{d \ln x}{dx} = \frac{1}{x}$$

$$\frac{de^x}{dx} = e^x$$

$$\frac{d \sin x}{dx} = \cos x$$

$$\frac{d \cos x}{dx} = -\sin x$$

$$\frac{d \tan x}{dx} = \sec^2 x$$

$$\frac{d \cot x}{dx} = -\csc^2 x$$

INDEFINITE INTEGRALS

$$\int x^n \, dx = \frac{x^{n+1}}{n + 1}$$

$$\int \frac{dx}{x} = \ln |x|$$

$$\int e^x \, dx = e^x$$

$$\int \frac{dx}{1 + x^2} = \arctan x$$

$$\int \frac{dx}{\sqrt{1 - x^2}} = \arcsin x$$

$$\int \frac{dx}{x\sqrt{x^2 - 1}} = \text{arcsec } x$$

$$\int \frac{1}{(x^2 + a)^{3/2}} = \frac{x}{a\sqrt{x^2 + a}}$$

$$\int \sin x \, dx = -\cos x$$

$$\int \cos x \, dx = \sin x$$

$$\int \tan x \, dx = \ln |\sec x|$$

$$\int \cot x \, dx = \ln |\sin x|$$

A.2 PROPERTIES OF LOGARITHMS AND EXPONENTIALS

If $x = \log y$, then $y = 10^x$

$\log xy = \log x + \log y$

$\log \dfrac{x}{y} = \log x - \log y$

$\log x^y = y \log x$

$10^{x+y} = 10^x 10^y$

$(10^x)^y = 10^{xy}$

All the above formulas remain valid if log is replaced by ln and 10 is replaced by $e = 2.718$.

$\log e = 0.4343$

$\ln 10 = 2.303$

$\ln x = 2.303 \log x$

A.3 PARTIAL DERIVATIVES

The partial derivative of $f(x, y)$ with respect to x is defined as the ordinary derivative of $f(x, y)$ with respect to x with y treated as a constant. For example,

$$\frac{\partial}{\partial x}(x^2y + y^2) = 2xy$$

The differential of a function of two variables is given by

$$df(x, y) = \frac{\partial f}{\partial x}\,dx + \frac{\partial f}{\partial y}\,dy$$

For example, if

$$f(x, y) = x^2y + y^2$$

then

$$df = 2xy\,dx + (x^2 + 2y)\,dy$$

A.4 COMPLEX NUMBERS

The complex number $z = x + iy$ may be pictured as a two-dimensional vector in a complex plane. Its components are the real part x and the imaginary part y (see the accompanying figure). The rule for multiplying two complex numbers follows directly from the fact that $i^2 = -1$. It is

$$(x_1 + iy_1)(x_2 + iy_2) = (x_1x_2 - y_1y_2)$$
$$+ i(x_1y_2 + y_1x_2)$$

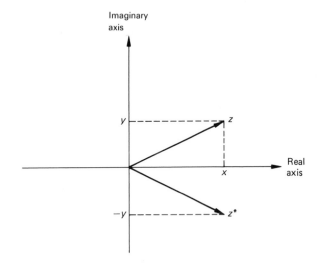

The complex conjugate of z is defined as $z^* = x - iy$. The square of the magnitude of z is given by

$$|z|^2 = x^2 + y^2 = z^*z$$

If we use the series expansion for e^x (given above) to evaluate $e^{i\theta}$, where θ is a real constant, we see that

$$e^{i\theta} = 1 + i\theta + \frac{i^2\theta^2}{2!} + \frac{i^3\theta^3}{3!} + \frac{i^4\theta^4}{4!} + \frac{i^5\theta^5}{5!} + \cdots$$

$$= 1 + i\theta - \frac{\theta^2}{2!} - i\frac{\theta^3}{3!} + i\frac{\theta^4}{4!} + i\frac{\theta^5}{5!} + \cdots$$

$$= \left(1 - \frac{\theta^2}{2!} + \frac{\theta^4}{4!} - \cdots\right) + i\left(\theta - \frac{\theta^3}{3!} + \frac{\theta^5}{5!} - \cdots\right)$$

$$= \cos\theta + i\sin\theta$$

The complex number $z = e^{i\theta}$ can be pictured as a vector of magnitude 1 in the complex plane (see the following figure). Thus

$$|e^{i\theta}|^2 = (e^{i\theta})^*(e^{i\theta}) = \cos^2\theta + \sin^2\theta = 1$$

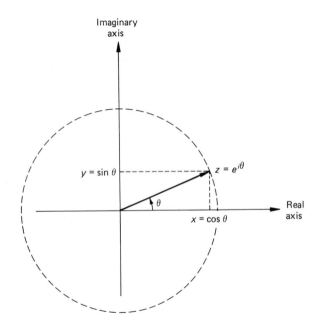

Appendix B

TRIGONOMETRIC FUNCTIONS

Angle	Sine	Cosine	Tangent	Angle	Sine	Cosine	Tangent
0°	0.000	1.000	0.000				
1°	.017	1.000	.017	46°	0.719	0.695	1.036
2°	.035	0.999	.035	47°	.731	.682	1.072
3°	.052	.999	.052	48°	.743	.669	1.111
4°	.070	.998	.070	49°	.755	.656	1.150
5°	.087	.996	.087	50°	.766	.643	1.192
6°	.105	.995	.105	51°	.777	.629	1.235
7°	.122	.993	.123	52°	.788	.616	1.280
8°	.139	.990	.141	53°	.799	.602	1.327
9°	.156	.988	.158	54°	.809	.588	1.376
10°	.174	.985	.176	55°	.819	.574	1.428
11°	.191	.982	.194	56°	.829	.559	1.483
12°	.208	.978	.213	57°	.839	.545	1.540
13°	.225	.974	.231	58°	.848	.530	1.600
14°	.242	.970	.249	59°	.857	.515	1.664
15°	.259	.966	.268	60°	.866	.500	1.732
16°	.276	.961	.287	61°	.875	.485	1.804
17°	.292	.956	.306	62°	.883	.469	1.881
18°	.309	.951	.325	63°	.891	.454	1.963
19°	.326	.946	.344	64°	.899	.438	2.050
20°	.342	.940	.364	65°	.906	.423	2.145
21°	.358	.934	.384	66°	.914	.407	2.246
22°	.375	.927	.404	67°	.921	.391	2.356
23°	.391	.921	.424	68°	.927	.375	2.475
24°	.407	.914	.445	69°	.934	.358	2.605
25°	.423	.906	.466	70°	.940	.342	2.747
26°	.438	.899	.488	71°	.946	.326	2.904
27°	.454	.891	.510	72°	.951	.309	3.078
28°	.469	.883	.532	73°	.956	.292	3.271
29°	.485	.875	.554	74°	.961	.276	3.487
30°	.500	.866	.577	75°	.966	.259	3.732
31°	.515	.857	.601	76°	.970	.242	4.011
32°	.530	.848	.625	77°	.974	.225	4.331
33°	.545	.839	.649	78°	.978	.208	4.705
34°	.559	.829	.675	79°	.982	.191	5.145
35°	.574	.819	.700	80°	.985	.174	5.671
36°	.588	.809	.727	81°	.988	.156	6.314
37°	.602	.799	.754	82°	.990	.139	7.115
38°	.616	.788	.781	83°	.993	.122	8.144
39°	.629	.777	.810	84°	.995	.105	9.514
40°	.643	.766	.839	85°	.996	.087	11.43
41°	.656	.755	.869	86°	.998	.070	14.30
42°	.669	.743	.900	87°	.999	.052	19.08
43°	.682	.731	.933	88°	.999	.035	28.64
44°	.695	.719	.966	89°	1.000	.017	57.29
45°	.707	.707	1.000	90°	1.000	.000	

COMMON LOGARITHMS

10^x	x	10^x	x	10^x	x
x	$\log x$	x	$\log x$	x	$\log x$
1.0	0.0000	4.0	0.6021	7.0	0.8451
1.1	.0414	4.1	.6128	7.1	.8513
1.2	.0792	4.2	.6232	7.2	.8573
1.3	.1139	4.3	.6335	7.3	.8633
1.4	.1461	4.4	.6435	7.4	.8692
1.5	.1761	4.5	.6532	7.5	.8751
1.6	.2041	4.6	.6628	7.6	.8808
1.7	.2304	4.7	.6721	7.7	.8865
1.8	.2553	4.8	.6812	7.8	.8921
1.9	.2788	4.9	.6902	7.9	.8976
2.0	.3010	5.0	.6990	8.0	.9031
2.1	.3222	5.1	.7076	8.1	.9085
2.2	.3424	5.2	.7160	8.2	.9138
2.3	.3617	5.3	.7243	8.3	.9191
2.4	.3802	5.4	.7324	8.4	.9243
2.5	.3979	5.5	.7404	8.5	.9294
2.6	.4150	5.6	.7482	8.6	.9345
2.7	.4314	5.7	.7559	8.7	.9395
2.8	.4472	5.8	.7634	8.8	.9445
2.9	.4624	5.9	.7709	8.9	.9494
3.0	.4771	6.0	.7782	9.0	.9542
3.1	.4914	6.1	.7853	9.1	.9590
3.2	.5051	6.2	.7924	9.2	.9638
3.3	.5185	6.3	.7993	9.3	.9685
3.4	.5315	6.4	.8062	9.4	.9731
3.5	.5441	6.5	.8129	9.5	.9777
3.6	.5563	6.6	.8195	9.6	.9823
3.7	.5682	6.7	.8261	9.7	.9868
3.8	.5798	6.8	.8325	9.8	.9912
3.9	.5911	6.9	.8388	9.9	.9956

EXAMPLES

(a) $\log 2.9 = 0.4624$

(b) $\log 29 = \log(10)(2.9) = \log 2.9 + \log 10 = 0.4624 + 1 = 1.4624$

(c) $\log 0.29 = \log\left(\dfrac{2.9}{10}\right) = \log 2.9 - \log 10 = 0.4624 - 1 = -0.5376$

(d) Let $e^{2.9} = y$; then $\log y = \log e^{2.9} = 2.9 \log e = 2.9(0.4343) = 1.26$. So $y = 10^{1.26} = 10^{(1+0.26)} =$ $(10)(10^{0.26}) = (10)(1.8) = 18$. Therefore $e^{2.9} = 18$.

Appendix C

ASTRONOMICAL DATA

Mass of earth $= 5.97 \times 10^{24}$ kg
Radius of earth (average) $= 6.3 \times 10^6$ m
Distance of earth from sun (average) $= 1.50 \times 10^{11}$ m

Mass of moon $= 7.34 \times 10^{22}$ kg
Radius of moon (average) $= 1.74 \times 10^6$ m
Distance of moon from earth (average) $= 3.8 \times 10^8$ m

Mass of sun $= 1.99 \times 10^{30}$ kg
Radius of sun $= 6.96 \times 10^8$ m

CHAPTER 1

1.A.1. 2.7 mi
1.A.3. 36°, 0.63 rad, 0.1 rev
1.A.5. (a) 8.7, 5.0
     (b) −5.0, 8.7
     (c) 5.0, −8.7
1.A.7. (a) −132, 188
     (b) 229.7 at 125°
1.B.5. 7.2 m

CHAPTER 2

2.A.1. 0.2 s
2.A.3. 44 ft
2.A.5. 1.5×10^8 cm/s^2, 1.5×10^5 g
2.A.7. (a) v/R
     (b) v
     (c) $2v$
     (d) $\sqrt{2}v$
     (e) v^2/R
     (f) v^2/R
2.B.1. 600 ft
2.B.3. (a) 80 ft/s
     (b) 128 ft
2.B.5. 4 ft
2.B.7. 27° south of east
2.B.9. 201 mi/h
2.C.1. 5 h, 1800 ft
2.C.3. 4.0 ft

CHAPTER 3

3.A.1. 60
3.A.3. (a) 6000 N
     (b) 1950 N
3.B.1. $s_1 = m_2 \ell / (m_1 + m_2)$
     $s_2 = m_1 \ell / (m_1 + m_2)$
3.B.3. (a) 24 min
     (b) 25.2 min
3.B.5. $(m/M)g$

CHAPTER 4

4.A.1. 16 ft/s²
4.A.3. 18 N, 12 N, 6 N
4.A.5. 167 lb, 133 lb
4.A.7. $v' = \sqrt{r/2xv}$
4.A.9. 5.6 ft/s²
4.B.1. 19.6 m/s²
4.B.3. 7.9 s
4.B.5. 0
4.B.7. (a) 5.6×10^{-3}
   (b) 2.2
4.B.9. 16.7 lb

CHAPTER 5

5.A.1. (a) 5 m/s
   (b) 5 m/s
5.A.3. 0
5.A.5. 4.0 m/s
5.A.9. (a) 0, -0.67
   (a) Stable, unstable
   (c) $-2axy$, $-ax^2$
5.B.3. 5 ft
5.B.5. (a) 0.97 ft
   (b) 17.9 ft/s
5.B.7. 17 lb

CHAPTER 6

6.A.1. More HCl and O_2 and less H_2O
   and Cl_2
6.A.3. 5.7 cal
6.A.5. 19,000 m
6.A.7. Increase
6.B.1. 3.8 days
6.B.3. 66% of the time

CHAPTER 7

7.A.1. (a) 46 N · m
   (b) 70 N · m
7.A.3. 1005 ft/s
7.A.5. $x = -0.75$
   $y = 0.75$
7.A.7. Same velocity
7.B.1. (a) 392 cm/s
   (b) 5800 dyn, 76 cm/s
7.B.5. $\frac{1}{2}g$ sin 37 = 9.6 ft/s²
7.B.9. $\frac{7}{4}$ ml²
7.C.3. 2000 r/min
7.C.5. 13.3 rad/s
7.C.7. 1.5 ft

CHAPTER 8

8.A.1. $\frac{1}{2}$ lb, $1\frac{1}{8}$ lb, 2.1 lb
8.A.3. $2.40
8.A.5. 0.8 ft
8.A.7. 644 lb, 633 lb
8.B.1. 4.3 ft
8.B.3. 1155 lb
8.B.5. 33.3 lb, 20 lb, 20 lb
8.B.7. 450 lb, 600 lb
8.C.1. (a) The right edge of the lowest
   board should be $\ell/4$ to the left of the
   right edge of the middle board.
   (c) 10 boards

CHAPTER 9

9.A.1. 0.31 Hz
9.A.3. (a) π s (π seconds)
   (b) $1/\pi$ Hz
   (c) 8 cm
   (d) 5 rad
   (e) 16 cm/s
   (f) 32 cm/s²
9.B.1. 0.45 s
9.B.3. $2\pi\sqrt{2\ell/3g}$

9.B.5. (a) $\sqrt{2gl}$
(b) $(\pi/2)\sqrt{gl}$
(c) 0.9
9.C.1. π s
9.C.3. (a) $\frac{2}{3}\pi$ s
(b) 1.1 m
9.C.5. $1/\sqrt{3}$

CHAPTER 10

10.A.3. (a) 71 mi/h
(b) 24 mi/h
10.A.5. 50 lb
10.B.1. 0.29 in
10.B.3. (a) $V_2 = 8V_1$, 8 times
(b) $A_2 = 4A_1$, 4 times

CHAPTER 11

11.A.1. 90.9 in²
11.A.3. $\frac{3}{4}$
11.A.5. 1.5×10^5 dyn/cm²
11.A.7. Unstable
11.A.9. 3.6 ft/s
11.B.1. (a) 0.98 J
(b) 0.98 W
(c) 2.2×10^9 J
(d) 1600 ft
11.B.3. 16 g/cm³
11.B.5. 11 lb
11.B.7. 23.7 ft/s
11.B.9. 6240 lb/ft²
11.C.1. $T = 0.10$ s
11.C.3. 7.3×10^7 dyn

CHAPTER 12

12.A.1. 29.7 cm
12.A.3. The stress is twice as great in the larger vessel.
12.A.5. 1.5×10^4 dyn

12.B.3. Eight times as long
12.B.5. 39 dyn
12.B.7. 0.59 cm
12.B.9. No
12.C.1. Same height as if the tube were straight and vertical

CHAPTER 13

13.A.1. None
13.A.3. 480 m/s, 2.9 h
13.A.5. 4.2
13.A.7. 190 m/s
13.B.1. 300 K
13.B.3. 31.5 lb/in²
13.B.5. 1.7¢/mol, 1.5¢/mol
13.B.7. 4×10^6 J, 4×10^4 s
13.C.1. $n = n_0 \exp(-\frac{1}{6} Avt/v)$

CHAPTER 14

14.A.1. 4.5 J
14.B.1. 7
14.B.3. Left
14.B.5. Left
14.C.1. (a) 90
(b) 60

CHAPTER 15

15.A.1. 10^4 J
15.A.3. 0.04 K, $p = 1$ N/m²
15.B.1. -0.97 J
15.B.3. (a) 784 N/m²
(b) 1.4°
15.B.5. $S = +2\sqrt{3E} - 4/v$
15.B.7. 600 J
15.B.9. 500 J
15.C.3. $T = 0.19$, $P = 3.8$

CHAPTER 16

16.A.1. $F = E - \frac{2}{3}E \ln(E^{3/2}V)$,
$G = \frac{5}{3}E - \frac{2}{3}E \ln(E^{3/2}V)$

16.B.1. $F = -2(nVT^3)^{1/2}$,
$G = -nT^3/p$

16.B.3. $F = \frac{3}{2}nRT - nRT \ln V - \frac{3}{2}nRT$
$\ln(\frac{3}{2}nRT)$
$G = \frac{5}{2}nRT - nRT \ln (nRT/p)$
$- \frac{3}{2}nRT \ln(\frac{3}{2}nRT)$

CHAPTER 17

17.A.1. $g' = 10.5$ g
17.A.3. 2.8×10^{-9}
17.B.1. (a) 2×10^4 m
(b) 0.003
(c) 1.2×10^5 m
(d) Yes
17.B.3. 0.99989, 0.33, 0.037
17.B.5. 1.3×10^5 g
17.B.7. 1.1×10^{-4} mol/m³
17.B.9. 0.85 mol/m³, 0.15 mol/m³
17.C.1. 0.30 mol

CHAPTER 18

18.A.1. 0.39
18.A.3. (a) 37.3 J
(b) 27.3 J
18.B.1. 0.6
18.B.3. One ball in each of the three lowest levels
18.B.5. Room will be heated more with the door open than with the door closed.
18.B.7. $e_{ref} = 1/e_{eng} - 1$
18.B.9. Q_{hot}(heat pump)
$= 12Q_{hot}$(radiant)

CHAPTER 19

19.A.1. $P = \alpha^3 bVT^3$
19.A.3. (a) 0.006 g
(b) 0.007 J
19.A.5. 97 W
19.A.7. 90°C
19.B.1. 10.5°C
19.B.3. 37.3°C
19.B.5. 4200 W
19.B.7. 1.3×10^3 N/m²
19.C.1. (a) No (b) No (c) Yes, 0.004°C

CHAPTER 20

10.A.1. 1.02 kcal/s
20.A.3. 5900 W
20.A.5. 6.3 W
20.A.7. 4.83×10^{-5} m
20.B.1. 20°C
20.B.3. 19°C
20.B.5. No
20.B.7. 1/size
20.C.1. 10.47 h
20.C.3. 703 ft²

CHAPTER 21

21.A.1. 7.9×10^3 N
21.A.3. 4.9×10^{-5} C
21.A.5. 1.42×10^{15} rad/s
21.A.7. 0.56 m from the 5-μC charge
21.A.9. (1) σ/ϵ_0 left
(2) $\sigma/2\epsilon_0$ left
(3) $\sigma/2\epsilon_0$ right
(4) σ/ϵ_0 right
21.B.1. 1.9°
21.B.3. 0.12 cm
21.B.5. 2.5×10^{-3} μC/m²
21.B.7. $4\sqrt{2}kq/a^2$
21.B.9. $2akq/x^3$, kqa/y^3
21.C.1. $\sigma^2/2\epsilon_0$
21.C.3. 2.5 N

CHAPTER 22

22.A.1. $V_A = 0 = V_B$
$V_C = (4kq/a)(1 - 1/\sqrt{5})$
22.A.3. 10^4 V/m
22.A.5. 5.9×10^5 m/s
22.A.7. $E_x = -2(x + y)$,
$E_y = -(2x + z)$, $E_z = -y$
22.B.1. (a) 1 J
(b) 1 J
(c) 1 J
(d) 1 J
22.B.3. $E_x = 200(x - 1)$
22.B.5. (a) $\dfrac{k}{x} + \dfrac{2k}{1 - x}$
(b) $\dfrac{k}{x^2} - \dfrac{2k}{(1 - x)^2}$
22.B.7. (a) 0
(b) $\dfrac{kq^2}{a}$
(c) $\dfrac{kq^2}{a}\left(1 + \dfrac{1}{\sqrt{2}}\right)$
(d) $\dfrac{kq^2}{a}\left(2 + \dfrac{1}{\sqrt{2}}\right)$
(e) $\dfrac{kq^2}{a}(4 + \sqrt{2})$
22.B.9. $\oint dV = 0$
22.C.1. (b) $2kp/r^3$
22.C.3. 0.41 m from the 1-C charge

CHAPTER 23

23.A.1. -2 μC, 1 μC
23.A.5. (a) All on outer surface
(b) -2 μC on the inner surface
and 6 μC on the outer surface
23.B.1. $\rho r/2\epsilon_0$
23.B.3. $V_{in} = -\dfrac{1}{2}\dfrac{kqr^2}{R^3} + \dfrac{\frac{3}{2}kq}{R}, V_{out} = \dfrac{kq}{r}$
23.B.5. (a) 1.8×10^7 V/m, 0,
7.2×10^5 V/m
(b) -40 μC/m²
(c) 9.9 μC/m²
23.C.1. 1.1×10^5 N · m²/C
23.C.3. $2\epsilon_0 x$
23.C.5. $\frac{1}{2}kq^2/R$
23.C.6. Perpendicular to the plane

CHAPTER 24

24.A.1. (a) 4 V, 2 V
(b) 2 μC
24.A.3. 0.57 V
24.A.5. 1.6 μF
24.B.1. $2\epsilon_0$ A/s
24.B.3. 2.2 V
24.B.5. 3.06 μF
24.B.7. 3.8 V
24.B.9. $q^2/2\epsilon_0$ A
24.B.11. 0.67
24.C.1. (a) $2\epsilon_0$ A/s
(b) $3\epsilon_0$ A/s
24.C.3. Connect n_2 capacitors in
series n_1 times. Next connect the n_1
banks of capacitors in parallel.
24.C.5. 9.3×10^{-11} F

CHAPTER 25

25.A.1. 5.4×10^{-6} m²
25.A.3. 6 h
25.A.5. (a) 125 Ω
(b) 250 V
25.B.1. (a) 4.0×10^{-6} m/s
(b) 1.2×10^5 m/s
25.B.3. (a) 24.6 Ω
(b) 16 min
25.C.1. $2\rho/\pi$
25.C.3. 0.22 A

CHAPTER 26

26.A.1. 3.11 Ω
26.A.3. (a) 10^4 r
(b) 10^8 r
26.A.5. 500 W
26.B.1. n_1 banks in series of n_2
resistors in parallel
26.B.3. 0.68 A
26.B.5. 10 A
26.B.7. (a) 1 Ω
(b) 6 V
26.B.9. 9×10^8 V
26.C.1. $\frac{5}{6}$ Ω

CHAPTER 27

27.A.1. (a) The particle will revolve counterclockwise in a circular orbit that drifts in the negative y direction.
    (b) Both the rotation and the drift would be reversed.

27.A.3. 0.3 N

27.A.5. 1.3 cm

27.A.7. To the left

27.B.1. A circular path that drifts in the negative y direction

27.B.3. 14.7 A

27.B.5. 200 m/s

27.B.7. 0.006 N · m

27.C.1. $\frac{1}{4}\sigma\omega R^4 B$

CHAPTER 28

28.A.1. Attract

28.A.3. (a) 8×10^{-7} T
    (b) 2×10^{-7} T

28.A.5. 2.0×10^{-3} m

28.A.7. $0, -\mu_0 i, \mu_0 i, -\mu_0 i$

28.B.1. (a) 0
    (b) $B_x = 2 \times 10^{-7}$ T, $B_y = -2 \times 10^{-7}$ T

28.B.3. (a) $\mu_0 i r / 2\pi R^2$
    (b) $\mu_0 i / 2\pi r$

28.B.5. 2×10^{-6} N

28.B.7. $B = \mu_0 Ni / 2\pi r$

28.C.1. $\dfrac{\mu_0 q \omega}{2\pi R^2}\left(\sqrt{R^2 + z^2} + \dfrac{z^2}{\sqrt{R^2 + z^2}} - 2|z|\right)$

CHAPTER 29

29.A.1. (a) Opposite to that of the stationary ring
    (b) In the same direction as the initial current in the stationary ring

29.A.3. 3.9×10^{-5} H

29.A.5. 0.47 A

29.B.1. $L\dfrac{di}{dt} + iR + 6t + 1 = 0$

29.B.3. $\mu_0 A N^2 / 2\pi R$

29.B.5. $\omega BA \sin \omega t$

29.B.7. $\mu_0 N_1 N_2 \alpha \pi r_1^2 / \ell$

29.B.9. 2×10^{-4} A

29.C.1. $\dfrac{\mu_0 i}{2\pi x}\ell v,\ \dfrac{\mu_0 i v}{2\pi}\ln\left(1 + \dfrac{\ell}{a}\right)$

29.C.3. 2.9×10^{-7} H

CHAPTER 30

30.B.3. $\dfrac{2\pi\epsilon_0}{\ln(r_2/r_1)}$

30.B.5. $\dfrac{\mu_0}{2\pi}\ln\left(\dfrac{r_2}{r_1}\right)$

30.B.7. 55 V/m, 1.8×10^{-7} T

CHAPTER 31

31.A.5. 0.048 cm

31.A.7. 0.05 kg/m

31.B.5. $u(x, t) = \dfrac{\sin(x - 3t)}{x - 3t}$

31.B.7. 0.2 W

31.B.9. Four times

31.B.13. 1.02 W

31.C.1. 3.5×10^{-4} in to 1.05 in

31.C.3. (a) 50 m/s
    (b) $u(x, t) = 0.03 \cos(4x - 200t)$
    (c) 0.016 m
    (d) 1.57 m, 0.063 s

31.C.5. $\mu\omega_n^2 A^2 L/4$, where $\omega_n = n\pi v/L$

31.C.9. $f = \dfrac{v}{[2L(1 - \alpha/\pi)}$

CHAPTER 32

32.A.1. 82.5 Hz
32.A.3. 1.44×10^{-3}
32.B.1. 2750 Hz
32.B.3. 1.7×10^{-6} m
32.B.5. 0.1 mm
32.C.1. 6.8×10^{-3} W
32.C.3. 6.06 rad

CHAPTER 33

33.A.1. $A = 67 \times 10^{-6}$ m,
$P = 1.05 \times 10^4$ W
33.A.3. $43.2°$
33.A.5. 0.01 rad
33.A.7. 1.36×10^8 m/s
33.B.1. 7.36×10^{-7} m, 22.1 km
33.B.3. $\lambda = 6 \times 10^{-7}$ m
33.B.5. 1.75 ft

CHAPTER 34

34.A.1. (a) 40 cm, real
(b) -16 cm, virtual
(c) -13.7 cm, virtual
(d) 6 cm, real
(e) -2.67 cm, virtual
(f) -16 cm, virtual
(g) 8 cm left of the right surface, virtual
34.B.1. (a) 6 cm left of right lens, virtual
(b) 2.69 cm left of right lens, virtual
(c) 2.4 cm right of right lens, real
34.B.3. $f_1 = 25$ cm
$f_2 = 66.7$ cm
$f_{12} = 18.2$ cm
34.B.5. 0.063 μm
34.C.1. (a) $f = -20$ cm
(b) -16.7 cm
(c) 10^{-2} rad
34.C.3. (a) 12.5
(b) 25 μm

CHAPTER 35

35.A.1 $y' = 8 - gt^2/2$, $x' = 7t$
35.A.5. $x = 1.48 \times 10^9$ m,
$t = 11.1$ s
35.A.7. $p_i = 3 \times 10^{-19}$ kg m/s,
$p_f = 2.4 \times 10^{-19}$ kg m/s
35.A.11. $\Delta m = 1.39 \times 10^{-18}$ kg
$= 1.5 \times 10^{12} m_e$
35.B.1. 0.025
35.B.7. $U'_t = 1.27c$, $U_x = 1.94c$,
$U_y = 0.6c$, $U_t = 2.27c$
35.B.9. 1.48×10^{14} metric ton/yr
35.B.11. 6.2×10^7 s
35.B.13. (b) 1.55×10^{-10} J
35.C.1. $\tan \theta = \gamma_v \tan \theta'$
35.C.7. (a) $v = gt/\sqrt{1 + \alpha^2 t^2}$,
$x/c = (\sqrt{1 + \alpha^2 t^2} - 1)/\alpha$,
$a = g/(1 + \alpha^2 t^2)^{3/2}$,
where $\alpha = g/c$
(b) $x = 2.91 \times 10^{17}$ m
(c) $x = 7.06 \times 10^{29}$ m
(d) 7.5×10^{13} yr
35.C.9. $E_{lab} = 999.5 \, mc^2$,
$E_{lab}/E_{c.m.} = 44.7$

CHAPTER 36

36.A.1. 6.36×10^{10} J/m³
36.A.3. Use the engine.
36.A.5. 6.63×10^{-29} J
36.A.9. $f_e = 3.63 \times 10^{16}$ Hz,
$f_n = 1.97 \times 10^{13}$ Hz,
$f_{ph} = 3 \times 10^{18}$ Hz,
$E = hf$
36.B.1. $C_v = (4\pi^2 k_B^4 T^3)/(15c^3\hbar^3)$,
$T = 5.9 \times 10^7$ K
36.B.3. 5.13×10^5 m/s
36.C.1. Density of photons $= 1/\pi^{-2} \int_0^\infty (e^{\beta\hbar ck} - 1)^{-1} k^2 \, dk$

CHAPTER 37

37.A.1. D

37.A.3. 9.7×10^{18} Hz,
6.2×10^{-12} m

37.A.5. $\lambda_{ph}/\lambda_{el} = 4.53$

37.A.7. 6.1×10^{-18} J,
2.44×10^{-17} J

37.B.1. 1.1×10^{6} W

37.B.5. 7.07×10^{-25} kg · m/s,
2.74×10^{-19} J

37.B.7. Use the engine.

37.C.1. $A^2 = \hbar/\sqrt{km}$,
$E = (\hbar \pi^2/8)\sqrt{k/m}$

37.C.5. $F = \pi^2 \hbar^2/mL^3$, same

CHAPTER 38

38.A.1. $-(\hbar^2/2m)(\partial^2 u/\partial x^2 + \partial^2 u/\partial y^2) + \frac{1}{2}kx^2 u = Eu$;
$u = e^{-\alpha x^2} e^{jky}$, where $\alpha = \sqrt{km}/2\hbar$

38.A.3. 27.5 eV, 371.2 eV

38.A.5. 2.1×10^{10} Hz

38.B.1. $(r, \theta, \phi) = (2.45, 65.9°, 63.4°)$

38.B.3. (a) $A^2 e^{-2r/a}$
$B^2(1 - r/2a)^2 e^{-r/a}$,
$c^2 r^2 e^{-r/a}\cos^2 \theta$,
$c^2 r^2 e^{-r/a}\sin^2 \theta$,
$c^2 r^2 e^{-r/a}\sin^2 \theta$
(b) All, first two.

38.B.5. $m^2 A$, 9.27×10^{-24}

38.C.1. $(x, y, z) = (1.22, 1.22, 1.00)$

38.C.3. 1.65×10^{4} T

CHAPTER 39

39.A.1. 1.49×10^{-14} m

39.A.3. m(calculated) $= 385.221 \times 10^{-27}$ kg,
m(actual) $= 385.232 \times 10^{-27}$ kg

39.A.5. 2.03×10^{-12} m

39.A.7. ^{80}Kr

39.B.1. 7.9%

39.B.3. 9.9×10^{-13} J

39.B.5.

	Surface
Be	1.12×10^{-11} J
Co	3.9×10^{-11} J
Bi	9.2×10^{-11} J

	Electrostatic
Be	8.8×10^{-13} J
Co	2.2×10^{-11} J
Bi	1.3×10^{-10} J

	Symmetry
Be	4.1×10^{-13} J
Co	1.6×10^{-12} J
Bi	3.3×10^{-11} J

39.B.7. 231 s

39.C.1. 3.7×10^{19} N/m²

39.C.3. $F = kZe^2/(1.2 \times 10^{-15}A^{1/3})^2$

INDEX

Abscissa, 2
Absolute rest frame, 606
Absolute zero, 308
Absorption spectrum, 679
Acceleration, 23
  angular, 30
  constant, 19
  direction of, 24
  in free fall, 19
  magnitude of, 24
  in one dimension, 17
  radial, 25, 31
  relative, 33
  tangential, 25, 31
Acoustic energy, 529
Action and reaction, 59, 127
Action at a distance, 348
Addition of velocity, relativistic, 622
Addition principle, 496
Adiabatic bulk modulus, 529
Adiabatic process, 261
Adsorption, 283
Air resistance, 44
Alpha decay, 728
Alpha-particle scattering, 651
Ampere (unit), 406
Ampere's law, 443, 454
Amplitude, vibration, 172
Amplitude, wave, 507, 539
Angles, 2
Angular acceleration, 30
Angular momentum, 124
  quantization of, 700
  quantum number, 703
Angular velocity, 30
Antineutrino, 730
Archimedes' principle, 200
Arc length, 4
Arteriosclerosis, 205
Aspirator, 205
Atmospheric hydrogen, 233
Atomic configuration, 711
Atomic radii, 717
  shells, 715
  structure, quantum theory of, 710
Auditory canal, 538
Auditory nerve, 538
Avogadro's number, 235

Back emf, 464
Barometric equation, 278
Basal metabolic rate, 113, 337
Basilar membrane, 535
Batteries, 411
  emf of, 412
  internal resistance of, 412
Bernoulli's theorem, 202
Beta decay, 728
Big bang theory, 301
Binding energy of nuclei, 724
Biological cells, size of, 181
Biot-Savart law, 451
Blackbody radiation, 333, 647
Blind spot, 583
Bohr
  orbits, 657
  radius, 658
  theory of hydrogen, 656
Boiling, 322
Boltzmann's constant, 234
Bubble chamber, 47
Bubbles, 216
Bulk modulus, adiabatic, 529
Buoyant force, 200

Calorie, 110, 318
  dietary, 111
Capacitance, 387, 492
Capacitor
  energy in, 476
  energy stored in, 397
  parallel plate, 389
  series and parallel, 391
  spherical, 391
Capillarity, 220
Cat's eyes, shining of, 583
Cavity radiation, 644
Celsius scale, 237
Center of gravity, 143
Center of mass, 118
Centimeter, 2
Centrifugal force, 55
Charge, motion of
  in nonuniform magnetic field, 426
  in uniform magnetic field, 423
Chemical bond, 712

Chemical potential, 274
   of dilute solute, 278
   in a gravitational field, 278
   of ideal gas, 277, 301
Chimney, falling, 147
Circular motion, 30
Cochlea, 538
Coefficient of friction
   kinetic, 71
   static, 72
Collisions, 40
Complex sounds, 539
Compton effect, 648
Compton wavelength, 650
Conductance, 407
Conduction, heat, 329
Conductors, 366
Configuration, atomic, 711
Conservation
   of electric charge, 453
   of energy, 88
   of momentum, 42, 43, 626, 630
   theorems, 77
Conservative forces, 85
Convection, heat, 329, 330
Converging lens, 578
Coordinate systems, 2
Coriolis force, 129
Coulomb's law, 51, 60, 344, 373, 382, 451
Countercurrent heat exchanger, 338
Covalent bond, 714
Critical point, 315
Cross product, 11
Cyclotron, 426
   frequency, 425
Cylindrical waves, 549

Dalton's law, 232
Davisson-Germer experiment, 661
de Broglie
   relations, 661
   theory, 659
   waves, 659
Decay constant, 731
Degeneracy of quantum states, 705
Dehydration, 188
Delayed effects, principle of, 478
Density, 193
   probability, 676

Derivative, 16
Determinism, 57
Dielectric constant, 399
Dielectrics, 397
Dietary calorie, 111
Diffraction
   by aperture, 555
   of electron waves, 667
   grating, 553
   single slit, 557
Dilute solute, 277
Direct current circuits, 411
Disorder and the velocity distribution, 299
Displacement, 5
Displacement current, 452, 456
Diverging lens, 578
Dot product, 10
Double string, imaginary, 504
Drag, 44
Drift velocity, 406
Drops, formation of, 217
Dynamics, 39
Dyne, 53

Ear
   anatomy of, 538
   physical description of, 534
Eddington, Sir Arthur, 49
Efficiencies of engines, 303
Elastic collisions, 107
Electric
   charge, conservation of, 453
   current, 405
   dipole, 346, 352, 370, 481
   field, 347
      calculating, 349
      of charged ring, 350
      in conductors, 381
      energy in, 475
      of infinite plane, 351
      lines, 357
   flux, 378
   permittivity, 399
   potential, 358
      of a charged ring, 361
      of a dipole, 370
      of a point charge, 360
   resistance, 406

Electromagnetic
  energy, 473, 475
  induction, 457
  quanta, 647
  waves, 478
    frequency spectrum of, 485, 553
    velocity of, 488
Electromotive force, 460, 464
  of a battery, 412
Electron scattering, quantum theory of, 670
Electron spin, 708
Electron wave diffraction, 667
Electrostatic energy of nucleus, 727
Electrostatic force, 343
Electrostatic potential, 95
Elementary particles, masses of, 636
emf, *see* Electromotive force
Emission spectrum of atoms, 653
Empirical mass formula, 725
Energy
  binding, of nuclei, 724
  in capacitor, 476
  conservation of, 88
  density, electromagnetic, 475
  in electric field, 474
  electromagnetic, 473
  exchange, 248
  flux, 531
  ground state, 654
  in inductor, 478
  inertia and, 101
  internal, 109, 244
  ionization, 715
  kinetic, 79, 99, 101
  levels of atoms, 654
  in magnetic field, 477
  metabolic, 110
  potential, 86
  quantization of, 653
  relativistic, 635
  of rotating body, 148
  spectrum of atoms, 653
  wave, *see* wave energy
Engines, efficiencies of, 303
Entropy, 246
  changes in, 255
  and disorder, 295
  of ideal gas, 265
  measurement of, 262

of the universe, 292
Equation of state, 313
Equilibrium of rigid bodies, 157
Equipotential surface, 365
Equivalent forces, 143
Erg, 80
Escape velocity, 97, 107
Ether, the, 491, 606
  velocity of, 607
Evaporation, 322
Evolution, law of, 296
  and second law of thermodynamics, 302
Expansion of the universe, 97
Eye, angular resolution of, 588
  human, 581
  insect, 590

Farad, 389
Faraday's law, 374
Field, 193
First law of dynamics, 43
First law of thermodynamics, 245
First-order spectrum, 554
Fission, nuclear, 476, 737
  neutron-induced, 743
  reactor, 745
Fixed end of string, 504
Flow rate, 376
Fluid flux, 376
Fluids, 192
Flux, electric, 378
  fluid, 376
  wave energy, 531
Flying, 186
Focal length, 577
Foot (unit), 2
Foot pound, 80, 103
Force, 50
  addition of, 60
  conservative, 85
  determination from potential, 98
  electric, 348
  gravitational, 51, 343
  magnetic, 421, 428, 429, 440
Forced oscillations, 534
Fourier amplitude, 542
  phase, 542
  series, 542
Four-momentum, 630

Franck-Hertz experiment, 655
Free energy
    Gibbs, 270
    Helmholtz, 271
Free expansion, 291
Free fall, 28, 66
Frequency, 172
    angular, of wave, 539
    discrimination by ear, 536
Friction, 70
    kinetic, 71
    static, 72
Fusion, nuclear, 737
    heat of, 321

Gamma decay, 728
Gamma rays, 486
Gauge pressure, 208
Gauss, 422
Gaussian surface, 378
Gauss's law
    for electric fields, 373, 377
    for fluids, 374
Geomagnetic drift, 426
Golf, 107
Gram, 2
Gravitation, law of, 51
Gravitational force, 51, 343
Gravitational potential energy, 93
Ground-state energy, 654

Half-life, 731
Hall effect, 434
Harmonic motion, 167
Harmonic oscillator, quantum, 688
Hearing, 534
Heat, 260
    conduction, 329
    convection, 329, 330
    of fusion, 321
    loss, 181
    pumps, 311
    radiation, 330
    specific, 318—319
    of vaporization, 321
Heat transfer, 329
    and physiology, 336
Heisenberg uncertainty relation, 683

Helecotrema, 535
Henry, 463
Hertz, 172
High jump, 91
Hooke's law, 52
Horsepower, 103
Huygens' principle, 555
Hydrogen, Schrödinger theory of, 704
Hydrogen atom, Bohr theory of, 656
Hydrostatic paradox, 198

Ideal gas, 230
    pressure, 231
    thermometer, 235
    translational energy, 232
Imaginary double string, 504
Impedance, wave, *see* Wave impedance
Index of refraction, 563
Inductance, 463
Inductor, energy in, 478
Inertia, 41, 49
    and energy, 101
Inertial frame, 49, 611
Insect eye, 590
Instantaneous velocity, 16
Interference, 549
    constructive, 550
    destructive, 550
Internal energy, 109, 244
Intrinsic properties, 387
Ionic bond, 714
Ionization, 283
Ionization energy, 715
Isentropic process, 261
Isothermal process, 261
Isotopes, 724

Jet engines, 44
Jet planes, ultimate velocity of, 46
Joule, 80
Jumping, 186
Junction rule, 415

Kelvin scale, 234
Kilogram, 2
Kinematics, 15
    and calculus, 21
    charges in an electric field, 363

of harmonic motion, 170
in two dimensions, 22
Kinetic energy, 79
physical significance of, 99
relativistic, 101
Kinetic theory of gases, 229
Kirchhoff's rules, 415
Kneecap, 162

Laplace's law, 215
Law of gravitation, 51
Law of large numbers, 244
Law of sines, 14
Length contraction, 619
Lens
converging, 578
diverging, 578
double surface, 576
focal length of, 577
single surface, 571
Lenz's law, 461
Lift, 207
Light
refraction of, 561
speed of, 448, 491, 563
Light rays, 562
Loop rule, 415
Lorentz-Fitzgerald contraction, 611, 619
Lorentz transformation, 614
equations, 618
Lorentz transformer, 625
Lungs, 211

Macrostate, 242
Magnetic bottle, 428
Magnetic circulation law, 442
Magnetic dipole moment, 431
Magnetic field, 422, 435
calculating, 439
circular current loop, 437, 440
current sheet, 438, 439, 445
energy, 477
motion of charge, 423, 426
solenoid, 437, 445, 460
straight wire, 436, 439
trajectory in a, 423
Magnetic flux, 458
Magnetic flux law, 442

Magnetic force, 421
between two parallel wires, 440
on a current, 428
current loop, 429
straight wire, 429
Magnetic induction, see Magnetic field
Magnetic materials, 446
Magnetic moment and the Zeemann effect, 707
Magnetic torque on a current loop, 430
Magnification, linear, 586
Magnifying glass, 584
Magnifying instruments, 583
Magnifying power, 584
Mass, 41
center of, 118, 120
and energy, 101
gravitational, 42
inertial, 42
Mass-energy conversion, 633
Mass formula, empirical, 725
Mass number, 724
Mass-zero particles, 636
Maxwell's equations, 349
Measurement process, 1
Membrane
basilar, 535
circular, 533
long strip, 533
normal mode vibrations of, 533
waves on, 532
Membrane diffusion, 284
Metabolism, 110
Meter, 2
Michelson-Morley experiment, 607
Microfarad, 389
Microscope
dark-field, 594
oil immersion, 593
phase-contrast, 594
resolution of, 591
simple, 586
ultraviolet, 594
Microstate, 242
Molar number, 236
Molecular weight, 235
Moment arm, 122
Moment of a force, 122
Moment of inertia, 130
physical significance of, 131

Moment of momentum, 124
Momentum, 42
    conservation of, 42, 43, 626, 630
    linear, 42
    moment of, 124
    physical significance, 48
    relativistic, 48, 101, 629
    time component of, 630
Moon, motion of, 38
Motion of the center of mass, 120

Near point, 581
Neutrino, 730
Neutron, 721
Neutron-induced fission, 743
Neutron star, 724
Newton (unit), 53
Newton's laws of motion, 61
Nonconservative forces, 85
Normalization of wave function, 682
Nuclear decay modes, 728
Nuclear fission and fusion, 737
Nuclear
    fluid, 723
    force, 347, 721
    physics, 721
    shells, 736
    stability, 732
Nucleon, 721
Nucleon-nucleon scattering, 721
Nucleus (nuclei)
    binding energy, 724
    surface energy, 727
    symmetry energy, 727

Ohm's law, 407
Oil-immersion objective, 593
Order and disorder, 295
Ordinate, 2
Oscillations, forced, 534
Ossicles, 535

Parallel axis theorem, 135
Parallel-plate capacitor, 389
Particle exchange, 251
Particle in a box, 677
Pascal's law, 195
Patella, 162

Pauli exclusion principle, 710
Pendulum, 89
    simple, 174
Period, 172
Periodic table, quantum theory of, 710
Phase, 171, 539
Phase equilibrium, 283
Phase transitions, 316
Photoelectric effect, 647
Photosynthesis, 297
Planck radiation formula, 646
Planck's constant, 646
Plane wave, 528
Poe, Edgar Allen, 37
Point sink, 374
    source, 374
Polar coordinates, 2
Polarization, 487
Pole vaulting, 90
Polygon method of vector addition, 6
Potential energy, 93
    minimum principle, 102
    physical significance, 99
    spring, 96
Pound, 53
Power, 103
Pressure, 193, 256, 259
Principle of delayed effects, 478
Principle of minimum potential energy, 102
Principle of work and energy, 79
Probability density, 676
Projectile motion, 28
Propulsion, 44
Proton, 721
Pseudoforces, 55

Quanta, electromagnetic, 647
Quantum and classical physics related, 671
Quantum mechanics, classical limit of, 685
Quantum numbers, 703
Quantum physics
    and classical physics, 671
    origins of, 643
Quantum states
    degeneracy of, 705
    of particle in a potential, 687
    in three dimensions, 701
Quantum theory, 668

Radian, 3
Radiation, 333
   blackbody, 333, 647
   cavity, 644
   heat, 330
Radiation field, 481
   of a dipole, 487
Radiation zone, 481
Radioactive decay, law of, 731
Radioactivity, 731
Radiometer, 298
Radio waves, 485
Radius of curvature, 26
Range, 29
Reaction equilibrium, 280
Rectangular coordinates, 2
Reference state, 87
Reflection, law of, 564
   total internal, 565
Reflection coefficient, 516
Reflection of waves, 503
   partial, 514
Refraction
   index of, 563
   of light, 561
Relativity
   Einstein's theory of, 611
   electromagnetic theory and, 605
   Newtonian principle of, 604
   theory of, 603
Resistance, 407
Resistivity, 407
Resistors
   energy dissipated in, 408
   in parallel, 413
   in series, 413
Resolution
   of the eye, 561
   of an optical system, 587
Rest energy, 101
Retina, 582
Reversed effective force, 56
Reversible engine, 304
Right-hand rule, 11, 123, 436
Rigid body, 130
   equilibrium, 157
Ritz combination principle, 655
Rockets, 44
Rotational invariance, 699

Rotation in a plane, 30
Running, 184
Rutherford scattering, 651

Satellites, 75
Saturated vapor pressure, 323
Saturn's rings, 128
Scalars, 5
Schrödinger equation
   derived, 673
   interpretation of, 675
   with a potential, 686
   without a potential, 675
   in two and three dimensions, 697
Second, 2
Second law of dynamics, 50
   and determinism, 57
   physical significance of, 59
Second law of thermodynamics, 241, 290
   and the death of the individual, 302
   and the efficiency of engines, 303
   and the fate of the universe, 301
   and the law of evolution, 296, 302
Screw jack, 93
Sedimentation, 278, 284
Self-inductance, 492
   of solenoid, 463
Shear, 192
Shells
   atomic, 715
   nuclear, 736
Sign conventions for lenses, 578
Simple harmonic motion, 167
Simple pendulum, 174
Size and function, 179
Slope, 16
Slug, 2
Snell's law, 561
Solenoid
   magnetic field of, 437, 445, 460
   self-inductance of, 463
Sound wave, 528
Spacetime, geometry of, 620
Spacetime plot, 15
Specific heat, 318
   at constant pressure, 319
   at constant volume, 319
Spectrometer, 680

Spectrum
  electromagnetic, 553
  energy, 653
Speed, 22
Spherical coordinates, 702
Spherical waves, 547
Spin of electron, 708
Stability, nuclear, 732
Standards, 1
  table of, 2
Standing wave, *see* Wave, standing
Standing-wave states in Schrödinger theory, 677
Statics, 69
  and anatomy, 160
Stationary states, 678
Steady state, 243
Stefan-Boltzmann constant, 333
Streamline flow, 201
Streamlines, 374
String waves, 511
Structural strength, 183
Sublimation, 315
Superposition principle, 496. *See also* Addition principle
Surface current density, 477
Surface energy, 212
  of nucleus, 727
Surface tension, 212
Surface-to-volume effects, 179
Symmetry energy of nucleus, 727

Tangent to curve, 16
Temperature, 234, 256, 265
  absolute zero of, 308
Tesla, 422
Thermal conductivity, 331
Thermal velocity, 406
Thermionic emission, 655
Thermography, 338
Tides, 68, 76
Time dilation, 618
Torque, 121
Total internal reflection, 565
Translational invariance, 698
Traveling wave, *see* Wave, traveling
Triple line, 315
Two-slit interference, 549

Ultracentrifuge, 31
Units, 2

Unit vector, 5
Universal gas constant, 236
Universal gravitational constant, 52
Universe
  entropy of, 292
  expansion of, 97
  fate of, 301

Vaporization, heat of, 321
Vapor pressure, 323
Vector analysis, 5
Vector product, 11
Vectors
  addition, 5, 6
  multiplication, 9, 10, 11
  resolution, 7
  subtraction, 7
Velocity, 15, 16, 22
  angular, 30
  direction of, 23
  field, 195
  instantaneous, 16
  magnitude, 23
  of light, 488, 491
  relative, 33
  relativistic, 628
  selector, 427, 652
  time component of, 628
Volume exchange, 251
Vortex, 442

Watt, 103
Wave, cylindrical, 549
Wave, harmonic
  defined, 506
  wave function of, 506
Wave, periodic
  definition, 506
  frequency of, 506
  period of, 506
  wavelength of, 506
Wave, sound
  described, 527
  plane, 528
  velocity of, 528
Wave, spherical, 547
Wave, standing
  defined, 508
  frequencies of, 511
  and normal modes, 510

Wave, string
  defined, 495
  wave function of, 500
Wave, traveling
  defined, 496
  wave function of, 500
Wave energy, 512
  in harmonic wave, 514
  in triangular wave, 513
  in sound wave, 529
Wave equation, derivation, 517
Wave frequency, angular, 539
Wave front, 489, 527
Wave impedance, 513
Wave function
  and addition principle, 500
  defined, 497
  example of, 498
  of harmonic wave, 506
  of nerve impulse, 501
  normalization of, 682
  of traveling wave, 500
Wave motion, 495

Wave packets, 659
Wave reflection
  coefficient of, 516
  at fixed end, 503
  partial, 514
Wave speed
  electromagnetic waves, 488
  sound waves, 528
  string waves, 511
Weight, 53
Weightlessness, 57
Weight lifting, 187
Whirlpool, 128
Work, 80
  and energy, principle of, 79
  function, 648
  path independence of, 84
Window, oval and round, 535

X-rays, 486

Zeemann effect, 706
Zero-mass particles, 636

PHYSICAL CONSTANTS

Symbol	Value	Units	Description
a_0	5.29177×10^{-11}	m	Bohr radius
c	2.9979246×10^8	m/s	Speed of light
e	1.60218×10^{-19}	C	Electron charge
g	9.8	m/s²	Gravitational acceleration
	32	ft/s²	
G	6.67×10^{-11}	N · m²/kg²	Gravitational constant
$\hbar$	1.05458×10^{-34}	J · s	Planck's constant
k	1.3807×10^{-23}	J/K	Boltzmann's constant
k	8.987552×10^9	N · m²/C²	Coulomb force constant ($1/4\pi\epsilon_0$)
m_e	9.1095×10^{-31}	kg	Electron mass
m_p	1.67265×10^{-27}	kg	Proton mass
m_n	1.67495×10^{-27}	kg	Neutron mass
N_A	6.0220×10^{23}	mol^{-1}	Avogadro's number
R	8.314	J/mol · K	Molar gas constant
ϵ_0	$8.8541878 \times 10^{-12}$	F/m	Permittivity of vacuum
μ_0	$4\pi \times 10^7$	H/m	Permeability of vacuum
σ	5.670×10^{-8}	W/m² · K⁴	Stefan-Boltzmann constant

UNIT ABBREVIATIONS

A	ampere	ft	foot	km	kilometer	nm	nanometer
Å	angstrom	G	gauss	keV	kiloelectron volt	rev	revolution
atm	atmosphere	g	gram	lb	pound	s	second
C	coulomb	H	henry	m	meter	T	tesla
°C	degree Celsius	h	hour	MeV	megaelectron volt	V	volt
cal	calorie	Hz	hertz	mi	mile	W	watt
cm	centimeter	in	inch	min	minute	Ω	ohm
eV	electron volt	J	joule	mm	millimeter		
°F	degree Fahrenheit	K	kelvin	msec	millisecond		
fm	fermi, femtometer	kg	kilogram	N	Newton		